W9-ARW-318

Microbial Cell Walls and Membranes

Microbial Cell Walls and Membranes

H. J. ROGERS
Head of the Department of Microbiology
National Institute for Medical Research, Mill Hill

H. R. PERKINS
Professor of Microbiology
University of Liverpool

J. B. WARD
Department of Microbiology
National Institute for Medical Research, Mill Hill

1980
LONDON NEW YORK
CHAPMAN AND HALL
150TH ANNIVERSARY

First published 1980 by
Chapman and Hall Ltd
11 New Fetter Lane, London EC4P 4EE

Published in the USA by
Chapman and Hall
in association with Methuen, Inc.
733 Third Avenue, New York NY 10017

Printed in Great Britain at the
University Press, Cambridge

British Library Cataloguing in Publication Data

Rogers, Howard John
Microbial cell walls and membranes.
1. Micro-organisms 2. Cell membranes
3. Plasma membranes 4. Plant cell walls
I. Title II. Perkins, Harold Robert
III. Ward, J B
576 QR77 80-40517

ISBN 0-412-12030-5

Contents

Preface

In 1968 when *Cell Walls and Membranes* was published it was still reasonable to attempt to write a book covering the whole subject. Accordingly this edition of the book had something to say about walls from micro-organisms and plants as well as about membranes from bacteria and animal cells. A decade later this is manifestly impossible. Knowledge about almost all the subjects has grown explosively, particularly about membranes and the biosynthesis of macromolecules. Moreover aspects of the subject that were still in a relatively primitive state ten years ago have grown into highly sophisticated subjects worthy of extended treatment. The result is that the present book has had to be confined to structures and functions relating to only one division of the biological kingdom, namely micro-organisms. Even then severe limitations have had to be made to keep the task within the time available to the authors and their expertise. A few of the titles of chapters such as those on the isolation of walls and membranes, the structure of the components of bacterial and micro-fungal walls and their biosynthesis remain from the earlier book. These chapters have been almost completely rewritten and a number of quite new chapters added on topics such as the action of the antibiotics that inhibit bacterial wall synthesis, on the function of bacterial membranes, and the bacterial autolysins. The vast majority of the work described in the book has been published in the last ten to fifteen years; and much of it in the last two or three years. An attempt has been made to summarize subjects in which there is by now some degree of consensus about the general structures or mechanisms involved. Other important areas, which some might have liked to have seen included such as, for example, the biosynthesis and export of proteins, particularly into the outer membrane of Gram-negative species of bacteria, or the inter-relations between surface growth, cell division and DNA replication in bacteria have been deliberately omitted. It was felt that these subjects are still at the growing points of the science and that it was impossible to make statements about them that would not be extremely evanescent.

Clearly the levels to which the various chapters in the book have been written are not the same. Some are much more detailed and deal very much more with the frontiers of knowledge than others. This is partly a reflection of the nature of the subjects, and partly of the nature of the authors and their professional expertise in the subjects. In general, the readers the authors have had in mind are final year undergraduate students in microbiology and postgraduate students working on the appropriate subjects. Some chapters, however, may be of interest to research workers, particularly those whose interests border the topics dealt with in the book.

The authors would like to thank all who have been kind enough to supply pictures for the text of the book and in particular, Dr. I. D. J. Burdett of the National Institute for Medical Research who was also kind enough to read and comment on the various drafts of the chapter on ultrastructure. They would also like particularly to thank Mrs. H. B. Sharp whose dedicated labours in disentangling the written manuscripts and the errors in referencing of two of the authors (H. J. R. and J. B. W.) and turning them into beautifully typed pages has added much to any value the book may have. We should also like to thank Miss J. White and Mrs. J. L. Marsh for doing the same services for the remaining author (H. R. P.).

1
Ultrastructure of bacterial envelopes

1.1 Introduction

The majority of the text of this book will be concerned with the chemistry, biochemistry and physiology of the envelopes and intracytoplasmic membranes of bacteria. As a background to this discussion, something needs to be said about the appearance and arrangement of these structures in the cell, even if this serves no other purpose than to allow people some idea of the complexity of present day aims in trying to understand the more complicated functions of bacteria, such as their ability to divide. This chapter should, however, be regarded only as a topological guide to what follows, not as a thorough review of the ultrastructure of bacteria.

The last two decades have seen steady progress in the resolution of the layers that surround bacteria and of the inclusions they contain. That bacteria had 'walls' was recognized long before the advent of the electron microscope. Even, however, with the design and application of special wall stains, the level of resolution under the light microscope was too low to allow reliable dimensions, much less infrastructures to be seen. The isolation of wall preparations [39, 45, 91, 111, 116, 150] and the ensuing chemical studies did little more from the ultrastructural point of view than to confirm the work with the light microscope. The preparations were monitored by metal shadowing under low-powered electron microscopes. Collapsed structures of the same shape as the original bacteria were seen, but these techniques yielded no information about internal wall structure and little about external structures. The preparation and examination of 'ultra-thin' sections of bacteria was in its infancy at this time [6, 30] and staining with molybdate and tungstate, the so-called negative stains, had not been applied. The development and application of methods for the preparation, embedding, fixing and staining of material, along with the increased resolving power of electron microscopes has revolutionized our understanding of the ultrastructure of biological material, including the layers of the envelopes of bacteria.

1.2 The Gram-positive cell wall

The structural and molecular differences between the outer layers of bacteria that retain the purple iodine–gentian violet complex, despite extraction with polar solvents such as alcohol and acetone, compared with those that do not do so, are usually quite clear and very few anomalies among true bacteria have been brought to light (eg. *Butyrivibrio fibrisolvens* [32, 158]), with the exception of the *Halobacteria* and the methanogenic bacteria which have envelopes fitting neither category. The mechanism of Gram's [76] empirical method of staining is still only partially understood [148]. The general, rather featureless appearance of the walls of Gram-positive bacteria was apparent at a relatively early stage of electron microscopic studies, whereas the complex nature of the Gram-negative envelopes demanded all the skill of biological workers in designing better fixation and staining conditions, and of physicists in designing electron microscopes giving greater resolution to solve their structure. Even as late as 1975, a considerable advance was still possible in our understanding of the envelope of the experimentally ubiquitous *Escherichia coli.*

The walls of most Gram-positive bacteria appear thick and relatively structureless in section (Fig. 1.1), irrespective of the methods for fixation and staining. Their thickness is capable of great variation, but during steady state growth or within the exponential phase in ordinary batch cultures these are rather slight. Wall synthesis is not coupled to protein or nucleic acid synthesis other than through the formation of the wall biosynthetic enzymes themselves and the supply of a very limited number of amino acids and nucleotides (see Section 8.2), so that variation in wall thickness, outside steady-state or nearly steady-state growth conditions is in no way surprising [73]. If protein synthesis by bacteria is deliberately stopped, either by the exhaustion of an essential amino acid or by the addition of antibiotics, such as chloramphenicol, wall synthesis is not necessarily affected and great thickening can occur [161]. For example, walls of *Staphylococcus aureus* have been increased in thickness from 30 nm to 100 nm by incubation in the presence of chloramphenicol [67]. Likewise, the thickness of walls of bacilli doubles when a tryptophan auxotroph of *Bacillus subtilis* is incubated in the absence of tryptophan for an hour at 37° C [98]. The precise measurement of wall thickness is not easy, however, [82, 120] since it demands that sections should be exactly median whether cut longitudinally or radially in rod-shaped organisms. Often the inner and outer edges of the wall are difficult to define with precision. The problem of measuring wall thickness precisely has been carefully studied [82, 83] using *Streptococcus faecalis* and *B. subtilis* [14]. In seven separate exponential phase cultures of *S. faecalis*, the thickness measured in sections of whole cells was found to vary between 26.7 and 28.3 nm with an average of 27.2 nm and with an error of about 10–15%. These authors' criterion, for a precisely antitangential section, was that the wall should appear sharply tribanded after glutaraldehyde–osmium tetroxide fixation and staining according to Ryter and Kellenberger [145] (see Fig. 1.2). The thickness of walls in exponentially growing cultures of *B. subtilis* was studied in cross-fractured cells in freeze-fracture preparations. The value obtained was 27.62 ± 2.75 nm. The thickness, however, of fragments of isolated walls varied considerably according to the staining procedure used. For example, treatment with uranyl acetate following glutaraldehyde fixation altered the mean value to 25.13 nm. The excellent use to which the biosynthetic process of wall thickening has been put in studying cell

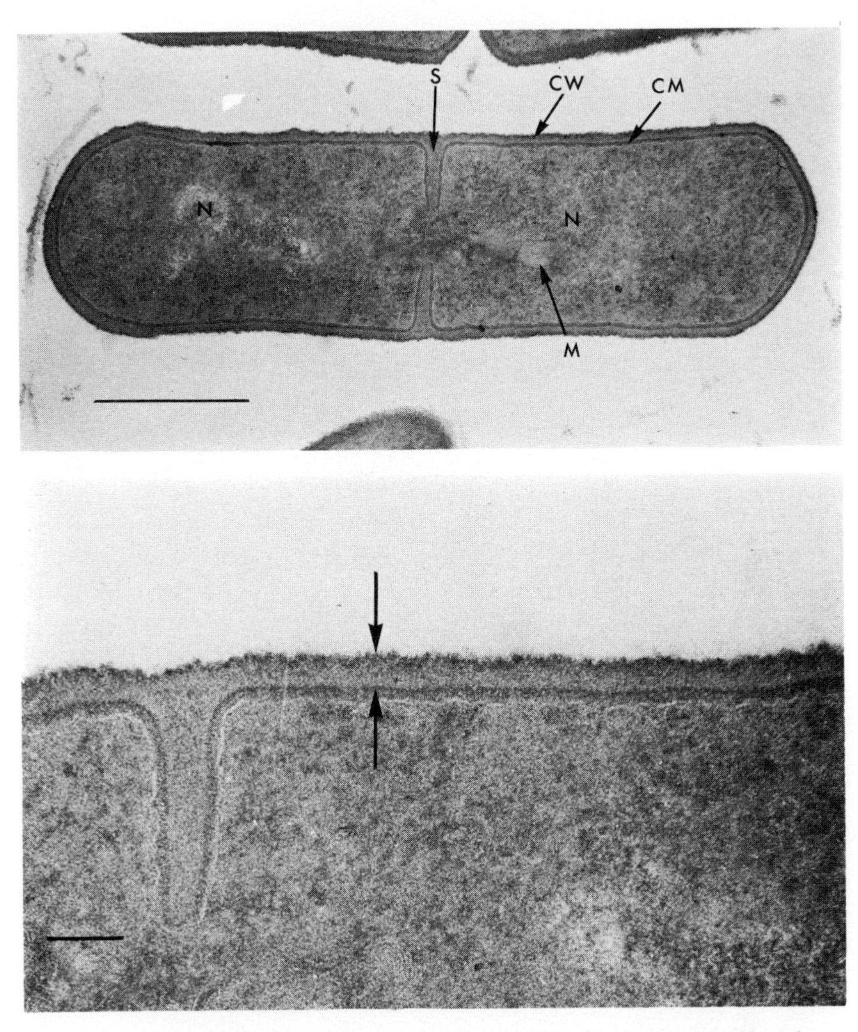

Figure 1.1 (a) A longitudinal section of *Bacillus subtilis* fixed and stained with glutaraldehyde and osmium, showing the absence of structure within the walls apart from their tribanded appearance. N, nuclear bodies; CM, cytoplasmic membrane; CW, cell wall; S, septum; M, mesosomes. The bar is equivalent to 0.5 μm. (We are indebted to Dr. I. D. J. Burdett of the National Institute for Medical Research for this photograph.) (b) A higher magnification where the bar represents 0.1 μm.

division of bacteria will be described in Chapter 15. Too few reliable measurements have otherwise been made of wall thickness in micro-organisms growing exponentially, or in a steady state, to comment on the results quoted. It would appear that many rapidly growing Gram-positive bacteria have walls that are usually of the order of 30 nm in thickness but values as high as 50–80 nm have been quoted [70, 149] and

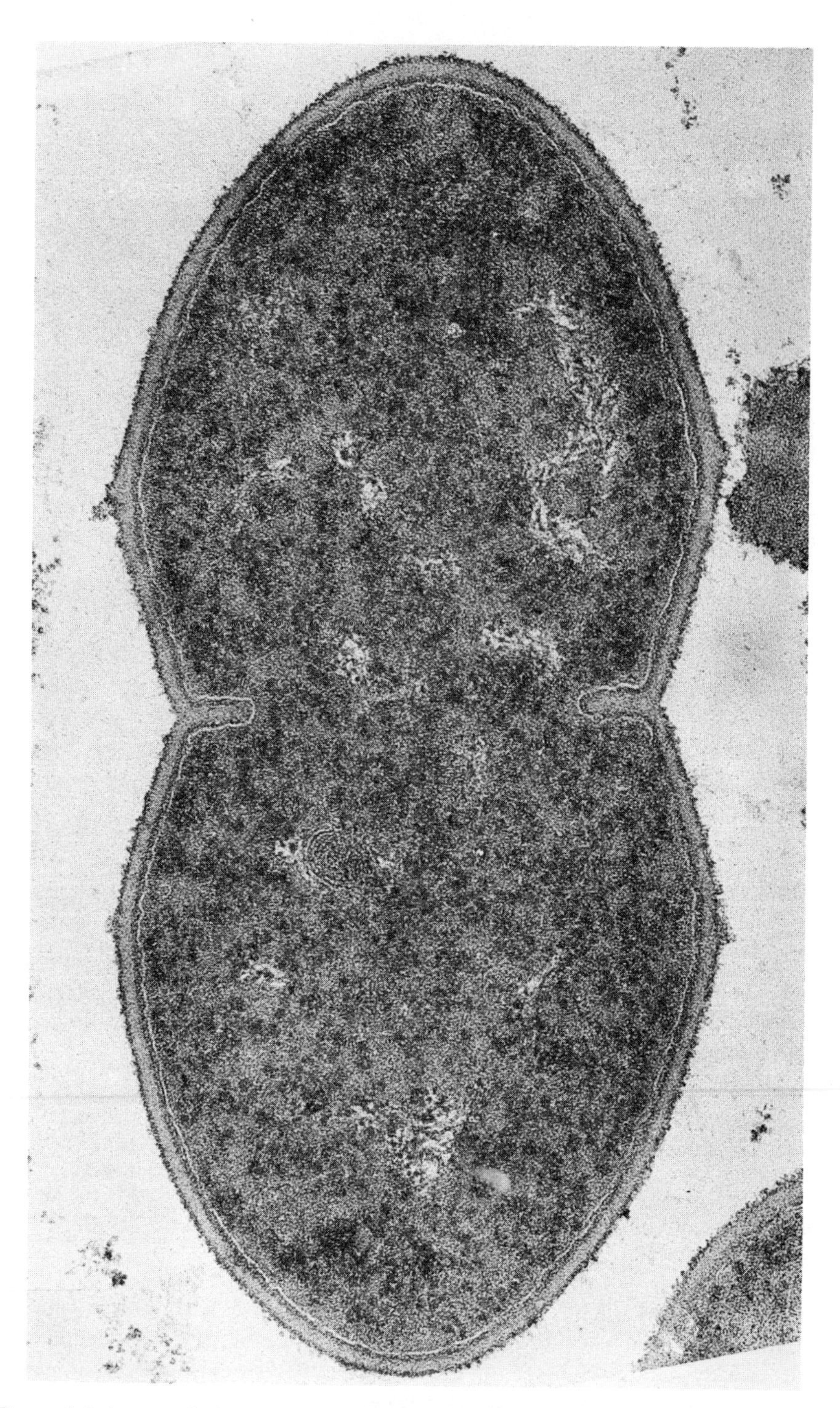

Figure 1.2 An exactly longitudinal section of *Streptococcus faecalis* showing the tribanded appearance of the wall. Magnification is 96 720. (We are indebted to Prof. M. Higgins of the Department of Microbiology, Temple University, Philadelphia, U.S.A. for this photograph.)

as low as 15.17 nm for *B. fibrisolvens* [32]. This latter organism although having a Gram-positive type wall stains Gram-negatively.

Claims that are made from time to time, for the presence of meaningful detailed structure in the depth of walls of Gram-positive species from the examination of very highly magnified sections of stained preparations or of negatively stained material, may be regarded with suspicion. The former would often appear to be due to chance distribution of stain in sections and the latter to the vagaries of the negative staining method itself. Repeated observations from many laboratories have nevertheless shown banding of walls from Gram-positive species in sections subjected to various fixing and staining procedures, and in freeze-fractured material. The sections usually have dark bands on the outer and inner limits of the wall and the effects of using a variety of fixation and post-staining methods on this phenomenon in the walls of *B. megaterium* and *B. subtilis* have been examined [14, 125]. In the former organism a triple-layered appearance was obtained when either isolated walls or whole organisms had been stained with osmium, permanganate or lead citrate. When uranyl acetate, phosphomolybdate or phosphotungstate had been applied the 'stains' were deposited evenly throughout the wall and no banding was seen. Unfortunately our knowledge of the physical and chemical reasons for the deposition of stain in biological material is not sufficiently detailed to allow deductions from this difference. It may nevertheless represent some real difference in properties between the innermost and outermost layers of the wall compared with that in its depth. When wall thickening had occurred during inhibition of protein synthesis in *B. subtilis,* the dark innermost layer seen after prefixation with glutaraldehyde, post-fixation with OsO_4 and staining with uranyl acetate, moved into the interior resulting in a striped appearance of the walls (Fig. 1.3) [98]. This suggests a chemical or physical distinction between the original innermost wall layer and the central region. In general, the low atomic number of the elements of greatest abundance in nature has precluded serious attempts to use unstained material for electron microscope examination. However, by adjusting the photographic recording process to give pictures of maximum contrast, Weibull [200] succeeded in obtaining rewarding pictures of the envelopes of bacteria that had been treated only with 4% glutaraldehyde at pH 7.0. A clearly triple or double-banded wall overlying the cytoplasmic membrane was seen in sections of *B. subtilis.* Results such as this suggested that the outer layer of the wall may differ in composition as far as the electron scattering power of its atoms is concerned and the results obtained with stained preparations reflect this difference. Since the atom with the highest atomic number in the wall is phosphorus, Weibull's [200] picture appeared to provide support for the suggestion that teichoic acid (see Section 7.1.1) is concentrated in the outer layers of the wall [125]. However, Millward and Reaveley [117] concluded that the trilamellar appearance of unstained walls of *B. licheniformis* and *S. aureus* was more likely to be due to variable packing of wall material than to the distribution of teichoic acid. They thought that the latter was, in fact, diffusely scattered through the thickness of the wall. This result is also supported by some aspects of a more complex study of the contrast patterns across bacterial walls produced by a variety of techniques [60]. Staining

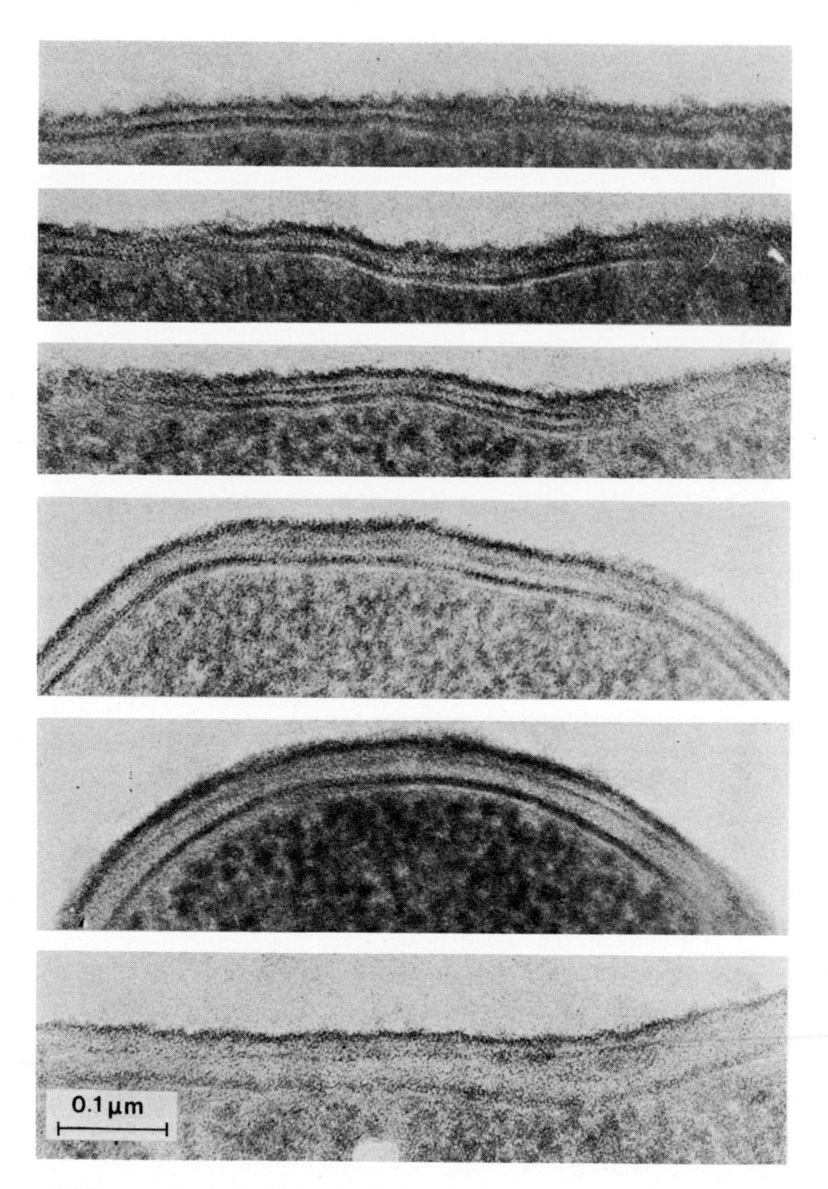

Figure 1.3 Progressive wall thickening in *B. subtilis trp* during incubation in the absence of tryptophan but presence of wall amino acids and glucose. The movement of the inner darkly staining band of the original wall into the interior, during the thickening process, can be seen (from [98]).

of osmium-fixed cells and walls of *S. aureus* with various combinations of lead citrate, uranyl acetate and ruthenium red were compared with pictures of sections of walls obtained after treatment with gold-coupled concanavalin A [59] and by a concanavalin A–peroxidase method. The concanavalin A studies were interpreted to show that teichoic acid was distributed throughout the thickness of the wall. This study illustrates some of the difficulties of reaching unambiguous conclusions about the distribution of teichoic acid in walls.

Evidence from the effects of extracting the accessory polymers from walls by reagents such as acids or formamide upon the tribanded appearance of stained preparations of walls from the same species of bacteria, has been conflicting. For example, Swanson and Gotschlich [177] found that the inner electron-dense layer of streptococcal walls was lost after treatment with HNO_2 and concluded that this was due to the loss of teichoic acid. Wagner *et al.* [192], on the other hand, extracted cells of Group A and Group C streptococci with eight different reagents that were found to remove proportions of the rhamnose-containing polysaccharides varying from 20–95%. The trilaminar appearance of the walls of the organisms was not altered after treatment with 10% trichloroacetic acid for 24 h at 4° C, despite a reduction in their thickness by 85%. Likewise, when wall preparations from *S. faecalis* and *Lactobacillus arabinosus* were treated with TCA, a very high proportion of the accessory polymers was removed and the walls were thinner but the trilaminar appearance persisted even though it was weaker [4, 186]. Thus, the peptidoglycan component of streptococcal walls, at least, can itself express the trilaminar appearance in sections after the application of standard Ryter–Kellenberger treatment. Wagner *et al.* [192] also showed that ferritin-labelled antibody produced against peptidoglycan would react with both sides of the wall. Antibodies raised against peptidoglycan will also react specifically with homologous organisms in a number of different species. Thus, even if any layering of the walls is present, peptidoglycan is accessible to antibody molecules, at least on the outer surface, and in streptococci apparently, on the inner surface as well. On the other hand a conditional mutant of *B. subtilis* with grossly reduced amounts of teichoic acid in its walls, did not have trilamellar walls when examined in section, whereas when grown under the appropriate conditions it regained both its wall teichoic acid and its wall staining pattern [38]. A different type of explanation of staining patterns has been suggested by considering the chemistry of the various stains used for electron microscopy [14]. The outer 'stained' bands might, according to this explanation, result from the mutual electrostatic charge properties of the wall and the stain. Some stains may have difficulty in entering the wall, and thus combine with the surfaces preferentially.

Although it is still not possible to be dogmatic about the relationship between the tribanded appearance of walls from Gram-positive bacteria and the distribution of different polymers within them, a simple layering seems highly unlikely. Nevertheless study of freeze-fracture material [186] again emphasizes that the stain distribution may correspond to some real differences possibly in the physical properties of the different parts of the wall. In whole cells, the walls appeared amorphous but in wall preparations first extracted with sodium dodecyl sulphate to remove membrane remnants and hydrophobic proteins, they appeared as "made up of irregular rows of 'globules' separated by a channel . . ." [186]. This appearance was only visible in cross-fractured walls, when freezing was fast enough to prevent separation of water and the cryostatic agent, glycerol. At slower rates of freezing the globules were less apparent but the walls looked 'much more like the tribanded structures seen in thin section' [186]. Removal of the accessory polymers again left intact the banded appearance in the freeze-fractured preparations.

It would seem that some infra-structure exists in walls, not dependent upon the continued presence of the accessory polymers, such that polyvalent substances such as OsO_4 or uranyl salts can either fix more firmly or enter more easily into the inner and outer layers. Moreover these layers are less transparent to electrons even in the absence of stains: freeze-fractured preparations confirm this heterogeneity. The relationship to molecular properties and architecture is quite uncertain. Evidence for layering in the walls of *Sarcina flava* has been obtained by studying sections of cells acted upon by lysozyme when two or three layers of the initially featureless wall separated [99]. The cells examined were, however, from resting phase cultures that had been incubated for 48 h and results for rapidly growing bacteria have not been quoted.

1.2.1 *Patterned outermost layers*

In a number of species of Gram-positive and -negative bacteria, an outermost, very regularly patterned layer can be seen after negative staining or in replicas from freeze-etching [72, 165, 178, 183]. These outer layers can be readily removed as sheets from the wall preparations of *Bacillus sphaericus, B. polymyxa, Clostridium thermosaccharolyticum* and *Cl. thermohydrosulfuricum* [43, 75, 96, 126, 163, 168] using mild treatments such as aqueous solutions of guanidine hydrochloride and urea. They are not, therefore, likely to be attached to the rest of the wall components by covalent bonds. Almost all these sheets of material consist of tetragonal arrays with centre to centre spacings varying from 7.5 nm to 19 nm. Examination of the layer from *B. polymyxa* by optical diffraction and optical filtering [51] showed that each unit was made of four sub-units 4–5 nm in diameter arranged to give a square pattern. Patterned layers have also been seen by use of the freeze-etching technique on a variety of other species of bacilli [93, 165].

The patterned layers consist of protein but carbohydrate has been reported in some [75, 78, 126, 165–166]. The patterned layers from *B. sphaericus* and from a mutant resistant to phage M have been examined chemically [96], and have been subject to detailed optical reconstruction [2]. The T (patterned) layer from the wild-type accounted for 30% of the protein extracted from whole cells by 8 M urea and about 16% of the total cell protein. A protein was purified from the extract and shown to be made of sub-units with a molecular weight of about 150 000. In polyacrylamide gel electrophoresis it ran as a single major band with a number of minor bands. Similar molecular weights of about 140 000 were obtained for the sub-units from the layers from two clostridia [168]. The purified T protein from *B. sphaericus*

Table 1.1. Inactivation of bacteriophage M by the purified T layer of *Bacillus sphaericus* Strain P-1 (from [96]).

Protein concentration as purified T layer (μg/ml)	*Bacteriophage titre* ($\times 10^{-4}$ PFU/ml)*
0	90.0
0.022	93.0
0.22	77.0
2.20	45.0
22.0	6.0
280.0	0.015

* The number of plaque forming units remaining after incubation at 37° C for 1 h of the bacteriophage M suspension with T layer protein in yeast extract medium.

inactivated phage M (Table 1.1). Sub-unit proteins were also isolated by a similar technique from a number of phage resistant mutants of the organism but their molecular weights were lower, the smallest being of the order of 80 000–90 000. The amino acid compositions of some of these proteins was similar. It would thus seem possible that one function of the patterned outer layer in this species is to provide attachment sites for bacteriophage. As the authors point out, however, no phage-resistant strains devoid of the layer were isolated. Detergents, such as sodium dodecyl sulphate, seem not only to remove the sheets but also to disaggregate the sub-units. The sub-units from *B. sphaericus* P1 and other bacteria can usually be made to reassemble into sheets. Those from two species of clostridia, for example, could be reversibly disassembled by lowering the pH to about 3.0 and assembled again by raising it to 7.0. The sub-units can also often be made to reassemble on the surface of micro-organisms previously stripped of their layers by treatment with reagents such as guanidine–HCl. This was done with the two species of clostridia and interestingly the reassembly process was not specific; the sub-units from one organism were able to attach to the other. For attachment to the surfaces a low pH buffer had to be used but the particles only rearranged to the characteristic pattern when the pH was raised [164]. Patterned layers may not be universally present on the surfaces of all species of bacilli or clostridia and have not so far been seen on the surface of most strains of common organisms such as *B. subtilis, B. licheniformis,* or *Cl. perfringens.* A single report of such a pattern on a strain of *B. subtilis* has appeared, however, [97] and caution over negative results is necessary. Nermut and Murray [126] for example, showed that repeated washing with saline of bacteria from young cultures of *B. polymyxa* revealed a pattern otherwise not visible, whilst Goundry *et al.* [75] revealed patterns by extraction with chloroform–methanol. Such experiments imply that buried pattern-forming material may occur in otherwise smooth surfaced bacteria. Difficulty has sometimes been experienced in seeing patterned layers in sections of bacteria, although it has been stated [96] that a layer, stained heavily with osmium, can be distinguished in sections of *B. sphaericus*, but is missing after extraction of the cells or their walls with 8 M urea. The regularly arranged layers on two clostridia could readily be seen in sections [168]. At present it may be more profitable to regard these patterned superficial layers as micro-capsules rather than as an essential part of the wall, but their evolutionary relationship to the outer track of the Gram-negative cell wall might be a subject for thought. Similar patterned layers have also been shown on the outside of Gram-negative bacteria attached to the outer membrane (see pp. 18–19).

1.3 The Gram-negative cell wall

In Gram-positive cells it is relatively easy, both in terms of morphology and chemistry, to be dogmatic about defining a 'wall' and a membrane even - as will be seen later (Section 15.1.3) – if this dogmatism becomes misleading when growth of the envelope is to be considered. In Gram-negative cells, such is the complexity of the outer layers of the cell that use of the term 'envelope' to describe the whole outer structure of the bacteria may be less confusing. Morphologically, at least three or four layers of the envelope can now be recognized, some themselves appearing in sections after the application of orthodox techniques of fixing and staining, as double-track membrane-like structures. Some of these layers can be isolated and shown to be chemically distinguishable.

In early work [103], four layers were distinguished in the walls of *E. coli*, an inner membrane and a triple layered wall. Present ideas would see the envelope of this and many Gram-negative species as built up of the following layers counting outwards from the cytoplasm of the cell:

(1) The plasma or cytoplasmic membrane appearing as a unit membrane of three layers,
(2) A thin peptidoglycan layer,
(3) A usually somewhat structureless zone,
(4) An outer membrane appearing as a second unit membrane of three layers,
(5) Lipopolysaccharide that can usually only be made visible by application of suitably labelled antibiodies, such as those conjugated with ferritin.

Differences in distances of separation between the layers, if not in the number of layers, exist in various Gram-negative species. Fig. 1.4 shows this general arrangement of the layers.

1.3.1 *The peptidoglycan layer*

Critical and extensive early studies of the layers in the walls of sections of *E. coli* were made by de Petris [47, 46] and by Murray *et al.* [121]. Detailed differences exist in interpretation by these authors, but they agreed that there was a densely osmium-staining layer of peptidoglycan (i.e. lysozyme sensitive) 2–3 nm thick, usually separated from the plasma membrane by an electron-transparent space. A densely staining, usually thin layer has been found in many if not all other Gram-

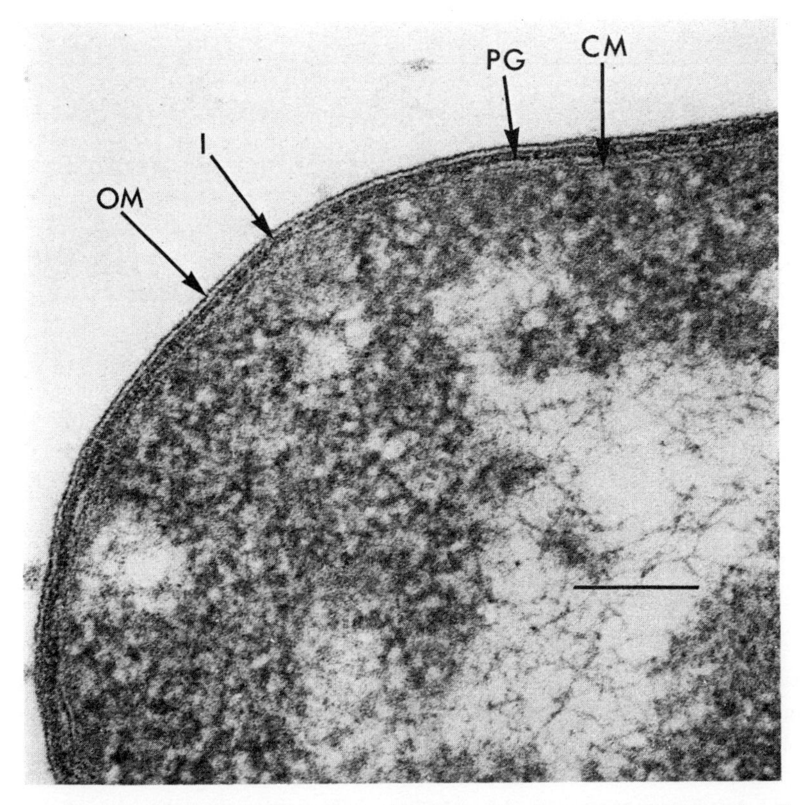

Figure 1.4 A longitudinal section of *E. coli* showing the characteristic layered walls of Gram-negatively staining bacteria. The close apposition of the inner layer of the wall and the cytoplasmic membrane may be noticed. No visible periplasmic space can be seen. OM, outer membrane; I, intermediate layer; PG, peptidoglycan layer; CM, cytoplasmic layer. The bar is equivalent to 0.1 μm (from [25] and we are grateful to the authors for providing the photograph).

negative bacteria, save the *Halobacteria* and the methanogenic bacteria, that varies in thickness from 2 to 10 nm (see Table 1.2). Whether or not this layer increases in thickness like the Gram-positive wall, when cells are incubated under circumstances such that protein biosynthesis cannot occur, is not clear. No increase in thickness of the peptidoglycan layer of *E. coli* occurred when the bacteria were incubated in the presence of chloramphenicol [67]. On the other hand, there was an increase in the thickness of this layer in *Acinetobacter* when similarly treated [169]. Whether or not this is a true species difference is as yet uncertain. It might be noted, however, in passing that in the *Cyanobacteria* which also have peptidoglycan layers, thickening takes place when they are incubated in the presence of chloramphenicol [100]. The greatest figure of 50 nm for the thickness of the peptidoglycan layer has been reported for *Veillonella parvula* in cultures that had been incubated for a considerable time. All the cells had similarly thick layers and had lost the

Table 1.2. Thickness and description of the layers in the envelope of Gram-negative bacteria as seen in transverse sections (after [73]).

Organism	*Peptidoglycan layer* (nm)	*Outer membrane* (nm)	*Additional external layers*
Ectothiorhodospira mobilis	5–7	ca. 7.5	+
Chlorobium thiosulfatum	ca. 4, dense	blebs	
Pelodictyon clathratiforme	3	7–8	+
Thiobacillus thiooxidans	ca. 3	8.5; wavy, loose-fitting	
Asticcacaulis C19	ca. 4; associated with PM	ca. 8.5 wavy	
Ferrobacillus ferrooxidans	3–5; dense	ca. 6	
Spirillum serpens	thin, taut, dense	7.5; wavy	+
Escherichia coli	2–3; see Fig. 1.2a	6–10	+
Moraxella duplex	8	ca. 10	
Veillonella parvula	10–50; thicker in older cells;	7.5–8; convoluted	
Thioploca ingrica	12–19; dense	ca. 8, wavy	+
Vitreoscilla	ca. 3; dense	5; folded crenated	

outer wavy membrane. Occasional cells of this nature were said also to be present in young cultures [18]. Other less well known species of bacteria, (e.g. some rumen bacteria) have been reported to have thick peptidoglycan layers but still retain the rest of the structure characteristic of the Gram-negative wall [40]. Evidence that the densely staining inner layer of the walls in a variety of species other than *E. coli* contains peptidoglycan, has been provided by sensitivity to lysozyme [18, 47, 57, 58, 121, 180, 201].

Although the peptidoglycan layer is the innermost staining layer of the wall in *E. coli*, a gap of 3–4 nm with the same electron transparency as the outside medium has been described between this layer and the cytoplasmic membrane [18, 121]. It is often referred to as the periplasmic space or zone [41] but its exact significant is hard to judge since any degree of plasmolysis could produce a similar gap, as might shrinkage during the dehydration of the cells prior to embedding for section cutting. More recent work using modified fixation procedures has produced sections that do not show the gap [25] (Fig. 1.5).

On the outer side of the dense peptidoglycan layer, considerable differences exist between species and between what has been seen by different authors looking at the same species. Undoubtedly, however, the so-called 'intermediate zone' would seem narrower in some strains of *E. coli* [73, 121, 122] than others [46, 47, 57, 133] and in representatives of other species, such as *Acinetobacter* [181]. The material in this zone, remaining relatively electron transparent after the application of different staining methods appears to contain protein, since proteolytic enzymes remove material from it [47, 181]. It is said to contain globular protein [11] and in

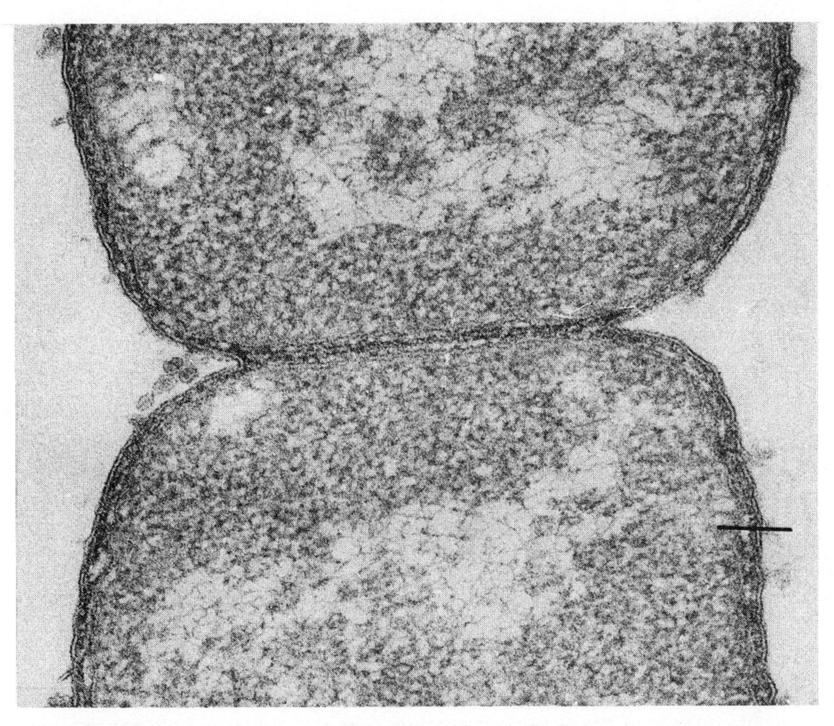

Figure 1.5 A tranverse section through the division plane of *E. coli* showing the nature of the septum. The bar is equivalent to 0.1 μm (from [25] and we are grateful to the authors for providing the photograph).

Chondrococcus columnaris fibrils have also been described as occurring in this region [133], connecting the peptidoglycan layer with the outer membrane. Whether this material in *E. coli* and other species represents the lipoprotein we now know to be attached to the peptidoglycan (see Chapter 7), is not known.

1.3.2 *The outer membrane*

All species of Gram-negative bacteria so far examined, again except the *Halobacteria,* have an outer membrane-like layer, which in sections of bacteria stained by different methods shows two dense lines separated by a transparent one with an overall thickness of about 7.5–8.0 nm (see Table 1.2). In other words, it has the characteristic appearance of a 'unit membrane'. The nature of this structure no longer needs speculative discussion, since it has been isolated as a separate membrane containing phospholipids and a limited number of proteins as well as lipopolysaccharide (see Section 2.3.3). It can be clearly distinguished from the cytoplasmic membrane on the basis of its relative lack of enzymic activity, the pattern of polypeptides shown after polyacrylamide electrophoresis and its content of lipopolysaccharide components. Extraction of cells with aqueous phenol removes the outer membrane [18]. It appears in sectioned cells as a wavy track that can be easily persuaded to shed 'blebs', an extreme example of which could be seen during the growth of a lysine requiring auxotroph of *E. coli* in medium containing limiting amounts of lysine [105, 205, 206]. The formation of blebs by the *Enterobacteriaceae* [101, 170] and by *Pseudomonas aeruginosa* [114] appears also related to the organization of the lipopolysaccharide, since rough mutants with incompleted side chains (see Section 7.2) form blebs particularly easily. Bleb formation from the outer membrane has also been seen in freeze-etched preparations of other species such as *Neisseria gonorrhoeae* [176] and *Acinetobacter* sp. [169]. Blebs have been described as frequently occurring near the point of cell division [25]. Negatively stained, whole Gram-negative bacteria have a highly wrinkled appearance [11, 12, 18, 46] presumably a three-dimensional picture of the waviness of the outer membrane seen in sections. Interestingly, isolated wall preparations do not show this wrinkling when negatively stained [94, 150, 151]. Freeze-etched preparations of whole organisms show a lumpy outer surface when they have been grown in media containing 20% (w/v) glycerol and glycerol is used subsequently as a cryostatic agent during freezing [123]. Only a smooth surface was observed when grown in the absence of glycerol, with or without the subsequent use of glycerol as a cryostatic agent [12]. In neither case, however, do the freeze-etch replicas give an appearance compatible with the extreme corrugations seen after negative staining. The waviness in thin sections and the corrugation seen after negative staining are more likely to be due to the shrinkage of the cell in the dehydration and staining procedures, which is known to take place [10, 12] leading to a partial collapse of the envelope; possibly loss of fluid from between the peptidoglycan layer and the outer membrane may also occur. The appearance of the negatively stained preparations is not due only to the application of the stain since high-resolution scanning pictures of *E. coli* show a similar wrinkling [3]. The negative staining method consists of mixing the bacteria with a solution containing atoms of greater scattering power than contained in the living micro-organism. The substances usually used are molybdate, silicotungstate or uranyl salts. The solution surrounding the bacteria is then dried under vacuum to form a mass relatively opaque when compared with the bacterium. Darker areas mean penetration of the stain into the bacterium. The reverse phenomenon, namely extreme swelling of the outer membrane as a result of placing a marine *Pseudomonad* in sucrose solutions, is to be seen in the work of MacLeod and his colleagues [56].

A study [188] of complementary freeze-fracture replicas of *E. coli* cells grown in 30% (v/v) glycerol showed that two fracture planes occurred, one through the plasma membrane and another through the outer membrane (see Fig. 1.6). In both, of course, the fracture plane is within the membranes and the true membrane surfaces are not visualized. This then demonstrated with some certainty that the flat elements (Fig. 1.7) of about 10 nm in diameter, were within the outer membrane. In the previous study [124] these had been thought to be associated with the inner peptidoglycan layer and perhaps to represent the lipoprotein known to be covalently attached to it. As a three-dimensional model of the envelope of *E. coli* the picture (see Fig. 1.7) shown by van Gool and Nanninga [188] agrees well with other

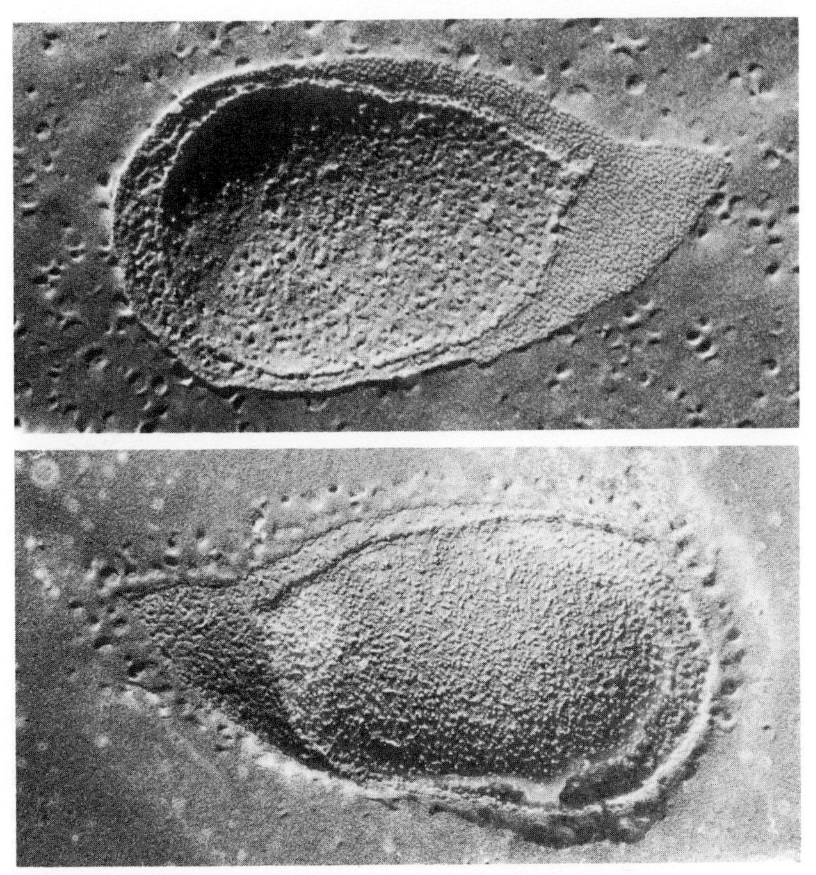

Figure 1.6 Complementary fracture faces in the cell envelope of *E. coli.* The central region in the converse face shows the particulate surface of the cytoplasmic membrane. The edges show the profiles through the wall layers and cytoplasmic membrane (from [188]. The authors are grateful to Dr N. Nanninga of the Laboratory of Electron Microscopy, University of Amsterdam, Amsterdam, The Netherlands for providing the photographs).

studies [12] and with sectioned material, it is probably definitive. A similar arrangement has been found for *Ps. aeruginosa* from freeze-etching studies [69]. A very careful study, using the freeze-fracture technique, has been made of an *Acinetobacter* sp. with essentially similar results, save that the patterned layer is the outermost layer of the cell [169]. However, it should be noted that some organisms, particularly among autotrophic bacteria (e.g. *Nitrosocystis oceanus*) have more complicated envelope structures (Fig. 1.8) as is illustrated from the work of Watson and Remsen [195] in which an outer easily-removed fibrous layer is deposited over a highly patterned layer (see pp. 18–19). Under these is the outer membrane with a globular layer and under this is the peptidoglycan that is separated from the cytoplasmic membrane by a further globular layer; a similar patterned

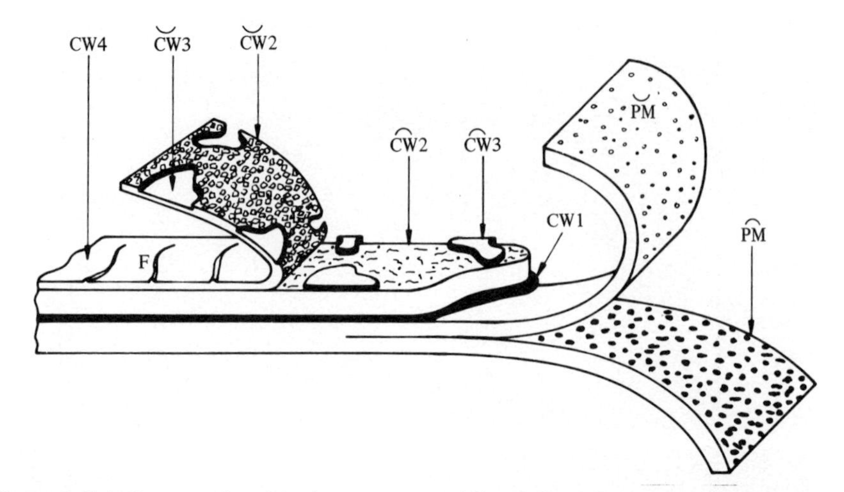

Figure 1.7 A diagram showing the structure of the wall of *E. coli* reconstructed from freeze-fractured preparations. The fracture-face dimensions are arbitrary. P̆M and P̑M are fracture-faces in the plasma membrane; CW1, presumed fracture profile of the peptidoglycan layer; C̆W2, C̑W2, C̆W3, C̑W3, fracture faces in the cell wall; CW4, outer surface of the wall revealed by etching (from [188]).

layer apparently associated with the cytoplasmic membrane has been seen in some *Halobacteria* [104] which of course, do not have a peptidoglycan layer. Some uncertainty necessarily still exists as to the location of the fracture planes in freeze-etch studies and it therefore is not entirely clear whether the globules are on or in membranous structures.

1.3.3 *Lipopolysaccharides*

Antibodies to lipopolysaccharides act as agglutinating antibodies for whole micro-organisms which shows that the antigenic determinants of these molecules protrude through the outer membrane of the envelope. Earlier work [47] showed that smooth strains of *E. coli* did not 'fuse', that is, the outer membrane tracks of contiguous micro-organisms did not touch but were separated by a relatively electron-transparent layer. Rough strains did not show this phenomenon. Specific visualization of the location of the antigenic part of the lipopolysaccharide was achieved by treatment of *E. coli, S. typhimurium* and *V. parvula* with ferritin-labelled antibody [10, 115, 156, 157]. The electron-dense particles of ferritin can be seen to extend as far as 150 nm outside the outer membrane of the envelope of *E. coli* and *S. typhimurium* in the form of a true micro-capsule [157]. Lipopolysaccharide is known to be removed from some bacteria by treatment with chelating agents such as EDTA [5, 42, 77, 109]. In some organisms a network of particles, seen under the electron microscope in freeze-etched preparations, is removed and can be found as rodlets 20–25 nm long in the supernatant fluid from such extractions, but whether these rodlets in fact represent mostly lipopolysaccharide may be doubted. When Mg^{2+} was

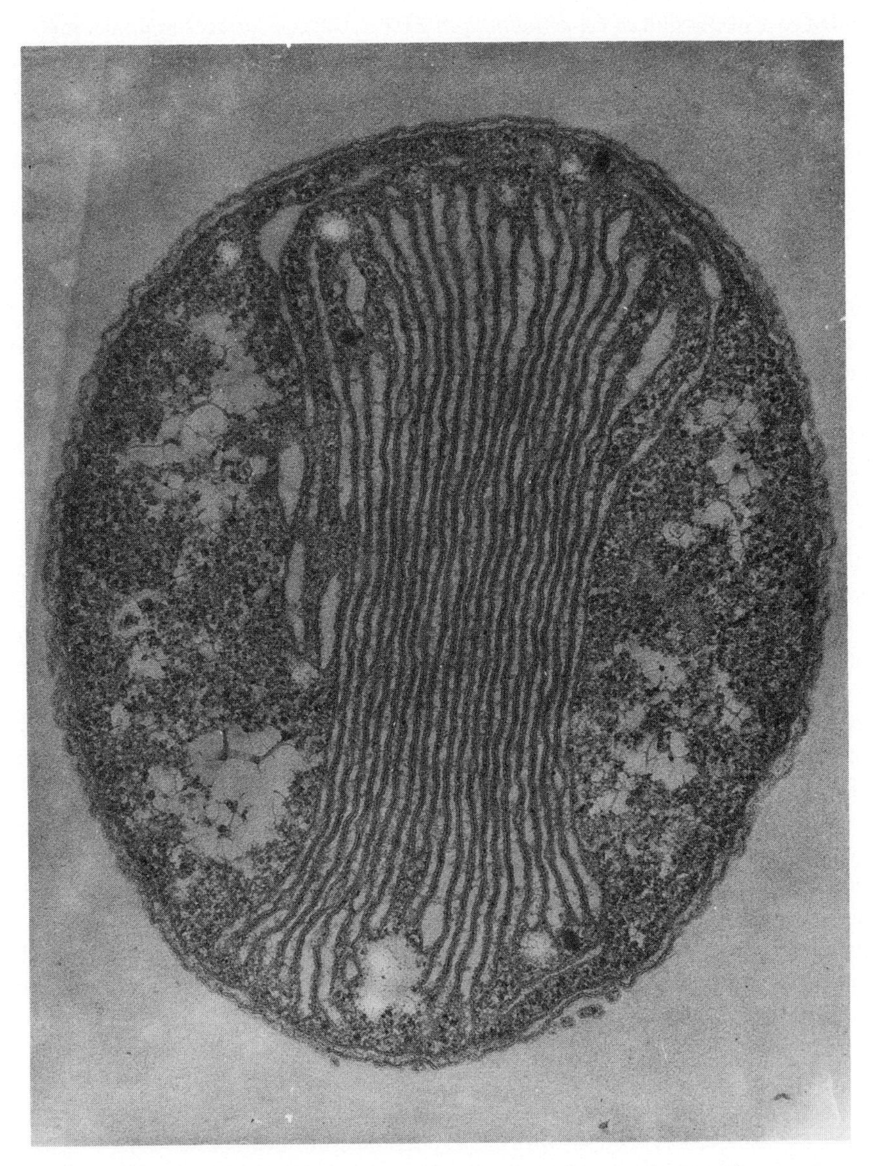

Figure 1.8 A section of *Nitrocystis oceanus* showing the complex envelope outer layer and the stacked membranes derived by infolding of the cytoplasmic membrane (from [122]. Photograph kindly supplied by Professor S. Watson, Woods Hole Oceanographic Institution, Woods Hole, Mass., USA).

added to a suspension of *Ps. aeruginosa* in EDTA, the cell surfaces again became coated with particles but it appeared that only protein was removed from the solution on to the cells [69]. This suggests that the lipopolysaccharide itself did not reassociate with the cell surface and, therefore, that the original particulate material may not yield reliable information about the distribution of the lipopolysaccharide.

1.3.4 *Superficial patterns on Gram-negative cells*

As with a number of Gram-positive species, Gram-negative bacteria also have arrays of ordered particles on the outer surface of their envelopes. Among the first of such patterns to be demonstrated was that on the envelopes of *Spirillum serpens* [94] and on *Halobacterium halobium* [95] but many species have since been found to have similar arrays [73, 139, 165, 178]. These patterns are best demonstrated by negative staining or freeze-etching, although they have been seen by metal shadowing and with care can be demonstrated in thin sections (eg [73, 120]). The sub-units making up the patterns on *Spirillum* and *Acinetobacter* have been investigated in detail [16, 22, 74, 120, 181, 183]. Three strains of *Spirillum* were investigated, two of which showed dissimilar patterns, whereas one showed no pattern at all. In *S. serpens* strain VHA, the centre-to-centre spacing of these particles in lightly shadowed wall fragments was 14.5 nm and they were situated on the outside of the wall. Beautiful proof of the exterior situation of similar particles on the surface of *Acinetobacter* is to be found in the work of Glauert and her colleagues [73, 169, 182]. Each particle of the material in *S. serpens* could be seen to consist of six sub-units hexagonally arranged. Each sub-unit was estimated to have a size of 2 × 6 nm. Spurs projected from each, connecting it with the next hexagonal array in the overall pattern. The patterned material was shown earlier to be removed by either phenol-water extraction or by treatment of the wall fragments with sodium dodecyl sulphate. These methods might be expected to disrupt the outer membrane of the organisms to which the particles are attached, as well as possibly to dissociate the sub-unit hexagonal structures if the coherence of these is dependent on hydrophobic interactions. In various species, however, the particles can also be removed by treatment with 8 M urea, 1.5 M guanidine-HCl, 10 mM EDTA, or by simply washing with 0.85% NaCl [22, 179, 180, 183]. Such gentle treatments thus give the opportunity to strip the outer surface of the bacteria of their patterned layer and to study reassembly of the material onto the denuded cell. The pattern seen on the surface of *Spirillum serpens* has been shown [22, 23] to consist of protein sub-units of molecular weight 140 000. These will reassemble to form the original pattern on cell wall fragments previously stripped by treatment with guanidine-HCl, provided that either Ca^{2+} or Sr^{2+} is present; Mg^{2+} cannot be substituted. The nature of the so-called backing layer or the outermost layer of the cell next to the patterned layer to which the sub-units bind has been examined [33] and shown to contain lipopolysaccharide and lipid. Both substances together with Ca^{2+} were necessary for sub-unit binding. The isolated units of the patterned material from the surface of *Acinetobacter* have a single protein with a molecular weight of about 67 000 [183], which again can reassemble under the appropriate conditions [74, 180]. The conditions necessary for the isolated sub-units to form sheets of ordered material away from the cell, ie self-assembly, can be distinguished from those necessary for reattachment. For self-assembly Cl^- or NO_3^- are necessary as well as 20 mM Mg^{2+}, whereas a much lower concentration of Mg^{2+} only is necessary for reattachment. It would appear [178, 180] that protein–protein interactions are involved not only in self-assembly but also in reattachment of the units into an orderly array on the outer face of the outer membrane of *Acinetobacter.* This conclusion is particularly interesting in relation to the finding that lipopolysaccharide and lipid as well as a different cation are involved in assembling the patterned layer on *S. serpens* [33]. Even greater morphological complexity of patterned outer layers has been reported for *S. putridiconchylium* and *S. metamorphum* [15, 16], the first of which has two ordered layers outside the outer membrane of the envelope and the second, two ordered layers and an amorphous one. In the former organism there is a linear array covering a tetragonal one of larger sub-units. In the latter, the differently constructed layers were held very loosely to the cell and tended to be shed towards the end of exponential growth. Particular precautions had to be taken to preserve them in place during preparation for electron microscopic examination. This complicated arrangement has been analysed by the aid of optical diffraction, image filtering, linear and rotary integration techniques. In many ways, this paper illustrates the wealth of information that can be obtained from electron microscope images. Treatment with proteolytic enzymes such as papain also removes the patterned material from the surface of *Acinetobacter* [181, 182]. At least in some organisms it would appear [140] from freeze-etch studies that a patterned layer can exist on the inner side of the outer layer of the wall, rather than on the outer surface.

Other apparently very specialized structures such as, for example, the sheath that surrounds and organizes groups of cells of *Lampropedia hyalina* [31, 131] and *Micrococcus radiodurans* with its sculptured wall (see [73] and [108] for the most recent observations) also have very complicated patterned layers that have not so far been fully resolved. Clearly, much remains to be learned about the physiological significance of these patterned self-assembling layers. Do they, for example, have antigenic or enzymic activity and do these particles traverse the depth of the wall if they are biosynthesized on the ribosomes in the cytoplasm, as they presumably are, being proteinaceous? Whatever their function and however formed, they are very common, having been recognized and examined as occurring on the surfaces of some 50 species of Gram-negative and 26 species of Gram-positive bacteria [165].

1.4 Membrane morphology

The membranes of bacteria are less diverse than those in even the least complicated eukaryotic cells. The bacterial 'nucleus' is not divided from the cytoplasm by nuclear membranes, there are no specialized sub-cellular bodies such as mitochondria, which themselves have two sets of membranes, and no internal membranes comparable in morphological complexity with the endoplasmic reticulum. Just because bacterial cell membranes are relatively functionally unspecialized, they must be in other senses more complex, in as much as the few differentiated types of membrane that are present must carry out many of the same functions as those carried out by the wealth of different membranes in the eukaryotic cell. For example, the plasma membrane of bacteria is involved with oxidative phosphorylation, transport of metabolites and biosynthesis of exocellular polysaccharide and wall materials (see Chapters 8 and 10). In the eukaryote, each of these functions is undertaken by a specialized organelle or set of membranes. In Gram-negative species the cytoplasmic membrane and the outer membrane of the envelope are clearly differentiated both chemically and functionally. The mesosome tubules isolated from Gram-positive bacteria can be differentiated chemically from the cytoplasmic membrane. In photosynthetic and nitrogen fixing micro-organisms more extensive, probably specialized internal membranes can be seen, of which more will be said later. Many, if not all, inclusion bodies are surrounded by chemically simple membranes. In general, nevertheless, the bacterial cytoplasmic membrane can be looked on as an undifferentiated multifunctional organ. Despite this, the morphological appearance of the membrane is scarcely different from, for example, cytoplasmic membranes surrounding mammalian cells.

1.4.1 *Cytoplasmic membranes*

Removal of the walls of suitable rod-shaped Gram-positive bacteria such as *B. megaterium* by digestion with egg-white lysozyme leads to the formation of spherical osmotically-sensitive protoplasts, providing the bacteria are first suspended in relatively high concentrations of slowly permeating solutes, such as sucrose [197]. This implies the presence in the cell of a selectively permeable membrane and protoplasts show most of the properties such as active transport, respiration, protein biosynthesis and even growth, shown by the original bacteria [112, 141, 197]. Osmotic lysis of these protoplasts leaves so-called ghosts with the appearance of membranes [198]. Despite this clear evidence and the introduction and application [146] of the Ryter and Kellenberger [145] technique, originally designed for fixing and staining preparations with buffered osmium tetroxide and uranyl acetate to preserve nuclear material, no distinct cytoplasmic membrane can be unequivocally distinguished in thin sections of whole bacteria made in the past [29, 30, 118, 119]. Paradoxically, internal membranes were seen then both in *B. subtilis* [146] and *Streptomyces coelicolor* [72]. The situation changed dramatically in the early 1960s when excellent and clear pictures of the cytoplasmic membrane in a variety of species of micro-organism were obtained, still by using the Ryter and Kellenberger [146] fixation method [71, 72, 189] but with improved embedding and sectioning techniques, together with electron microscopes of higher resolution. The membranes, after fixing with OsO_4 and staining with either uranyl or lead salts, resembled in most respects those from animal cells, consisting of dark inner and outer layers, separated by a much less darkly staining, middle layer. The total thickness of the structure was 7–8 nm. Difficulties have often been encountered subsequently in distinguishing the outer darker staining layer of the cytoplasmic membrane from the inner, also darkly staining layer of the cell wall of Gram-positive species. The membrane and the wall are indeed very closely applied in these species and plasmolysis is difficult. Claims have been made [162] for an asymmetry of the membrane in a number of bacteria with a relatively thicker dense layer on the outer, wall side, of the membrane than on the inner. Some credence was given to this view by the reversal of the asymmetry in the internal membranes of the mesosomes (see pp. 23-30). Omission of ions from the Ryter–Kellenberger fixative led to all the membranes staining symmetrically. It is difficult to appraise these observations on the cytoplasmic membrane because of the frequent lack of resolution between the outer edge of the membrane and the inner edge of the wall, but it may be that there is a greater concentration of some materials fixing uranyl ions or osmium on one face of the membrane than on the other. Because of the paucity of knowledge about the basic chemistry of the fixation and staining reactions, this conclusion from morphological observations should be regarded as most tentative.

All earlier authors remarked on this close apposition of the membrane and wall to be seen in sections of Gram-positive species. Even where gaps have been seen, these have been described as containing bridges of material running between the outer side of the membrane and the inner side of the wall. Pegs have also been described which appeared to run from the membrane into the depth of the wall [26]. In freeze-etched preparations from many species but possibly not all, strips or fibres of material run from the membrane to the wall (eg in *Sarcina ureae, B. anthracis*). Ring-like discs embedded in a sort of 'cloddy' material have been seen on the outer side of the cytoplasmic membranes of protoplasts of staphylococci after freeze-etching and are claimed to exist between the wall and the membrane [68]. Thus it seems that the cytoplasmic membrane and walls of Gram-positive bacteria may well be physically anchored together. In Gram-negative species, on the other hand, although also anchored together, attachment points seem to be much less frequent. When *E. coli* is plasmolysed by transfer into concentrated solutions, areas of close contact between the wall and membrane remain [8, 9], or long thin fibrils of cytoplasm extend between outer envelope and shrunken cytoplasm according to the degree of plasmolysis [17]. Evidence has been obtained suggesting that the areas of contact are also areas at which bacteriophages fix to the outside of the cell [9] and at which lipopolysaccharides are synthesized [10]. In freeze-etched preparations small islands of membrane have been found still associated with the wall, after the bulk of the latter had been sheared away [12]. In other Gram-negative species (eg *Rhizobium meliloti*) no attachments may exist between cytoplasmic membrane and wall [113]. Other workers have seen fibrils running between wall and membrane in freeze-etched preparations of unplasmolysed cells [50, 69, 110] but the latter two groups of workers favour the possibility that these are artefacts. Clearly, more work is needed to explore the extent and firmness of links between the wall and the plasma membrane in both Gram-positive and Gram-negative species. It becomes very important to say whether or not these structures are firmly fixed together when thinking about the growth and division of individual bacterial cells.

1.4.2 *Negative staining of cytoplasmic membranes*

Negative staining of mitochondrial membranes has shown the presence of stalked particles attached to the inner membrane. These particles have subsequently been shown to be enzyme proteins which can be isolated and purified. Abram [1] negatively stained membranes from *B. stearothermophilus* with phosphotungstate at pH 7.0–7.2. The membranes from this organism, isolated by a variety of methods including lysozyme action, heat shock and ultrasonic treatment, had the unusual property of partially retaining the rod shape of the original organism, even though the wall had been removed or separated in the absence of osmotic support. The cytoplasmic membranes were studded with particles about 6.5–8.5 nm in diameter that were deduced to be on the inward face of the membrane. At the highest magnifications it seemed possible that the particles were attached to the membrane by fine stalks of the order of 5 nm long. The attached particles could be particularly clearly seen at the folded edges of reversed fragments of membrane. Similar particles were seen on the membranes of *B. pumilis, B. licheniformis, B. brevis, B. circulans, E. coli, Proteus vulgaris* and *Shigella dysenteriae* Y6R. Subsequently, similar particles in the membranes of *S. faecalis* and *Micrococcus luteus* were found, stripped from the membranes and purified. They have been shown to be made of aggregates of sub-units of the enzyme ATPase in these organisms. Ways of inducing the enzyme to reattach to the membranes have been found (see Chapter 4.5.1 for further description of this work).

1.4.3 *Halobacteria*

These bacteria pose to modern scientists some of the same problems as were represented by fossils to eighteenth and nineteenth century geologists. They grow best in 5 M NaCl, fall to pieces in more reasonable environments, have patterned internal membranes, and produce a purple membrane containing only one ordered pigmented protein, very nearly related to the mammalian visual pigment, rhodopsin. Their walls contain no detectable peptidoglycan and seem composed mostly of proteins. Ultrastructural examination naturally posed some difficulties [21]. These were overcome by modification of orthodox procedures, such as having 25% NaCl present during prefixing with formaldehyde and post-fixing with $KMnO_4$ followed by staining with uranyl acetate again in the presence of 25% NaCl [173]. Alternatively, the bacteria could be prefixed with glutaraldehyde and postfixed with 1% OsO_4 in the presence of 25% NaCl, after which the structures were stable [34]. Like other bacteria, a bounding plasma membrane about 7.5–8 nm thick was found, overlaid by a wall. Unlike other Gram-negative species of bacteria, however, the ultrastructure of the wall was relatively homogeneous, though the outermost boundary stained rather more densely than the areas next to the cytoplasmic membrane. Negative staining [94], shadowing [107, 173] and freeze-etching [104] have all shown the presence of a highly ordered patterned layer with a periodicity of 12–15 nm on the outside of the walls. In some instances, this persists in isolated envelope preparations. Intracytoplasmic inclusion membranes have also been found; like gas vacuoles from other species [193, 194] they are patterned by parallel ribs 4 nm apart after negative staining and freeze-etching [173, 174]. These membranes are indeed collapsed gas-vacuoles [174]. Freeze-fracture of *H. halobium* [104] revealed a further regular pattern internally in the envelope that appeared to represent one of the two surfaces of the plasma membrane. A light-diffraction study suggested that this pattern was hexagonal.

Striated intracytoplasmic membranes were seen by negative staining or by freeze-etching. Lowering the NaCl concentration in which envelope preparations were suspended from 5 M to 1.0–1.2 M for *H. halobium* or to 2.2 M for *H. salinarium* led to a partial dissociation of the wall leaving a furry-appearing membrane [172, 173]. Complete removal of the salt by dialysing the envelopes against water left clean cytoplasmic membranes. In the supernatant after the cytoplasmic membranes of *H. halobacterium* had been removed, the purple membrane remained. This was subsequently isolated as circular or oval sheets showing no pattern after negative staining [174]. The envelopes of a wide variety of species of *Halobacteria* all appear to have a similar structure. No intracytoplasmic membranes could be detected in *H. salinarium* and no membranes, other than the cytoplasmic membrane itself, could be isolated from the supernatant fluids remaining after dissociation by distilled water [172]. There are indications that other organisms, for example, *Sulfolobus acidocaldarius* [203] that live in extreme environments may also have walls with quite different arrangements from the common run of Gram-positive and Gram-negative bacteria. So far, exploration of the envelopes of such species is limited.

1.5 Internal membranes

1.5.1 *Mesosomes*

The term mesosome was coined by Fitz-James [52] to describe the internal membranous bodies seen in *Bacilli* and, although other names such as chondrioids and plasmalemmasomes have been used, this term is now universal. The whole of the structures, whether demonstrated in thin sections or by negative staining, are usually referred to as mesosomes. These bodies have the form of either doughnuts [48] (American type!) or goblets [26] which are filled with tubules, vesicles or sheets of membrane (Fig. 1.9). No evidence yet exists to differentiate the outer wrapping membrane of the mesosome from the plasma or cytoplasmic membrane. The bag of the mesosomes is presumed to become part of the boundary membrane of the protoplast when the internal membranes are ejected. No fractionation procedure has yet been designed to separate the bag from the rest of the cytoplasmic membrane, once the protoplasts are made.

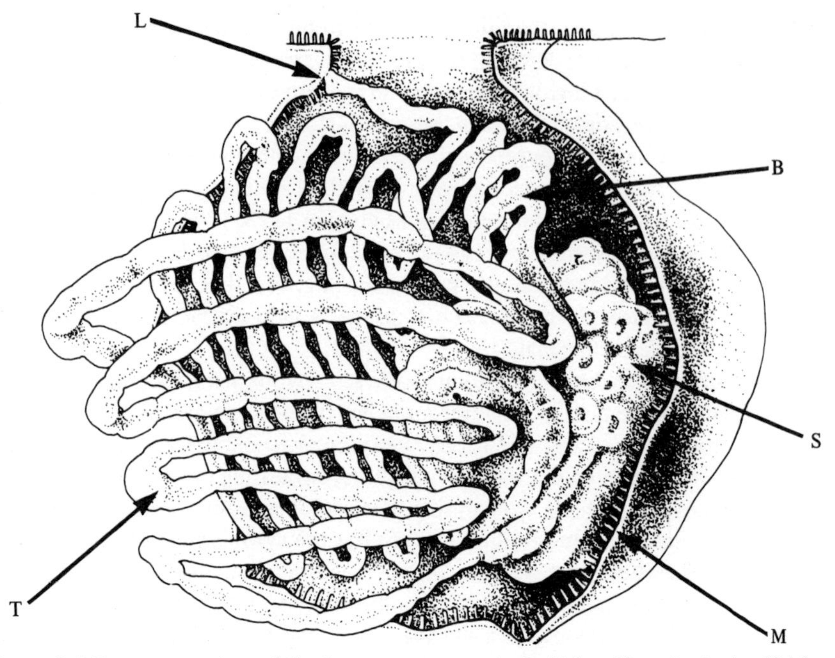

Figure 1.9 Reconstruction of the large mesosome in *B. licheniformis* strain 6346 made from serial sections and negatively-stained preparations. L, linkage of tubules (T) to the rough surfaces cytoplasmic membrane (M); B, a branched tubule; S, sheets of partially tubulated membrane (from [27]).

As has been pointed out already, internal membranes were seen in sections of *Bacilli* at a very early stage of the subject [30, 145] and very beautiful and well resolved pictures of them were published by van Iterson [189, 190]. Despite this, however, there is still some uncertainty about their structure, more about their development and almost complete ignorance about their true function. There is also considerable reason to doubt whether they exist at all in the growing organisms! As originally described [189], they appeared in section to be, more or less, large round and, therefore presumably, spherical bodies, usually attached to the septum as were the bodies first seen by Chapman and Hillier [30] and they were buried in the nuclear material. Serial sectioning [26] showed them indeed to be deformed spheres, at least in *Bacilli.* The inside of the sac formed by invagination of the cytoplasmic membrane, is often packed with what appear to be vesicles about 24 nm in diameter, each surrounded by a double membrane. When bacteria are placed in hypertonic solutions, these vesicles are extruded into a shallow depression of the cytoplasmic membrane under the wall [53, 143, 189, 190, 199]. If the walls are subsequently removed by digestion with an appropriate enzyme, such as lysozyme, the vesicles extend out from the protoplasts like long strings of pearls [147] or as tubules [26]. The number of mesosomes per cell in bacilli where they have been most studied was, and still is, open to much debate and varies from one [87, 88] to five [144]. A few factual comments may be made, which, although not settling the arguments about the structure and enumeration of mesosomes, at least point to the reasons for the difficulties. One problem frequently raised is whether the internal membranes exist as sheets of membranes arranged as concentric whorls [87, 152], as vesicles [147, 189] or as both [138]. The form in which the membranes are seen seems to depend strongly upon the ionic strength or osmolarity and the presence or absence of divalent cations in the fixative, at least when osmium tetroxide is used to study *B. licheniformis* or *B. subtilis* [26, 162]. If the amount of salt present is very low, or if Ca^{2+} is omitted, then sheets of whorled membrane are seen whereas with higher salt concentrations, or if sucrose is added instead and Ca^{2+} is present, tubules or vesicles are seen. The vesicles are in fact probably sections of tubes which are expanded at regular intervals along their length (see Fig. 1.10). As far as number is concerned, most investigators would agree that this depends upon the fixation and staining processes and whether the mesosomes are counted in serial sections or in negatively stained preparations. Examination of single, even exactly medial longitudinal sections, is of little value owing to the frequent grossly asymmetric positioning and shape of the mesosome bodies within the cell. Where serial sections have been made [27, 87, 88, 144] the number per cell is small. In negatively stained preparations, the situation is much more confusing. Again, a small number of large sized mesosomes is seen and if the internal structure of these is carefully compared with the arrangement of the tubules constructed from serial sections [27] there is good agreement. As well as the presence of large mesosomes however, numbers of other small mesosome-like areas frequently appear as a result of negative staining (see Fig. 1.11), and sometimes these can even make the cell look as though it is smitten by a pox [65]. What we have to ask ourselves is whether these bodies exist as such as in the cell or whether some of them may be the result of the staining or fixation procedures used. Nanninga [123, 124] for example, found that if he grew *B. subtilis* in glycerol, or if he first fixed with osmium tetroxide before freeze-etching, vesicular mesosomes of the sort previously demonstrated in this organism by both freeze-etching and by sectioning could be found. If, however, he grew the cells without glycerol and neither treated with osmium tetroxide fixative before freezing nor added glycerol as a cryostatic agent during freezing, he did not find significant numbers of mesosomes. Similar results for chemical fixation have been found for *S. faecalis* [85], *S. aureus* [55] and *B. licheniformis* strains 749c and 749 [64]. This apparent paradox may have been resolved by an experiment [86] in which the appearance of mesosomes was studied in cultures of *S. faecalis* growing at 37° C to which 2.5% glutaraldehyde (a common fixing agent) had been added. After various intervals, bacteria in the cultures were rapidly concentrated by filtration, frozen in liquid N_2 and freeze-fractured. In a parallel series of samples, radioactively-labelled amino acid was added to the cultures before the addition of glutaraldehyde. The frequency of appearance of mesosomes in the freeze-fractured material increased exactly in proportion to the binding of the amino acid, presumably by covalent cross-linkage to cell structures, by the glutaraldehyde. Before addition of glutaraldehyde, only about 2% of the cells showed the presence of mesosomes. Even without the addition of glutaraldehyde simply cooling the culture and filtering

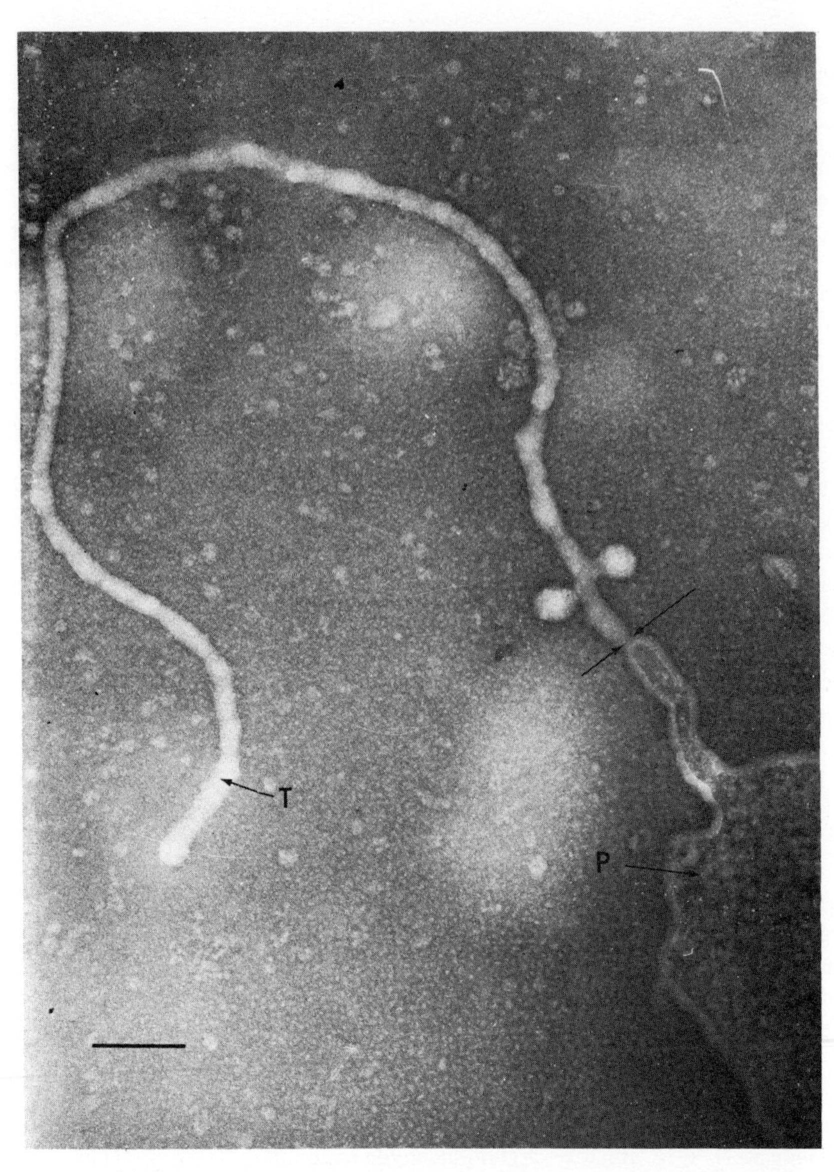

Figure 1.10 (a) Mesosomes extruded from protoplasts as tubules constricted at regular intervals. The preparations have been negatively stained with ammonium molybdate. The bacteria were from growth on a solid agar medium. The smooth surfaced tubule is seen still attached to the exceedingly rough cytoplasmic membrane in a protoplasm lysate. P, protoplast membrane; T, mesosome membrane.

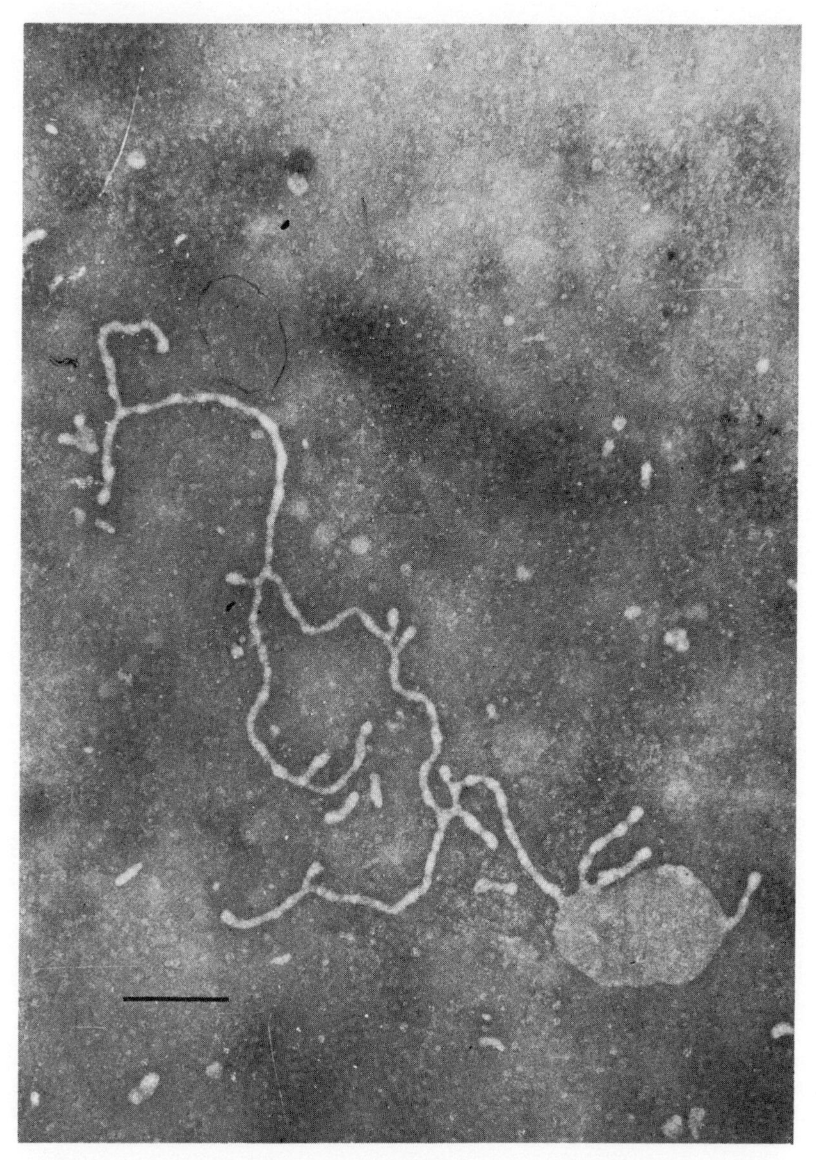

The arrow shows the junction between the two sorts of membrane. The bar is equivalent to 0.1 μm. (b) Showing the length and branching of the mesosomal tubules. The bar represents 0.2 μm. (This photograph was kindly supplied by Dr. I. D. J. Burdett, National Institute for Medical Research, Mill Hill, London).

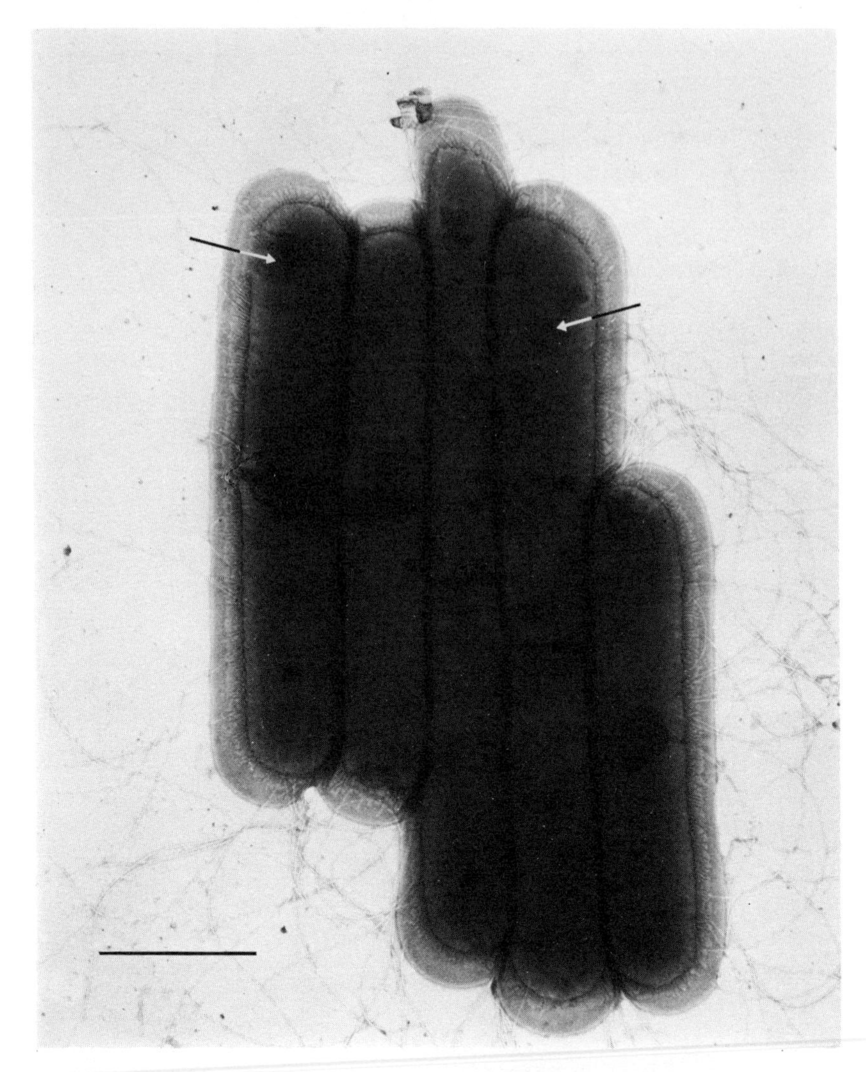

Figure 1.11 A 'raft' of *B. licheniformis* bacteria after negative staining with phosphotungstic acid. The multiple mesosomes are darker than the background and two are indicated by arrows. The bar is equivalent to 1 μm. (We are indebted to Dr. I. D. J. Burdett for this photograph).

off the cells increased this proportion to 15–20%. After 10 min treatment at 37° C or 60–120 min at 3° C the proportion increased to 70–80%. The implication of this work is that the mesosomes appear as a result of even mild insults to the cell. Nevertheless, they always occur at the same location and have about the same morphology. Thus they may be perhaps regarded as flags marking some abnormality of the membrane in the region of the forming septum that will eventually divide the cell, this being the situation in which the mesosome forms or is found. Thus, even if they consist of membrane components generated from the existing cytoplasmic membrane by the fixing techniques, they may still mark the position of specialized membrane.

In cocci, such as *S. faecalis* [83, 84], *S. aureus* or *Streptococcus pneumoniae* [184, 185] it is generally accepted that only one mesosome per cell exists although the structure of this would seem to be somewhat different from those in *Bacilli*. When orthodox means of fixing, staining and sectioning bacteria are used, mesosomes are found in representatives of a very wide range of bacterial species. Their presence has been recognized and their form described in over 40 species belonging to 20–25 genera. Gram-positive species have the most clearly demonstrable ones, but the statement that they either do not occur or are rare in Gram-negative species is scarcely true since descriptions exist for some 13 species of such bacteria. Indeed, in some, such as *Caulobacter, Chondrococcus, Chromobacterium violaceum* [142] and *Cytophaga aquatilis* [175] they are nearly as elaborate as in *Bacilli*. In a variety of other Gram-negative bacteria less orderly accumulations of membrane or membrane-like material have been reported. For example, using new modified fixation and staining procedures, various inclusions including multilayered, vesicular and tubular material of a possibly membranous nature were seen in *Ps. aeruginosa* [28, 90]. Polar accumulations of membrane occur at the end of the exponential phase of growth in *Glucenobacter*, and *Acetobacter suboxidans* [7, 35]. Large amounts of intra-cytoplasmic membrane accumulate as multilayered whorls in some temperature-sensitive mutants of *E. coli*, such as strain 0111a. These have been studied and their appearance reported both in sections, after negative staining and after freeze-etching [202].

1.5.2 *Comparative ultrastructure of cytoplasmic and mesosomal membranes*

A difference exists, at least in some species, between the appearance of the cytoplasmic membrane and the internal membranes of the mesosomes eg in *B. licheniformis*, after negative staining, freeze-etching or sectioning. The negatively-stained cytoplasmic membrane is studded with particles of about 11–11.5 nm in diameter, whereas the mesosomal membrane usually appears smooth and without particles (Fig. 1.12) [27]. This distinction can be seen very beautifully when protoplasts, together with their strings of extruded mesosomal tubules, are examined. Similar results have been found for a number of species such as *B. subtilis* [54, 147] *Listeria monocytogenes* [61] and *M. luteus* [130]. Conflicting results have been reported from the examination of replicas of freeze-etched preparations. In *B. licheniformis* they appeared smooth with occasional particles [63]; in *Nocardia* [13] they were usually smooth, although some particles could be seen in some of the vesicles; in *B. subtilis*, vesicular mesosomal membrane was smooth, although the lamellar membrane in this type of mesosome showed particles [138]. The vesicular mesosomes of *M. luteus* and the cytoplasmic membrane of the bacteria had a full complement of particles [130]. These differences may be of importance if the particles seen in the replicas of membranes after freeze-fracturing are indeed proteins as has been suggested, since it may indicate that the functions of mesosomes are not the same in all

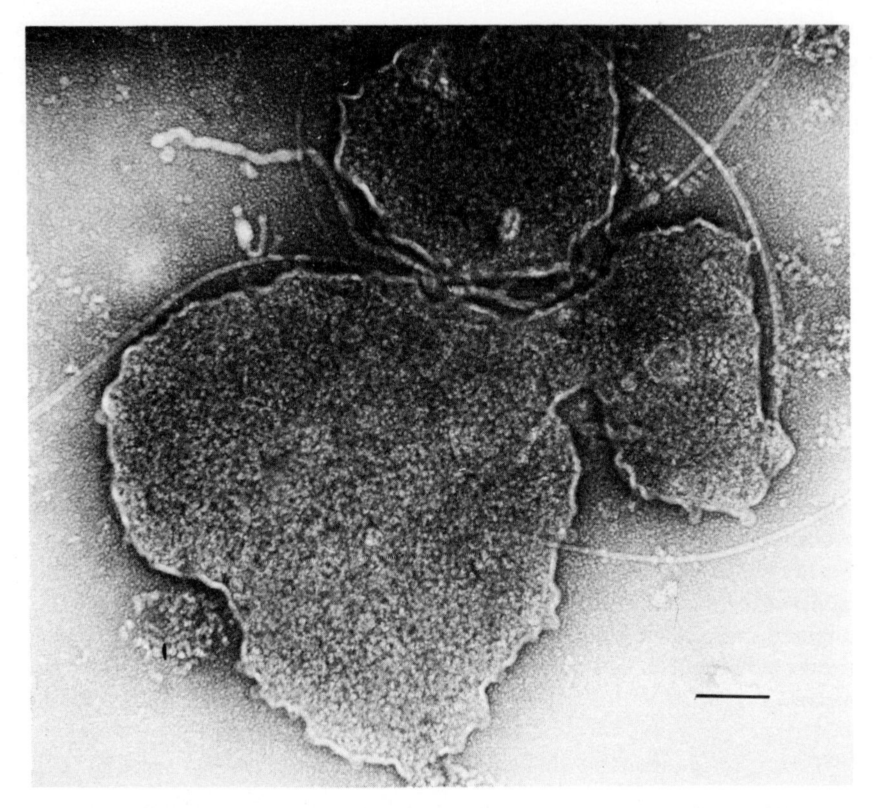

Figure 1.12 Tubular smooth surfaced mesosomal membrane, flagellae and the very rough cytoplasmic membrane as they appear after negatively staining a protoplast lysate from *Bacillus licheniformis* strain 6346. The bar is equivalent to 0.2 μm. (We are indebted to Dr. I. D. J. Burdett for this photograph).

organisms. Those mesosomal membranes that do not show particles in freeze-fractured preparations may indeed have rather few proteins buried within the lipid bilayer (see Chapter 4.9). Two comprehensive reviews on mesosomes have been published and can be consulted by those wishing for further information [62, 79].

1.6 Specialized membrane systems

Extensive internal membrane systems are present in some groups of micro-organisms such as the photosynthetic bacteria, nitrogen fixers, and methane oxidizing micro-organisms. Specialized membranes are also developed by sporulating bacteria. In some instances the appearance of these membranes is known to be correlated with the switching on of the particular function involved. In sporulation, for example, this goes without saying but a necessary connection is less obvious in processes such as nitrogen fixation or photosynthesis. In many ways the organisms in which function and ultrastructure are correlated are useful potential models for the study of membrane biogenesis.

1.6.1 *Photosynthetic bacteria*

Under conditions where photosynthesis occurs during growth, all members of the Chromatiaceae (purple sulphur bacteria) and the Rhodospirillaceae (non-sulphur purple bacteria) develop characteristic intracytoplasmic membrane systems [128]. The relationship between some of these systems that arise as invaginations of the cytoplasmic membranes of the micro-organisms, and the mesosomes just discussed is an exciting realm for speculation. It is tempting to wonder whether the mesosomes may not be vestigial evolutionary structures in view of their apparent lack of functional characteristics. In the green sulphur bacteria and the cyanobacteria, the photosynthetic apparatus does not appear to be associated either structurally or developmentally with intracytoplasmic membranes ultimately derived from the cytoplasmic membrane. A wide variety of organization in the photosynthetic apparatus has been described (see Fig. 1.13). One species investigated in great detail is *Rhodospirillum rubrum.* Vatter and Wolfe's [191] early pictures of sections of this organism growing photosynthetically, showed it to be packed with circular membranous profiles 50–100 nm in diameter. Bacteria grown heterotrophically in the dark did not show such profiles. These were presumably the particulate photosynthetic material that had already been isolated from sonically disrupted cells by Schachman *et al.* [153]. Further examination [92] suggested that the circular profiles originally seen, represented the cross-sections of tubules constricted at intervals, Fig. 1.14 shows a reconstruction of the ideas put forward by Holt and Marr [92]. Interconnection between the vesicles was also supported by Hickman and Frenkel [80, 81]. When ghosts are formed from spheroplasts, the intracytoplasmic membrane apparatus has adherent groups of vesicles [19] which contain most of the photosynthetic pigments, thus explaining the physiological activity of whole ghosts [187]. The vesicles or tubules it should be noted, remain on the inside of the cytoplasmic membrane in the ghosts and are not extruded when spheroplasts are made. Thus they behave differently from the mesosomes of Gram-positive bacteria which are extruded from protoplasts (see p. 24). Interconnection between the vesicles cannot be seen in section but this is perhaps not surprising. Although the hypothesis that the chromatophores arise from the cytoplasmic membrane, an idea originally put forward by Cohen-Bazire and Kunisawa [36], appears to be generally acceptable and well supported [80, 81], considerable arguments have been advanced for either tubule-like structures, interconnected vesicles or quite separate vesicles [66]. So far no complete decision between these possibilities can be made, more especially in the light of the known ability of membranous bodies to adhere or dissociate under, for example, the influence of altered divalent cation concentration. Arguments have also been raised against the bounding membrane of chromatophores being of normal dimensions of about 7–8 nm. A value of 4 nm has been found by some workers [19], who refer to it as a non-unit membrane, whilst others (eg [128])

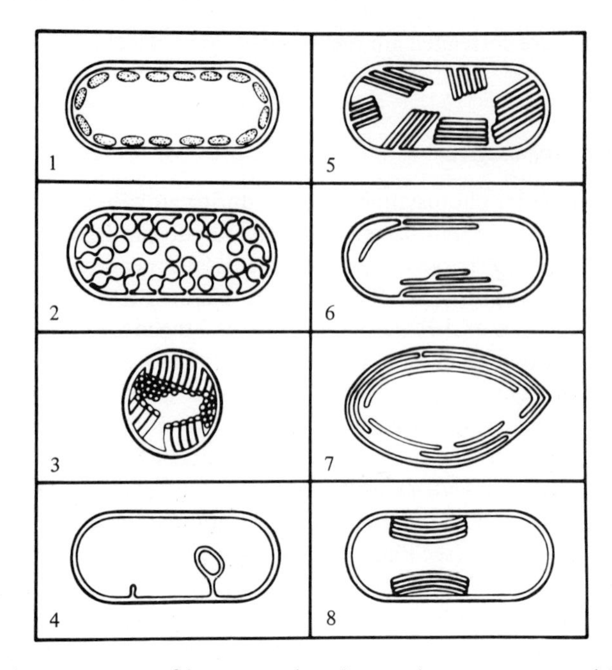

Figure 1.13 Arrangements of intracytoplasmic membrane structures (chromatophores or thylakoids) of photosynthetic bacteria. (1) Chlorobium vesicles in the green sulphur bacteria *Chlorobium limicola, C. thiosulphatophilum, C. phaeobacteroides, C. phaeovibrioides, Pelodictyon clathratiforme, P. aggregatum, Chloropseudomonas ethylicum* and *Prothecochloris aestuarii.* (2) Vesicular membrane systems in *Rhodospirillum rubrum, R. spheroides, R. capsulatei,* Chromatium strain D, *C. okenii, C. weissei, Thiospirillum jenense* and *T. roseopersicina.* (3) Tubular membrane systems in *Thiocapsa pfennigii* (*Thiococcus* sp). (4) Single small membrane invaginations in *Rhodopseudomonas gelatinosa* and *Rhodospirillum tenue.* (5) Stacks of short double membranes bound to the cytoplasmic membrane in *Rhodospirillum molischianum, R. fulvum, Ectothiorhodospira* and *R. photometricum.* (6) Parallel lamellae underlying and continuous with the cytoplasmic membrane in *Rhodopseudomonas palustris, Rps. acidophila.* (7) Concentric layers of double membranes in cells of *Rhodomicrobium vannielii.* (8) Grana-like stacks of double membrane in *Rps. viridis* (from [128]).

suggest more normal values of 7 or 8 nm.

As can be seen from Fig. 1.14, the arrangement of the intracytoplasmic membrane system in *Rh. rubrum* and *Rh. spheroides* is not shared by numbers of other photosynthetic bacteria. For example, the obligately photosynthetic *Rhodomicrobium vannielii* forms looped lamellae of membranes which run around the cell, presumably as shells underneath the wall [20] ; these loops are open at the ends of the cells. The membranes have a thickness of 6.5–8.7 nm and the photosynthetic pigments are fixed to them. This organism also has the characteristic of forming filaments connecting the cells together with septa cutting off the cells; the photosynthetic membranes do not seem to extend into the filaments. When isolated, the stacks of

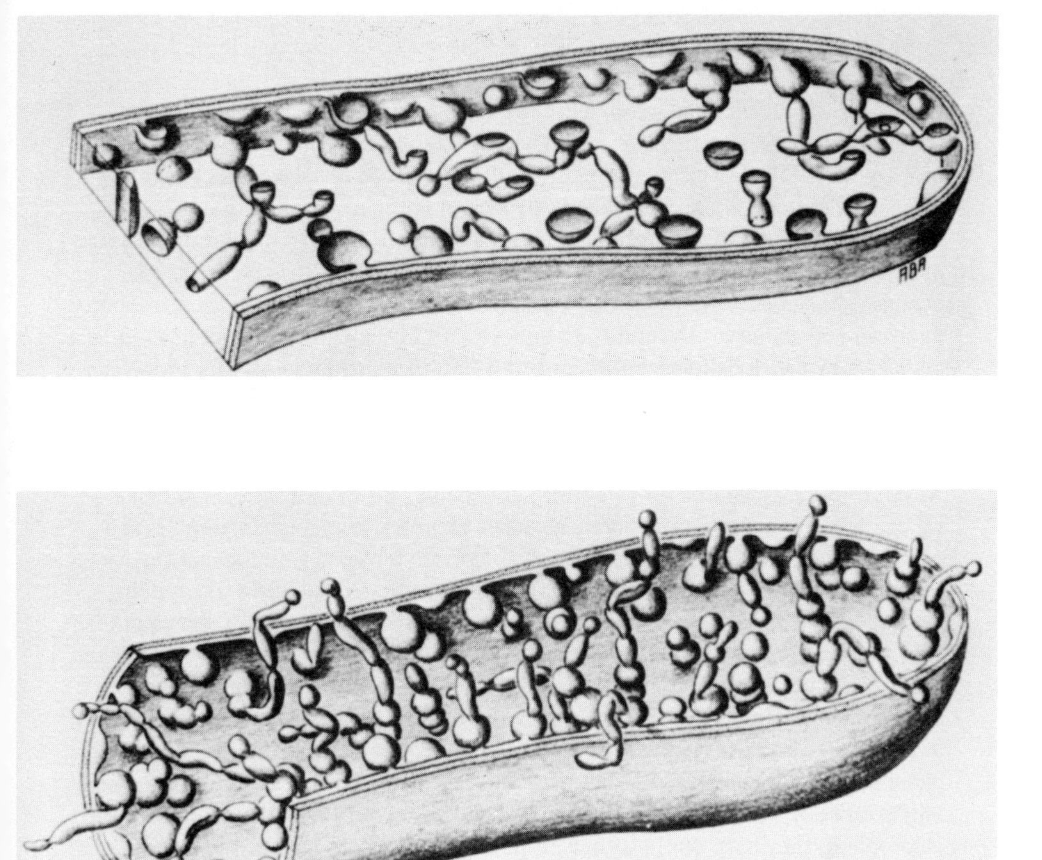

Figure 1.14 Hypothetical three-dimensional representation of the internal membrane systems of *Rh. rubrum* (from [92]).

membrane appear as various sized large vesicles up to 0.5–1.0 nm in diameter. Other species and genera form grana-like stacks of membrane. The growth of such stacks that originate from the cytoplasmic membrane has been studied in *Rhodospirillum molischianum* [81]. In cells from young cultures only, single internal membranes could be seen and these increased as the cultures grew older. It is interesting to note that lysed cultures showed no stacked membrane and only variable sized vesicles, a point to be remembered in interpreting the studies of the nitrifying organism *Azotobacter agilis* (see below). The photosynthetic apparatus in the green sulphur bacteria, the *Chlorobiaceae* and the *Chlorofexaceae*, is organized somewhat differently with chlorobium vesicles firmly attached to the cytoplasmic membrane [37]. The cytoplasmic membrane itself contains a bacteriochlorophyll called a_p whilst the chlorobium vesicles are the major organelles for gathering light-energy in the cell and contain bacteriochlorophylls *c, d* or *e* [134]. Since the vesicles and the membrane to which they are attached contain different pigments, their interrelation 'ontologically' is obviously of great interest.

The cyanobacteria (blue-green algae) however, constitute according to expert witnesses 'the largest, most diverse and widely distributed group of photosynthetic prokaryotes', [171]. This group traditionally assigned to the algae has come to be recognized structurally and biochemically as more properly prokaryotic being essentially photosynthetic bacteria but with a degree of internal organization that is more complicated than that in most other bacteria. The wall structure of these organisms is similar to those of Gram-negative bacteria, with an inner layer of peptidoglycan and an outer membrane. The photosynthetic apparatus is contained in bodies known as thylakoids which consist of flattened membranous sacs and which are the sites of chlorophyll and carotenoids. The major photopigments, the phycobiliproteins, however, are on the outside of the thylakoids and are contained in organelles known as phycobilisomes, attached in regular arrays to their surfaces. Apart from the apparatus for gathering light-energy, the cyanobacteria contain a variety of other inclusions including glycogen granules, cyanophycin granules and carboxysomes, while some also have poly-β-hydroxybutyrate granules and gas-vacuoles. All these inclusions are likely to have membrane structures surrounding them (see pp. 37–38). The total organization of most cyanobacteria is shown in Fig. 1.15 taken from Stanier and Cohen-Bazire [171]. The individual cells of some cyanobacteria can be arranged in groups, called trichomes, surrounded by an external sheet which is often fibrous in nature. Some cyanobacteria can form heterocysts and in these there is very extensive structural and molecular reorganization when compared with the vegetative cell. Their envelopes are completely resistant to lysozyme unlike those of the vegetative cell, which are sensitive. Although they contain some chlorophyll, they have no thylakoids or phycobilisomes or many of the inclusions already mentioned. Those interested in further information about cyanobacteria should consult the review by Stanier and Cohen-Bazire [171].

1.6.2 *The nitrogen-fixing and nitrifying organisms*

Increased attention to the distribution of the ability to fix free N_2 among bacterial species has greatly widened our horizons. A list of names [135] covers some 20 genera from species of *E. coli,* through *Bacilli* and photosynthetic bacteria to the more commonly regarded nitrogen-fixers, such as *Azotobacter, Nostocaceae,* and the symbiotic *Rhizobia.* Whether micro-organisms such as *Bacilli* or *E. coli* develop specialized membranes when growing under conditions favouring the fixation of N_2 is not known, at least to the present authors. Since this chapter is to be regarded only as a guide for the reader of the remainder of this book, we shall confine ourselves to describing the ultrastructure of a few of the more widely recognized nitrogen-fixing bacteria.

Early pictures [132] of whole and partially disrupted cells of *A. agilis* suggested

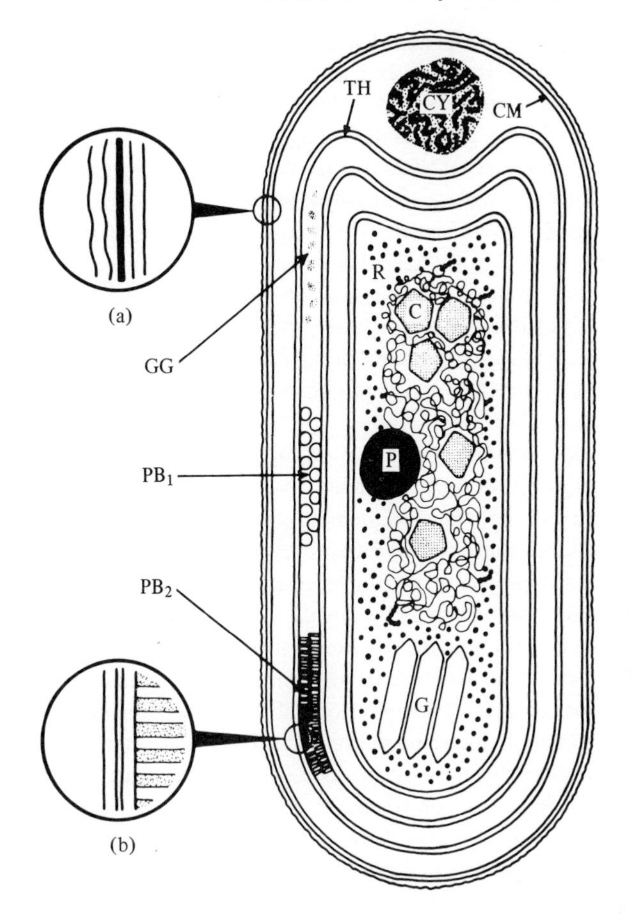

Figure 1.15 A diagram of the vegetative cell of cyanobacteria. CM, cytoplasmic membrane; TH, thylakoid; PB_1, PB_2, front and side views of two rows of phycobilisomes attached to thylakoids; GG, glycogen granules, CY, cyanophycin granule, P, polyphosphate granule, R, 70S ribosomes, C, carboxysome surrounded by nucleoplasm, G, gas vacuole. (a) Enlarged view of cell envelope showing the outer membrane, peptidoglycan layer and cytoplasmic membrane. (b) Enlarged view of thylakoid showing paired unit membranes with attached phycobilisomes (from [171]).

the presence of large numbers of vesicles and tubules. Unfortunately, clear pictures of the membranes could only be obtained after the cytoplasmic contents had been partially emptied by rupture of the envelope. Later studies [89, 129] have confirmed and extended this early work. Some difficulty was still experienced due to the electron opacity of the cytoplasm, but Oppenheim and Marcus [129] showed pictures of sections of the bacteria grown both with gaseous N_2 and with three sources of fixed nitrogen. Of these only the bacteria growing on gaseous N_2 showed the presence of large numbers of small, oval or round bodies empty of cytoplasm.

Osmotically shocked spheroplasts of such bacteria were filled with round and deformed empty membranous bodies [89]. In bacteria grown on nitrate, none of these intracytoplasmic inclusions were present and in those grown on ammonia [89, 129] or casein hydrolysate [129] there were far fewer. Whether these latter vesicles had been present in the original cells or were formed by the well recognized abilities of other membranes in the cells, such as the cytoplasmic membrane, to vesiculate, is not clear from the pictures. Studies of particulate nitrogenase from the bacteria grown under different conditions makes it likely that the membranes are related to the nitrogenase activity rather than to the respiratory activity [89]. The vesicles have many similarities in appearance to those in *Rh. rubrum,* some strains (at least) of which also fix N_2 as well as fixing CO_2 by photosynthesis [135]. Whether in *Azotobacter* they are derived from and attached to the cytoplasmic membrane as Oppenheim and Marcus [129] believe is not, however, clear.

Among the nitrifying organisms that oxidize ammonia to nitrite the internal membrane system of *N. oceanus* has been described with particular clarity [122, 139, 195]. Stacks of triple layered membranes could be seen in sections (Fig. 1.7). Examination of these stacks showed that they were likely to consist of large collapsed vesicles, each lamella of the stacks consisting, therefore, of two normal unit membranes closely opposed. It seems highly likely that these large vesicles are derived from the cytoplasmic membrane by invagination. On the other hand, they form a highly specialized structure which divides when the cells divide. Organisms involved in the next logical stage of autotrophic nitrogen assimilation, such as *Nitrobacter agilis* that oxidizes nitrite to nitrate, also have stacked membranes, though in a somewhat different arrangement (Fig. 1.16) being closely opposed to the cytoplasmic membrane, and to each other over the poles of the cell [122]. Essential aspects of the oxidative systems are membrane bound in these organisms but definitive proof that the intracytoplasmic membranes are specialized for this task has not yet been obtained.

1.6.3 *Hydrocarbon-utilizing bacteria*

The arrangement of the internal membranes of the methane utilizing bacteria [44, 136, 196, 204] seems remarkably similar to that found in the nitrifying *Nitrosomonas* group (see p. 000). Each 'membrane' in the stack consists of two closely opposed unit-membranes, probably resulting from the collapse of a vesicle, rather than being made from two open sheets. Evidence from freeze-etched preparations [196] suggests that as with *Nitrosomonas oceanicus* the stack of membranes divides with the cell, implying a perpetuated structure rather than, as with the photosynthetic species, renewed development in the freshly born cell. The origin and significance of the internal structures seen [44] in negatively-stained preparations of organisms using the higher hydrocarbons (eg propane) as a carbon source would seem to be more problematical. They are rather similar to the mesosomes of Gram-positive bacteria (see pp. 23–29) and the apparent membrane invaginations are reminiscent of those shown by Ghosh *et al.* [63] in *B. licheniformis.* Sections of organisms

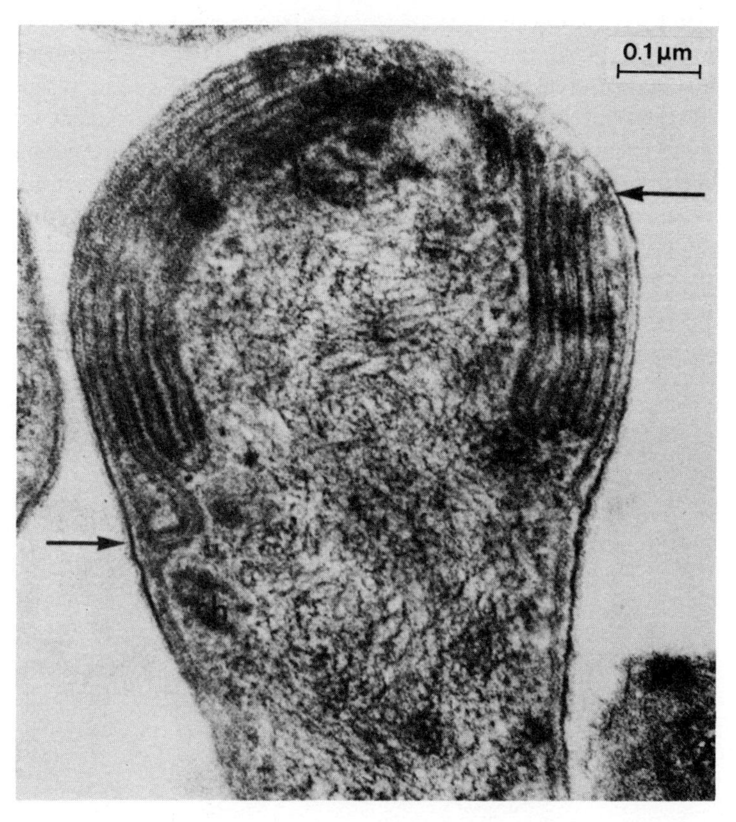

Figure 1.16 Transverse section of *Nitrobacter agilis* showing the infolded plasma membrane so that it forms layers over the end of the bacteria (from [122]).

grown on still longer chain hydrocarbons showed the presence of no internal membranes.

1.6.4 *Membrane-enclosed inclusion bodies*

Inclusions within bacteria, apart from those already discussed, are varied and widespread. Among them may be mentioned the polyglucoside amylopectin granules found in a number of species of *Clostridia*, poly-β-hydroxybutyrate granules in numbers of genera including *Bacilli*, sulphur globules in the *Thiorhodaceae*, polygonal carboxysomes containing ribulose-5-phosphate carboxylase in *Thiobacillus neapolitanus* [159] and gas-vacuoles in aquatic prokaryotes including extreme halophiles and some cyanobacteria [193, 194]. Work on all these bodies has been summarized [160]. In many instances sectioned bacteria show that the included material is packaged in thin membranous material of the order of 2–4 nm thick, compared with the thickness of 7–8 nm for normal bilayer unit membranes. When these thin membranes have been isolated and studied (see Chapter 2.4.2), they have a different composition from the thicker 'true' membranes from other situations within the cell. For example, those from the sulphur particles in *Chromatium vinosum* seem to consist of a single protein possibly together with a very small amount of lipid [155]. Empty ghosts left after the sulphur had spilt out of them appeared made up from globular components 2.5 nm in diameter [127]. The thin membranes forming the gas vacuoles in five species of prokaryote have been shown [194] to consist of a single protein. In only one species of *Halobacterium* is the presence of more than one protein proven [49]. There have also been claims for the presence in the vesicle coat of small amounts of phosphorus and carbohydrate [106].

1.6.5 *Membrane and spore formation*

A number of genera, prominent among them being *Bacilli* and *Clostridia,* form heat-

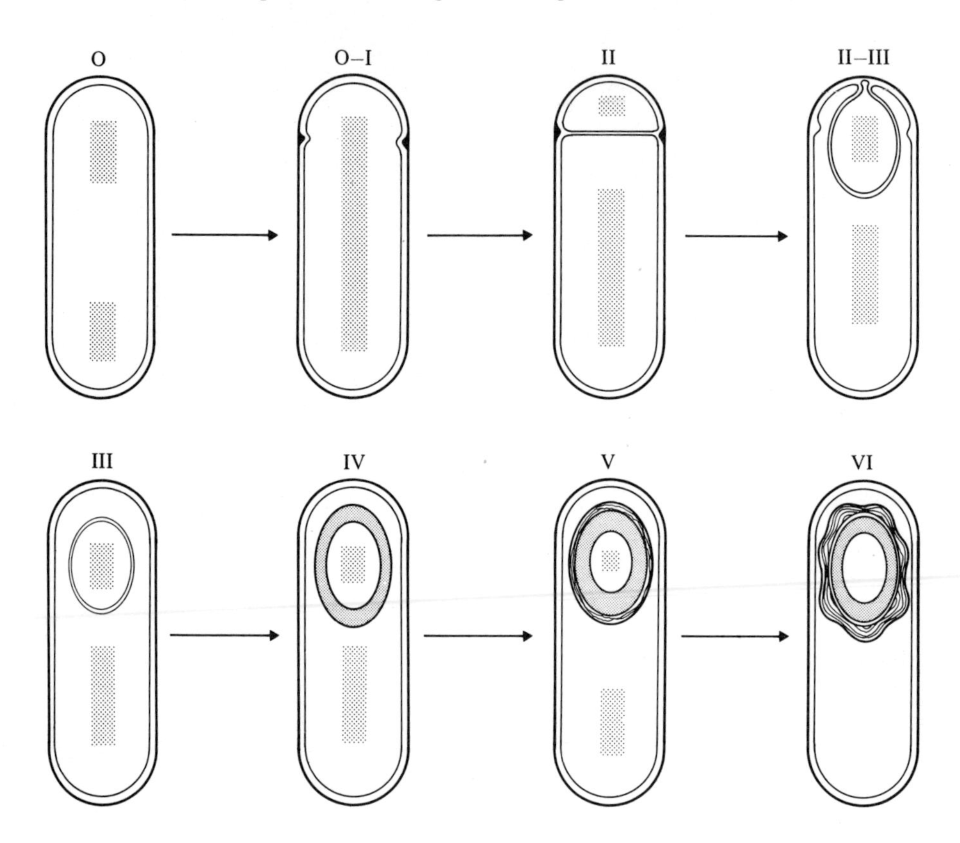

Figure 1.17 Generalized diagram of the major morphological events occurring during sporulation. The stages are indicated by Roman numerals. The shaded areas are to show the disposition of the nuclear material (from [47a]).

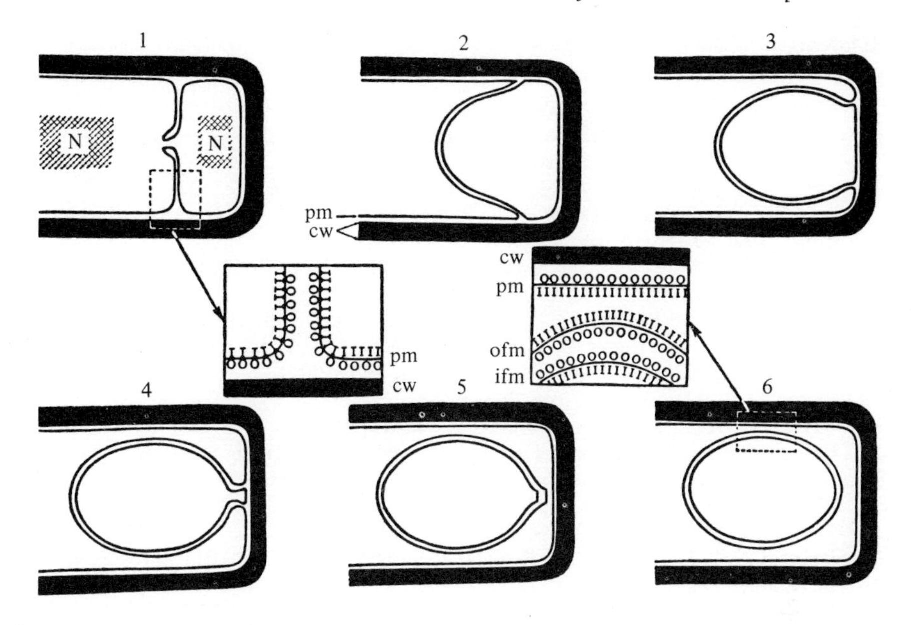

Figure 1.18 The formation of the forespore during stages II, III of sporulation (see Fig. 1.17). The inset diagrams illustrate the 'sidedness' of the membranes (O, outside; I, inside). In the main diagram N, nuclear material; PM, cytoplasmic membrane; CW, cell wall; OFM, outer forespore membrane; IFM, inner forespore membrane (from [47a]).

resistant spores. The spores are known to have specialized peptidoglycan and coats of protein as well as membranes. The recognised stages in sporulation are shown in Fig. 1.17 [154]. At an early stage (stage II) the formation of a double membranous septum near a ring-like intrusion of the wall to one end of the cell, looking like spikes in section, occurs at the edge of the membrane. The double membrane appears closely associated with a large mesosome of the type commonly found in *Bacilli* [137] and a similar situation has been found in *Bacillus cereus* [207] although the mesosome is less easy to see in these latter pictures. The initial spore septum or membrane extends by what appears to be a process of ballooning towards the centre of the cell carrying with it a mesosome. Finally, this double membrane is detached from the ring of wall and cytoplasmic membrane and becomes a separate oval shaped prespore (Fig. 1.18) with a mesosome under it. The material between the double membranes grows in thickness to form the spore cortex, known to contain a high proportion of peptidoglycan of unique structure. Meanwhile, the membrane around the outside of the forming spore grows in thickness and begins to show layering and to become very dense, eventually forming the spore coat. The original inside membrane which surrounded the prespore becomes very indistinct and difficult to see. Eventually, both the outer spore membrane and the inner layer adopt the angular corrugated appearance of the fully mature spore.

References

1. Abram, D. (1965) *J. Bact.* **89**, 855–73.
2. Aebi, U., Smith, P. R., Dubochet, J., Henry, C. and Kellenberger, E. (1973) *J. supramolec. Struct.* **1**, 498–522.
3. Amako, K. and Umeda, A. (1977) *J. gen. Microbiol.* **98**, 297–9.
4. Archibald, A. R., Armstrong, J. J., Baddiley, J. and Hay, J. B. (1961) *Nature, Lond.* **191**, 570–2.
5. Asbell, M. A. and Eagon, R. G. (1966) *Biochem. biophys. Res. Commun.* **22**, 664–71.
6. Baker, R. F. and Pease, D. C. (1949) *Nature, Lond.* **163**, 282.
7. Batzing, B. L. and Claus, G. W. (1973) *J. Bact.* **113**, 1455–61.
8. Bayer, M. E. (1968a) *J. gen. Microbiol.* **53**, 395–404.
9. Bayer, M. E. (1968b) *J. Virol.* **2**, 346–56.
10. Bayer, M. E. (1974) *Ann. N. Y. Acad. Sci.* **235**, 6–28.
11. Bayer, M. E. and Anderson, T. F. (1965) *Proc. natn. Acad. Sci. U.S.A.* **54**, 1592–9.
12. Bayer, M. E. and Reinsen, C. C. (1970) *J. Bact.* **101**, 304–313.
13. Beamen, B. L. and Shankel, D. M. (1969) *J. Bact.* **99**, 876–84.
14. Beveridge, T. J. (1978) *Can. J. Microbiol.* **24**, 89–104.
15. Beveridge, T. J. and Murray, R. G. E. (1974) *J. Bact.* **119**, 1019–31.
16. Beveridge, T. J. and Murray, R. G. E. (1975) *J. Bact.* **124**, 1529–44.
17. Birdsell, D. C. and Cota-Robles, E. H. (1967) *J. Bact.* **93**, 427–37.
18. Bladen, H. A. and Mergenhagen, S. E. (1964) *J. Bact.* **88**, 1482–92.
19. Boatman, E. S. (1964) *J. Cell Biol.* **20**, 297–311.
20. Boatman, E. S. and Douglas, H. G. (1961) *J. biophys. biochem. Cytol.* **11**, 460–80.
21. Brown, A. D. and Shorey, C. D. (1963) *J. Cell Biol.* **18**, 681–9.
22. Buckmire, F. L. A. and Murray, R. G. E. (1973) *Can. J. Microbiol.* **19**, 59–66.
23. Buckmire, F. L. A. and Murray, R. G. E. (1976) *J. Bact.* **125**, 290–99.
24. Burdett, I. D. J. and Murray, R. G. E. (1974) *J. Bact.* **119**, 303–324.
25. Burdett, I. D. J. and Murray, R. G. E. (1974) *J. Bact.* **119**, 1039–56.
26. Burdett, I. D. J. and Rogers, H. J. (1970) *J. Ultrastruct. Res.* **30**, 354–67.
27. Burdett, I. D. J. and Rogers, H. J. (1972) *J. Ultrastruct. Res.* **38**, 113–33.
28. Carrick, L. and Berk, R. S. (1971) *J. Bact.* **106**, 250–56.
29. Chapman, G. B. (1959) *J. Bact.* **78**, 96–104.
30. Chapman, G. B. and Hillier, J. (1953) *J. Bact.* **66**, 362–73.
31. Chapman, J. A., Murray, R. G. E. and Salton, M. R. J. (1963) *Proc. R. Soc. Lond., Ser B* **158b**, 498–513.
32. Cheng, K-J. and Costerton, J. W. (1977) *J. Bact.* **129**, 1506–512.
33. Chester, I. R. and Murray, R. G. E. (1978) *J. Bact.* **133**, 932–41.
34. Cho, K. Y., Doy, C. H. and Mercer, E. H. (1967) *J. Bact.* **94**, 196-201.
35. Claus, G. W., Batzing, B. L., Baker, C. A. and Goebel, E. M. (1975) *J. Bact.* **123**, 1169–83.
36. Cohen-Bazire, G. and Kunisawa, R. (1963) *J. Cell Biol.* **16**, 401–16.
37. Cohen-Bazire, G., Pfennig, G. and Kunisawa, R. (1964) *Arch. Mikrobiol.* **72**, 115–34.
38. Cole, R. M., Popkin, T. J., Boylan, R. J. and Mendelson, N. H. (1970) *J. Bact.* **103**, 793–810.
39. Cooper, P. D., Rowley, D. and Dawson, I. M. (1949) *Nature, Lond.* **164**, 842–3.
40. Costerton, J. W., Damgaard, H. N. and Cheng, K-J. (1974) *J. Bact.* **118**, 1132–43.
41. Costerton, J. W., Ingram, J. M. and Cheng, K-J. (1974) *Bact. Rev.* **38**, 87–110.
42. Cox, S. T. and Eagon, R. G. (1968) *Can. J. Microbiol.* **14**, 913–22.

43. Crowther, R. A. and Sleytr, U. B. (1977) *J. Ultrastruct. Res.* **58**, 41–9.
44. Davies, S. L. and Whittenbury, R. (1970) *J. gen. Microbiol.* **61**, 227–52.
45. Dawson, I. M. (1949) *Symp. Soc. gen. Microbiol.* 119–28.
46. De Petris, S. (1965) *J. Ultrastruct. Res.* **12**, 247–62.
47. De Petris, S. (1967) *J. Ultrastruct. Res.* **19**, 43–83.
47a. Ellar, D. J. (1978) In *Relations between Structure and Function in the Prokaryotic Cell,* 28th Symp. Soc. Gen. Microbiol., eds. Stanier, R. Y., Rogers, H. J. and Ward, J. B., pp. 295–325, Cambridge University Press.
48. Ellar, D. J., Lundgren, D. G. and Slepecky, R. A. (1967) *J. Bact.* **94**, 1189–1205.
49. Falkenberg, P. (1974) *Thesis Institut for Teknisk Biokjemi Norges, Tekiske Hogskole Universitetet, Trondheim.*
50. Fiil, A. and Branton, D. (1969) *J. Bact.* **98**, 1320–27.
51. Finch, J. T., Klug, A. and Nermut, M. V. (1967) *J. Cell Sci.* **2**, 587–90.
52. Fitz-James, P. C. (1960) *J. biophys. biochem. Cytol.* **8**, 507–528.
53. Fitz-James, P. C. (1964) *J. Bact.* **87**, 1483–91.
54. Fitz-James, P. C. (1968) *Microbial Protoplasts, Spheroplasts and L-forms* ed. Guze, L. B., p. 124. Baltimore: Williams and Wilkins, Co.
55. Fooke-Achterrath, M., Lickfield, K. G., Reusch, V. M. Jr. Aebi, U., Tschöpe, U. and Menge, B. (1974) *J. Ultrastruct. Res.* **49**, 270–85.
56. Forsberg, C. W., Costerton, J. W. and MacLeod, R. A. (1970) *J. Bact.* **104**, 1338–53.
57. Frank, H. and Dekegel, D. (1965) *Zentbl. Bakt. ParasitKde. Abt.* **198**, 81–3.
58. Frank, H., Lefort, M. and Martin, H. H. (1962) *Z. Naturf. Teil B* **17B**, 262–8.
59. Garland, J. M. (1974) In *Electron Microscopy and Cytochemistry,* eds. Wisse, E. and Daems, W. Th. pp. 303–309, Amsterdam: Elsevier.
60. Garland, J. M., Archibald, A. R. and Baddiley, J. (1975) *J. gen. Microbiol.* **89**, 73–86.
61. Ghosh, B. K. and Murray, R. G. E. (1969) *J. Bact.* **97**, 426–40.
62. Ghosh, B. K. (1974) *Sub-Cell Biochem.* **3**, 311–67.
63. Ghosh, B. K., Lampen, J. O. and Remsen, C. C. (1969) *J. Bact.* **100**, 1002–1009.
64. Ghosh, B. K. and Nanninga, N. (1976) *J. Ultrastruct. Res.* **56**, 107–120.
65. Ghosh, B. K., Sargent, M. G. and Lampen, J. O. (1968) *J. Bact.* **96**, 1314–28.
66. Gibson, J. D. (1965) *J. Bact.* **90**, 1059–72.
67. Giesbrecht, P. and Ruska, H. (1968) *Klin. Wschr.* **11**, 575–82.
68. Giesbrecht, P., Wecke, J., Reinicke, B. and Tesche, B. (1977) *Arch. Mikrobiol.* **115**, 25–35.
69. Gilleland, H. E., Stinnett, J. D., Roth, I. L. and Eagon, R. R. (1973) *J. Bact.* **113**, 417–32.
70. Glauert, A. M. (1962) *Br. med. Bull.* **18**, 245–50.
71. Glauert, A. M. and Hopwood, D. A. (1959) *J. biophys. biochem. Cytol.* **6**, 515–16.
72. Glauert, A. M. and Hopwood, D. A. (1960) *J. biophys. biochem. Cytol.* **7**, 479–86.
73. Glauert, A. M. and Thornley, M. J. (1969) *A. Rev. Microbiol.* **23**, 159–98.
74. Glauert, A. M. and Thornley, M. J. (1973) *John Innes Symp.* **1**, 297–305.
75. Goundry, J., Davison, A. L., Archibald, A. R. and Baddiley, J. (1967) *Biochem. J.* **104**, 1c.
76. Gram, C. (1884) *Fortschr. Med.* **ii**, 185–90.
77. Gray, G. W. and Wilkinson, S. G. (1965) *J. appl. Bact.* **28**, 153–67.
78. Gray, G. W. and Wilkinson, S. G. (1965) *J. gen. Microbiol.* **39**, 385–99.
79. Greenawalt, J. W. and Whiteside, T. L. (1975) *Bact. Rev.* **39**, 405–463.
80. Hickman, D. D. and Frenkel, A. W. (1965) *J. Cell Biol.* **25**, 261–78.

81. Hickman, D. D. and Frenkel, A. W. (1965) *J. Cell Biol.* **25**, 279–91.
82. Higgins, M. L. (1976) *J. Bact.* **127**, 1337–45.
83. Higgins, M. L. and Shockman, G. D. (1970) *J. Ultrastruct. Res.* **31**, 247–59.
84. Higgins, M. L. and Daneo-Moore, L. (1970) *J. Ultrastruct. Res.* **31**, 260–71.
85. Higgins, M. L. and Daneo-Moore, L. (1974) *J. Cell Biol.* **61**, 288–300.
86. Higgins, M. L., Tsien, H. C. and Daneo-Moore, L. (1976) *J. Bact.* **127**, 1519–23.
87. Highton, P. J. (1970) *J. Ultrastruct. Res.* **31**, 247–59.
88. Highton, P. J. (1970) *J. Ultrastruct. Res.* **31**, 260–71.
89. Hill, S., Drozd, J. W. and Postgate, J. R. (1972) *J. appl. chem. Biotechnol.* **22**, 541–58.
90. Hoffmann, H. P., Geftic, S. G., Heymann, H. and Adair, F. W. (1973) *J. Bact.* **114**, 434–8.
91. Holdsworth, E. S. (1952) *Biochim. biophys. Acta* **8**, 110–12.
92. Holt, S. C. and Marr, A. G. (1965) *J. Bact.* **89**, 1402–1412.
93. Holt, S. C. and Leadbetter, E. R. (1969) *Bact. Rev.* **33**, 346–78.
94. Houwink, A. L. (1953) *Biochim. biophys. Acta* **10**, 360–8.
95. Houwink, A. L. (1956) *J. gen. Microbiol.* **15**, 146–150.
96. Howard, L. and Tipper, D. J. (1973) *J. Bact.* **113**, 1491–1504.
97. Hughes, R. C. (1970) *Biochem. J.* **119**, 849–60.
98. Hughes, R. C., Tanner, P. J. and Stokes, E. (1967) *Biochem. J.* **120**, 159–70.
99. Hunter, M. I. S., Muir, D. D. M. and Thirkell, D. (1973) *J. Bact.* **116**, 483–7.
100. Ingram, L. O., Thurston, E. L. and van Baalen, C. (1972) *Arch. Mikrobiol.* **81**, 1–12.
101. Irvin, R. T., Chatterjee, A. K. and Sanderson, K. E. (1975) *J. Bact.* **124**, 930–41.
102. Kay, D. and Warren, S. C. V. (1968) *Biochem. J.* **109**, 819–24.
103. Kellenberger, E. and Ryter, A. (1958) *J. biophys. biochem. Cytol.* **4**, 323–5.
104. Kirk, R. G. and Ginzburg, M. (1972) *J. Ultrastruct. Res.* **41**, 80–94.
105. Knox, K. W., Vesk, M. and Work, E. (1966) *J. Bact.* **92**, 1206–1217.
106. Krantz, M. J. and Ballou, C. E. (1973) *J. Bact.* **114**, 1058–67.
107. Kushner, D. J., Bayley, S. T., Boring, J., Kates, M. and Gibbons, N. E. (1964) *Can. J. Microbiol.* **10**, 483–97.
108. Lancy, P. and Murray, R. G. E. (1978) *Can. J. Microbiol.* **24**, 162–76.
109. Leive, L. (1965) *Biochem. biophys. Res. Commun.* **21**, 290–6.
110. Lundgren, D. G. and Remsen, C. (1966) *J. Bact.* **91**, 2096–2102.
111. McCarty, M (1952) *J. exp. Med.* **96**, 569–80.
112. McQuillen, K. (1960) In *The Bacteria I*, eds. Gunsalus, I. C. and Stanier, R. C. pp. 249–359. New York and London Academic Press.
113. Mackenzie, C. R., Vail, W. J. and Jordan, D, C. (1973) *J. Bact.* **113**, 387–93.
114. Meadow, P. M., Wells, P. L., Salkinoja-Salonen, M. and Nurmiaho, E-L. (1978) *J. gen. Microbiol.* **105**, 23–8.
115. Mergenhagen, S. E., Bladen, H. A. and Hsu, K. C. (1966) *Ann. N. Y. Acad. Sci.* **133**, 279–91.
116. Mitchell, P. and Moyle, J. (1951) *J. gen. Microbiol.* **5**, 981–92.
117. Millward, G. R. and Reaveley, D. A. (1974) *J. Ultrastruct. Res.* **46**, 309–326.
118. Murray, R. G. E. (1957) *Can. J. Microbiol.* **3**, 531–2.
119. Murray, R. G. E. (1960) In *The Bacteria I*, eds. Gunsalus, I. C. and Stanier, R. Y. p. 35, New York and London: Academic Press.
120. Murray, R. G. E. (1963) *Can. J. Microbiol.* **9**, 381–92.
121. Murray, R. G. E., Steed, P. and Elson, H. E. (1965) *Can. J. Microbiol.* **11**, 547–60.
122. Murray, R. G. E. and Watson, S. W. (1965) *J. Bact.* **89**, 1594–1609.

123. Nanninga, N. (1969) *J. Cell Biol.* **42**, 733–44.
124. Nanninga, N. (1970) *7th Congress Internat. de Microscop. Electronique, Grenoble* **111**, 349–50.
125. Nermut, M. V. (1967) *J. gen. Microbiol.* **49**, 503–512.
126. Nermut, M. V. and Murray, R. G. E. (1967) *J. Bact.* **93**, 1949–65.
127. Nicolson, G. L. and Schmidt, G. L. (1971) *J. Bact.* **105**, 1142–8.
128. Oelze, J. and Drews, G. (1972) *Biochem. biophys. Acta* **265**, 209–239.
129. Oppenheim, J. and Marcus, L. (1970) *J. Bact.* **101**, 286–91.
130. Owen, P. and Freer, J. H. (1972) *Biochem. J.* **129**, 907–917.
131. Pangborn, J. and Starr, M. P. (1966) *J. Bact.* **91**, 2025–36.
132. Pangborn, J., Marr, A. G. and Robrish, S. A. (1962) *J. Bact.* **84**, 669–78.
133. Pate, J. L. and Ordal, E. (1969) *J. Cell Biol.* **35**, 37–51.
134. Pfennig, N. (1977) *Adv. Microbiol.* **31**, 275–90.
135. Postgate, J. R. (1974) In *Evolution in the Microbial World,* 24th Symp. Soc. Gen. Microbiol., eds. Carlile, M. J. and Skehel, J. J. pp. 263–92, Cambridge University Press.
136. Proctor, H. M., Norris, J. R. and Ribbons, D. (1969) *J. appl. Bact.* **32**, 118–23.
137. Remsen, C. C. (1966) *Arch. Mikrobiol.* **54**, 266–75.
138. Remsen, C. C. (1968) *Arch. Mikrobiol.* **61**, 40–47.
139. Remsen, C. C. and Watson, S. W. (1972) *Int. Rev. Cytol.* **33**, 253–96.
140. Remsen, C. C., Valois, F. W. and Watson, S. W. (1967) *J. Bact.* **94**, 422–33.
141. Roth, G. S. and Daneo-Moore, L. (1971) *J. Bact.* **108**, 980–85.
142. Rucinsky, T. E. and Cota-Robles, E. H. (1974) *J. Bact.* **118**, 717–24.
143. Ryter, A. (1968) *Bact. Rev.* **32**, 39–54.
144. Ryter, A. and Jacob, F. (1966) *Ann. Inst. Past.* **107**, 384–400.
145. Ryter, A. and Kellenberger, E. (1958) *J. biophys. biochem. Cytol.* **4**, 671–8.
146. Ryter, A., Kellenberger, E., Birch-Anderson, A. and Maaløe, O. (1958) *Z. Naturf.* **13B**, 597–605.
147. Ryter, A., Frehel, C. and Ferrandes, B. (1967) *C. r. Acad. Sci. Paris* **265**, 1259–62.
148. Salton, M. R. J. (1961) *Bact. Rev.* **25**, 77–99.
149. Salton, M. R. J. (1964) In *The Bacterial Cell Wall,* Amsterdam: Elsevier.
150. Salton, M. R. J. and Horne, R. W. (1951) *Biochim. biophys. Acta* **7**, 19–42.
151. Salton, M. R. J. and Williams, R. C. (1954) *Biochim. biophys. Acta* **14**, 455–8.
152. Salton, M. R. J. and Chapman, J. A. (1962) *J. Ultrastruct. Res.* **6**, 489–98.
153. Schachman, Hu., Pardee, A. B. and Stanier, R. Y. (1952) *Archs. Biochem. Biophys.* **38**, 245–60.
154. Schaeffer, P. (1969) *Bact. Rev.* **33**, 48–71.
155. Schmidt, G. L., Nicolson, G. L. and Kamen, M. D. (1971) *J. Bact.* **105**, 1137–41.
156. Shands, J. W. (1965) *J. Bact.* **90**, 266–70.
157. Shands, J. W. (1966) *Ann. N. Y. Acad. Sci.* **133**, 292–8.
158. Sharpe, E. M., Brock, J. H. and Phillips, B. A. (1975) *J. gen. Microbiol.* **88**, 355–63.
159. Shively, J. M., Ball, F., Brown, D. H. and Saunders, R. E. (1973) *Science* **182**, 584–6.
160. Shively, J. M. (1974) *A. Rev. Microbiol.* **28**, 167–87.
161. Shockman, G. D. (1965) *Bact. Rev.* **29**, 345–58.
162. Silva, M. T. (1971) *J. Microscopy* **93**, 227–32.
163. Sleytr, U. B. (1975) *Nature, Lond.* **157**, 400–402.
164. Sleytr, U. B. (1976) *J. Ultrastruct. Res.* **55**, 360–77.
165. Sleytr, U. B. (1978) *Int. Rev. Cytol.* **53**, 1–64.
166. Sleytr, U. B. and Glauert, A. M. (1976) *J. Bact.* **126**, 869–82.

167. Sleytr, U. B. and Thornley, M. J. (1973) *J. Bact.* **116**, 1383–97.
168. Sleytr, U. B. and Thorne, K. J. I. (1976) *J. Bact.* **126**, 377–83.
169. Sleytr, U. B., Thornley, M. J. and Glauert, A. M. (1974) *J. Bact.* **118**, 693–707.
170. Smit, J., Kamio, Y. and Nikaido, H. (1975) *J. Bact.* **124**, 942–58.
171. Stanier, R. Y. and Cohen-Bazire, G. (1977) *A. Rev. Microbiol.* **31**, 225–74.
172. Steensland, H. and Larsen, H. (1969) *J. gen. Microbiol.* **55**, 325–36.
173. Stoeckenius, W. and Rowen, R. (1967) *J. Cell Biol.* **34**, 365–93.
174. Stoeckenius, W. and Kunau, W. H. (1968) *J. Cell Biol.* **38**, 337–57.
175. Strohl, W. R. (1979) *J. gen. Microbiol.* **112**, 261–8.
176. Swanson, J. (1972) *J. exp. Med.* **136**, 1258–71.
177. Swanson, J. and Gotschlich, E. C. (1973) *J. exp. Med.* **138**, 245–58.
178. Thorne, K. (1977) *Biol. Rev.* **52**, 219–34.
179. Thorne, K. J. I., Thornley, M. J. and Glauert, A. M. (1973) *J. Bact.* **116**, 410–17.
180. Thorne, K. J. I., Thornley, M. J., Naisbitt, P. and Glauert, A. M. (1975) *Biochim. biophys. Acta* **389**, 97–116.
181. Thornley, M. J. and Glauert, A. M. (1968) *J. Cell Sci.* **3**, 273–94.
182. Thornley, M. J., Glauert, A. M. and Sleytr, U. B. (1973) *J. Bact.* **114**, 1294–1308.
183. Thornley, M. J., Thorne, K. J. I. and Glauert, A. M. (1974) *J. Bact.* **118**, 654–62.
184. Tomasz, A., Jamieson, J. D. and Ottolenghi, E. (1964) *J. Cell Biol.* **22**, 453–64.
185. Tomasz, A., Westphal, M., Briles, E. and Fletcher, P. (1975) *J. Supramolec. Struct.* **3**, 1–16.
186. Tsien, H. C., Shockman, G. D. and Higgins, M. L. (1978) *J. Bact.* **133**, 372–86.
187. Tuttle, A. L. and Gest, H. (1959) *Proc. natn. Acad. Sci. U.S.A.* **45**, 1261–9.
188. Van Gool, A. P. and Nanninga, N. (1972) *J. Bact.* **108**, 474–81.
189. Van Iterson, W. (1961) *J. biophys. biochem. Cytol.* **9**, 183–92.
190. Van Iterson, W. (1965) *Bact. Rev.* **29**, 299–325.
191. Vatter, A. E. and Wolfe, R. S. (1958) *J. Bact.* **75**, 480–88.
192. Wagner, M., Wagner, B. and Ryc, M. (1978) *J. gen. Microbiol.* **108**, 283–94.
193. Walsby, A. E. (1972) *Bact. Rev.* **36**, 1–32.
194. Walsby, A. E. (1978) In *Relations between Structure and Function in the Prokaryotic Cell*, 28th Symp. Soc. Gen. Microbiol., eds. Stanier, R. Y., Rogers, H. J. and Ward, J. B. pp. 327–58, Cambridge University Press.
195. Watson, S. W. and Remsen, C. C. (1970) *J. Ultrastruct. Res.* **33**, 148–60.
196. Weaver, T. L. and Dugan, P. R. (1975) *J. Bact.* **121**, 704–710.
197. Weibull, C. (1953) *J. Bact.* **66**, 688–95.
198. Weibull, C. (1956) *Expl. Cell Res.* **10**, 214–21.
199. Weibull, C. (1965) *J. Bact.* **89**, 1151–7.
200. Weibull, C. (1973) *J. Ultrastruct. Res.* **43**, 150–59.
201. Weidel, W., Frank, H. and Martin, H. H. (1960) *J. gen. Microbiol.* **22**, 158–66.
202. Weigand, R. A., Holt, S. C., Shively, J. M., Decker, G. L. and Greenawalt, J. W. (1973) *J. Bact.* **113**, 433–44.
203. Weiss, R. L. (1974) *J. Bact.* **118**, 275–84.
204. Whittenbury, R., Phillips, K. C. and Wilkinson, J. F. (1970) *J. gen. Microbiol.* **61**, 205–227.
205. Work, E. (1967) *Folia microbiol., Prague* **12**, 220–6.
206. Work, E., Knox, K. W. and Vesk, M. (1966) *Ann. N. Y. Acad. Sci.* **133**, 438–49.
207. Young, F. E. (1964) *J. Bact.* **88**, 242–54.

2 Isolation of walls and membranes

2.1 Introduction

As has been described in Chapter 1, the structure of the outer layers of Gram-positive bacteria is reasonably simple as far as can be seen by present techniques. They consist of walls of variable, but always considerable, thickness and of no very clearly defined internal structure. In some species, these walls may bear on their outer surfaces patterned layers of protein which are easily removed, each unit of which consists of a number of sub-units. Underlying the wall and probably physically anchored to it, is the cytoplasmic membrane. This may be invaginated at a limited number of places to form the mesosomes. These sacs or goblets are filled with membranous tubules which are extruded between the wall and membrane when the bacteria are placed in hypertonic solutions. Gram-negative cells, on the other hand, differ considerably in the complexity of their envelopes. Several layers and two membranes are present and the mesosomal membranes are usually less well developed. Among such bacteria are those more specialized micro-organisms that can fix inorganic nitrogen, photosynthesize and use such unlikely food as gaseous H_2 and methane. In these groups well developed membrane systems are present, often as packed lamellae or vesicles. Mutants of very common Gram-negative bacteria such as *Escherichia coli* have also been isolated that under appropriate conditions, seem to accumulate large amounts of extra internal membrane.

Methods for preparation of membranes and walls have undoubtedly been dictated by these differences in structure of organisms but also by the use to which the preparations are to be put. It would seem obvious, for example, that the most expedient method for obtaining preparations of cytoplasmic membrane from lysozyme-sensitive Gram-positive bacteria would be to first remove the walls with lysozyme or some other suitable enzyme preparation, either in an osmotically protective solution so that protoplasts are first formed, or in its absence when insoluble membrane particles result. In fact membranes isolated by these means are not infrequently deficient in enzymic functions present in those made by mechanically disrupting the bacteria and isolating membrane particles by differential centrifugation. Indeed, in studying the biosynthesis of cell wall polymers by some species, only the fragments of membrane still firmly adhering to the wall seem fully functional. Protoplast membranes sometimes need a special 'conditioning' treatment to restore some of their activities. For this, and other reasons, particulate membranous material from whole bacteria has usually been used in biosynthetic studies with Gram-positive bacteria even though easy means of hydrolysing away wall material were available. For chemical and many other studies, on the other hand, it has been usual first to strip away the wall.

Methods for the preparation of bacterial walls, have nevertheless been dictated largely by the differences in the organization of the Gram-positive and Gram-negative envelopes. From Gram-positive species the procedures can be relatively straightforward once enough is known to avoid the pitfalls of damage by the action of autolytic enzymes, on the one hand, and by hydrolysis of bonds between polymers, on the other, resulting from over vigorous treatment with acidic or alkaline solutions. For Gram-negative species, the evolution of procedures for the separation of the various layers outside the cytoplasmic membrane has been slower and only recently achieved with apparent success and reproducibility. Good methods now exist for isolating the outer membrane and the peptidoglycan layer but other components have not yet been so clearly isolated, and methods for obtaining the cytoplasmic membrane are not entirely satisfactory.

2.2 Isolation of walls and membranes from Gram-positive species

2.2.1 *Disruption processes and methods*

The processes for breaking open micro-organisms by exposing them to sonic or ultrasonic energy, grinding with sand or alumina, shearing by forcing suspensions through slits and orifices or shaking them with small glass beads were, of course, first developed in order to study and isolate enzymes and components of the cytoplasm. In order to do this, the suspensions of broken cells were centrifuged and the insoluble material thrown away as 'debris'. With the availability of the electron microscope and means for the chemical analysis of small amounts of material, a number of workers [21, 57, 67, 88, 89] examined this 'debris'. All of the groups of workers ruptured the bacteria by shaking suspensions of them with small glass beads of 0.1 to 0.15 mm in diameter, sometimes known as ballotini. This method was first shown to kill bacteria by King and Alexander [52].

Many of these methods have also been found to be applicable to eukaryote micro-organisms such as microfungi and much of what will now be said about bacteria can be applied to these. After the micro-organisms in a suspension have been ruptured in this way, walls are, of course, still mixed with all the cytoplasmic contents, including the membranes, and with any capsular substances not removed by previously washing the whole bacteria. The pictures taken by Dawson [21] of the whole deposit centrifuged down from ruptured suspensions suggested fairly extensive contamination of the cell walls. Salton and Horne [89] separated the walls by first removing denser, more bulky, whole cells by centrifuging the suspension at low speed (about 1000 *g*) for 10 min. The insoluble material not deposited at this stage was then obtained by further centrifuging at about 10 000 *g* for 10 min. This deposit was then washed repeatedly, first with either 0.1 M phosphate buffer or

M NaCl solution and then with water. From some Gram-positive bacteria at least this procedure can yield cell walls that are relatively free from cytoplasmic and membrane material, as judged by chemical analysis and appearance under the electron microscope. This early method for preparing bacterial walls is described because, although numerous modifications have been introduced, it remains at the basis of most of those used subsequently.

In the original method, as described by Salton and Horne [89], the bacteria were shaken with ballotini in a Mickle [62] disintegrator. This machine vibrates at the frequency of the alternating mains, and uses two small bottles that can contain about 10 ml of suspension and 10 ml of ballotini beads. The suspension can be at a density of about 10 to 20 mg dry weight/ml, so that about 0.2 to 0.4 g dry weight of bacteria can be dealt with at one time. No cooling is provided and the machine has to be run in a cold room; even then, relatively short bursts of a few minutes are desirable if the temperature of the suspensions is to be kept near to 0° C.

Several of the modifications that have been introduced are designed to increase the mass of bacteria that can be dealt with, to decrease the time involved and improve the control of temperature of the suspension during disruption. Two pieces of apparatus can be mentioned. One is the shaker centrifuge head designed by Shockman *et al.* [97]. This apparatus is made to be fitted to the spindle of a refrigerated International Centrifuge, and can take up to 6 g dry wt of bacteria. It consists of four steel pots, fitted to horizontal arms, and when the apparatus is fitted over the centrifuge spindle and the centrifuge run at a speed of about 2000 rev/min a fast, complicated gyratory-vertical motion is given to the pots. A suspension of *Streptococci*, for example, can be almost completely disrupted in about 10 min. Some bacteria may require very much longer. A major disadvantage of this apparatus is the damage to the centrifuge. Modifications to the original design have been suggested [96] to reduce this disadvantage but the method has never been very widely used.

Another piece of apparatus that is widely used was first designed by Merkeschlager *et al.* [61], and is manufactured by B. Braun Melsungen Apparatebau Melsungen, West Germany. In this technique, too, the bacteria are shaken with ballotini, but the suspension is contained in a single glass bottle that is held in a horizontal position and vibrated in a rotary fashion on its axis by a powerful motor. Cooling is achieved by a flow of liquid carbon dioxide, which evaporates between concentric metal tubes around the bottle. In a careful study by Huff *et al.* [43], *Staphylococci* were found to be over 90% disrupted during a 3 min treatment and completely broken in 5 min (see Table 2.1). The temperature did not rise above 4° C. This machine requires a certain amount of artistry in adjusting the flow of liquid carbon dioxide so that the contents of the bottle do not freeze. Otherwise the apparatus is very satisfactory.

A different type of energy has been imparted to mixtures of glass beads and bacteria in suspensions, by putting the mixture in a Waring blender. This method has been used by a number of workers and is useful when large masses of wall are required for certain purposes, but has the disadvantage that walls are frequently contaminated with particles of metal from the blades of the blender.

In all the methods in which glass beads are used, care must be taken over the

Table 2.1 The disruption of *Staphylococcus aureus* during shaking with glass beads in a Braun disintegrator (after [43]).

Shaking time (min)	*Percentage*			
	Viable counts	*Cells remaining*	*Walls released*	*Optical density*
0	100	100	0	100
0.6	1.8	32	53	50
1.5	1.4×10^{-2}	15	83	27
2.6	3×10^{-4}	4.1	93	13
4.0	1×10^{-4}	1.1	96	7.7
5.0	7×10^{-5}	0.2	100	6.3

nature of the beads. Some samples of glass liberate considerable amounts of alkali, producing suspensions, after bacterial disruption, with pH values as high as 10 [103]. High as well as low pH values can solubilize specific components from walls. It has been recommended [103] that the glass from which the beads are made should have a refractive index of 1.68. Alternatively, the use of styrene-divinylbenzene copolymer beads has been suggested for rupture of *Streptococcus faecalis* [98]. Less heat is generated during disruption using such plastic beads and no pH change is observed. Whatever beads are used, finely powdered bead material is produced during disruption which can contaminate the wall preparations. Glass powder considerably upsets chemical analysis not only by its weight but by its content of heavy metals and silica, particularly important if interest is centred on the ionic composition of the walls.

Energy for disruption has also been supplied by sonic or ultrasonic waves to suspensions of micro-organisms, sometimes mixed with glass beads, early examples of such procedures are: Salton [85], Bosco [5], Roberson and Schwab [81] and Ikawa and Snell [47]. This method is obviously attractive for the breakage of large masses of bacteria. It suffers rather more, perhaps, than other methods from the disadvantage that the walls when emptied of their cytoplasmic contents become further fragmented. Most methods of disrupting bacteria lead to the formation of some 'soluble' wall constituents either because the walls are mechanically broken down so that the smaller pieces cannot be deposited at the centrifugal forces used for the preparation of the wall or because lytic enzymes act sufficiently on the walls, despite the use of low temperatures. Ultra-sound seems to be particularly effective at 'solubilizing' walls.

All the methods so far described have depended upon exposure of all the particles in suspension to the kinetic energy imparted to the system for the whole time of treatment. Thus the wall from an organism, disrupted in the first second of the process, will continue to be exposed to the same energy throughout the treatment, whereas walls from organisms, disrupted immediately before the energy supply is disconnected, will not. Two groups of methods that have been used avoid this disadvantage. Both depend on the sheer forces arising from the passage of the suspension through slits or orifices.

The first group includes the presses designed by Hughes [44] and by Edebo [24]. In the former a cell paste, usually frozen solid, is forced through a slit between two blocks of metal by a series of heavy blows, usually applied by a fly-press, hitting the top of a piston working in a cylinder containing the paste. In the latter a stiff liquid paste of the micro-organisms at about 0° C is forced through an orifice by hydraulic pressure, again applied to a piston working in a cylinder. This press is so designed that the material can be repeatedly passed back and forth through the orifice. The Hughes [44] press may break the cell without stripping off the cytoplasmic membrane from the walls, at least in some organisms [46, 45] as judged by the retention of enzymes in wall preparations from Pseudomonads. The Edebo or so-called X-press [24], has been used successfully by a number of workers to prepare walls from a variety of Gram-positive and Gram-negative organisms.

The second group of methods that forces the suspension of organisms through an orifice uses a pressure cell in which a relatively thin cell suspension (1-3 mg/ml dry weight) is forced under pressure through an adjustable needle valve. Various fairly elementary modifications of the so-called French pressure cell have been used, if not always described in the literature. All of these pieces of apparatus are based on the original described by Milner *et al.* [63], but a very much more elaborate apparatus was designed by Ribi *et al.* [80] which can break fairly large masses of bacteria although the rather dilute suspension employed sets limits on its use; the expense of the commercially available apparatus sets others. The suspension is forced, under a pressure of 35 000–40 000 lb/in^2 ($1\ lb/in^2 = 6894.76\ Pa$), through a needle valve cooled to 2° C by CO_2. These pressure cell methods seem to have special advantages over others for preparing walls from some genera of micro-organisms such as the *Mycobacteria,* and for making so-called wall membrane preparations (see Section 8.6.1) in which membrane is left attached to the wall. Another method which is particularly satisfactory for the latter purpose is to grind a stiff paste of bacteria in a cold agate pestle and mortar with Al_2O_3 [58].

Other methods for the disruption of bacteria, such as rapid heating, cell autolysis, osmotic lysis and internal pressure disruption have not been described in detail because, although they have been used, they are either limited in use to a few species of bacteria (e.g. osmotic lysis and rapid heating), or have grave and obvious disadvantages (e.g. cell autolysis) or they have not yet been tried sufficiently to pronounce upon (eg internal pressure disruption); a few words might be said about the method of cell autolysis. The method of incubating a suspension of micro-organisms under toluene for a time and then isolating the insoluble material was used in some of the early work on walls. As knowledge has progressed, however, it has become clear that, from the point of view of interpretation, this is a highly dangerous procedure. Many, perhaps all, bacteria contain lytic enzymes that bring about changes, particularly in the peptidoglycan part of the wall (see Section 11.2). Thus, material isolated by this method is almost certainly considerably changed from the initial material present in the cell. As knowledge has increased, efforts have moved in quite the opposite direction, so that considerable precautions are usually taken to inactivate the lytic enzymes at as early a stage as is practicable during the isolation of wall material. This can be done by a variety of methods discussed below.

2.2.2 *Isolation of the walls*

Once the organisms have been broken open, the problem arises of separating the wall material from soluble material and fragments of the remainder of the cell. This raises the question of what we mean by 'bacterial cell wall'. If we apply too drastic conditions for separation, we may remove substances that should properly be regarded as part of the wall. On the other hand, if too little treatment is done, contaminating substances from other parts of the cell may appear in the wall fraction. There is no ideal answer to this dilemma, since there is no absolute definition of a cell wall. Some would say that the outer layers of the cell are organized continuously from the outer edge of any capsule surrounding the cell to the inner edge of the cytoplasmic membrane, perhaps even including the mesosomes. In all organisms, this situation can be simplified by regarding the wall as an organelle that has the same shape as the original cell and which, in section, corresponds in appearance with the region of the original undisrupted cell called the wall. By modern critical methods, the wall and the underlying cytoplasmic membrane can usually be distinguished fairly clearly in sections of the whole organism. In most Gram-positive organisms, the chemistry of the materials is also characteristic (see Chapters 6 and 7.1), although this can become a circular argument if ideas about the chemistry become too firmly entrenched in the worker's mind. Finally, in those Gram-positive cells sensitive to lytic enzymes such as egg-white lysozyme, that are known to remove the wall by specifically attacking certain bonds in the peptidoglycan (see Section 11.2), the wall and cytoplasmic membrane can be differentiated clearly.

The early method already described that involves shaking suspensions of bacteria with glass beads and washing the walls, is still among the most satisfactory when one is interested in obtaining fairly large masses of wall material in as nearly complete a state as possible, providing certain precautions are taken. One of the most important of them has already been mentioned, that is to avoid the action of the autolytic enzymes nearly always present in cell wall preparations. Satisfactory yields of wall material from many bacteria can be obtained by the unmodified original procedure, providing the cell suspension is cooled to 0–4° C before being disrupted and all subsequent work is then done in a cold room at 0–4° C, including filtration and washing procedures. If it is to be kept, the final product should be dried from the frozen state immediately, or frozen to –20° C––50° C. Although wall preparations from many bacteria may carry such very powerful lytic enzymes that they are completely solubilized by incubating a suspension at pH 7.0 and 35° C for a few hours, the use of this simple method is still very important if the enzymic content is to be preserved, as for example, in studies concerned with the activity and distribution of the autolytic enzymes. Also, crude wall preparations are now known [64, 65, 66, 109] to be able to complete peptidoglycan synthesis, whereas membrane preparations from some organisms cannot do so (see Section 8.5). When the walls are required only for chemical studies, the autolytic enzymes can be inactivated at an early stage and two methods have been commonly used to do this. One that is probably somewhat damaging to the preparations, is to heat the suspension of walls at pH 7.0 to 100° C for a few minutes immediately after the first differential spinning of the disrupted cell suspension. However, the method that has become most widely used, was first introduced by Shockman and his colleagues [14, 98]. The crude wall preparation obtained by high speed centrifugation of the disrupted bacteria is suspended in a 2% solution of sodium decyl sulphate. The decyl sulphate is recommended, rather than the more common dodecyl sulphate, on the basis of its greater solubility. Treatment with anionic detergent not only inactivates the autolytic enzymes present but also removes traces of membranous materials that otherwise adhere firmly to the walls. The detergent is ultimately removed from the preparation by repeated washing with buffer containing 1 M NaCl and with water. This method can be modified still further for bacteria producing very active autolytic enzymes, by suspending the bacteria directly in the detergent before disruption. When this is done, however, the method of disruption must be carefully chosen to avoid methods leading to foam formation. The French pressure cell and its modifications become the methods of choice. A clear example of the results of treating the walls with detergent is to be found in work with *Bacilli*. For example earlier workers [48] had found that small amounts (5–10% by weight) of insoluble protein were always present in wall preparations from *Bacillus subtilis* and *Bacillus licheniformis.* When detergent treatment was used, this protein was no longer present presumably because it was an hydrophobic protein, removed by the detergent treatment. It should be emphasized that when walls are required to retain their autolytic or other enzymic activity, they must be prepared at low temperature in the absence of detergent.

Most walls, from Gram-positive bacteria, even when prepared by these methods and hydrolysed by acids still yield traces of a range of amino acids, although the peptidoglycan amino acids predominate enormously. The other amino acids are usually regarded as 'contaminants' and, in order to avoid the difficulty of interpreting the picture, various processes for 'cleaning' the wall preparations were designed in early studies and are occasionally still used. One from which most subsequent methods stem, is that introduced by McCarty [57] and developed further by Cummins and Harris [17, 18]. In the latter authors' method the walls are first separated by differential centrifugation, as in the Salton and Horne [89] procedure; they are then subjected to the successive action of trypsin and ribonuclease acting at pH 7.6 followed by pepsin acting in 0.01 M-HCl. In applying this method it is even more important to inactivate any lytic enzymes in the preparations, since some of these are very active at pH 7.6. The disadvantage of such a method only became apparent as understanding of the nature of bacterial cell walls deepened and it became clear that some of the polymers such as the teichoic and teichuronic acids could be partly removed during its application. It is probable that the worst offender in the process is the dilute acid used to allow the pepsin to work. Dilute acids are commonly used in the dissection of walls into their component polymers, but it is also possible, and in some instances certain, that proteases themselves will remove wall components.

For example, the M protein produced by some haemolytic *Streptococci* is firmly fixed to the wall, of which it may therefore be part, but is removed by treatment with proteolytic enzymes.

A number of workers [2, 43, 81, 123] have examined the possibilities of distributing the particles in wall preparations in density gradients of sucrose or caesium chloride. In this work it has often been observed that the material distributes itself in three places in the gradients. In some work [2] there was a deposit of a small number of unbroken cells, a band that appeared to be wall material together with substances that have not usually been regarded as part of the wall, and a less dense band of pure wall. In other work [81] the middle band appeared to be walls mixed with a few whole cells. Yoshida *et al.* [123] on the other hand, found that wall preparations from only *Staphylococcus aureus* and *S. faecalis* gave three bands, whilst those from *Bacillus megaterium, E. coli* and *Bordatella pertussis* gave a single band of cell-wall material and a deposit of whole cells. Answers to the apparent discrepancies between the findings obtained by the different workers may lie with the use of different methods to disrupt the cells, the growth phase and species of the organisms disrupted and the very small total gravitational field used, that may have been too small to allow equilibration. Huff *et al.* [43] found that they obtained only two bands when disruption of *Staphylococci* was incomplete. After complete disruption a single band was obtained. This band was, however, polydisperse in the sense that if different parts of it were isolated and retested in a fresh gradient, they were found in the same positions in the new gradient (ie they did not redistribute to give the broad band found in the first gradient). The authors suggest that the different pieces were produced during the disruption of the bacteria, rather than by subsequent disintegration of the empty walls.

One of the difficulties of the gradient technique for preparing walls for chemical work is that the amounts that can be handled and appear in the wall band are usually small. This method applied to walls from Gram-positive organisms needs further examination along the lines used with *Staphylococci* by Huff *et al.* [43]. If the latter's conclusions can be applied to other organisms, then it would seem that the critical factor is to obtain as complete a rupture as possible of the organisms in the first place. If this is obtained, then little is to be gained by the use of density-gradient centrifugation simply to purify the walls. If, however, it is found that significance can be attached to the polydispersity found in the 'pure' wall band, this may prove a useful method for the further fractionation of cell-wall preparations obtained by other methods. Density gradients are used extensively in separating the components of the Gram-negative bacterial envelope (see p. 60).

2.2.3 *Isolation of cytoplasmic membranes*

The isolation and 'purification' of bacterial membranes from all species is necessarily a different type of problem from that of isolating the walls. As defined, walls have a limited number of molecular components arranged in a definite order and although, as will be seen later (Section 10.1.6), their composition can change with circumstances they, nevertheless, have unique polymers and components that make their recognition relatively easy and provide criteria for their purity. Membranes have few unique characters to distinguish their sources or to define their characteristics. They are labile fragile structures which can fragment, fuse together and differ according to the growth conditions used (eg [103]). They consist of lipid bilayers with proteins and enzymes associated with them to degrees varying from those which can be removed very easily, by for example, chelating agents or buffers of low ionic strength to those which are difficult or even impossible to obtain completely free of lipid. Thus 'purified' tends sometimes rather to be a shortened description of the technology used in the isolation procedure than a meaningful term of approbation. The cytoplasmic membrane in bacteria performs many of the functions undertaken by the mitochondria, the endothelial reticulum, the Golgi apparatus, and the cytoplasmic membrane in the cells of higher plants and animals. Its importance and complexity are, therefore, not to be doubted.

A description of the isolation of the membranes from Gram-positive bacteria is being included before turning to the isolation of any of the components of envelope of the Gram-negative species, because of the greater structural simplicity of the former bacteria. This simplicity allows the design of relatively straightforward procedures for the isolation of the cytoplasmic membrane. As was said in Section 2.1, the methods have been decided to a considerable extent by the usage envisaged. The preparation of what are undoubtedly membrane particles for much biochemical, and in particular biosynthetic work, has been relatively uncontrolled from the 'membranologists' viewpoint. Function has been the sole criterion, the particles being treated as insoluble enzymes or enzyme complexes. Therefore, it would not be particularly useful to detail the methods that have been used. In general, the bacteria have been disrupted by methods which are best for preserving the enzymic function: ultra-sound has often been used. Unbroken cells and, presumably, walls have been removed by low speed centrifugation, and the pellet obtained frequently called debris. The membrane particles have then been deposited by centrifuging at high speeds, of the order of 100 000 *g*. The resulting preparation has rarely been examined in a way that would define it as consisting of fragmented membranes. The work on isolation, which will be described here, has usually been undertaken either to allow chemical analysis of the membranes or for other purposes for which relatively defined and complete preparations were required.

2.2.4 *Preparation of protoplasts*

A variety of enzymes that hydrolyse walls (see Chapters 11.2 and 11.3) have been purified or partially purified and are, therefore, theoretically available for removing walls to make protoplasts. Few of these, however, have attained the state of purity or accessibility of egg-white lysozyme. As a result the detailed studies of membranes have tended to centre around those from lysozyme sensitive species, such as *Micrococcus* (*lysodeikticus*) *luteus*, *Sarcina lutea*, *B. megaterium*, *B. stearothermophilus* and *B. subtilis* or its near neighbours like *B. licheniformis*. Early work

[110-112] with *B. megaterium* showed that when the walls were hydrolysed from bacteria suspended in 1.2 M sucrose or 7.5% polyethylene glycol, the resulting osmotically fragile bodies respired as actively as the original bacteria. Since we now know that the protoplast membrane is the source of a number of the electron transport components, this observation is of value in indicating the preservation of a degree of organization in the membrane of the protoplast. However, since we now also know (see Chapter 4.6) that particles and variously prepared vesicles of membrane smaller than the whole protoplast membrane, can carry out functions such as active transport, respiration and many stages of wall biosynthesis, the original arguments for the intermediate preparation of protoplasts are less compelling.

In the preparation of protoplasts a variety of stabilizing agents, concentrations of lysozyme, buffers and conditions of incubation have been used. Particularly some cocci such as *Staphylococci* and *Micrococci* have a rather high internal osmotic pressure which has to be equalized and the viscosity of sufficiently concentrated sucrose solutions is inconvenient. To overcome this, M NaCl has been successfully used [68, 69, 77]. The potentials of a wide variety of solutes to stabilise the protoplasts of *B. megaterium* was studied by Corner and Marquis [15] and Marquis and Corner [59]. The simple calculated osmolality of the solutions is not the only factor involved in the stabilization of a population of protoplasts (see also [113]). For example, at equal osmolality, the higher molecular weight sugar, raffinose, is more effective than sucrose which, in turn, is considerably better than ribose. Peptides again are more effective than sugars at equal osmolality. The authors suggest that the diffusion constants and hence the shape of the molecules are involved in the stabilization process. Whether or not protoplasts lyse is also profoundly influenced by the rate of alteration of the concentration of the external solution. Slow dilution of the suspensions leads to great expansion of the protoplast membrane without rupture, whereas rapid dilution leads to a process of brittle fracture. The greater the expansion of the protoplast membrane the larger the stabilizing molecule has to be, whilst distinction between the efficiency of solutes is largely obliterated by rapid dilution. The importance of divalent cations to the integrity of protoplasts and membranes was also recognized in Weibull's [111-113] early work. If protoplasts were allowed to lyse in the absence of Mg^{2+} a small number of membranous vesicles, usually 3–4, was produced from each protoplast, whereas in the presence of 2 mM Mg^{2+} one ghost was formed from each. Subsequent work [82, 74] has shown that the concentration of divalent cations is important in reducing the escape of cytoplasmic constituents from protoplasts (see Table 2.2). Whether such escape is due to leakage through the membrane or more likely to the frank lysis of some individuals in the population is uncertain. The required concentration of divalent cation seems to differ with bacterial species and also probably according to the growth conditions.

One factor in deciding whether the intermediate preparation of protoplasts is necessary before isolating membranes from Gram-positive species is provided by whether or not the mesosomal membrane should be distinguished. When protoplasts are formed, mesosomes are extruded and can be left behind in the supernatant fluid

Table 2.2 The effect of Mg^{2+} upon the stability of protoplasts (after [74, 82] and H. J. Rogers, unpublished work).

Micro-organism	Mg^{2+} *concentration* (mM)						
	0	1	5	10	15	20	40
M. luteus	19.0	6.1	5.9	5.8	–	5.7	5.7
B. subtilis 168	–	–	–	7.0	–	6.0	3.0
B. licheniformis 749	–	–	–	3.5	–	10.0	3.0
B. licheniformis 6346	87.0	–	48.0	–	46	25–35	5.0

The protoplasts were made using either 0.8 M sucrose for the micrococci or 20% polyethylene glycol (average MW 400) for the bacilli as stabilizing agents, the concentrations of Mg^{2+} shown in the table were also present. They were removed from suspension by centrifuging and the extinction of the supernatant fluid at 260 nm measured. The value was compared with that of a lysate of the protoplasts and stated as a percentage. The values obtained give an indication of the soluble material liberated.

by deposition of the protoplasts at low centrifugal speeds. Whether mesosome vesicles adhere to the protoplasts and are deposited or are left free in the supernatant fluid depends upon the Mg^{2+} concentration [25, 27, 74, 77, 83, 84]. Fig. 2.1 shows a quantitative study of this phenomenon for *M. lysodeikticus.* So far no specific positive functions of mesosomal membrane have been discovered, although certain quantitative differences in protein and enzymic composition have been described (see Chapter 4.9). Since the amount of mesosomal membrane liberated from the protoplasts under the most advantageous circumstances yet described does not amount to more than 10–15% of the total cell membrane, it is probably justifiable for many purposes to use methods which do not separate mesosomal from cytoplasmic membranes.

2.2.5 *Preparation of membranes from protoplasts or cells*

When protoplasts are prepared they are normally deposited by low speed centrifugation and washed at least once with a solution of the same composition as that used for making them. The deposit is then usually resuspended in the same solution, lacking the osmotically protective solute, and a little deoxyribonuclease is often added at this time to reduce the viscosity of the lysate. If Mg^{2+} is omitted from the solution used for lysis, fragmentation of membrane and the loss of easily dissociable enzymes such as ATPase (see Chapter 4.5.1) occurs. However, the presence of this ion often causes adherence of rather large numbers of ribosomes [82]. Thus the conditions chosen for the lysis of protoplasts or cells must be a compromise guided by the purposes for which the membranes are required. Very little controlled study has been made of the effect of different conditions used for lysing protoplasts and cells on the composition and properties of the membrane obtained. Different times, temperatures, buffers and Mg^{2+} concentrations have been used and no

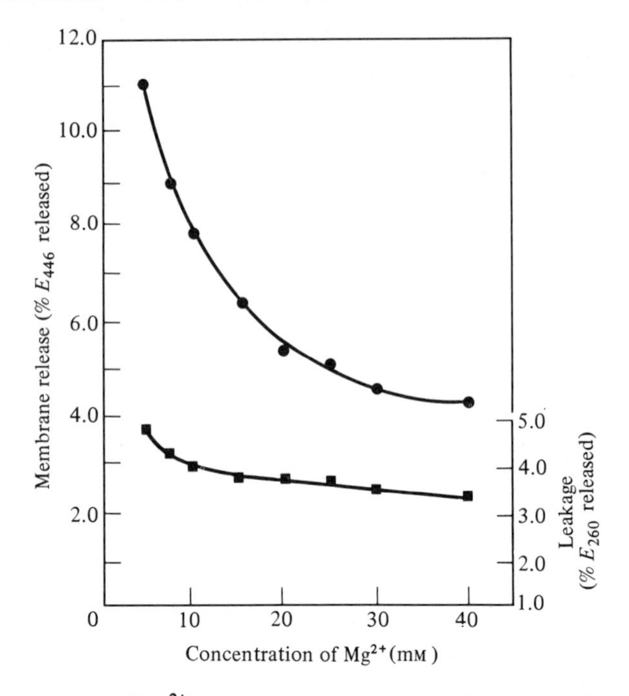

Figure 2.1 The effect of Mg^{2+} concentration during the formation of protoplasts from *Micrococcus luteus* upon the release of mesosomal membrane and the stability of the protoplasts. The bacteria had first been plasmolysed by suspension for 1.5 h at room temperature in 2.0 M sucrose containing the various Mg^{2+} concentrations. The protoplasts were then kept for 2 h at 30° C. Membrane not sedimentable by centrifuging at 12 000 *g* for 2 h was measured by the E_{466} (●) of the supernatant fluid and taken as a measure of mesosomes. The E_{260} (■) of the same supernatant fluid was used as evidence of protoplast instability. (From [74]).

comparisons of a wide range of conditions upon an equally wide range of membrane properties have been made. This is particularly disturbing in view of the known dissociation and association of proteins with membranes and the presence of phospholipases and proteases. It is by now clear that, although hydrophobic components such as ubiquinone, carotenoids, and many cytochromes, together with enzymes such as succinic dehydrogenase can be quantitatively recovered from bacteria in the membranes almost irrespective of the method used for lysis of the cells or protoplasts, the recovery of these substances is not an adequate criterion for the intactness of a membrane preparation or for the quantitative yield of the total membranous material from the cells, since much other less firmly fixed material may be lost.

A potentially useful way of monitoring the degree of contamination of membrane preparations was described by Salton [86] who reacted antibodies made against soluble cell components with the washings from the membrane preparation. Bacteria suspended in 0.1 M tris buffer, pH 7.5, were lysed by incubation with lysozyme and the viscosity of the lysate was reduced with deoxyribonuclease. The membranes were deposited by centrifuging and washed a number of times with 0.1 M tris buffer, pH 7.5. Antiserum produced in rabbits against soluble cell contents (supernatant from the original lysate) was then placed in the centre well of a double diffusion plate and successive washing fluids put in the surrounding wells; the fourth washing and the succeeding ones showed no reaction with the 'protoplasmic' antibody. Likewise, antibodies were made to the membranes, and sonicated membranes after successive washes were tested against them in double diffusion plates. Washings and membranes were also examined by polyacrylamide electrophoresis. A test carried out in this way clearly demands a number of arbitrary decisions about what is protoplasm and what is membrane. However, it would seem worthwhile to re-examine this type of approach now that our understanding of the various degrees of association of proteins with membranes is better. At worst, it should provide reliable criteria for the 'purity' of a 'membrane' defined as consisting of those components most firmly fixed to it.

2.2.6 *Fractionation of cytoplasmic membranes*

A number of authors [26, 82, 86] have noted that bacterial membrane preparations dissociate into discrete bands when centrifuged through sucrose gradients. Examination of material from such bands, however, has failed to show any remarkable differences in ultrastructure, composition or enzymatic function. The number of bands and the distribution of the amount of material in the bands is greatly influenced by the number of washes and composition of the solutions used for washing the original membranes. There is also variation according to the conditions that have been used for growing the bacteria [82]. It would seem most likely that the behaviour of membranes on gradients is more a reflection of different degrees of vesiculation and association of the vesicles with each other and possibly with ribosomes than an indication of different species of membrane. A rather long fractionation procedure involving changes of pH and Mg^{2+} concentration was claimed to yield fractions of different composition, one of which contained little RNA and represented about 40% of the original membrane [121]. A method [20] of fractionation, which was new in principle, grew out of studies of the association of DNA with membranes. Trembley *et al.* [104] first found that part of the membrane associated with DNA, adsorbed to insoluble Mg-lauryl sarcosinate. A band of this 'complex', the so-called M band, could be recognized in gradients. Daniels [20] modified the method by adsorbing membranes, prepared from protoplasts of *B. subtilis,* to forming crystals of cadmium lauryl sarcosinate. The crystals were deposited on a disc of Whatman No. 41 filter paper, and fractions were eluted from them by a series of KCl and sodium deoxycholate solutions. Nine fractions differing in their ratios of protein to lipid were obtained and the patterns from isolated cytoplasmic and mesosomal membranes were different. Some evidence was obtained for the separation into different fractions of those enzymes that were stable enough to withstand the treatment. This novel preliminary study has not so far been repeated.

2.2.7 *Separation of mesosomal membranes*

As has already been said, mesosomal membrane can be found suspended in the supernatant fluid from protoplasts which have been removed by centrifuging at relatively low speeds providing the Mg^{2+} concentration in the environment is suitable (see p. 55). Mesosomes, however, as seen in fixed and stained sections of whole bacteria, consist of an infolding of the cytoplasmic membrane to form something like a pleomorphic goblet which is filled with turbules or sheets of membrane. The membrane remaining in the supernatant after the removal of carefully prepared protoplasts represents only the internal membrane of these structures. The infolding of the cytoplasmic membrane that formed the walls of the goblets has presumably been stretched out to make part of the cytoplasmic membrane when the protoplasts are formed. Conditions for liberating mesosomal membrane from cells, apart from the Mg^{2+} concentration, have been studied very little, partly because no positive markers exist for the mesosomal membrane and differences in composition are too small to provide criteria. In one very thorough study, Owen and Freer [74] examined the effects of plasmolysis, temperature and agitation, as well as Mg^{2+}, on the liberation of mesosomal membrane from protoplasts of *M. lysodeikticus.* Of these factors, only plasmolysis and Mg^{2+} concentration had important effects. The yield of mesosomal membrane could be doubled by first treating the bacteria with 2 M sucrose before diluting to 0.8 M and adding lysozyme. The supernatant fluid from protoplasts contains not only the mesosomal membrane but also all the wall digestion products and superficial structures, such as flagella. It would, therefore, seem desirable that some method in addition to depositing the membrane by high speed centrifugation and washing, should be used to separate the mesosomal membrane from other insoluble material. Examination of the deposit on either sucrose [26, 77] or caesium chloride [82] gradients has shown multiple bands containing membrane material. Often this multiplicity is in the form of two bands very close together on the gradients. Gradient separation of mesosomal membrane seems more meaningful than similar treatment of cytoplasmic membranes. In two instances [82, 101] the top band of mesosomal membrane consisted of material closely similar to that to be expected from the mesosomes seen in sections of micro-organisms and was relatively free from nucleic acids and where relevant from flagella. The lower band on the other hand was heavily contaminated with vesicles of cytoplasmic membrane, nucleic acids and with flagella when prepared from *Bacilli.* Some workers [74, 79] have, nevertheless, obtained preparations of mesosomal membrane omitting gradient fractionation, that have the expected negative properties such as low activities of succinic dehydrogenase, NADH oxidase, and cytochromes which are characteristic of the material (see Chapter 4.9) but which are not grossly contaminated by nucleic acids. Other workers [76] have found multiple bands on sucrose gradients consisting of two slowly sedimenting peaks similar to those recognized by others and a rapidly sedimenting peak but claimed that there were no differences between them. Since no attempt was made to separate the two slowly sedimenting peaks the conclusion drawn is probably unjustified. On balance, it would appear desirable to purify mesosomal membrane by gradient centrifugation for more critical work, if only to remove pinched-off vesicles of cytoplasmic membrane and flagella. For contrary arguments, the reader is referred to Reusch and Burger [79].

2.3 Separation of the components of the wall from Gram-negative species

2.3.1 *Separation of the peptidoglycan layer*

The known presence in the envelope of Gram-negative bacteria of two membranes together with a layer of peptidoglycan covalently combined with a lipoprotein, let alone a number of other less explored components, has presented biochemists and microbiologists with a considerable problem. As might be expected, the peptidoglycan layer was the first to be dissected out, if only because it was sufficiently recalcitrant to withstand chemical treatments to which the remaining layers yielded. The first preparation, a 'rigid' or R-layer from Gram-negative cells, was obtained from *E. coli* strain B by Weidel *et al.* [115]. The procedure for preparing walls included extraction of the cells with 0.03 M NaOH followed by suspension and disruption in 0.4% sodium dodecyl sulphate. This treatment had to be repeated frequently before empty structures were obtained. These were then extracted four times with 90% phenol and repeatedly washed to remove remaining lipopolysaccharide. The insoluble material was suspended again in 0.4% sodium dodecyl sulphate and vibrated with glass beads. Finally, the detergent was removed by washing, to leave thin, rod-shaped, slightly flexible structures covered with raised lumps, the whole amounting to about 10% of the weight of the original walls. Acid hydrolysates of such preparations contained the expected peptidoglycan constituents (see Chapter 6) but also contained quite large amounts of eight other amino acids. We can now deduce, of course, that these almost certainly had their origin in the lipoprotein now known to be covalently attached to the peptidoglycan [7]. Lysozyme or the lysin from coliphage T2 disintegrated the R-layers 'by disengaging from one another the tiny spheres that gave the R-layer its characteristic structure'. When the above procedure was replaced [60, 114, 115] by one that involved repeated treatment with 2% and 4% sodium dodecyl sulphate followed by digestion with either trypsin or pepsin, thin, smooth, featureless, collapsed sausages were obtained from either *E. coli* or *Salmonella gallinarum.* These analysed to show the presence of only peptidoglycan constituents. It took a number of years to realize [7] that an important covalently linked constituent of the rigid layer, namely a lipoprotein, had been removed by the proteolytic enzymes. The isolation of an intact peptidoglycan layer from Gram-negative organisms had important consequences at the time, in answering critics of the idea of the presence of an entire layer in these organisms, chemically similar to the very much thicker and morphologically simpler wall in Gram-positive cells [116]. It had been observed, for example, that treatment of Gram-negative bacteria with alkalis or anionic detergents led to lysis of the cells as measured by the optical density of suspensions. This, Weidel and Pelzer [116] were able to argue, was due to damage to the underlying cytoplasmic membrane with consequent leakage of internal cell contents through the holes of the thin layer of peptidoglycan, rather than to the absence of such a layer around the cell. Other phenomena were likely to be due to the activation of enzymes hydrolysing linkages in the peptidoglycan layer. The proposal [116] that the peptidoglycan of the envelopes of Gram-negative bacteria is one huge 'bag-shaped' molecule, is still a subject for discussion but might now be less readily accepted in the form put forward originally than it was a few years ago. The original method used to isolate the rigid layer from Gram-negative bacteria has been modified in minor ways but has remained over the years the basic method of choice.

The isolation of the very much thinner layer of peptidoglycan from the marine *Pseudomonad* B16 has also been described [28–30]. By manipulating the environment of this micro-organism it was found possible to remove the outer membrane leaving naked, so-called mureinoplasts. This theoretically made it possible to prepare all three of the commonly accepted layers of the envelope in one operation. Attempts to prepare the peptidoglycan layer from the mureinoplasts, however, even when they were first treated with sodium dodecyl sulphate, were at first unsuccessful and only disaggregated fragments could be isolated. To obtain whole sacculi, retaining the shape of the mureinoplasts, bacteria from exponentially growing cultures had to be treated immediately with boiling 0.5 M NaCl, followed by sodium dodecyl sulphate. Subsequent treatment with trypsin to remove, presumably, denatured protein, led to the isolation of the very thin sacculi. So far no evidence has been given as to whether a lipoprotein was also removed by this procedure, as it was from *E. coli*, but the presence of a number of non-peptidoglycan amino acids in acid hydrolysates of the sacculi before trypsin treatment makes its presence likely.

2.3.2 *Separation of outer and the cytoplasmic membranes*

In the above procedures for the isolation of the peptidoglycan layer, both the outer membrane of the 'wall' and the underlying cytoplasmic membrane were dissolved away. The converse procedure in which these were preserved at the expense of the peptidoglycan layer was first designed by Miura and Mizushima [70]. They made use of the observation [4] that when spheroplasts were made from *E. coli* by the combined action of EDTA and lysozyme at pH 8.0 in tris buffer [78] the outer membrane of the wall was still present on the surface. Such spheroplasts were burst in the presence of 5 mM $MgCl_2$ and after washing with more $MgCl_2$ solution, were suspended in and dialysed against 3 mM EDTA at pH 7.0. The resulting membranes were layered on to a linear 35% to 50% sucrose gradient, and centrifuged for 4 h at 140 000 *g*. Three bands were obtained. The middle band was re-run on the same type of sucrose gradient and again three bands were obtained. The denser and lighter bands from the two runs were also re-examined. The upper band was run on a 30% to 45% sucrose gradient and the lower one on a 40% to 55% gradient, each yielding a single band. Morphological examination of the upper and the lower bands suggested that the lower, more dense band might consist of the outer membrane, while the upper one was composed of cytoplasmic membrane. Biochemical evidence fully confirmed this supposition. The membrane in the less dense band was rich in ATPase, succinic dehydrogenase, NADH dehydrogenase, and cytochromes, characteristic of cytoplasmic membranes, whilst these were either missing or present in much lower activity in the denser band. On the other hand, the latter contained a relatively high concentration of carbohydrate, presumably from the associated lipopolysaccharide. As might be predicted from its fractionation behaviour, the middle band behaved as a mixture of the other two. This work has been described in some detail because it has formed the basis for many other later modifications [31, 73, 95, 105]. Another method that differs in an important aspect from that of Miura and Misushima [70] was designed by Schnaitman [92]. This method does not arrive at the membrane preparations via the formation of spheroplasts, rather the cells are broken by two passages through a French pressure cell after first blending the suspension in a Sorvall Omnimixer to remove flagella, pili and capsular material. After removal of larger material from the disintegrated bacteria by low speed centrifugation, the crude envelope fraction is deposited and, after resuspension, fractionated on a continuous sucrose gradient to yield cytoplasmic membrane and outer membrane presumably together with peptidoglycan. Since the lysozyme treatment is omitted the peptidoglycan layer has not been deliberately removed or disrupted. Subsequent treatment of the original crude envelopes with Triton X-100 specifically solubilizes the cytoplasmic membrane, leaving partially intact the outer membrane (from the point of view of its integral proteins; see Chapter 3.4.2) although half the lipopolysaccharide and some of the phospholipid is also removed [93, 94]. This method allowed the initial examination of the protein composition of the outer membrane. Modifications of this method have been used by a number of workers studying *E. coli* [41, 53, 56]; and *Caulobacter crescentus* [1]. Modifications of both the original methods of Miura and Mizushima [70] and that of Schnaitman [92] have been applied to other micro-organisms such as *Neisseria gonorrhoeae* [49, 108].

Osborn *et al.* [73] were able to label the lipopolysaccharide specifically with ^{14}C-galactose and the lipids with ^{3}H-glycerol by using the appropriate mutant. Four membrane bands were obtained on EDTA-sucrose gradients at densities of 1.14, 1.16, 1.19 and 1.22. The most dense band contained almost all the ^{14}C-galactose (i.e. the lipopolysaccharide) and accounted for 40–60% of the total membrane proteins. The two lightest bands contained $>$90% of the electron transport enzymes and components. The outer membrane band was thought to be 2–3% contaminated by cytoplasmic membrane, whilst the membrane in the upper bands was contaminated to the extent of not more than 10% by outer membrane. This apparent contamination of the cytoplasmic membrane fraction, of course, is measured by the presence of ^{14}C-galactose. Since lipopolysaccharide is biosynthesized on the cytoplasmic membrane and transported to the outer membrane, the figure of 10% contamination may partially represent the presence of incomplete or untransferred lipopolysaccharide, rather than contaminating outer membrane. The importance of the use of chelating agents in the process was emphasized in this work, since the presence of Ca^{2+} ions led to aggregation and failure to resolve the outer and inner membranes. Fox *et al.* [31] used sucrose gradients containing 2-mercaptoethanol, the modification being introduced so that cytoplasmic membranes from auxotrophs that had respectively incorporated oleic and bromostearic acid could be separated by isopycnic centrifugation in continuous sucrose gradients. The separation of the cytoplasmic and outer membranes did not seem as satisfactory by this technique as by that of Osborn *et al.* [73].

A modification of the spheroplast–sucrose gradient technique which is simpler to use on a large scale and is said to be more reproducible than other methods has been described by Yamamato *et al.* [122] and used for *E. coli* by some workers [8, 99]. By this technique spheroplasts are made at high cell density (10^{11} cells/ml) with lysozyme acting at 0°–4° C in the presence of EDTA. They are then broken by a single passage through a French pressure cell. The ionic strength is subsequently low and 0.3 mM EDTA is always present. The washed membranes in suspension are centrifuged after layering over a 44% (w/v) sucrose solution. An upper layer of inner or cytoplasmic membrane is formed. The suspended pellet of membrane that has passed through the 44 % (w/v) sucrose is layered on to a discontinuous sucrose gradient with 52% (w/v) and 56% (w/v) steps. The material passing through the 52% (w/v) layer and resting on top of the 56% (w/v) layer consists of outer membranes.

2.3.2 *Preparation of outer membranes*

A lot of attention has been paid to the specific isolation and study of the outer membrane, particularly of *E. coli*. The somewhat unusual behaviour of the outer membrane towards detergents [93, 94] and pH makes the isolation of this structure easier than the isolation of other membranes. The presence of lipopolysaccharide components and latterly the recognition of its relatively simple protein composition have made its identification more sure. The first isolation of the outer membrane in a reasonably homogeneous state was by De Pamphilis and Adler [23] and Schnaitman [93, 94]. The methods were either first to make spheroplasts by the action of lysozyme in the presence of EDTA or to disintegrate the cells and prepare crude envelopes. In the former type of method sufficient $MgCl_2$ was added to make the molar ratio of Mg^{2+} to EDTA, 6:1. After 30 min incubation at 30° C, Triton X-100 was added. Under these conditions the cytoplasmic membrane was dissociated but not the outer membrane. The lysate from spheroplasts was fractionated with $(NH_4)_2SO_4$, a floating layer from this treatment was centrifuged over sucrose layers and finally fractionated in a CsCl gradient. The work of De Pamphilis and Adler [23] was specifically directed towards studying the organization of flagella rather than of the membranes and it might be noted that the outer membrane fraction obtained contained various ratios of flagella to membranes. De Pamphilis [22] showed that L-membranes (i.e. outer membranes) purified by the method already described [23] could be dissociated by combined treatment with EDTA and Triton X-100, but that they could be reassembled by dialysis against solutions of $MgCl_2$. A major contribution from Schnaitman's [92] early work, already mentioned was recognition that the protein component of the outer membrane was likely to be a great deal simpler than that in the cytoplasmic membrane. At this time only one structural protein was recognized, whereas, of course, 30–40 bands can be seen in gels after polyacrylamide electrophoresis of solubilized cytoplasmic membranes. A number of observations drew attention, however, to some of the difficulties involved in studying membrane structures. For example, lipopolysaccharide itself can form membrane-like structures which dissociate and re-associate under much the same conditions as outer membrane preparation. EDTA, as is well known [55], removes LPS from the *E. coli* surface as well as small amounts of protein and phospholipid. Reassembled membrane preparations were found [23] to contain 25% less protein than 'native' membrane preparations. These observations caution against drawing conclusions that are too dogmatic about the composition of the outer membrane, as it exists *in vivo*, from the results for isolated preparations, without close scrutiny of the methods used for isolation. Outer membrane isolated from spheroplast membranes by treatment with Triton X-100 in the presence of Mg^{2+} and purified as by De Pamphilis and Adler [23] on CsCl gradients has been compared with sequential sodium dodecyl sulphate extracts from Triton X-100 treated cell envelopes [6]. The fractions containing outer membrane both differed from cytoplasmic membrane, in the bands shown after polyacrylamide gel electrophoresis but whilst it was clear that more than one protein was present in the outer membrane preparations, it was equally clear that the composition was comparatively simple with a few major bands predominating. Both Wu [120] and Ames *et al.* [3] prepared outer membranes from *E. coli* and *Salmonella typhimurium* respectively by the Osborn *et al.* [73] method. Ames *et al.* [3] examined the polyacrylamide gel electrophoresis patterns produced by samples taken from the sucrose gradients in which the outer and inner membranes had been separated. This picture (Fig. 2.2) showed that there were a limited number of predominating bands in the region of the gradient corresponding to the relatively heavy outer membrane with a background of a very large number of lesser staining bands. The cytoplasmic membrane region of the gradient separation showed the presence of a very large number of protein bands all staining much more equally. In the work of Ames *et al.* [3], Bragg and Hou [6] and Wu [120] the importance of the conditions used for solubilizing the isolated membrane with anionic detergent as a preliminary to gel electrophoresis was emphasized. In particular, the resulting pattern was dependent on the temperature used. A similar picture for the comparative protein composition of inner and outer membranes from *E. coli* was obtained by Sekizawa and Fukui [95] using the original Miura and Mizushima [70] method of membrane fractionation. The composition of the outer membrane will be described in greater detail in Chapter 3. A very simple method for the preparation of outer membrane from *E. coli* has been suggested [117]. This consists simply of first shearing off flagella, pili, capsular material etc. by brief treatment of the organisms suspended in tris buffer, pH 7.8, and 1 mM EDTA in an Omnimixer. The organisms are then suspended in 30% sucrose, containing 10 mM EDTA, and treated with lysozyme, after 10 min at 22° C, $MgCl_2$ to a final concentration of approximately 20 mM, ribonuclease and deoxyribonuclease are added. The suspension is then centrifuged and the outer membrane remaining suspended in

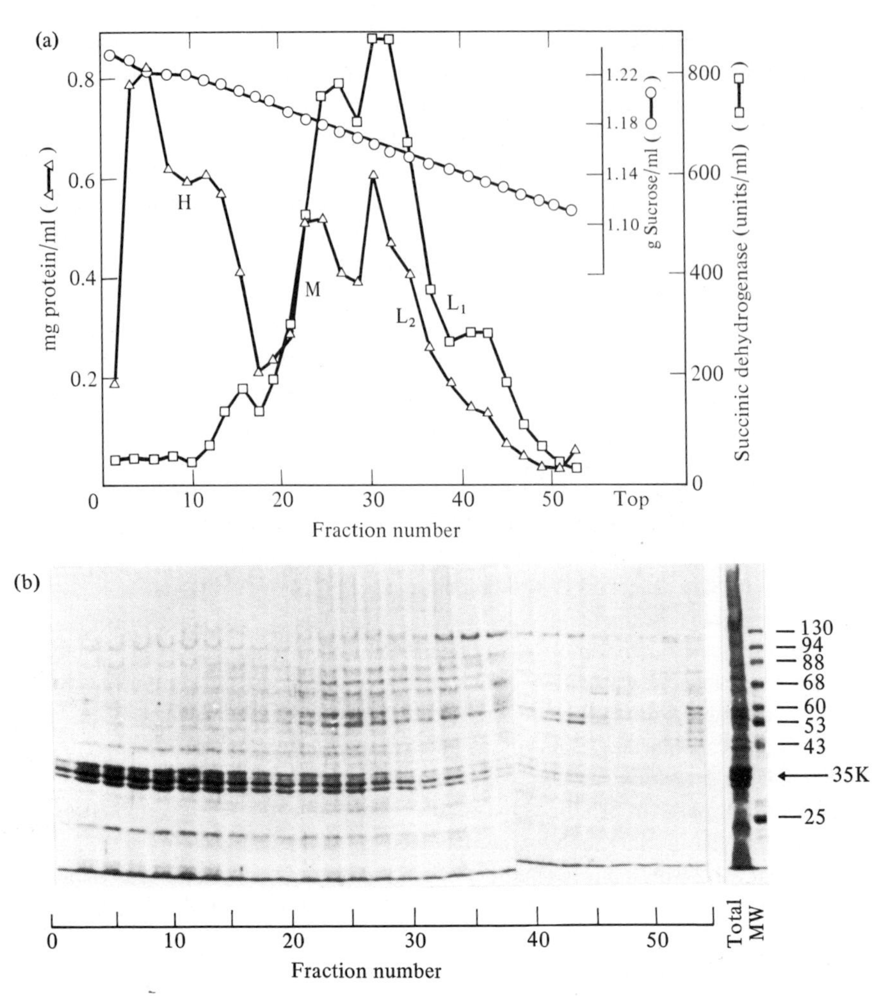

Figure 2.2 The separation of inner and outer membranes from *Salmonella typhimurium.* (a) Examination of samples for protein content and succinic dehydrogenase (SDH) activity from a sucrose density gradient, showing the outer membrane band, H, and the cytoplasmic membrane bands, L_1 and L_2, having SDH activity. (b) The results of SDS-polyacrylamide gel electrophoresis of fractions across the gradient. The dominant 33 000–36 000 mol. wt proteins characteristic of outer membranes can be seen on the samples from band H. (From [3].)

the supernatant fluid is aggregated by lowering the pH to 5.0 with HCl. This work is based on the reverse of the deduction from Birdsell and Cota-Robles' study [4], drawn by Miura and Mizushima [70]. Nevertheless, the authors show that the material obtained sediments at about the same density as the heavy band obtained by Osborn *et al.* [73], and is not likely to be contaminated by cytoplasmic membrane. The preparation contained about 50% protein, 17–19% LPS and 6–8% lipid. The pattern obtained after polyacrylamide gel electrophoresis, however, appeared very much more complicated than those shown by Wu [120] or Ames *et al.* [3]. Problems of relative loading of the gels in the various studies, may complicate comparisons, as well as the temperatures of treatment with anionic detergents used to solubilize the membranes before electrophoretic examination. A relatively simple method was also described [102] for the preparation of outer membrane from *Acinetobacter* sp. Bacteria from late exponential phase cultures were broken by the use of a French pressure cell and the walls prepared by differential centrifugation and repeated washing in the cold, as described for the preparation of walls from Gram-positive bacteria. The resulting envelope material appeared in section to have lost its cytoplasmic membrane. These 'walls' were then treated successively by lysozyme and pepsin. Analyses for lipids and lipopolysaccharide constituents were published. Diaminopimelic acid, used as an indicator of the presence of peptidoglycan, was shown to be absent. Despite the rather large number of suggestions for separating the outer membrane from Gram-negative species of bacteria, a preponderance of studies has used the convenient modification of the Miura and Mizushima [70] method designed by Osborn *et al.* [73].

The properties of a marine *Pseudomonad* B16 allowed a different and elegant method for the preparation of the outer membrane and other material on the outside of the cell [28, 29]. This organism is grown in a balanced salt mixture containing 0.22 M NaCl, 0.026 M $MgCl_2$ 0.01 M KCl and 0.1 mM $FeSO_4(NH_4)_2SO_4$. When the harvested cells are washed several times with 0.5 M NaCl and then suspended in 0.5 M sucrose, the outer membrane swells, bursts and can be collected from the supernatant fluid remaining after the naked mureinoplasts (see p. 60) have been removed. The preliminary NaCl washes remove material which is presumably outside the outer membrane and when negatively stained, appears to be of a plaque-like nature. Soluble material is also liberated when the outer membrane bursts. If the NaCl washings were omitted and the cells washed with 0.5 M sucrose and suspended again in sucrose, the outer membrane swelled but did not burst until the cells were repeatedly washed with 0.5 M sucrose. A similar picture of grossly swollen outer membrane has been seen in *E. coli*, after the action of lysozyme at 0–4° C on a very high density of cells in the presence of EDTA [122]. After removal of the outer membrane, from the marine *Pseudomonad*, the peptidoglycan layer could be hydrolysed by the action of lysozyme and the cytoplasmic membrane then prepared from the protoplasts. Membranes have been prepared from another marine *Pseudomonad* after lysing the cells by metabolic swelling, an unusual procedure. The crude membranes were purified on CsCl gradients. It was claimed that higher specific activities for several cytoplasmic membrane enzymes were obtained by this method [33].

2.4 Preparation of specialized intracytoplasmic membranes

2.4.1. *Photosynthetic vesicles–chromatophores*

These bodies were first isolated in the classical study by Schachman *et al.* [109] as vesicles of about 60 nm in diameter. They were actually seen under the electron microscope as collapsed discs of much larger diameter. The original study had been on the patterns obtained in the ultracentrifuge from several species of bacteria, including *Rhodospirillum rubrum* after breaking by a variety of means. The photosynthetic bacterium was broken by grinding with alumina [62] and a fast sedimenting fraction isolated that contained all the photosynthetic pigments of the bacteria in the vesicles mentioned. Also of historical interest in this paper, is the observation that a wide variety of bacteria contained material sedimenting at 4° C which was particulate and contained all the RNA of the cells – this was before our sophisticated approach to ribosomes! Particles (chromatophores) were isolated soon after from *Rh. rubrum* that had been disrupted by either sonic oscillation or by grinding with alumina using differential centrifugation or sucrose gradients [10, 34, 35, 36, 51, 71, 119]. These spherical particles which were of two sizes could undertake photo-activated phosphorylation, had succinic dehydrogenase activity and cytochromes. In order to avoid contamination with membrane fractions other than those concerned with photosynthetic reactions, it was found necessary to avoid very drastic methods for breaking the cells and to fractionate on sucrose gradients. When this was done, the photosynthetically active fraction could be isolated as homogeneous spherical vesicles that when sectioned were seen to be bounded by single unit membranes of the usual appearance [9]. The facts that all the bacteriochlorophyll is bound to membrane and that vesicles of a similar nature can be seen in whole light grown micro-organisms (see Chapter 1.6.1), are strong indications that the isolated particles in fact represent the photosynthetic apparatus of the cell. As might be expected, the membranes of the photosynthetic vesicles can be disrupted by organic solvents, anionic or neutral detergents, and they can be ruptured osmotically whereupon part or all of the bacteriochlorophyll components are liberated as soluble protein pigment complexes (see [75] for refs).

The chromatophores of *Chromatium* species have been carefully isolated and studied [19]. The bacteria were disrupted by the use of a Ribi press (see p. 49) and the chromatophores first separated from the ribosomes which were deposited through a 35% solution of RbCl by centrifuging the disrupted cells at 30 000 *g*. The floating chromatophores were then subjected to centrifugation through sucrose density gradients. A light and a heavy fraction were obtained as with the chromatophores from the *Rhodospirilla*. The light fraction was homogeneous. However, the significance of these apparent separations of chromatophores into two fractions is questionable [13, 37, 38, 51, 72]. For example, chromatophores could be isolated as a single band from *Rhodopseudomonas spheroides* by zonal gradient centrifugation, or after sucrose density gradient centrifugation following chromatography on Sepharose 2B columns [32]. Interaction between chromatophores and envelope components from *Rhodospirillum rubrum* and *Rh. spheroides* was found to occur [37, 38, 51, 72] unless the concentration of Mg^{2+} was maintained at a low level. Such an interaction is a likely explanation of the heavy, somewhat morphologically heterogeneous band containing bacteriochlorophyll observed by others. The cytoplasmic membrane, together probably with other envelope components in such bands, could be obtained almost free of bacteriochlorophyll. Purified cytoplasmic membrane and outer membranes have also been prepared [12] from aerobic cultures of this organism and chromatophores from anaerobic phototropically grown cultures [13]. Various methods of disrupting the organisms grown with light were compared and, from reconstruction type experiments, it was concluded that chromatophores could be obtained that gave a single band after sucrose gradient isopycnic centrifugation and that were free of contamination with cytoplasmic membrane. This was confirmed by their appearance after negative staining. After lysis of the cells by successive treatment with lysozyme, EDTA and Brij 58 and isolation in the absence of the detergent, about 50% of the bacteriochlorophyll was in the chromatophore fraction. Membrane fractions of the chromatophores appeared similar to membranes generally. They contained both succinic dehydrogenase and cytochromes; their SDS-polyacrylamide gel patterns of polypeptides and proteins were quite distinct, however, from those obtained either from the total envelope fraction or the cytoplasmic membrane. The suggestion that, in phototrophically grown organisms, the chromatophore membranes are a major repository of succinic dehydrogenase, is obviously of great interest [13, 118].

In as much as 50% of the chromatophores of *Rh. rubrum* can be isolated from cells gently disrupted by lysozyme in the absence of detergents, one is tempted to conclude that at least this proportion of the organelles is free in the cytoplasm and not tightly linked to other membranous structures. However, this problem resembles that of deciding upon the original location of proteins such as cytochromes, some of which are found soluble in disrupted preparations but might be expected to be membrane bound in the original living cells.

The green photosynthetic bacteria have vesicles that are easier to distinguish from those formed by comminution of other membranes during cell disruption and are therefore easier to isolate with confidence. They are cylindrical in shape and surrounded by a thin 'non-unit' membrane. They were isolated from *Chloropseudomonas ethylicum* and from *C. thiosulphatofilium* [11] on density gradients of sucrose [42] after the cells had been broken by the French pressure cell. These vesicles had the expected shape and contained the photosynthetic pigments. The chemistry of these so-called chlorobium vesicles had been examined by Cruden and Stanier [16]. The vesicles as a whole consist of 50% lipid, 30% protein and 15% carbohydrate. Of the lipid 75% is in the chlorophyll containing fraction, the remaining 25% contains monogalactosyl diglyceride which does not occur in the cytoplasmic membrane. No phospholipids are present. These preliminary indications confirm that the membrane around the vesicles is chemically, as well as ultrastructurally, different from other membranes in the cell.

Part of the photosynthetic apparatus of the cyanobacteria has been isolated, [39, 40] namely the phycobilisomes (see p. 34), by breaking the cells in solutions of high ionic strength and somewhat bizarrely again in the presence of detergent. The bodies can then be purified by sucrose gradient centrifugation. Obviously, interesting questions about the nature of these bodies are raised by their apparent resistance to detergents at high ionic strength. If the ionic strength of the buffer is lowered after cell disintegration, the phycobilisomes disintegrate and a crude membrane fraction containing thylakoids (see p. 34) along with membrane and wall fragments can be obtained by centrifuging the suspension. A purified preparation of thylakoids has not so far been obtained [100].

2.4.2 *Isolation of other inclusions and their membranes*

Sulphur granules, poly-β-hydroxybutyrate granules and other inclusions have been isolated from disrupted cell suspensions usually by differential centrifugation and to an extent purified and analysed [91], but the small amounts of material available have not usually allowed extensive examination of the homogeneity of these preparations, or of the 'membranes' surrounding them. The membranes surrounding gas vacuoles are an exception [106]. The vesicles can be isolated relatively easily providing they are not exposed to pressures causing them to collapse. Bacteria containing them can be lysed by chemical [54] osmotic [107] or enzymatic [50] means. The vesicles float to the top of such cell-disintegrates during centrifugation. Since by their nature and function these vesicles are empty of solutes in solution, the membranes can be readily examined without fear of contamination from the inside. Contamination of the outside of the vesicles by cytoplasmic contents of the cells might have been thought to provide a problem. However, this does not seem to be so and examination of the protein components of gas vesicle membranes has shown only a single protein to be present. This work is summarized by Walsby [106].

References

1. Agabian, N. and Unger, B. (1978) *J. Bact.* **133**, 977–94.
2. Allsop, J. and Work, E. (1963) *Biochem. J.* **87**, 512–9.
3. Ames, G. F-L., Spudich, E. N. and Nikaido, H. (1974) *J. Bact.* **117**, 406–416.
4. Birdsell, D. C. and Cota-Robles, E. H. (1967) *J. Bact.* **93**, 427–37.
5. Bosco, G. (1956) *J. infect. Dis.* **99**, 270–4.
6. Bragg, P. D. and Hou, C. (1972) *Biochim. biophys. Acta* **274**, 478–88.
7. Braun, V. (1975) *Biochim. biophys. Acta* (Rev. in Biomembranes) **415**, 335–77.
8. Chopra, I., Howe, T. G. B. and Ball, P. R. (1977) *J. Bact.* **132**, 411–18.
9. Cohen-Bazire, G. (1963) In *Bacterial Photosynthesis*, eds. Gest H., san Pietro, A. and Vernon, L. P., pp. 89–110, Yellow Springs, Ohio: Antioch Press.
10. Cohen-Bazire, G. and Kunisawa, R. (1963) *J. Cell Biol.* **16**, 401–16.
11. Cohen-Bazire, G., Pfennig, N. and Kunisawa, R. (1964) *J. Cell. Biol.* **22**, 207–218.
12. Collins, M. L. P. and Niederman, R. A. (1976) *J. Bact.* **126**, 1316-25.
13. Collins, M. L. P. and Niederman, R. A. (1976) *J. Bact.* **126**, 1326–38.
14. Conover, M. J., Thompson, J. S. and Shockman, G. D. (1966) *Biochem. biophys. Res. Commun.* **23**, 713–19.
15. Corner, J. R. and Marquis, R. E. (1969) *Biochim. biophys. Acta* **183**, 544–58.
16. Cruden, D. and Stanier, R. Y. (1970) *Arch. Mikrobiol.* **72**, 115–34.
17. Cummins, C. S. and Harris, H. (1956) *J. gen. Microbiol.* **14**, 583–600.
18. Cummins, C. S. and Harris, H. (1958) *J. gen. Microbiol.* **18**, 173–89.

19. Cusanovich, M. A. and Kamen, M. D. (1968) *Biochim. biophys. Acta* **153**, 376–96.
20. Daniels, M. J. (1971) *Biochem. J.* **122**, 197–207.
21. Dawson, I. M. (1949) In *The Nature of the Bacterial Surface*, eds. Miles A. A. and Piris, N. W., pp. 119–25, Oxford: Blackwell.
22. De Pamphilis (1971) *J. Bact.* **105**, 1184–99.
23. De Pamphilis, M. L. and Adler, J. (1971) *J. Bact.* **105**, 396–407.
24. Edebo, L. (1960) *J. biochem. microbiol. technol. Eng.* **2**, 453.
25. Ellar, D. J. and Freer, J. H. (1969) *J. gen. Microbiol.* **58**, vii.
26. Ferrandes, B., Frehel, C. and Chaix, P. (1970) *Biochim. biophys. Acta* **223**, 292–308.
27. Fitz-James, P. C. (1967) *Proteids of Biological Fluids, Proceedings of the 15th Colloquium*, ed. Peeters., H., pp. 289–301, Amsterdam: Elsevier.
28. Forsberg, C. W., Costerton, J. W. and MacLeod, R. A. (1970) *J. Bact.* **104**, 1338–53.
29. Forsberg, C. W., Costerton, J. W. and MacLeod, R. A. (1970b) *J. Bact.* **104**, 1354–68.
30. Forsberg, C. W., Rayman, M. K., Costerton, J. W. and MacLeod, R. A. (1972) *J. Bact.* **109**, 895–905.
31. Fox, C. F., Law, J. H., Tsukagoshi, N. and Wilson, G. (1970) *Proc. natn. Acad. Sci. U.S.A.* **67**, 598–607.
32. Fraker, P. J. and Kaplan, S. (1971) *J. Bact.* **108**, 465–73.
33. Franklin, R. M., Datta, A., Dahlberg, J. E. and Braunstein, S. N. (1971) *Biochim. biophys. Acta* **233**, 521–37.
34. Frenkel, A. W. (1956) *J. biol. Chem.* **222**, 823–34.
35. Frenkel, A. W. and Hickman, D. A. (1959) *J. biophys. biochem. Cytol.* **6**, 285–9.
36. Geller, D. M. and Lipman, F. (1960) *J. biol. Chem.* **235**, 2478–84.
37. Gorchein, A. (1968) *Proc. R. Soc. Lond. Ser. B* **170**, 247–97.
38. Gorchein, A., Neuberger, A. and Tait, G. H. (1968) *Proc. R. Soc. Lond. Ser. B.* **170**, 229–46.
39. Gray, B. H. and Gantt, E. (1975) *Photochem. Photobiol.* **21**, 121–8.
40. Gray, B. H., Lipschultz, C. A. and Gantt, E. (1973) *J. Bact.* **116**, 471–8.
41. Henning, U. and Haller, I. (1975) *FEBS Letts.* **55**, 161–4.
42. Holt, S. C., Conti, S. F. and Fuller, R. C. (1966) *J. Bact.* **91**, 349–55.
43. Huff, E., Oxley, H. and Silverman, C. S. (1964) *J. Bact.* **88**, 1155–62.
44. Hughes, D. E. (1951) *B. J. exp. Path.* **32**, 97–109.
45. Hughes, D. E. (1962) *J. gen. Microbiol.* **29**, 39–46.
46. Hunt, A. L., Rodgers, A. and Hughes, D. E. (1959) *Biochim. biophys. Acta* **34**, 354–72.
47. Ikawa, M. and Snell, E. E. (1960) *J. biol. Chem.* **235**, 1376–82.
48. Janczura, E., Perkins, H. R. and Rogers, H. J. (1961) *Biochem. J.* **80**, 82–93.
49. Johnston, K. H. and Gotslich, E. C. (1974) *J. Bact.* **119**, 250–57.
50. Jones, D. D. and Jost, M. (1970) *Arch. Mikrobiol.* **70**, 43–64.
51. Ketchum, P. A. and Holt, S. C. (1970) *Biochim. biophys. Acta* **196**, 141–61.
52. King, H. K. and Alexander, H. (1948) *J. gen. Microbiol.* **2**, 315–24.
53. Koplow, J. and Goldfine, H. (1974) *J. Bact.* **117**, 527–43.
54. Larsen, H., Omang, S. and Steensland, H. (1967) *Arch. Mikrobiol.* **59**, 197–203.
55. Leive, L. (1974) *Ann. N. Y. Acad. Sci.* **61**, 1435–9.
56. Lutkenhaus, J. F. (1977) *J. Bact.* **131**, 631–7.
57. McCarty, M. (1952) *J. exp. Med.* **96**, 569–80.
58. McIlwaine, H. (1948) *J. gen. Microbiol.* **2**, 288–91.

59. Marquis, R. E. and Corner, T. R. (1976) In *Microbiology and Plant Protoplasts*, eds Peberdy J. F., Rose, A. H., Rogers, H. J. and Cocking, E. C., pp. 1–22 New York and London: Academic Press.
60. Martin, H. H. and Frank, H. (1962) *Z. Naturf.* **17b**, 190–6.
61. Merkeshlager, M., Schlossmann, K. and Kurz, W. (1957) *Biochem. Z.* **329**, 332.
62. Mickle, H. (1948) *J. R. mic. Soc.* **68**, 10.
63. Milner, H. W., Lawrence, N. S. and French, C. S. (1950) *Science*, **111**, 633–4.
64. Mirelman, D., Bracha, R. and Sharon, N. (1972) *Proc. natn. Acad. Sci. U.S.A.* **69**, 3355–9.
65. Mirelman, D. and Sharon, N. (1972) *Biochem. biophys. Res. Commun.* **46**, 1909–917.
66. Mirelman, D., Shaw, D. R. D. and Park, J. T. (1971) *J. Bact.* **107**, 239–44.
67. Mitchell, P. and Moyle, J. (1951) *J. gen. Microbiol.* **5**, 981–93.
68. Mitchell, P. and Moyle, J. (1956) *Symp. Soc. Gen. Microbiol.* **6**, 150–80.
69. Mitchell, P. and Moyle, J. (1956) *J. gen. Microbiol.* **15**, 512–20.
70. Miura, T. and Mizushima, S. (1968) *Biochim. biophys. Acta* **150**, 159–61.
71. Newton, J. W. and Kamen, M. D. (1957) *Biochim. biophys. Acta* **25**, 462.
72. Niederman, R. A. (1974) *J. Bact.* **117**, 19–28.
73. Osborn, M. J., Gander, J. E., Parisi, E. and Carson, J. (1972) *J. biol. Chem.* **247**, 3962–72.
74. Owen, P. and Freer, J. H. (1972) *Biochem. J.* **129**, 907–17.
75. Parson, W. W. (1974) *A. Rev. Microbiol.* **28**, 41–59.
76. Patch, C. T. and Landman, O. E. (1971) *J. Bact.* **107**, 345–57.
77. Popkin, T. J., Theodore, T. S. and Cole, R. M. (1971) *J. Bact.* **107**, 907–17.
78. Repaske, R. (1958) *Biochim. biophys. Acta* **30**, 225–32.
79. Reusch, V. M. and Burger, M. M. (1973) *Biochim. biophys. Acta* **300**, 79–104.
80. Ribi, E., Perrine, T., List, R., Brown, W. and Goode, G. (1959) *Proc. Soc. Exp. Biol. Med.* **100**, 647–9.
81. Roberson, B. S. and Schwab, J. A. (1960) *Biochim. biophys. Acta* **44**, 436–44.
82. Rogers, H. J. and Reaveley, D. A. (1969) *Biochem. J.* **113**, 67–79.
83. Rogers, H. J., Reaveley, D. A. and Burdett, I. D. J. (1967) *Proteids of Biological Fluids, Proceedings of the 15th Colloquium*, ed. Peeters H., p. 803, Amsterdam, London: Elsevier.
84. Ryter, A., Frehel, C. and Ferrandes, B. (1967) *C. R. Acad. Sci. Ser. D.* **26**, 1259–62.
85. Salton, M. R. J. (1953) *J. gen. Microbiol.* **9**, 512–23.
86. Salton, M. R. J. (1967) In *The specificity of cell surfaces*, eds. Davis, B. D. and Warren L., p. 71, Englewood Cliffs. N. J.: Prentice-Hall.
87. Salton, M. R. J. (1967) *Trans. N. Y. Acad. Sci.* **29**, 764–81.
88. Salton, M. R. J. and Horne, R. W. (1951) *Biochim. biophys. Acta* **7**, 19–42.
89. Salton, M. R. J. and Horne, R. W. (1951) *Biochim. biophys. Acta* **7**, 177–91.
90. Schachman, H. K., Pardee, A. B. and Stanier, R. Y (1952) *Archs. Biochem. Biophys.* **38**, 245–60.
91. Schmidt, G. L., Nicolson, G. L. and Kamen, M. D. (1971) *J. Bact.* **105**, 1137–41.
92. Schnaitman, C. A. (1970) *J. Bact.* **104**, 890–901.
93. Schnaitman, C. A. (1971) *J. Bact.* **108**, 545–52.
94. Schnaitman, C. A. (1971) *J. Bact.* **108**, 553–63.
95. Sekizawa, J. and Fukui, S. (1973) *Biochim. biophys. Acta* **307** 104–117.
96. Shockman, G. D. (1962) *Biochim. biophys. Acta* **59**, 234–5.
97. Shockman, G. D., Kolb, J. J. and Toennies, G. (1957) *Biochim. biophys. Acta* **24**, 203–4.

98. Shockman, G. D., Thompson, J. S. and Connover, M. J. (1967) *Biochemistry* **6**, 1054–65.
99. Smyth, C. J., Siegel, J., Salton, M. R. J. and Owen, P. (1978) *J. Bact.* **133**, 306–319.
100. Stanier, R. Y. and Cohen-Bazire, G. (1977) *A. Rev. Microbiol.* **31**, 225–74.
101. Theodore, T. S., Popkin, T. J. and Cole, R. M. (1971) *Prep. Biochem.* **1**, 233.
102. Thorne, K. J. I., Thornley, M. J. and Glauert, A. M. (1973) *J. Bact.* **116**, 410–17.
103. Toennies, G., Bakay, B. and Shockman, G. D. (1959) *J. biol. Chem.* **234**, 3269–75.
104. Trembley, G., Daniels, M. J., and Schaechter, M. (1969) *J. molec. Biol.* **40**, 65–76.
105. Tsukagoshi, N. and Fox, C. F. (1971). *Biochemistry* **10**, 3309–13.
106. Walsby, A. E. (1978) In *Relations between Structure and Function in the Prokaryotic Cell*, 28th Symp. Soc. for Gen. Microbiol., eds. Stanier, R. Y., Rogers, H. J. and Ward, J. B., pp. 327–357, Cambridge University Press.
107. Walsby, A. E. and Buckland, B. (1969) *Nature, Lond.* **224**, 716–17.
108. Walstad, D. L., Guymon, L. F. and Sparling, P. F. (1977), *J. Bact.* **129**, 1623–7.
109. Ward, J. B. (1974) *Biochem. J.* **141**, 227–41.
110. Weibull, C. (1953) *J. Bact.* **66**, 696–702.
111. Weibull, C. (1953) *J. Bact.* **66**, 688–95.
112. Weibull, C. (1956) *Expl. Cell Res.* **10**, 214–21.
113. Weibull, C. (1956) In *Bacterial anatomy*, eds. Spooner, E. T. C. and Stocker, B. A. D., pp. 111–126, Cambridge University Press.
114. Weidel, W., Frank, H. and Leutgeb, W. (1963) *J. gen. Microbiol.* **30**, 127–30.
115. Weidel, W., Frank, H. and Martin, H. H. (1960) *J. gen. Microbiol.* **22**, 158–66.
116. Weidel, W. and Pelzer, H. (1964) In *Advances in Enzymology*, ed. Nord, F. F., pp. 193–233, New York and London: Interscience Publishers.
117. Wolf-Watz, H., Normark, S. and Bloom, G. D. (1973) *J. Bact.* **115**, 1191–7.
118. Woody, B. R. and Lindstrom, E. S. (1958) *J. Bact.* **69**, 353–6.
119. Worden, P. B. and Sistrom, W. R. (1964) *J. Cell. Biol.* **23**, 135–50.
120. Wu, H. C. (1972) *Biochim. biophys. Acta* **290**, 274–89.
121. Yamaguchi, T., Tamura, G. and Arima, K. (1967) *J. Bact.* **93**, 483–9.
122. Yamamato, I., Anraku, A. and Hirosawa, K. (1975) *J. Biochem., Tokyo* **77**, 705–718.
123. Yoshida, A., Heden, C. G., Cedergren, B. and Edebo, L. (1961) *J. biochem. microbiol. technol. Eng.* **3**, 151.

3

Membrane structure and composition in micro-organisms

3.1 General ideas of membrane structure

As we have already seen, ultrastructure proves of comparatively little help in distinguishing between membranes with manifestly different functions. This statement excludes highly specific membranes, such as the purple membrane of the *Halobacteria* with its array of protein molecules or the disc membranes of retinal-rod outer segments. Otherwise, all the modern techniques of examination, from freeze-etching to the examination of simple sections, lead only to the visualization of rather small differences that are often hard to interpret. This similarity in ultrastructure, together with those of composition and properties, has led to a consideration of a general structure for membranes. Such general ideas are quite vital as guidelines to projecting further experimentation and thinking. Nevertheless, their acceptance must not be allowed to prevent us remembering the vast range of detailed differences in function that exist. Indeed, a few general differences such as the absence of sterols from bacteria and their universal presence in membranes from eukaryotes may suggest that the hypothetical general structures are as yet only beginning to embrace the complexities and niceties necessary to explain the functioning of any given membrane.

When first writing *Cell Walls and Membranes* ten years ago, the underlying rhythm of ideas about general structure was still dominated by the bilayer 'unit membrane' hypothesis originally suggested by Davson and Danielli [27]. Alternative different models had been proposed on the basis of various experimental approaches and have been summarized by Mahundra and White (see Fig. 2) [80] and by Rothfield and Finkelstein [116]. Since this time, however, one different model has become outstandingly attractive. This is the fluid mosaic model proposed by Singer and Nicholson [139]. It was proposed primarily on thermodynamic grounds, being the arrangement of proteins and lipids that possessed minimum free energy, a criterion not met by the original Davson-Danielli model. In membranes, the proteins which usually account for 60–70% of the dry weight of the preparations and the bulk of the lipids which account for most of the rest of the material, are not covalently linked together. This is not to lose sight of the small amounts of lipids that are known to be linked to some of the proteins isolated from membranes, and which have sometimes been shown to be necessary for enzymic function. The major interactions between the lipids and proteins are then non-covalent and involve large numbers of small forces working co-operatively together, rather than fewer very strong ones, as with covalent linkage in polymers. The most important non-covalent linkages are *hydrophobic* and *hydrophilic.* Hydrophobic interaction is a thermodynamic concept embodying the forces that sequester non-polar groups away from water. Contrary-wise, hydrophilic forces are those which summarize the preference of ionic and polar groups for an aqueous environment.

Before describing the Singer–Nicholson model, it is desirable to appreciate that isolated membranes from most sources have proteins associated with them in two different ways. These have been called 'peripheral' or 'extrinsic' proteins and 'integral' or 'intrinsic' proteins [137, 138] and are distinguished operationally by the methods that can be used to separate them from the membranes. Peripheral proteins can be removed easily by altering the ionic strength of the suspending fluid, by the use of chelating agents such as EDTA, or even by repeatedly washing with buffers of low ionic strength. Peripheral proteins associated with microbial membranes will be dealt with later. Is is clear that they are attached principally by ionic bonds probably with the involvement of divalent cations, such as Mg^{2+}. They are not only attached differently but also their mode of biogenesis has been shown to differ in some instances from the integral proteins. As implied by their name, integral proteins cannot be easily removed from the membranes. Indeed, relatively few of them have been isolated and purified. Most of these proteins are at the stage of having been recognized by application of gel electrophoresis techniques. To examine membranes by this latter process they are first disaggregated by treatment with anionic detergents, such as sodium dodecyl sulphate, and then moved through a porous gel of polyacrylamide under the influence of an electric current. Since the proteins carry an overwhelming charge conferred by their association with the acidic detergent, they are separated by virtue of their different sizes which gives an indication of their molecular weight. Table 3.1 summarizes the properties of 'peripheral' and 'integral' proteins.

The fluid-mosaic model was reasonably designed only to take account of the integral proteins. The peripheral proteins may be, and in some instances are known

Table 3.1. Criteria for distinguishing peripheral and integral membrane proteins (from [138]).

Property	*Peripheral proteins*	*Integral proteins*
Treatments necessary for dissociation from membrane	High ionic strength, chelating agents	Reagents breaking hydrophobic bonds eg detergents, some organic solvents, chaotropic agents.
Association with lipids when soluble	0	+
Solubility after dissociation in a neutral aqueous environment	+	0 or aggregated

to be, vitally important to function, but their attachment to the structure once formed is very much less of a problem. Among the integral proteins are dehydrogenases, cytochromes, and other components of the electron and metabolic transport systems.

Returning to the membrane model and in particular the arrangement of the lipids, much physical evidence has accumulated to show that these are arranged, at least over a high proportion of the surfaces, as a bilayer. The upper limit suggested [69] for such a bilayer is 70–80% of the membrane area. Accepting this, we have to arrange the amphipathic lipids and phospholipids in such a way as to minimize the free energy in the structure. (The term amphipathic is applied to structurally asymmetric molecules with one end highly polar and the other strongly non-polar.) The only way to do this in a bilayer structure existing in an aqueous environment, is to have the hydrocarbon tails of the fatty acids facing inwards towards each other, and the polar fatty acid and zwitterionic groups, facing outwards (see Fig. 3.1). A most important property of the lipid bilayer is the fluidity of its interior made up of the non-polar fatty acid chains, the nature of which controls the degree of fluidity. However, such a model does not suggest whether or not the two leaflets making up the bilayer are the same or different. Evidence now strongly suggests [16, 41] that they are in fact different; that is the lipid bilayer is asymmetric. Some of the most convincing experiments done with bacteria are contained in the work of Rothman and Kennedy [118]. These authors treated *Bacillus megaterium* with reagents that reacted specifically with amine groups but which under the conditions could not permeate through membranes. After treatment only about one third of the phosphatidylethanolamine in the cells became rapidly labelled. When conditions were changed so that the reagents could slowly penetrate through the membranes, the remaining phosphatidylethanolamine was also slowly labelled. When the experiments were repeated with inverted membrane vesicles (see Chapter 4.6), two-thirds of the phospholipid reacted quickly under the non-penetrating conditions. Thus, it would appear that in the membrane of the bacterium the outer leaflet has only about half the phosphatidylethanolamine of the inner layer. A summary of other methods used to investigate the transverse distribution of membrane lipids is given by de Pierre and Ernster [28]. If the lipid bilayer were covered with protein, as was suggested by the original Davson–Danielli model, a number of problems would arise. The polar groups of the lipids would be involved, presumably in combining with the proteins, again presenting a non-polar surface to the aqueous surroundings. This would not correspond to a situation of minimum free energy, a problem resolved if the proteins are thought of, not as stretched out on the surface, but as inserted into the bilayer. The conformation of the proteins can then be such that the non-polar groups face outwards, whilst the polar groups are inwards. Examination, for example, by circular dichroism has shown that a proportion, in some instances up to 40%, of the proteins in many membranes has an α-helical structure [60, 69, 159] so that the evidence would support a globular form for the proteins, rather than the extended conformation, to be expected if they were stretched out on interfaces. Proteins thus inserted as 'globules' also explain rather nicely the particles seen in freeze-etched replicas of

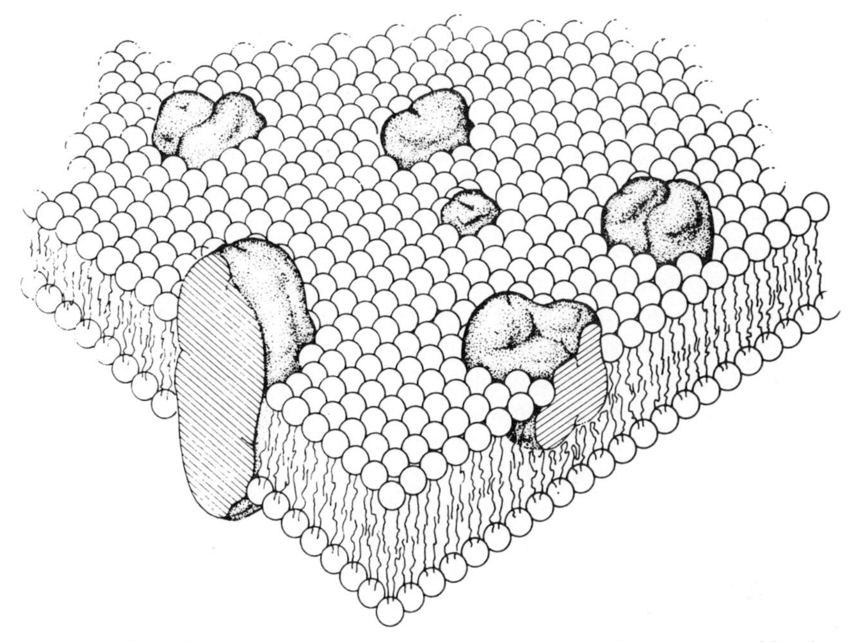

Figure 3.1 The fluid-mosaic model for membrane structure. The matrix of lipid bilayer is shown with the stippled bodies inserted into it as it is imagined integral membrane proteins exist (from [139]).

membranes. The plane of cleavage being within the non-polar heart of the bilayer exposes the embedded proteins as particles on one fracture face and indentations on the other. Some proteins pass completely through the bilayer and are exposed to reagents on both sides of the membrane. Fig. 3.1 shows a diagram of the fluid mosaic or 'iceberg' arrangement taken from Singer and Nicholson [139]. The whole subject of the long range order in biomembranes and the various models that have been suggested has been discussed quite recently [80].

3.2 Some physical properties of membranes

We should then, if this model is correct, be able to regard membranes as viscous lipid seas with protein molecules floating in them to various depths. The physical properties of membranes have been intensively examined over the years with this picture in mind. The methods that have been used with great success to examine the viscosity of the lipid phase, and to show an internal gradient of viscosity within the hydrophobic heart of the lipid bilayer include spin label esr, the use of fluorescent probes, high resolution nmr, X-ray diffraction and freeze-etching electron microscopic techniques. For those interested in pursuing in more detail this approach to membranes, reviews may be consulted [5, 20, 25, 32, 68, 82]. The results of such investigations have shown that both the lipids and the proteins are capable of rather rapid lateral movements. There is general agreement that lipids can redistribute over the surface of a membrane at a rate of the order of 10^{-8} cm^2/sec, in other words a molecule could travel the length of an *Esherichia coli* or *Bacillus subtilis* rod in about 1–2 sec. Although the most extensive work both by high resolution nmr [68] and by esr [36] has been done with artificial membrane systems rather than with isolated cell membranes, the lesser number of measurements using biological material would suggest values of the same order. The membrane proteins can also move laterally at speeds about 100 times less than those for the lipids [123]. The slower rate is largely to be accounted for by the molecular weight and size difference. They can also rotate on an axis perpendicular to the plane of the membrane. This sort of movement has been very beautifully demonstrated by making use of the dichroic properties of the rhodopsin chromophore, II cis-retinal [18, 22]. Calculations from the rate of rotation of the rhodopsin molecules in the visual receptor membrane suggest values for the micro-viscosity of the lipid layers that correspond rather well with those obtained from the other techniques. Although lateral movements cannot be doubted, the situation may be complicated in bacterial membranes that are particularly rich in embedded proteins. Even in membranes from mammalian sources complications arise. For example, investigation [146] of the cytochrome P_{450}–cytochrome P_{450} reductase enzyme system involved in hydroxylation reactions in liver microsomal membranes, has shown the proteins to be surrounded by a halo of segregated lipids with a different temperature for phase transition from the remainder. Clearly the general dynamic properties of the membrane will depend on the extent of this sort of organization of the lipid, the nature of the fatty acid tails of the lipids, the degree to which their hydrophilic head groups react with divalent cations and other potentially cross-linking reagents, and the presence or absence of proteins (such as spectrin of erythrocyte membranes) that traverse the membrane.

One movement of both lipids and proteins that may be very slow or even forbidden is that of traversing the membrane – a so-called 'flip-flop' movement. Measurement of the rate of such movement in artificial phosphatidylcholine vesicles has given half times of 6.5 h at 30° C. Although some faster movements have been suggested such as 72 min at 30° C for the half-time for flip-flop of a fluorescent sterol probe in liposomes [70, 141], the general opinion is that in natural membranes, the rate is much slower or infinitely slow. If indeed the slowness or absence of flip-flop movement is correct, the origin of asymmetric membranes with different lipid inner and outer faces becomes easier to understand.

From some view points these physical investigations are still in their early stages and do not go very far towards providing explanations for functions of the membranes in living cells. Nevertheless, the consistency of the results obtained already with the suggested general model for membrane structure is very reassuring. Eventually it would seem possible that their further application and development will help provide explanations of the multifarious functions of the many lipid and protein components in membranes.

3.3 Composition of microbial membranes

Like membranes from all sources, those from bacteria contain a variety of phospholipids and a very large variety of proteins but, unlike cytoplasmic membranes from mammalian cells, usually lack glycoproteins although exceptions have been described. Bacterial membranes also have a number of lipid peculiarities that distinguish them from those of other micro-organisms and multi-cellular creatures. They have no sterols and the range of phospholipids is very much more limited. However, they have a wide range of glycolipids and *O*-acylated lipid moieties, with a limited range of amino acids. Yeasts on the other hand, have a wide range of sterols in their membranes and a phospholipid composition much nearer that of membranes from mammalian sources [109]. Moreover bacterial membranes contain a considerably higher proportion of protein than do membranes from say microfungi. Table 3.2 [124] shows that whereas the membranes from microfungi usually contain about 40–45% protein and 40–45% lipid, bacterial cytoplasmic membranes are composed of from 50–70% protein and 20–30% lipid. How far this difference stems from the lower degree of differentiation of the bacterial membrane systems is not clear. In the prokaryotes all the membrane functions of the cell must be carried out by the single cytoplasmic membrane together with the small amount of mesosomal membrane if this exists in the living organism as a real entity (see Chapter 1.5.1). In the eukaryote micro-organisms a range of types of specialized membrane exists. Presumably a greater variety of proteins must be packed into the single prokaryotic membrane. This, together with generally rapid metabolism and fast growth rates, may account for the necessity for a higher proportion of membrane protein. Evidence exists [127] that very little of the phospholipid is covalently attached to protein.

The whole subject of membranology has grown explosively over the last 5–10 years and there is no possibility within the covers of a single book devoted to the whole subject of microbial walls and membranes to give a detailed and complete review of the subject. In some instances, such as for example, the lipid components, not only has there been an explosive growth in the number of components known to be present in membranes, but our knowledge of the function of this exponentially

Table 3.2. Chemical composition of membranes from Gram-positive bacteria and micro-fungi

Organism	*Protein* (%)	*Total lipid* (%)	*Hexose* (%)	*RNA* (%)	*Reference*
Bacillus megaterium	58–75	20–28	0.2–8	1.2–5.1	123a
Micrococcus luteus	52–68	23–28	16–19	2.3	
Staphylococcus aureus	69–73	30	1.7	2.4	
Saccharomyces cerevisiae	46–47.5	37.8–45.6	3.2	6.7	51a
Candida utilis	38.5	40.4	5.2	1.1	
Candida albicans (mycelial form)	45.0	31.0	25	0.5	115a
Fusarium culmorum	25.0	40.0	30.0	–	

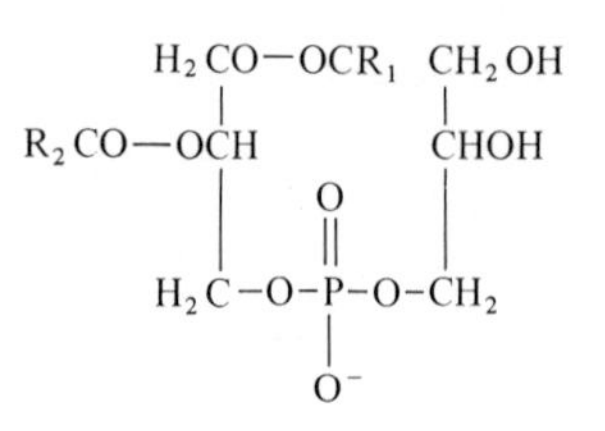

Phosphatidylglycerol

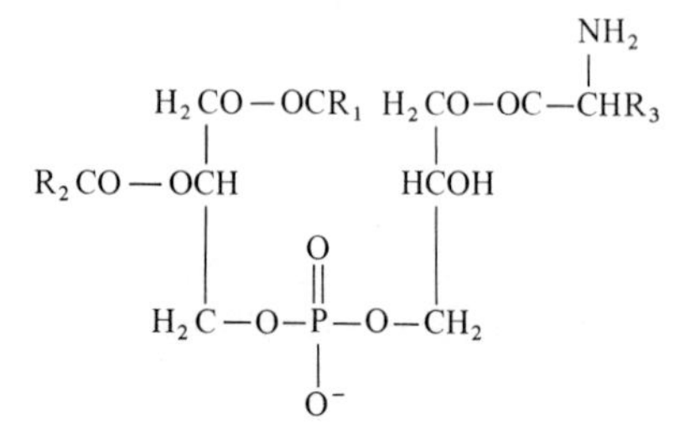

Amino-acyl ester of phosphatidylglycerol

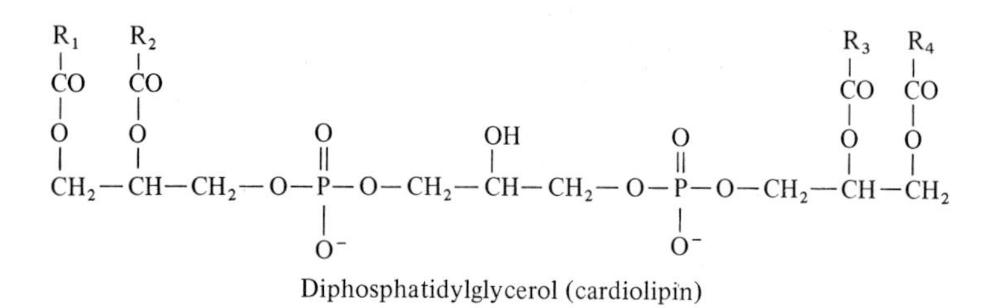

Diphosphatidylglycerol (cardiolipin)

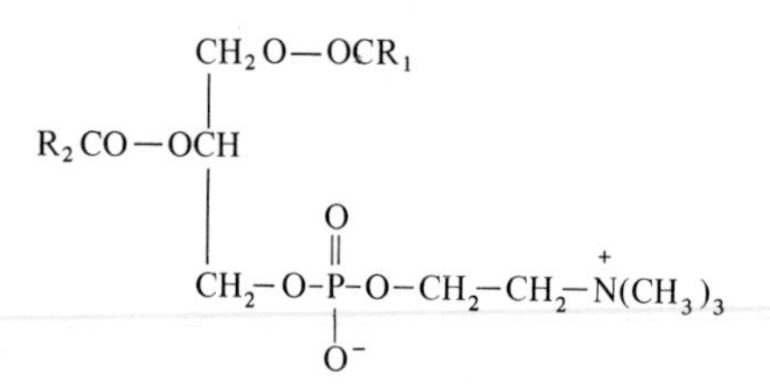

Phosphatidylcholine (lecithin)

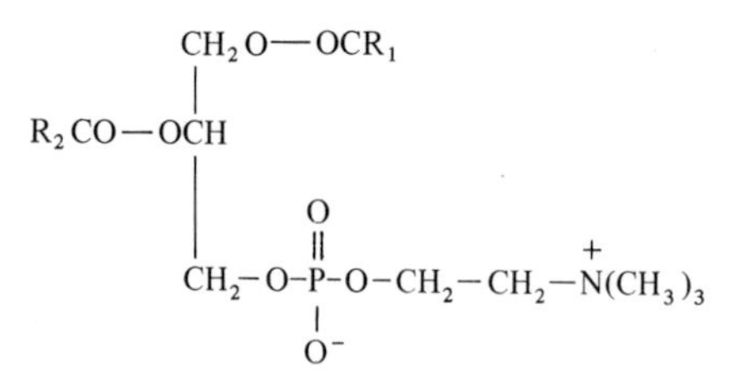

Phosphatidylethanolamine

Figure 3.2 Structures of the phospholipids in microbial membranes.

Table 3.3. Phospholipids in Gram-positive bacteria (from [40])*

Organism	*Phosphatidyl-ethanolamine*	*Phosphatidyl-glycerol*	*O-amino acylphos-phatidyl-glycerol*	*Cardio-lipin (Diphos-phatidyl-glycerol)*
Micrococcaceae				
Micrococcus luteus	–	72	–	4
Staphylococcus aureus	–	10–60	18–80 (lys)	0–20
	–	63	16–18 (lys)	11–14
Sarcina lutea	–	14	–	17
Streptococaeae				
Diplococcus pneumoniae		c25	–	c70
Streptococcus faecalis		++	++	+
Bifidobacterium				
Bifidobacterium bifidum		8.6–19	c3 (ala)	42–52
Propionibacterium				
Propionibacterium shermanii		++		++
Corynebacteriaceae				
Listeria monocytogenes		++		++
Microbacterium thermo-sphactum	++	++		++
Arthrobacter simplex		c85		
Bacillaceae				
Bacillus cereus	46	35	8 (orn)	+
Bacillus cereus ATCC4342	45–64	25–32	1–5 (ala)	5–25
Bacillus licheniformis	23	c60	2 (lys)	2–12
Bacillus megaterium				
MK10D pH 5.0†	+	5–10	+	
MK10D pH 7.0	36–45	35–45	8–14 (lys)	
Bacillus natto (log phase)	22	27		34
Bacillus subtilis	40	13	10 (lys)	38
Bacillus stearothermophilus	14–32	22–30		45–57
Clostridia				
Clostridium welchii				
Clostridium butyricum	14‡	26‡	++ (lys, ala)	+‡
SPIROCHAETALES				
Treponema pallidum§ kaza-5	5–10			
Treponema pallidum (Reiter)**,††		7		4
Leptospira canicola	60–65			
Leptospira patoc	80–90	5–10		1–5

* The values are given as the percentage of lipid phosphorus unless otherwise stated.
† 30–35% – glucosaminylphosphatidylglycerol.
‡ 38% phosphatidylmethylethanolamine is present of which 78% is plasmalogen, 55% of the phosphatidylethanolamine and 9% of the phosphatidylglycerol also of the plasmalogen form.
§ 30–40% phosphatidylcholine present and 25% monogalactosyldiglyceride.
** 41% phosphatidylcholine present and 37% glucosyldiglyceride.
†† Values given as the percentage of total lipid by weight.

Table 3.4. Phospholipids of Gram-negative bacteria (from [40]).*

Organism	*Phosphatidyl-ethanolamine*	*Phosphatidyl-N-methyl-ethanolamine*	*Phosphatidyl-choline*	*Phosphatidyl-glycerol*	*Cardiolipin (Diphosphatidyl-glycerol)*
Azotobacteraceae					
Azotobacter agilis (log phase)	64	5	1	27	2.4
Azotobacter agilis (stationary phase)		5	2.4	13	2.3
Azotobacter vinelandii	53			←7→	
Rhizobiaceae					
Agrobacterium tumefaciens (log phase)	45	14	7	29	
Agrobacterium tumefaciens (stationary phase)	18	16	28	13	19
Chromobacterium violaceum	77			18	4.6
Enterobacteriaceae					
Escherichia coli (log phase)	76			20	1.1
Escherichia coli (stationary phase)	77			11	6.6
Serratieae					
Serratia marcescens	66			14	17
Proteeae					
Proteus vulgaris (log phase)	63	4.0		17	2.9
Proteus vulgaris (stationary phase)	63	12.1		5.8	14.8
Salmonelleae					
Salmonella typhimurium	78			18	3.2
Brucellaceae					
Brucella abortus Bang 1119†	←27→		37	16	5.8
Haemophilus parainfluenzae	78	0.4	0.4	18	3.0

Chromatium strain D	[illegible]			39	5
Anthiorhodaceae					
Rhodopseudomonas spheroides	41		19	36	
Rhodopseudomonas capsulata	46		13	41	
Rhodospirillum rubrum	19		10		5
Nitrobacteraceae					
Nitrocystis oceanus	67		3	←28→	
Nitromonas europaea	78			←17→	
Thiobacteriaceae					
Thiobacillus neapolitanus†	42–45	20–27		11–15	17–23
Thiobacillus thioparus	65			24	11
Thiobacillus intermedius†	55–60	12–15		2–20	10–26
Thiobacillus thio-oxidans (log phase)	20	36		37	7
Thiobacillus thio-oxidans (stationary phase)	4	53		27	16
Thiobacillus novellus†	23–27	3–11	33–37	24–30	5–7
Pseudomonadaceae					
Pseudomonas aeruginosa	69		1.4	15	9.4
Siderocapsaceae					
Ferrobacillus ferro-oxidans	20	42	1.5	23	13
Hypermicrobiales					
Hyphomicrobium vulgare NQ521	23		29	10	
Rhodomicrobium vannielii	4.5		27	10	

* The values are given as the percentage of lipid phosphorus

† Ranges for different media.

increasing array of different molecules has not kept pace. The result, therefore, of any attempt to include everything would rapidly reduce this chapter to little more than a catalogue.

3.3.1 *Phospholipids in bacterial membranes*

In those instances examined some 90% or more of the total bacterial cell content of phospholipids and glycolipids is found in the membranes. In Gram-positive species this is found in the cytoplasmic and mesosomal membranes, and in Gram-negative species it is found in the outer membrane as well. The latter also contains, of course, the lipid A molecules as part of the lipopolysaccharide. By far the commonest phospholipids are phosphatidylethanolamine (frequently predominant in Gram-negative species) phosphatidylglycerol and diphosphatidyl glycerol or cardiolipin (see Fig. 3.2). Phosphatidylcholine, universally present as the dominant phospholipid in membranes from animals, plants and fungi, is a more rare component of bacteria. It is absent altogether from Gram-positive species and has been found principally in *Agrobacteria, Nitrocystis oceanus, Thiobacillus novellus, Ferrobacillus ferro-oxidans* and some photosynthetic bacteria (see Table 3.3). Likewise sphingolipids are usually absent and phosphatidylinositol, a common component of membranes from other sources is rare in bacteria except in the *Mycobacteria, Propionibacteria* and *Corynebacteria,* although it has been detected in membranes from *Micrococcus luteus* [30, 78, 98]. Some of the quantitative values for contents of phospholipids are set out in Tables 3.3 and 3.4. Noteworthy are the very large differences in relative proportions of the three commonest compounds between different species. For example the proportion of phosphatidylglycerol varies from 72% of the total in *M. luteus* to 13% in *Bacillus subtilis* and 5-6% in stationary-phase cultures of *Proteus vulgaris.* The result for the latter bacterium calls attention to the variations in phospholipid composition that can occur when organisms are grown under different conditions, since, for example, exponentially growing bacteria contain 17% of phosphatidylglycerol. *B. megaterium* strain NK10D grown at pH 5.0 had 5-10% of this phospholipid whereas at pH 7, the proportion rose to 35-45% [40, 93]. In the Gram-positive cocci and in some *Bacilli* a relatively high proportion of the phosphatidylglycerol exists as the amino-acyl derivative, the amino acids most commonly involved being lysine and alanine. Again the proportion of these compounds varies considerably according to the growth phase of the cultures and their pH [51]. Even more dramatic changes can be achieved by growing organisms in continuous culture. For example, when *Pseudomonas aeruginosa* or *Ps. fluorescens* were grown under phosphate limitation, ornithine amide lipids came to dominate the picture [86, 87]. The occurrence of ornithine lipids in bacterial membranes had been previously described [17, 61, 165]. Strictly anaerobic bacteria such as the *Clostridia* have a further peculiarity in that the phospholipids occur in their aldehyde forms as plasmalogens [57] (see Fig. 3.3). In *Cl. butyricum* for example, these are ethanolamine, *N*-methylethanolamine, and phosphatidylglycerol

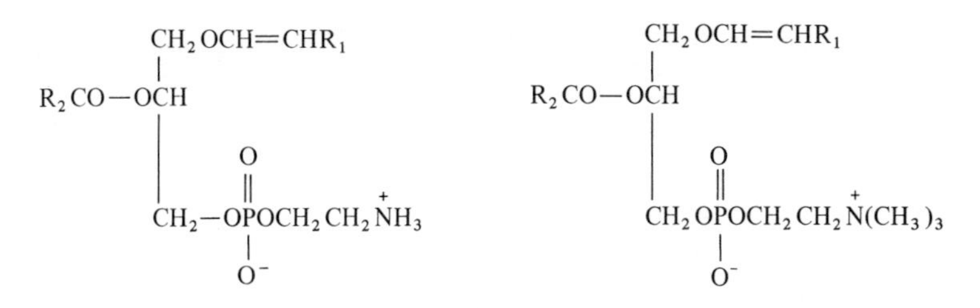

Figure 3.3 Structures of the ethanolamine and choline plasmalogen phospholipids in microbial membranes.

Table 3.5. Phospholipid content of some anaerobic Gram-negative cocci (from [39]).*

	Megasphaera elsdenii	*Veillonella parvula*	*Selenomonas ruminantium*	*Anearovibrio lipolytica*
Diacylphosphatidyl-serine	10	19	17.6	19.8
Serine plasmalogen	25.8	7.5	6.6	3.5
Diacylphosphatidyl-ethanolamine	7.8	28.5	27.2	44.8
Ethanolamine plasmalogen	54.3	45.1	45.0	29.8
Unknown diacyl-phospholipid	2.0	Trace	3.6	2.1

* Values are expressed as the percentage of total lipid phosphate

compounds in the plasmalogen form, making up respectively 78%, 55% and 9% of the total phospholipid. In the anaerobic Gram-negative cocci, however, a further novelty is found [39] where a high proportion of what is normally regarded as the precursor of phosphatidylethanolamine namely phosphatidylserine occurs in the form of plasmalogen (Table 3.5). As in a number of other anaerobes *O*-alkyl acetyl-phosphatide was also present among the phospholipids of this group of bacteria.

In Gram-negative bacteria the phospholipid content of the cytoplasmic membrane differs from that of the outer membrane. In general the same compounds are present although the contents of phosphatidylglycerol and cardiolipin are lower in the outer membrane whereas the proportion of phosphatidylethanolamine is increased (see Table 3.6). The amount of total phospholipid per unit mass of membrane in the outer layer is also considerably lower than in the cytoplasmic membrane. Examination [120] of the viscosity of the hydrophobic heart of the inner and outer membranes of, for example, *P. vulgaris,* showed that the latter had a much higher viscosity. Evidence would suggest [140] that almost all the phospholipids of the outer membrane are confined to the inner leaflet, their place being taken in the outer leaflet by the lipid A

Table 3.6. Composition of inner and outer membranes from Gram-negative species.

Organism	*Phospholipid*	*Membrane*	
		Cytoplasmic	*Outer*
*Salmonella typhimurium**	Total phospholipid (PL) (mg PL/mg protein)	0.53–0.61	0.3
	Phosphatidylglycerol (% total PL as glycerol)	33	17
	Diphosphatidylglycerol (% total PL as glycerol)	6.8	1.7
	Phosphatidylethanolamine as ratio	1	1.3
Escherichia coli†	Total phospholipid (mg/g cells)	9.1 ± 0.4	3.4 ± 0.2
Proteus mirabilis‡	Total phospholipid (% of membrane dry wt)	38	18
	Phosphatidylglycerol, diphosphatidylglycerol (% total polar lipid)	30	15–20
	Phosphatidylethanolamine (% total polar lipid)	55–60	75–80

* [96].
† [88].
‡ [120].

part of the lipopolysaccharides. This association of the fatty acids with the largely exteriorly directed lipopolysaccharide, may account for the higher viscosity demonstrated [120].

3.3.2 *The fatty acids of the phospholipids*

Just as the proportions of the phospholipids in bacterial membranes differ from those in other living cells, so do the fatty acids in them. Whereas in the membranes of other sorts of cell, polyunsaturated acids are common whilst branched chain and cyclopropane acids are missing, the reverse is true for bacteria. Polyunsaturated fatty acids are not present in bacteria, except for *Mycobacterium phlei* [7]. The common fatty acids present have chains 10–20 carbons long with 15–19 carbon chains predominating, and are of four types: straight chain saturated, straight chain mono-unsaturated, branched chain commonly iso- and anteiso-, and lastly cyclopropane. In the latter the single unsaturated bond of a mono-unsaturated acid is satisfied by a methylene group. Examples of branched and cyclopropane fatty acids are illustrated in Fig. 3.4. Branched chain fatty acids are much more common among Gram-positive species, where they may be always present. Among Gram-negative species there is much greater variation and branched chain acids may be missing altogether in some instances [58]. In some extreme halophiles the fatty acids are replaced by ether-

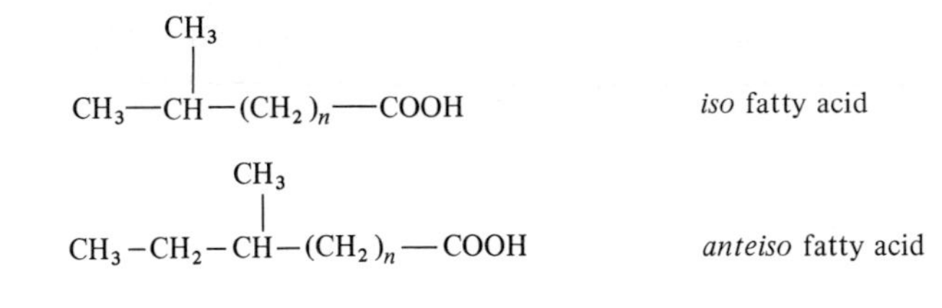

$CH_3-(CH_2)_{n_1}HC \overset{CH_2}{\triangle} CH(CH_2)_{n_2}-COOH$ cyclopropane fatty acid

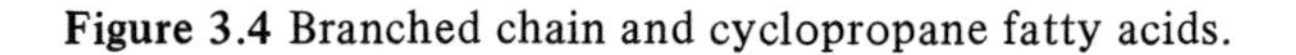

Figure 3.4 Branched chain and cyclopropane fatty acids.

$CH_3-CH(CH_3)-CH_2-CH_2-(CH_2-CH(CH_3)-CH_2-CH_2)_3-OCH$

$CH_2-CH_2(CH_3)-CH_2-CH_2-(CH_2-CH(CH_3)-CH_2-CH_2)_3-OCH$

$CH_2OPO_3^{-}-CH_2-CHOH-CH_2OPO_3^{2-}$

Figure 3.5 Structure of the dihydrophytyl ether analogue of phosphatidylglycerol-phosphate.

substituted alkyl residues [59, 133], the major alcohol being dihydrophytyl alcohol (see Fig. 3.5). Lipids containing glycerol ether-linked residues rather than ester linkages, are also present in the membranes of *Thermoplasma acidophilum.* For those needing further detailed information about the fatty acid composition of bacterial lipids, a number of tables of analyses exist [58, 115]. It must, however, be pointed out that these values must be closely correlated with the conditions used for growing the organisms, since variations in pH, growth phase, medium constituents and temperature can all cause very considerable variation in the proportions of the classes of fatty acid involved. Two well known examples are that the proportion of unsaturated fatty acids in the lipids of a Gram-negative organism such as *E. coli* can increase in an inverse relationship to the growth temperature, over a range between 10° C and 43° C [81]. Another well known change is that of the increase in the proportion of cyclopropane fatty acids when growth of an aerobic organism is limited by the supply of oxygen [1, 66].

Over and above such changes introduced by nutritional factors, use can be made of the large number of mutants conditionally blocked at various stages in the synthesis of fatty acids [23, 136]. Some of these mutants are listed in Table 3.7. Considerable and deliberate manipulation of the fatty acid content of membranes can be achieved using them. Similar changes can also be brought about in the membranes of

Table 3.7. Mutants of *E. coli* disturbed in lipid biosynthesis (from [136] and [108a]*).

Mutants	*Phenotype*	*Biochemical defect*	*Map position* (min)
Fatty acid biosynthesis			
fab A	Unsaturated fatty acid auxotroph	β-Hydroxydecanoyl thi thioester dehydrase	22
fab B	Unsaturated fatty acid auxotroph	β-ketoacyl ACP synthetase II	44
cvc	Reduced *cis*-Δ^{11}-18:1 synthesis	β-ketoacyl ACP synthetase II	?
fab D	Temperature-sensitive fatty acid synthesis	Malonyl CoA–ACP transacylase	24
Phospholipid biosynthesis			
gps A	L-glycerol-3-phosphate auxotroph	L-glycerol-3-phosphate dehydrogenase	79
pls B	L-glycerol-3-phosphate auxotroph	L-glycerol-3-phosphate acyltransferase (altered K_M for substrate)	69
pls A	Temperature-sensitive phospholipid synthesis	L-glycerol-3-phosphate acyltransferase	13
psd	Temperature-sensitive	Phosphatidylserine decarboxylase	95
pgs	No obvious phenotype	Phosphatidylglycerol phosphate synthetase	41
cls	None	Cardiolipid synthetase	28
dgk	Osmotic fragility	Diglyceride kinase	91

* For references to the original studies this review should be consulted.

Acholeplasma by simply growing the organisms with different fatty acids [111, 112], for example changes in viscosity and packing of the lipid tails in the lipid bilayer. Evidence for the dependence of the function of membrane enzymes in relation to the fatty acid composition has been obtained, but a surprising degree of variation is sometimes possible without obvious effects. For example, mutants with the lesions affecting the enzyme that transfers the methyl carbon of *S*-adenosylmethionine to the double bond of a fatty acid making cyclopropane phospholipid, synthesized <0.1% of the normal amount but showed no change in physiological phenotype. One very general property of the lipids found essential for normal growth of *E. coli* has now been determined by three laboratories using different techniques and manipulations to vary the degree of unsaturation of the fatty acids [53, 151, 155]. All three groups of workers concluded that about half the lipid must be in a disordered non-crystalline form. A very similar result had been reported for *Acholeplasma laidlawii* [75, 76]. Thus definite, if rather broad, limits are set for the acceptable degree of the fluidity of membrane lipids at any given growth temperature. However, it must be pointed out that growth of the *Acholeplasma* only stopped completely when less than 10% of the lipid remained in the fluid state.

As has been pointed out before, bacteria contain no sterols unless grown in their presence [110]. In mammalian membranes cholesterol is thought to play an important part in regulating the fluidity of the lipid bilayer. The *Mycoplasmas,* excepting the genus *Acholeplasma,* require sterols for growth and incorporate them into their membranes [112]. By manipulating a strain of *Mycoplasma mycoides* var. *capri* an adapted variety was established that contained $<3\%$ cholesterol, compared with 22–25% in the parent [121, 122]. With the lower sterol content, phase transitions in the membrane lipids with varying temperature could be clearly demonstrated, a result agreeing with studies of phosphatidylcholine liposomes, in which the presence of large amounts of cholesterol suppressed phase transitions. The reduced sterol content of the *Mycoplasma* led, as would be expected from what was said earlier about membrane fluidity, to failure to grow at low temperatures. The cells were also more osmotically fragile and Arrhenius plots of the membrane ATPase and α-methyl glucosidase activities showed discontinuities corresponding with the transition point of the lipids at 25° C.

3.3.3 *Glycolipids and glycophospholipids*

Glycosyldiglycerides

The first recognition of this class of membrane components in bacteria was by Macfarlane [79] who isolated a compound from the protoplast membranes of *M. luteus* that on acid hydrolysis of the water soluble saponification product gave equivalent amounts of glycerol and mannose. Such compounds were well recognized in plants and are abundantly present in their chloroplast membranes. Since this time a large number (see Table 3.8) of such compounds has been isolated from bacteria. They are carbohydrate derivatives of 1, 2 diacyl-*sn*-glycerol with the carbohydrates joined to the 3-hydroxyl of the glycerol moiety by a glycoside bond. The carbohydrate can be a mono-, di-, tri- or tetrasaccharide. Table 3.8 lists the compounds that have been isolated and examined structurally. There is a considerable number of other organisms where evidence exists for the presence of glycolipids but the compounds have not been fully examined. An amino sugar containing diglyceride, 1-(*O*-β-glucosaminyl-2, 3-diglyceride, has been isolated from *B. megaterium* as a minor constituent [106]. *Acholeplasma modiculum* contains [84, 85] a pentaglycosyl diglyceride, the carbohydrate residues being two molecules of galactose, one of mannoheptose and two of glucose. This compound is the major glycolipid component of the membranes of this species, but also present are the more common diglucosyl and monoglucosyldiglyceride. Thus, organisms may contain more than one glycolipid and these are major lipid components amounting in some instances to 50% of the total lipid in the cell. Where complex glycolipids co-exist in the membrane with simpler ones it is reasonable to suppose that the latter may be accumulations of intermediates for the synthesis of the oligosaccharide containing ones. Biosynthesis is by the simple process of addition from nucleoside diphosphate precursors [134]. There is evidence that the glycolipid composition of membranes, like that of other cell components,

Table 3.8. The glycosyl diglycerides of known structure in bacteria (from [134, 135]).

Carbohydrate residues	*Organism*
Monoglycosyl	
Glcα1 →	*Acholeplasma laidlawii*
	Acholeplasma modicum
Glcβ1 →	*Mycoplasma neurolyticum*
Galα1 →	*Treponema pallidum*
Galβ1 →	*Arthrobacter* sp.
Galfβ1 →	*Mycoplasma mycoides*
	Bifidobacterium bifidum
GlcAα1 →	*Pseudomonas diminuta*
GlcAβ1 →	*Pseudomonas nibexcus*
GlcNβ1 →	*Bacillus megaterium*
Diglycosyl	
Glcα1 → 2Glcα1 →	*Streptococcus lactis*
	Streptococcus haemolyticus
	Streptococcus faecalis
	Acholeplasma laidlawii
Glcβ1 → 6Glcβ1 →	*Staphylococcus lactis*
	Staphylococcus saprophyticus
	Staphylococcus aureus
	Bacillus subtilis
	Mycoplasma neurolyticum
	Pseudomonas iodinium
Galβ1 → 6Galβ1 →	*Arthrobacter crystallopoietes*
	Arthrobacter pasceus
	Arthrobacter globiformis
Galβ1 → 2Galβ1 →	*Bifidobacterium bifidum*
Galfβ1 → 2Galfβ1 →	*Bifidobacterium bifidum*
Manα1 → 3Manα1 →	*Micrococcus luteus*
	Arthrobacter sp. (see above)
Galα1 → 2Glcα1 →	*Diplococcus pneumococcus* type 1
	Diplococcus pneumococcus type XIV
	Lactobacillus casei
	Lactobacillus buchneri
	Lactobacillus plantarum
	Lactobacillus helveticus
	Lactobacillus acidophilus
	Listeria monocytogenes
Glcβ1 → 4GlcAα1 →	*Pseudomonas diminuta*
Glcα1 → 4GlcAα1 →	*Streptomyces LA7017*
Triglycosyl	
Glcα1 → 2Glcα1 → 2Glcα1 →	*Streptococcus haemolyticus*
Glcα1 → 6Galβ1 → 2Glcα →	*Lactobacillus casei*
	Lactobacillus helveticus
	Lactobacillus acidophilus
3′–SO_3–Galβ1 → 6Manα1 → 2Glcα1 →	*Halobacterium cutirubrum*
Galβ1 → 2Galβ1 → 2Galβ1 →	*Bifidobacterium bifidum*
Tetraglycosyl	
Glc1 → 6Glcα1 → 6Galα1 → 2Glcα1 →	*Lactobacillus acidophilus*
Galf1 — 2Gal1 — 6GlcNHWR1 → 2Glc1 →	*Flavobacterium thermophilum*

can be altered by the conditions under which micro-organisms are grown. One of the most dramatic results was that, when *Pseudomonas diminuta* was grown under phosphate limitation in continuous culture, a high proportion of the acid phospholipids was converted to uronic acid containing glycolipids [88]. It will be remembered (p. 82) that *Pseudomonads* similarly treated replaced phosphatidylethanolamine with an ornithine-amide lipid. Too little has been done to be able to say how far deliberate nutritional limitations affect the glycolipid or phospholipid composition of the membranes of other species.

Acylated sugar derivatives

A variety of compounds of acylated sugars exist in bacteria but rather little attention has been paid to their topological location. In some instances such as that of the dirhamnose-β-hydroxydecanoyl-β-hydroxdecanoate isolated from *Pseudomonas aeruginosa* [33], it would seem likely that they are indeed membrane components. On the other hand, the well known 6, 6-dimycolates of α, α-D-trehalose (cord factors) of the mycobacteria are treated as wall components and the arabinogalactans linked to mycolic acid (wax D) are thought to be actually covalently attached to the peptidoglycan [104].

Glycophospholipids

This group of compounds like the glycolipids consists of sugars attached to glycerol acylated in the 1–2 positions but instead of a direct glycosidic linkage, a phosphate diester group is interposed. Thus they form an intermediate group between the phospholipids and the glycolipids. The earliest members of the group to be discovered were a family of phosphatidylinositol mannosides in *Mycobacteria* which were isolated in the 1930s but the structures of which were not proposed until 1955 [67]. Mannosides of this type are illustrated in Fig. 3.6. The tri- and tetramannosides are made by extending the mannosyl residue on the six position of inositol by α-1, 6-linkages. In the pentamannoside, the terminal mannose has an α-1,2-linkage. These types of glycophospholipid have been found in the *Actinomycetales, Nocardia, Streptomycetes* and *Corynebacteria.* Phosphatidylinositol has been reported in *Arthrobacter* and a diacylinositol mannoside in *Propionibacteria* [135]. Phosphatidylglycerol β-glucosaminides have been isolated from *B. megaterium.* In one, the glycosidic linkage is to the three position of glycerol, in the other to the C_2 [74]. An α-glucosaminide with linkage to the C_2 of glycerol has been isolated from *Pseudomonas ovalis* [74, 95]; a glucosaminyl phospholipid had been isolated earlier from *Bacillus* membranes [93, 94]. Phosphatidyldiglucosyldiglyceride and glycerolphosphoryldiglucosyldiglyceride (Fig. 3.7) have both been isolated from *Streptococcus faecalis* [3, 34, 35, 135]. Analogous lipids have also been found in *Acholeplasma laidlawii,* species of *Cellulomonas, Leuconostoc mesenteroides, Listeria monocytogenes* and *Pseudomonas diminuta.* The function of this large family of glycolipids and glycophospholipids in the membrane of bacteria and in chloroplast and mitochondrial membranes from more complex forms of life, is quite unclear. In a few instances among bacteria the lipoteichoic acids (see Section

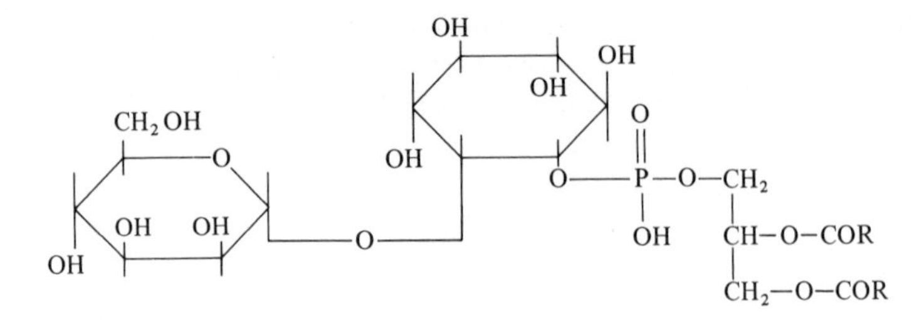

(a) Phosphatidylinositolmonomannoside

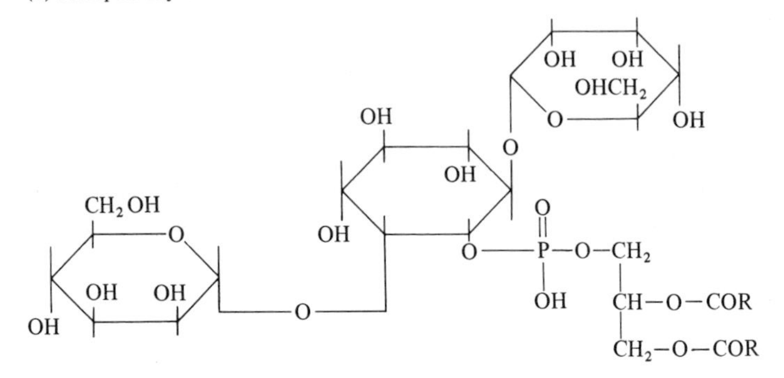

(b) Phosphatidylinositoldimannoside

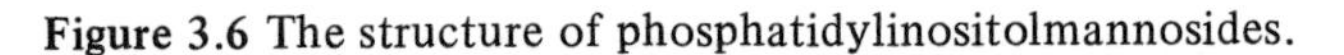

Figure 3.6 The structure of phosphatidylinositolmannosides.

7.2.1) are known to be 'anchored' to the membranes via a membrane glycolipid [19, 42, 154, 164], which may either be one of the normal glycolipids or a special one not found free in the membrane preparations.

3.3.4 *Hydrocarbons, quinones and isoprene alcohols*

Carotenoids

The carotenoids are ubiquitous in nature and they occur widely as membrane components of bacteria. Those in the photosynthetic bacteria have been subject to particular attention and Fig. 3.8 shows the structure and source of some carotenoids. The chemistry and biosynthesis of these and other carotenoids has been reviewed [71]. Apart from the photosynthetic bacteria the carotenoid content of *Flavobacterium dehydrogenans, Halobacterium salinarum* and *Corynebacterium poinsettiae* have been studied each yielding a number of carotenoids and acyclic precursors.

Benzoquinones and naphthoquinones

The widespread occurrence of small amounts of quinone substances with attached polyisoprenoid side chains was the origin of the trivial name 'ubiquinone' for the

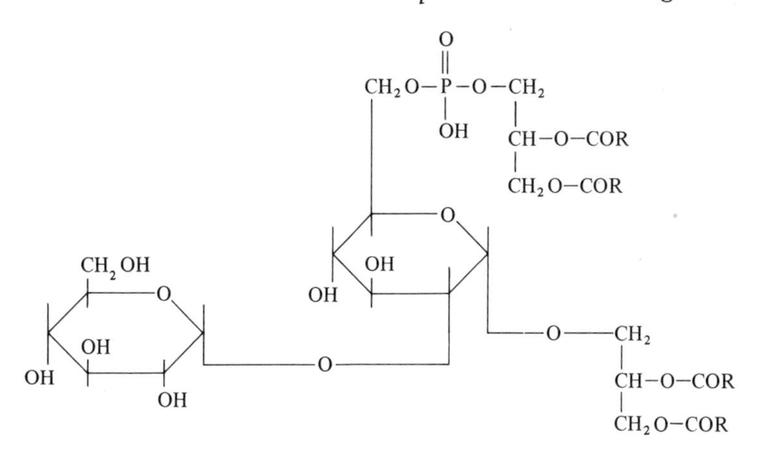

(a) Phosphatidyldiglucosyldiglyceride

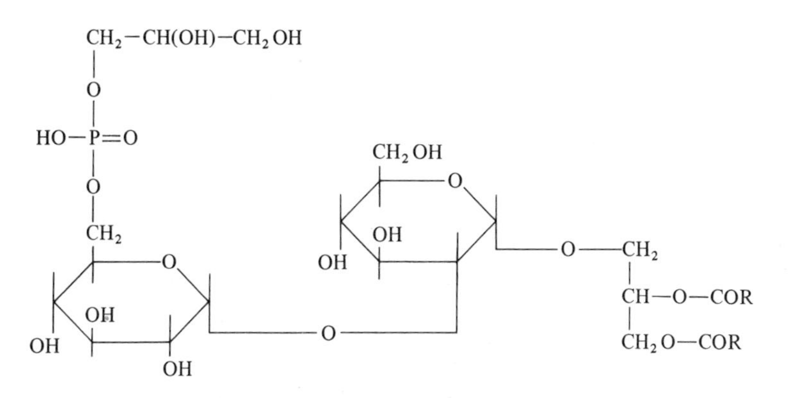

(b) Glycerylphosphoryldiglucosyldiglyceride

Figure 3.7 Structures of the phosphoglycolipids isolated from *Streptococcus faecalis.*

benzoquinone series [89]. They consist of two series, the 5,6-dimethoxyquinones and the menaquinones or K vitamins. The isoprene chains are of variable length according to the source of the quinone [152]. They are highly lipophilic and are dissolved in the membranes where they play a vital role in electron transport (see Chapter 4.3.3). The quantitative distribution of the quinones in bacteria has been examined [10] (see Table 3.9). Whereas Gram-positive species grown aerobically have principally or only menaquinones, Gram-negative species may contain both, or only the ubiquinone series, anaerobes such as the *Clostridia* contain neither; a single strain of *Staphylococcus aureus* grown anaerobically contained neither, although under aerobic conditions it contains naphthoquinone. A survey of the isoprenoid quinones in Coryneform bacteria and related species, has been undertaken [21]. Menaquinones were found in 85 out of 95 Coryneforms as the only quinone.

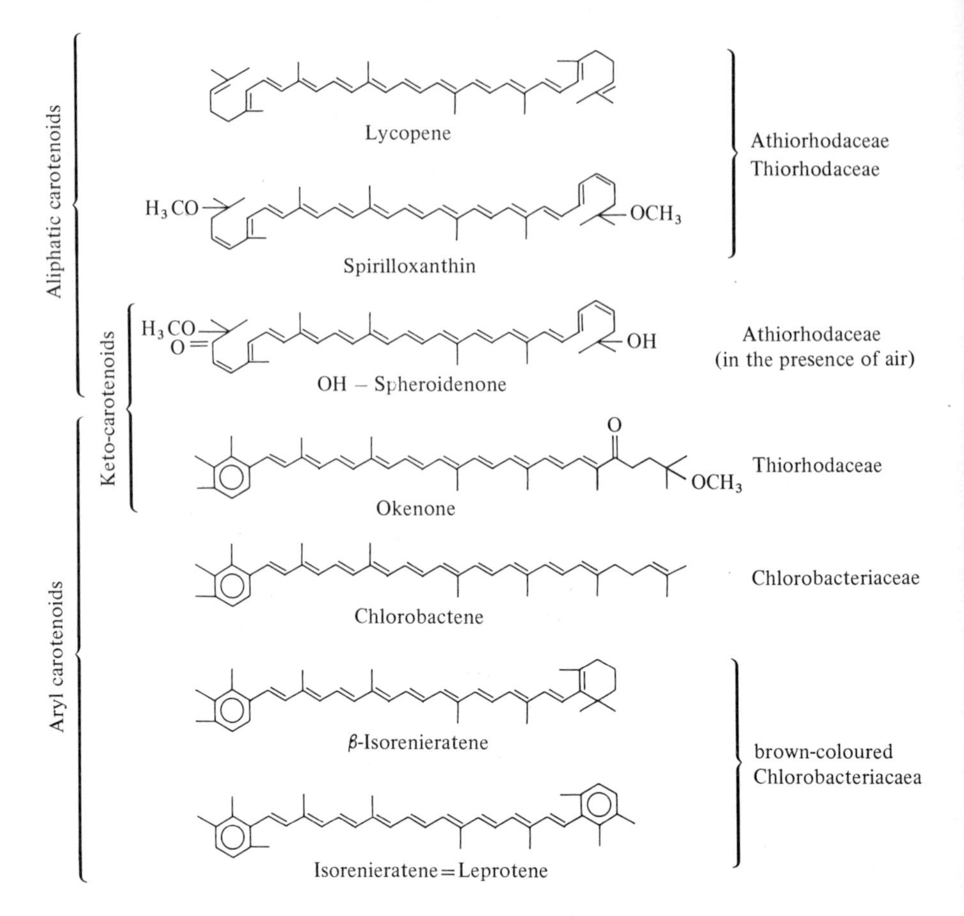

Figure 3.8 Structures of some of the carotenoids from bacteria together with their source of origin (from [105]).

Isoprenoid alcohols

The functional importance of phosphorylated C_{55} isoprenoid alcohol as a carrier lipid in the biosynthesis of peptidoglycan, lipopolysaccharides, teichoic acids and other extracellular hydrophilic substances by bacteria is by now well known (see Sections 8.3 and 10.3). Like the quinones, isoprene alcohols of various chain lengths seem ubiquitous and much evidence would point to an export function for them. For example compounds with chain lengths of from C_{90} to C_{105} have been isolated from bakers yeast [31] and polyprenols are known to act as carriers in the synthesis of wall mannan [149] (see Section 14.3). A series of hexahydroprenol esters of palmitic, stearic, palmitoleic, oleic and linoleic acids have been isolated from *Aspergillus fumigatus* [147, 148].

Table 3.9. Distribution of ubiquinone and vitamin K in micro-organisms (from [10]).

Organism	*Cultural conditions*	*Total lipid* (% dry wt)	*Ubiquinone* (μmol/g dry wt)	*Vitamin K_2* (μmol/g dry wt)
Gram-positive				
Bacillus subtilis	aerobic	1.1	<0.001	0.7
B. subtilis (spores)	aerobic	0.2	<0.001	<0.001
B. megaterium (NCTC9848)	aerobic	1.9	<0.001	0.66
Lactobacillus casei	aerobic	1.5	<0.001	<0.001
Staphylococcus aureus	aerobic	3.8	<0.001	1.4
	anaerobic	3.0	<0.001	<0.01
Sarcina lutea	aerobic	1.8	<0.001	1.8
Clostridium sporogenes	anaerobic	2.0	<0.001	<0.001
Corynebacterium PW8 diphtheriae	aerobic (low Fe)	–	<0.001	1.3
Gram-negative				
Azotobacter chroococcum (NC1B 8003)	aerobic	1.7	0.40–0.48	<0.05
Escherichia coli	aerobic	2.0	0.32–0.41	0.32
	anaerobic	2.3	0.33–0.40	0.28
Proteus vulgaris	aerobic	5.4	0.62–0.67	0.62
	anaerobic	5.0	0.51–0.61	0.51

Other isoprenoid hydrocarbons with less obvious functions have also been isolated from bacteria such, for example, as 15,16-dihydrosqualene from *S. aureus* and di- and tetra-hydrosqualenes from *Halobacterium cutirubrum* [153]. These too are presumably membrane components considering their hydrophobicity.

3.3.5 *The lipomannan of Micrococcus luteus*

Whilst Gram-positive organisms generally contain a teichoic acid covalently-linked to a membrane glycolipid, *M. luteus* does not. It had been known for many years [38, 77] that its protoplast membranes contained a very large proportion of carbohydrate as mannose and this sugar was ultimately shown [99–101, 107, 108, 128] to be present as a mannan of a size corresponding to about 50 monomer residues. Substitution by succinyl groups make it acidic and it is probably attached to a membrane glycolipid, accounting for its membrane location. In some ways this substance is analogous to the lipoteichoic acids in other organisms. It appears to be accumulated preferentially in the mesosomal vesicle of the organism [101].

3.4 Proteins in membranes

As already pointed out, the undifferentiated cytoplasmic membranes of bacteria are relatively rich in proteins compared with the more highly differentiated structures present in the cells of eukaryotes, apart from that is, the inner membrane of mitochondria. The total content of protein amounts to 60–70% of their dry weight. However, it is probably true to say that we still cannot be sure just how many proteins are present in any membrane. The commonest method of tackling this problem is to disaggregate washed membranes with solutions of neutral or anionic detergents such as Triton X-100 or sodium dodecyl sulphate (SDS) respectively, followed by application of the 'solution' to a tube or sheet of polyacrylamide gel made to contain a known degree of cross-linkage. The gel is then subjected to electrophoresis in a suitable apparatus. When an anionic detergent like SDS is used it couples tightly with the hydrophobic membrane proteins, so that all of them have a strongly negative charge and move distances in the gels that are related to their molecular size and the degree of cross-linking of the polyacrylamide. A number of factors influence the qualitative results obtained;

(1) The method of preparing the membrane (e.g. removal of peripheral proteins or ribosomes),
(2) The degree of cross-linkage of the polyacrylamide,
(3) The length of storage of the membrane preparation before disaggregation by SDS; proteolytic enzymes present in the preparations can hydrolyse the proteins during storage,
(4) The temperatures at which the membranes are treated with SDS solution to disaggregate them.

When suitable precautions have been taken and membranes are examined by this method, some 30–40 bands can be visualized in the gels by first staining them with a suitable dye such as Coomassie Blue, followed by destaining of the background with acetic acid. Alternatively the bands may be detected by autoradiography if the organisms are first grown in the presence of a suitable radioactively-labelled precursor. Judging by the distances travelled in SDS-containing gels, the molecular weights of the proteins vary from about 16 000 to more than 100 000. Scanning either the gels themselves after staining, or photographing them can give some rather uncertain indications of the relative amounts of proteins in each of the bands. However, it is clear that the results obtained are only crude. The proteins in the bands react to different extents with the dyes and complete resolution of the bands cannot be obtained. Attention is drawn to the limits of resolution of undirectional polyacrylamide gel electrophoresis of complex protein mixtures when they are examined by two-dimensional methods. Proteins can be separated by their charge in one direction and by their molecular size in the other. Very many more can then be shown to be present in the mixture. The originator [92] of the first satisfactory method for two-dimensional gel electrophoresis, for example, used it to examine the total proteins of *E. coli*. When unidirectional gradient polyacrylamide gel electrophoresis was undertaken, 100 bands were recognized. When, however, this method of separation was combined with iso-electric focussing in the second direction, 1100 spots were found. Since 70 proteins were resolved using the latter method alone, the two-dimensional technique 'should resolve some 7000 components' [92]. However, the chromosome of *E. coli* is thought to code for only some 4000 proteins and, as the author says, the theoretical resolving power of the system may not have been fully saturated. A further group of methods which adds specificity to electrophoretic analysis of membrane proteins is immunoelectrophoresis. In these methods the membrane proteins are drawn by electrophoresis through gels containing homogeneously distributed antibody. A specialized form of this is called crossed or rocket-immunoelectrophoresis [11, 63–4, 158, 161]. By this technique the membranes are first disaggregated by neutral detergents and then electrophoresed, first in agarose gel, and then in a second direction into a block of the same gel containing homogeneously distributed antibodies made against the membrane proteins. Rocket-like or arc-shaped tracks are seen in the second gel after staining with suitable dyes. Some of the advantages of this method are firstly that the areas underneath the stained arcs are directly related to the quantity of the particular protein. Secondly, since only neutral detergents need be involved, enzymes are not inactivated and should their actions not be neutralized by combination with antibody, their presence in the gels can be identified by specific staining [103].

3.4.1 *Cytoplasmic and mesosomal membranes*

Consideration of the range of enzyme functions undertaken by bacterial membranes would suggest the presence of at least an equally large number of proteins. When the membranes are disaggregated by heating to 100 °C for 1 min in 1% SDS solution and applied to a polyacrylamide gel made up to give about 10–13% cross-linking, about 30–40 bands can be visualized as has been said (Fig. 3.9). These bands vary in width and in depth of staining and the pattern produced is characteristic of the micro-organism. Indeed gel patterns from unidirectional electrophoresis of membrane proteins have been suggested [119] as sufficiently characteristic to be used as 'finger prints' in the classification of *Mycoplasmas*. When however, cytoplasmic membranes are disaggregated and examined by the O'Farrell [92] two-dimensional electrophoresis, a very much larger number of spots can be seen in membrane preparations from *Acholeplasma laidlawii* about 140 have been found [6]. Even so it may still be doubted whether enzymes represented by only a limited number of copies in the membrane would be present in sufficient quantity to be registered by either of these techniques.

Crossed immunoelectrophoresis has so far been used by rather few groups of workers to study microbiological membranes [54, 55, 99, 102, 103, 125, 126, 142, 143]. Salton and his co-workers have nevertheless made great progress in identifying enzymes after application of the crossed-immunoelectrophoresis or so-called rocket-electrophoresis analytical method. The resolving power of the technique is, however, only about the same as the unidirectional method and so far 46 antigens have been recognized in the plasma membrane of *E. coli* [143] and 27 antigens in that of *M. luteus* [102]. A major factor limiting the analysis of membrane antigens is undoubtedly the spreading of the proteins in the first direction of electrophoresis. This leads to subsequent arcs with very wide bases during the running in the antibody-containing gel. Some of the major antigens give arcs, the bases of which occupy about half or more of the total distance run by any of the entities. Nevertheless, the

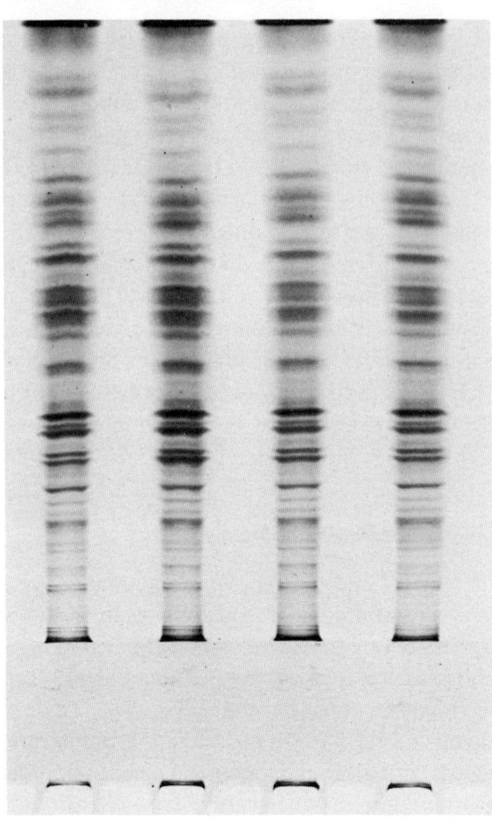

Figure 3.9 Uni-directional polyacrylamide gel electrophoresis of a membrane preparation from *Bacillus subtilis* disaggregated in sodium dodecyl sulphate. The degree of cross-linking of the gel was adjusted to 13%. The gel was stained with Coumassie Blue. The three runs were from the same membrane and are shown to demonstrate the excellent reproducibility of the method (Rogers and Thurman, unpublished work).

ability to quantitate the immuno-precipitates and to detect specific enzymes more than justifies use of the technique in its present form. Salton and his co-workers have succeeded in identifying eleven active membrane enzyme–antibody immuno-precipitates in the gels to date. These include dehydrogenases utilizing succinate, 1,6-phosphogluconate, dihydro-orotate, two utilizing NADH, malate, glycerol-3-phosphate, lactate and glutamate as substrates; an adenosine triphosphatase and a protease. More will be said about membrane enzymes in the chapter on membrane function.

Examinations of the mesosomal membranes from *B. licheniformis* [113], *M. luteus* [97] and *S. aureus* [150] by unidirectional polyacrylamide gel electrophoresis after solubilization in sodium dodecyl sulphate solutions have shown that their proteins are not identical to those of the cytoplasmic membranes. The mesosomal membrane from *B. licheniformis* was considerably enriched in three proteins while mesosomes from *M. luteus* appeared to differ even more from the cytoplasmic membranes than did those from the *Bacillus*: a number of high molecular weight proteins present in the cytoplasmic membranes were either greatly reduced or missing in the mesosomes. There also seemed to be some enrichment in lower molecular weight proteins. Staphylococcal mesosomes, on the other hand, were specifically enriched in a protein of molecular weight of about 35 000. Unambiguous results showing a reduction in the amounts of a number of enzymically active proteins in mesosomes have been obtained by crossed-electrophoresis [126]. Greatly reduced arcs corresponding to the dehydrogenases and ATPase were clearly demonstrated in mesosomal membranes of *M. luteus*. The arc corresponding to the membrane mannan in mesosomes was increased confirming chemical estimations [97, 100]. The demonstration that certain enzyme proteins were present in greatly reduced amounts in mesosomes is particularly important and gratifying. Much earlier work (see Chapter 4.9) had demonstrated the reduced enzymic activity present, but it was always possible that this was due either to the enzymes being inaccessible to their substrates in mesosomes or that the enzyme protein was present but in an inactive conformation.

3.4.2 *The outer membrane of Gram-negative bacteria*

The outer membranes like the cytoplasmic membranes are composed of phospholipids and proteins but unlike the latter also have the lipid A portion of the lipopolysaccharide immersed in the bilayers. Indeed the phospholipid of the outer membrane consisting mainly of phosphatidylethanolamine, phosphatidylglycerol and cardiolipin [24, 96, 162] appears almost completely confined to the inner leaflet of the bilayer [140]. The molecular composition of the membranes is such that 1 μm^2 contains 10^5 molecules both of lipopolysaccharides, and of proteins together with 10^6 molecules of phospholipid [14], the contribution of fatty acids from phospholipids and lipopolysaccharides being about the same. The ratio of protein to phospholipid in the membrane is obviously considerably higher than that in the cytoplasmic membranes being about 2.2:1 instead of 1.5:1. The only enzyme so far detected in the outer membrane is a phospholipase [2, 9, 163]. As was noted in Chapter 2.3.2 it was at first thought [129] that the majority (about 70%) of the protein present was represented by a single molecular species of molecular weight of 44 000. However, subsequent work showed that, although the number of major proteins was limited, the situation was more complicated than was at first thought. There would be no virtue in reproducing all the arguments which led to the present recognition of the presence of four or possibly five major proteins. Some five groups of workers contributed to this picture and unfortunately each chose a somewhat different nomenclature to designate the entities identified. A useful summary is given in Table 3.10, which also can be used as a guide to the historical evolution of the subject. Most of the reasons for confusion and differences of opinion over the numbers of bands to be demonstrated by unidirectional electrophoresis are to be found in the different techniques used by the various authors. The causes of variation were outlined previously as factors affecting the results of gel electrophoresis (p. 94).

Table 3.10. Designations used by different authors for bands seen after polyacrylamide gel electrophoresis of SDS disagregated outer membranes (from [29]).

De Rienzo *et al.* [29]	Inouye [52]	Bragg [13]	Schnaitman [130, 8]	Henning *et al.* [37, 50, 132] (a)	Henning *et al.* [37, 50, 132] (b)	Mizushima [90, 156]	Lugtenberg [73]
Matrix protein							
1a	Peak 4	A_1 A_2	1	1	1a	σ–9	b
1b					1b	σ–	c
Protein 2*	–	–	2	–	–	–	–
TolG protein	Peak 7	B	3a	II*	–	σ–10	d
Protein 3†	(Peak 6)†	nd‡	3b	nd‡	–	(σ–11)†	a
Lipoprotein	Peak 11	F	nd	IV	–	(σ–18)†	nd‡

* Present only in strains lysogenic for phage PA-2.
† The correlation of these bands with those of other authors is unclear.
‡ nd means not detected.

So far no two-dimensional gel examination of the outer membrane proteins has been reported although Smyth *et al.* [143] have detected five major antigens and six or seven lesser ones by rocket-electrophoresis. The established presence of the four or five major proteins shown by unidirectional examination should not be allowed to diminish the importance of some or all of the 10–20 minor bands also to be seen by this analytical technique. Many of these are likely to be involved in important functions such for example as phage receptors, and in the uptake of nutritional substances [14] (see Chapter 4.10.3).

The two larger proteins, 1a or 1b in the classification of Schmitges and Henning [132] have been christened matrix proteins or porins partly because they bind by non-covalent bonds to form an ordered pattern on the isolated peptidoglycan layers from the envelope providing that the bound lipoprotein is also present. When associated, which they do optimally at pH 8.0 and in the presence of 5 mM Mg^{2+} [45, 166, 167], they form a hexagonal lattice with a 7.7 nm repeat [117, 145]. Although it has been suggested that the binding to peptidoglycan involves 'the D-diaminopimelic acid residue' of peptidoglycan (S. Mizushima, personal communication quoted in [29]) it should be noted that almost the same ultrastructure was formed when the proteins bound to lysozyme spheroplasts [145]. The whole formation of arrays on bacterial surfaces by subunit proteins is reminiscent of the formation of the outermost layer patterns already described (Sections 1.2.1 and 1.3.4). One of the matrix proteins (1a) has been purified to homogeneity [117]: its molecular weight determined by several methods is 36 000 and it appears to be a single polypeptide of 336 amino acid residues; of which a short *N*-terminal sequence has been specified.

The relative amounts of proteins 1a and 1b are greatly influenced by the medium used to grow the organisms [8, 45, 73a, 131], particularly by the availability of readily fermentable carbon sources such as glucose or lactate. When these are present, the amount of 1b is enhanced at the expense of 1a. An increase in osmolarity of the growth medium has the same effect. Since the alteration can be achieved by supplementing a yeast extract medium with sucrose as well as with NaCl and KCl [157], it is not unlikely to be entirely a function of the ionic strength.

A high proportion of both proteins 1a and 1b exists in the extended β-conformation [90, 117] in contrast to many intrinsic membrane proteins which have large regions arranged in the more common α-helical conformation (see p. 74). This may have a profound influence on the structural contribution of the matrix proteins or porins to the membrane as well as their functions in it. For example it is suggested that they span the membrane [145], a possibility that both supports and makes it easy to understand the suggested triplet association of the proteins to form pores through the membrane, and which provide the indentations of 2 nm diameter described [145] as present in freeze-etched preparations.

Apart from the matrix proteins the other outer membrane protein that has been isolated and studied is the tolG protein (Protein II* of [50]) (see Table 3.10). This protein can be distinguished from the others by two characteristics;

(1) it is the only major outer membrane protein susceptible to the action of trypsin or pronase,

(2) its molecular weight is 'heat modifiable'.

When heated above 50 °C in SDS, its weight increases [13, 52, 130, 156] but in the absence of SDS its molecular weight decreases [90, 116]. The increase in the presence of SDS seems due rather to a change in the conformation of the protein than to any form of polymerization or to the binding of greater amounts of SDS [114]. Various results for the molecular weight of the tolG protein have been suggested varying from 27 000–38 000 in the unheated state to 33 000–48 000 after heating.

A further major protein component found in the outer membrane is the smallest and therefore runs furthest in polyacrylamide gels of SDS disaggregated material. This is the lipoprotein of Braun and Rehn [15]. The structure and properties of this will, however, be discussed in Section 7.2, as a component of the envelope also found covalently attached to the peptidoglycan in *E. coli* and other Gram-negative species.

The majority of the evidence about the protein composition of the outer membrane of Gram-negative bacteria, reviewed above has come from work with *E. coli* and some from *Salmonella typhimurium*. As already pointed out, the proportions of the proteins can be altered by the conditions used for growing the organisms [72, 131] and by the organization of the lipopolysaccharide as in deep rough mutants [4, 62]. Moreover mutants of *E. coli* lacking all the major outer envelope proteins except the lipoprotein have been isolated and are reported to grow normally [49]. On the other hand mutants such as *omA lpp* lacking both the lipoprotein and the tolG protein required very high (30 mM) concentrations of either Mg^{2+} or Ca^{2+} compared with the wild-type (1 mM) to support rapid growth and avoid lysis [144]. These mutants, moreover, always grew as cocci rather than rods. Further, transductants lacking the matrix proteins too were still found to grow well, even though as cocci. These observations might lead one to suspect that different species of micro-organisms could afford to evolve different sorts of outer membrane. However examination of *Ps. aeruginosa* [12, 44, 83], *Proteus mirabilis* [26, 91] and *Neisseria gonorrheae* [43, 46–8, 56, 160] suggest that the organization of the outer membranes may be very similar in these disparate species, there always being three or four major proteins present.

References

1. Abel, K., de Schmartzing, H. and Peterson, J. L. (1963) *J. Bact.* **85**, 1039–44.
2. Albright, F. R., White, D. A. and Lennarz, W. J. (1973) *J. biol. Chem.* **248**, 3968–77.
3. Ambron, R. T. and Pieringer, R. A. (1971) *J. biol. Chem.* **246**, 4216–25.
4. Ames, G. F-L., Spudich, E. N. and Nikaido, H. (1974) *J. Bact.* **117**, 406–483.
5. Andersen, H. C. (1978) *A. Rev. Biochem.* **47**, 359–83.
6. Archer, D. B., Rodwell, A. W. and Rodwell, E. S. (1978) *Biochim. biophys. Acta* **513**, 268–83.
7. Asselineau, C., Montrozier, H. and Primé, J. C. (1969) *Eur. J. Biochem.* **10**, 580–84.
8. Bassford, P. J. jr., Diedrich, D. L., Schnaitman, C. A., Reeves, P. (1977) *J. Bact.* **131**, 608–622.
9. Bell, R. M., Mavis, R. D., Osborn, M. J. and Vagelos, P. R. (1971) *Biochim. biophys. Acta* **249**, 628–35.
10. Bishop, D. H. L., Pandye, K. P. and King, H. K. (1962) *Biochem. J.* **83**, 606–614.
11. Bøg-Hansen, T. C. (1973) *Analyt. Biochem.* **10**, 358–61.
12. Booth, B. R. and Curtis, N. A. C. (1977) *Biochem. biophys. Res. Commun.* **74**, 1168–76.
13. Bragg, P. D. and Hou, C. (1972) *Biochim. biophys. Acta* **274**, 478–88.
14. Braun, V. (1978) In *Relations between Structure and Function in the Prokaryotic Cell*, 28th Symp. Soc. Gen. Microbiol., eds. Stanier, R. Y., Rogers, H. J. and Ward, J. B., pp. 111–38, Cambridge University Press.
15. Braun, V. and Rehn, K. (1969) *Eur. J. Biochem.* **10**, 426–38.
16. Bretscher, M. S. (1973) *Science* **181**, 622–9.
17. Brooks, J. and Benson, A. A. (1972) *Archs. Biochem. Biophys.* **152**, 347–55.
18. Brown, P. K. (1972) *Nature, New Biol.* **236**, 35–8.
19. Button, D. and Hemmings, N. L. (1976) *Biochemistry* **15**, 989–95.
20. Chapman, D. and Wallach, D. F. H. (1973) *Biological Membranes, Vols. I and II,* New York and London: Academic Press.
21. Collins, M. D., Goodfellow, M. and Minnikin, D. E. (1979) *J. gen. Microbiol.* **110**, 127–36.
22. Cone, R. A. (1972) *Nature, New Biol* **236**, 39–43.
23. Cronan, J. E. (1978) *A. Rev. Biochem.* **47**, 163–89.
24. Cronan, J. E. jr. and Vagelos, P. R. (1972) *Biochem. biophys. Acta* **265**, 25–60.
25. Cronan, J. E. jr. and Gelman, E. P. (1975) *Bact. Rev.* **39**, 232–56.
26. Datta, D. B., Kramer, C. and Henning, U. (1976) *J. Bact.* **128**, 834–41.
27. Davson, H. and Danielli, F. G. (1943) In *The permeability of natural membranes,* Cambridge University Press.
28. De Pierre, J. W. and Ernster, L. (1978) *A. Rev. Biochem.* **46**, 201–262.
29. Di Rienzo, J., Nakamura, K. and Inouye, M. (1978) *A. Rev. Biochem.* **47**, 481–532.
30. De Siervo, A. J. and Salton, M. R. J. (1973) *Microbios.* **8**, 73-8.
31. Dunphy, P. J., Kerr, J. D., Pennock, J. F., Whittler, K. J. and Feeney, J. (1967)

Biochim. biophys. Acta **136**, 136–47.
32. Edidin, M. (1974) *A. Rev. Biophys. and Bioeng.* **3**, 179–201.
33. Edwards, J. R. and Hayashi, J. A. (1965) *Archs. Biochem. and Biophys.* **111**, 415.
34. Fischer, W. (1970) *Biochem. biophys. Res. Commun.* **41**, 731–6.
35. Fischer, W., Ishizuka, I., Landgraf, H. R. and Herrman, J. (1973) *Biochim. biophys. Acta* **296**, 527–45.
36. Gaffney, B. J. and Chen, S-C. (1977) In *Methods in membrane biology*, ed. Korn, E. D., pp. 291–358; New York and London: Plenum Press.
37. Garten, W., Hindennach, I. and Henning, U. (1975) *Eur. J. Biochem.* **59**, 215–21.
38. Gilby, A. R., Few, A. V. and McQuillen, K. (1958) *Biochim. biophys. Acta* **29**, 21–9.
39. Golde van, L. M. G., Akermans-Kruyswijk, W., Franklin-Klein, W., Lankhorst, A. and Prins, R. A. (1975) *FEBS Letts.* **53**, 57–60.
40. Goldfine, H. (1972) *Adv. microb. Physiol.* **8**, 1–58.
41. Gordesky, S. E. and Marinetti, G. V. (1973) *Biochem. biophys. Res. Commun.* **50**, 1027–31.
42. Granfield, M-C. W. and Pieringer, R. A. (1975) *J. biol. Chem.* **250**, 702–709.
43. Guymon, L. F., Walstad, D. L. and Sparling, P. F. (1978) *J. Bact.* **136**, 391–401.
44. Hancock, R. E. W. and Nikaido, H. (1978) *J. Bact.* **136**, 381–90.
45. Hasagawa, Y., Yamada, H. and Mizushima, S. (1976) *J. Biochem., Tokyo* **80**, 1401–409.
46. Heckels, J. E. (1977) *J. gen. Microbiol.* **99**, 333–41.
47. Heckels, J. E. (1978) *J. gen. Microbiol.* **108**, 213–19.
48. Heckels, J. E. and Everson, J. S. (1978) *J. gen. Microbiol.* **106**, 179–82.
49. Henning, U. and Haller, I. (1975) *FEBS Letts.* **55**, 161–4.
50. Henning, U., Hohn, B. and Sonntag, I. (1973) *Eur. J. Biochem.* **47**, 343-52.
51. Houtsmuller, U. M. T. and van Deenen, L. L. M. (1964) *Biochim. biophys. Acta* **84**, 96–8.
51a. Hunter, K. and Rose, A. H. (1971) P.211 In *The Yeasts*, eds. Rose A. H., and Harrison J. S., London and New York: Academic Press.
52. Inouye, M. and Yee, M. (1973) *J. Bact.* **113**, 304–312.
53. Jackson, M. B. and Cronan, J. E. jr. (1977) *Biochim. biophys. Acta* **512**, 472–80.
54. Johansson, K-E. and Hjerkú, S. (1974) *J. molec. Biol.* **86**, 341–8.
55. Johansson, K-E. and Wroblewski, H. (1978) *J. Bact.* **136**, 324–30.
56. Johnston, K. H. and Gotschlich, E. C. (1974) *J. Bact.* **119**, 250–57.
57. Kamio, Y., Kim, K. C. and Takahashi, H. (1969) *J. gen. appl. Microbiol.* **15**, 439.
58. Kaneda, T. (1977) *Bact. Rev.* **41**, 391–418.
59. Kates, M., Palameta, B., Joo, C. N., Kushner, D. J. and Gibbons, N. E. (1966) *Biochemistry*, **5**, 4092–9.
60. Ke, B. (1965) *Archs. Biochem. Biophys.* **112**, 554.
61. Knoche, H. W. and Shively, J. M. (1972) *J. biol. Chem.* **247**, 170–78.
62. Koplow, J. and Goldfine, H. (1974) *J. Bact.* **117**, 527–43.
63. Laurell, C-B. (1965) *Scand. J. clin. lab. Invest.* **17**, 271.
64. Laurell, C-B. (1965) *Analyt. Biochem.* **10**, 358–61.
65. Laurell, C-B. (1972) *Scand. J. clin. lab. Invest.* **29**, 124.
66. Law, J. H., Zalkin, H. and Kaneshireo, T. (1963) *Biochim. biophys. Acta* **70**, 143–51.
67. Kee, Y. C. and Ballou, C. E. (1965) *Biochemistry* **4**, 1395–404.

68. Lee, A. G., Birdsall, N. J. M. and Metcalf, J. C. (1974) In *Methods in Membrane Biology* **2**, ed. Korn E. D., pp. 1–56, New York and London: Plenum Press.
69. Lenard, J. and Singer, S. J. (1966) *Proc. Natn. Acad. Sci. U.S.A.* **56**, 1828–35.
70. Lenard, J. and Rothman, J. E. (1976) *Proc. Natn. Acad. Sci. U.S.A.* **73**, 391–5.
71. Liaaen-Jensen, S. and Andrews, A. G. (1972) *A. Rev. Microbiol.* **26**, 225–48.
72. Lugtenberg, B., Peters, R., Bernheimer, H. and Berendsen, W. (1976) *Mol. gen. Genet.* **147**, 251–62.
73. Lugtenberg, B., Meijers, J., Peters, R., van der Hoek, and van Alphen, L. (1975) *FEBS Letts.* **58**, 254–8.

73a. Lugtenberg, B., Peters, R., Bernheimer, H. and Berendsen, W. (1976) *Molec. gen.Genet.* **147**, 251–62.

74. McDougall, J. C. and Phizackerley, P. J. R. (1969) *Biochem. J.* **114**, 361–7.
75. McElhaney, R. N. (1974a) *J. Supramolec. Struct.* **2**, 617–28.
76. McElhaney, R. N. (1974b) *J. molec. Biol.* **84**, 145–58.
77. McQuillen, K. (1960) In *The Bacteria I* eds. Gunsalus, I. C. and Stanier R. Y., pp. 249–359, New York and London: Academic Press.
78. Macfarlane, M. G. (1961a) *Biochem. J.* **79**, 4P–5P.
79. Macfarlane, M. G. (1961b) *Biochem. J.* **80**, 45P.
80. Mahendra, K. J. and White, H. B. (1976) *Adv. Lipid Res.* **15**, 1–60.
81. Marr, A. G. and Ingraham, J. L. (1962) *J. Bact.* **84**, 1260–7.
82. Marsh, D. (1975) *Essays in Biochemistry* **11**, 139–174.
83. Matsushita, K. O., Adachi, E., Shinagawa, E. and Ameyawa, M. (1978) *J. Biochem.* **83**, 171–81.
84. Mayberry, W. R., Smith, P. F. and Langworthy, T. A. (1974) *J. Bact.* **118**, 898–904.
85. Mayberry, W. R., Langworthy, T. A. and Smith, P. F. (1976) *Biochim. biophys. Acta* **441**, 115–22.
86. Minnikin, D. E., Abdolrahimzadeh, H., Baddiley, J. (1974) *Nature, Lond.* **249**, 268–9.
87. Minnikin, D. E. and Abdolrahimzadeh, H. (1974) *FEBS Letters.* **43**, 257–60.
88. Miura, T. and Mizushima, S. (1968) *Biochim. biophys. Acta* **150**, 159–61.
89. Morton, R. A. (1961) *Vitamin and Hormone* **19**, 1–37.
90. Nakamura, K. and Mizushima, S. (1976) *Biochem., Tokyo* **80**, 1411–22.
91. Nixdorff, K., Fitzer, H., Gmeiner, J. and Martin, H. H. (1977) *Eur. J. Biochem.* **81**, 63–9.
92. O'Farrell, P. H. (1975) *J. biol. Chem.* **250**, 4007–4021.
93. Op den Kamp, J. A. F., Houtsmuller, U. M. T. and van Deenen, L. L. M. (1965) *Biochim. biophys. Acta* **106**, 438–41.
94. Op den Kamp, J. A. F., van Iterson, W. and van Deenen, L. L. M. (1967) *Biochim. biophys. Acta* **135**, 862–84.
95. Op den Kamp, J. A. F., Bonsen, P. P. M., and van Deenen, L. L. M. (1969) *Biochim. biophys. Acta* **176**, 298–305.
96. Osborne, M. J., Gander, J. E., Parisi, E. and Carson, J. (1972) *J. biol. Chem.* **247**, 3962–72.
97. Owen, P. and Freer, J. H. (1972) *Biochem. J.* **129**, 907–917.
98. Owen, P. and Salton, M. R. J. (1975) *Proc. Natn. Acad. Sci. U.S.A.* **73**, 3711–15.
99. Owen, P. and Salton, M. R. J. (1975) *Biochim. biophys. Res. Commun.* **406**, 875–80.
100. Owen, P. and Salton, M. R. J. (1975) *Biochim. biophys. Acta* **406**, 235–47.
101. Owen, P. and Salton, M. R. J. (1975) *Biochem. biophys. Acta* **406**, 214–34.

102. Owen, P. and Salton, M. R. J. (1977) *J. Bact.* **132**, 974-985.
103. Owen, P. and Smyth, C. J. (1977) In *Immunochemistry of Enzymes and their Antibodies,* ed. Salton, M. R. J., pp. 147-202, New York and London: Wiley and Sons.
104. Petit, J-F. and Lederer, E. (1978) In *Relations between Structure and Function in the Prokaryotic Cell,* 28th Sym. Soc. Gen. Microbiol., eds. Stanier, R. Y., Rogers, H. J. and Ward, J. B., pp. 177-99, Cambridge University Press.
105. Pfenning, N. (1967) *A. Rev. Microbiol.* **21**, 285-324.
106. Phizackerley, P. J. R., MacDougall, J. C. and Moore, R. A. (1972) *Biochem. J.* **126**, 499-502.
107. Pless, D. D., Schmit, A. S. and Lennarz, W. J. (1975) *J. biol. Chem.* **250**, 1319-27.
108. Powell, D., Duckworth, M. and Baddiley, J. (1974) *FEBS Letts.* **41**, 259-63.
108a. Raetz, C. R. H. (1978) *Microbiol. Rev.* **42**, 614-59.
109. Rattray, J. B., Schibeci, A. and Kidby, D. K. (1975) *Bact. Rev.* **39**, 197-231.
110. Razin, S. (1975) *J. Bact.* **124**, 570-72.
111. Razin, S. (1975) In *Progress in Surface and Membrane science* **9**, eds. Danielli, J. F., Rosenberg, M. D. and Cadenhead, D. A., pp. 257-312, New York and London: Academic Press.
112. Razin, S. (1978) *Microbiol. Revs.* **42**, 414-70.
113. Reaveley, D. A. (1968) *Biochem. biophys. Res. Commun.* **30**, 649-55.
114. Reithmeier, R. A. and Bragg, P. D. (1977) *Arch. biochem. Biophys.* **178**, 527-34.
115. Rogers, H. J. and Perkins, H. R. (1968) In *Walls and membranes*, pp. 362, London: E. & F. N. Spon, Ltd.
115a. Rose, A. H. (1976) In *The Filamentous Fungi, 2 Biosynthesis and Metabolism*, ed. Smith J. E. and Berry, R., p. 308, London: Edward Arnold.
116. Rothfield, L. and Finkelstein, A. (1968). *A. Rev. Biochem.* **37**, 463-96.
117. Rosenbusch, J. P. (1974) *J. biol. Chem.* **249**, 8019-8029.
118. Rothman, J. E. and Kennedy, E. P. (1977) *Proc. Natn. Acad. Sci. U.S.A.* **74**, 1821-5.
119. Rottem, S. and Razin, S. (1967) *J. Bact.* **94**, 359-64.
120. Rottem, S., Hasin, M. and Razin, S. (1975) *Biochim. biophys. Acta* **375**, 395-405.
121. Rottem, S., Yashouv, Y., Ne'eman, Z. and Razin, S. (1973) *Biochim. biophys. Acta* **323**, 495-508.
122. Rottem, S., Cirillo, V. P., de Kruyff, B., Shinitzky, M. and Razin, S. (1973) *Biochim. biophys. Acta* **323**, 509-519.
123. Sackman, E. H., Trauble, H., Galla, H. J. and Overath, P. (1973) *Biochemistry* **12**, 5360-69.
123a. Salton, M. R. J. (1967) *A. Rev. Microbiol.* **21**, 417-42.
124. Salton, M. R. J. (1978) *J. gen. Physiol.* **52**, 227s-252s.
125. Salton, M. R. J. (1978) In *Relations between Structure and Function in the Prokaryotic Cell,* 28th Cymp. Soc. Gen. Microbiol., eds Stanier, R. Y., Rogers, H. J. and Ward, J. B., pp. 201-222, Cambridge University Press.
126. Salton, M. R. J. and Owen, P. (1976) *A. Rev. Microbiol.* **30**, 451-82.
127. Salton, M. R. J. and Schmidt, M. D. (1967) *Biochem. biophys. Res. Commun.* **27**, 529-34.
128. Scher, M. and Lennarz, W. J. (1969) *J. biol. Chem.* **244**, 2777-89.
129. Schnaitman, C. A. (1970) *J. Bact.* **104**, 890-901.
130. Schnaitman, C. A. (1974) *J. Bact.* **118**, 442-53.
131. Schnaitman, C. D. (1974) *J. Bact.* **118**, 454-64.

132. Schmitges, C. J. and Henning, U. (1976) *Eur. J. Biochem.* **63**, 47–52.
133. Sehgal, S. N., Kates, M. and Gibbons, N. E. (1962) *Can. J. Biochem. Physiol.* **40**, 69–81.
134. Shaw, N. (1970) *Bacteriol. Rev.* **34**, 365–77.
135. Shaw, N. (1975) *Adv. Microbiol. Physiol.* **12**, 141–167.
136. Silbert, D. F. (1975) *A. Rev. Biochem.* **44**, 315–39.
137. Singer, S. J. (1971) In *Structure and Function of Biological Membrane*, ed. Rothfield L. I., p. 145, New York and London: Academic Press.
138. Singer, S. J. (1974) *A. Rev. Biochem.* **43**, 805–833.
139. Singer, S. J. and Nicholson, G. L. (1972) *Science* **175**, 720–31.
140. Smit, J., Kamio, Y. and Nikaido, H. (1975) *J. Bact.* **124**, 942–58.
141. Smith, R. J. M. and Green, C. (1974) *FEBS Letts.* **42**, 108–111.
142. Smyth, C. J., Friedman-Kien, A. E. and Salton, M. R. J. (1976). *Infect. Immun.* **13**, 1273–88.
143. Smyth, C. J., Siegel, J., Salton, M. R. J. and Owen, P. (1978) *J. Bact.* **133**, 306–319.
144. Sonntag, I., Schwarz, H., Hirota, Y. and Henning, U. (1978) *J. Bact.* **136**, 280–5.
145. Steven, A. C., Ten-Heggeler, B., Muller, B., Kistler, J., Rosenusch, J. P. (1977) *J. Cell Biol.* **72**, 292–301.
146. Stier, A. and Seckman, E. (1973) *Biochim. biophys. Acta* **311**, 400–408.
147. Stone, K. J. and Hemming, F. W. (1968) *Biochem. J.* **109**, 877–82.
148. Stone, K. J., Butterworth, P. H. W. and Hemming, F. W. (1967) *Biochem. J.* **102**, 443–55.
149. Tanner, W., Jung, P. and Behrens, N. H. (1971) *FEBS Letts.* **16**, 245–8.
150. Theodore, T. S. and Panos, C. (1973) *J. Bact.* **116**, 571–6.
151. Thilo, L. and Overath, P. (1976) *Biochemistry* **15**, 328–4.
152. Thomson, R. H. (1971) In *Naturally Occurring Quinones*, New York and London: Academic Press.
153. Tornabene, T. G., Kates, M., Gelpi, E. and Oro, J. (1969) *J. Lipid Res.* **10**, 294–303.
154. Toon, P., Brown, P. E. and Baddiley, J. (1972) *Biochem. J.* **127**, 399–409.
155. Uehara, K., Akutsu, H., Kyogoku, Y. and Akamatsu, Y. (1977) *Biochim. biophys. Acta* **466**, 393–401.
156. Uemura, J. and Mizushima, S. (1975) *Biochim. biophys. Acta* **413**, 163–76.
157. Van Alphen, W. and Lugtenberg, B. (1977) *J. Bact.* **131**, 623–30.
158. Verbruggen, R. (1975) *Clin. Chem.* **21**, 5–43.
159. Wallach, D. F. H. and Zahler, P. H. (1966) *Proc. Natn. Acad. Sci. U.S.A.* **56**, 1552–9.
160. Walstad, D. L., Guyman, L. F. and Sparling, P. F. (1977) *J. Bact.* **129**, 1623–7.
161. Weeke, B. (1973) *Scand. J. Immunol.* **2**, (Suppl. 1) 37–46.
162. White, D. A., Lennarz, W. J. and Schnaitman, C. A. (1972) *J. Bact.* **109**, 686–90.
163. White, D. A., Albright, F. R., Lennarz, W. J. and Schnaitman, C. A. (1971) *Biochim. biophys. Acta* **249**, 636–42.
164. Wicken, A. J. and Knox, K. W. (1970) *J. gen. Microbiol.* **60**, 293–301.
165. Wilkinson, S. G. (1972) *Biochim. biophys. Acta* **270**, 1–11.
166. Yamada, H. and Mizushima, S. (1977) *J. Biochem, Tokyo* **81**, 1889–99.
167. Yu, F. and Mizushima, S. (1977) *Biochem. biophys. Res. Commun.* **74**, 1397–402.

4
Membrane functions

4.1 Active components and functions of bacterial cell walls

The relative simplicity of the structures of most bacteria, combined with their ability to grow very rapidly and to adapt themselves physiologically to a wide variety of environments, means they must pack a large number of functions into few structures. In many eukaryotic cells, for example, the mitochondrion can be looked upon as the power house, the cytoplasmic membrane as the molecular sieve for small metabolites and the Golgi apparatus and endoplasmic reticulum as the means of synthesizing and exporting large molecules like proteins and polysaccharides. In bacteria, all these functions must be undertaken by the available membranes, which ultimately means the cytoplasmic membrane. Mesosomes may or may not exist as such in the growing cell (see Chapter 1.5.1) and the outer membrane of Gram-negative species, although now recognized as functionally important, does not seem directly concerned with either the cell's energy supply or with the passage of most small metabolites in and out of the cytoplasm under normal growth conditions. Where other functions such as photosynthesis and nitrogen fixation are served by specialized membrane structures, these can frequently be shown also to have their origin in the cytoplasmic membrane and to have many of the same enzymes involved with energy supply, such as the dehydrogenases.

Outside the cytoplasmic membrane, and biosynthesized by it is the peptidoglycan layer with, in Gram-positive species, teichoic and teichuronic acids, polysaccharides and in some organisms such as staphylococci and streptococci, immunologically specific proteins attached to it. Gram-negative species have lipoproteins attached and lipopolysaccharides fixed to the outer membrane. Finally the bacteria may be embedded in capsules or micro-capsules of polysaccharides. All of these macromolecules are synthesized from intermediates, soluble in the cytoplasm, by enzymes fixed in or on the cytoplasmic membrane. The biosynthetic processes involved are described elsewhere in this book (Chapters 9 and 10). Not only do the cytoplasmic membranes have a major role in the synthesis of exo-cellular macromolecules, but many of the enzymes concerned with the synthesis of the lipid bilayer itself are also fixed within the layers; the membranes are to this extent self-generating. Thus the cytoplasmic membranes of bacteria are truly multi-functional.

It is not, therefore, surprising that a strong wall layer has been evolved by bacteria to protect such an all important cellular organelle as the cytoplasmic membrane.

Undoubtedly this is a major function of the peptidoglycans in the Gram-positive wall and of the Gram-negative sacculus (also known as the peptiodoglycan layer or 'rigid layer'). Inter-relationships between the evolution of transport systems and of regulation of biosynthesis, leading to high concentrations of low molecular weight metabolites in the cytoplasm and the strong external wall, are matters for interesting speculation. Clearly the large pools to be found particularly in Gram-positive species are of great value to the bacteria in allowing ready mobilization of intermediates for macromolecular synthesis. The presence of this large pool gives rise to high osmotic pressures when the organisms are growing in the usual rather dilute media. The delicate cytoplasmic membrane could only be protected against the damaging consequences of this internal pressure by being encased in a wall of considerable tensile strength. The effects of damaging the wall can readily be seen when it is attacked by autolytic or externally added lytic enzymes. Nevertheless, wall-less prokaryotes exist and have been classified separately from the eubacteria into a new class, the *Mollicutes* [253], perhaps currently more readily recognizable, as the *Mycoplasmas*. It was possible until fairly recently to regard the success of these apparently unprotected forms as due to their adaptation to an entirely parasitic life either in plants or animals. However, wall-less forms similar to mycoplasmas, have been isolated from self-heated coal refuse piles [54]. Prokaryotic microbes can therefore exist successfully in nature without walls. At present only relatively few (about 60) species of *Mollicutes* have been recognized which following some deductions might suggest that their success is comparatively limited. However, it would be rash to be dogmatic and many more so far unrecognized examples may yet be discovered. Other wall-less multiplying prokaryote forms such as the L-forms grow only on laboratory media providing adequate osmotic protection; their ability to live in the environment outside the laboratory is uncertain and their ability to grow as parasites much debated.

4.2 Functions of the cytoplasmic membrane

It is clearly impossible in a single chapter to give a detailed and complete picture of the present state of knowledge of all the functions of the cytoplasmic membranes of bacteria. A truly overwhelming amount of information is available on each aspect. Yet generalization is often made difficult or impossible because of the specialization of the systems in the different species. As we shall see later, electron transport chains in the membrane are a good example of diversity; the components differing from species to species. The regulation of their biosynthesis varies likewise according to the supply of electron donors, in particular oxygen for those organisms capable of using it. A simple but effective general summary of the bioenergetic membrane functions of a bacterium like *Escherichia coli* has been provided [86] and is reproduced here as Fig. 4.1.

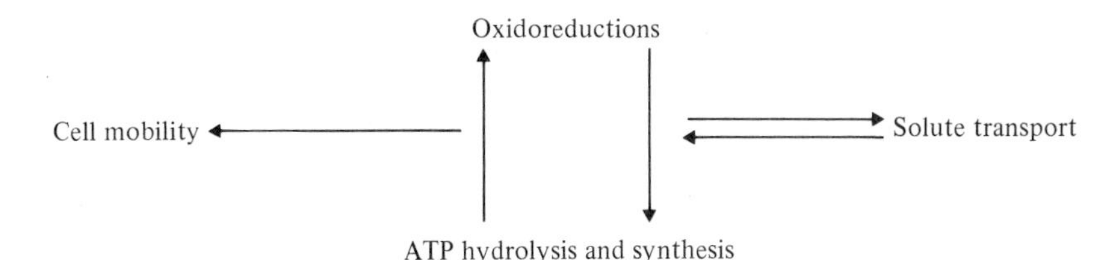

Figure 4.1 Bioenergetic functions of the cytoplasmic membrane in micro-organisms such as *E. coli* (from [86]).

4.2.1 *The chemiosmotic hypothesis – one likely generalization*

Although so few generalizations about components responsible for membrane functions are possible, one hypothesis relating to energy supply now seems generally accepted. Like the fluid-mosaic model for membrane structure (see Chapter 3.1) the chemiosmotic hypothesis has come to dominate thinking to the exclusion of numbers of rivals with which it co-existed for so long. First proposed by Mitchell [202–4] the idea of the anisotropic organization of membrane components so that the passage of electrons across the system would lead to the production of H^+ on one side of the membrane and its disappearance with the appearance of OH^- on the other, grew out of ideas first put forward by Lundegardh [183] and developed by a number of other workers to explain acid secretion by gastric mucose (see [264] for a review). As applied to bacteria the chemiosmotic theory has been repeatedly reviewed fairly recently [86, 106, 114].

The central ideas of the chemiosmotic hypothesis are:

(1) An asymmetric arrangement of the membrane-bound enzymes in such a way that they can catalyse vectorial reactions resulting in the translocation of ions, molecules or chemical groups across the membrane which has a low permeability to protons and ions generally.

(2) That some of these reactions lead to separation of charged entities within the membrane.

The recombination of such charges provides the energy for osmotic, chemical and mechanical work. To achieve these conditions it is suggested the electron transport chains, consisting of dehydrogenases, cytochromes, ferredoxins and quinols, are arranged as a series of loops spanning membranes that are essentially impermeable to most ions including H^+ and OH^-. These co-exist with an ATPase which can translocate protons according to the equation

$$ATP^{4-} + H_2O + xH^+_R \rightleftharpoons ADP^{3-} + PO_4^{3-} + xH^+_L \qquad (1)$$

where the symbols L and R represent the left and right hand sides of the membrane which by convention are synonymous with the extra-cellular and cytoplasmic space. Since this reaction is reversible, energy can either be accumulated as ATP synthesized, or expended by its hydrolysis, during H^+ translocation. We may then represent this very simplified version of the chemiosmotic hypothesis as in Fig. 4.2.

The end result of either the transport of electrons or hydrolysis of ATP in a system arranged asymmetrically, as suggested, is that one side of the membrane

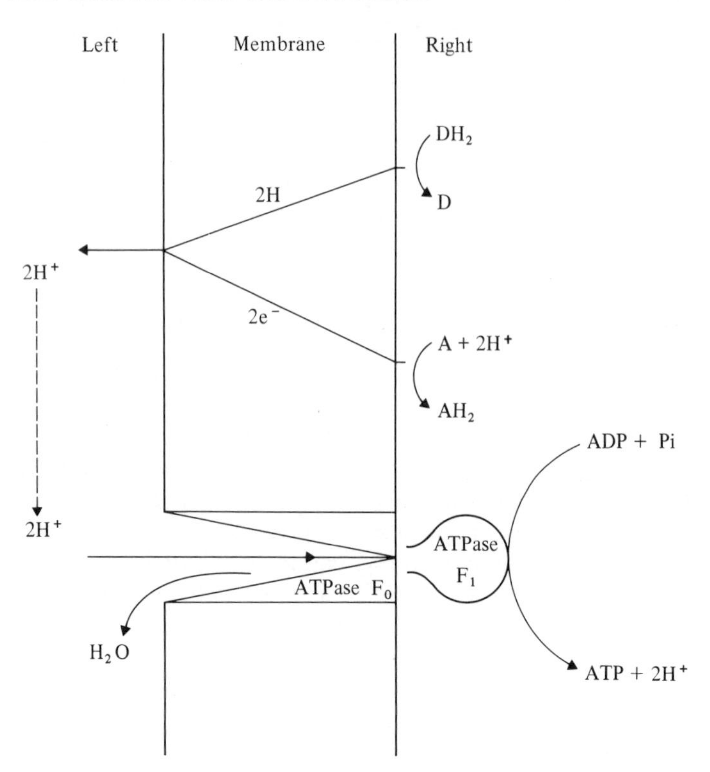

Figure 4.2 Schematic representation of a proton-translocating oxidoreduction segment of the electron transport chain and of the proton translocating ATPase. D is an electron donor, A is an acceptor, whilst F_1 and F_0 are parts of the membrane ATPase discussed later (p. 133) (from [104]).

becomes alkaline and negatively-charged. Thus a membrane potential ($\Delta\psi$), and a pH gradient (Δ pH) are set up by the movement of H^+ across the membrane by either electron transport or ATP hydrolysis. The electrochemical activity of the protons in the phases on either side of the membrane will differ and the difference is known as the proton motive force or Δp. The interrelationship between $\Delta\psi$, ΔpH and Δp is given by the equation

$$\Delta p = \Delta\psi - \left(\frac{2.3RT}{F}\right)\Delta\text{pH} \qquad (2)$$

where R is the molecular gas constant, T the absolute temperature, and F the Faraday Constant. The value of the constant, 2.3 RT/F is 59.2 mV at 25° C or 60.1 at 30° C.

One important aspect of the proton motive force Δp is that it opposes the ATPase-driven translocation of protons, a situation illustrated by Fig. 3.2. If the ATPase reaction is allowed to proceed to equilibrium the proton motive force will eventually be poised against that set up by the equilibrium ATP/(ADP + PO_4^{3-}). The force of this couple can be expressed in millivolts as the phosphate potential ΔGp and at equilibrium

$$x\Delta p = \Delta Gp \tag{3}$$

where x is the same as in Equation 1. From this the relationship between the number of protons translocated and the molecules of ATP synthesized can be deduced.

The above summarizes the way in which it is suggested that the oxido-reduction chains in the membrane are coupled to energy supply in the form of ATP synthesis, in other words, the mechanisms underlying oxidative-phosphorylation. The hypothesis then can be expanded to account for the transport of cations on monofunctional carriers, or uniports, in response to the membrane potential, inside negative. Anions, on the other hand, enter on bifunctional carriers, or proton symports, in response to the pH gradient, inside alkaline; neutral entities can also enter on proton symports under the influence of both the potential and pH gradients. The various aspects of the hypothesis will be discussed again at appropriate points in this chapter.

That bacteria do indeed generate proton motive forces, made of potential and pH gradients between the inside and outside of the cells, seems beyond dispute and a considerable number of determinations have been made. For example, values of 230 mV for Δp, 132 mV for $\Delta\psi$ and a ΔpH of 1.65 have been found for *E. coli* and 211 mV, 134 mV and 1.3 for *Staphylococcus aureus* [44]. A review has been published on both the methods of determination and the results for transmembrane proton gradients particularly in mitochondria and chloroplasts [272]. At a relatively early point in the development of the idea outward translocation of protons coupled to respiration in photosynthetic bacteria and in *Micrococcus denitrificans* were observed [300, 301]. It is of interest that the chromatophores in the former translocate H^+ in the reverse direction that is, from the cell cytoplasm into the organelle interior (see also [198a]).

4.3 Components of the electron transport chain

Aerobically grown, respiring bacteria and even obligate anaerobic ones such as *Desulfovibrio*, or photo-anaerobes such as *Chlorobium limicola* that reduce inorganic compounds, make use of the cytochromes for the transfer of electrons from oxygen, or materials in highly oxidized states to substrates of lower oxidation levels with concomitant storage of useful free energy. Likewise those species that have been examined also have quinones and iron sulphur ferredoxin-like proteins. Exceptions to the general rule that energy storage is mediated by electron flow down the cytochrome chain are bacteria such as the streptococci, and some lactobacilli that use exclusively flavin mediated systems and obligate anaerobes that derive energy exclusively from fermentation. In principle the mechanism of transfer of electrons by the cytoplasmic membranes of bacteria is similar to that by the inner membranes of mitochondria isolated from eukaryotic micro-organisms or from plant and animal tissues. Here, however, the analogy stops because as soon as the systems are looked at in detail it becomes apparent that the eubacteriales differ from mitochondrial electron transport in their numbers of cytochromes, which have different absorption spectra and properties towards inhibitors, and which vary in amounts according to the growth conditions. Only a few species of bacteria, such as *Paracoccus denitrificans* and *Alcaligenes eutrophus* have components similar to those in mitochondria [139] but it has been stated [131] 'no single demonstration has yet been successful in establishing complete identity between any of the bacterial cytochrome components and other mitochondrial analogues'. A full complement of cytochrome types has been found in organisms such as *Mycobacterium phlei, Corynebacterium diphtheriae, Micrococcus luteus, M. denitrificans, Azotobacter agilis, Bacillus subtilis* and *Haemophilus parainfluenzae* but many others contain rather few. Reviews of bacterial cytochromes are to be found [131, 142, 149, 173, 197, 282]. The omission of cytochromes is common; for example, bacilli may have only three namely b, aa_3 and o with the latter present in low concentration, compared with the six present in the mitochondrion-like *Paracoccus*. Again looking at the various cytochromes one finds that those of bacterial origin have slightly different spectra from those of mitochondria. Whereas mitochondria use ubiquinones exclusively, bacteria, particularly Gram-positive species, also have substituted menaquinones.

The components of the bacterial electron transport chain cannot be regarded as constant in proportion or even always present in the organism. For example, when subjected to oxygen limitation, or grown in the presence of cyanide, a number of organisms modify their cytochromes and quinones presumably in order to combat diminution of respiratory activity. If *E. coli* is oxygen limited, cytochrome oxidase aa_3 is decreased and cytochrome oxidase o increased. If it is grown anaerobically the aa_3 component disappears completely and, if grown in the presence of 150 μM cyanide, aa_3 disappears, o is increased and cytochrome oxidase d appears [139]. Likewise variation of the quinones present has been rather dramatically recorded for *Pasteurella multocida* growing with either O_2 or fumarate as electron acceptors [160]; when it was grown with O_2 the ratio of desmethylmenaquinone to ubiquinone was 0.44:1 whereas during growth with fumarate it was 5.67:1.

4.3.1 *Respiratory pathways*

Bacteria differ from the generally accepted picture for the ordering of components in the respiratory electron transport chain in mitochondria (see Fig. 4.3) not only

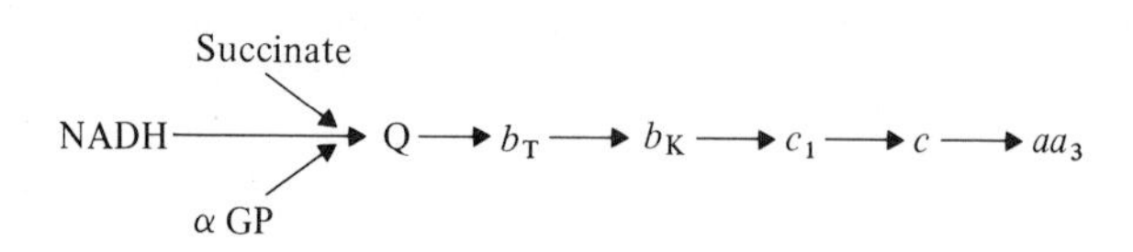

Figure 4.3 Linear respiratory chain in mitochondria. Q is ubiquinone; b_T, b_K, c, c_1 are cytochromes; and aa_3 cytochrome oxidase.

by the nature of the cytochromes and quinones involved but also by the complexity of their organization. Whereas the electron transfer chain in mitochondria is linear, bacteria, apart from the two species already mentioned and *E. coli* when grown highly aerobically, have branched chains. The use of the term 'linear' excludes, of course, branching at the level of primary dehydrogenases which is present to greater or lesser extents in all respiratory systems. Some examples [139] are shown to illustrate the meaning of the term 'branched'. In Fig. 4.4 the chains are branched only at the level of cytochrome oxidase, whereas in organisms such as *Azotobacter vinelandii* branching is more complex (see Fig. 4.5), occurring at *b* and/or *c* type

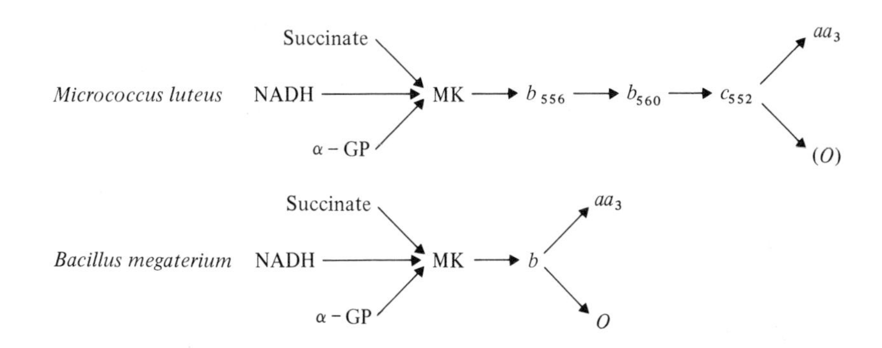

Figure 4.4 Respiratory pathways at the level of the cytochrome oxidases. MK is menaquinone; *o* is a cytochrome; and aa_3 cytochrome oxidase, as in Fig. 4.3.

cytochromes as well as at the cytochrome oxidases. One of the terminal branchways is less sensitive to cyanide than the other, and the sensitive pathway catalyses the oxidation of artificial substrates such as the couple ascorbate-2,6 dichlorophenol-indophenol whilst the resistant pathway is mainly concerned with oxidizing natural (physiological) substrates.

In the intensively investigated micro-organism *E. coli*, no fewer than nine different cytochromes have been detected [102, 308]. However, not all of these are likely to be of membrane origin and involved in the energetically important transfer of electrons. In as much as the chemiosmotic hypothesis explains the coupling of respiration to the storage of useful energy as ATP, the cytochromes thus involved must be associated with the membranes. A commonly quoted cytochrome involved in transfer but not likely to be membrane linked, is cytochrome c_{550} that is present in a form that can be washed from sphaeroplasts and can therefore be regarded as periplasmic [79, 80]. Whether this enzyme is involved with formate dehydrogenase or with nitrate reductase is as yet still warmly discussed; however neither of these pathways is necessarily coupled to ATP synthesis. Cytochrome c_{550} is one example of a soluble cytochrome, a number of others are to be found soluble after various forms of cell disintegration. Whether or not these are truly soluble in the cytoplasm of the living cell or are really membrane peripheral proteins that are easily solubilized (see Chapter 3.1) is not always clear.

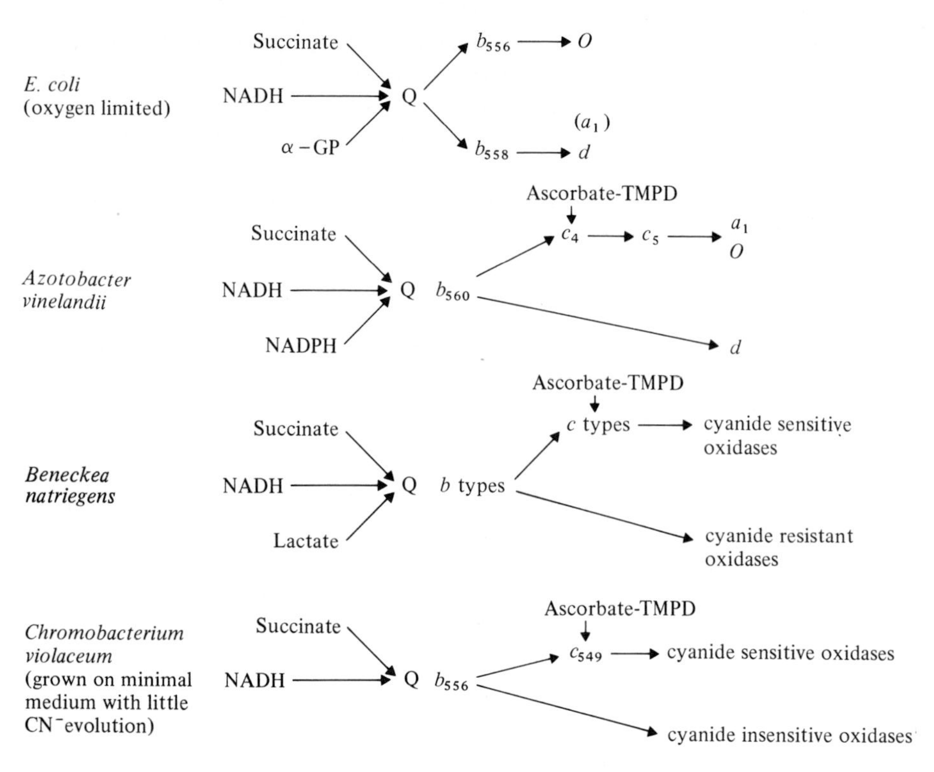

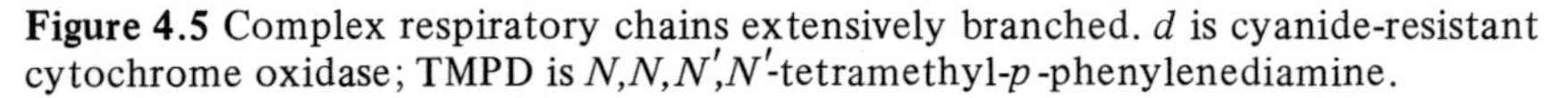

Figure 4.5 Complex respiratory chains extensively branched. *d* is cyanide-resistant cytochrome oxidase; TMPD is *N,N,N′,N′*-tetramethyl-*p*-phenylenediamine.

4.3.2 *Organization of the electron transport chain*

Whilst it is now almost a platitude to say that the apparatus for bacterial respiration is fixed into the cytoplasmic membrane, historically this has been only recently recognized as a dogma. Much earlier confusion and some not so early, arose from the ability of shattered membranes to form closed vesicles. This, together with the impermeable nature of bacterial membranes to metabolites and in particular to nucleotides, led to much controversy about the nature of function of 'particles' found after bacterial cell disruption. The use of so-called 'vesicles' in the study of metabolite transport, however, led to urgent consideration of the 'sidedness' of these bodies. As a result of this work it became possible to re-interpret much of the earlier work with so-called electron transport particles and insoluble and soluble fractions obtained after manipulation of membranous material.

The chemiosmotic hypothesis demands that the membrane electron transport chain should be organized as a series of loops traversing the membrane. A variety of arrangements have been suggested such as the one shown in Fig. 4.6. Although these schemes have a strongly speculative element in them, evidence has accumulated [114] for asymmetric functioning of the inner membrane of mitochondria

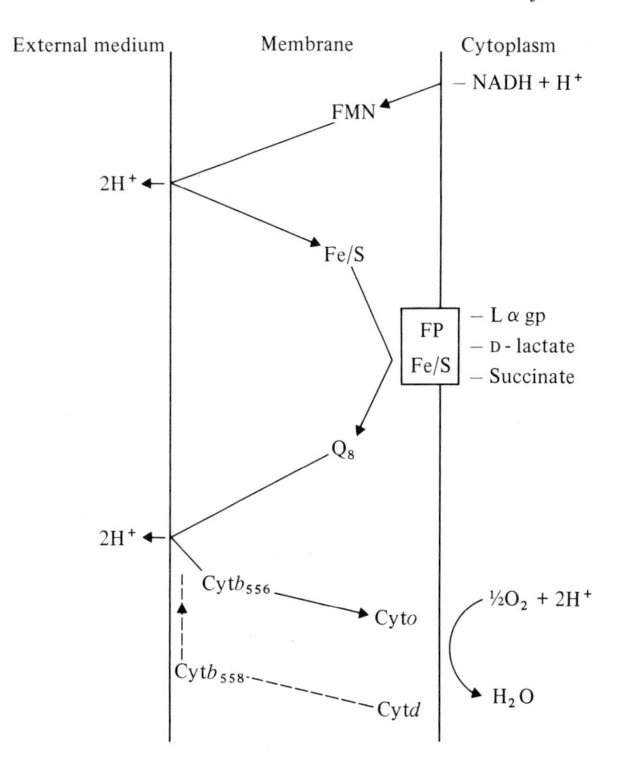

Figure 4.6 Functional organization of the electron transport chain on aerobically grown *E. coli* (from [104]). FMN and FP are flavoproteins; Fe/S is a ferredoxin-type protein; Q_8 is ubiquinone, an isoprenoid with a side-chain 8 carbons long; Cytb_{556}, Cyt_{558}, Cyt*o* and Cyt*d* are cytochromes; and Lαgp is *L*-α-glycerophosphate.

that is consistent with these ideas and in particular with the spanning of the membrane by the cytochromes [123]. Evidence, for example, is available to show that the cytochrome-nitrate reductase system of *E. coli* spans the cytoplasmic membrane structurally and functionally [25, 141]. The methodological simplicity of this demonstration is appealing and makes its extension seem likely. The essence of the method is to use the low-potential electron acceptor–donor, bipyridynium compounds known as methyl, ethylene and benzyl viologens: methyl viologen is the well known herbicide Paraquat whilst the ethylene compound is Diquat. The oxidized radicals of none of these dyes can cross the cytoplasmic membrane. The unoxidized form of the benzyl compound can, however, do so but those of the methyl derivative, and Diquat cannot. Thus, if the cells are allowed to take up reduced benzylviologen and then induced with NO_3^- to form nitrate reductase, electrons pass from outside the cells to the viologen inside, giving the coloured oxidized radical which cannot escape. With the wild-type strain, electrons could be transferred from Diquat (outside the cell) to nitrate whereas with a haem-deficient mutant this could not happen. The simplest interpretation is that in the wild-type electrons were transferred via the cytochrome chain from the outside to the nitrate reductase–nitrate couple on the cytoplasmic side of the membrane. This result is in good agreement with ^{125}I-lactoperoxidase labelling of spheroplasts and inverted vesicles in which the sub-unit polypeptides of the nitrate reductase were labelled in the inverted vesicles whereas that of the cytochrome b_{556} was labelled in the spheroplasts [25].

Further evidence for which entities in the cytoplasmic membrane are exposed on the culture medium side of *M. luteus* has come from the application of immunological methods [229, 230]. By examining membranes solubilized by neutral detergents using crossed immuno-electrophoresis (see Chapter 3.4.1), it has been possible not only to follow the behaviour of the individual proteins but also to identify the enzymic activities of many of them. Antisera were made to the whole membrane fraction from *M. luteus* including the mesosomes. Then, by absorbing such antisera with suspensions of undamaged protoplasts, the various antigens and enzymes exposed on the outer face of the membrane surrounding the protoplasts could be distinguished. A total of 27 antigens were identified in the membrane preparations of which the surprisingly large proportion of 12 were exposed on the outer face of protoplasts and these same antigens were also exposed on the surface of whole cells. Among 14 antigens expressed only on the cytoplasmic side of the membrane, five were identified as succinate and malate dehydrogenases together with two NADH dehydrogenases and ATPase [284]. Thus, consistently and reasonably, the enzymes responsible for substrate level oxidation in the electron transport chain are situated on the cytoplasmic side of the membrane.

4.3.3 *Dissection of the electron transport chain*

The twin concepts of integral and peripheral membrane proteins (see Chapter 3.1) and that of sidedness of fragmented membranes in vesicles have both helped to make easier an understanding of the results of attempts to dissect the electron transport chain. The ready removal of some enzymes when membrane preparations are treated with chelating agents, or in some instances with buffers of low ionic strength, can now be appreciated as typical behaviour of peripheral proteins – the ATPases in organisms such as *Micrococcus luteus*, Streptococci and *E. coli* are, of course, classical examples (see p. 126). NADH dehydrogenase, as measured either by the reduction of dyes such as 2,6-dichlorophenolindophenol or phenazine methosulphate or by the reduction of potassium ferricyanide, is another. The sidedness of the vesicles formed by disrupting membranes is important because substrates such as NADH can only reach enzymes exposed on their outer faces. Therefore, if NADH oxidase activity is sought, the results are likely to be meaningful only if all the membrane has been inverted during disruption. Many essential enzymes are positioned on the inner cytoplasmic side of the membrane in the living cell and are inaccessible to substrates such as NADH as we have seen. The stages in attempts to dissect apart the various entities and then reconstitute them into a functional membrane have been well set out by Razin [252] as:

(1) The solubilization of the biomembrane to its building blocks,
(2) The biochemical and biophysical characterization of the solubilization products,
(3) Their reassembly to a membrane identical to the membrane in structure and function.

Despite much effort devoted to attempts to achieve these goals with both bacterial and acholeplasmal membranes, in the late 60s and early 70s, comparatively little progress was made. Whilst it proved relatively easy to remove peripheral enzymes such as ATPase and add them back to membranes (see p. 125), it was difficult to dissociate the hydrophobic integral enzymes by means sufficiently gentle not to damage them functionally.

Early work on the mitochondrial electron transport chain allowed the separation of four functional complexes [116]. The membranes were solubilized with bile salts and fractionated with $(NH_4)_2SO_4$. Membranous structures could be formed by the reassociation of these fractions occurring when the concentration of surface active substances and the ionic strength were lowered. The functions of the reassociation of various mixtures of these four complexes was studied [337]. Complex IV was cytochrome oxidase, a mixture of Complexes I and III acted as NADH–cytochrome *c* reductase, whilst a mixture of Complexes II and III was succinate–cytochrome *c* reductase, II being succinate–ubiquinone reductase, and I, III and IV acted together as NADH oxidase.

Attempts to disaggregate the functions of the bacterial electron transport system have been more difficult although partial success has been achieved with membranes from *B. megaterium* [65, 66], and *M. luteus* [63, 64, 88]. In these experiments the membranes were 'disrupted' with 0.3% deoxycholate which was subsequently removed by passage through a column of Sephadex G25. L-Malate and NADH dehydrogenases (using 2,6-dichlorophenolindophenol as acceptor) were solubilized, and the insoluble residues contained cytochromes *b*, *c* and *a* types and menaquinones. When a mixture of soluble dehydrogenases and the insoluble part of the membrane was treated with 30 mM Mg^{2+}, the L-malate and NADH oxidase activities were partially restored. A study [88] of the effect of a variety of detergents without reconstitution experiments had earlier led to similar results for the solubilization of malate and NADH dehydrogenase by Triton X-100 and cholate, but cytochrome b_{556} and cytochrome b_{560} were also solubilized. The solubilized enzymes could still transfer electrons from substrates to the cytochrome *b*. By considering the effects of the various detergents, it was possible to order tentatively the electron transport chain. Dissection of succinoxidase of *E. coli* into three components has been achieved [256] by deoxycholate extraction of the membranes and subsequent fractionation of the extract. Two of the soluble fractions were succinic dehydrogenase–cytochrome *b*, and cytochrome oxidase complexes. When these were mixed in the presence of the detergent and the latter then removed, succinoxidase activity was obtained. The activity was, however, increased about 100% by the addition of a factor containing phospholipid. The reconstituted material was vesicular in form. The electron transport chain in *Mycoplasma laidlawii* membranes would seem to be simpler and more robust than that in bacteria. Cytochromes are not involved and the NADH oxidase activity is rather resistant to detergents (even sodium dodecyl sulphate). The oxidase activity could be solubilized with 50 mM deoxycholate in the presence of Mg^{2+} and reconstituted on to membranes [215]. Some evidence was obtained that the reaggregated membrane was not present as sealed vesicles. Both the ATPase and the NADH oxidase were shown to be on the cytoplasmic side of the original membranes. More recently succinic dehydrogenase has been purified from the chromatophore of *Rhodospirillum rubrum* [55a] and from the cytoplasmic membranes of *Bacillus subtilis* [117a]. Both were complexes consisting of unequal subunits, the former contained two (with molecular weights of 60 000–70 000 and 25 000–27 000) and the latter three (with molecular weights of 65 000, 28 000 and 19 000). The *B. subtilis* enzyme contained cytochrome probably of type *b* and the 19 000 mol.wt subunit appeared to be its *apo*-protein. It also contained a molar proportion of flavin associated with the 65 000 mol.wt. subunit. The suggested organization of the enzyme on the membrane is for cytochrome *b* to be the specific binding site with which the 28 000 mol.wt. subunit combines, the 65 000 mol.wt. subunit then combining with it to form the enzyme proper. A number of succinic dehydrogenase negative mutants (*cit* F) that did not produce the 28 000 mol.wt. unit had the 65 000 mol.wt. subunit as a component of the cytoplasm. The analogies between this isolated enzyme and the mitochondrial complexes is considerable.

The role of quinones in electron transport can be shown chemically in two ways. They can be extracted from freeze-dried preparations with pentane [326] leading to gross reduction in the NADH oxidase activity, for example, of the membranes from *Azotobacter vinelandii*. When ubiquinone is added back to the preparation its activity is restored. A second method makes use of the sensitivity of quinones to so-called 'black light' of 360 nm wavelength. Again, when the appropriate quinone (either one of the ubiquinone or menaquinone families) is added back to the preparation, electron transport activities are restored [64, 80]. Black light treated membrane vesicles from *B. subtilis* had transport activities restored when menaquinones or ubiquinone were added back to them [185].

4.3.4 *Flavodoxins, ferredoxins and other iron–sulphur proteins*

Nothing has so far been said about the iron–sulphur and flavin-containing proteins but as can be seen from Fig. 4.6 they occur in addition to the cytochromes, at critical points along the electron transport chains of both mitochondria and bacteria, as for example in the enzymic transfer of electrons to ubiquinone and between the cytochrome *b* components and NADH. One of the triumphs of biochemistry and chemistry over the last few years has been to isolate a series of small proteins of rather unusual amino acid composition which have linked to them a number of molecules of FeS in the ferredoxins, and a molecule of flavinmononucleotide in the flavodoxins. Several of the proteins have been crystallized, the sequence of their amino acids determined and their three-dimensional structures determined by X-ray crystallography [136, 227]. The first ferredoxins were isolated from *Clostridium pasteurianium* by Mortenson *et al.* [207] in 1962. Since this time they have been obtained from a rather wide variety of micro-organisms (Table 4.1). The molecular weights of the *apo*-proteins vary from 6000–14 400 and the number of iron and sulphur atoms from 2–8 (Table 4.2). They appear to have peptide cage structures with two cores of iron and sulphur atoms co-ordinately linked through the S atoms to the cysteine residues of the peptide [136].

Table 4.1. Distribution of bacterial ferredoxins (from [363]).

Fermentative bacteria	
Clostridium pasteurianium	*Cl. thermocetum*
Cl. acidi-urici	*Methanobacillus omelianskii*
Cl. butyricum	*Methanosarcina barkeri*
Cl. tetanomorphum	*Peptostreptococcus elsdenii*
Cl. tartarivarum	*Desulfovibrio desulphuricans*
Cl. thermosaccharolyticum	*Micrococcus lactilyticus*
Cl. kluyveri	*Butyribacterium rettgeri*
Cl. significans	*Diplococcus glycinophilus*
Cl. lactoacetophilum	*Sarcina maxima*
Cl. sporogenes	*Bacillus polymyxa*
Cl. sticklandii	*Escherichia coli*
Photosynthetic bacteria	
Chromatium D	*Chlorobium thiosulphatophilum*
Chloropseudomonas ethylicum	*Rhodospirillum rubrum*
Aerobic bacteria	
Azotobacter vinelandii	*Rhizobium*
Azotobacter chroococcum	*Pseudomonas putida*

Ferredoxins are known to be involved as electron carriers in a considerable number of reactions in fermentative bacteria and, perhaps most relevantly, they have been shown to be involved [151] in the heterocyclic hydroxylation in which electrons are transferred to cytochrome p_{450} in *Pseudomonas putida*:

$$\text{NADH} \rightarrow \text{Reductase} \rightarrow \text{Ferredoxin (Putidaredoxin)} \rightarrow \text{Cytochrome } p_{450}.$$

Likewise they are involved in nitrogen fixation and the nitrogenase reaction as performed by *Azotobacter vinelandii* according the the scheme [19]:

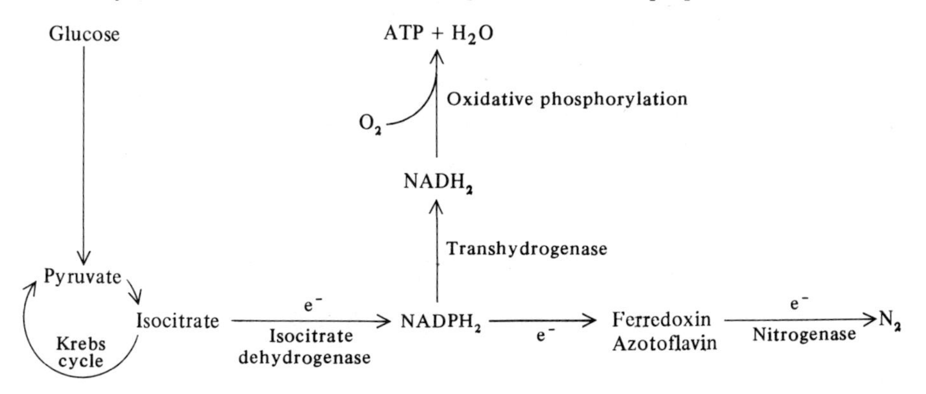

Table 4.2. Chemical properties of some ferredoxins (from [363]).

Micro-organisms	*Molecular weight* (10^{-3})	*Oxidation reduction potential* (mV)	*Iron–Sulphide* (atom/mol)
Clostridium pasteurianium	6	−390	8
Escherichia coli	12.5	—	2
Bacillum polymyxa	9	−390	4
Diplococcus gigas	6.6		4
Azotobacter vinelandii	14.4	−420	8
Rhodospirillum japonicum	9.4	—	—*
Pseudomonas putida	12.0	−250	2
Chromatium	16.0	−490	8
Rhodospirillum rubrum Ed. I	8.7	—	6
Rhodospirillum rubrum Ed. II	7.5	—	2

*Fe and S, 0.31 and 0.35 μmoles/mg protein

These and the other reactions involving the ferredoxins are reviewed in [19] and [363]. The ferredoxins shown in Table 4.2 have been isolated from the soluble fraction of disrupted bacteria. Other iron–sulphur proteins that are also involved in electron transport undoubtedly occur as intrinsic proteins in the membranes. Whether the known isolated ferredoxins and flavoproteins are truly soluble in the cytoplasm of the living cell, or whether they are peripheral proteins that have been dissociated from the membranes during either cell disruption or possibly at the early steps of purification is not easy to say. Evidence for the presence of firmly fixed iron- and flavin-containing proteins has come from examination of preparations of the intrinsic membrane enzyme, succinic dehydrogenase, after it has been solubilized from the inner membrane of beef heart mitochondria. At least three so-called HiPIP–iron–sulphur centres involved in the electron transport chain have been recognized. Fractionation of the enzyme preparation with the chaotropic agent, sodium trichloroacetate, gave two proteins; one with a molecular weight of 97 000 containing flavin, iron and sulphur in the ratios of 4:1:1, and another with a molecular weight $<$48 000 free of flavin but containing iron and sulphur [55]. Chaotropic agents function by breaking up the ordered structure water which surrounds lipophilic molecules, thus reducing the energy involved in the reaction of the molecule with random water molecules in the aqueous surroundings. Examination [18a, 222, 223] of a number of other soluble preparations of succinic dehydrogenase by electron paramagnetic resonance has allowed three functional FeS protein (HiPIP) centres to be recognized. One is involved with the succinate-ubiquinone dehydrogenase side of the chain and the other with the transfer of electrons from succinate to the cytochrome b_1-cytochrome *c* complex. There seems little doubt that these iron-sulphur centres in solubilized preparations of an intrinsic protein obtained by the use of chaotropic agents, or alcohol, represent intrinsic ferredoxins. The smaller protein obtained by trichloroacetate treatment of the succinic dehydrogenase may represent another intrinsic membrane ferredoxin. So far HiPIP centres have been recognized in hydrogenase preparations from a variety of bacteria including *Rhodospirillum rubrum* [3a]. Hydrogenase is the enzyme that catalyses the reversible activation of hydrogen. As more intrinsic enzymes concerned with electron transport are purified and studied, it is expected that more active FeS centres will be recognized.

4.3.5 *Mutants with disturbances in electron transport linked to ATP synthesis*

The use of mutants provides a further and powerful tool for dissecting and analysing the relative importance of different aspects of the flow of electrons through the electron transport chain and its coupling to the synthesis of available energy in the form of ATP. The different general methods used to isolate such mutants have been summarized [101] as follows:

(1) Differential ability to utilize alternative carbon sources;
(2) Differential ability to utilize alternative terminal electron acceptors for growth;
(3) Inhibitor resistance;

(4) Antibiotic resistance;
(5) Search for auxotrophic mutants;
(6) Use of intrinsic and extrinsic chromophores.

As might be expected the majority of the work has been done with *E. coli* because of its well-understood genetic system and the vast amount of information available. Some less complete work has, however, been done with *B. subtilis, Pseudomonas aeruginosa* and *Salmonella typhimurium*. Regarding isolation of the mutants from *E. coli*, Method 1 is by far the most common in which strains have been sought that are capable of growth by fermentation on carbon substrates such as glucose but which are unable to grow on non-fermentable sources, such as succinate, D-lactate, glycerol, malate or acetate. The rationale of this method is that an organism unable to effect electron transport-mediated synthesis of ATP and H^+ translocation because of mutations, will still be able to carry out these essential functions by fermentation of glucose using substrate level phosphorylations. Alternatively if the ATPase is non-functional because of an appropriate genetic lesion, the organism will grow providing the electron transport chain is still able to play its part in H^+ translocation and in maintaining the membrane potential gradient. Method 1 has been used to isolate mutants in the biosynthesis of quinones (*ubi, men* mutants) (see Tables 4.3 and 4.4) and dehydrogenase deficient mutants (*dld, sdh, glpD*) as

Table 4.3 Concentrations of components of the electron transport chain in the wild-type and a ubiquinone deficient mutant of *E. coli* (after [51]).

	Concentration (nmol/mg protein)	
Component	*Revertant wild-type*	*Ubiquinone deficient mutant*
Total flavin	0.25	0.39
Cytochrome b_1	0.19	0.19
Cytochrome a_2	0.027	0.047
Cytochrome o	0.073	0.04
Cytochrome a_1	+	+
Ubiquinone	4.7	<0.05
Vitamin K_2	0.67	2.7

Table 4.4 Oxidase activities in membrane preparations from the wild-type and the effect of adding quinone to a ubiquinone-deficient mutant of *E. coli* [51]. Quinones added a final concentration of 32 μM.

Substrate	*Quinone added*	O_2 *uptake* (nmoles/min/mg membrane protein)	
		Revertant wild-type	*Ubiquinone deficient mutant*
NADH	—	230	25
	Ubiquinone	430	560
	Menaquinone	230	82
D-lactate	—	68	19
	Ubiquinone	150	210
	Menaquinone	77	30
D-malate	—	31	5
	Ubiquinone	50	42

well as mutants disturbed in ATPase functions (the *unc* mutants). Method 2 involves looking for mutants that cannot grow anaerobically by using alternative electron acceptors to O_2, such as nitrate or fumarate. The search for chlorate resistant mutants illustrates Method 3 best. The rationale of this method is that chlorate induces nitrate reductase, which in turn reduces the chlorate to the toxic chlorite. Therefore, organisms deficient in nitrate reduction appear as resistant to chlorate and grow by fermentation of glucose. A number of these mutations are pleiotropic (e.g. *chlA, chlB, chlD*) showing phenotypic absence of a number of enzymes. Analysis of this pleiotropy has shown that the mutants lack proteins involved with the incorporation of molybdenum into the *apo*-reductases. The study of chlorate resistance has already provided and still continues to provide a fascinating probe into interactions between the membrane components at the low potential end of the electron transport chain. The antibiotics used in searching for mutants (Method 4) have been the amino-glycosides (streptomycin, neomycin, and kanamycin). This method has been successful, particularly in providing mutants deficient in haem synthesis and ATPase function. Unfortunately, its rationale is quite unclear, although very early work suggested some interference by streptomycin in the electron transport chain. The search for auxotrophic mutants (Method 5) is an obvious method when looking for mutants lacking the ability to synthesize prosthetic groups such as flavins or haems, as well as those unable to make quinones. It has been limited by the inability of many of the precursors to penetrate the envelope of Gram-negative species. The use of chromophores (Method 6), though not common for the isolation of the sorts of mutants we are discussing, involves the most time-honoured approach to the recognition of distinctive species and strains in classical bacteriology when the organisms are grown as colonies on plates. Colonies of haem-synthesizing bacteria can be distinguished from non-porphyrin accumulating ones by their red–brown colour, and inclusion of suitable redox dyes such as benzyl viologen (see p. 113) into the agar medium allows coloured colonies to be distinguished from the white colonies of cells lacking the particular dehydrogenase for the acceptor in the medium. For this method to work, of course, the dyes must not be toxic and must be able to penetrate the cells.

Some of the mutants isolated from *E. coli* together with their phenotypes are summarized in Table 4.5 modified from [101] and [104].

4.4 The coupling of energy flow to phosphorylation

As one of the most distinguished workers on the electron transport chain said '. . . the mechanism of membrane-bound energy-coupling processes remains one of the most challenging and as yet unsolved problems of modern biology . . .' [41]. The chemiosmotic hypothesis as we have seen supposes the linkage to involve an interaction or interactions between electron transport and ATPase systems through a proton flux across the membrane [205]. However, other hypotheses have been vigorously followed and are still possible. Direct intervention of a conventional high

Table 4.5 Mutants of *E. coli* disturbed in their ability to synthesize components of the electron transport chain (after [101, 104]).

Mutation	*Gene*	*Defective functions*	*Phenotypic restoration*	*References*
Ubiquinone biosynthesis	*ubiA*-H	Synthesis of ubiquinone from chorismic acid (structural genes)	Addition of appropriate ubiquinone or in some instances intermediates	89, 90, 51 103, 244
Menaquinone biosynthesis	*men*A	1,4-dihydroxy-2-naphthoic acid accumulated	Addition of appropriate menaquinone	366, 217
	*men*B	2-succinylbenzoic acid accumulated		
Haem biosynthesis	*hem*A (popC)	5-aminolaevulinic acid synthetase	Addition of 5-amino-laevulinic acid to non growing cells or haem-atin + ATP to membrane	246, 362, 100, 103, 265, 102, 289
	*hem*B (popD)	Porphobilinogen synthetase	—	246
	*hem*C (popE)	Uroporphyrinogen I synthetase	—	246
	*hem*D (hemC)	Uroporphyrinogen cosynthetase	—	186
	*hem*E	Uroporphyrinogen decarboxylase	—	290
	*hem*F (popB)	Coproporphyrinogen III oxidase	—	246
	*hem*G (popA)	Ferro-chelatase	—	47
Anaerobic electron	*chl*A	Pleiotropic. Likely two complementation		338, 190,

reduction		groups. Suggested lesion in synthesis of factor(Fa) for attachment or insertion of Mo-cofactor.		338, 190, 191, 259
	*chl*C	Nitrate reductase? α subunit (structural gene)	—	190, 96
	*chl*D	Suggested for processing Mo	Excess Mo	91
	*chl*E	Pleiotropic. Possibly two complementation groups? γ subunit nitrate reductase	—	190, 191
	*chl*F	Suggested formate dehydrogenase	—	91
	*chl*G	Unknown	—	91
Dehydrogenase activities	*dld*	D-lactate dehydrogenase	—	312
	sdh	Succinic dehydrogenase	—	130
	frd	Fumarate reductase	—	321, 322
	glpD	L-glycerophosphate dehydrogenase (membrane-bound variety)	—	155
	nut	Energy linked transhydrogenase	—	90
	ndh	NADH dehydrogenase	—	367
Additional mutants	*hyd*	Hydrogenase	—	236
	*nir*A, *nir*B	Nitrite reductase	—	152, 43

energy coupling phosphorylated compound has not yet been ruled out. Energy linked conformational changes in proteins may be critical as some believe [26] and indeed evidence for relevant changes has been obtained. Nevertheless, it is now noticeable that many workers previously favouring the latter hypotheses have introduced a large measure of thinking consistent with the chemiosmotic hypothesis [27].

The flow of electrons from the substrate to either O_2 or some other terminal electron acceptor via the chain of dehydrogenases, cytochromes, flavoproteins, iron-proteins and quinones must be coupled (converted) in some way to the synthesis of chemically useful energy usually in the form of ATP. The electrons appear to be 'bled off' and coupled to the process of synthesis at a number of points, thus giving the impression that the electron transport chain is organized in segments (labelled 0 to 3 in bacteria). These correspond to chemically realizable fractions in mitochondria. Supposedly such segments have a relation to the loops necessary for the chemiosmotic hypothesis. The hypothesis (see p. 107) suggests that ATPase together with the charged impermeable membrane constitutes the essential coupling factor for these segments. The search for coupling factors has a history of 20 or 30 years, most of the work being done with mitochondria, and those interested are advised to turn to the copious literature on the subject. A brief review is given by Garland [86] whilst protagonists of the several hypotheses have been brought together to express their views in the multi-author article already mentioned [27].

Evidence for such a central role of ATPase in energy coupling in bacteria has been greatly helped by the possibility of applying the techniques and reasoning of genetics. A major problem however in considering energy conservation in bacteria has always been the difficulty of obtaining adequate measurements from membrane preparations. These difficulties are to be expected perhaps because of the impermeability of membranes to nucleotides, their ability to vesiculate when isolated and the siting of the ATPase on the cytoplasmic side of the membrane. Direct measurement of the P:O ratio in membrane preparations, using substrates such as lactate, leads to very low values of 0.1–0.3, [14, 38, 242, 336]. For oxidation of NADH they are higher, of the order of 1.0–1.5, especially when chemolithotrophs are involved [88, 141]. Doubt about the usefulness of such absolute values has led to the preference for P:2e ratios during transfer of electrons between substrates of different redox potential, e.g. NADH or reduced cytochrome *c* as the electron donor and O_2 or oxidised cytochrome *c* as acceptor. With growing support for the idea of chemiosmosis, however, it is becoming increasingly common to measure the energy conservation, using whole bacteria, in terms of proton translocation. If an $\rightarrow H^+$:P ratio of 2 moles H^+/mole of ATP [32, 208, 349] is assumed, the measured value for $\rightarrow H^+$:O gives a good indication of the potential number of coupling sites in the electron transport chain.

4.4.1 *Uncoupled mutants*

Examination of bacterial mutants uncoupled between electron flow and ATP synthesis – the so-called *unc* mutants – has provided evidence that ATPase is indeed a vital coupling factor. Although the absolute values of P:O for the membrane preparations, and even for the whole bacteria of the wild-type, were often low, there was no doubt about the very much lower values obtained from the *unc* mutants. A series of such mutants of *E. coli* has been isolated [30, 37, 97, 218, 312, 336] usually by Method 1 (see p. 119): i.e. the organisms would not grow on succinate or glucose anaerobically and gave poor growth yields on glucose aerobically. Membrane preparations made from them had wild-type levels of NADH and D-lactate oxidases, cytochromes b_1, a_2, o and aa, ubiquinone, menaquinone and flavin. They had, however, only very low levels of Ca^{2+}–Mg^{2+} activated ATPase and were unable to form esterified phosphate using D-lactate as substrate. This class of mutants was called *unc*A and the gene mapped at 73.5 min on the chromosome. Other classes of uncoupled mutants have since been isolated as *unc*B [38, 48, 52], *unc*C [90a], *unc*D [329] and *uncE* [61b]. The *unc*C mutants are particularly interesting because, although they are uncoupled, membrane preparations still have functional ATPase activity. Since ATPase is a peripheral membrane protein (see Section 4.5.1) it can be removed by very gentle methods such as washing with buffers of low ionic strength, or dilute solutions of chelating agents, without damaging the function of the integral proteins. This has made possible experiments in which the ATPase is removed from membranes and then reassociated with them whilst indicators of coupling such as NADH transhydrogenase are monitored. When the stripped membranes of the wild-type are reassociated with isolated ATPase in the presence of Mg^{2+} and Ca^{2+} the ATP-driven transhydrogenase activity is restored [30]:

$$NADP^+ + NADH \xrightleftharpoons{ATP} NADPH + NAD^+.$$

When membranes from the *uncA* and *uncB* mutants were compared, those from *uncA* responded to the added ATPase after they had been stripped but not before. Moreover, the wash fluid from *uncB* membranes also restored the ATP-driven transhydrogenase of stripped membranes isolated from either the *uncA* strain or the *uncB* strain [48, 52]. It seems likely from these experiments that since the ATPase-inactive membranes from *uncA* mutants would not reassociate with added ATPase until they were stripped, the sites on the membrane were already occupied by defective ATPase [52]. Other *unc* mutants have been described in which the membranes appear not to have aggregate ATPase on them [49, 334, 335] and yet others have been described in which association between the ATPase and membrane appears to be the unsatisfactory aspect [150, 314]. In one such strain, over 90% of the enzyme appeared in the supernatant fraction during membrane preparation. Examination of the membrane-proteins in another strain not fixing ATPase showed the loss of one of about 54 000 mol.wt. and the gain of another of 25 000 mol.wt. This mutant was deficient in ATP-driven transhydrogenase and it seemed likely that the altered membrane proteins led to poor attachment of the ATPase, possibly due to a lesion in the F_0 region rather than the F_1 region solubilized as ATPase (see p. 126). Yet another so-called *etc* mutant had only slightly less ATPase than the wild-type but was deficient in ATP-driven transhydrogenase and active transport [31, 130]. Examination of the protein subunits of the ATPase (see p. 000) showed a defective γ subunit [31]. It is a matter of some interest that all the *unc* mutants whether phenotypically ATPase$^-$ or ATPase$^+$ all map at a position of 73.5 min in the *E. coli* chromosome. Those interested in reading more about the *unc* and other uncoupled mutants are referred to reviews by Haddock and Jones [104], Simoni and Postma [313], and Downie *et al.* [61a].

4.5 Isolation and properties of Mg^{2+}–Ca^{2+} ATPase

4.5.1 *Purification of F_1*

A Ca^{2+}-Mg^{2+} activated ATPase is associated with membranes prepared from all bacteria so far examined. A high proportion of the enzyme is present in the form of a multi-subunit protein that can be removed from the membranes either by washing them with low ionic strength buffers in the absence of divalent cations or by the use of chelating agents such as EDTA. Thus, the enzyme falls clearly into the category of membrane peripheral proteins. Electron microscopic examination of membranes by negative staining shows arrays of particles apparently fixed to the cytoplasmic side of the membranes [1, 68]. Examined in detail these particles appear hexagonal and to consist of six globules arranged around a central hole. In the enzyme from *S. faecalis*, for example, the globules making up the hexagon have a long axis of 4 nm and the hexagon is 15 nm across [225, 298]. The proof that they are ATPase rests on the identity of the conditions that remove both the enzymic activity and the particles, together with the fact that the particles can be reassembled by treating extracted membranes with isolated purified enzyme. The soluble isolated enzyme also has a similar morphology to that associated on the membrane surface. The location of the ATPase activity on the cytoplasmic side of the plasma membrane has been best demonstrated in *Micrococcus lysodeikticus* by two methods. Antiserum prepared against purified ATPase reacted with isolated membrane preparations and inhibited their ATPase activity, whereas the antibodies failed to react with intact protoplasts [225, 284]. When ferritin labelled antibodies were used, the distribution of ferritin particles, after reaction with the membranes, was very similar to the pattern of particles seen by negative staining on the unreacted membranes [225]. Likewise, the ATPase was only accessible to ^{125}I in the lactoperoxidase ^{125}I system in isolated membrane preparations and not in intact protoplasts [287]. The enzyme has now been isolated and purified from a considerable number of different species of bacteria (Table 4.6). The method of isolation has almost always been to wash membrane preparations first with buffer containing Mg^{2+} to rid them of cytoplasmic cell consituents, and then repeatedly with low concentration buffers not containing Mg^{2+}, with or without EDTA present. Monitoring the washing fluids for the presence of ATPase has shown that the enzyme is removed rather cleanly in a few such treatments. Subsequent purification has usually been by column chromatography on DEAE-cellulose columns, often

Table 4.6 Sub-unit size and composition of the F_1 ATPases (modified from [104])

Micro-organism	*Molecular weights* ($\times 10^{-3}$) Enzyme	α	β	γ	δ	ϵ	*Subunit composition*	*Reference*
Rat liver mitochondria	340–380	53–62.5	50–57	25–36	12–12.5	7.5	$\alpha_3\beta_3\gamma\delta\epsilon$	140
Yeast mitochondria	340	58	54	38	38	12	—	140
Escherichia coli	340–400	56–60	52–56	32–35	21	11–13	$\alpha_3\beta_3\gamma\delta\epsilon$*	140
Salmonella typhimurium		57	52	31	21	13	$\alpha_3\beta_3\gamma\delta\epsilon$	140
Alcaligenes faecalis	350	59	54	43		12	—	140
Thermophilic bacterium	380	56	53	32	11	15	—	140
Streptococcus faecalis	345	60	55	37	20	12	—	140
Bacillus megaterium	399	68	65	—	—	—	$\alpha_3\beta_3$	140
Micrococcus luteus	345	52–60	47–60	41	28		$\alpha_3\beta_3\gamma(\delta\epsilon)$†	140
Bacillus subtilis	315	59	57	—	—	—	$\alpha_3\beta_3$	307

*One claim for a composition of $\alpha_2\beta_2\gamma_2\delta_{1-2}\epsilon_2$ [340]

†Presence of δ and ϵ not yet settled

followed by columns of Sepharose or Sephadex. The best preparations obtained have had specific activities of the order of 100–130 units/mg protein where a unit represents the hydrolysis of 1 μmol of ATP/min to give ADP and P_i. This high specific activity has been achieved with the enzyme extracted from *E. coli* which has been subjected to more intensive study by more groups of workers than the enzyme extracted from other sources. The specific activity of enzyme extracted from other sources, such as *M. lysodeikticus* is usually lower. These enzymes seem pure and the difference may indicate a truly different catalytic activity by the proteins. When examined by polyacrylamide gel electrophoresis (see p. 94) in the absence of reagents known to dissociate multimer proteins (e.g. sodium dodecyl sulphate, high concentrations of guanidine HCl or urea) a single band is obtained but, when run on gels containing such agents, a number of bands are seen. In all the ATPases examined two major bands, arbitrarily called α and β, appeared which ran more slowly than the remaining bands which correspond to subunits present in smaller amounts and which have been called γ, δ, and ϵ. The α and β subunits carry the major part of the ATPase enzyme activity. The first recognition that a separate protein subunit in crude ATPase was concerned with attachment of the enzyme to membranes, came from work with the enzyme from *Streptococcus faecalis.* It was named 'nectin' or the γ subunit [2] and was isolated in a purified state [18] as an enzymically inactive protein with a molecular weight of 37 000. It induced the α and β subunits to reaggregate on membranes from which ATPase had already been removed [298]. The aggregate had the composition $\alpha_3 \beta_3 \gamma$ [2]. Since this time, it has become clear that the δ and ϵ subunits of ATPase are both concerned with attaching the enzyme to membranes. The early hints with the enzymes from both *M. lysodeikticus* [285] and *E. coli* [31] were that enzymes prepared in certain ways showed fewer subunits and were not able to reassociate with membranes stripped of their ATPase. The story for the enzyme from *E. coli* is now rather more complete than from other sources. Bragg *et al.* [31] found that when they extracted ATPase from polyacrylamide gels run at pH 8.7 in the absence of dissociating agents, it was not capable of restoring the ATP-driven transhydrogenase reaction to membranes stripped of other ATPase, whereas if they prepared the enzyme by centrifugation in sucrose gradients [29] it could do so. Examination of the enzyme from the gels showed that the δ subunit was missing. Likewise a method used [111, 216] for purification of ATPase from *E. coli* gave rise to preparations that would not reassociate with membranes and which dissociated to only four subunits. Another method, however, [85] led to enzyme consisting of five subunits and which would associate. Treatment of the five subunit enzyme with pyridine, a step in the production of the four unit enzyme led to the solubilization of two subunits, the δ and ϵ units [316]. The addition of the minor subunit fraction to the deficient enzyme restored its ability to attach to membranes and enabled them to undertake oxidative phosphorylation. Subsequent fractionation [318] of the pyridine extract containing both δ and ϵ subunits suggested that only the δ subunit was concerned with attachment. Nevertheless, other work [324] suggests that both δ and ϵ subunits are necessary. A further important fact was established during this work, namely one of the functions of the ϵ subunit. It was found that this subunit, which is hydrolysed by trypsin, inhibits the ATPase activity of the aggregate that remains after the removal of the δ and ϵ subunits. However, an aggregate consisting of only the major α and β subunits obtained by trypsin digestion was not inhibited, which suggests that in some way the remaining γ subunit is necessary for the inhibitory action of ϵ on the $\alpha\beta$ aggregate. A very active soluble ATPase can be reconstituted from isolated α, β and γ subunits when the mixture is dialysed against buffer at pH 6 containing ATP, all three units being necessary [83]. A very suggestive piece of work [28, 30a] on the possible arrangement of the subunits in the hexagonal arrays that have been seen both in preparations of the isolated enzyme and on the inner surface of the plasma membrane, involved the use of dithio*bis*succinimidyl propionate (DSP), a well known protein cross-linking reagent. When reacted with DSP neither the α nor the β subunits were found homogeneously linked together. Rather α and β were linked together and a fragment containing α, β, and γ was also isolated. The suggested arrangement on the basis of this work is as shown in Fig. 4.7. The δ and ϵ subunits can also be shown to form a cross-linked pair [15] and, since δ and ϵ both appear to be concerned with membrane attachment, one might suggest the arrangement shown in Fig. 4.8.

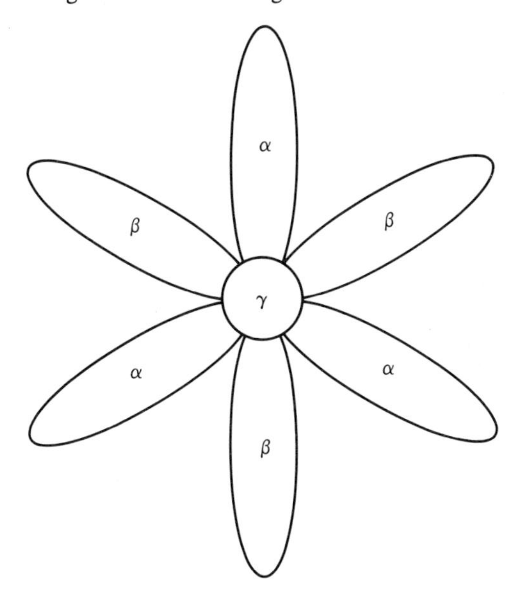

Figure 4.7 A suggested structure for ATPase F_1 (from [28]).

As can be seen from Table 4.6 the stoichiometry of the aggregate ATPase from a number of bacteria is $\alpha_3\ \beta_3\ \gamma\ \delta\ \epsilon$ as in *E. coli.* The foregoing all applies to the analysis of the enzymically active ATPase that can be solubilized as a peripheral membrane enzyme. In order for it to fulfil its role in coupling electron transport and protein translocation, there must exist some specific property of the membrane at the location occupied by the ATPase complex. This is frequently referred to as a specific channel or pore, and has been named [239] the F_0 part of the ATPase complex with the solubilizable subunit proteins of the active enzyme already discussed, being called F_1. If the F_1 complex sits at the mouth of a channel (see Figs 4.2 and 4.8), then its removal from the membrane should cause an increase in proton conductance of the membrane. Evidence from mitochondria would suggest that this is true; for further discussion of this matter the reader is referred to Haddock and Jones [104].

The function of all the minor subunits of ATPase from other organisms is at present less well defined, although it would appear that the δ subunit in a number of organisms is involved in attachment of the remaining subunits to the membrane [3, 82, 317]. For example, homogeneous enzymes, prepared from *M. luteus* by washing membranes with low ionic strength buffers and examined by polyacrylamide gel electrophoresis in the presence of dissociating agents, show the presence of two major bands and from one to three lesser ones according to the preparation [10, 258, 285, 286]. A study of trypsin action on a purified preparation containing only α, β and γ subunits, however, found inactivation of the enzyme to occur with a half-life of 3 h and the production of two bands of multimer enzyme. It is suggested that the α subunit peptide had been shortened [128]. The *M. luteus* enzyme will reassociate with membranes in the presence of Mg^{2+} and is then activated by treatment with trypsin [285, 286]. After subsequent extraction

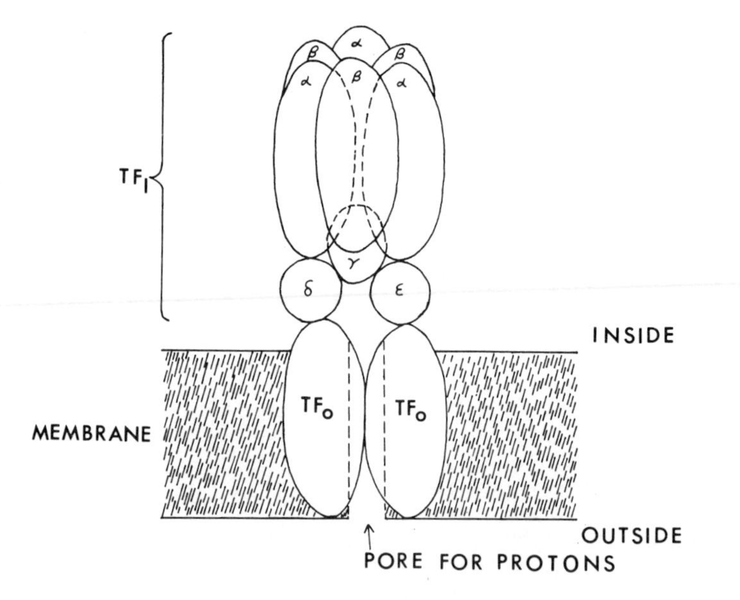

Figure 4.8 A possible arrangement of the TF_1 and TF_0 regions of ATPase in the membrane. The arrangement of subunits of TF_0 to form a pore penetrable by protons is hypothetical, whilst considerable evidence exists for the arrangement of the subunits of TF_1. After [363a].

of the membrane with aqueous *n*-butanol, only two subunits are found and the enzyme will not reassociate with the membranes and is not activated by trypsin treatment [286]. An obvious interpretation would appear to be that the minor subunits including ϵ had been digested during the trypsin treatment leaving only the major α and β units bearing the enzyme activity. The authors, however, pointed out that the molecular weights of the enzymes, prepared by these two methods appeared to be scarcely significantly different being 352 000 and 345 000 respectively. Nevertheless, the fact that trypsin treatment of the ATPase from *E. coli* also destroyed the minor subunits and left only the α and β subunits [216] strongly suggests that something similar may have happened to the *M. luteus* enzyme. Antibodies prepared against the α or β units of the ATPase from *E. coli* completely inhibited the ATP-linked transhydrogenase activity of the membranes [150, 216] again supporting a major catalytic role for these subunits. The isolated multimer enzyme itself, from various sources, shows similar, though apparently not identical, properties. For example, the molecular weight varies from 315 000 to 400 000 (see Table 4.6). Some differences exist in substrate specificity and in the range of divalent cations capable of giving activation. Almost all the isolated enzymes have been found to be sensitive to cold, in contrast to the enzyme *in situ* on the membrane. Investigation [340] of the enzyme from *E. coli* showed that freezing preparations to $-70°$ C and subsequent thawing dissociated the aggregate into two particles which were different from each other and both still multimers. These could be reassociated into active enzyme under the correct conditions of ionic strength and pH. The enzymes from *Mycobacterium phlei* [148] and a thermophilic bacterium [99] are said to be insensitive to cold. All the isolated ATPases studied are strongly inhibited by ADP and are activated by Mg^{2+}. Although the enzyme is often referred to as Mg^{2+}–Ca^{2+} ATPase to distinguish it from the well known enzyme from mammalian sources that is activated by Na^+-K^+, preparations from some sources such as thermophilic bacteria [99] are rather poorly activated by Ca^{2+} compared with Mg^{2+} or divalent cations such as Mn^{2+} and Cd^{2+}. In some examples the reverse is also true and Ca^{2+} gives a greater degree of activation than Mg^{2+} [201]. Nevertheless all the enzymes studied are activated by Mg^{2+} and the activation obtained bears a definite relationship to the ratio of the concentration of the ion to the substrate, ATP. Binding to membranes also shows different specificities towards divalent cations whilst, with enzymes from most sources Mg^{2+} is effective, the enzyme from *B. megaterium* binds only in the presence of Ca^{2+} [201]. The substrate specificities of the different enzymes also appear to show quantitative differences particularly towards pyrimidine triphosphate nucleotides such as CTP and UTP. In general they act better on the purine nucleotides. Strongly enzyme bound ADP and ATP have been described in the ATPase isolated from *Streptococcus faecalis* [2]. The nucleotide bound in the enzyme when isolated from the bacteria cannot be displaced by exogenous nucleotide although this too, is strongly bound.

Table 4.7 Inhibition of ATPases from the cytoplasmic membranes of *Micrococcus luteus* (from [296]).

Agent	*Concentration*	*% Inhibition*		
	(M × 10^4)	F_1-*ATPase*	F_0–F_1-*ATPase*	*Membrane-bound ATPase*
Rutamycin	4.6	—	77	50
Dio-9	(20 μg/ml)	—	49	34
Quercetin	0.2	100	48	11
Oligomycin	4.6	—	43	48
Botrycidin	(20 μg/ml)	—	42	53
E frapeptin	0.6	—	35	14
Leucinostatin	1.3	—	28	34
Valinomycin	0.5	—	28	9
Venturicidin	(20 μg/ml)	—	24	49
N,N'-dicyclohexylcarbodiimide (DCCD)	4.0	8	80	67
Dinitrophenyl	3.0	—	49	16
Triethyltin-Cl	2.9	13	—	62
Tributyltin-Cl	2.9	24	—	40

4.5.2 *Purification of F_0-F_1*

As has been already said, it was hypothesized that the soluble multimeric enzyme isolated and studied above must be situated on specific points on the membrane surface organized to act as channels for protons. Membrane bound ATPases are sensitive to dicyclohexylcarbodiimide (DCCD) and other carbodiimides, whereas the isolated preparations of F_1 are not. A number of attempts have, therefore, been made to isolate a DCCD-sensitive ATPase complex from bacterial cytoplasmic membranes [112, 218, 295, 319, 364, 365]. ATPase with the properties of the membrane-bound enzyme was first isolated from *M. luteus* cytoplasmic membranes [210], by extraction with Triton X-100 in the presence of Mg^{2+}, and identified as an antigen and enzyme that was not expressed on the surface of undamaged protoplasts [229]. The isolated purified 'membrane-bound' enzyme from *E. coli* was also obtained and studied [295]. It reacted identically to F_1 with F_1 antibodies but unlike the water soluble multimer F_1, the F_0-F_1 complex was as strongly inhibited by a wide range of substances including DCCD and dinitrophenol as was the original membrane preparation (see Table 4.7). The chromatophore membranes of *Rhodospirillum rubrum* have been shown to have ATPase on their outward (cell cytoplasmic) surfaces [21, 209]. This enzyme has been purified first in the water soluble F_1 form [138, 158, 241], then in an oligomycin-sensitive form using Triton X-100 extraction of the membrane [226, 299]. Again there was immunological identity between membrane-bound ATPase, F_1 ATPase and F_0-F_1 ATPase [209]. This F_1 ATPase can also be dissociated into the usual five subunits [137, 138, 241], there again being some evidence [209] that the δ subunit is involved in attachment of the F_1 multimer to the F_0 region of the membrane. Having isolated F_0-F_1 enzymes the matter of great interest is to know, of course, the nature of the F_0 proteins. In some of the F_0-F_1 ATPases a small number of subunits over and above the five of the F_1 enzyme can be recognized. For example, from an unidentified thermophilic bacterium, PS3, three extra bands after polyacrylamide gel electrophoresis of F_0-F_1 were found with molecular weights of 5400, 13 500 and 19 000 [224, 319]; these three extra bands corresponded with those in isolated F_0 preparations made from membranes successively extracted with cholate and Triton X-100 [364, 365]. The F_0 preparation rendered F_1 preparations sensitive to inhibitors and when incorporated into a phospholipid preparation rendered the liposomes conductive to protons [224]. A very exciting system capable of light-induced ATP synthesis has been reconstructed with an F_0-F_1 preparation, phospholipids and purple membranes from *Halobacterium halobium* and has been shown to have a light-dependent ATP synthesis that is sensitive to uncoupling agents [365].

Thus we are close to showing that membrane Mg^{2+}-Ca^{2+} activated ATPase is indeed, if not the coupling factor, one very closely involved with energy conservation in membranes. This is at one level of explanation but clearly many years of fascination await scientists investigating the precise mechanism of action of the multimeric F_1 ATPase and its interaction with hydrophobic membrane proteins to form the F_0-F_1 enzymes. Studies on F_1 and F_0-F_1 have been recently reviewed [61c].

4.6 Vesiculation of membranes

Discussion of the behaviour of membranes when they are disrupted is interjected at this point, before discussing analysis of their transport functions, because the use of membrane vesicles has become so important in this field of study. Much of the earlier work done with so-called membrane particles can be interpreted more easily when it is realized that the particles are enclosed vesicles and there have been a number of references, earlier in the text, to membrane vesiculation. When membranes, free of wall material, are disrupted they do not exist as sheets available to all molecules at both surfaces but as resealed small closed vesicles. Naturally occurring membranes, however, are not symmetric either in the distribution of lipids or of enzymes (see Chapters 3.1 and 4.3.2). It is, therefore, of importance whether the vesicles are formed with membrane in the same or a different orientation from that in the living cell. In other words, the sidedness of the membrane has to be considered. Many methods and arguments have been used in attempts to decided membrane orientation unambiguously. Some of these methods are summarized in Table 4.8. They may be briefly categorized as:

(1) The use of antibodies to membrane enzymes with the reasonable assumption that antibodies cannot penetrate membranes. The antibodies are frequently coupled with ferritin to allow electron-microscopic examination of the preparation.
(2) The availability or otherwise of membrane enzymes to non-penetrating substrates. The presence of the active enzyme or enzyme system in the preparation may then be checked by using a penetrating substrate. Potassium ferricyanide as a non-penetrating substrate, and phenazine methosulphate as a penetrating substrate have frequently been used as a pair of electron donors in oxidation–reduction reactions. Results obtained using potassium ferricyanide should, however, be accepted with great caution, in view of the results [22] for membrane vesicles from *B. subtilis* which seem to suggest that ferricyanide can penetrate membranes after all.
(3) Electron microscopic examination using negative staining to look for the characteristically particulate ATPase (see Section 1.4.2) or freeze–fractured preparations for membrane particles.
(4) Examination of the direction of translocation of H^+. Fluorescent organic anions have their fluorescence quenched within vesicles such as chloroplasts or liposomes [58, 302, 303]. Their distribution in a vesicle preparation can then be readily followed and used to calculate the ΔpH across the membrane [57].

Doubts have been raised about the validity of some of the methods for determining membrane orientation, due to possible migration of some enzymes, (e.g. ATPase) but not others, from one side of the membrane to the other during the process of vesiculation [6, 309]. Ignoring this possibility for the moment, it would seem unlikely that any vesicle preparation is made up of either wholly inverted or wholly uninverted membrane. Gentle lysis of lysozyme-produced spheroplasts of *E. coli* seems to produce a population which contains not less than about 50% of the vesicles with the membrane orientated as it was in the original bacteria, and under some circumstances similar results are obtained with *B. subtilis*. Sonication, or passage through the French Pressure Cell, of bacteria, spheroplasts or protoplasts produce a high proportion of vesicles with inverted membrane. An unexpected complication is that in some instances, the cultural conditions for growing the micro-organism can be important in defining the membrane orientation in vesicles. When *Paracoccus denitrificans* was grown with succinate as carbon source and nitrate as electron acceptor, the vesicles resulting from the lysis of lysozyme-produced protoplasts appeared to have inverted membranes. When the bacteria were grown with CO_2 as the carbon source, H_2 as the reductant and O_2 as the electron acceptor, the membrane of the vesicles was as in the original bacteria. This difference did not seem to be related to the slow growth rate obtainable under auxotrophic conditions, since bacteria growing rapidly in rich media also liberated uninverted vesicles [36].

One of the dangers of working with preparations consisting of two populations of vesicles with membranes of opposite orientations is that the results may need careful interpretation. If two functions are measured, such for example as NADH dehydrogenase and metabolite transport, each may be carried out by a separate part of the population. Therefore, causal relationships between the two should not be deduced. Methods involving gradient centrifugation have been devised in attempts to separate the two populations of vesicles prepared from mammalian cell membranes [323, 368] and antibodies to ATPase have also been used for bacteria [113]. A particularly ingenious method for vesicles prepared from bacterial membranes employs antibodies to the coat protein of the filamentous bacteriophage M13 [353]. Evidence had been produced [352] that the protein was present only on the outside of the membrane of spheroplasts of infected *E. coli* and that antibody would not react with vesicles produced by sonication. In order to obtain agglutination of the vesicles they had first to be treated with the coat antibody raised in rabbits, washed and treated with anti-rabbit antibodies raised in goats. Under these conditions about 80% of the vesicles produced by osmotic lysis of lysozyme–EDTA-induced spheroplasts were agglutinated. Very disturbingly, there was no difference between the agglutinated and non-agglutinated populations as far as ATPase and NADH oxidase activities were concerned (neither substrate can, of course, penetrate membranes). The authors suggest that either the M13 coat protein, or the enzymes had been redistributed during formation of the vesicles. This problem clearly needs further study, particularly since the results with sonically produced vesicles had been clear-cut in showing no reaction with anti-coat antibody [352]. In a very thorough examination of the situation with respect to vesicles produced from membranes of *E. coli* strain K12-7 by the lysozyme–EDTA and French Pressure Cell methods, Adler and Rosen [4] have raised a number of points, hitherto over-

Table 4.8 The orientation of bacterial membranes in vesicles

Micro-organism	*Method of preparation*	*Method of examination*	*Orientation deduced*	*Reference*
E. coli	Lysozyme–EDTA	Activities of succinic and glycerol 3-phosphate dehydrogenases.	c. 50% inverted	344
		ATPase and reaction with antibody NADH-$K_3Fe(CN)_6$ * reductase.	c. 50% inverted	82
	Lysozyme–EDTA	D-lactate dehydrogenase. Reaction with antibody.	85% uninverted	84
	Lysozyme–EDTA	Ultrastructure. Section and freeze-fracture	uninverted	145
	Lysozyme–EDTA	Crossed immuno-electrophoresis	99% uninverted	228a
	Lysozyme–EDTA	Ultrastructure. Freeze-etching.	uninverted	309
	Lysozyme–EDTA	Uptake of [^{3}H]-vinylglycollate	uninverted	310
	Lysozyme–EDTA, freezing and thawing	Ultrastructure. Freeze-etching	25% inverted	6
	EDTA–lysozyme	Antibody to ATPase	50% uninverted	113
	Lysozyme–EDTA and sonication	D-lactate dehydrogenase. Reaction with antibody.	> 70% inverted	309
	Lysozyme–EDTA	ATPase, NADH dehydrogenase. ATPase antibody.	Enzyme translocation (see text)	4
	French Pressure Cell	ATPase. Reaction with antibody	60–100% inverted	82

	Sonication	antibody. Ultrastructure. Freeze-etching ATPase antibody reaction. $NADH_2$–$K_3FE(CN)_6$ reductase.	all inverted	309
Mycobacterium phlei	Lysozyme	Cryptic oxidative phosphorylation	uninverted	13
	Sonication	Expressed oxidative phosphorylation	inverted	13
Bacillus subtilis	Lysozyme	Ultrastructure.† Freeze-etching	uninverted	159
Micrococcus lysodeikticus	Lysozyme	ATPase reaction with antibody and [^{125}I]-lactoperoxidase.	at least partly inverted	225
	Lysozyme, Sonication	Accumulation of organic anions and K^+	mixture of inverted and uninverted	93
Bacillus caldolyticus	Sonication	H-pump.	inverted	57

* $K_3Fe(CN)_6$ was used as an impenetrable anionic electron acceptor. It was not reduced in the presence of NADH by protoplasts or spheroplasts unless these were treated with either toluene or Triton X-100. When vesicles behaved similarly they were assumed to be surrounded by membrane of the same orientation as that on the protoplasts i.e. uninverted.

† Vesicles within vesicles were seen and of these about 85% were uninverted.

looked. Firstly, they have pointed out that vesicles formed from the outer membrane with strains of *E. coli* such as K12-7, are likely to form about 50% of the vesicle population, while strain ML308 [145, 228a] may shed its outer membrane in some way without producing vesicles. Outer membrane vesicles have no enzymic activity. Secondly, in the total population of vesicles about 60% of the NADH dehydrogenase and ATPase were expressed compared with toluene-treated vesicles as the 100% control. In vesicles produced by the French Pressure Cell, 100% of both activities was expressed. None of the D-lactate dehydrogenase is available to antibody in the lysozyme–EDTA type vesicles [82, 309]. Thus it would seem that some enzymes, such as ATPase and NADH dehydrogenase, appear on both sides of the membrane whereas others stay in their original position. This migration of some enzymes and not others produces, it is suggested, a mosaic membrane. The consequences for the polarity of the membrane with various electron donors are curious and results of transport studies, for example, need very careful interpretation. With NADH as donor, Ca^{2+} ions were accumulated by the vesicles, suggesting that the everted NADH dehydrogenase still acted as part of its electron transporting complex and pumped protons out and Ca^{2+} in on the anti-porter (see [335] and p. 109) with a negative alkaline gradient to the outside of the membrane. D-Lactate, on the other hand, did not serve as the energy source for Ca^{2+} uptake by the vesicles because the polarity of the proton gradient gave positive and acid on the outsides, negative and alkaline on the insides. The reverse situation would be true for proline uptake by the vesicles which is energized by a gradient acid-positive outside. Re-examination [228a] of the vesicle population produced from *E. coli* ML308, on the other hand, by crossed immunoelectrophoresis of 14 recognizable immunogens showed that 11 exhibited minimal expression in EDTA–lysozyme produced material and that migration of enzymes from one side of the membrane to the other could not be greater than 10% for any one of the immunogens. Thus, although the use of vesicles is widespread and extremely valuable in studies of membrane function, much care must be exerted in the interpretation of results.

4.7 Transport of metabolites and ions

The passage of substances into bacterial cells is usually considered under three or four main headings representing fundamentally different processes. These are group translocation, simple diffusion, facilitated diffusion, and active transport. Brief descriptions of these processes will precede their detailed discussion.

Group translocation

This process is distinguished from the other three by chemical modification of the molecule transported, presumably at the interface of the membrane. This means that translocation itself does not take place against a concentration gradient, the concentration of the molecule before modification always being low on the cytoplasmic side of the membrane. The best known example of group translocation is the phosphotransferase system for sugar transport.

Simple diffusion

This needs no description and of course, for neutral metabolites always proceeds along a downhill concentration gradient, the cytoplasmic membrane having a similar role to that of dialysis sac. Whether such a simple process ever occurs through living membranes may be questioned.

Facilitated diffusion

This also occurs downhill but at a much faster rate than can be explained by simple diffusion and involves specific carrier molecules in the membrane. The rate is not strictly proportional to the concentration gradient and the system appears saturatable. It is also stereospecific unlike simple diffusion.

Active transport

The substrate molecule is not modified but is driven along an uphill chemical gradient so that high concentrations of substances are accumulated inside cells. The literature on the process is voluminous and still disputatious. Three aspects of the problem can be distinguished:

(1) The nature of the specific protein carriers both in the membrane and as periplasmic proteins, which are genetically specified and regulated.
(2) The way in which energy derived from the electron transport chain is transferred so that it can drive the uphill transport.
(3) The actual process of translocation of the molecules across the membrane.

Much of the argument and interest over the last few years has been centred on problems under (2). Very little is known about (3).

It is important to realize that a micro-organism may have the ability to take up sugars or other metabolites not by one of these methods alone, but by several of them. For example, *E. coli* and *Salmonella typhimurium* transport sugars by facilitated diffusion, group translocation, and active transport, whereas *S. aureus* appears to transport all its sugars by group translocation. In *E. coli* four separate galactose permease systems have been described and many Gram-negative organisms appear to take up glucose, mannose, fructose, *N*-acetylglucosamine, glucosamine, *N*-acetylmannosamine, mannosamine, β-glucosides, hexitols, and possibly galactose by the phosphotransferase system. On the other hand, disaccharides, pentoses and glycerol, are either actively transported or taken up by facilitated diffusion.

4.7.1 *Phosphotransferase system (PTS) for sugar transport*

The overall reaction involved in the translocation of carbohydrates through the membranes by this system is:

$$\text{Sugar} + CH_2{=}\underset{\displaystyle O{-}PO_3^{2-}}{\underset{|}{C}}{-}CO_2^- \xrightarrow[\text{PTS}]{Mg^{2+}} \text{Sugar}{-}PO_3^{2-} + CH_3CO{-}CO_2^-$$

This overall reaction has turned out to consist of a complex series of steps, even now not fully dissected. It has been analysed largely through the work of Roseman and his colleagues [267, 268] into the following:

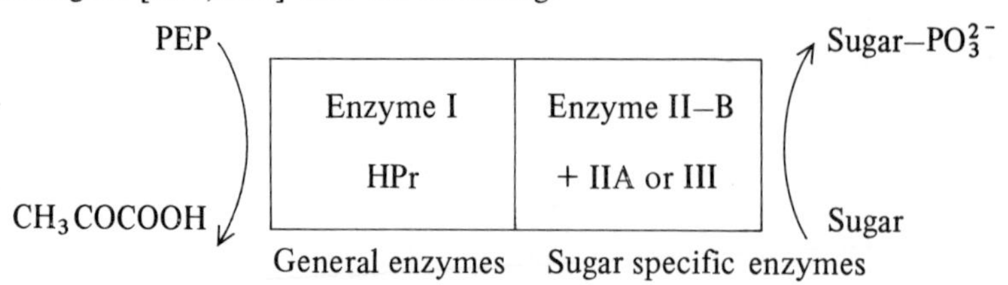

Thus, the simplest analysis of the complex says it has two parts: the non-specific enzymes important for the transfer of all sugars and which always appear soluble in the cytoplasm, and the highly carbohydrate specific enzyme II and III complexes some parts of which are membrane-bound. Two enzymes in this complex are required for the transport of any given sugar. The soluble and membrane-bound enzymes are interrelated through a soluble low molecular weight protein (HPr) that is reversibly phosphorylated on the *N*-1 positions of the imidazole ring of a histidyl residue. HPr proteins from *S. aureus*, *S. typhimurium* and *E. coli* have been isolated in homogeneous forms, crystallized and studied. That from *S. aureus* has one residue of histidine, whereas that from *E. coli* has two. Thus the scheme above can be expanded to:

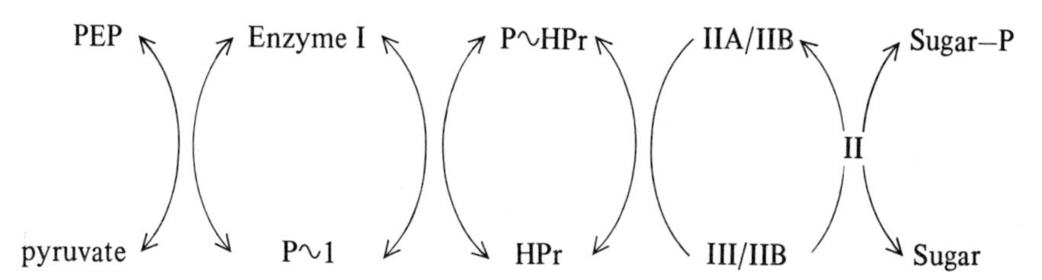

Enzyme I like the HPr protein is also phosphorylated as a first step in the total reaction, the phosphate again being attached to a histidyl residue. The transfer of the phosphate from PEP to enzyme I requires Mg^{2+} unlike the reaction transferring phosphate from protein-I to HPr. The equilibrium constant for the overall reaction (1) has been determined as 10 ± 5:

$$\text{PEP} + \text{HPr} \rightleftharpoons \text{P-HPr} + \text{pyruvate} \tag{1}$$

It would be deduced from this that the energy of the P ~ HPr bond is greater than the pyrophosphate bond of ATP and only slightly less than the phosphate bond in PEP.

As has been said, Enzyme I and HPr both appear in the soluble fraction of disintegrated bacteria. Whether, however, this means that they are truly cytoplasmic or have been displaced from the membrane during cell disintegration is once more not known. Enzyme I has been purified from *E. coli* and *S. typhimurium.* There is some evidence that it may be dissociated into subunits [267, 268]. This part of the phosphotransferase system is generally thought to be constitutively formed and is not specific. Mutants lacking or defective in Enzyme I or HPr are not able to transport any of the sugars using the phosphotransferase system [45, 70, 280, 281]. The map positions of the genes coding for these two components have been determined in both *E. coli* and *S. typhimurium*. They occur very close together and constitute the *pts* operon which also includes a promoter [45, 46, 70, 281].

Apart from these non-specific constitutive proteins, a pair of sugar specific proteins is required which, together with lipid and divalent cation, constitute Enzyme II. Each component of the Enzyme II complex is divided into membrane-bound (IIB) and phosphoryl transfer proteins which may be either membrane-bound (called IIA) or soluble (called III). The rather complex classification of these enzymes according to the sugar involved is set out in a review [245].

The need for several proteins can be demonstrated using *E. coli* membranes. When the organism is grown on a minimal salt medium with glucose as the sole carbon source, the membrane fraction from the cells will transfer phosphate from P ~ HPr to glucose, mannose or fructose. If the membranes are extracted with *n*-butanol and urea, the resulting soluble fraction and the residue are both necessary for phosphorylation of the three sugars. From the soluble fraction three proteins can be purified, each specific for one of the sugars. These proteins have been designated the II proteins. From the pellet the IIB protein can be obtained by deoxycholate extractions [161, 267, 268]. IIB appears to be a single homogeneous protein and if so it must be shared by multiple IIA proteins. Genetic evidence supports the idea of IIB as a single protein [245]. Phosphatidylglycerol and divalent ions are necessary for the activity of the IIA–IIB complex. The III Glc/IIB Glc system will also phosphorylate glucose and III Glc has been isolated and purified to homogeneity. It can be dissociated into 3–4 subunits each of 5000–7000 mol.wt. [267, 268]. Curiously, the homogeneous preparation shows a very potent phosphatase activity, specific for hexoses of the D-gluco or D-manno configurations but it is thought that this is due to contamination [245]; the phosphatase can be inactivated by heating without damaging the transferase function but the III Glc protein is dissociated into its subunits [268]. III Glc can be phosphorylated by P ~ HPr but in contrast to other proteins in this system, can also be phosphorylated by ATP [268]. The IIB Glc protein which is different from the IIB protein has not been purified. It should be noted that whilst IIB is an enzyme, the proteins HPr and IIA (or III) are essentially substrates acting as phosphate acceptors. The soluble protein III lac involved in the transport of lactose into *S. aureus* has been purified [117] and shown to consist of three identical subunits each of which when present in the multimer enzyme can accept a phosphoryl group on to the *N*-3 position of histidine.

The phosphotransferase system would seem to be involved in a more generally regulatory function in bacterial cells, than is indicated by the sugars known to be transported by it. Sugars not transported into organisms such as *E. coli* by the system are accumulated by active transport and, as is now generally thought, move along a gradient of cations into the cell, i.e. by the symport mechanism of the chemiosmotic hypothesis. Mutants defective in enzyme I and HPr are hypersensitive to repression of the permeases necessary for the active transport to take place, particularly by traces of the carbohydrates normally transported by the phosphotransferase system. This effect itself can be abolished by another mutation, close to the phosphotransferase operon, known as *crr* [45, 267, 268].

The involvement of the phosphotransferase system in catalysing uptake of sugars by membrane vesicles prepared from *E. coli*, *S. typhimurium* and *B. subtilis* has been very elegantly demonstrated [342]. Particularly impressive was the demonstration that when the vesicles had been pre-washed with ^{14}C-glucose under conditions where no phosphorylation would occur, they showed absolute preference for ^{3}H-glucose supplied along with phosphoenolpyruvate in the external environment. There was no evidence of any mixing of the pool ^{14}C-glucose into the ^{3}H-glucose phosphate that had been transported into the cells. A potentially valuable inhibitor of Enzyme I of the phosphotransferase system in membrane vesicles has been found in vinylglycolate [342]. It is thought that this substance is taken up by membrane vesicles via the lactate transport system [193] and oxidized by lactic dehydrogenase on the inner surface of the membranes to 2-keto-3-butenoate which then reacts with enzyme I. It does not inactivate HPr.

Evidence for the functioning of all the components of the transferase system in *E. coli* and *S. aureus* [314a, b] has been obtained. One broader survey [266] suggested that it was not, however, universally present as a method for transporting glucose. Activity was present in 8 out of 13 species of different bacteria and whilst most of these were facultative anaerobes, *B. subtilis*, a strict aerobe also had a very active system. This is clearly a highly complex method of transport; one has to ask whether other transport systems will be found equally complex when the right keys are found and turned with the fortitude and persistence of those who have investigated the phosphotransferase system. Those interested in further information about the phosphotransferase system, are urged to consult the review by Postma and Roseman [245] and the references therein.

Another possible type of group translocation cycle is illustrated by the uptake of purines by bacteria. Preliminary work showed that the concentrative uptake of these bases by *Bacillus subtilis* was controlled by purine nucleoside pyrophosphorylases [20]. Subsequently [126], it was found that the uptake of adenine by membrane vesicles prepared from *E. coli* was stimulated by P-ribose-PP and that it was converted to AMP during transport. Adenine phosphoribosyl transferase was isolated from the membranes and shown to be identical to that isolated from the soluble fraction from sonically disrupted cells. The adenine taken from the medium accumulated inside the vesicles as AMP. About 70% of the phosphoribosyl transferases could also be released from whole cells of *E. coli* by the osmotic shock procedure [125, 127]. Thus it would seem likely that the enzymes were on the outside of the cytoplasmic membrane as well as possibly in the periplasmic space. The former location would seem necessary for at least some of the enzyme since vesicles also transported the bases, and these have lost much of their truly periplasmic enzyme complexity. The role of the purine phosphoribosyl transferases as necessary components in transport was supported by the observation [134] that, whilst membrane vesicles from a mutant of *S. typhimurium* lacking the guanine phosphoribosyl transferase were not stimulated in the uptake of guanine by P-ribose-PP, the wild-type was. Hypoxanthine uptake was stimulated in vesicles from both the mutant and the wild-type. Thus adenosine, guanosine and hypoxanthine appear to be converted to mononucleotides probably at the outer membrane first, and translocated in this form. Moreover evidence has been found [125] that, when presented with a nucleoside such as adenosine, the P-ribose phosphate is split off and accumulated inside vesicles as P-ribose-PP whilst it is renewed from exogenous P-ribose-PP. Thus for the uptake of nucleosides the reactions suggested are:

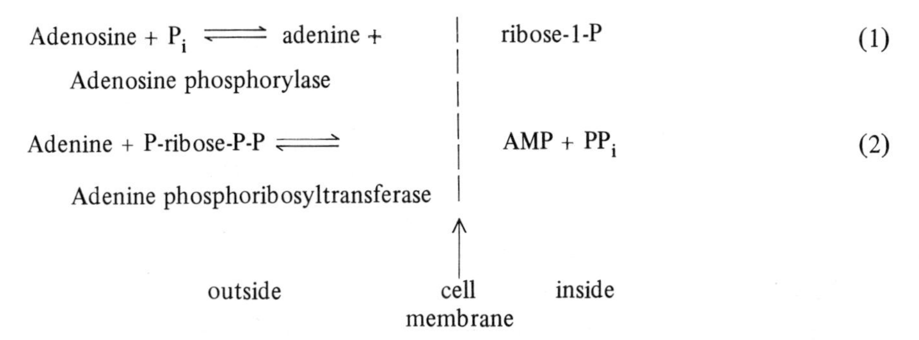

This interesting system needs further investigation. The possible role and cellular location of these very clearly superficial enzymes is of considerable importance. It must also be noted that there is disagreement with this whole interpretation of the transport of purines and arguments have been put forward in favour of specific active transport systems [272a].

4.7.2 *Active transport*

The process of active transport of substances into cells, bacterial or otherwise is distinguished by three criteria:

(1) The substance appears inside the cell in a chemically unchanged form.
(2) Accumulation takes place against a concentration gradient of the substance (outside $\ll$ inside).
(3) energy is expended, to achieve this accumulation.

We know by now that bacterial cells actively transport a large range of metabolites from inorganic ions to sugars and amino acids. The process is probably the principal way by which bacteria ensure an adequate readily available supply of nutrients to feed rapid growth and flexibility of synthesis such as is involved, for example, in the inductive ability to use substrates. Indeed, steps in the active transport processes themselves can quite commonly be induced, as for example in the uptake of β-galactosides or of dicarboxylic acids. Unambiguous study of the problem is difficult and intensive study over many years still leaves many unsolved problems. A first requirement has to be a system in which the substance taken up is not further metabolized. Although it is true that the uptake of utilizable metabolites by wild-type cells has been successfully studied, special techniques and very short time intervals have to be employed. Three ways have commonly been used to avoid the difficulties:

(1) The use of chemical analogues of the substrate which although not metabolized by the cells are well transported and accumulated. The use of methyl-β-D-thiogalactoside in studying the transport of β-galactosides by *Escherichia coli* is a classical example of this approach.

(2) The use of mutants blocked in an early, or preferably the first, step in the metabolism of a substrate but in which the accumulation system is unaffected. An example of this approach is provided by the uptake of arabinose by a mutant *E. coli* deficient in arabinose isomerase which is the first step in metabolism of the pentose [118].

(3) The use of membranous vesicles, prepared from disrupted micro-organisms. In preparing the vesicles essential parts of the cytoplasmic machinery are lost with the result that further metabolism of the substrate is completely or almost completely lost. This method was pioneered by Kaback and his colleagues [143, 144, 146, 156, 157]. We have already discussed some of the problems about the sidedness of the membrane of such vesicles, but great gains in knowledge have resulted from their use and for many purposes this has clearly become the method of choice for studying aspects of active transport. The binding proteins (see Section 4.8) are lost during the preparation of vesicles and when these form necessary components of the total transport process by whole cells, vesicles will be deficient. For the uptake of many substances binding proteins are not obligatory components and membrane vesicles have been shown to be able to concentrate a wide range of substances when suitably energized.

Whichever system is used to study active transport the resulting problem is susceptible to three types of analysis:

(a) What are the carriers? The process of transport is specific and under genetic control and it is usually assumed that a range of specific proteins is involved in carrying substances across the membrane.

(b) What modifications of location, conformational or other changes in the protein, persuade the carrier to move the substrate across the membrane from the external environment into the cytoplasm. We have no effective knowledge of these processes and are still in the realms of speculation.

(c) How is the energy from the electron transport chain linked into the transport process to fuel it? This is the area into which most thinking went for a number of years. It is also an area that generated much energy in the form of heat! One of the penalties of this discussion, which has by no means yet been brought to a generally acceptable conclusion, is that whatever the authors of this book say they are liable to bring down wrath from the clouds upon their heads. However, this is a risk they will have to take! Fortunately there is by now a shift towards the chemiosmotic hypothesis as a general explanation (see Section 4.2.1) and even the most active proponents of the more direct hypotheses have now accepted this thermodynamically orientated one [250, 251]. It will be remembered that the proton motive force resulting from the translocation of protons and made up of the membrane potential ($\Delta\psi$) and the transmembrane pH gradient (ΔpH), is thought to drive ATP synthesis, *via* the ATPase. Imposed artificial proton and pH gradients have been repeatedly shown to lead to nutrient accumulation as would be expected from the hypothesis. Likewise stoichiometric proton flux associated with uptake of neutral substrates such as β-galactosides by *E. coli* was demonstrated long before general acceptance of the chemiosmotic hypothesis; it was found that there was a strictly 1:1 ratio between protons translocated and molecules of sugar transported [347–48, 349]. Moreover transport negative mutants (facilitated diffusion was still present) have been found to have lost their ability to co-translocate protons [350]. Thus it would appear that a number of substances can be transported actively into the cell as a direct result of the $\Delta\psi$ and the ΔpH. This can happen either directly, by proton symport, by antiport, or by symport with some other cation such as Na^+, as it does in one of the Na^+-dependent glutamate–aspartate uptake systems [105, 115, 200]. It has been stated [359] that about 40% of the transport systems of *E. coli* are of this type, directly explicable by the chemiosmotic theory. Another 40% however, involved periplasmic binding proteins, that are lost when vesicles are made. It is thought that these systems are not driven *directly* by the proton motive force but by ATP or some other high energy compound. Since, of course, ATP is likely to be synthesized during electron transport via ATPase, the proton motive force is, in these instances, sometimes working indirectly to drive transport rather than directly. In other instances ATP may be synthesised fermentatively by substrate level phosphorylation.

Recognition of specific transport proteins

In discussing this subject, one of the difficulties met is that, in much of the most interesting modern work devoted to sorting out the likely specific entities involved in the transport into bacterial cells of substances such as amino acids, distinction between mechanisms is not always possible with certainty. In some instances, for example, strong genetic and kinetic evidence has been produced for involvement of more than one 'permease' for a given substance. Sometimes no evidence is available as to whether both these systems can be regarded as active transport processes or whether facilitated diffusion may not account for one or the other. Both these processes show saturation kinetics and both are stereospecific. In other words, both involve specific carrier functions. It may not matter for the purposes in hand of unravelling the complexity of some of the systems, but one supposes that since fuelling drives accumulation against a concentration gradient in one but not the other processes, differences must exist in the properties and behaviour of the carrier proteins. However, in what follows no clear distinction is possible and the carriers in what is probably usually active transport but may sometimes be facilitated diffusion will be discussed together.

Two classes of entities involved in transport can be recognized. Firstly, we can recognize the proteins that behave as genetically specified integral membrane proteins. A few such carriers have been isolated and studied; more have been deduced from genetic studies and from the kinetics of transport processes. Secondly, the so-called binding proteins, which behave as periplasmic proteins and can therefore be isolated relatively easily as water soluble entities from the shock fluids from Gram-negative species of bacteria, can be recognized.

Although it is clear that the most important aspect of transport lies in the nature of membrane proteins our knowledge of them and their behaviour is still scarce. This reflects both the difficulty of isolating and purifying integral hydrophobic proteins and the difficulty of recognizing specific relevant properties of such 'permeases' when isolated, other than by their ability to combine with the metabolite transported.

The *lac* permease – the classical β-galactoside transport system of *E. coli* controlled by the *lac* operon – is by now too well known to require detailed description. One gene, the y gene of the operon specifies the active transport of β-galactosides, including analogues such as methyl-β-thiogalactoside, into the cells and only one protein is therefore, likely to be involved. In early work [77] the M protein concerned with this system was specifically labelled and partially purified, by making ingenious use of the competition between inhibitor and 'substrate' and the inducibility of the transport system. Both uninduced and induced cells were treated with unlabelled *N*-ethylmaleimide in the presence of isopropylthiogalactoside. *N*-ethylmaleimide is an –SH reagent known to inhibit transport but it was also known that substrates for the transport system effectively competed with it and prevented inactivation. The unlabelled excess *N*-ethylmaleimide was destroyed. The uninduced cells were then treated with ^{3}H-*N*-ethylmaleimide whilst ^{14}C-*N*-ethylmaleimide was used for the induced ones. The labelled cells were mixed in exactly equal amounts and an envelope fraction made from the mixture was extracted with Triton X-100 and the extract chromatographed on a column of ECTEOLA. In later work, sodium dodecyl sulphate was found to be a better extractant [140]. A product was obtained, the M protein, that was assessed to be 60–70% pure from its ratio of ^{14}C:^{3}H. Its formation was shown to be regulated by the y gene [78]. This method of labelling, recognizing and isolating the M protein has been described in some detail partly because of its historical importance. A more direct method of labelling part of the M protein with ^{14}C-methylthiodigalactose has been designed since [154]. It has been purified so that only a single band is present after polyacrylamide gel electrophoresis and the molecular weight both from this method and from gel filtration is about 30 000 [140, 339]. Cell free synthesis of the y gene product has been achieved using DNA from a constructed plasmid [62a]. The protein obtained was of the same molecular weight and had the same amino acid sequence at its *N*-terminus as that isolated from membranes. The plasmid DNA itself has been sequenced [35a] and from this the full sequence of the M protein deduced. The calculated molecular weight was 46 500, which was significantly higher than the molecular weight of 30 000 measured by gel electrophoresis of the isolated protein. The M protein represents about 4% of the total membrane protein [140, 304] and is a component of the cytoplasmic membrane. If the membranes are sonicated however, it can be randomized to the outer membrane [339]. So far, purified M protein has not been successfully used in reconstitution experiments, but by using extracts from vesicles prepared from a y^+ strain of *E. coli*, transport could be restored to those prepared from a y^- strain [6]. The y^+ vesicles were extracted with the aprotic solvent, hexamethylphosphorictriamide, and the extract added to y^- vesicles. Success was only met, however, when the extract and vesicles were sonicated together. It may be remembered that sonication under some conditions leads to vesicles with inverted membranes (see p. 135). The accumulation of lactose by the reconstituted vesicles was either energy dependent or could be driven by an artificial applied membrane potential, negative inside. Both processes were dependent upon the extract being present and did not occur with sonicated y^- vesicles Clearly the aprotic solvent extracts from y^+ membranes many proteins besides the M protein but further study may allow a dissection of the membrane components necessary for accumulation of galactosides. A second protein in the cytoplasmic membrane of 15 000 mol.wt. has also been found to be induced along with the M protein but its function is unknown [339].

The use of galactosides such as the *p*-nitrophenyl and dansyl derivatives has enabled demonstration of some of the complexities of the uptake system [255, 273-304]. These derivatives combined with the transport proteins but were not themselves thought to be accumulated [273, 304, 305]. The fluorescence of the dansyl compounds is greatly altered when they combine specifically with the transport proteins. There appeared to be an energy dependent component and a smaller energy independent component of binding. The latter appeared to represent 20% of the carrier molecules and had high affinity for substrates whereas the former had a low affinity unless it was activated by a proton motive force. In the absence of energy, both forms of the carrier appeared to have identical properties on either side of the membrane: that is, they appeared symmetrically arranged. When a proton motive force was set up both forms appeared asymmetric with a high affinity and low affinity aspect on opposite sides of the membrane according to the direction of the proton motive force. More details and references for these conclusions can be found in a review by Wilson [359] and in a paper by Schuldiner *et al.* [305a]. A totally different interpretation of these experiments is, however, now available [227b], namely that the dansyl- and nitrophenyl galactosides combine stoichiometrically with the lactose carrier protein at a molecular ratio of 1:1, even in the absence of an electrochemical gradient. When the latter is applied the derivatives are in fact transported into the vesicles but because of their hydrophobicity they associate non-specifically with the membranes.

Ingenious use of the dansyl substrate analogues has been made to measure the distance of the carrier protein from the hydrophobic membrane surface [305a]. The basis of this method is that, if the galactosyl end of the molecule (see Fig. 4.9) combines with the carrier molecule as would be expected, then the distance of the dansyl group from the carrier protein will vary according to the length of the methylene chain joining the galactosyl and naphthyl parts of the molecule. The fluorescence of the dansyl group changes according to whether its environment is

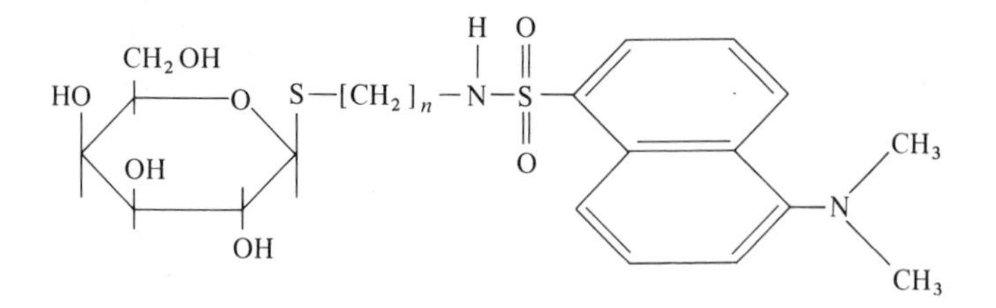

Figure 4.9 Structure of dansylthiogalactoside used to study the disposition of carrier proteins in the membranes.

hydrophobic or hydrophilic. Thus, by increasing the length of the methylene chain (n in Fig. 4.9) from 2 to 6 carbons and by making certain assumptions about the conformation of the molecule, it was deduced that the binding site in the *lac* carrier protein is approximately 50–60 nm from the aqueous solvent at the surface of the membrane. The nature of the environment of the dansyl groups was confirmed by studying the effect of an aqueous soluble agent that quenched its fluorescence when in an hydrophilic environment. In view of the experiments of Overath *et al.* [227b] these data may need reinterpretation. In chemiosmotic terminology, the galactoside uptake system in *E. coli* is a symport system [203] : that is protons and galactosides enter the cell together. The movement of a molecule of lactose into energy deficient *E. coli* removes one proton from the environment [347, 349]. An ingenious experiment using an ATPase negative mutant and *N,N'*-dicyclohexylcarbodiimide (DCCD) (see p. 000) to block proton permeability in the mutant showed that accumulation of thiomethylgalactoside could be driven by a pH gradient set up by reducing the external pH from 8.0 to 6.0 by the additon of HCl [75]. Thus, the most apt brief summary of the galactoside permease system would seem that energy is derived from the proton motive force and that this, in some unknown way, is used to alter or modify a specific carrier protein within the membrane. One aspect of this alteration may be demonstrated by the asymmetric arrangement of the protein in the membrane.

Study of mutants of *E. coli* deficient in the transport of dicarboxylic acids demonstrated the presence of at least three lesions [179, 180] one of these, the *cbt* gene, was shown to specify a periplasmic binding protein with unique properties [175]. The other two genes *dctA* and *dctB* specify two integral membrane proteins [178]. They can be extracted with the non-ionic detergent Lubrol 17A-10 and fractionated on an affinity column of Sepharose 4 B–aspartate, from which they can be eluted with 0.2 M succinate. The whole fractionation was carried out in the presence of 0.5% of the detergent. The binding constants for succinate by the two proteins are very different being 23 μM and 2.3 μM. The K_M for transport of succinate is about 20 μM at Ph 7.4 [98]. The binding of succinate by both proteins is strongly inhibited by fumarate and malate, though not by a range of other acids including malonate and D-lactate [176] and the latter competes rather ineffectively for the isolated proteins. The inducible transport pathway for succinate into *E. coli* is shared by fumarate and malate [153] and appears complex. In whole cells, transport is inhibited by -SH reagents such as *p*-chloromercuribenzoate and *N*-ethylmaleimide [179] as is the uptake of β-galactosides on the M carrier protein. On the other hand, the binding of malate, succinate and fumarate by membrane vesicles made from the organism is inhibited only by *N*-ethylmaleimide, yet transport using D-lactate as electron donor is immediately blocked by *p*-chloromercuribenzoate and any material already accumulated is effluxed. Transport into vesicles is not affected by *N*-ethylmaleimide but efflux is blocked [180, 181]. Both the -SH reagents reversibly inhibit the binding of succinate to the two isolated carrier proteins, although to rather different extents [176, 177]. Examination [178] of the K_m values for combination with succinate by the two isolated proteins and comparison with the constants for inverted and uninverted vesicles, led to the conclusion that the two proteins were probably on opposite sides of the membrane.

The dicarboxylic acid transport system is energized by a proton gradient [98, 251]. One molecule of succinate is symported into the cells along with two protons and the K_m for proton uptake is, like the uptake of succinate, about 20 μM at pH 7.4 and is strongly pH dependent as would be expected for a symporter mechanism.

Fractionation of cholate disrupted membranes from *Halobacterium halobium* on an agarose column led to the recognition of a glutamate carrier protein [167]. The molecular weight of this protein was about 50 000. When added to liposomes it promoted transport by what appeared to be facilitated diffusion. Only the binding protein eluted from the agarose column was active in the liposome system. Competition studies with the glutamate analogues, kainic acid, α-methylglutamic acid and *N*-methylglutamic acid showed a consistent pattern using either intact envelopes or protein binding or the liposome system.

The energetics of glutamate transport are particularly interesting in vesicles prepared from *H. halobium* membranes since bacteriorhodopsin is present and upon illumination protons are ejected. It is suggested that the Na^+/H^+ antiport is electrogenic with $H^+/Na > 1$ and resulting in a large Na^+ gradient (outside ≫ inside) which, together with the induced membrane potential, will drive the transport of 18 amino acids. The accumulation, however, of glutamate is said to be indifferent to $\Delta\psi$ and is transported by symport with sodium in an electrically neutral manner depending entirely on the chemical gradient.

Like the transport processes for β-galactosides on the M protein, the accumulation and efflux of amino acids such as proline into membrane vesicles from *E. coli* is inhibited by reagents combining with -SH groups; this inhibition is reversible [17, 147]. When treated with Brij 36T about 20% of the vesicle membrane protein can be solubilized and by fractionation on Sephadex G100, one peak with high binding capacity for L-proline, and two with much poorer ability have been recognized [92]. Extraction of *E. coli* membranes with acidified *n*-butanol is more satisfactory, however, and removes less (2%) of the proteins along with binding proteins. Some purification can then be achieved by chromatography on Sephadex LH-20. Reconstitution of this material with phospholipids from the organism led to liposomes with the ability to accumulate proline when driven by valinomycin-induced K^+ efflux. The crude extract similarly treated, led to liposomes capable of also transporting glutamate and cysteine [7]. A protein that specifically binds proline has been purified from the membranes of *Mycobacterium phlei* [168], as well as a second protein that binds several amino acids. Only the former was active when reconstituted into liposomes or into *M. phlei* vesicles after detergent extraction. Highly hydrophobic proteins binding folate [119] and thiamine [120] have been extracted and purified from the membranes of *Lactobacillus casei*.

4.8 Binding proteins

The transport abilities of Gram-negative bacteria are sometimes greatly reduced by treating them with the osmotic shock procedure for liberating periplasmic proteins. When the shock fluid is examined, proteins can be recognized that combine strongly with the metabolites, the transport of which has been affected. The dissociation constants of the protein–metabolite complexes frequently reflect the K_m for the transport by whole cells. Vesicles prepared from Gram-negative species are ineffective in transporting such metabolites. For example, *E. coli* cells subjected to the osmotic shock procedure transport glutamine at only 10% of the rate of the untreated bacteria and membrane vesicles prepared from the organism show only a marginal stimulation of uptake by energy sources [144, 271]. The transport of other amino acids such as proline and glycine are relatively unaffected. Both of these latter amino acids are transported by membrane vesicles, proline by active transport and glycine probably by facilitated diffusion. Generalizations about which amino acids are transported by whole cells and not by vesicles should be cautious and closely referred to particular strains. For example, vesicles prepared from *E. coli* ML308-225 actively transport both L-glutamate and L-aspartate [182] whereas when prepared from *E. coli* strain K12, the glutamate transport ability is almost completely lost [16] and when prepared from *E. coli* strain D_2W is partially lost whilst that for aspartate is retained [358]. Some of the best evidence for the involvement of binding proteins in transport comes from genetic studies showing coregulation of transport and binding protein synthesis, both for amino acids and carbohydrates.

Table 4.9 shows a list of recognized binding proteins together with some of their properties. The probability that they play a role in the passage of some metabolites from the environment into Gram-negative cells would seem strong. For example, osmotic shock causes a 90% reduction in the uptake of glutamine or leucine without seriously affecting that of other amino acids, such as alanine, glycine or proline [243, 345]. If, therefore, the explanation for the reduction in uptake of some amino acids were attributed to cellular damage, this must be highly specific. Nevertheless, attempts at reconstruction experiments have met with variable success; it would be very strong evidence for the role of binding proteins if one of the purified proteins could be reproducibly used to restore the specific transport activity of a depleted

Table 4.9 Properties of periplasmic binding proteins

Micro-organism	*Metabolite*	*Molecular weight* ($\times 10^{-4}$)	*Dissociation constant K_d* ($\times 10^8$)	*K_m for transport by whole cells* ($\times 10^8$)	*Reference*
Salmonella typhimurium LT_2	SO_4^{2-}	3.2	2	3.6×10^3	232, 233, 62, 166
Escherichia coli str. AB3311	PO_4^{3-}	4.2	80	80	195, 196
E. coli str. K_{12}	D-galactose	3.6	10 and 10^3	50	11, 12, 23, 24
		(3.6)	10^2 and 10^4	50	235
E. coli str. K_{12}	leucine, isoleucine, valine	2.4–3.6	24	10–15	240, 11, 12
	(for leucine)		13	10–15	243
E. coli str. 7	leucine (specific)	3.6	70	200	81
E. coli str. 7	glutamine	2.3	30	8	345
E. coli str. K_{12}	glutamate–aspartate	3.1	70		
			6.7×10^2	*	345, 357, 358
			1.2×10^2		16, 271
E. coli str. K_{12} W	cysteine, 2,6-diaminopimelic acid	2.8	1	30	172, 122
E. coli str. K_{12} W	arginine	2.8	3	2.6	270
	lysine, arginine, ornithine	2.6	20		269, 270
E. coli str. K_{12}W3092	ribose	2.9	13	30	356
S. typhimurium	histidine (specific)	2.5	10	2.6	9
E. coli str. B/r	L-arabinose	3.8	28	8.3×10^2	174, 35, 234, 235
Pseudomonas sp.	phenylalanine	2.4	10	2×10^3	163
Neurospora crassa	tryptophan		8×10^3	c8×10^3	354
E. coli str. K12	fumarate	1.5	5.5×10^3		177, 179
	malate		3.4×10^3		
	lactate		1.3×10^5		
	succinate		4.0×10^3	1.2×10^3	
Halobacterium halobium	glutamate	5.0	6.0	—	167

*The K_m for glutamate transport in *E. coli* K_{12} is strongly Na^+ dependent [105].

cell. Failure in such ventures is not altogether surprising and the most reasonable explanation for the liberation of binding proteins along with other periplasmic proteins by osmotic shock is that damage to the outer membrane of the cell envelope allows them to leak out. Unless a binding protein can be introduced into the envelope and the outer membrane resealed, its retention and function may not be possible. It is probably significant that soluble binding proteins have only been isolated from Gram-negative bacterial species, such as *E. coli*, *Salmonellae* and *Pseudomonads* and not from true Gram-positive species. This suggests that they are primarily concerned with the movement of the metabolite between the outer membrane and the cytoplasmic membrane rather then in actual translocation across the latter structure. It has been suggested [315] that the periplasmic binding proteins may be in fact peripheral membrane proteins associated with integral outer membrane proteins such as the porins, or structural proteins (see Chapter 3.4.2) so that the binding sites are exposed *via* the pores through the outer membrane formed by the porins (see Section 4.10.1). A conformational change in the binding protein would then bring the substrate to face the cytoplasmic membrane. Such a function raises the teasing question of why this device should be necessary for some amino acids and not others. In a number of instances, both specific and relatively non-specific transport systems involving binding proteins have been recognized. For example, two binding proteins for leucine have been isolated [11, 12, 81, 240, 243]. One is highly specific for leucine whereas the second combines equally well with leucine, isoleucine or valine. These two transport systems for leucine have been recognized by a quite different method [81] in cells of *E. coli* strain 7 by making use of the observation that trifluoroleucine inhibits the binding of leucine to the specific protein but not to the less specific one. It was found that the transport of L-leucine into cells was not completely inhibited by the addition of isoleucine and that the residual transport was very sensitive to trifluoroleucine. Likewise, the trifluoro compound did not completely inhibit leucine transport in the absence of the iso compound but the residual transport was very sensitive to the latter amino acid. The kinetics of transport of leucine into the cells indeed showed evidence of two processes with K_m values of 2×10^{-7} and 2×10^{-6}. Similar situations seem to apply to the transport of arginine, cystine and histidine by *E. coli* and *S. typhimurium* [39, 122, 248, 269, 270, 360] and for sugars such as L-arabinose by *E. coli* [35] for Mg^{2+} by *B. subtilis* [355]. The system for histidine transport has been examined in detail for its genetic control [9, 162, 174]. In this organism the *hisJ* gene codes for a histidine binding protein. A class of promoter mutations called, *dhuA* which were derived from *hisF* mutants unable to use D-histidine to supply L-histidine, produce more of the binding protein, whilst mutants in the *hisJ* gene, produce none or modified protein [162] according to the type of mutation [8]. Examination of the K_m for the transport of histidine by cells of these mutants gave the results shown in Table 4.10. The level of histidine binding protein has a marked effect on the kinetics of the transport of amino acid into the cells. Its elimination, however, still leaves a very effective uptake system. A further *hisP* gene controlled this remaining high affinity system and mutants damaged in this gene were unable to grow on histidine. The remaining

Table 4.10 Affect of altered levels of histidine binding protein on histidine transport by *Salmonella typhimurium*

Strain	K_m *transport* (M) ($\times$ 10^8)	*Histidine-binding protein*
wild-type	2.6	Control level
*dhuA**	0.66	Increased five-fold
hisJ†	20	Absent
hisP‡	100	Normal level

*Produce more binding protein
†No binding protein or modified protein
‡Lacking the membrane-bound high affinity transport system

transport of histidine in hisP^- mutants had the very high K_m of 10^{-6} compared with 6.6×10^{-9} for the *dhu* mutants, and was completely inhibited by the presence of the aromatic amino acids, phenylalanine, tyrosine and tryptophan. It therefore appears that the high affinity specific system consists of at least two proteins: namely, the gene products of *hisJ* and *hisP*, the former being the binding protein. There was no evidence of missing periplasmic proteins from the shock fluid of *hisP* mutants [8]. Further *hisJ* mutants, selected by their ability to grow in the presence of azaserine, which uses the same transport system as histidine, have been studied [162] and shown to have a modified histidine binding protein. Both spontaneous and mutagenized cultures were used and about 25% of the former and 65% of the latter still produced a protein, seen on polyacrylamide gels in the position of the J protein. The original strain used for these experiments was a *dhu* mutant so that more of the J gene product was formed and could be easily recognized. Isolation of the periplasmic proteins from these mutants showed the presence of a false J protein with almost normal binding properties for histidine but which was not able to promote transport of the amino acid in the bacteria. Clearly, this first indication of the presence of two functional sites on binding proteins, one for binding and the other for transport is of great importance and needs further exploration. The binding protein appears to have the effect in this system of a very effective helper, greatly increasing the affinity of the membrane system for histidine by a factor of ten, rather than having an essential role as in the transport of glutamine or diaminopimelic acid. Another example of a binding protein acting as helper in transport is that for galactose. Strains with mutations in the structural gene for the protein still transport substrates but with a K_m value increased by a factor of a thousand [262, 263]. An example of another periplasmic protein that appears essential is that involved in the transport of *sn*-glycerol-3-phosphate by *E. coli* [311]. This protein is under the regulation of the *glpT* gene. Mutations in this gene lead to resistance to phosphonomycin (see Chapter 9), loss of transport of *sn*-glycerol-3-phosphate and sometimes to the disappearance of the periplasmic protein as demonstrated by two dimensional polyacrylamide electrophoresis. When revertants were selected by growth on glycerol-3-phosphate all three properties (i.e. sensitivity to phosphonomycin, transport and periplasmic protein) return. Transport activity in the revertants was related to the amount of periplasmic protein they contained.

Whether or not this protein can bind glycerol-3-phosphate is unknown. A number of amino acids, apart from histidine, appear to have multiple transport systems. Glutamate and aspartate, for example, have been claimed to have five [291]:

(1) A binding-protein independent, sodium independent, glutamate–aspartate system inhibited by β-hydroxyaspartate or cysteate.
(2) A binding-protein dependent, sodium independent, cysteate inhibited glutamate–aspartate system.
(3) A binding-protein independent, sodium dependent, α-methylglutamate inhibited, glutamate specific system.
(4) A binding protein-independent, aspartate specific system.
(5) A dicarboxylic acid transport system also transporting aspartate.

Mutants have been obtained affecting the levels of each of these systems.

To return for a moment to any attempt to assign a role to the binding proteins in the total transport process, it is moderately clear, even from the work surveyed above and that summarized in Table 4.9 that all the binding proteins have remarkably similar sizes and binding constants for their specific metabolites; only that for the dicarboxylic acids departs considerably from the rest and this protein has other remarkable features. One of the urgent problems for decision is the exact location in the cell of these and other periplasmic proteins. The variable loss of transport activity when different strains of the same bacterial species are converted to membranous vesicles, raises the question of whether periplasmic proteins including binding proteins may not be peripheral membrane proteins that are held on the outer surface of the cytoplasmic membrane by very weak bonds of strength differing according to the strain and the particular protein (see above). It seems unlikely that binding proteins are concerned directly and necessarily only with passage of small metabolites through the outer membrane of the Gram-negative bacterial envelope; such proteins are known and are integral outer membrane proteins (see Section 4.10.1). Also, the loss of transport activity on shocking the cells or on treating with lysozyme and EDTA is hardly compatible with such a function, since gross damage to the outer membrane happens during these procedures. The evidence that there are two sites on the binding proteins, one for binding and the other for transport, does not seem to favour an alternative suggestion that the binding proteins simply increase the concentration of the ligands near the outside of the cytoplasmic membrane. If this were true one might expect the binding site to be the relevant one for transport as well. If on the other hand the transport site were concerned with the loose attachment of the protein to the outside of the cytoplasmic membrane, like the minor subunits of ATPase for example, this would be more reasonable. the conformation of the binding proteins is clearly important since almost all of them cease to bind in the presence of strong urea or guanidine salts. It may be important that at least in one instance [356],–SH groups have been revealed during this treatment. The arabinose binding protein has a known amino acid sequence [129] and its three-dimensional structure has been worked out with a resolution of 0.28 nm [249], whilst preliminary crystallographic data have been published for the leucine–isoleucine–valine binding protein [194]. A clear short review of the whole subject of the transport of organic solutes by bacteria has been published [333] as have a number of other more detailed or specialized ones [106, 107, 114, 313, 359].

4.9 Mesosomal membrane

As stated in Section 1.5.1, when many species of Gram-positive bacteria are suspended in concentrated solutions of poorly penetrating solutes such as sucrose, the mesosomes are extruded between the cytoplasmic membrane and the wall. When the wall is removed enzymatically they appear as tubules and strings of small globules. Then, by suitably adjusting the concentration of Mg^{2+} and differentially centrifuging they can be isolated as collections of tubules or vesicles. As has also been pointed out (Section 1.5.1), their exact status is uncertain and it is by no means clear that they even exist, as such, in living, dividing bacteria. No explanations for the mechanism of their appearance in sections of 'ill-treated' cells has been advanced and, as we shall see, no function has so far been unequivocally proven for them when isolated. However, a number of functions of the cytoplasmic membrane are partially or completely missing from them. Long and very thorough reviews of mesosomes have been published [95, 257].

On the basis of cytochemistry [169], it was at first thought that the coiled internal membranes seen in sections of bacilli and other bacteria were mitochondrion-like structures in which the oxidative apparatus of bacteria was packed – and hope comes to die slowly [95]. Indeed an initial claim [71] was made that the cytochromes were concentrated in the mesosomal membrane from *B. subtilis*. However, in the light of other work [254], this claim was withdrawn [72]. It would indeed have been embarrassing if the initial claim had been supported, since almost all workers who have studied the problem agree that mesosomal membrane is grossly deficient in other functions of the electron transport chain [72, 228, 254, 328]. Among the deficiencies, are NADH oxidase and succinic dehydrogenase, both dependent upon integral membrane proteins. When the cytochrome contents have been looked at critically, as they have in mesosomes from *M. luteus* [228], they too have been very low or absent except, curiously, cytochrome b_{556}. Other much less exacting work with *B. subtilis* found little difference in the cytochrome content of cytoplasmic and mesosomal membrane [237, 254]. Membrane peripheral enzymic proteins, however, appear to be present. For example, NADH-dehydrogenase, measured using either potassium ferricyanide, dichlorophenol-indophenol or cytochrome *c* as electron acceptors, has the same or even slightly higher activity in the mesosomal compared with that in the cytoplasmic membrane.

Table 4.11 summarizes the quantitative results for comparisons between the activities of cytoplasmic and mesosomal membrane preparations from a number of species for succinic dehydrogenase, NADH oxidase and NADH dehydrogenase, with one exception the two former activities are low or very low in the mesosomes. The exceptional products from *L. monocytogenes* were separated by a different method and the exact relationship of the mesosomal membrane obtained by this method to that obtained by the more usual one of preparing protoplasts, needs further examination. Table 4.12 summarizes qualitative results for a wider range of enzymes and components of the electron transport chain. With hindsight, it would indeed have been strange if mesosomes had carried a major part of the electron transport chain since very early work showed that protoplasts were as fully competent to respire as the bacteria from which they were derived, and moreover, had a full complement of cytochromes [189, 343]. The protoplasts studied had presumably already been separated from the mesosomes in this early work since they had been deposited by centrifuging at only slow speeds.

A further, major function suggested for the mesosomes was that they were involved in wall synthesis. This would have been particularly appropriate since they universally occur at the site of cell septation, where for organisms such as streptococci the peripheral wall is manufactured (see Chapter 15.1). However, again essential enzymes for synthesis of wall polymers are missing. Nevertheless, it must be pointed out that more subtle regulatory functions may still be possible as a role for the mesosomal membrane in helping to form the septum. Other functions in nuclear segregation, transformation, export of proteins and orderly cell division have been suggested but are so far unproven. These various suggestions are summarized in Table 4.13 along with an assessment of their likely verity.

4.10 Outer membrane of Gram-negative bacteria

When the earlier book, *Cell Walls and Membranes*, was published in 1968, little was known about the outermost membrane in Gram-negative bacteria, other than that it existed. It could be seen in sections under the electron microscope but had not been isolated or studied as a separate entity. Increased attention has been paid to it over the last few years so that we know that it has important functions as well as knowing about its biochemical and biophysical structure. It has three main types of function, *vis-à-vis* the micro-organism. Firstly and probably most important, it is a selective permeability barrier, keeping out hydrophilic substances above a certain size and a wider range of hydrophobic ones that would otherwise be injurious to the bacteria. It probably also keeps active proteins concentrated in the periplasmic space that would otherwise be exported from the cell and diluted in the medium. Secondly, it contains proteins responsible for the specific uptake of some metabolites and the adsorption of bacteriophages. Thirdly, the lipid A ends of the lipopolysaccharides or O-antigens are anchored in it and contribute part of its lipid bilayer. Thus it is a primary organ in controlling the interaction of Gram-negative bacteria with their external environment.

It will be remembered from earlier chapters in this book that the outer membrane differs in a number of purely 'structural' characteristics from the inner cytoplasmic membrane. It is, for example, relatively rich in proteins and, whereas the quantities of the large variety of molecular species in the inner membrane are of the same order of magnitude, four or five species dominate the picture in the outer membrane. It demonstrates the asymmetry of structure now recognized for all membranes to a very much larger degree than is usual, and the phospholipids are likely to be confined to the inner leaflet, whilst the lipid component of the outer leaflet is provided by the lipid A of the lipopolysaccharide. Whereas components of inner cytoplasmic membranes in common with membranes from other sources move laterally with considerable freedom, examination of components of the outer membrane suggests that lateral movement is very much slower. Whether this 'rigidity' is due simply to the large proportion of protein in the membrane or to specific protein–protein, protein–lipopolysaccharide or protein–lipopolysaccharide–divalent cation interactions cannot be said yet with certainty. Finally, it will be remembered that the lipoprotein, fixed to the peptidoglycan, almost certainly extends from this layer to the outer membrane, where its lipophilic substituent is buried. The so-called matrix proteins of the outer membrane have a special relationship with the peptidoglycan layer and its fixed lipoprotein constituent which is dependent on the latter being present. These structural considerations should be born in mind as we talk about outer membrane functions.

4.10.1 *Permeability to hydrophilic substances*

Evidence that the outer membrane of *E. coli* was likely to control entrance to and exit from the cell of certain molecules came from three independent approaches. It was found [238] that the organisms would not grow on peptides having greater than a certain molecular volume. These larger molecules did not compete with smaller peptides and it was therefore concluded that the cell wall excluded them from reaction with specific transport sites in the cytoplasmic membrane. Secondly, examination [170, 171] of the effect of chelating agents such as EDTA upon *E. coli* showed that part of the lipopolysaccharide was removed and the bacteria were then more sensitive to some reagents; other work [108, 109] showed that Gram-negative organisms were made more sensitive to penicillin when treated with EDTA. Thirdly, the kinetics of exit of β-thiogalactoside from bacteria after it had been transported, suggested a further permeability barrier outside the cytoplasmic membrane [261]. Isolation and analysis of outer membranes along with the isolation of mutants deficient in major proteins [221] allowed a more thorough analysis of this problem. Examination [59] of a range of saccharides and polyethylene glycols of increasing molecular weight showed very clearly that saccharide molecules above a molecular weight of about 600–700 were excluded from the cytoplasm, by an outer permeability barrier in four different organisms, namely *E. coli*, *Salmonella typhimurium*, *Pseudomonas aeruginosa* and *Alcaligenes faecalis.* The polyethylene glycols had a somewhat higher exclusion limit of about 1500 but the discrepancy could have been due either to polydispersity of the compounds or to the very different conformation of such thread-like molecules compared with the saccharides. Nevertheless, microorganisms, other than the enterobacteria, almost certainly have higher exclusion limits for the permeability of hydrophilic molecules (see Section 9.7.8). It was suggested that the passage and exclusion results could best be explained by the presence of water-filled pores in the outer layers of the cell – not of course a new

Table 4.11 Distribution of three electron transport activities (as specific activities) between mesosomal and cytoplasmic membranes. Latter arbitrarily taken as 100.

Micro-organism	*Succinic dehydrogenase*	*NADH Oxidase*	*NADH dehydrogenase*	*Reference*
Bacillus subtilis 168	44	—	0.5–300	237
Bacillus subtilis 172	1–10	0	—	H. J. Rogers and S. M. Fox, unpublished work
Bacillus subtilis Marburg	4*	5*	—	72
Bacillus licheniformis 6346	4–7	21	66–200	254
Bacillus megaterium KM	6	—	67	D. J. Ellar, personal communication
Micrococcus luteus	10	6	—	228
Staphylococcus aureus	38*	24*	—	328
Listeria monocytogenes†	180	36	280	94

*Measured by O_2 uptake.
†May not be representative of the whole mesosome fraction from the bacteria.

Table 4.12 Components of the electron transport systems absent or poorly represented in mesosomal membranes

Enzyme	*Micro-organisms*	*Reference*
Succinic dehydrogenase	*Bacillus subtilis*, *B. licheniformis*, *B. megaterium*, *Staphylococcus aureus*, *Micrococcus luteus*	254, 228, 72, 69, 328
NADH oxidase	As for succinic dehydrogenase also *Listeria monocytogenes*	69, 94, 228
Malate dehydrogenase	*B. megaterium*, *M. luteus*, *S. aureus*	69, 228, 328
	S. aureus	
α-glycerophosphate dehydrogenase	*S. aureus*	328
Cytochromes	As for succinic dehydrogenase	69, 72, 228, 328
L-lactate dehydrogenase	*B. megaterium*, *M. luteus*, *S. aureus*	69, 228, 328

Table 4.13 Ideas about the functions of mesosomes.

Function	*Evidence*	*Reference*	*Status*	*Evidence*	*Reference*
Mitochondrial analogues i.e. organelles for electron transport	Deposition of tellurium crystals and formazon vital staining methods	71, 73, 132, 133, 169, 306	Not acceptable	Cytochromes and de-hydrogenases missing. Deposition of tellurium artifactual.	254, 228, 72, 69, 328
Essential for wall bio-synthesis	Morphological position	42, 73, 67, 283	Not acceptable	a) Important enzymes missing. b) L-forms regenerate.	257, 165
Precursors of membrane lipids	Some pulse experiments	74	Not acceptable	Pulse and other experi-ments using better defined material.	53, 69, 199, 206, 330
Orderly cell division	Morphological correlation	Numerous authors	Possible	Doubts raised on basis of growth of bacilli on the surface of 25% gelatin without meso-somes.	237
Nuclear segregation and separation	Morphological association between nucleus and mesosome.	135, 276, 278, 279	Possible	Further experimental evidence would be desirable.	
	Membrane growth experiments.	274, 275, 276			
DNA uptake during transformation	Some early evidence for association of DNA enter-ing cell in the region of the mesosome.	361, 5	Doubtful	Evidence available is inadequate or alterna-tive explanations available.	
	Reduction of competence in the 'absence' of mesosomes	331	Doubtful	Evidence available is inadequate or alterna-tive explanations available.	
Export of enzymes and proteins in particular β-lactamase	Enrichment of mesosome vesicles with β-lactamase	288	Possible but not as simple secretory organs	Protoplasts without mesosome tubules secrete at normal rates.	164, 164a

suggestion for the permeability of membranes generally by hydrophilic substances [56]. The analysis of the control of the permeability of walls to hydrophilic substances proceeded [213] by studying plasmolysed cells which had their peptidoglycan layers severely damaged either by lysozyme or by the natural autolysins acting during penicillin treatment. This work allowed the conclusion that the peptidoglycan layer of the wall did not form a permeability barrier with the molecular weight limits found. Indeed the limit of exclusion for Gram-positive cells with their very much thicker layers of peptidoglycan is of the order of molecular weight of 10^5 [292, 341]. Convincing evidence for the role of outer membrane proteins in permeability phenomena came from studying vesicles made from components of the isolated membrane [211-214]. When the phospholipids, lipopolysaccharides, or the protein fraction alone were used, either vesicles were not formed at all or no penetration took place. When outer membrane proteins were included with the first two components and the vesicles that formed were treated in a rather complex manner with trypsin, temperature and Mg^{2+}, their permeability limits were similar to those of whole plasmolysed cells. Separation of the proteins present showed that one 'porin' [211] was responsible in *E. coli* and none other would do, whereas in vesicles prepared from *Salmonella typhimurium* three were necessary [212]. The isolation of mutants deficient in the porins [221] allowed further and convincing proof of their importance for smaller molecular weight solutes to penetrate the membrane. In a particularly beautiful study [220], the presence and absence of the porin proteins in mutants of Salmonellae was correlated with the activity of β-lactamase, coded for by introduced plasmids, using cephaloridine as substrate (see Table 4.14). It can be seen that the permeability coefficient for the antibiotic decreased by a factor of ten when the amounts of the outer membrane proteins with molecular weights of 36 000 and 34 000 were greatly reduced. The presence or absence of protein with a molecular weight of 33 000 was irrelevant. Meanwhile, however, one problem had been raised and another settled by the isolation of porin deficient mutants. It had been found that such strains grew as normal rods at a normal rate on ordinary broth medium [121]. This observation disposed of the suggestion that the outer membrane proteins formed a self-assembling shape determining layer and that any morphogenetic properties of the wall resided in this layer; this problem has however now returned in a more sophisticated and realistic form [320] (see Chapter 15.5). The problem raised, rather than disposed of, was that the permeability of the outer membrane, even in the absence of the porin proteins, to normal hydrophilic metabolites appeared unaffected, since the organisms grew normally. This problem was answered by two pieces of work [184, 220] in which it was shown that the growth rate of the mutants was only normal when high concentrations of nutrients were present in the medium – the specific nutrients studied were isoleucine and histidine using the appropriate auxotrophs. At low concentrations, growth was very slow and the K_m for the uptake of these amino acids was found much reduced. Very interestingly, mutants resistant to Cu^{2+} were found deficient in outer membrane porins. These mutants also incorporated radioactive methionine at very much reduced rates when the concentration of the amino acid was low. Thus, the original

Table 4.14 Evidence for the role of outer membrane porin proteins in permeability to cephaloridine by *Salmonella typhimurium* (from [220]).

Strain number	*Porins* (mol.wt. × 10^{-3})				*β-lactamase* nmol/min/mg cells			
	36	35	34	33	*Intact cells* (a)	*Sonicated extract* (b)	*Ratio* a:b	*Permeability coefficient* (× 10^6)
SH5014*	+++	+	+++	+++	44	108	0.41	9.3
SH5551*	+++	+	±	+++	21	97	0.27	5.0
SH6017*	±	+	+++	+++	18	105	0.17	3.1
SH6260*	±	+	±	+++	3.4	121	0.028	0.57
SH6261*	±	+	±	+++	8.0	91	0.088	1.34
SH6263*	±	+	±	+++	4.2	99	0.042	0.69
SL1917*	+++	+	+++	—	35	64	0.55	9.0
SH5014†	+++	+	+++	+++	95	1850	0.051	15.4
SH6260†	±	+	±	+++	<10	1840	<0.005	<1.3

*Strains bearing the R1 plasmid

†Strains bearing the R47 1a plasmid

report of normal growth of porin deficient mutants was due to their being grown in a rich medium containing high concentrations of nutrients. The function of the porins in regulating the permeability of the outer membrane to hydrophilic substances can be assumed by other proteins. A number of mutants (see for example [40, 76, 247, 297] have been described arising from strains already deficient in porins and which showed the expected slow growth rates under appropriate conditions. The mutants have gene products in the outer membrane that have molecular weights differing from the already recognized porin proteins. The phenotype of these mutants is that of the wild-type organisms. Therefore these different proteins also have a porin-like action.

Although strict proof of the organization of the hydrophilic porin pores is lacking, suggestions have been put forward that seem eminently reasonable. The very strong association of the matrix proteins with the peptidoglycan is dependent upon both the fixed and possibly the free lipoprotein [60], and occurs neither with *lpo* mutants [124] which do not form lipoprotein, or with trypsin-treated walls from which the lipoprotein has been removed but in which the matrix proteins remain unaltered. It is suggested [188] that the fixed lipoprotein adopts a double helical structure. Three different approaches have led to the conclusion that in the outer membrane of Salmonellae, three porin molecules are organized to give the functional pore. One approach has been by image analysis of whole outer membranes when a three-fold symmetry was found [325], another by applying protein cross-linking reagents both to whole outer membranes and to porin aggregates in solution [231] and a third, by isolating porin aggregates by salt–SDS solutions that are active in the vesicle systems already described. The molecular weight of these aggregates was carefully measured and the monomers, when separated from them again, measured. The conclusion from these studies was unambiguously that trimers were involved [332]. One suggestion for the total pore structure is that shown in Fig. 4.10 [61]. Other properties of the mutants deficient in major membrane proteins are less easy to understand. Some of these have been summarized by Braun [33] (see Table 4.15).

Although the matrix proteins or porins have a major non-specific role in the exclusion and passage of hydrophilic molecules according to size through the outer membrane, they are not the only permeability regulators present.

Table 4.15 Properties of *E. coli* mutants lacking major outer membrane proteins (from [33]).

Protein				*Behaviour of cells*
1a	1b	II$^+$	Lp	
+	+	+	–	EDTA (1 mM) hypersensitive, leakage of periplasmic enzyme
+	+	–	+	Conjugation-deficient recipient cell
+	+	–	–	Require 30 mM $MgCl_2$ in nutrient both for growth
–	–	+	+	EDTA (0.5 M) sensitive
–	–	+	–	Like mutant lacking only lipoprotein
–	–	–	+	Phospholipase suicide, killed on freezing in 30% glycerol

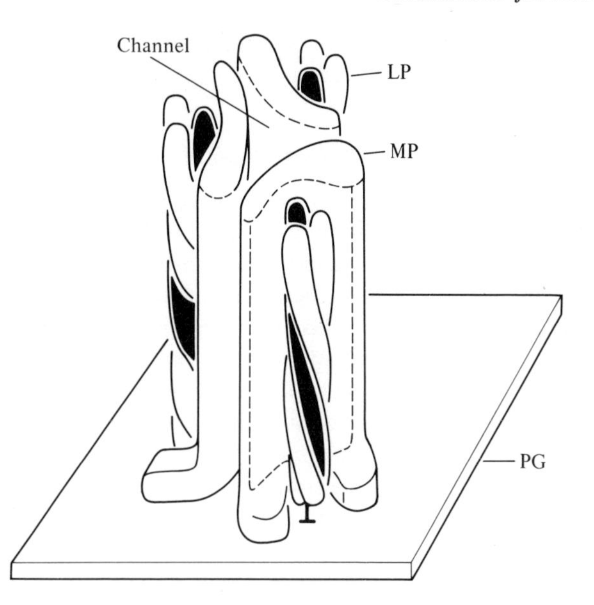

Figure 4.10 An idea for the organization of the hydrophilic pore in the outer membrane of Gram-negative bacteria. LP is lipoprotein; MP is matrix protein (porins) and PG is peptidoglycan (from [61]).

4.10.2 *Permeability to hydrophobic substances*

The passage of hydrophobic substances is not mediated by the porins. Earlier observations [260, 293–295] showed that the sensitivity of Salmonellae to a range of antibiotics and dyes, was greatly increased in deep rough mutants compared with the wild-types (see Chapter 7). For this to happen 80–90% of the polysaccharide chain of the lipopolysaccharide had to be missing. Analysis of this phenomenon [219] showed that there was a good correlation between the hydrophobicity of compounds as measured by their partition between 1-octanol and 0.05 M phosphate buffer pH 7.0, and the ratio of their penetration into deep rough mutants compared with that into the wild-type. This meant that substances such as actinomycin D, novobiocin, and nafcillin penetrated the rough mutants with relative ease whereas those such as benzylpenicillin and ampicillin did not. Other work [369] has also suggested the penetration of penicillin antibiotics into wild-type organisms is inversely related to their hydrophobicity.

An obvious explanation that the polysaccharide part of the lipopolysaccharides forms a hydrophilic zone around the wild-type organisms, thus holding at bay hydrophobic compounds, is not acceptable since great shortening of the saccharide chain has no effect and penetration only occurs in the true deep rough mutants which have lost 80–90% of their polysaccharide. The loss of the LPS-polysaccharide chains, however, also leads to loss of outer membrane proteins with compensating appearance in the outer leaflet of phospholipid. It is suggested [219] that it is this which leads to the permeability change towards hydrophobic compounds in the deep rough mutants.

4.10.3 *Phage receptors and specialized permeability*

A variety of much more specific mechanisms for the penetration of hydrophilic substances of varying molecular weight, greater than the critical 600–700, is also present. Early examples indicating such mechanisms have already been noted, such as that of the passage of maltodextrins through the membrane which is switched on when the *lamB* gene is induced with maltose. This outer membrane protein also serves as the λ bacteriophage receptor, a situation of interrelationships between permeability proteins and bacteriophage receptors that is rather common. Another example is the finding that high concentrations of nucleosides are not necessary to ensure rapid growth of porin-deficient strains of *E. coli*. Investigation of this problem further demonstrated the complexities of the function of outer membrane proteins. A gene mapping at 9 min on the *E. coli* chromosome which controlled the uptake of nucleosides was called *nup* [110, 187] and a mutant called *tsx* mapping at 10 min on the chromosome was resistant to T6 and colicin K. The gene product was known to be a protein [198, 346]. It is probable that the *nup* and *tsx* are one and the same gene since the latter mutants also take up the nucleosides very poorly when they are present at low concentration, and lack an outer membrane protein of 25 000 mol.wt. [110, 192]. Thus, this outer membrane protein which is not one of the recognized porins, has specificity for bacteriophage T6, colicin K and allows the passage of nucleosides, although it may be noted that the porins themselves also act as bacteriophage receptors, for example for T2 or T4 phages. A number of other outer membrane proteins have now been recognized that serve as permeability factors as well as receptors for bacteriophages and colicins. Some of these are summarized in Table 4.16. The problem has been reviewed by Braun [33]. The apparently multifunctional nature of these outer membrane proteins presents a fascinating problem. Again it may be worth emphasizing that, as with the porins, the permeability proteins for low molecular weight metabolites such as maltose, or nucleosides, are only required when these are present at very low concentrations (i.e. of the order of micromolar); at millimolar concentrations growth of the deficient mutants was as fast as that of the wild-type organisms. That different sites of the membrane proteins are likely to be associated with different functions, is suggested by the observation that mis-sense *lamB* mutants showed greatly reduced growth on maltotriose but still had normal amounts of λ receptor protein [34, 327]. Again neither maltose nor nucleosides prevent binding of the bacteriophages to their respective receptor proteins. The *tonB* gene product is not fully understood but mediates the transport of a variety of substances across the outer membrane and in the irreversible binding of phages. Studies of Fe^{3+}–enterochelin transport and binding of bacteriophages T1 and ϕ80 has led to the suggestion [33] that it may in some way serve as an energy coupling device between the inner and outer membranes.

Table 4.16 Multifunctional gene products in the outer membrane of *E. coli* (after [33]).

Permeability function	*Phage receptor function*	*Colicin receptor function*	*Outer membrane gene product*
Fe^{3+}–ferrichrome	T5	M	*tonA*
Albomycin	T1 ϕ80	–	*tonA tonB*
Fe^{3+}–enterochelin	–	B	*feuB tonB*
–	–	J	*cir tonB*
Fe^{3+} citrate	–	–	*cit tonB*
Vitamin B_{12}	BF_{23}	E group	*bfe tonB*
Nucleoside	T6	K	*tsx (nup)*
Maltose, Maltodextrins	λ	–	*lamb*

References

1. Abrams, D. (1965) *J. Bact.* **89**, 855–61.
2. Abrams, A., Jensen, C. and Morris, D. H. (1975) *J. supramolec. Struct.* **3**, 261–74.
3. Abrams, A., Jensen, C. and Morris, D. H. (1976) *Biochem. biophys. Res. Commun.* **69**, 804–811.

3a. Adams, M. W. W. and Hall, D. O. (1979) *Archs. Biochem. Biophys.* **195**, 288–99.

4. Adler, L. W. and Rosen, B. P. (1977) *J. Bact.* **129**, 959–66.
5. Akrigg, A., Ayad, S. R. and Blamire, J. (1969) *J. theor. Biol.* **24**, 266–72.
6. Altendorf, K. H. and Staehelin, L. A. (1974) *J. Bact.* **117**, 888–99.
7. Amanuma, H., Motojima, K., Yamaguchi, A. and Anraku, Y. (1977) *Biochem. Biophys. Res. Commun.* 366–73.
8. Ames, G. F-L. (1974) *J. biol. Chem.* **249**, 634–44.
9. Ames, G. F. and Lever, J. (1970) *Proc. Natn. Acad. Sci. U.S.A.* **66**, 1096.
10. Andreu, J., Albendea, J. A. and Muñoz, E. (1973) *Eur. J. Biochem.* **37**, 505–517.
11. Anraku, Y. L. (1968a) *J. biol. Chem.* **243**, 3116–22.
12. Anraku, Y. L. (1968b) *J. biol. Chem.* **243**, 3123–7.
13. Asano, A., Cohen, N. S., Baker, R. F. and Brodie, A. F. (1978) *J. biol. Chem.* **248**, 3386–97.
14. Baillie, R. D., Hou, C. and Bragg, P. D. (1971) *Biochim. biophys. Acta* **234**, 46–56.
15. Baird, B. A. and Hammes, G. G. (1977) *J. biol. Chem.* **252**, 4743–8.
16. Barash, H. and Halpern, Y. S. (1971) *Biochem. biophys. Res. Commun.* **45**, 681–8.
17. Barnes, E. M. jr. and Kaback, H. R. (1971) *J. biol. Chem.* **246**, 5518–22.
18. Barn, C. and Abrams, A. (1971) *J. biol. Chem.* **246**, 1532–44.

18a. Beinert, H., Ackrell, B. A. C., Keamy, E. G. and Singer, T. P. (1974) *Biochem. biophys. Res. Commun.* **58**, 564–72.

19. Benemann, J. R. and Valentine, R. C. (1971) *Adv. microbial Physiol.* **5**, 135–72.
20. Berlin, R. D. and Stadtman, E. R. (1966) *J. biol. Chem.* **241**, 2679–86.
21. Berzborn, R. J., Johansson, B. C. and Baltscheffsky, M. (1975) *Biochim. Biophys. Acta* **396**, 360–70.
22. Bisschop, A., de Hong, L., Lima-Costra, M. E. and Konings, W. N. (1975) *FEBS Letts.* **60**, 11–15.

23. Boos, W. (1969) *Eur. J. Biochem.* **10**, 66–73.
24. Boos, W., Gordon, A. S., Hall, R. E. and Price, A. D. (1972) *J. biol. Chem.* **247**, 917–24.
25. Boxer, D. H. and Clegg, R. A. (1975) *FEBS Letts.* **60**, 54–7.
26. Boyer, P. D. (1977) *A. Rev. Biochem.* **46**, 957–66.
27. Boyer, P. D., Chance, B., Ernster, L., Mitchell, P., Racker, E. and Slater, E. C. (1977) *A. Rev. Biochem.* **46**, 955–1026.
28. Bragg, P. D. (1975) *J. supramolec. Struct.* **3**, 297–303.
29. Bragg, P. D. and Hou, C. (1972) *FEBS Letts.* **28**, 309–312.
30. Bragg, P. D. and Hou, C (1973) *Biochem. biophys. Res. Commun.* **50**, 729–36.
30a. Bragg, P. D. and Hou, C. (1975) *Archs. Biochem. Biophys.* **167**, 311–321.
31. Bragg, P. D., Cavies, P. L. and Hou, C. (1973). *Archs. Biochem. Biophys.* **159**, 664–70.
32. Brand, M. D., Reynafarje, B. and Lehninger, A. L. (1976) *Proc. Natn. Acad. Sci., U.S.A.* **73**, 437–41.
33. Braun, V. (1978) *Soc. gen. Microbiol. Symp.* **28**, 110–38.
34. Braun, V. and Krieger-Brauer, H. J. (1977) *Biochim. biophys. Acta* **469**, 89–98.
35. Brown, C. E. and Hogg, R. W. (1972) *J. Bact.* **111**, 606–13.
35a. Buchel, D. E., Gronenborn, B. and Müller-Hill, B. (1980) *Nature (London)*, **283**, 541–5.
36. Burnell, J. N., Johns, P. and Whatley, L. (1975) *Biochem. J.* **150**, 527–36.
37. Butlin, J. D., Cox, G. B. and Gibson, F. (1971) *Biochemistry* **124**, 75–81.
38. Butlin, J. D., Cox, G. B. and Gibson, F. (1973) *Biochim. biophys. Acta* **292**, 366–75.
39. Celis, T. F. R., Rosenfeld, H. J. and Maas, W. K. (1973) *J. Bact.* **116**, 619–26.
40. Chai, T-J. and Foulds, J. (1978) *J. Bact.* **135**, 164–70.
41. Chance, B. (1977) *A. Rev. Biochem.* **46**, 967–80.
42. Chapman, G. B. and Hillier, J. (1935) *J. Bact.* **66**, 362–73.
43. Cole, J. A. and Ward, F. B. (1973) *J. gen. Microbiol.* **76**, 21–9.
44. Collins, S. H. and Hamilton, W. A. (1976) *J. Bact.* **126**, 1224–31.
45. Cordaro, J. C. and Roseman, S. (1972) *J. Bact.* **112**, 17–29.
46. Cordaro, J. C., Anderson, R. P., Grogan, E. W., Wenzel, D. J., Engles, M. and Roseman, S. (1974) *J. Bact.* **120**, 245–52.
47. Cox, R. and Charles, H. P. (1973) *J. Bact.* **113**, 122–32.
48. Cox, G. B., Gibson, F. and McCann, L. (1973a) *Biochem. J.* **134**, 1015–21.
49. Cox, G. B., Gibson, F. and McCann, L. (1974) *Biochem. J.* **138**, 211–15.
50. Cox, G. B., Newton, N. A., Butlin, J. D. and Gibson, F. (1971) *Biochem. J.* **125**, 489–93.
51. Cox, G. B., Newton, N. A., Gibson, F., Snoswell, A. M. and Hamilton, J. A. (1970) *Biochem. J.* **117**, 551–62.
52. Cox, G. B., Gibson, F., McCann, L. M., Butlin, J. D. and Crane, F. L. (1973b) *Biochem. J.* **132**, 689–95.
53. Daniels, M. J. (1971) *Biochem. J.* **122**, 197–207.
54. Darland, G., Brock, T. D., Samsonoff, W. and Conti, S. F. (1970) *Science* **170**, 1416–18.
55. Davis, K. A. and Hatefi, Y. (1971) *Biochemistry* **10**, 2509–16.
55a. Davis, K. A., Hatefi, Y., Crawford, I. P. and Baltsceffsky, H. (1977) *Archs. Biochem. Biophys.* **180**, 459–64.
56. Davson, H. and Danielli, J. F. (1943) *The Permeability of Natural Membranes* New York: Macmillan.
57. Dawson, A. G. and Chappell, J. B. (1978) *Biochem. J.* **170**, 395–405.
58. Deamer, D. W., Prince, R. C. and Crofts, A. R. (1972) *Biochim. biophys. Acta* **274**, 323–35.

59. Decad, G. M. and Nikaido, H. (1976) *J. Bact.* **128**, 325-36.
60. De Martini, M. and Inouye, M. (1976) *J. Bact.* **133**, 329-35.
61. Di Rienzo, J. M., Nakamura, K. and Inouye, M. (1978) *A. Rev. Biochem.* **47**, 481-532.
61a. Downie, J. A., Gibson, F. and Cox, G. B. (1979) *A. Rev. Biochem.* **48**, 103-31.
61b. Downie, J. A., Senior, A. E., Gibson, F. and Cox, G. B. (1979) *J. Bact.* **137**, 711-18.
62. Dreyfus, J. (1964) *J. biol. Chem.* **239**, 2292-7.
62a. Ehrling, R., Beyreuther, K., Wright, J. K. and Overath, P. (1980) *Nature (London)* **283**, 537-40.
63. Eisenberg, R. C. (1971) *J. Bact.* **108**, 964-72.
64. Eisenberg, R. C. (1972) *J. Bact.* **112**, 445-52.
65. Eisenberg, R. C., Yu, L. and Wolin, M. J. (1970) *J. Bact.* **102**, 172-7.
66. Eisenberg, R. C., Yu, L. and Wolin, M. J. (1970) *J. Bact.* **102**, 161-71.
67. Ellar, D. J., Lundgren, D. G. and Slepecky, R. A. (1967) *J. Bact.* **94**, 1189-205.
68. Ellar, D. J., Munoz, E. and Salton, M. R. J. (1971a) *Biochim. biophys. Acta* **225**, 140-50.
69. Ellar, D. J., Thomas, T. D. and Postgate, J. A. (1971b) *Biochem. J.* **122**, 44P-45P.
70. Epstein, W., Jewett, S. and Fox, C. F. (1970) *J. Bact.* **104**, 793-7.
71. Ferrandes, B., Chaix, P. and Ryter, A. (1966) *C.r. Acad. Sci., Paris* **263**, 1632-5.
72. Ferrandes, B., Fréhèl, C. and Chaix, P. (1970) *Biochim. biophys. Acta* **223**, 292-303.
73. Fitz-James, P. C. (1960) *J. biophys. biochem. Cytol.* **8**, 507-528.
74. Fitz-James, P. C. (1967) In *Protides of the Biological Fluids*, ed Peters, H., pp. 289-301, Amsterdam: Elsevier Co.
75. Flagg, J. L. and Wilson, T. H. (1976) *J. Bact.* **125**, 1235-6.
76. Foulds, J. and Chai, T-J. (1978) *J. Bact.* **133**, 1478-83.
77. Fox, C. F. and Kennedy, E. P. (1965) *Proc. Natn. Acad. Sci. U.S.A.* **54**, 891-9.
78. Fox, C. F., Carter, J. R. and Kennedy, E. P. (1967) *Proc. Natn. Acad. Sci. U.S.A.* **57**, 698-705.
79. Fugita, T. (1966) *J. Biochem., Tokyo* **60**, 329-34.
80. Fugita, T. and Sato, R. (1966) *J. Biochem. Tokyo* **60**, 568-77.
81. Furlong, C. E. and Weiner, J. H. (1970) *Biochem. biophys. Res. Commun.* **38**, 1076-83.
82. Futai, M. (1974a) *J. Memb. Biol.* **15**, 15-18.
83. Futai, M. (1977) *Biochem. biophys. Res. Commun.* **79**, 1231-7.
84. Futai, M. and Tanaka, Y. (1975) *J. Bact.* **124**, 470-75.
85. Futai, M., Sternweis, P. C. and Heppel, L. A. (1974) *Proc. Natn. Acad. Sci U.S.A.* **71**, 2725-9.
86. Garland, P. B. (1977) *Soc. Gen. Microbiol. Symp.* **27**, 1-21.
87. Gel'man, N. S., Lukoyanova, M. A. and Ostroskii, D. W. (1975) In *Biomembranes* **6**, ed. Manson, L. A., New York and London: Plenum Press, pp. 129-209.
88. Gel'man, N. S., Tikkonova, C. V., Simakova, I. M., Lukoyanova, M. A., Taptykova, S. D. and Mikelsaar, H. M. (1970) *Biochim. biophys. Acta* **223**, 321-31.
89. Gibson, F. (1973) *Biochem. Soc. Trans.* **1**, 317-26.
90. Gibson, F. and Cox, G. B. (1973) *Essays in Biochemistry* **9**, 1-29.
90a. Gibson, F., Cox, G. B., Downie, J. A. and Radik, J. (1977) *Biochem. J.* **164**, 193-8.
91. Glaser, J. H. and de Moss, J. A. (1972) *Mol. gen. Genet.* **116**, 1-10.

92. Gordon, A. S., Lombardi, F. J. and Kaback, H. R. (1972) *Proc. Natn. Acad. Sci. U.S.A.* **69**, 358–62.
93. Gorneva, G. A. and Ryaborc, I. D. (1974) *FEBS Letts.* **42**, 273–4.
94. Gosh, B. K. and Murray, R. G. E. (1969) *J. Bact.* **97**, 426–40.
95. Greenwalt, J. W. and Whiteside, T. L. (1975) *Bact. Rev.* **39**, 405-463.
96. Guest, J. R. (1969) *Mol. gen. Genet.* **105**, 285-97.
97. Gutrick, O. L., Kanner, B. I. and Postma, P. W. (1972) *Biochim. biophys. Acta* **283**, 217–22.
98. Gutowski, S. J. and Rosenberg, H. (1975) *Biochem. J.* **152**, 647–54.
99. Hachimori, A., Muramatsu, N. and Nosok, Y. (1970) *Biochim. biophys. Acta* **206**, 426–7.
100. Haddock, B. A. (1973) *Biochem. J.* **136**, 877–84.
101. Haddock, B. A. (1977) *Soc. Gen. Microbiol. Symp.* **27**, 95–120.
102. Haddock, B. A. and Schairer, H. U. (1973) *Eur. J. Biochem.* **35**, 34–45.
103. Haddock, B. A. and Downie, J. A. (1974) *Biochem. J.* **142**, 703–706.
104. Haddock, B. A. and Jones, C. W. (1977) *Bact. Rev.* **41**, 47–89.
105. Halpern, Y. S., Varash, H., Dover, S. and Druck, K. (1973) *J. Bact.* **114**, 53–8.
106. Hamilton, W. A. (1975) *Adv. microbial Physiol.* **12**, 1–53.
107. Hamilton, W. A. (1977) In *Microbial Energetics, 27th Symp. Soc. gen. Microbiol.*, eds. Haddock, B. A. and Hamilton, W. A., Cambridge University Press, pp. 185–216.
108. Hamilton-Miller, J. M. T. (1965) *Biochem. biophys. Res. Commun.* **20**, 678–81.
109. Hamilton-Miller, J. M. T., Smith, J. T. and Knox, R. (1965) *Nature, Lond* **208**, 235–7.
110. Hantke, K. (1976) *FEBS Letts.* **70**, 109–12.
111. Hanson, R. L. and Kennedy, E. P. (1973) *J. Bact.* **114**, 772–81.
112. Hare, J. E. (1975) *Biochem. biophys. Res. Commun.* **66**, 1329–37.
113. Hare, J. F., Olden, K. and Kennedy, E. P. (1974) *Proc. Natn. Acad. Sci. U.S.A.* **71**, 4843–6.
114. Harold, F. M. (1972) *Bact. Rev.* **36**, 172–230.
115. Hasam, S. M. and Tsuchiya, T. (1977) *Biochem. biophys. Res. Commun.* **78**, 122–8.
116. Hatefi, Y., Haavik, A. G., Fowler, L. R. and Griffiths, D. E. (1962) *J. biol. Chem.* **237**, 2661–70.
117. Hays, J. B., Simoni, R. D. and Roseman, S. (1973) *J. biol. Chem.* **248**, 941–56.
117a. Hederstedt, L., Holmgren, E. and Rutberg, L. (1979) *J. Bact.* **138**, 370–76.
118. Henderson, P. J. F. and Kornberg, H. L. (1975) In *Energy Transformation in Biological Systems* (Ciba Foundation) pp. 243–269, Amsterdam, Elsevier.
119. Henderson, G. B., Zevely, E. M. and Huennekens, F. M. (1977a) *J. biol. Chem.* **252**, 3760–65.
120. Henderson, G. B., Zevely, E. M., Kadner, R. J. and Huennekens, F. M. (1977b) *J. supramolec. Struct.* **6**, 239–47.
121. Henning, U. and Haller, I. (1975) *FEBS Letts.* **55**, 161–4.
122. Heppel, L. A., Rosen, B. P., Friedberg, I., Berger, E. and Weiner, J. H. (1972) In *The Molecular Basis of Biological Transport*, eds. Woessner, J. F. and Huijing, F., New York and London: Academic Press, pp. 139–49.
123. Hinckle, P. and Mitchell, P. (1970) *J. Bioenerg.* **1**, 45–60.
124. Hirota, Y., Suzuki, H., Nishimura, Y. and Yasuda, S. (1977) *Proc. Natn. Acad. Sci. U.S.A.* **74**, 1417–20.

125. Hochstadt-Ozer, J. (1972) *J. biol. Chem.* **247**, 2419–26.
126. Hochstadt-Ozer, J. and Stadtman, E. R. (1971a) *J. biol. Chem.* **246**, 5304–11.
127. Hochstadt-Ozer, J. and Stadtman, E. R. (1971b) *J. biol. Chem.* **246**, 5312–20.
128. Hockel, M., Hulla, F. W., Risi, S. and Dose, K. (1976) *Biochim. biophys. Acta* **429**, 1020–28.
129. Hogg, R. W. and Hermondson, M. A. (1977) *J. biol. Chem.* **252**, 5135–41.
130. Hong, J. S. and Kaback, H. R. (1972) *Proc. Natn. Acad. Sci. U.S.A.* **69**, 3336–40.
131. Horio, T. and Kamen, M. D. (1970) *A. Rev. Microbiol.* **24**, 399–428.
132. Iterson, W. van (1965) *Bact. Rev.* **29**, 299–325.
133. Iterson, W. van and Letne, W. (1964) *J. Cell Biol.* **20**, 361–77.
134. Jackman, L. E. and Hochstadt, J. (1976) *J. Bact.* **126**, 312–26.
135. Jacob, F., Ryter, A. and Cuzin, F. (1966) *Proc. R. Soc. B* **164**, 267–78.
136. Jensen, L. H. (1974) *A. Rev. Biochem.* **43**, 461–74.
137. Johansson, B. C. and Baltscheffsky, M. (1975) *FEBS Letts.* **53**, 221–4.
138. Johansson, B. C., Baltscheffsky, M., Baltscheffsky, H., Baccarini-Melandri, A. and Melandri, B. A. (1973) *Eur. J. Biochem.* **10**, 109–117.
139. Jones, C. W. (1977a) *Symp. Soc. Gen. Microbiol.* **27**, 23–59.
140. Jones, T. H. D. and Kennedy, E. P. (1969) *J. biol. Chem.* **244** 5981–7.
141. Jones, R. W. and Garland, P. B. (1977) *Biochem. J.* **164**, 199–211.
142. Jurtshuk, P., Mueller, T. J. and Acord, W. C. (1975) *Crit. Rev. Microbiol.* **3**, 399–468.
143. Kaback, H. R. (1971) In *Methods in Enzymology* **22**, ed. Jackoby, W. B., pp. 99–120, New York and London: Academic Press.
144. Kaback, H. R. (1973) In *Bacterial Cell Walls and Membranes*, ed. Leive, L., pp. 241–92, New York: Dekker Inc.
145. Kaback, H. R. (1974) *Science* **186**, 882–92.
146. Kaback, H. R. and Stadtman, E. R. (1966) *Proc. Natn. Acad. Sci.* **55**, 920–7.
147. Kaback, H. R. and Barnes, E. M. (1971) *J. biol. Chem.* **246**, 5523–31.
148. Kalra, V. K., Lee, S-H., Ritz, C. J. and Brodie, A. F. (1975) *J. supramolec. Struc.* **3**, 231–41.
149. Keman, M. D. and Horio, T. (1970) *A. Rev. Biochem.* **39**, 673–700.
150. Kanner, B. I., Nelson, N. and Gutnick, D. L. (1975) *Biochim. biophys. Acta* **396**, 347–59.
151. Kabagin, M., Ganguli, B. W. and Gunsalus, I. (1968) *J. biol. Chem.* **243**, 3543–6.
152. Kavanagh, B. M. and Cole, J. A. (1976) *Proc. Soc. Gen. Microbiol.* **3**, 84.
153. Kay, W. W. and Kornberg, H. L. (1971) *Eur. J. Biochem.* **18**, 274–81.
154. Kennedy, E. P., Rumley, M. K. and Armstrong, J. B. (1974) *J. biol. Chem.* **249**, 33–9.
155. Kistler, W. S. and Lin, E. C. C. (1971) *J. Bact.* **108**, 1224–34.
156. Konings, W. N. and Freese, E. (1971) *FEBS Letts.* **14**, 65–8.
157. Konings, W. N. and Freese, E. (1972) *J. biol. Chem.* **247**, 2408–18.
158. Konings, A. W. J. and Guillory, R. J. (1973) *J. biol. Chem.* **248**, 1045–50.
159. Konings, W. N., Bisschop, A., Veenhuis, M. and Vermeulen, C. A. (1973) *J. Bact.* **116**, 1456–65.
160. Kroger, A. (1977) *Symp. Soc. Gen. Microbiol.* **27**, 61–93.
161. Kundig, W. and Roseman, S. (1971) *J. biol. Chem.* **246**, 1407–18.
162. Kustu, S. G. and Ames, G. F-L. (1974) *J. biol. Chem.* **249**, 6976–83.
163. Kuzuya, H., Bromwell, K. and Guroff, G. (1971) *J. biol. Chem.* **246**, 6371–80.
164. Lampen, J. O. (1974) In *Transport at the Cellular Level, Symp. Soc. Exp. Biol.* **28**, eds. Sleigh, M. A. and Jennings, D. H., pp. 357–74, Cambridge University Press.

164a. Lampen, J. O. (1978) In *Relations Between Structure and Function in the Prokaryotic Cell, 28th Symp. Soc. Gen. Microbiol.*, eds. Stanier, R. Y., Rogers, H. J. and Ward, J. B., pp. 231–44, Cambridge University Press.
165. Landman, O. E., Ryter, A. and Frehel, C. (1968) *J. Bact.* **96**, 2154–70.
166. Langridge, H., Shinagawa, H. and Pardee, A. B. (1970) *Science* **169**, 59–61.
167. Lanyi, J. K. (1977) *J. supramolec. Struct.* **6**, 169–77.
168. Lee, S-H., Cohen, N. S., Jacobs, A. J. and Brodie, A. F. (1978) *J. supramolec. Struct.* **7**, 111–17.
169. Leene, W. and Iterson van W. (1966) *J. Cell Biol.* **20**, 361–75.
170. Leive, L. (1965) *Biochem. biophys. Res. Commun.* **21**, 290–6.
171. Leive, L. (1968) *J. biol. Chem.* **243**, 2373–80.
172. Leive, J. and Davis, B. D. (1965) *J. biol. Chem.* **240**, 4362–76.
173. Lemberg, R. and Barrett, J. (1973) In *Cytochromes*, New York and London: Academic Press, pp. 217–326.
174. Lever, J. E. (1972) *J. biol. Chem.* **247**, 4317–26.
175. Lo, T. C. Y. and Sanwal, B. D. (1975a) *J. biol. Chem.* **250**, 1600-602.
176. Lo, T. C. Y. and Sanwal, B. D. (1975b) *Biochem. biophys. Res. Commun.* **63**, 278–85.
177. Lo, T. C. Y. and Sanwal, B. D. (1975c) *J. biol. Chem.* **250**, 1600–1602.
178. Lo, T. C. Y. and Bewick, M. A. (1978) *J. biol. Chem.* **253**, 7826–31.
179. Lo, T. C. Y., Rayman, M. K. and Sanwal, B. D. (1972a) *J. biol. Chem.* **247**, 6323–31.
180. Lo, T. C. Y., Rayman, M. K. and Sanwal, B. D. (1972b) *J. biol. Chem.* **247**, 6332–9.
181. Lo, T. C. Y., Rayman, M. K. and Sanwal, B. D. (1974) *Can. J. Biochem.* **52**, 854–66.
182. Lombardi, F. and Kaback, H. R. (1972) *J. biol. Chem.* **247**, 7844–57.
183. Lundegardh, H. (1945) *Ark. Bot. 32A*, **12**, 1–18.
184. Lutkenhaus, J. F. (1977) *J. Bact.* **131**, 631–7.
185. McCleod, R. A., Thurman, P. and Rogers, H. J. (1972) *J. Bact.* **113**, 329–40.
186. McConville, M. and Charles, H. P. (1975) *Proc. Soc. Gen. Microbiol.* **3**, 14–15.
187. McKeown, M., Kahn, M. and Hanawalt, P. (1976) *J. Bact.* **126**, 814–822.
188. McLachlan, A. D. (1978) *J. molec. Biol.* **121**, 493–506.
189. McQuillen, K. (1960) In *The Bacteria Vol. 1,* eds. Gunsalus, I. C. and Stanier, R. Y., New York and London: Academic Press.
190. MacGregor, C. H. (1975) *J. Bact.* **121**, 1117–21.
191. MacGregor, C. H. and Schnaitman, C. A. (1973) *J. Bact.* **114**, 1164–76.
192. Manning, P. A. and Reeves, P. (1976) *Biochem. biophys. Res. Commun.* **71**, 466–71.
193. Martin, A. and Konings, W. N. (1973) *Eur. J. Biochem.* **34**, 58–67.
194. Meador, W. E. and Quiocho, F. A. (1978) *J. molec. Biol.* **123**, 499.
195. Medveszky, N. and Rosenberg, H. (1969) *Biochim. biophys. Acta* **192**, 369–71.
196. Medveszky, N. and Rosenberg, H. (1970) *Biochim. biophys. Acta* **211**, 158–68.
197. Meyer, D. J. and Jones, C. W. (1973) *Int. J. Syst. Bact.* **23**, 459–67.
198. Michael, J. G. (1968) *Proc. Soc. Exp. Biol. and Med.* **128**, 434–438.
198a. Michels, A. M. and Konings, W. N. (1978) *Eur. J. Biochem.* **85**, 147–55.
199. Mindich, L. and Dales, S. (1972) *J. Cell Biol.* **55**, 32–41.
200. Miner, K. M. and Frank, L. (1974) *J. Bact.* **117**, 1093–8.
201. Mirsky, R. and Barlow, V. (1971) *Biochim. biophys. Acta* **241**, 835–45.
202. Mitchell, P. (1961) *Nature, Lond.* **191**, 144–8.
203. Mitchell, P. (1963) *Biochem. Soc. Symp.* **22**, 142–68.
204. Mitchell, P. (1966) *Chemiosmotic Coupling and Energy Transduction*, Glynn Research Ltd. Bodwin, England.

205. Mitchell, P. (1977) *A. Rev. Biochem.* **46**, 956–1026.
206. Morrison, D. C. and Morowitz, H. J. (1970) *J. molec. Biol.* **49**, 441–59.
207. Mortenson, L. E., Valentine, R. C. and Carnahan, J. E. (1962) *Biochem. biophys. Res. Commun.* **17**, 448–52.
208. Moyle, J. and Mitchell, P. (1973) *FEBS Letts.* **30**, 317–20.
209. Muller, H. W., Schmitt, M., Schneider, E. and Dose, K. (1979) *Biochim. biophys. Acta* **545**, 77–85.
210. Muñoz, E., Salton, M. R. J., Ng, M. H. and Schor, M. T. (1969) *Eur. J. Biochem.* **7**, 490–501.
211. Nakae, T. (1976a) *Biochem. biophys. Res. Commun.* **71**, 877–84.
212. Nakae, T. (1976b) *J. biol. Chem.* **251**, 2176–8.
213. Nakae, T. and Nikaido, H. (1975) *J. biol. Chem.* **250**, 7359–65.
214. Nakamura, K. and Mizushima, S. (1975) *Biochim. biophys. Acta* **413**, 371–93.
215. Ne'em, Z. and Razin, S. (1975) *Biochim. biophys. Acta* **375**, 54–68.
216. Nelson. N., Kanner, B. I. and Gutnick, D. L. (1974) *Proc. Natn. Acad. Sci. U.S.A.* **71**, 2720–24.
217. Newton, N. A., Cox, G. B. and Gibson, F. (1971) *Biochim. biophys Acta* **244**, 155–66.
218. Nieuwenhuis, F. J. R. M., Kanner, B. I., Gutnick, D. L., Postma, P. W. and van Dams, K. (1973) *Biochim. biophys. Acta* **325**, 62–71.
219. Nikaido, H. (1976b) *Biochim. biophys. Acta* **433**, 118–32.
220. Nikaido, H., Shaltiel, S. A. S. and Nurminen, M. (1976) *Biochem. biophys. Res. Commun.* **76**, 324-30.
221. Nurminen, M., Lounotmaa, K., Sarvas, M., Makela, P. H. and Nakae, T. (1976) *J. Bact.* **127**, 941–55.
222. Ohnishi, T., Winter, D. B., Lim, J. and King, T. E. (1973) *Biochem. biophys. Res. Commun.* **53**, 231–7.
223. Ohnishi, T., Winter, D. B., Lim, J. and King, T. E. (1974) *Biochem. biophys. Res. Commun.* **61**, 1017–25.
224. Okamoto, H., Sone, N., Hirata, H., Yoshida, M. and Kagawa, Y. (1977) *J. biol. Chem.* **252**, 6125–31.
225. Oppenheim, J. D. and Salton, M. R. J. (1973) *Biochim. biophys. Acta* **298**, 297–322.
226. Oren, R. and Gromet-Elhanan, Z. (1977) *FEBS Letts.* **79**, 147–50.
227. Orme-Johnson, W. H. (1973) *A. Rev. Biochem.* **42**, 159–204.
227a. Orme-Johnson, N. R., Orme-Johnson, W. H., Hansen, R. E., Beiner, H. and Hatefi, Y. (1971) *Biochem. biophys. Res. Commun.* **44**, 446–52.
227b. Overath, P., Teather, R. M., Simoni, R. D., Aichele, G. and Wilhelm, U. (1979) *Biochemistry* **18**, 1–11.
228. Owen, P. and Freer, J. H. (1972) *Biochem. J.* **129**, 907–917.
228a. Owen, P. and Kaback, H. R. (1979) *Biochemistry* **18**, 1422–6.
229. Owen, P. and Salton, M. R. J. (1975) *Proc. Natn. Acad. Sci. U.S.A.* **72**, 3711–15.
230. Owen, P. and Salton, M. R. J. (1977) *J. Bact.* **132**, 974–85.
231. Palva, E. T. and Randall, L. L. (1978) *J. Bact.* 133, 279–83.
232. Pardee, A. B. (1966) *J. biol. Chem.* **241**, 5886–92.
233. Pardee, A. B. (1967) *Science* **156**, 1627–8.
234. Parsons, R. G. and Hogg, R. W. (1974a) *J. biol. Chem.* **249**, 3602–607.
235. Parsons, R. G. and Hogg, R. W. (1974b) *J. biol. Chem.* **249**, 3608–614.
236. Pascal, M. C., Casse, F., Chippaux, M. and Lepelletier, M. (1975) *Mol. gen. Genet.* **141**, 173–9.
237. Patch, C. T. and Landman, O. E. (1971) *J. Bact.* **107**, 345–57.

238. Payne, J. W. and Gilvarg, C. (1968) *J. biol. Chem.* **243**, 6291–9.
239. Pedersen, P. L. (1975) *J. Bioenerg.* **6**, 243–75.
240. Penrose, W. R., Nichoalds, G. E., Piperno, J. R. and Oxender, D. L. (1968) *J. biol. Chem.* **243**, 5921–8.
241. Philiosoph, S., Binder, A. and Gromet-Elhanan, Z. (1977) *J. biol. Chem.* **252**, 9747–52.
242. Pinchot, G. B. (1953) *J. biol. Chem.* **205**, 65–74.
243. Piperno, J. R. and Oxender, D. L. (1966) *J. biol. Chem.* **241**, 5732–4.
244. Poole, R. K. and Haddock, B. A. (1974) *Biochem. J.* **144**, 77–85.
245. Postma, P. W. and Roseman, S. (1976) *Biochim. biophys. Acta* **457**, 213–57.
246. Powell, K. A., Cox, R., McConville, M. and Charles, H. P. (1973) *Enzyme* **16**, 65–73.
247. Pugsley, A. P. and Schnaitman, C. (1978) *J. Bact.* **133**, 1181–9.
248. Quay, S. and Christensen, H. N. (1974) *J. biol. Chem.* **249**, 7011–17.
249. Quiocho, F. A., Gilliland, G. L. and Phillips, G. N. jr. (1977) *J. biol. Chem.* **252**, 5142–9.
250. Ramos, S. and Kaback, H. R. (1977a) *Biochemistry* **16**, 848–54.
251. Ramos, S. and Kaback, H. R. (1977b) *Biochemistry* **16**, 854–8.
252. Razin, S. (1972) *Biochim. biophys. Acta* **265**, 241–196.
253. Razin, S. (1978) *Microbiol. Rev.* **42**, 414–70.
254. Reaveley, D. A. and Rogers, H. J. (1969) *Biochem. J.* **113**, 67–79.
255. Reeves, J. P., Shechter, E., Weil, R. and Kaback, H. R. (1973) *Proc. Natn. Acad. Sci. U.S.A.* **70**, 2722–6.
256. Reddy, J. L. P. and Hendler, R. W. (1978) *J. biol. Chem.* **253**, 7972–9.
257. Reusch, V. M. and Burger, M. M. (1973) *Biochim. biophys. Acta* **300**, 79–104.
258. Risi, S., Höckel, M., Hulla, F. W. and Dose, K. (1977) *Eur. J. Biochem.* **81**, 103–109.
259. Rivière, C., Giordano, G., Pommier, J. and Azoulay, E. (1975) *Biochim. biophys. Acta* **389**, 219–35.
260. Roantree, R. J., Kuo, T., MacPhee, D. G. and Stocker, B. A. D. (1969) *Clin. Res.* **17**, 157.
261. Robbie, J. P. and Wilson, T. H. (1969) *Biochim. biophys. Acta* **173**, 234–44.
262. Robbins, A. R. and Rotman, B. (1975) *Proc. Natn. Acad. Sci. U.S.A.* **72**, 423-7.
263. Robbins, A. R., Guzman, R. and Rotman, B. (1976) *J. biol. Chem.* **251**, 3112-16.
264. Robertson, R. N. (1960) *Biol. Rev.* **35**, 231–64.
265. Rockey, A. E. and Haddock, B. A. (1974) *Biochem. Soc. Trans.* **2**, 957–60.
266. Romano, A. H., Eberhard, S. J., Dingle, S. L. and McDowell, T. D. (1970) *J. Bact.* **104**, 808–813.
267. Roseman, S. (1972) In *Metabolic Pathways* (3rd Edn.), ed. Hokin, L. E., New York and London: Academic Press, pp. 41–89.
268. Roseman, S. (1975) *Ciba Found. Symp.* **31** (new series) 225–41.
269. Rosen, B. P. (1971) *J. biol. Chem.* **246**, 3653–62.
270. Rosen, B. P. (1973) *J. biol. Chem.* **248**, 1211–8.
271. Rosen, B. P. and Heppel, L. A. (1973) In *Bacterial Membranes and Walls*, ed. Leive L., New York: Marcel Dekker Inc.
272. Rottenberg, H. (1975) *J. Bioenerg.* **7**, 61–74.
272a. Roy-Burman, S. and Visser, D. W. (1975) *J. biol. Chem.* **250**, 9270–75.
273. Rudnick, G., Schuldiner, S. and Kaback, H. R. (1976) *Biochemistry* **15**, 5126–31.
274. Ryter, A. (1967) *Folia microbiol.* **12**, 283–300.
275. Ryter, A. (1968) *Bact. Rev.* **23**, 39–54.

276. Ryter, A. (1969) *Current Topics. Microbiol. and Immunol.* **48**, 151–177.
277. Ryter, A (1971) *Annls. Inst. Past.* **131**, 271–88.
278. Ryter, A. and Jacob, F. (1964) *Annls. Inst. Past.* **107**, 389–400.
279. Ryter, A. and Jacob, F. (1966) *Annls. Inst. Past.* **110**, 801–812.
280. Saier, M. H. jr. and Roseman, S. (1972) *J. biol. Chem.* **247**, 972–8.
281. Saier, M. H. jr., Simoni, R. D. and Roseman, S. (1970) *J. biol. Chem.* **245**, 5870–73.
282. Salemmi, F. R. (1977) *A. Rev. Biochem.* **46**, 299–329.
283. Salton, M. R. J. (1956) *Symp. Soc. Gen. Microbiol.* **6**, 51–100.
284. Salton, M. R. J. (1978) In *Relations Between Structure and Function in the Prokaryotic Cell*, 28th Symp. Soc. Gen. Microbiol. eds. Stanier, R. Y., Rogers, H. J. and Ward, J. B., Cambridge University Press, pp. 201–33.
285. Salton, M. R. J. and Schor, M. T. (1972) *Biochem. biophys. Res. Commun.* **49**, 350–57.
286. Salton, M. R. J. and Schor, M. T. (1974) *Biochim. biophys. Acta* **345**, 74–82.
287. Salton, M. R. J., Schor, M. T. and Ng, M. H. (1972) *Biochim. biophys. Acta* **290**, 408–413.
288. Sargent, M. G., Gosh, B. K. and Lampen, J. O. (1968) *J. Bact.* **96**, 1329–38.
289. Săsărman, A., Surdeanu, M., Szegli, G., Horadniceanu, T., Greeanu, V. and Dumistrescu, A. (1968) *J. Bact.* **96**, 570–72.
290. Săsărman, A., Chartrand, P., Proscher, R., Desrochers, M., Tardif, D. and Lapointe, C. (1975) *J. Bact.* **124**, 1205–212.
291. Schellenberg, G. D. and Furlong, C. E. (1977) *J. biol. Chem.* **252**, 9055–64.
292. Scherer, R. and Gerhardt, P. (1971) *J. Bact.* **107**, 718–35.
293. Schlecht, S. and Schmidt, G. (1969) *Zentbl. Bakt. ParasitKde. Orig.* **212**, 505–11.
294. Schlecht, S. and Westphal, O. (1970) *Zentbl. Bakt. ParasitKde. Orig.* **213**, 356–80.
295. Schmidt, G., Schlecht, S. and Westphal, O. (1970) *Zentbl. Bakt. ParasitKde. Orig.* **212**, 88–96.
296. Schmitt, M., Rittinghaus, K., Scheurich, P., Schwulera, U. and Dose, K. (1978) *Biochim. biophys. Acta* **509**, 410–8.
297. Schnaitman, C., Smith, D. and De Salsas, M. F. (1975) *J. Virol.* **15**, 1121–30.
298. Schnebli, H. P., Vatter, A. E. and Abrams, A. (1970) *J. biol. Chem.* **245**, 1122–7.
299. Schneider, E., Schwuléra, U., Müller, H. W. and Dose, K. (1978) *FEBS Letts.* **87**, 257–60.
300. Scholes, P. and Mitchell, P. (1970) *J. Bioenergetics* **1**, 309–23.
301. Scholes, P., Mitchell, P. and Moyle, J. (1969) *Eur. J. Biochem.* **8**, 450–4.
302. Schuldiner, S. and Avron, M. (1971) *FEBS Letts.* **14**, 233–6.
303. Schuldiner, S., Rottenberg, H. and Avron, M. (1972) *Eur. J. Biochem.* **25**, 64–70.
304. Schuldiner, S., Kerwar, G. K., Weil, R. and Kaback, H. R. (1975b) *J. biol. Chem.* **250**, 1361–70.
305. Schuldiner, S., Kung, H. F., Kaback, H. R. and Weil, R. (1975a) *J. biol. Chem.* **250**, 3679–82.
305a. Schuldiner, S., Weil, R., Robertson, D. E. and Kaback, H. R. (1977) *Proc. Natn. Acad. Sci. U.S.A.* **74**, 1851–4.
306. Sedar, A. W. and Burde, R. M. (1965) *J. Cell Biol.* **27**, 53–66.
307. Serrahima-Zieger, M. and Monteil, H. (1978) *Biochim. biophys. Acta* **502**, 445–57.
308. Shipp, W. S. (1972) *Arch. Biochem. Biophys.* **150**, 459–72.

309. Short, S. A., Kaback, H. R. and Kohn, L. D. (1975) *J. biol. Chem.* **250**, 4291–6.
310. Short, S. A., Kaback, H. R., Kaczorowski, G., Fisher, J., Walsh, C. T. and Silvertein, S. C. (1974) *Proc. Natn. Acad. Sci. U.S.A.* **71**, 5032–6.
311. Silhavy, T. J., Hartig-Becken, I. and Boos, W. (1976) *J. Bact.* **126**, 951–8.
312. Simoni, R. D. and Shallenberger, M. K. (1972) *Proc. Natn. Acad. Sci. U.S.A.* **69**, 2663–7.
313. Simoni, R. D. and Postma, P. W. (1975) *A. Rev. Biochem.* **44**, 523–54.
314. Simoni, R. D. and Shandell, A. (1975) *J. biol. Chem.* **250**, 9421–7.
314a. Simoni, R. D., Smith, M. F. and Roseman, S. (1968) *Biochem. biophys. Res. Commun.* **31**, 804–11.
314b. Simoni, R. D., Nakazawa, T., Hays, J. B. and Roseman, S. (1973) *J. Biol. Chem.* **248**, 932–40.
315. Singer, S. J. (1974) *A. Rev. Biochem.* **43**, 805–833.
316. Smith, J. B. and Sternweis, P. C. (1975) *Biochem. biophys. Res. Commun.* **62**, 764–71.
317. Smith, J. B. and Sternweis, P. C. (1977) *Biochemistry* **16**, 306–311.
318. Smith, J. B., Sternweis, P. C. and Heppel, L. A. (1975) *J. supramolec. Struct.* **3**, 248–65.
319. Sone, N., Yoshida, M., Hirata, H. and Kagawa, Y. (1975) *J. biol. Chem.* **250**, 7917–23.
320. Sonntag, I., Sechwarz, H., Hirota, Y. and Henning, U. (1978) *J. Bact.* **136**, 280–85.
321. Spencer, M. E. and Guest, J. R. (1973) *J. Bact.* **114**, 563–70.
322. Spencer, M. E. and Guest, J. R. (1974) *J. Bact.* **117**, 947–53.
323. Steck, T. L., Weinstein, R. S., Strauss, J. H. and Wallach, D. F. M. (1970) *Science* **168**, 255–7.
324. Sternweis, P. C. (1978) *J. biol. Chem.* **253**, 3123–8.
325. Steven, A. C., Heggeler, B. T., Müller, R., Kistler, J. and Rosenbusch, J. P. (1977) *J. Cell Biol.* **72**, 292–301.
326. Swank, R. T. and Burris, R. H. (1969) *J. Bact.* **98**, 311–3.
327. Szmelcman, S. and Hofnung, M. (1975) *J. Bact.* **124**, 112–28.
328. Theodore, T. S. and Weinbach, E. C. (1974) *J. Bact.* **120**, 562–4.
329. Thipayathana, P. (1975) *Biochim. biophys. Acta* **408**, 47–57.
330. Thomas, T. D. and Ellar, D. J. (1973) *Biochim. biophys. Acta* **316**, 180–95.
331. Tichy, P. and Landman, O. E. (1969) *J. Bact.* **97**, 42–51.
332. Tokunaga, M., Tokunaga, H., Okajima, Y. and Nakae, T. (1979) *Eur. J. Biochem.* **95**, 441–8.
333. Tristram, H. (1978) In *Companion to Microbiology*, eds. Bull, A. T. and Meadow, P., pp. 297–370, London: Longman.
334. Tsuchiya, T. and Rosen, B. P. (1975a) *J. biol. Chem.* **250**, 8409–915.
335. Tsuchiya, T. and Rosen, B. P. (1975b) *J. biol. Chem.* **250**, 7687–92.
336. Turnock, G., Erickson, S. K., Ackrell, B. A. C. and Birch, B. (1972) *J. gen. Microbiol.* **70**, 507–515.
337. Tzagoloff, A., MacLennan, D. H., McConnell, D. B. and Green, D. E. (1967) *J. biol. Chem.* **242**, 2051–61.
338. Venables, W. A. (1972) *Mol. gen. Genet.* **114**, 223–31.
339. Villarejo, M. and Ping, C. (1978) *Biochem. biophys. Res. Commun.* **82**, 935–42.
340. Vogel, G. and Steinhart, R. (1976) *Biochemistry* **15**, 208–216.
341. Wallach, D. F. H. and Kamat, V. B. (1964) *Proc. Natn. Acad. Sci. U.S.A.* **52**, 721–8.

342. Walsh, C. T. and Kaback, H. R. (1974) *Ann. N. Y. Acad. Sci.* **235**, 519–41.
343. Weibull, C. (1953) *J. Bact.* **66**, 688–712.
344. Weiner, J. H. (1974) *J. Membrane Biol.* **15**, 1–14.
345. Weiner, J. H. and Heppel, L. A. (1971) *J. biol. Chem.* **246**, 6933–41.
346. Weltzieu, M. U. and Jesaitis, M. A. (1971) *J. exp. Med.* **133**, 534–53.
347. West, I. C. (1970) *Biochem. biophys. Res. Commun.* **41**, 655–61.
348. West, I. C. and Mitchell, P. (1972) *Bioenergetics* **3**, 445–62.
349. West, I. C. and Mitchell, P. (1973) *Biochem. J.* **132**, 587–92.
350. West, I. C. and Wilson, T. H. (1973) *Biochem. biophys. Res. Commun.* **50**, 551–8.
352. Wickner, W. (1975) *Proc. Natn. Acad. Sci. U.S.A.* **72**, 4749–53.
353. Wickner, W. (1976) *J. Bact.* **127**, 162–7.
354. Wiley, W. R. (1970) *J. Bact.* **103**, 656–62.
355. Willecke, K., Gries, E. M. and Oehr. P. (1973) *J. biol. Chem.* **248**, 807–813.
356. Willis, R. C. and Furlong, C. E. (1974) *J. biol. Chem.* **249**, 6926–9.
357. Willis, R. C. and Furlong, C. E. (1975a) *J. biol. Chem.* **250**, 2574–80.
358. Willis, R. C. and Furlong, C. E. (1975b) *J. biol. Chem.* **250**, 2581–6.
359. Wilson, D. B. (1978) *A. Rev. Biochem.* **47**, 933–65.
360. Wilson, O. H. and Holden, J. T. (1969) *J. biol. Chem.* **244**, 2743–9.
361. Wolstenholme, D. R., Vermuelen, C. A. and Venema, G. (1966) *J. Bact.* **92**, 1111–21.
362. Wulff, D. L. (1967) *J. Bact.* **93**, 1473–4.
363. Yoch, D. C. and Valentine, R. C. (1972) *Recent Adv. Microbiol.* **26**, 139–62.
363a. Yoshida, M., Okamoto, H., Sone, N., Hirata, H. and Kagawa, Y. (1977) *Proc. Natn. Acad. Sci. U.S.A.* **74**, 936–40.
364. Yoshida, M., Sone, N., Hirata, H. and Kagawa, Y. (1975a) *J. biol. Chem.* **250**, 7910–16.
365. Yoshida, M., Sone, N., Hirata, H., Kagawa, Y., Takeudi, Y. and Ohno, K. (1975b) *Biochem. biophys. Res. Commun.* **67**, 1295–1300.
366. Young, I. G. (1975) *Biochemistry* **14**, 399–406.
367. Young, I. G. and Wallace, B. J. (1976) *Proc. Aust. Biochem. Soc.* **9**, 66.
368. Zachowski, A. and Paref, A. (1974) *Biochem. biophys. Res. Commun.* **57**, 787–92.
369. Zimmerman, W. and Rosselet, A. (1977) *Antimicrobial Agents and Chemotherapy* **12**, 368–72.

5 Membranes of bacteria lacking peptidoglycan

5.1 Introduction

Bacteria lacking peptidoglycan come into four main categories: Mycoplasmas, bacterial L-forms, halobacteria and methanogenic bacteria. The last-named were only recently identified as falling in this class, when Kandler and Hippe [14] showed that the walls of *Methanosarcina barkeri* and some other species lacked muramic acid, glucosamine and D-glutamic acid. The structural component of the wall of *M. barkeri* consisted of galactosamine, uronic acids, glucose and a little galactose and thus resembled that of *Halococcus morrhuae* (see below) except that it was not sulphated. Further evidence has now indicated that many of the Methanobacteria have walls containing 'pseudomurein' (see Chapter 6) and that they may be considered as a group along with the Halobacteria, *Thermosplasma* and *Sulfolobus* to form a separate kingdom of prokaryotes designated Archaebacteria [66].

5.2 Mycoplasmas

These very small prokaryotes, some of them pathogenic, have no cell walls and are bounded only by their cytoplasmic membrane [9]. They fall into two genera, Mycosplasma which require sterols for their growth and Acholeplasma which do not (although their membranes may contain sterols if they happen to be present in the medium). Although they lack a rigid wall, the mycoplasmas are quite resistant to osmotic lysis during most of their growth cycle. In fact the mycoplasma membrane is more resistant to mechanical shock, such as sonication, than are the cytoplasmic membranes of walled bacteria, perhaps because of stabilization with cholesterol, which is missing in other prokaryotes.

Mycoplasma membranes can be isolated after osmotic lysis [40, 41], in the absence of divalent cations, which affords a very gentle method of disruption. Some strains are resistant to lysis under these conditions, but can be induced to lyse by first being suspended in 2 M glycerol before being placed in hypotonic medium. Alternatively, lysis can be procured with the aid of digitonin, which acts upon membranes that contain cholesterol. The isolated membranes contain about 50–60% protein, 30–40% lipid, and small amounts of RNA and DNA. (Table 5.1). The membranes can be seen in electron micrographs of thin sections, in which they

Table 5.1 Composition of Mycoplasma membranes (after [42]).*

Source of membrane	*Protein*	*Total lipid*	*Cholesterol*	*Carbohydrate*	*RNA*	*DNA*
Acholeplasma laidlawii B	57	32	1.3	0.5	2.0	1.0
Mycoplasma bovigenitalium	59	37	8.9	2.2	2.5	0.8
Mycoplasma mycoides var. *capri*	50	40	12.0	2.0	4.5	1.1
Mycoplasma mycoides var. *mycoides*	51	39	n.d.	3.2	5.6	0.8
Mycoplasma hominis	57	41	15.2	n.d.	0.8	0.6
Mycoplasma pneumoniae	37	58	n.d.	3.2	2.1	n.d.

*The values represent the percentage of the dry weight. n.d. Not determined.

appear as typical bilayered structures.

Although it was reported that glycoproteins had not been found on mycoplasma membranes [42], there was evidence for various carbohydrate components in particular species, probably as glycolipids. An extensive study employing agglutination by plant lectins provided evidence that many strains had specific sugar residues on their surfaces [50] (Table 5.2). In some instances, such as *M. pneumoniae* and *M. mycoides* var. *mycoides* the carbohydrates were presumably bound to lipids, since pronase treatment did not affect agglutinability. This result agreed with the known presence of di- and tri-galactosyldiglycerides and corresponding structures with glucose as well as galactose as glycolipids of the membranes of *M. pneumoniae* [55]. Others, however, showed Pronase-susceptible agglutinability (Table 5.2), which implied that the surface carbohydrates were bound to protein. This result contrasted with the supposed absence of glycoproteins from Mycoplasma membranes. Surface location of carbohydrate residues was demonstrated under the electron microscope by a cytochemical procedure involving fixation of concanavalin A, which in turn fixed a horseradish peroxidase [51].

A thermophilic Mycoplasma *Thermoplasma acidophilum*, now considered to be a member of the Archaebacteria [66], contains as part of its membrane an unusual lipopolysaccharide, consisting of mannose, glucose, glycerol and ether-linked long-chain alkyl residues. Its proposed structure is $(\alpha\text{-}1, 2\text{-Man})_8$-1,3-Glc-glyceroldiether [30]. It was shown to behave morphologically, in negative stained electron microscope preparations, very similarly to the more conventional lipopolysaccharides of Gram-negative bacilli [31].

Apart from expected phospholipids, such as phosphatidylglycerol which occurs in all mycoplasmas so far examined, *Acholeplasma laidlawii* has a phosphoglycolipid, glycerylphosphoryldiglucosyl diglyceride [54]. It has also been found that *Acholeplasma axanthum* synthesizes a sphingolipid as its main phospholipid [38]. On the other hand phosphatidylethanolamine, which is the main phospholipid in many bacteria, is rare in Mycoplasmas. Many of them, however, contain large amounts of free fatty acid in their membranes [49].

The membrane of *A. laidlawii* has altogether about 25 to 30% phosphatidylglycerol and diphosphatidyl glycerol, about 60% of glucolipids and phosphoglucolipids and less than 10% of neutral lipids, so that a very large proportion of the total contains bound carbohydrate [43].

There is some evidence that the lipid components of mycoplasma membranes, in common with those of other biological membranes, are unequally distributed between the two halves of the bilayer. Thus immunological methods have shown that the glycolipids of *M. pneumoniae* and the phosphoglucolipid of *M. mycoides* subsp. *capri* appear on the outside of the cells [44, 53]. In the latter organism, binding of polycationic ferritin to the lipid phosphate groups showed that only the outside surface could bind in this way, whereas in *M. hominis* these groups had to be unmasked by Pronase before they could be shown, but they then appeared on both sides of isolated membranes [52]. Cholesterol, on the other hand, appears to occur on both sides of the membrane, even though the fact that it is exogenous

Table 5.2 Agglutination of mycoplasma species by lectins, where ++ is strong and + is weak agglutination and where the lectin and sugar specificity is shown in brackets (from [50]).

Mycoplasma species	*Archis hypogaea* (*β-D-galactose*)	*Vicia cracca α-N-acetyl-D-* (*galactosamine*)	*Ricinus communis* (*β-D-galactose*)	*Phaseolous vulgaris* (*N-acetyl-D-galactosamine*)	*Canavalia ensiformis* (*α-D-glucose D-mannose*)
M. pneumoniae	0	0	++	0	+
M. fermentans	0	0	0	0	0
A. laidlawii	0	0	0	0	0
M. neurolyticum	0	0	0	0	++
M. mycoides var. *mycoides*	0	0	+	0	+
M. hominis	0	0	0	0	0
M. pulmonis	0	++*	++*	++*	+
M. mycoides var. *capri*	0	++*	++*	+	++*
M. gallisepticum	0	++*	++*	++*	+
M. gallinarum	0	++*	++*	++*	+

*Pronase treatment of Mycoplasmas decreased this agglutination.

necessitates its original access from the outside. Filipin binding studies showed that in *M. gallisepticum* and *A. laidlawii* cholesterol is found equally on both sides of the membrane, whereas in *M. capricolum* about two thirds of the unesterified sterol resides in the outer half [1, 8].

Just as lipids are unevenly distributed across the membrane, so are proteins. Mycoplasma membranes, in common with plasma membranes generally, probably have most of their proteins facing inwards [43]. Thus antisera to purified membranes reacted with only about a quarter of the total membrane proteins in intact *A. laidlawii* [13]. As in other systems, the proteins move across the membrane with difficulty, and their mobility within the particular layer in which they find themselves is affected by the physical state of the membrane lipids and hence by the amount of cholesterol present [46, 62].

The membranes of mycoplasma have occasioned a great deal of interest because they have the property of permitting themselves to be disrupted and then reconstituted as recognizable membranes (e.g. [45]).

In the presence of ionic detergents such as sodium dodecyl sulphate or cetyltrimethylammonium bromide the membranes can be completely dissolved, and the former reagent has been mainly used as the solvent prior to reconstitution studies. In these experiments the detergent solution is dialysed at 4 °C against a large volume of 30 mM NaCl containing 2.5 mM Tris, 0.5 mM 2-mercaptoethanol and 20 mM $MgCl_2$. After some days, centrifugation of the dialysis sac contents yields a pellet of reconstituted membranes, generally consisting of small vesicles. Thin sections of these look exactly like the original native membranes from which they were ultimately derived. The close resemblance between the composition of native membranes and those reconstituted in 20 mM Mg^{2+} can be seen from Table 5.3.

There are suggestions that the reconstituted mycoplasma membranes are built up by a multi-step assembly process, in which a lipid-rich membrane is formed first, followed by more protein depending upon the Mg^{2+} concentration [42]. In some cases the reconstituted membranes superfically resemble the native membranes, but in fact have far more of their protein on the surface.

Table 5.3 Composition of native and reconstituted *Mycoplasma laidlawii* membranes (from [43]).

Preparation	*Mg^{2+} in dialysis buffer* (mM)	*Total lipid* (% dry wt.)	*Lipid/protein* (*cpm per mg protein*)	*Density* (g/cm³)
Reconstituted membranes	5	49.0	51 800	1.140
Reconstituted membranes	20	33.2	26 800	1.170
Native membranes	–	32.0	25 400	1.172

Membranes containing lipids labeled with ³H-oleic acid were solubilized by 20 mM SDS and dialyzed in the cold for 72 h against dilute buffer containing Mg^{2+}.

5.3 Bacterial L-forms

For many years there was confusion over the possible connection between bacterial L-forms (L-phase variants) and what were known as pleuropneumonia-like organisms (PPLO), subsequently called Mycoplasmas. It is now clear that although the colonial morphologies of these two groups are alike (the so-called 'fried-egg' appearance), their only other similarity is that they lack a rigid wall containing peptidoglycan. The properties of the retaining membranes of L-forms are described below.

In 1935 Klieneberger-Nobel originally described a growth form of *Streptobacillus moniliformis* that differed from normal vegetative cells, and since then L-forms have been isolated by suitable treatment of many Gram-negative and Gram-positive bacteria. Characteristically, these wall-less forms will grow on the surface of soft agar medium supplemented with serum and they have most often been obtained by methods involving agents that interfere with cell wall integrity e.g. penicillin, lytic enzymes such as lysozyme or high glycine concentrations. The individual cells within a single culture differ greatly in size, as seen under phase contrast or by electron microscopy of ultra-thin sections (e.g. [67]), but in each case they lack a rigid cell wall and are bounded by a typical bilayer membrane resembling the cytoplasmic membrane of vegetative cells. Once induced by various methods the growing L-forms may readily revert to the normal vegetative bacteria, and often many passages in the presence of antibiotic are required before the L-form culture can be considered stable. For Gram-positive organisms such as Bacilli and Streptococci treatment with lysozyme to produce protoplasts (vegetative cells stripped of their walls), followed by plating on a suitable support medium can lead to almost one-for-one conversion to the L-form [10, 18, 21]. The ease with which stabilized L-forms will revert to vegetative cells depends on the support medium, e.g. increase in the agar concentration or transfer to the surface of a 25% gelatin medium will both favour reversion. De Castro-Costa and Landman [7] have proposed that in stable L-forms of *B. subtilis*, produced by mass conversion via protoplasts, reversion is normally prevented by a reversion inhibitory factor that has the properties of a protein. It is suggested that in fact this protein may be the autolysin of the vegetative cells, which continues to be produced by the L-forms and thus inhibits the re-establishment of peptidoglycan and hence of the whole wall with its ancillary polymers. Anything that destroys the autolysin (e.g. trypsin) or prevents its production (e.g. chloramphenicol) favours reversion. The same authors invoke for these processes a control function by lipoteichoic acid, which is known to exert an inhibitory effect on lytic enzymes that attack peptidoglycan (see Chapter 11).

The chemical composition of the membranes of the L-forms of Gram-positive bacteria resembles in many respects that of the membranes of the parent cells. Thus in *Staphylococcus aureus* there was, perhaps, some shift from protein towards lipid in going from the membranes of vegetative cells to L-forms but the overall composition of the two types was grossly similar, except for the carbohydrate content (Table 5.4). Decreased proportions of protein relative to lipids have also been found in L-forms from *Streptococcus pyogenes* the protein–lipid ratio being 1.7:1

Table 5.4 Staphylococcal protoplast and L-form membranes (from [64]).

Source of membrane	*Protein*	*Lipid*	*Carbohydrate*	*Phosphorus*	*RNA*
S. aureus 100 protoplasts	65.2	23.3	0.83	1.15	2.86
Derived L-form L_1	63.4	27.7	1.45	1.07	2.27
Derived L-form L_{20}	58.8	32.0	1.66	1.00	2.52
S. aureus H protoplasts	66.7	23.3	0.88	1.13	4.61
Derived L-form HL	59.0	30.3	2.55	0.84	3.48

*The values give the percentage of the dry weight of membrane.

compared with about 4:1 in protoplast membranes [12, 36]. The difference in carbohydrate content in *S. aureus* L-forms was related to the presence in the lipid fraction of the L-forms of amounts of glycolipid increased 2 to 3 fold at the expense of the phospholipids [64]. There was also some difference in the particular type of phospholipid, in that the major class in protoplast membranes was phosphatidylglycerol, whereas in the L-forms it was disphosphatidylglycerol. The relationship between these differences and the state of being an L-form rather than a protoplast is not known. Possibly the relatively high glycolipid contents might relate to the presence of lipoteichoic acid, but this component was not specifically looked for in the experiments with *S. aureus.*

In the L-form membranes of *Strep. pyogenes* there were some differences of detail in the relative amounts of particular unsaturated fatty acids compared with the corresponding protoplast membranes, but the overall proportion of unsaturated acids compared to saturated ones was little changed [36]. The trends in glycolipid contents seen in *S. aureus* were also observed in *Strep. pyogenes.* It seems, therefore, that such alterations in the chemical composition of membranes may well be essential to the development of growing L-forms from Gram-positive cocci, as opposed to their mere persistance as protoplasts.

In several investigations, Panos has sought for features in the L-form membrane that might account directly for their biological difference from protoplast membranes, but no factor unequivocally responsible for the difference has emerged. Thus membrane proteins differed in quantity rather than in kind [37] and a lipoprotein that might contain isoprenoid alcohols was present in ten-fold lower amount in L-forms than in parent cells [20]. Similarly both membranes were capable of synthesizing the rhamnose-rich wall-located polymer from ^{14}C-rhamnose, although only in the vegetative cells was the polymer attached to peptidoglycan components (*N*-acetylmuramic acid and *N*-acetylglucosamine). This result parallels the original observation of Chatterjee, Ward and Perkins [3] that the membranes of staphylococcal L-forms were capable of synthesizing peptidoglycan from its amino sugar nucleotide precursors, even though the L-forms themselves contained no peptidoglycan or precursors. Ward [63] has since shown that L-forms of *B. subtilis* and *B. licheniformis* also do not synthesize peptidoglycan, but accumulate either complete or truncated UDP-*N*-acetylmuramyl-pentapeptide precursors, or lack the membrane-bound translocase that converts the complete precursor to the undecaprenolypyrophosphate linked intermediate. Membrane preparations from some of the L-forms were also able to synthesize the secondary polymers glycerol teichoic acid, teichuronic acid and a polymer of *N*-acetylglucosamine.

Unstable L-forms from the Gram-negative *Proteus mirabilis*, which can be propagated indefinitely in the presence of penicillin and yet can revert in its absence, synthesize an apparently normal peptidoglycan [15, 28], which has about the same degree of cross-linking as in the vegetative cells. Stable L-forms from the same parent strain lack peptidoglycan completely, and yet they contain in their membranes enzymes such as D,D-carboxypeptidase and L,D-carboxypeptidase that have as substrates either completed peptidoglycan or its biosynthetic intermediates [29]. Thus in *P. mirabilis*, as in *S. aureus, Strep. pyogenes* and *B. licheniformis*, L-forms continue to synthesize enzymes for the production of non-existent cell walls. Martin *et al.* [29] suggested that demonstration of such enzymes could serve as a highly specific technique for the identification as bacteria of L-forms of wall-less prokaryotic organisms of unknown origin.

5.4 Archaebacteria

The Archaebacteria were first recognized as a phylogenetically distinct group of prokaryotes distinguishable from the classical bacteria on the basis of their 16S-ribosomal RNA constitution [65]. At that time the methanogenic bacteria were the only species known to belong to this group. Later the same technique was applied to *Halobacterium halobium* [26] and the conclusion arose that Halobacteria are also Archaebacteria. It has since become apparent that the membranes of Archaebacteria also have important features in common and some of these are discussed below.

5.4.1 *Extreme halophiles*

Some bacteria that can tolerate high NaCl concentration are Gram-positive and yet lack peptidoglycan (e.g. *Halococcus morrhuae*) [2, 47]. These organisms have thick walls, in which the structural component is a sulphated polysaccharide containing neutral sugars, amino sugars and uronic acids [56]. Other salt-dependent halophiles such as *Halobacterium halobium, H. cutirubum* and *H. salinarium* are Gram-negative and have no peptidoglycan or other carbohydrate polymer in its place. Thus for instance, no diaminopimelic acid was found [57]. The latter are more sensitive to osmotic damage than members of the genus *Halococcus*, which have thick walls. Both types of halophile have high intracellular salt concentration [5] and the lipids of their cell envelopes contain unique types of lipid with ether linked alkyl groups instead of the ester-bound fatty acids found normally [13]. An extensive review of the physiology of Halobacteria was written by Dundas [9].

The halophiles that lack a rigid wall (i.e. those other than the Halococci) are particularly sensitive to lowered salt concentration. Not only does osmotic damage occur, but the membrane becomes highly sensitive to physical challenge [25]. As in some non-halophiles, the surface of some halobacteria shows a regular hexagonal pattern under the electron microscope, which can persist even in the closed vesicles that form after the disruption of cells [6]. Although *H. salinarium* had in its envelope glycoproteins containing amino sugars, no peptidoglycan could be found. In addition to glycoproteins, sulphated acidic polysaccharides have also been described [19]. There is some evidence that the glycoproteins are involved in the cell morphology, since Mescher and Strominger [33] showed that low concentrations of bacitracin, which interferes with the semi-dephosphorylation of polyprenol-pyrophosphate (Chapter 9), transformed normal rod-shaped cells into spheres. This glycoprotein is highly acidic, having 33 mole % excess of acidic over basic amino acid residues, and is located on the outer surface of the membrane. If it is removed by proteolytic enzymes the cells become round but remain viable. Hence it appears that the glycoprotein is responsible for maintaining the rod-shape of the normal cells [32].

There is a close relationship between the salt concentration in the medium and the integrity of cell envelope proteins. Thus Lanyi [22] showed that as the sodium chloride concentration was gradually decreased, various proteins were selectively released from the cell membranes. Furthermore the proteins first set free were those for which retention by the envelope was most specific for the presence of sodium chloride *per se*, rather than just high salt concentration. This observation is related to the whole question of the salt-dependent proteins from extreme halophiles. Lanyi [23] proposed that high salt concentrations would give rise to new hydrophobic interactions within protein molecules and that halophilic proteins have high hydrophobicity as compared with protein from non-halophiles.

As mentioned earlier, the lipids of Halobacteria differ markedly from those of non-halophiles in having ether-linked long-chain polyisoprenoid alcohols instead of fatty acids. Kates *et al.* [17] showed that the structure of these alkyl groups resembled phytanyl (Fig. 5.1). Kates [16] suggested that the membrane lipids could help to give structural stability, since the ether-linked analogue of phosphatidylglycerophosphate binds Mg^{2+} strongly. The same substitution also rendered the phospholipid, when incorporated into model membranes, much more selective for the transport of K^+. It is also known that these membranes contain substantial amounts of

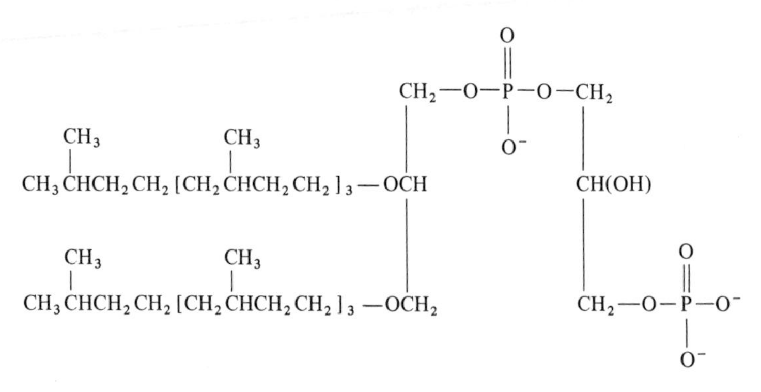

Figure 5.1 The major phosphatide of *H. cutirubrum* (after [17]).

squalenes known in other species as precursors for sterol biosynthesis. The relationship between these lipids and those of methanogenic bacteria is discussed below.

H. halobium contains a 'purple membrane' [58] which contains as its only protein component bacterio-rhodopsin [35]. This membrane component is involved in photophosphorylation, so that the low ATP concentration attained under anaerobic conditions can be restored either by aeration or by illumination and thus represents a new biological system for the conversion of light energy [34]. Reconstituted phospholipid vesicles containing the purple membrane fraction from *H. halobium* and also mitochondrial ATP-ase were able to perform photophosphorylation [39].

The purple membranes were shown by spin-label studies to be rigid structures that could not transport protons by protein mobility, but rather by a pore mechanism [4]. Indeed, detailed electron micrographic study and model building led to the proposal that each molecule of bacteriorhodopsin consists of seven α-helical sub-units arranged almost perpendicular to the plane of the membrane [11]. Light energy conversion in *H. halobium* has been reviewed in detail by Lanyi [24].

5.4.2 *Methanogenic bacteria*

Investigations of the membrane lipids of methanogenic bacteria are comparatively recent. Tornabene *et al.* [61] showed in 1978 that the chloroform-soluble lipids of the thermophilic chemolithotroph *Methanobacterium thermoautotrophicum* consisted of polar and non-polar lipids in a ratio of about 4:1. As in Halobacteria, the lipids in both categories consisted of dialkyl ethers of glycerol or its derivatives, in which the alkyl groups were isoprenoids with varying degrees of unsaturation. In the neutral lipid fraction additional components were squalene and its partially or totally hydrogenated derivatives. Thus the overall picture was extremely close to previous observations on Halobacteria.

It had been thought at one time that the presence of ether-linked lipids, rather than the ester-linked analogues of classical bacteria and eukaryotes, might be a response to existence in an extreme environment. However, further investigations among the methanogenic bacteria showed that mesophilic species had membranes made up of similar components [27, 59]. The proportions of diphytanyl diethers and dibiphytanyl tetraethers (Fig. 5.2) varied in the different species studied, thus bridging the gap between *Halobacterium*, which has phytanyl diethers and the thermoacidophilic *Sulfolobus* and *Thermoplasma*, which have biphytanyl tetraethers [59]. In this respect the coccal methanogens resembled the halophiles in lacking the more complex lipids. The presence of the isoprenoid lipids in relation to the sub-groups of methanogenic bacteria has been summarized by Tornabene *et al.* [60] (Table 5.5).

An additional peculiarity of the membranes of methanogenic bacteria concerns the phospholipids. There is evidence that in *M. thermoautotrophicum* and *M. formicicum*, but not in *M. hungatii*, some 20–30% of the total phospholipid is stable to acid hydrolysis, probably because the phosphorus is present as the phospho-

Table 5.5 Distribution of lipid and cell wall components and relationship of methanogenic bacteria (from [60]).

Dendrogram based on 16S-rRNA	Cell wall composition		Isoprenoid hydrocarbons	Major lipid components: Isopranylglycero ethers
M. strain AZ	ND	Polypeptide sacculus	$C_{30}H_{50}$	$C_{20}+C_{40}$ ethers
M. arbophilicum	Modified polypeptides		ND	ND
M. ruminantium PS			$C_{30}H_{50}$; $C_{30}H_{52}$	$C_{20}+C_{40}$ ethers
M. ruminantium M-1			$C_{30}H_{50}$; $C_{30}H_{52}$; $C_{30}H_{54}$; $C_{30}H_{56}$	$C_{20}+C_{40}$ ethers
M. strain M.o.H.	Polysaccharides in addition		$C_{30}H_{50}$; $C_{30}H_{52}$	$C_{20}+C_{40}$ ethers
M. formicicum			ND	$C_{20}+C_{40}$ ethers
M. thermoautotrophicum			$C_{30}H_{50}$; $C_{30}H_{52}$; $C_{30}H_{54}$; $C_{25}H_{50}$	$C_{20}+C_{40}$ ethers
Cariaco Isolate JR-1	Outer layer of protein subunits		ND	ND
Black Sea Isolate JR-1			ND	ND
M. hungatii	Protein sheaths		$C_{30}H_{50}$; $C_{30}H_{52}$	$C_{20}+C_{40}$ ethers
M. barkeri	Polysaccharide sacculus		$C_{25}H_{46}$; $C_{25}H_{48}$; $C_{25}H_{50}$; $C_{25}H_{52}$	C_{20} ethers

Dendrogram scale: 0.2 0.4 0.6 0.8 1.0 S_{AB}

ND: not determined

Dibiphytanyl diglycerol tetraether

Diphytanyl glycerol diether

Squalene

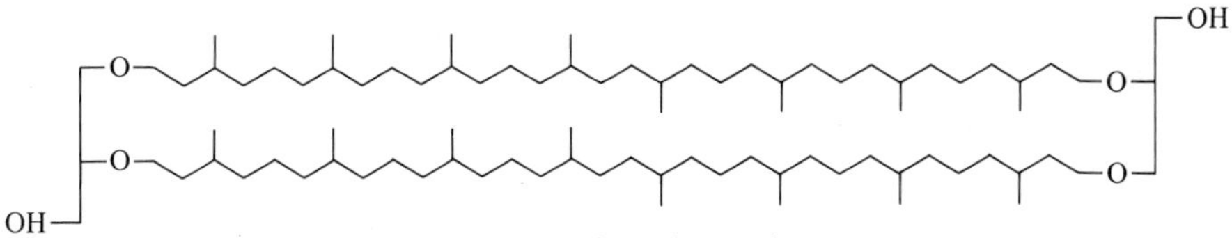

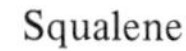

Figure 5.2 Structures of neutral lipids found in Archaebacteria.

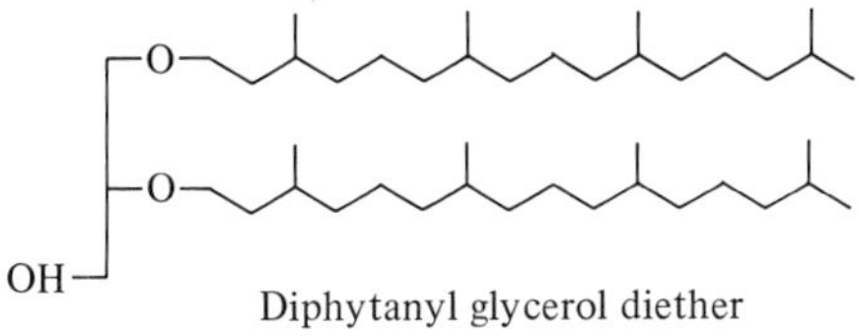

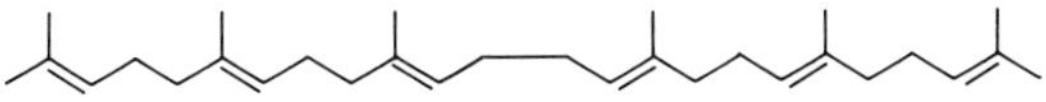

nate (i.e. with a direct C–P bond) rather than as a phosphate ester [27]. Of this phosphonolipid 70% occurs in the tetraether and 30% in the diether fraction.

A role in lipid stabilization has been suggested for the biphytanyl tetraethers, by which it is proposed that these compounds could span the complete bilayer of the membrane and have one hydrophilic moiety on each face [48].

It seems evident from various sources of evidence involving 16S-rRNA, cell wall composition and membrane lipids that the Archaebacteria form a distinct channel of evolution of great antiquity. No doubt their structure and physiology will hold more surprises in the near future.

References

1. Bittman, R., and Rottem, S. (1976) *Biochem. biophys. Res. Commun.* **71**, 318–24.
2. Brown, A. D. and Cho, K. Y. (1970) *J. Gen. Microbiol.* **62**, 267–70.
3. Chatterjee, A. N., Ward, J. B. and Perkins, H. R. (1967) *Nature, Lond.* **214**, 1311–14.
4. Chignell, C. F. and Chignell, D. A. (1975) *Biochem. biophys. Res. Commun.* **62**, 136–43.
5. Christian, J. H. B. and Waltho, J. A. (1962) *Biochim. biophys. Acta* **65**, 506–8.
6. D'Aoust, J. Y. and Kushner, D. J. (1972) *Canad.J.Microbiol.* **18**, 1767–8.
7. De Castro-Costa, M. R. and Landman, O. E. (1977) *J.Bact.* **129**, 678–89.
8. DeKruijff, B., Gerritsen, W. J., Oerlemans, A., van Dijck, P. W. M., Demel, R. A. and van Deenen, L. L. M. (1974) *Biochim. biophys. Acta* **339**, 44–56.
9. Dundas, I. E. D. (1977) *Adv. Microb. Physiol.* **15**, 85–120.

10. Gooder, H. (1968) In *Microbial protoplasts, Spheroplasts and L-forms*, ed. Guze, L. B., pp. 40–55, Baltimore, Md.: Williams and Wilkins.
11. Henderson, R. and Unwin, P. N. T. (1975) *Nature, Lond.* **257**, 28–32.
12. James, A. M., Hill, M. J. and Maxted, W. R. (1965) *Antonie van Leeuwenhoek*, **31**, 423–32.
13. Johansson, K. E. and Hjerten, S. (1974) *J.Mol.Biol.* **86**, 341–8.
14. Kandler, O. and Hippe, H. (1977) *Arch.Microbiol.* **113**, 57–60.
15. Kandler, O., Hund, A., and Zehender, C. (1958) *Nature, Lond.* **181**, 572–3.
16. Kates, M. (1972). *Ether lipids, chemistry and biology* ed. Snyder, F. pp. 351–98, New York: Academic Press.
17. Kates, M., Yengoyan, L. S. and Sastry, P. S. (1965) *Biochim. biophys. Acta* **98**, 252–68.
18. King, J. R. and Gooder, H. (1970) *J. Bact.* **103**, 686–91.
19. Koncewicz, M. (1972) *Biochem. J.* **130**, 40P–41P.
20. Lacave, C. and Panos, C. (1973) *Biochim. biophys. Acta.* **307**, 118–32.
21. Landman, O. E. and Halle, S. (1963) *J. Mol. Biol.* **7**, 721–38.
22. Lanyi, J. K. (1971) *J. biol.Chem.* **246**, 4552–9.
23. Lanyi, J. K. (1974) *Bact. Rev.* **38**, 272–90.
24. Lanyi, J. K. (1978) *Microbiol. Rev.* **42**, 682–706.
25. Larsen, H. (1973) *Antonie van Leeuwenhoek*, **39**, 383–96.
26. Magrum, L. J., Luehrsen, K. R. and Woese, C. R. (1978) *J.Mol.Evol.* **11**, 1–8.
27. Makula, R. A. and Singer, M. E. (1978) *Biochem.biophys.Res. Commun.* **82**, 716–22.
28. Martin, H. H. (1964) *J.Gen.Microbiol.* **36**, 441–50.
29. Martin, H. H., Maskos, C. and Burger, R. (1975) *Eur. J.Biochem.* **55**, 465–73.
30. Mayberry-Carson, K. J., Langworthy, T. A., Mayberry, W. R. and Smith, P. F. (1974) *Biochim,biophys.Acta* **360**, 217–29.
31. Mayberry-Carson, K. J., Roth, I. L. and Smith, P. F. (1975) *J.Bact.* **121**, 700–3.
32. Mescher, M. F. and Strominger, J. L. (1976) *Proc. Natn.Acad.Sci. U.S.A.* **73**, 2687–91.
33. Mescher, M. F. and Strominger, J. L. (1975) *J.Gen.Microbiol.* **89**, 375–8.
34. Oesterhelt, D. (1975) *CIBA Foundation Symposium on Energy Transformations and Biological Systems* pp. 147–165, Amsterdam: Elsevier.
35. Oesterhelt, D. and Stoeckenius, W. (1971) *Nature, New Biol.* **233**, 149–52.
36. Panos, C. (1968) *Microbial Protoplasts, Spheroplasts and L-forms*, ed. Guze, L. B., pp. 154–162. Baltimore, Md.: Williams and Wilkins.
37. Panos, C., Fagan, G. and Zarkadas, C. G. (1972) *J.Bact.* **112**, 285–90.
38. Plackett, P., Smith, P. F. and Mayberry, W. R. (1970) *J.Bact.* **104**, 798–807.
39. Racker, E. and Stoeckenius, W. (1974) *J.biol.Chem.* **249**, 662–3.
40. Razin, S. (1963) *J.gen.Microbiol.* **33**, 471–5.
41. Razin, S. (1964) *J.gen.Microbiol.* **36**, 451–9.
42. Razin, S. (1973) *Adv.microb.Physiol.* **10**, 1–80.
43. Razin, S. (1978) *Microbiol.Rev.* **42**, 414–70.
44. Razin, S., Prescott, B., and Chanock, R. M., (1970) *Proc.Natn.Acad.Sci.U.S.A.* **67**, 590–7.
45. Razin, S. and Rottem, S. (1974) *Meth.Enzymol.* **32B**, 459–68.
46. Razin, S., Wormser, M., and Gershfeld, N. C. (1974) *Biochim. biophys. Acta.* **352**, 385–96.
47. Reistad, R. (1972) *Arch.Microbiol.* **82**, 24–30.
48. Rohmer, M., Bouvier, P. and Ourisson, G. (1979) *Proc. Natn. Acad. Sci. U.S.A.* **76**, 847–51.
49. Romano, N., Smith, P. F., and Mayberry, W. R. (1972) *J.Bact.* **109**, 565–9.

50. Schiefer, H. G., Gerhardt, U., Brunner, H. and Krüpe, M. (1974) *J.Bact.* **120**, 81–8.
51. Schiefer, H. G., Krauss, H., Brunner, H. and Gerhardt, U. (1975) *J.Bact.* **124**, 1598–600.
52. Schiefer, H. G., Krauss, H., Brunner, H. and Gerhardt, U. (1976) *J. Bacteriol.* **127**, 461–8.
53. Schiefer, H. G., Gerhardt, U. and Brunner, H. (1977) *Zentr. Bakt. Parasit. Infekt. Hyg. Abt. I Orig.* A **239**, 262–9.
54. Shaw, N., Smith, P. F., and Verheij, H. M. (1970) *Biochem.J.* **120**, 439–41.
55. Smith, P. F. (1973) *J.Infect.Dis.* **172**, Suppl. 8–12.
56. Steber, J. and Schleifer, K. H. (1975) *Arch.Microbiol.* **105**, 173–7.
57. Steensland, H. and Larsen, H. (1969) *J.gen.Microbiol.* **55**, 325–36.
58. Stoeckenius, W. and Kunau, W. H. (1968) *J.cell.Biol.* **38**, 337–57.
59. Tornabene, T. G. and Langworthy, T. A. (1978) *Science* **203**, 51–3.
60. Tornabene, T. G., Langworthy, T. A., Holzer, G. and Oro, J. (1979) *J.mol.Evol.* **13**, 73–83.
61. Tornabene, T. G., Wolfe, R. S., Balch, W. E., Holzer, G., Fox, G. E., and Oro, J. (1978) *J.mol.Evol.* **11**, 259–66.
62. Wallace, B. A., Richards, F. M., and Engelman, D. M. (1976) *J.mol.Biol.* **107**, 255–69.
63. Ward, J. B. (1975) *J.Bact.* **124**, 668–78.
64. Ward, J. B. and Perkins, H. R. (1968) *Biochem.J.* **106**, 391–400.
65. Woese, C. R. and Fox, G. E. (1977) *Proc. Natn. Acad. Sci. U.S.A.* **74**, 5088–90.
66. Woese, C. R., Magrum, L. J. and Fox, G. E. (1978) *J.mol.Evol.* **11**, 245–52.
67. Wyrick, P. B. and Rogers, H. J. (1973) *J.Bact.* **116**, 456–65.

6
Structure of peptidoglycan

6.1 Introduction

The fundamental polymer that is a common component of the cell walls of Gram-positive and Gram-negative bacteria, Rickettsiae and blue-green bacteria is called peptidoglycan (formerly mucopeptide or murein). As its name implies, it consists of glycan chains with peptide substituents, and in all examples that have been studied the peptide subunits are cross-linked so that the overall structure is a network that surrounds the cell. This network seems responsible for the integrity of the shape of Gram-positive bacteria, and at least partially of Gram-negative bacteria as well. Certainly when the peptidoglycan is degraded, as for instance by lysozyme, the bacterium tends to lose its characteristic shape and to form a spherical body known as a spheroplast, which usually needs to be maintained in a hypertonic medium if it is not to burst because of the high osmotic pressure within it and the lack of external support. The chemical composition of peptidoglycan has been established over the period since the early 1950s, when Salton [40] first showed that the cell walls prepared from Gram-positive organisms were of a comparatively simple amino acid composition, although both he and Weidel [56] found that the walls of Gram-negative species were more complex.

The glycan part of the molecule consists of alternating *N*-acylated residues of glucosamine and its 3-*O*-D-lactyl ether derivative, muramic acid, which was originally discovered by Strange and Dark [43]. In the large majority of peptidoglycans the amino groups are in fact acetylated but in some organisms such as Mycobacteria [1, 3], *Nocardia kirovani* [23, 49] and *Micromonospora* [51] muramic acid is present as its *N*-glycolyl derivative (CH_3CO– is replaced by $HOCH_2CO$–). In *Bacillus cereus* many of the glucosamine residues are not acetylated but have free amino groups [2]. The muramic acid is always of the D-gluco configuration [57, 58]. Estimates of the length of these glycan chains have been somewhat contradictory and there seems little doubt that many of the wall preparations used had been exposed to the risk of possible glycosidase action. In several instances a large number of reducing groups attributable to *N*-acetylglucosamine were found. This result is at variance with the known biosynthetic mechanism for peptidoglycan (see Chapter 8), which would lead to terminal reducing groups (or potential reducing groups) of muramic acid rather than glucosamine. Ward [53] showed that in a non-lytic mutant of *Bacillus licheniformis* the average glycan chain length in peptidoglycan, as originally biosynthesized, was about 140 disaccharide units, compared with 180 in *Staphylococcus aureus* Copenhagen and 100 in *Bacillus subtilis* 168. The proportion of additional reducing groups of glucosamine varied with the species, being about one per chain for the Bacilli and twenty for *S. aureus*. The latter result agreed well with the overall degree of polymerization observed for *S. aureus* [48].

A chain length of the glycan as biosynthesized, say 150 disaccharide units, would represent an available length of about 150 nm [39]. Thus a bacterium of 1.5–3 μm in length and 0.6 μm in diameter would require 10–20 extended glycan chains to encompass its length, or about 13 chains to enclose its girth. There is now some evidence to show that the glycan chains lie parallel to the cell surface [9, 32] (see below). The relation to the other structures present is discussed below.

Apart from the glycan, the remainder of the peptidoglycan consists of the peptide chains that are joined to the carboxyl groups of muramic acid residues, a proportion of which are commonly cross-linked one to another, so that in this way the glycan chains are also joined. The peptide chains consist of two parts, one obligatory, consisting of the primary chain linked to muramic acid during the build up of the precursor nucleotide (Fig. 6.1), and the other optional, representing amino

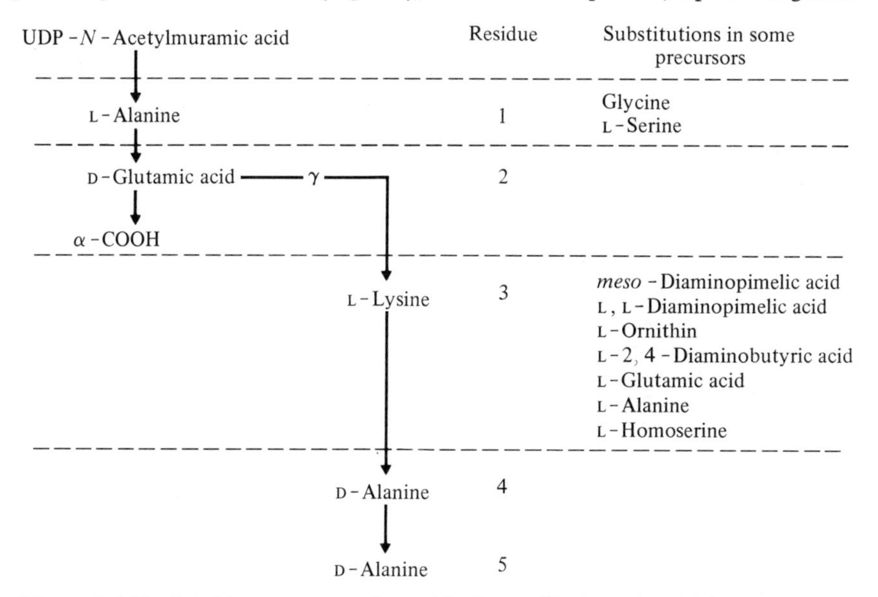

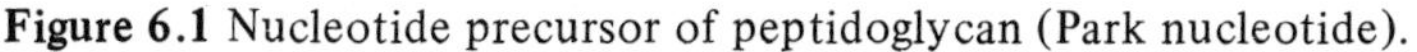

Figure 6.1 Nucleotide precursor of peptidoglycan (Park nucleotide).

acids added later, some of which may ultimately form the bridge between cross-linked chains. As Fig. 6.1 shows, the configurations of the amino acids in the primary chain extending from the muramic acid carboxyl group are L (or glycine), D, L, D, D and the linkages are α-bonds except for the one between the second and third residues, where the bond is formed by the γ-carboxyl group of D-glutamic acid.

Just as variations exist in the amino acids found in various species to make up the primary peptide chain so, many varieties of cross-linkage exist, ranging from those like the peptidoglycans of Gram-negative bacteria, in which there is no bridging amino acid at all, to those in many Gram-positive organisms where a peptide of up to three or four different amino acids may form the bridge.

As discussed in Chapter 8.6 cross-linkages are formed by a transpeptidation reaction, in which the terminal D-alanine of the primary chain is replaced by an amino group belonging directly or indirectly to another primary chain. Very often, primary peptide chains that do not become involved in cross-linkages lose either one or both of their terminal D-alanine residues by the action of carboxypeptidases, so that the final peptidoglycan rarely contains on average more than one D-alanine per D-glutamic acid residue, and frequently has less.

So far there have been two major attempts to classify the peptidoglycans according to their chemical structures, the first by Ghuysen [20] and the second by Schleifer and Kandler [41]. Both systems regard the type of cross-linking as being of paramount importance, but the latter is more complete. It is not, perhaps, surprising that variations in peptidoglycan composition and structure, first extensively studied by Cummins and Harris [11], have been shown to have taxonomic implications and for that reason a sound basis for the classification of peptidoglycan structure is a prerequisite for its use in the classification of bacteria.

The peptidoglycan classification of Schleifer and Kandler is essentially a tridigital system, in which the first digit, a Roman capital letter, represents the major class of cross-linking. Nearly all such links involve the carboxyl group of D-alanine residue 4 in one primary chain, and this may be linked either directly or indirectly to a second amino group on residue 3 of another chain (e.g. ε-amino group of L-lysine) (Group A) or indirectly to the α-carboxyl group of the D-glutamic acid residue 2 of another chain (Group B).

The second digit, a number, refers to the type of bridge, or lack of it, involved in forming the cross-links. The third digit, a Greek letter, indicates the amino acid formed at position 3 in the primary chain. As set out by Schleifer and Kandler, the meaning of the second digit is not absolute, but depends upon a knowledge of the first (A or B) and similarly the meaning of the third digit is dependent upon that of the first two. Ultimately it might be preferable to have a system in which the meaning of each digit was uniquely defined. A summary of the classification system of Schleifer and Kandler is set out in Table 6.1 and examples are given (Fig. 6.2).

It is apparent from this that the basic biological problem of linking one glycan strand to another by means of peptide cross-linking has been solved in many different ways in different species. Schleifer and Kandler [41] gave attention to the value of peptidoglycan structure as a taxonomic marker, and concluded that it met the following four criteria:

(1) Widespread distribution. Peptidoglycan is a characteristic constituent of almost all prokaryotic cells. Furthermore, higher organisms such as fungi show an even more simplified wall structure, in which the muramic acid has lost its 3-*O*-lactyl ether group (and hence its peptides) so that the resulting glycan becomes converted to chitin, and in higher plants even the 2-acetamido groups of chitin are lost, so that cellulose results.

Table 6.1 Tridigital Peptidoglycan classification system of Schleifer and Kandler [41]

Anchorage point of the cross-linkage	*Type of intervening bridge*	*Amino acid at position 3*	*Examples*
A Cross-linkage between positions 3 and 4 of two peptide sub-units	1 None	α L-lysine	*Gaffkya homari*
		β L-ornithine	*Spirochaeta stenostrepta*
		γ *meso*-diaminopimelic acid	*E. coli, B. megaterium, C. diphtheriae*
	2 Polymerized sub-units	α L-lysine	*Micrococcus luteus, Sarcina flava*
	3 Monocarboxylic L-amino acid or glycine or oligo-peptides thereof	α L-lysine	*Staph. aureus, Leuconostoc mesenteroides*
		β L-ornithine	*Micrococcus radiodurans*
		γ L,L-diaminopimelic acid	*Propionibacterium petersonii, Streptomyces albus*
	4 Contains a dicarboxylic amino acid	α L-lysine	*Streptococcus faecium, Sporosarcina ureae*
		β L-ornithine	*Lactobacillus cellobiosus*
		γ *meso*-diaminopimelic acid	*Arthrobacter duodecadis*
		δ L-diaminobutyric acid	*Arthrobacter* sp. Ar 22
B Cross-linkage between positions 2 and 4 of two peptide sub-units	1 Contains an L-amino acid	α L-lysine	*Microbacterium lacticum*
		β L-homoserine	*Brevibacterium imperiale*
		γ L-glutamic acid	*Arthrobacter* J39
		δ L-alanine	*Erysipelothrix rhusiopathiae*
	2 Contains a D-diamino acid	α L-ornithine	*Butyribacterium rettgeri*
		β L-homoserine	*Corynebacterium poinsettiae*
		γ L-diaminobutyric acid	*Corynebacterium insidiosum*

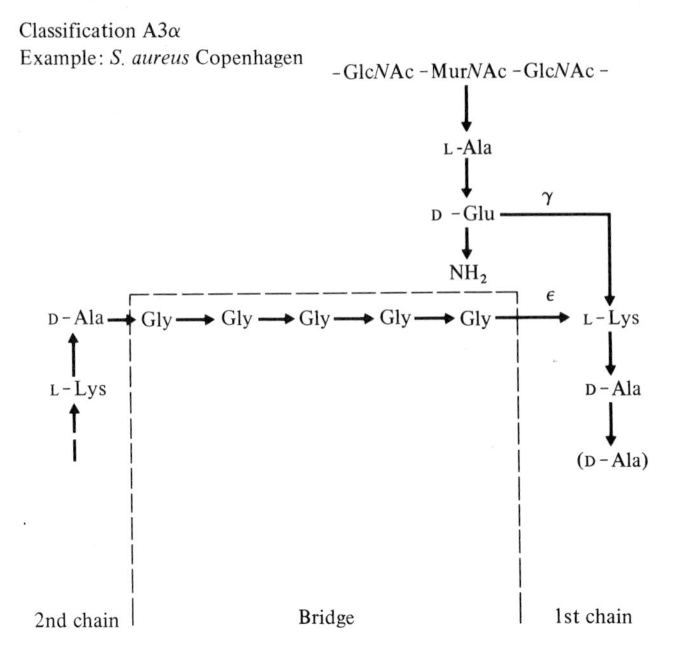

Figure 6.2 Examples of peptidoglycan structure classified according to Schleifer and Kandler [41].

(2) Multiple gene involvement. The formation of peptidoglycan structure clearly requires the action of many genes, since at least 20 different enzymes are involved in its biosynthesis. Furthermore, the diverse peptidoglycan types differ from each other by more than one gene, so that given peptidoglycan structures tend to be stable, and are unlikely to have arisen by convergence.

(3) Possibility of change without loss of viability. Although peptidoglycan structure is stable during periods of investigation, nevertheless in Gram-positive species numerous solutions to the basic structural problem have been achieved, all of which are compatible with survival.

(4) Changes should be recognizable as tending to be either more, or less, highly-developed; in other words, an evolutionary direction should be deducible. Although this is naturally the most debatable criterion, Schleifer and Kandler proposed a possible evolutionary sequence in which the multilayered peptidoglycans of Gram-positive bacteria are taken to be the most primitive, and to have the greatest variation of cross-links, whereas the most developed are the monolayered peptidoglycans of Gram-negative bacteria and Spirochaetes, in which bridging amino acids have disappeared from the cross-links (Fig. 6.3).

The above outline of cross-linking structures in bacterial peptidoglycans does not deal with the problem of the extent to which these links are made in the walls of living cells. The structural examples given in Fig. 6.2 in each case represent only a single link from one glycan strand to another (or, theoretically at least, to a point

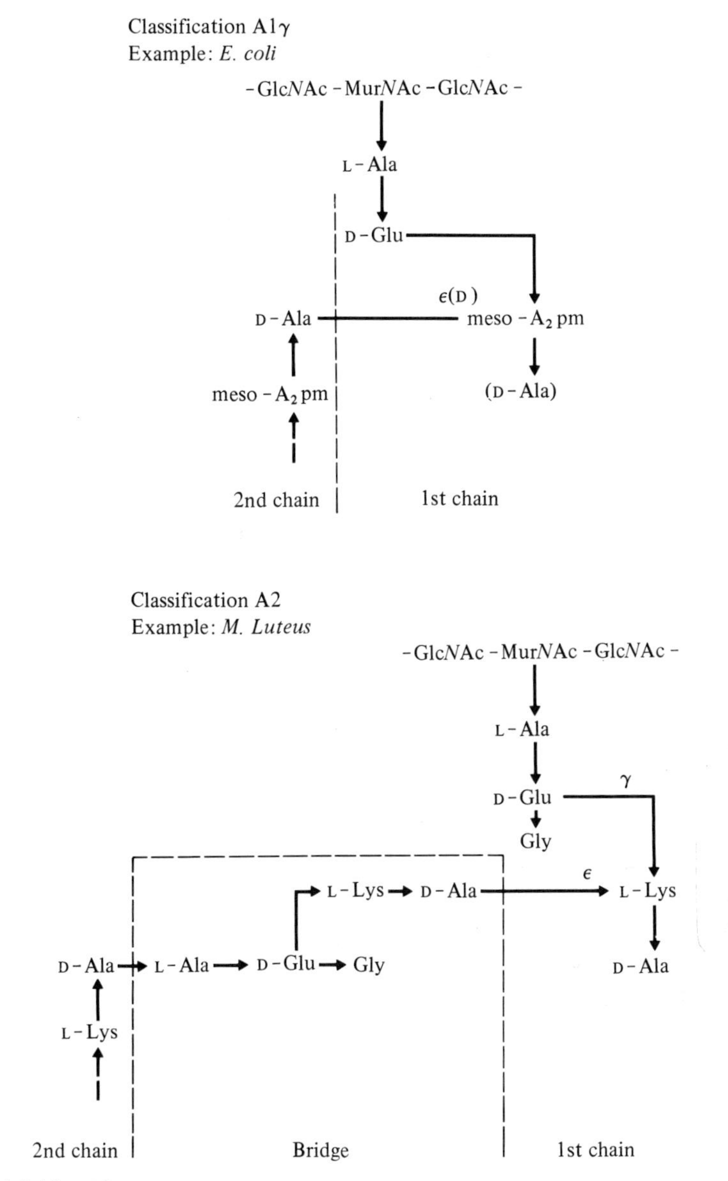

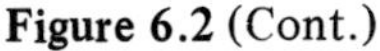

Figure 6.2 (Cont.)

further along the same strand since chemically there is no difference). In fact since all the links are made between identical units, each with a linkage initiation point (the carboxyl group of the D-alanine at residue 4) and a linkage closure point (either an amino group on residue 3, as in Class A, or the α-carboxyl group on the D-glutamic acid at residue 2, as in Class B) it is formally possible to make a continuous chain of cross-linkages (Fig. 6.4). The extent to which such multiple cross-linkages are formed

Classification A3β
Example: *Bifidobacterium longum*

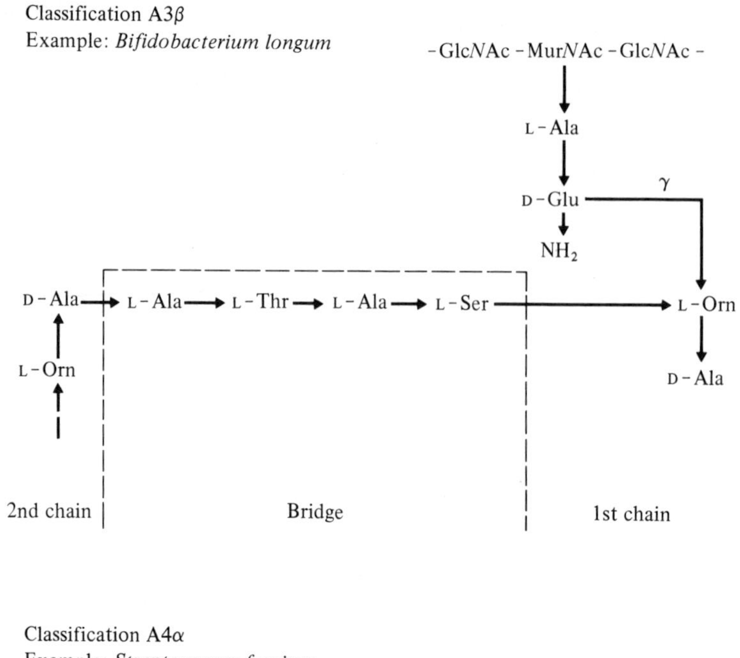

Classification A4α
Example: *Streptococcus faecium*

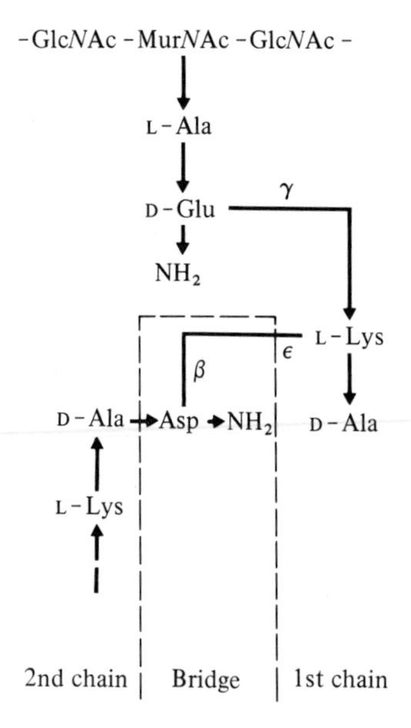

Figure 6.2 (Cont.)

has been examined, either directly or indirectly, in several species. An indirect expression of the degree of cross-linking can be made in the form of the index used by Fordham and Gilvarg [16]. These authors looked at the peptidoglycan of *B. megaterium* and measured the proportion of the total diaminopimelic acid that could be

Classification B1α
Example: *Microbacterium lacticum*
(in some conditions of growth, the
D-glutamic acid becomes converted
to threo-3-hydroxy glutamic acid)

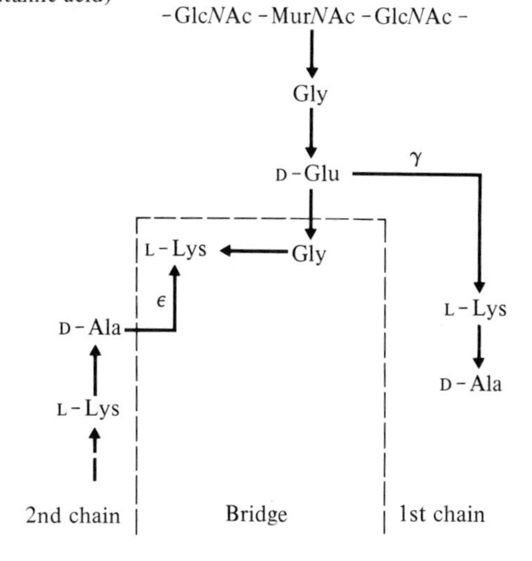

Classification B2β
Example: *Corynebacterium poinsettiae*

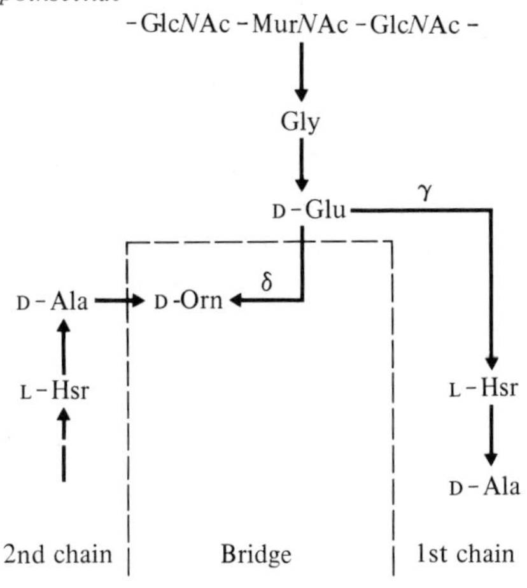

Figure 6.2 (Cont.)

converted to mono-hydroxymonoaminopimelic acid by the action of nitrous acid. Any diaminopimelic acid that reacted must have had one amino group free (in fact a D-centre) and therefore could not have been involved in forming a cross-link. The cross-linking index was defined as the amount of diaminopimelic acid *not* attacked by nitrous acid divided by the total amount originally present. Thus in Fig. 6.4 a simple dimer which was not further cross-linked would have one free amino group

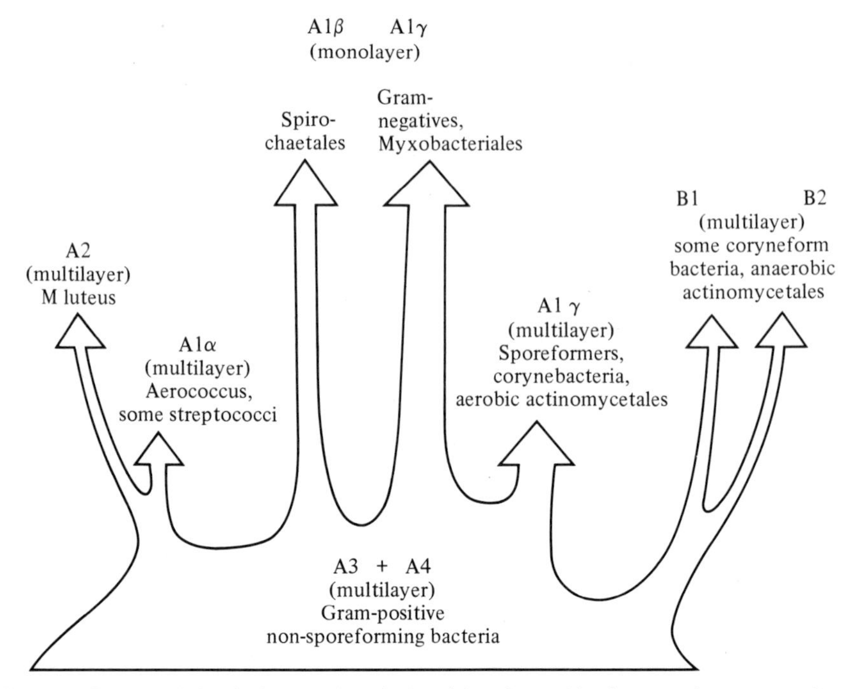

Figure 6.3 A possible phylogenetic relationship of peptidoglycan subgroups and variations (from [41]).

for two diaminopimelic acid residues and a peptidoglycan constructed entirely of of such units, whereby each side chain is directly linked to only one other, would give a degree of cross-linking of 50%. Values of less than this imply that some side-chains are not cross-linked at all, while higher values mean that at least some oligomers greater than dimers are present. The validity of this type of structural analysis depends on the assumption that none of the potentially cross-linking amino groups (e.g. in Bacilli, Gram-negative bacteria or those at the D-centre of *meso*-diaminopimelic acid) is blocked in any other way. In some species such as *Corynebacterium aquaticum* amino groups of L-diaminobutyric acid at residue 3, which would appear apt for cross-linking, are not so employed and in others such as *C. insidiosum* and *C. michiganense* they are also blocked by acetyl groups [14, 37].

The results of investigations of the degree of cross-linking in various species are summarized in Table 6.2. The cross-linking index varies from about 20% in *E. coli* growing exponentially in rich medium to over 93% in *S. aureus.* The former low value means that only 40% of the total primary side-chains are found in cross-linked dimers, whereas in *S. aureus* almost all the side-chains participate in cross-linked polymers of considerable size, containing on average about 15 repeating units. The latter result led to the proposal that the overall structure of the peptidoglycan of *S. aureus* consisted of parallel chains of glycan totally cross-linked at right-angles by parallel peptide chains [22].

Further complications in methods of cross-linking occur in certain peptidoglycans

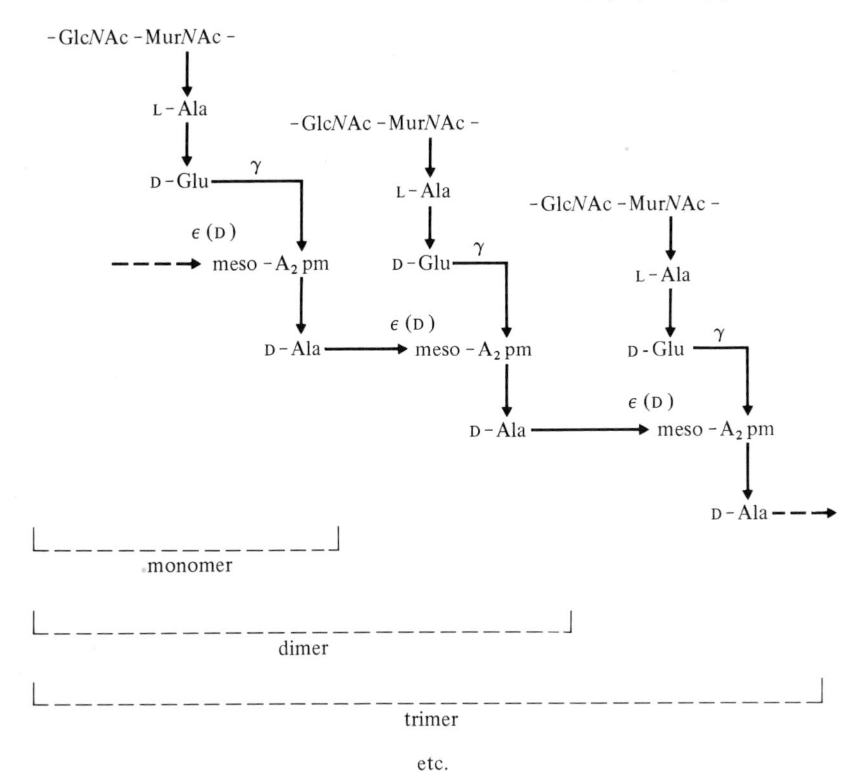

Figure 6.4 The possibility of multiple cross-linking in peptidoglycan. (Type A1 is used as an example.)

containing diaminopimelic acid. It is known that, in addition to a majority of the meso-isomers, the wall of *B. megaterium* contains L,L- and D,D-diaminopimelic acid [6, 12]. Since position 3 of the primary peptide chain has always been found to be occupied by an L-centre of asymmetry, it is easy to account for a proportion of the L,L- isomer. The D,D- isomer, however, presents a more difficult problem and it seems likely that in fact it participates in a different type of cross-linking unit, in which two primary chains are each bound by their D-alanine residue at position 4 to one end of a single molecule of D,D-diaminopimelic acid [50, 59] (Fig. 6.5). The D,D-isomer has also been observed in *Micromonospora* [25] and in *Bacillus cereus* [12]. Another type of cross-linking that would not be expected from the simple mechanism by which the D-alanine at position 5 of one primary chain is removed and replaced by an amino group belonging to another chain has been found in Mycobacteria [60]. Both *M. smegmatis* and *M. tuberculosis* BCG were found to contain sequences in which two or even three molecules of *meso*-diaminopimelic acid were linked together. These sequences represented about one-third of the total cross-links in *M. smegmatis*, the others being of the normal D-alanyl-D-*meso*-diaminopimelic acid type, and the structure proposed for them is shown in Fig. 6.6.

Table 6.2 Cross-linking in peptidoglycans*

Species	Type	Percentage of total primary chains found in: Monomer	Dimer	Trimer	Tetramer	Pentamer	Hexamer	Heptamer	Polymer	Cross-linking index (%)
E. coli	A1γ									33
E. coli	A1γ	54.5	45.5							22.8
		45.4	54.8							27.4
		59.3	40.6							20.4
		38.4	61.6							30.8
		40.3	59.5							29.8
P. vulgaris	A1γ	35	40	16						33.7
S. aureus	A3α								majority	93.5
M. tuberculosis H37 Rv	A1γ	50% of total amino groups is approx =								81
P. mirabilis	A1γ									33
C. diphtheriae	A1γ									67
M. smegmatis	A1γ	1.4	5.6	51			42			73.4
M. tuberculosis BCG AL1 enzyme (soluble peptides rep. 60% of whole peptidoglycan)	A1γ	15	27	32	26					59
B. megaterium	A1γ									
	Exp.									50
	Stat.									56
B. megaterium KM	A1γ									
Part with	D,D-Dap		35 }							} 30.6
Part without	D,D-Dap	28	36.7 }							
Streptomyces albus G		5.6	15.7	13	others of average = trimer 65.7 →					59
Cl. perfringens										50
Bacillus sphaericus	A4α	14.5	31.7	20.7	oligomers 12.8 →				20.4	54
L. casei	A4α	14.4	18.4	20.0	13.6	6.0	3.6	2.0		56.5
M. lysodeikticus†	A2	6	4.8	17.9			71.4			20.2
Bacillus cereus	A1γ	mean result from cell-walls of 4 strains →								61
B. subtilis	A1γ	20.8	72	7.2						41
Lactobacillus acidophilus (exponential phase)	A4α	10	37	30	remaining 23% unattributable →					50
Aerococcus viridans 201	A1α	34	50	remaining 16% unattributable →						30
Gaffkya homari	A1α	34	40	remaining 26% unattributable →						27

*The cross-linking index [16] represents the R_3 residues with bound amino group divided by the total R_3 residues.
†With type A2 peptidoglycan (primary side-chains detached to form multiple bridging peptides) the oligomers represent these links and not a multiplicity of glycan chains joined together.

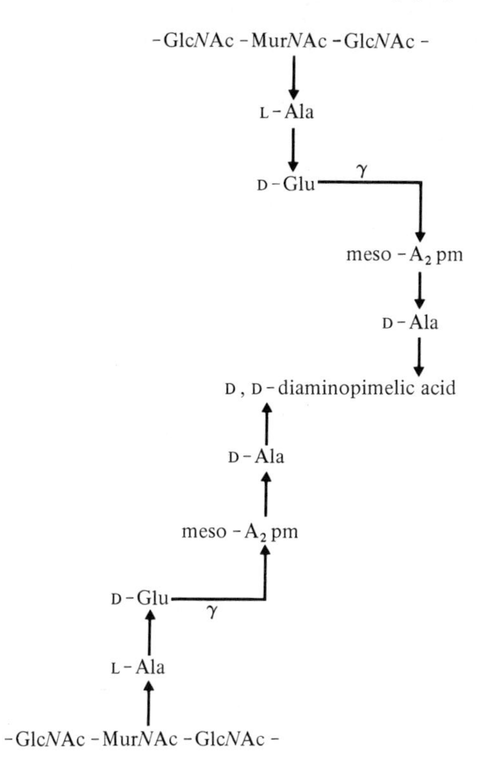

Figure 6.5 The proposed structure of the abnormal cross-linkage of peptidoglycan in *B. megaterium* [50, 59].

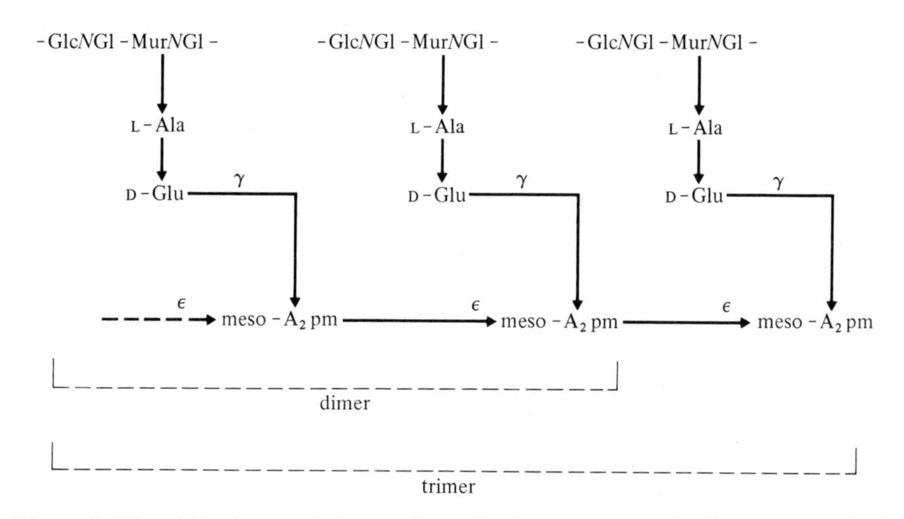

Figure 6.6 Specialized cross-linkages of peptidoglycan reported in Mycobacteria [60].

6.2 Modification of the basic peptidoglycan structure

The first type of modification involves substitution of the basic peptidoglycan network. The most common of these is amidation of certain carboxyl groups, namely the α-carboxyl groups of the D-glutamic acid at residue 2, the D-centre of *meso*-diaminopimelic acid and the D-aspartic acid that is present as a bridging amino acid in some species e.g. *Streptococcus faecium* (Fig. 6.2, structure A4α). Examples of some species in which peptidoglycan amidation has been demonstrated are shown in Table 6.3. These substitutions will obviously make the peptidoglycan less acidic,

Table 6.3 Amidation of peptidoglycans

			Bridging amino acids	
Species	D-*glutamyl* α-$CONH_2$	*meso-diaminopimelic acid*-$CONH_2$	D-*aspartyl* α-$CONH_2$	D-*glutamyl* α-$CONH_2$
S. aureus	+			
C. diphtheriae	+	+		
C. perfringeus	+			
Streptomyces albus	+			
E. coli	None			
B. megaterium	None			
B. subtilis		+		
Streptococcus faecium			+	
L. plantarum	+	+		
Cellulomonas biazotea				+

although the terminal carboxyl groups belonging to each primary peptide chain have never been reported to be blocked by amide groups. So far no functional or biochemical consequences have been ascribed to the presence of peptidoglycan amide groups, except for the observation that transpeptidation of the type involved in cross-link formation, when performed *in vitro* by an enzyme from *Actinomyces* R39, is highly susceptible to the concentration of a peptidoglycan peptide in which the α-carboxyl group of D-glutamic acid is amidated [21]. The peptidoglycan of *S. aureus* is exceptional in that, apart from the small proportion of residual uncross-linked primary pentapeptide side-chains, it has no carboxyl groups, since its D-glutamic acid α-carboxyl groups are amidated and most of its original terminal carboxyl groups are cross-linked to L-lysine ϵ-amino groups *via* pentaglycine bridges (see Fig. 6.2).

Staphylococcus aureus also provides an example of the second type of peptidoglycan modification, the presence of *O*-acetyl substituents on the *N*-acetylmuramic acid residues. In this organism about 50% of the muramic acid residues are present as the 4-*N*,6-*O*-diacetylderivative [45, 47], which also occurs in *Proteus vulgaris, Neisseria perflava, Movaxella glucidolytica* and *Pseudomonas alcaligenes* [15, 34a]. *O*-acetylation of muramic acid residues reaches nearly 70% in *L. acidophilus* [10]. This substitution has the effect of making the peptidoglycan resistant to the muramidase egg-white lysozyme [7]. The opposite type of modification, loss of acetyl groups, also occurs. In the peptidoglycan of some strains of *B. cereus* the

N-acetylglucosamine is deacetylated by as much as 60–70% [2]. The presence of excess free amino groups also inhibits the action of egg-white lysozyme [36].

Another alteration of the simple cross-linked peptidoglycan structure also occurs as a characteristic feature of the class A2 of Schleifer and Kandler [41], typified by *Micrococcus luteus* (*lysodeikticus*). Here a large proportion of the peptide side-chains become detached from the muramic acid residues and form pentapeptide cross-bridges (tetrapeptides L-Ala-D-isoGlu-L-Lys-D-Ala with a glycine residue added at the α-carboxyl group of the glutamic acid) between a D-alanine residue at position 4 of one primary chain and the ε-amino group of the lysine residue of another. These bridges often contain several pentapeptides joined end-to-end, and as many as half the muramic acid residues lose their side-chains [22] (Fig. 6.2 and Table 6.2).

A further modification of peptidoglycan occurs in bacterial spores. Warth and Strominger [54, 55] demonstrated that the spores of *B. subtilis* contained a peptidoglycan in which about 55% of the muramic acid residue had lost both peptide and acetyl group and formed a lactam. Lysozyme digests contained tetrasaccharides such as the one shown in Fig. 6.7. On reduction with sodium borohydride the lactam yielded mainly a cyclic secondary amine, which was identified and gave a clue to the presence of the lactam. Other differences from the peptidoglycan of vegetative cells were also observed:

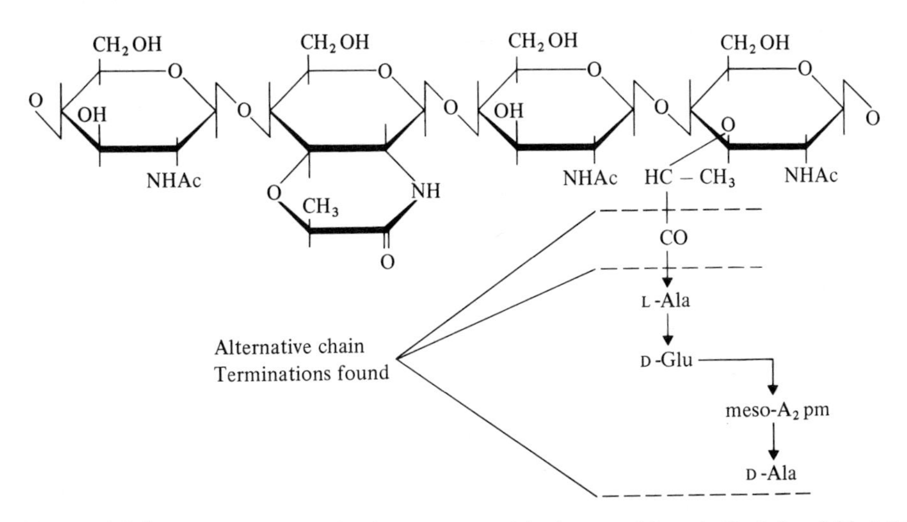

Figure 6.7 Structures present in the spore peptidoglycan of *B. subtilis* (after [54, 55]).

(1) The cross-linking index was reduced from 41% in vegetative cells (Table 6.2) to 19% in spores.
(2) Most uncross-linked side chains had lost both D-alanine residues in the vegetative cells, whereas one terminal D-alanine residue remained in the spores.
(3) The *meso*-diaminopimelic acid residues were mostly amidated in the vegetative cell walls, but not in the spores.
(4) Many of the muramic acid residues in spores that were not lactamized nevertheless did not have tetrapeptide side chains, but instead a single residue of L-alanine.

Additions to the peptidoglycan of Gram-positive bacteria also occur in the form of covalently-linked secondary polymers, such as teichoic acids or teichuronic acids, discussed in Chapters 7.1.1 and 7.1.2. These often seem to be attached by way of phosphodiester linkages on C_6 of some of the muramic acid residues. Some examples of the presence of muramic acid phosphate are given in Table 6.9 [33]. In *E. coli*, on

Table 6.4 Muramic acid phosphate content of Gram-positive bacterial cell walls (from [33]).*

Wall		*% Muramic acid as phosphate*
Group C Streptococcus	A	17.5
	B	13.7
Group A Streptococcus	S43	12.8
	T25	18.1
	B220	10.3
Group E Streptococcus	K129	18.4
Micrococcus luteus		7.1
Staphylococcus aureus	NYH6	7.7
Mycobacterium butyricum		5.1

*The values represent the percentage of total muramic acid present as the phosphate, each having been corrected for losses during hydrolysis.

the other hand, a different type of attachment occurs, in which some of the *meso*-diaminopimelic acid carboxyl groups at the L-centre (i.e. the one also linked directly to the primary peptide chain) are linked to a special type of lipoprotein via a lysine residue. This protein is discussed in Chapters 1 and 7.

6.3 Three-dimensional structure of peptidoglycans

Although the chemical structure of many peptidoglycans has been worked out, determination of their spatial arrangement has proved rather intractable. Wall polymers of other organisms, such as the cellulose of higher plants or the chitin of fungi, have yielded X-ray diffraction patterns sufficiently sharp to permit an accurate assignment of their three-dimensional structure. In addition it has often proved possible to observe distinct fibrils with the electron microscope, thus showing how the detailed arrangement of molecules found by X-ray methods has been assembled into larger structures. Attempts to follow these procedures with bacterial cell walls have met with limited success, for which two possible causes spring to mind. The first is that the chains of glycan or cross-linking peptide may be relatively short, so that extensive areas of ordered structure are not to be found, and hence X-ray diffraction pictures are likely to be poor. The second is that, whereas cellulose or chitin contain linear polymers of only one kind (poly-β-1,4-glucose or poly-β-1,4-*N*-acetylglucosamine respectively) peptidoglycan consists of glycan that is itself a heteropolymer, poly-β-1,4-(4-β-*N*-acetylglucosaminyl-*N*-acetylmuramic acid) and which is furthermore partially cross-linked in its peptide moieties (Fig. 6.2 and Table 6.2). Some success in using physical methods was achieved by Formanek *et al.* [17, 18], who used X-ray diffraction of dried foils of peptidoglycans from *Spirillum serpens* (Gram-negative) and *Lactobacillus plantarum* (Gram-positive), as well as measurements of density and infrared absorption. The infrared spectra showed the same amide I and II bands observed in the spectrum of chitin, and isopycnic density gradient centrifugation yielded a value of 1.46 g/cm^3 for the density of peptidoglycan. The Debye–Scherrer rings obtained by X-ray diffraction indicated repetitions of structure at 0.93–1 nm and at 0.44 nm. In a structure of β-1,4-linked glucose residues (and both chitin and peptidoglycan represent substituted versions of the same cellulose polymer) a periodicity of about 1 nm can only occur if the sugar residues form a two-fold screw axis [38]. Hence it was proposed that the glycan chain of peptidoglycan probably resembles cellulose and chitin, which both have that structure with a repeating distance of 1.03 nm. The two-fold screw axis of the glycan, stabilized by hydrogen bonds, would have the effect of directing all the peptide chains to the same side. The same authors suggested a structure for the peptide chains that would fit into the space allowed by the packed glycans [17]. The criteria were (a) that the peptide chains must be very flat in order to fit into the available space and (b) that they should be stabilized by as many hydrogen bonds as possible. The first model proposed that met these conditions was that of Kelemen and Rogers [29] who on the basis of model building suggested that a straight pleated sheet structure for the peptide part of peptidoglycan would fit in with a chitin-like glycan. However, Formanek's results did not show the infrared absorption maxima typical of pleated sheet, and similar objections were raised to the model of Higgins and Shockman [24], in which separated sheets of flat peptide and of glycan chains lay one above the other. The structure proposed by Formanek [17] to fit in with a chitin-like glycan was a variation of the 2.2 helix proposed by Donohue [13].

A somewhat different structure for peptidoglycan was developed on theoretical grounds by Oldmixon *et al.* [35]. The basic idea was that there should be a sheet of glycan chains, much as in Formanek's model, but that beneath it and at right angles should be a plane of peptide chains that could also be cross-linked to considerable distances, the whole making a covalent sheet joined by glycan chains in one direction and by peptide chains at 90° to it. The structures they proposed, based on different hydrogen bonding from that suggested by Formanek, resulted in glycan chains separated by a minimum of 1.14 nm within one sheet. However, Braun *et al.* [5] measured the size of isolated cell envelopes of *E. coli* and determined the diaminopimelic acid content of the same preparation. Hence they calculated the surface area per molecule of diaminopimelic acid, the average value being 12.9 nm^2. Again assuming the chitin-like repeating unit for the glycan chains (1.03 nm) their results yield an interchain distance of 1.25 nm.

The models for peptidoglycan based on a chitin-like glycan [17, 29, 35] have been subsequently discounted by two groups of workers on the basis of X-ray and other evidence. Burge *et al.* [8, 9] looked at the X-ray diffraction patterns of the cell walls and extracted peptidoglycans of *Micrococcus luteus*, *M. roseus*, *Bacillus licheniformis*, *Proteus vulgaris* and two strains of *Staphylococcus aureus*. Two methods were used. In the first aqueous suspensions of the chosen samples were slowly dried in the cold on silicone-treated glass slides to produce orientated specimens of layered cell walls. Strips of deposited material were piled up and examined at different humidities either 'edge on' or 'face on' [9]. In the second, amorphous scattering of X-rays by freeze-dried preparations pressed into blocks about 1 mm thick was compared with the results for the orientated specimens [8]. The former method yielded orientated diffraction patterns from layered specimens of peptidoglycan only from the three species in which it is loosely cross-linked, namely *P. vulgaris*, *B. licheniformis* and *M. luteus*. The sharp meridional reflexion varied with water content (0.98 nm 'wet' and 0.94 nm 'dry') and was identified with the axial advance of the glycan helix. The broad equatorial reflexion, on the other hand (1.9 nm 'wet' and 0.9 nm 'dry') was correlated to the distance between the glycan chains. The authors comment that the attached peptide chains provide special constraints on the glycan conformation, particularly in regard to its response to water content. The fact that the 0.98 nm sharp reflexion was meridional allowed the conclusion that the glycan chains lay in the plane of the peptidoglycan layers, while the differences of the equatorial reflexion indicated that the glycan chains were not in a precisely parallel arrangement.

The lack of orientation achieved with the tightly cross-linked peptidoglycans of *M. roseus* and *S. aureus* was attributed to built-in structural irregularities and distortions that prevented any coherent X-ray diffraction from being attainable in deposited layers. Model building studies and Ramachandran plots led to the conclusion that hydrogen bonds in the Tipper position [44] and as in chitin could occur at adjacent GlcNAc→MurNAc and MurNAc→GlcNAc linkages respectively, although the Knox and Murthy [30] bond (Fig. 6.8) was not excluded as an alternative. Structure building with space-filling atomic models showed that an axial repeating distance of 0.94–0.98 nm was incompatible with an α-chitin-like helix, because of the bulky D-lactyl group of muramic acid. In preference, a right-handed helix with four disaccharides per turn in an axial distance of about 4 nm was considered (Fig. 6.9). One attractive feature of this model was the fact that the arrangement of disaccharide units at roughly 90° to each other along the glycan axis would allow the cross-linked peptide chains to lie either in one plane (if linked to every other muramic acid) as perhaps required for the thin peptidoglycan layers of Gram-negative bacteria, or to provide three dimensional cross-linking as required in the thick layers of Gram-positive bacteria.

The second attack on a chitin-like structure for peptidoglycan also came from X-ray work, this time on the peptidoglycan of *M. luteus*, *S. aureus* and *Escherichia coli* [32]. Again both dried foils and powders were used, and in all cases X-ray diffraction showed two sharp peaks corresponding to 0.7 and 0.94 nm periodicities. In addition the Gram-positive bacteria showed a weak oriented reflexion at 4.2 nm. Comparison of these results with the intensity calculations derived from the models having chitin-like glycan and 2.2_7 helical peptide chains [17], ruled out the latter as

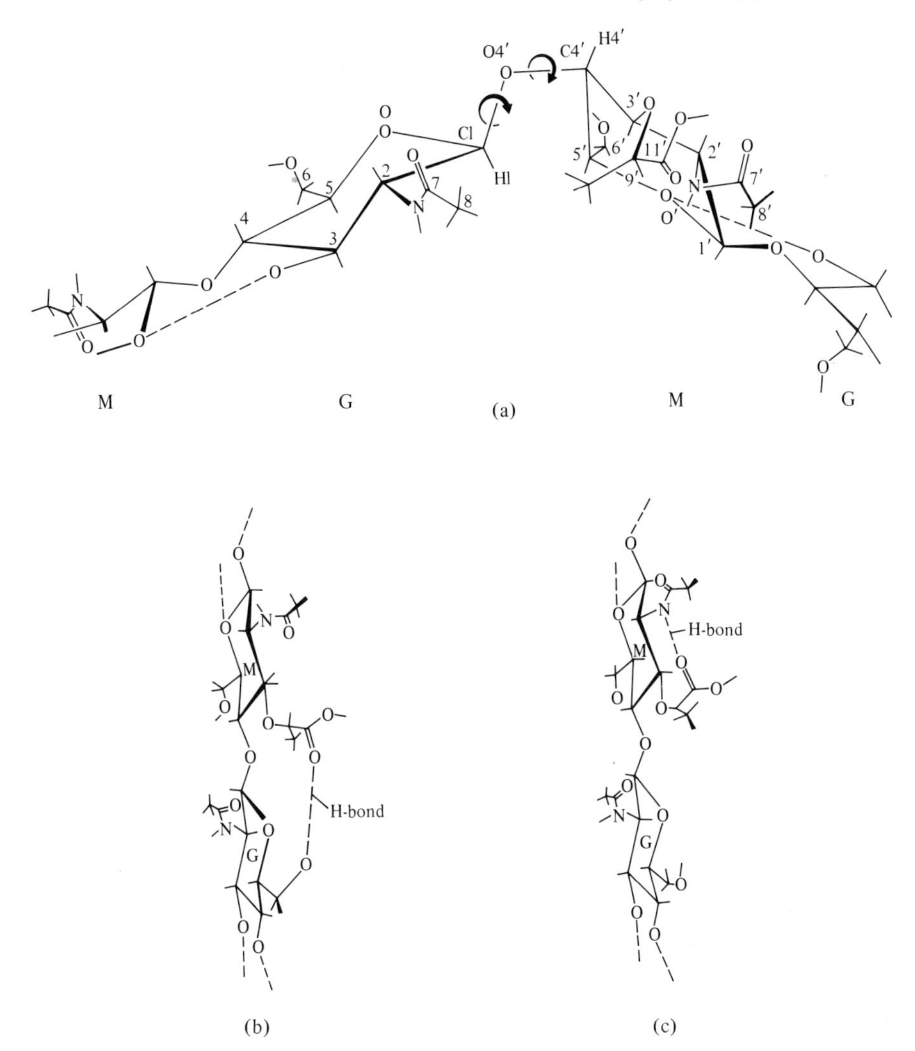

Figure 6.8 (a) Stereochemistry of the glycan chain of peptidoglycan. The Glc*N*Ac–Mur*N*Ac linkage is shown with the conformation $\psi = 4.0°$. The chitin-type hydrogen bonds are marked. (b) The Tipper [44] hydrogen bond proposal. (c) The Knox and Murthy [30] hydrogen bond proposal (from [9]).

a feasible structure. In addition the halo at 0.45 nm was discounted as being derived from 'chitin' chain separation, because the lactic acid side chains prevented such close proximity. It was concluded that the glycan chains do not possess a two-fold screw axis [9, 17] and that a greater value for the repeat distance would lead to the peptide chains being directed at a number of different angles, as described above. However, Labischinski *et al.* [32] comment that the attribution of the 0.94–0.98 nm to the disaccharide repeating distance, as by Burge *et al.* [8, 9] is only stereochemi-

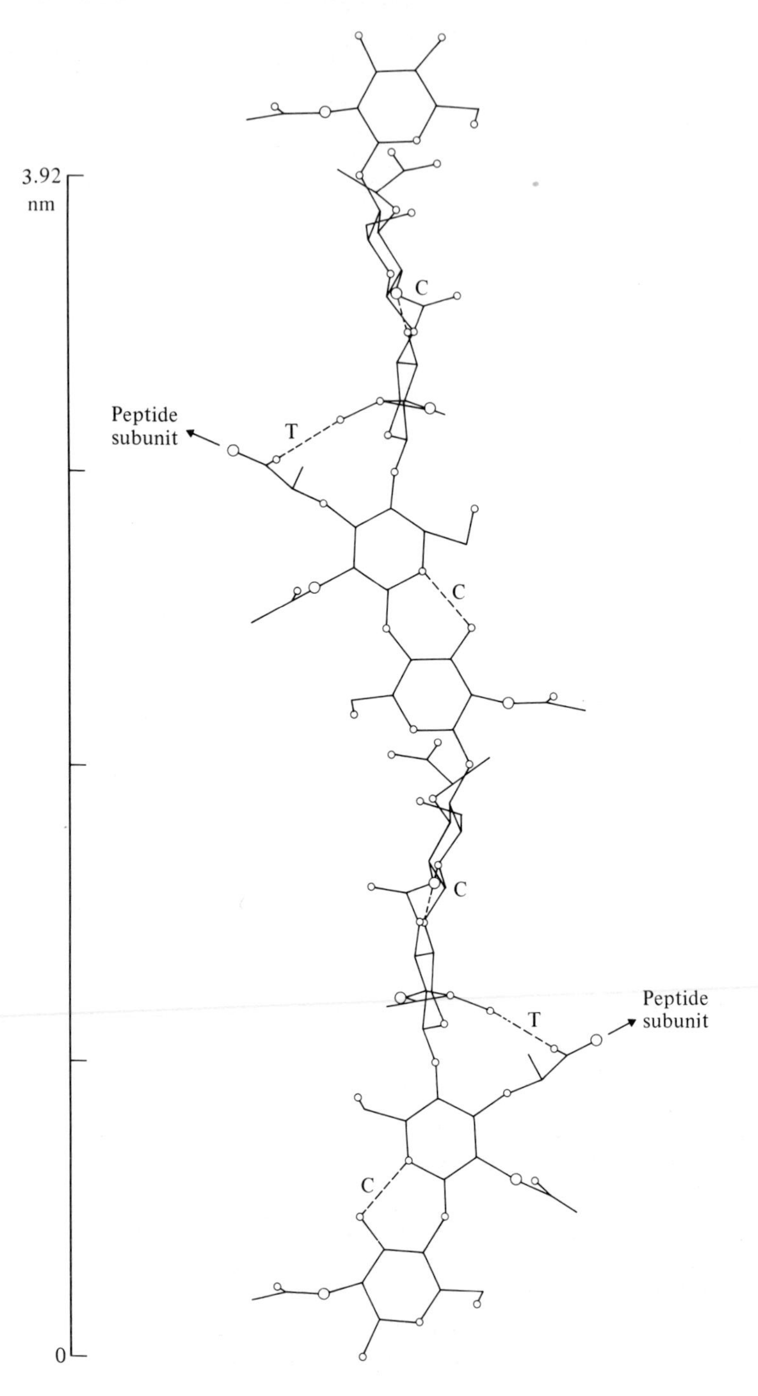
3.92
nm
0
C
Peptide
subunit
T
C
C
T
Peptide
subunit
C

cally possible if extreme limits of contact distances and glycosidic bridge angles are postulated. The former authors prefer to attribute all reflexions to the peptide portion except the weak oriented reflexion at 4.2 nm, which is present in Gram-positive bacteria, and is taken to represent the distance between the sheets of glycan chains. These sheets were supposed to have variable orientation relative to one another. These authors have furthermore proposed a structure on the basis of conformational energy calculations for a model peptide *N*-acetyl-L-alanyl-D-γ-glutamyl-L-lysyl-D-alanyl-D-alanine [4]. Working on the assumption that energetically favoured conformations should be chosen on the basis of a similarity for both the primary pentapeptide chain and the truncated one that results after cross-linking, they arrived at a ring-shaped conformation for the primary peptide, stabilized by a hydrogen bond between the -NH group of L-alanine and the -CO group of the penultimate D-alanine. Like Burge *et al.* [8, 9] they conclude that β-pleated sheet [29] and flat 2.2_7 helical structures [17] are not favoured energetically but point out that alternating L- and D-amino acid configurations are intrinsic to their model, unlike those of others [8, 9, 29]. The conformation that Barnickel *et al.* [4] propose is shown in Fig. 6.10. It is perhaps worth noting here that the conformational studies [4] did not support the analogy of β-lactam antibiotics for acyl-D-Ala-DAla [46, 52] or indeed for any other part of the peptide structure.

A significant property considered by Oldmixon *et al.* [38] and by Burge *et al.* [8, 9] was the ability of peptidoglycans to change in volume with the ionic conditions [34]. An extended model was proposed for peptidoglycans in their low-density form. This structure was sufficiently flexible to allow peptides from glycan in one plane to cross-link with those of glycan in another, probably an inevitable requirement to allow for the modification of wall structure that must necessarily occur during growth and cell division. Some bacteria, such as *B. megaterium, B. subtilis, L. acidophilus* and *S. aureus* lose peptidoglycan during exponential growth ('turnover') whereas others, such as *S. faecalis* and *E. coli* do not. This phenomenon of turnover, therefore, although it has great significance for the structure of peptidoglycan in certain species cannot be an essential process in all bacteria, and some other mechanism of wall modification must be sought. There is no reason why peptidoglycan fragments should not be re-distributed without any of them necessarily being lost from the wall, for instance by a series of concerted and successive transpeptidations in which one cross-link at a time along the length of a glycan chain is broken and re-attached to another site.

Figure 6.9 Computer plot of peptidoglycan chain with four disaccharides per turn in an axial distance of 3.92 nm. Only the D-lactyl residues of the peptide subunits in the plane of the paper are shown, so only every other Tipper [44] hydrogen bond is marked. All chitin-type hydrogen bonds are shown. C atoms are at line junctions, O atoms are small circles, N atoms larger circles (from [9]).

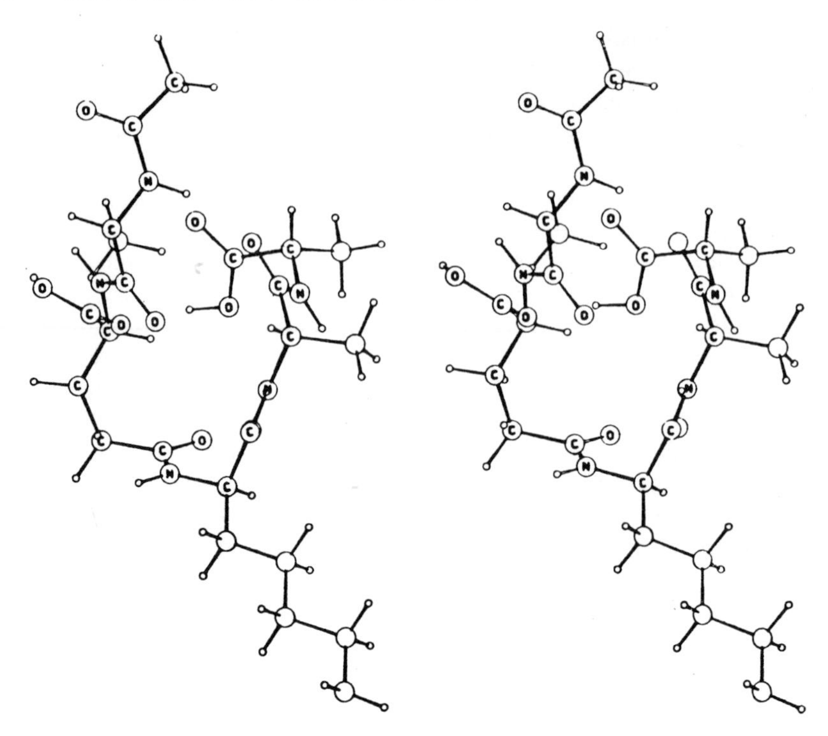

Figure 6.10 Stereoscopic view of the conformation corresponding to the global minimum obtained for Ac-L-Ala-D-Glu(L-Lys-D-Ala-D-Ala) (from [4]).

6.4 Cell walls of prokaryotes without peptidoglycan

For a long time it was thought that any prokaryotes that had a rigid wall were bound to contain peptidoglycan, because this cross-linked heteropolymer was the only constructional material used in this way. However, it has now become apparent that there are exceptions to this rule. The first examples were the Halococci (described elsewhere in this book) and *Halococcus morrhuae* was shown to have a sulphated heteroglycan as its major cell wall polymer [42]. The carbohydrates in this polymer were glucose, mannose, galactose, glucosamine, galactosamine, glucuronic acid, galacturonic acid and the unusual component gulosaminuronic acid. Such a polymer could clearly fulfill the functions normally attributed to Gram-positive wall material, since it obviously has a complex and hence probably branched structure, and has acidic groups in the form of sulphate and carboxylates that mimic the presence of carboxyl and phosphate groups in the more familiar organisms that have peptidoglycan and teichoic acids. In fact attempts to segregate specific polymers from the sulphated heteroglycan were unsuccessful and it may be that a single large molecule contains all the required functions.

More recently the methanogenic bacteria have been identified as another large group that can have rigid walls without peptidoglycan [27]. Whereas *Methanosarcina* has a rigid wall consisting of a heteropolysaccharide [26], the walls of methanobacteria have a polypeptide substituted with an amino sugar and associated with a polysaccharide made up of neutral and amino sugars. As in peptidoglycan the polypeptides probably comprise covalently linked sub-units, each of which has L-glutamic acid, L-lysine and L-alanine in the proportions 2:1:1 and which mostly have attached to them a residue of either *N*-acetylglucosamine or *N*-acetylgalactosamine [28]. Though many methanobacteria have the same peptide sub-units (some of which may carry an extra L-alanine residue) some have substitutions, such as L-threonine for L-alanine in *M. ruminantum.* In *Methanospirillum* the somewhat less rigid wall component is different, consisting of a sodium dodecyl sulphate and trypsin resistant protein. Kandler and König [27] propose that the Methanobacteria have, as it were, developed their own system of cell envelopes during the course of evolution quite independently of the mainstream of prokaryotes that answered the same general biological problems by making peptidoglycan. These ideas are set out in Fig. 6.11.

More recently the term pseudomurein (i.e. pseudopeptidoglycan) has been coined for the wall polymers of the Methanobacteria [31]. In the case of *M. thermoautotrophicum* the peptide units consist of short cross-linked chains as in eubacterial

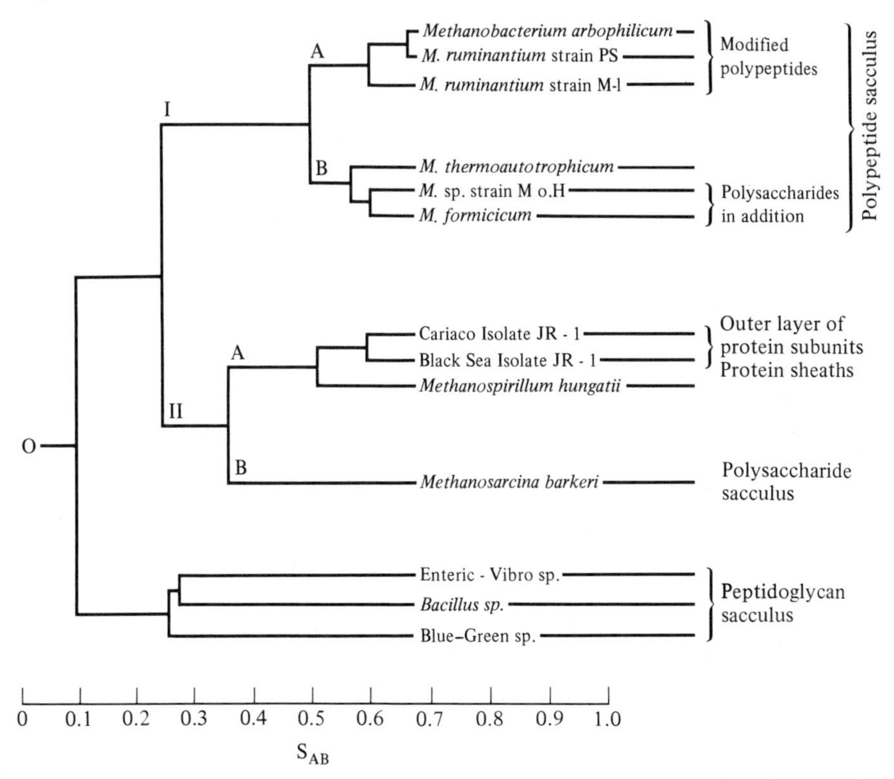

Figure 6.11 Distribution of cell wall components and relationship of methanogenic and classical bacteria according to the dendrogram based on comparison of 16S ribosomal RNA [19] (from [27]).

peptidoglycan, only in this case only L-amino acids are involved. The structures proposed are shown in Fig. 6.12. These short peptides are linked to the *N*-acetylhexosamine glycan chain via a newly-identified aminohexuronic acid, *N*-acetyltalosaminuronic acid, presumably by way of an amide linkage to its carboxyl group [31].

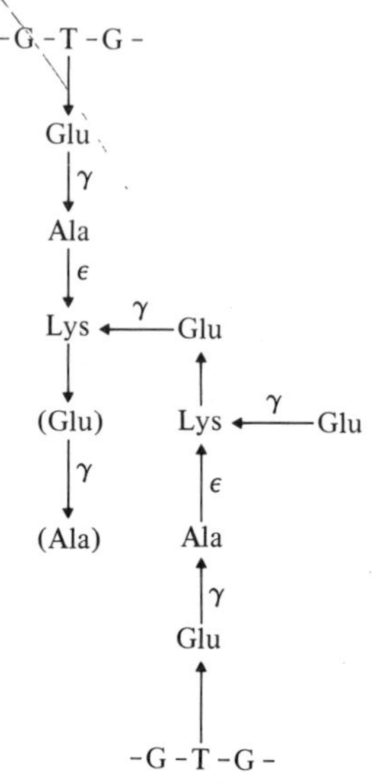

Figure 6.12 Structure proposed for the peptidoglycan (pseudomurein) of *Methanobacterium thermoautotrophicum*, G, *N*-acetylglucosamine; T, *N*-acetyltalosaminuronic acid; (Glu) and (Ala) represent residues that are partially removed in the final polymer (from [31a]).

References

1. Adam, A., Petit, J. F., Wietzerbin-Falszpan, J., Sinaÿ, P., Thomas, D. W. and Lederer, E. (1969) *FEBS Letts.* **4**, 87–92.
2. Araki, Y., Nakatani, T., Nakayama, K. and Ito, E. (1972) *J. biol. Chem.* **247**, 6312–22.
3. Azuma, I., Thomas, D. W., Adam, A., Ghuysen, J-M., Bonaly, R., Petit, J-F. and Lederer, E. (1970) *Biochim. biophys. Acta.* **208**, 444–51.
4. Barnickel, G., Labischinski, H., Bradaczek, H. and Giesbrecht, P. (1979) *Eur.J. Biochem.* **95**, 157–65.
5. Braun, V., Gnirke, H., Henning, U. and Rehn, K. (1973) *J.Bact.* **114**, 1264–70.
6. Bricas, E., Ghuysen, J-M. and Dezélée, P. (1967) *Biochemistry* **6**, 2598–607.
7. Brumfitt, W., Wardlaw, A. C. and Park, J. T. (1958) *Nature, Lond.* **181**, 1783–4.
8. Burge, R. E., Adams, R., Balyuzi, H. H. M. and Reaveley, D. A. (1977) *J.molec. Biol.* **117**, 955–74.
9. Burge, R. E., Fowler, A. G. and Reaveley, D. A. (1977) *J.molec.Biol.* **117**, 927–53.
10. Coyette, J. and Ghuysen, J-M. (1970) *Biochemistry* **9**, 2935–43.
11. Cummins, C. S. and Harris, H. (1956) *J.gen.Microbiol.* **14**, 583–600.

12. Day, A. and White, P. J. (1977) *Biochem.J.* **161**, 677–85.
13. Donohue, J. (1953) *Proc.Natn.Acad.Sci.USA.* **39**, 470–78.
14. Fiedler, F. and Kandler, O. (1973) *Arch.Mikrobiol.* **89**, 51–66.
15. Fleck, J., Mock, M., Minck, R. and Ghuysen, J-M. (1971) *Biochim.biophys. Acta.* **233**, 489–503.
16. Fordham, W. D. and Gilvarg, C. (1974) *J.biol.Chem.* **249**, 2478–82.
17. Formanek, H., Formanek, S. and Wawra, H. (1974) *Eur.J.Biochem.* **46**, 279–94.
18. Formanek, H., Schleifer, K. H., Seidl, H. P., Lindemann, R. and Zundel, G. (1976) *FEBS Letts.* **70**, 150–54.
19. Fox, G. F., Magrum, L. J., Balch, W. E., Wolfe, R. S. and Woese, C. R. (1977) *Proc.Natn.Acad.Sci.USA.* **74**, 4537–41.
20. Ghuysen, J-M. (1968) *Bact.Rev.* **32**, 425–64.
21. Ghuysen, J-M., Leyh-Bouille, M., Campbell, J. N., Moreno, R., Frere, J-M., Duez, C., Nieto, M. and Perkins, H. R. (1973) *Biochemistry* **12**, 1243–51.
22. Ghuysen, J-M., Strominger, J.L. and Tipper, D. J. (1968) In *Comprehensive Biochemistry 26A*, ed. Florkin, M. and Stotz, E. H., pp. 53–104, New York: American Elsevier.
23. Guinand, M., Vacheron, M. J. and Michel, G. (1970) *FEBS Letts.* **6**, 37–9.
24. Higgins, M. L. and Shockman, G. D. (1971) *Crit. Rev.Microbiol.* **1**. 29–71.
25. Hoare, D. S. and Work, E. (1957) *Biochem.J.* **65**, 441–7.
26. Kandler, O. and Hippe, H. (1977) *Arch.Microbiol.* **113**, 57–60.
27. Kandler, O. and König, H. (1978) *Arch.Microbiol.* **118**, 141–52.
28. Kandler, O. and König, H. (1978) *Hoppe-Seyler's Z.Physiol.Chem.* **359**, 282–3.
29. Kelemen, M. V. and Rogers, H. J. (1971) *Proc.Natn.Acad.Sci.USA.* **68**, 992–6.
30. Knox, J. R. and Murthy, N. S. (1974) *Acta crystallog. (B).* **30**, 365–71.
31. König, H. and Kandler, O. (1979) *Arch.Microbiol.* **121**, 271–5.
31a. König, H. and Kandler, O. (1979) *Arch.Microbiol.* **123**, 295–9.
32. Labischinski, H., Barnickel, G., Bradaczek, H. and Giesbrecht, P. (1979) *Eur.J. Biochem.* **95**, 147–55.
33. Liu, T-Y., and Gotschlich, E. C. (1967) *J.biol.Chem.* **242**, 471–6.
34. Marquis, R. E. (1968) *J.Bact.* **95**, 775–81.
34a. Martin, J.-P., Fleck, J., Mock, M. and Ghuysen, J-M. (1973) *Eur.J.Biochem.* **38**, 301–6.
35. Oldmixon, E. H., Glauser, S. and Higgins, M. L. (1974) *Biopolymers* **13**, 2037–60.
36. Perkins, H. R. (1965) *Biochem.J.* **95**, 876–82.
37. Perkins, H. R. (1971) *Biochem.J.* **121**, 417–23.
38. Ramachandran, G. N., Ramakrishnan, C. and Sasisekhavan, V. (1963) In *Aspects of Protein Structure*, ed. Ramachandran, G. N., pp. 121–35, New York and London: Academic Press.
39. Rogers, H. J. (1974) *Ann. N.Y. Acad.Sci.* **235**, 29–51.
40. Salton, M. R. J. (1953) *Biochim.biophys.Acta.* **10**, 512–23.
41. Schleifer, K. H. and Kandler, O. (1972) *Bact.Rev.* **36**, 407–477.
42. Steber, J. and Schleifer, K. H. (1975) *Arch.Microbiol.* **105**, 173–7.
43. Strange, R. E. and Dark, F. A. (1956) *Nature, Lond.* **177**, 186–88.
44. Tipper, D. J. (1970) *Int.J.syst.Bacteriol.* **20**, 361–77.
45. Tipper, D. J., Ghuysen, J-M. and Strominger, J. L. (1965) *Biochemistry* **4**, 468–73.
46. Tipper, D. J. and Strominger, J. L. (1965) *Proc.Natn.Acad.Sci.USA.* **54**, 1133–41.
47. Tipper, D. J. and Strominger, J. L. (1966) *Biochem.biophys.Res.Commun.* **22**, 48–56.

48. Tipper, D. J., Strominger, J. L. and Ensign, J. C. (1967) *Biochemistry* **6**, 906–920.
49. Vacheron, M-J., Guinand, M., Michel, G. and Ghuysen, J-M. (1972) *Eur.J.Biochem.* **29**, 156–66.
50. Van Heijenoort, J., Elbaz, L., Dezélée, P., Petit, J-F., Bricas, E. and Ghuysen, J. M. (1969) *Biochemistry* **8**, 207–211.
51. Vilkas, E., Massot, J. C. and Zissmann, E. (1970) *FEBS Letts.* **7**, 77–9.
52. Virudachalam, R. and Rao, V. S. R. (1977) *Int.J.Peptide Protein Res.* **10**, 51–9.
53. Ward, J. B. (1973) *Biochem.J.* **133**, 395–8.
54. Warth, A. D. and Strominger, J. L. (1969) *Proc.Natn.Acad.Sci.USA* **64**, 528–35.
55. Warth, A. D. and Strominger, J. L. (1972) *Biochemistry* **11**, 1389–96.
56. Weidel, W. (1950) In *Viruses 1950*, ed. Delbrück, M., pp. 119–21, California Institute of Technology.
57. Wheat, R. W. and Ghuysen, J-M. (1971) *J. Bact.* **105**, 1219–21.
58. Wheat, R. W., Kulkarni, S., Cosmatos, A., Scheer, E. R. and Steele, R. S. (1969) *J.biol.Chem.* **244**, 4921–5306.
59. Wickus, G. G. and Strominger, J. L. (1972) *J.biol.Chem.* **247**, 5297–306.
60. Wietzerbin, J., Das, B. C., Petit, J-F., Lederer, E., Leyh-Bouille, M. and Ghuysen, J-M. (1974) *Biochemistry* **13**, 3471–6.

7
Additional polymers in bacterial walls

7.1 Gram-positive bacteria

7.1.1 *Teichoic acids*

In addition to peptidoglycan, the walls of most Gram-positive bacteria contain other polymers, frequently with repeating acidic groups. Often these groups are phosphodiesters, which generally occur in polymers called teichoic acids. The teichoic acids were discovered in 1958, as the result of a search for a role for the CDP-glycerol and CDP-ribitol that had been identified in Gram-positive bacteria [4]. As originally defined, they consist of polymerized polyol phosphates, often with side-chains of mono- or oligo-saccharide units, and also ester-linked D-alanine residues. These ester linkages have exceptional lability, occasioned by the presence of vicinal hydroxyl or phosphate groups [75]. In some cases, such as the membrane teichoic acids of group D streptococci, the D-alanyl groups are attached to D-glucose residues rather than to the polyol part of the molecule, and in these instances the ester linkages are appreciably more stable to alkali [86]. As more information has accumulated it has become clear that many variations on the basic structural pattern of teichoic acids exist, although the various polymers presumably serve the same function in the cell. Some characteristic structures of teichoic acids are set out in Fig. 7.1 and of related structures in Fig. 7.2. An exception to the general rule that teichoic acids occur only in Gram-positive bacteria came with the description of lipoteichoic acid in the Gram-negative rumen bacterium *Butyrivibrio fibrisolvens* [36]. However, ultrastructural study showed that the walls were of the Gram-positive morphological type, but exceptionally thin. The thinness was probably the cause of the apparent Gram-negativity [11].

Originally, teichoic acids were extracted from bacteria or their isolated cell walls by treatment with cold dilute trichloroacetic acid. The degree of polymerization of such samples tended to be rather small, often only some 10 to 20 repeating units. However, a careful study of the degree of degradation that might be caused under these conditions suggested that little breakdown occurred in an overnight extraction [33]. Subsequently, potentially less degradative methods of extraction have been employed and greater chain lengths have been proposed.

The development of the understanding of wall teichoic acids has been somewhat confused because closely-related compounds also occur in the cytoplasmic membranes – often in the same organism. These compounds, originally called 'intra-

(a)

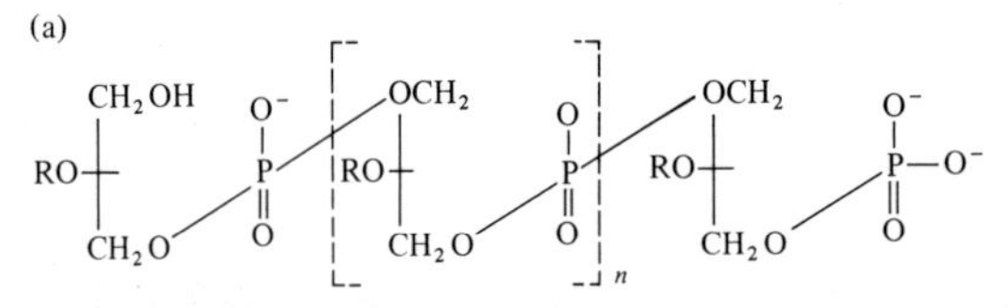

R = H, D-alanyl or glycosyl

Most glycerol teichoic acids are 1,3-linked .

(b)

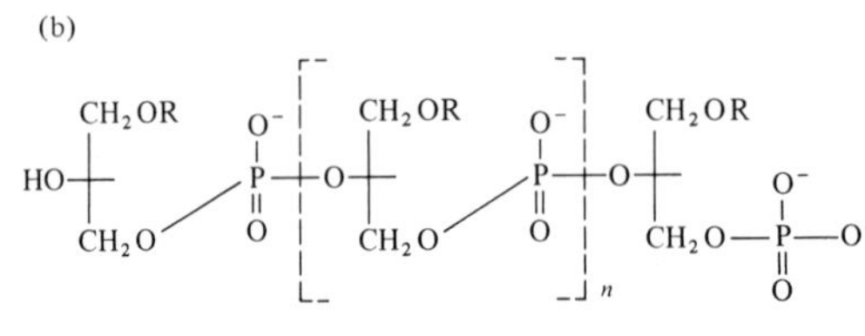

R = β-glucosyl (*B. subtilis* var. *niger* WM)
  = α-glucosyl (*B. stearothermophilus*)
  = *O*-α-D-galactopyranosyl-(1, 4) - α-D - *N*-acetylgalactosaminyl (*Actinomyces antibioticus*)

(c)

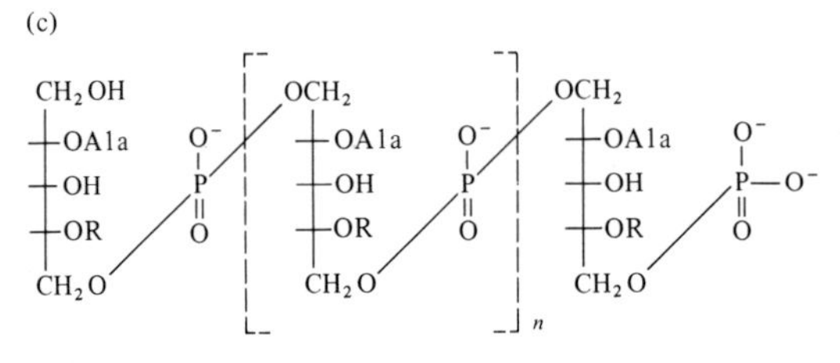

R = glycosyl

(d)

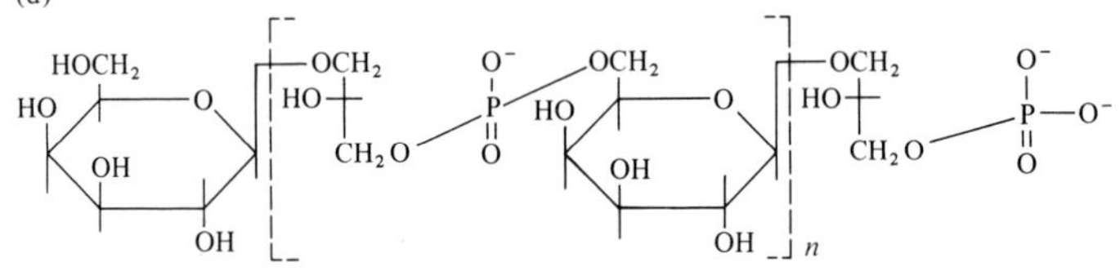

Example with galactosylglycerolphosphate repeating unit (*Bacillus licheniformis* ATCC 9945)

(e)

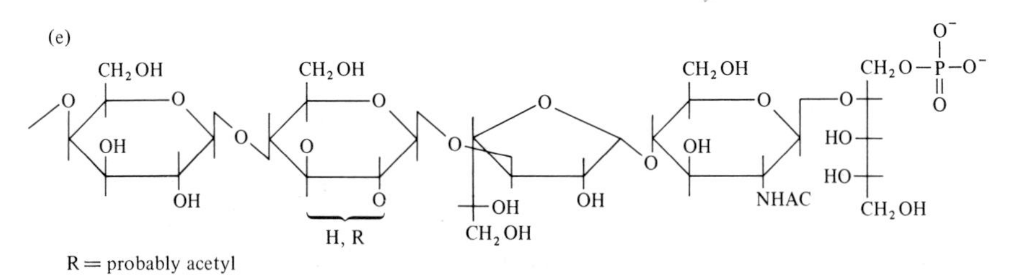

R = probably acetyl

Figure 7.1 Structure of the teichoic acids. (a) and (b) Glycerol teichoic acids. Most glycerol teichoic acids are 1,3-linked. (b) Glycerol teichoic acids – variant structure. (c) Ribitol teichoic acids. (d) and (e) Teichoic acids with intercalated glycosyl unit. The example shown has a galactosylglycerol phosphate repeating unit (*Bacillus licheniformis* ATCC 9945). (e) Teichoic acid of Type 13 pneumococcal capsules. This material has an oligoglycosylribitol phosphate repeating unit.

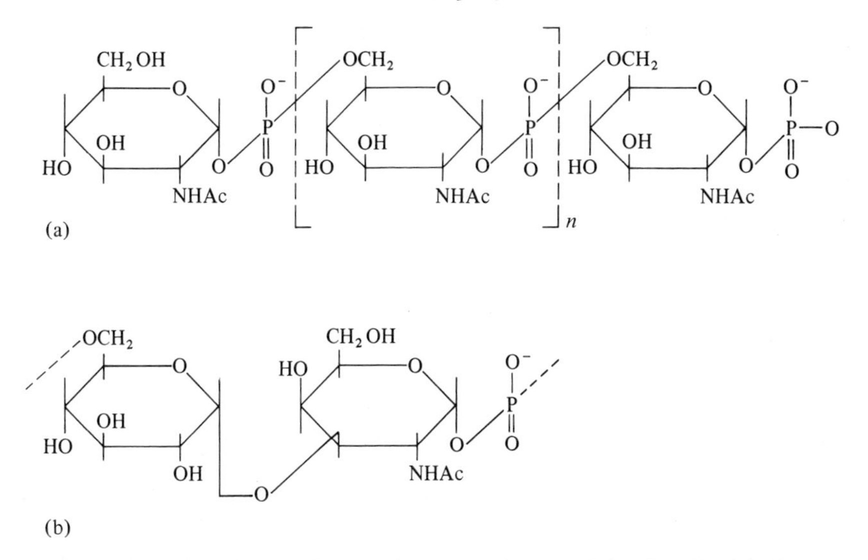

Figure 7.2 Sugar phosphate polymers that occur in bacterial cell walls. (a) *Micrococcus* sp. (*Staphylococcus lactis*) 2102. The repeating unit is *N*-acetylglucosamine-1-phosphate linked 1,6. (b) *Micrococcus* sp. A1. The repeating unit is 3-O-α-D-glucopyranosyl-*N*-acetylgalactosamine-1-phosphate linked 1,6.

cellular' and later 'membrane' teichoic acids, have gradually come to be recognized as lipoteichoic acids. This name refers to the fact that lipoteichoic acids consist of a typical glycerol teichoic acid linked at one end to a glycolipid, as shown in Fig. 7.3. It seems evident that the lipid groups serve to provide lipophilic attachment of one end of the molecule within the generally hydrophobic cytoplasmic membrane, leaving the other hydrophilic portion free to extend through the wall or at least to act as a surface antigen [87]. The occurrence and function of membrane teichoic acids has been reviewed [48].

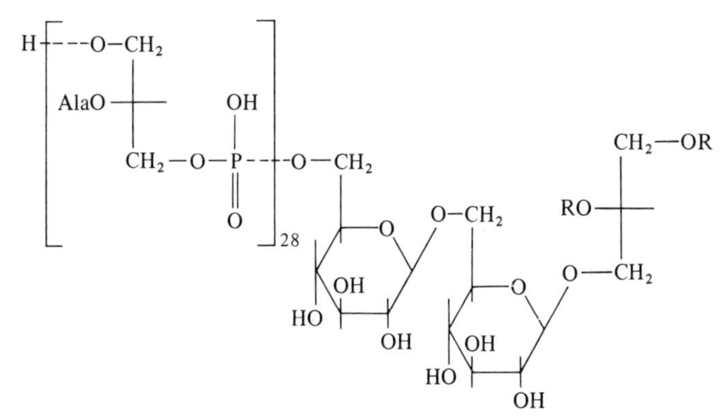

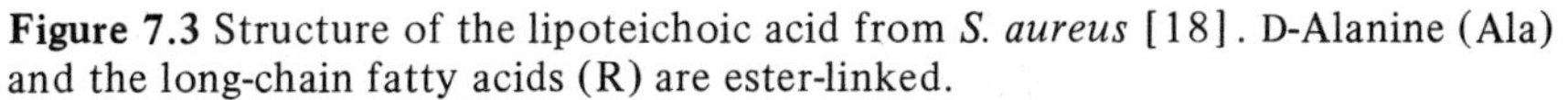

Figure 7.3 Structure of the lipoteichoic acid from *S. aureus* [18]. D-Alanine (Ala) and the long-chain fatty acids (R) are ester-linked.

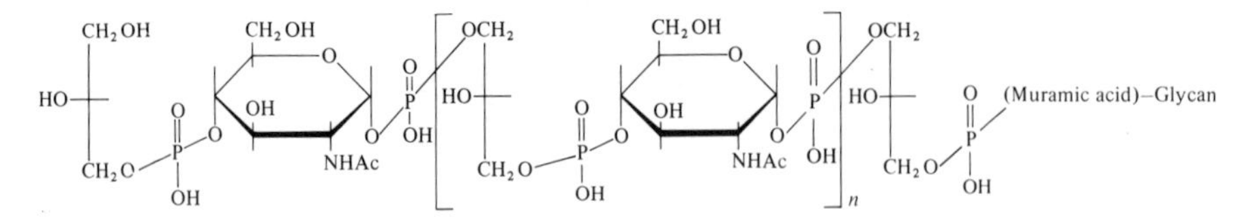

Figure 7.4 Proposal for the linkage to peptidoglycan of the wall teichoic acid of *Staph. lactis* I3 [10].

What is the relationship of the wall and membrane teichoic acids to each other and to the rest of the cell surface environment in which they find themselves? Let us first consider the wall teichoic acids. The very fact that degradative methods had to be used to remove them implied that they were covalently linked to the wall structure, and evidence has accumulated for a linkage to peptidoglycan. First, the presence of muramic acid phosphate in acid hydrolysates of cell walls, originally observed by Ågren and de Verdier [1], gradually came to be equated with a linkage to secondary polymers such as teichoic acids or polysaccharides [53]. The other end of the presumed phosphodiester linkage proved more difficult to establish. In the single instance of the cell walls of a strain of *Staphylococcus lactis*, Button *et al.* [10] obtained some evidence for a direct phosphodiester linkage between the teichoic acid and C_6 of muramic acid, as shown in Fig. 7.4. Latterly, work with other organisms has led to the conclusion that a common method for joining teichoic acids to peptidoglycan is by means of a link-piece consisting of three units of glycerol phosphate attached to *N*-acetylglucosamine, which in turn is linked by its reducing group to a phosphate on C_6 of muramic acid (Fig. 7.5) [13, 89]. The presence of a sugar-1-phosphate would easily account for the great acid lability of the linkage, so that teichoic acids are easily set free as, for instance, by cold trichloroacetic acid. Because of the frequent occurrence of phosphate bonds vicinal to hydroxyl groups, teichoic acids are easily degraded by mild alkali, though for the same reason this treatment should leave the linking *N*-acetylglucosamine attached to the muramic acid. In fact the bond linking the first glycerolphosphate to the *N*-acetylglucosamine is particularly alkali-labile. The sequence of biosynthetic events by which teichoic acids become polymerized and attached to peptidoglycan is discussed in Chapter 10.1.

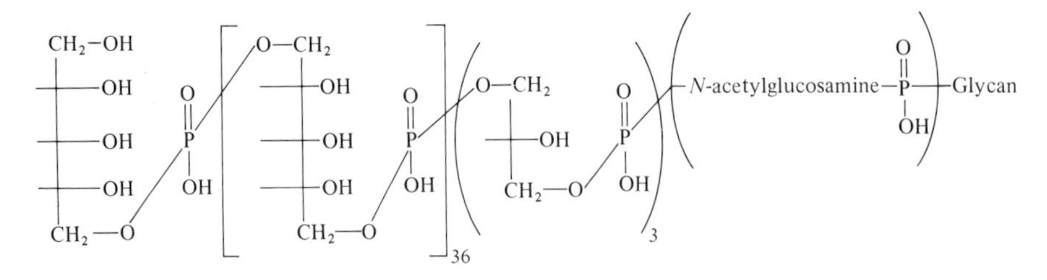

Figure 7.5 Linkage to peptidoglycan of the wall teichoic acid of *Staph. aureus* H mutant lacking glycosidic substituents [13].

Functions of teichoic acids

The place of teichoic acids in the bacterial economy is uncertain. Their most likely function seems to be as mediators for the access to the cell of divalent cations (particularly Mg^{2+}) as proposed by Heptinstall, *et al.* [35]. A polymer with so many equally spaced phosphate groups would be expected to have an affinity for magnesium ions and these authors were able to obtain evidence that the presence or absence of teichoic acid in the cell wall affected the cell's avidity for Mg^{2+}. Additional evidence for a role for wall teichoic acids comes from the work of Ellwood and Tempest [19]. These workers showed that when *Bacillus subtilis* var. *niger* was grown in continuous culture, if the cells were Mg^{2+}-limited then the cell wall contained teichoic acid, whereas if phosphate were the limiting nutrient then the wall teichoic acid was replaced by another polymer, teichuronic acid, where the repeating acidic functions were the carboxyl groups of glucuronic acid residues alternating with *N*-acetylgalactosamine. This was the same polymer originally identified in *Bacillus licheniformis* 6346 [44]. Similar results were obtained with several other Gram-positive cocci and bacilli. The close association with the availability of Mg^{2+} is evident and Ellwood and Tempest [20] proposed that wall teichoic acids are involved in the process of cation assimilation. Thus cells limited for phosphates, but with ample magnesium, often produce wall teichuronic acids that replace teichoic acids; when Mg^{2+} uptake is limited, then teichoic acid, which has a much greater affinity for Mg^{2+}, is once more formed. Incidentally, these authors also found that Na^{+} competed with Mg^{2+} in their systems, and that under Mg^{2+}-limitation lowering the pH value of a continuous culture during growth of *B. subtilis* W23 from 7.5 to 5.5 procured the appearance of some teichuronic acid in the walls and a decrease in their teichoic acid content. At the same time the ester alanine content of the teichoic acid increased threefold. Thus there is ample evidence for the response of the acidic surface polymers of Gram-positive bacteria (teichoic or teichuronic acids) to the conditions prevailing in the cell's environment, and an implication of an important function. As Ellwood and Tempest [20] pointed out, even in phosphate-limited cells the lipoteichoic acids continued to be synthesized, and hence it seems likely that these components are indispensable for the cells that bear them.

Certain Gram-positive cocci appear to lack teichoic acids altogether. In 1975 it was shown that *Micrococcus luteus, M. flavus* and *M. sodonensis* have no lipoteichoic acid in their membranes, but that its place is taken by a lipomannan [64, 67, 68]. This compound consists of an α-linked mannan of some 50–70 residues with some branch points, linked glycosidically to a diglyceride of which the fatty acids were similar to those of the whole cell lipids. In addition succinic acid residues are present as *O*-acyl substituents on about one in four of the mannose residues, the terminal carboxyl groups of the succinic acid providing the whole polymer with a considerable number of acidic functions (Fig. 7.6). The lipomannan, therefore, has functional components that resemble lipoteichoic acids: a lipophilic region and a hydrophilic portion with frequent acidic groups. *M. luteus* is known not to have a teichoic acid in its walls, but rather a teichuronic acid [66] and *M. sodonensis* and *M. flavus* have also been shown to have walls lacking teichoic acids [65].

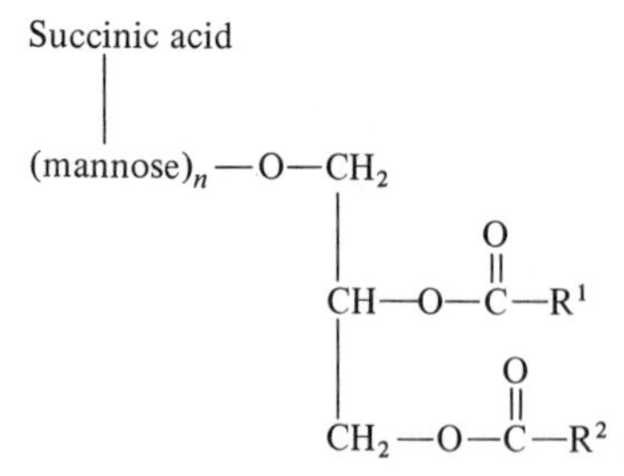

Figure 7.6 Structure of the lipomannan from *Micrococcus luteus* [68]. R^1 and R^2 are fatty acids, n = 52–70. The ratio of succinic acid to mannose is 1:4.

Another function of teichoic acids within the bacterial cell is their effect on autolytic enzymes. In *Streptococcus pneumoniae* the lytic enzyme, an *N*-acetylmuramic acid-L-alanine amidase, requires the presence of choline on the wall teichoic acid before it can act [38]. Cells grown in a medium in which choline is replaced by ethanolamine have walls that are not susceptible to autolysin, either endogenous or exogenous, because the wall teichoic acid also has an ethanolamine replacement. The choline–teichoic acid apparently has an additional effect in that it allows active 'C-type' autolysin to be produced instead of inactive 'E-type' enzyme. Although other organisms such as *Streptococcus faecalis* have lytic enzymes that bind strongly to the cell wall, there is no clear evidence that teichoic acids are involved.

Whereas the autolysin of *Streptococcus pneumoniae* requires wall-bound choline-bearing teichoic acid in order to cleave the bond in peptidoglycan between the *N*-acetylmuramic acid and L-alanine, the closely related Forssman antigen of the same organism (a lipoteichoic acid-like version of the wall component that occurs in the membrane) inhibits the action of the autolysin [38]. Similar inhibitory actions upon autolytic enzymes were also observed when lipoteichoic acids from *S. faecalis*, *Lactobacillus casei*, *L. fermentum* and *Streptococcus lactis* were added to the lytic enzymes of *S. faecalis* or *L. acidophilus*. However, deacylated lipoteichoic acid was without effect, and pneumococcal Forsmann antigen and the lipomannan from the membranes of *Micrococcus luteus* were also without inhibitory effect [12]. This lack of 'cross-inhibition' was also observed in reverse, since the lipoteichoic acids from the lactic acid bacteria were without inhibitory effect on pneumococcal autolysin, and indeed even produced some enhancement of enzyme activity.

The relationship between lipoteichoic acids and autolytic enzymes (also discussed in Chapter 11) has prompted suggestions that in growing cells lytic action is normally under negative control, only being released from inhibition by lipoteichoic acids when wall-modification demands it. There is also evidence that in suitable cells penicillin action is accompanied by the release of lipoteichoic acid into the medium, with the result that the autolysin, freed from inhibition, has caused lysis of the cells [82].

Why the action of penicillin (or other inhibitors of peptidoglycan synthesis and assembly, e.g. fosfomycin, D-cycloserine, vancomycin, β-chloro-D-alanine) [81] should cause the release of lipoteichoic acid (Forssman antigen) from the membrane and the cell remains to be explained. Another problem is that apparently only the acylated form of the lipoteichoic acid can inhibit the autolysin [12] and the acylated, lipophilic part of the molecule would be expected to occur in the cytoplasmic membrane. The autolysin, on the other hand, to exercise its function would have to reside in the wall, where it would seem that only the hydrophilic part of the lipoteichoic acid could reach. Hence inhibition of autolysis by lipoteichoic acid under normal conditions is difficult to understand at the molecular level.

Teichoic acids, apart from their functions in the internal economy of the cell, also affect the relationship of the bacteria with the external world. Many of them are responsible for the immunological properties of bacteria, examples of which are given in Table 7.1 [47].

Table 7.1 Teichoic acids as bacterial antigens (after [47]).

		Teichoic acid components		
Organism	*Group*	*Location*	*Polyol*	*Sugars*
Streptococcus	D	Membrane	Glycerol	Glucosyl-α-(1,2)-glucose
	N	Membrane	Glycerol	Galactose-phosphate
Streptococcus mutans	I	Membrane	Glycerol	Glucose, galactose
	II	Wall	Glycerol	Galactose
Staphylococcus aureus	Strain Wood 46	Wall	Ribitol	β-*N*-acetylglucosamine
	Strain 263	Wall	Ribitol	α-*N*-acetylglucosamine
Staphylococcus epidermidis	Strain T1	Wall	Glycerol	α-Glucose
	Strain T2	Wall	Glycerol	β-Glucose
Lactobacillus	A	Membrane (possibly wall)	Glycerol	α-Glucose
	D	Wall	Ribitol	α-Glucose
	E	Wall	Glycerol	α-Glucose
	F	Membrane	Glycerol	Galactosyl-α-(1,2)-Glucose

Sometimes they are in fact the group antigens, as for instance in some streptococci (Groups D and N, and *S. mutans*), staphylococci and particularly in lactobacilli, when they play an important part in serological classification [47]. Lipoteichoic acids have also been implicated in the binding of streptococci to host cells [63]. Certain teichoic acids serve as phage receptors, as in *Staphylococcus aureus* [14, 60] and *B. subtilis* W23 [2, 24]. There is evidence that a functional phage receptor requires the presence of teichoic acid organized in a particular way with the peptidoglycan, although the exact structural requirement has not yet been resolved. As mentioned earlier, *B. subtilis* grown in K^+-limited chemostat culture produces wall teichoic acid, whereas in PO_4^{3-}-limited culture it does not. Archibald and Coapes [3] used this fact, coupled with adsorption of phage SP50, to show that teichoic acid synthesized at the membrane took some time to react with phage, presumably because it required time to become accessible at the cell surface. It became available more rapidly along the length of the cell cylinder than at the cell poles.

7.1.2 *Teichuronic acids*

During the discussion of teichoic acids we mentioned that in many Gram-positive bacteria grown in PO_4^{3-}-limited media, wall teichoic acid disappears and becomes replaced by other acidic polymers known as teichuronic acids. The first example of these to be characterized was the teichuronic acid of *Bacillus licheniformis* 6346, which was shown to be a polymer consisting of equimolar proportions of *N*-acetylgalactosamine and D-glucuronic acid [44]. Since then polysaccharides with repeating acidic groups have been discovered in the cell walls of many Gram-positive bacteria, and some of these teichuronic acids are listed in Table 7.2. It is likely that reports of these compounds will continue to accumulate, particularly when bacteria are examined that have been grown under phosphate limitation, or where phage-resistant mutants are isolated.

The linkage of some teichuronic acids to the peptidoglycan has been studied, and involvement of a phosphodiester linkage to the 6-position of muramic acid has been proposed (Table 7.2). Thus the teichuronic acid from *B. licheniformis* is easily removed from the peptidoglycan by mild acid with liberation of a reducing group, identified as a hexosamine. Since the polymer consisted of glucuronic acid and *N*-acetylgalactosamine the latter was presumed to represent the sugar linked to the peptidoglycan, and the next sugar was shown to be glucuronic acid [39]. The most detailed information of linkage of a teichuronic acid has come from studies of *Micrococcus lysodeikticus* (*luteus*). Elegant work by Anderson and his colleagues on biosynthesis of the polymer by membrane preparations has shown that a polyprenol phosphate lipid intermediate has linked to it first *N*-acetylglucosamine, followed in succession by two residues of *N*-acetylmannosaminuronic acid [72]. Thereafter glucose and *N*-acetyl-mannosaminuronic acid are added alternately to complete the polymer [79], which is presumably extended at the non-reducing end and still attached to the carrier lipid. It can then be transferred to the peptidoglycan, where it is thought to be linked by a phosphodiester group to the C_6 hydroxyl group of a muramic acid residue. Some chemical evidence for the linkage was also obtained by Hase and Matsushima [32], although their observation that release by mild acid of nearly all the *N*-acetylglucosamine reducing groups was not accompanied by concomitant liberation of phosphate as monoester remains unexplained.

Apart from probable involvement in the divalent cation economy of the cell, as mentioned earlier in connection with teichoic acids, teichuronic acids have been shown to have some connection with cell separation. Thus morphological mutants of *M. luteus*, which grew in large regular cell packets rather than small groups or short chains, required a much higher concentration of Mg^{2+} for growth than the parent cells. These mutant cells lacked teichuronic acid, which implies that the presence of this polymer on the surface of the wild-type cells in some way promoted cell separation [90]. A similar effect was produced by sub-lethal concentrations of novobiocin or EDTA [91] and likewise a novobiocin-resistant mutant of *B. licheniformis* also grew in chains and lacked teichuronic acid [71]. It was shown that in this mutant deficiency in teichuronic acid alone led to a *lyt*$^-$ phenotype and that activity and binding of autolysin(s) depended on teichuronic acid but not teichoic acid.

Table 7.2 Teichuronic acids of the walls of Gram-positive bacteria

Organism	*Components of polymer*	*Repeating unit*	*Mode of attachment to peptidoglycan*	*References*
Bacillus licheniformis 6346	*N*-acetylgalactosamine D-glucuronic acid	→ 4)-*D*glucuronosyl (1,3)-*N*-acetylgalactosaminyl-(1 →	Acid labile, probably phosphodiester	39, 44
Bacillus subtilis var. *niger*	As above	Not known	Not known	20
Micrococcus lysodeikticus (*luteus*) NCTC 2665	D-glucose *N*-acetyl-D-mannosaminuronic acid	→ 4)-*N*-acetyl-D-mannosaminuronosyl-β(1,6)-D-glucosyl-α(1→	di-*N*-acetylmannosaminuronosyl-*N*-acetylglucosaminyl-1-phosphoryl-	29, 30, 31, 66, 72
Staphylococcus aureus T	*N*-acetyl-D-fucosamine *N*-acetyl-D-mannosaminuronic acid	Not known	Not known	88
Bacillus megaterium M46	D-glucose Glucuronic acid Rhamnose	Not known	Not phosphodiester. Glycosidic?	43, 85
Corynebacterium poinsettiae	Rhamnose Glucuronic acid Galactose Mannose Pyruvic acid	Not known	Not known	16
C. betae	Rhamnose Glucuronic acid Fucose Mannose	Not known	Not known	

7.1.3 *Neutral polysaccharides*

The cell walls of many Gram-positive bacteria contain neutral polysaccharides covalently bound to the peptidoglycan, examples being the Streptococci and Lactobacilli. Evidence for phosphodiester linkages has been produced [28, 46], and in some instances, as with the teichoic acids and teichuronic acids, the linking sugar has been shown to be a hexosamine. In others, however, release of the polymer by mild acid was not accompanied by the appearance of a reducing group of hexosamine, e.g. the polysaccharide responsible for *rhamnosus* specificity in *L. casei* var. *rhamnosus*, which consists of rhamnose, glucose and galactose and no hexosamine [28]. A neutral polymer with similar components is also attached to the peptidoglycan of *L. acidophilus* [15].

Among the streptococci, wall polysaccharide antigens are known to be important. Group A streptococci contain a carbohydrate that consists of only L-rhamnose and *N*-acetylglucosamine (in a molar proportion of 2:1), and a group A variant strain produces a polymer that contains almost entirely rhamnose but little amino sugar. This result, combined with other evidence from enzymic studies, shows that the *N*-acetylglucosamine residues occur as non-reducing terminals β-linked to a polyrhamnose core. Rhamnose units linked 1,3 [37] have been demonstrated. Altogether the evidence suggests that the group A and A-variant polysaccharide side chains consist of 1,3 linked rhamnose residues, and in group A these have additional non-reducing *N*-acetylglucosamine terminals [59].

Group C streptococcal wall polysaccharide resembles that of group A, except that *N*-acetylgalactosamine replaces *N*-acetylglucosamine as the immunodominant sugar. Immunological evidence suggests that the structure of the rhamnose part of the polymer is probably very similar to that found in group A [59].

As with the teichoic and teichuronic acids, ample evidence has accumulated that the neutral polysaccharides of the walls of the Gram-positive bacteria are linked to peptidoglycan through phosphodiester bonds to the C_6 of muramic acid. Liu and Gotschlich [53] were able to show for a number of cocci and *Mycobacterium butyricum* that between 5% and 18% of the wall muramic acid was probably originally present as the 6-phosphate, and similar evidence has indicated the same kind of linkage for the neutral polysaccharide of *Lactobacillus acidophilus* [15]. In this case the polysaccharide consisted of equimolar amounts of glucose, rhamnose and galactose, with one phosphate bridge to muramic acid C_6 for each 70 molar proportions of the neutral sugars.

It is worth mentioning at this point that Liu and Gotschlich [53] were unable to find any muramic acid-6-phosphate in isolated peptidoglycan, or in whole cells, of Gram-negative bacteria (they looked at *E. coli* B and *Branhamella catarrhalis*). Probably this kind of linkage is confined to Gram-positive bacteria. Linkages to peptidoglycan of other polymers in the walls of Gram-negative species are discussed below.

There seems to be some evidence that polymers found bound to peptidoglycans of Gram-positive bacteria may often also occur as capsules not covalently bound to the cell wall, or be set free into the medium. Thus the teichuronic acids of the plant pathogenic Corynebacteria are very similar in composition to the polysaccharides that these organisms produce in such large amounts, and which are implicated in their pathogenicity. Similarly, an encapsulated strain of *Lactobacillus casei* var. *rhamnosus* contained the same rhamnose-rich polysaccharide in its capsule that other strains contain in their cell walls [45].

7.1.4 *Proteins*

The question whether proteins occur in the cell walls of Gram-positive bacteria has often been obscured by the methods used for isolating cell wall fractions. Certainly the immunologically recognizable M proteins of Streptococci are found in the cell walls, and it seems likely that certain enzymes of wall metabolism, such as autolysins or transpeptidases, must occur within the wall of the living cell if they are to exert their characteristic actions. Generally speaking, however, few proteins have been identified in wall fractions. An exception is a teichoic acid-less mutant of *Bacillus subtilis.*

This spontaneous mutant (*Cbl1*) of *B. subtilis* 168 was resistant to phages that infect the wild-type, because it had no wall teichoic acid [34]. In place of teichoic acid the walls contained a protein of high molecular weight (255 600 by equilibrium sedimentation), which could be extracted with 6 M lithium chloride and separated by sodium dodecyl sulphate or urea into two products of molecular weight 105 000 and 155 000 as shown by gel electrophoresis [50].

Another protein that sometimes occurs covalently bound to peptidoglycan, but may also be excreted in the medium, is the Protein A of *Staphylococcus aureus* [78]. This protein of molecular weight 42 000 has an extended shape, and is isolated by digestion of the cell walls with lysostaphin before purification by affinity chromatography [5, 77]. Unlike the protein of *B. subtilis*, it is not extracted by lithium chloride. There is evidence that this protein consists of repetitive regions that probably lie outside the wall proper and which will bind the Fc fragments of immunoglobulins, and a different wall-bound region that may also have short repeating sequences [76]. The exact linkage to the peptidoglycan has not been elucidated.

The M proteins of group A streptococci are important in the virulence of this group, serving to protect the bacteria from phagocytosis, and were clearly found to be associated with the wall [73]. Various methods ranging from acid-extraction, alkali-extraction to enzymic procedures have been used to isolate these proteins, and they can be destroyed *in situ* by proteolytic enzymes such as trypsin, but the exact mode of linkage to the cell wall (and presumably directly or indirectly to the peptidoglycan) is not known. Depending upon the method of extraction, M proteins have been reported to have molecular weights of between 32 000 and 180 000, and it seems that at least the larger isolated forms contain multiple sub-units [21]. They have an iso-electric point of about pH 5.5 and are precipitated by concentrations of ethanol exceeding 50%. There is some evidence that the presence of M proteins increases the hydrophobicity of *S. pyogenes* [83] but that lipoteichoic acid may be more important in causing the streptococci to bind to epithelial cells [63].

The walls of streptococci contain proteins other than the M protein, since T and R proteins have also been described [49]. Protein antigens have been found not only in group A streptococci but also in membrs of groups B, C and G [59].

7.2 Gram-negative bacteria

Apart from peptidoglycan, which is of the simple directly cross-linked type (Chapter 6.1), the cell walls of Gram-negative bacteria have a more complex structure with more components than are found in Gram-positive bacteria. Outside the cytoplasmic membrane (or 'inner membrane') is a region that contains the peptidoglycan and the so-called periplasmic space, and beyond that is the outer membrane, with associated lipopolysaccharides [62]. These structures are discussed in Chapter 1. In *E. coli* there is in the outer membrane a specific lipoprotein of unusual composition, about one-third of which is covalently linked to the peptidoglycan [8]. This protein was first identified in its peptidoglycan-linked form [6] and later found also to exist free as an outer membrane component [40, 41]. In a series of elegant studies Braun elucidated the complete sequence of the lipoprotein, and showed that its lipophilic properties were attributable to a novel type of masked *N*-terminal cysteine, with fatty acyl residues present on the amino group and also on a diglyceride linked as a thioether, (Fig. 7.7) [8]. The carboxyl terminus of the 58-amino acid sequence is a lysine residue, and it is this amino acid which is linked by an amide bond from its ϵ-amino group to the α-carboxyl group of the L-centre of a *meso*-diaminopimelic acid residue that forms the C-terminal of a tripeptide linked in the usual way to a muramic acid residue in the peptidoglycan. The fatty acids found on the glyceryl cysteine were mainly palmitic acid (45%), palmitoleic acid (11%), *cis*-vaccenic acid (24%) cyclopropylenehexadecanoic acid (12%) and cyclopropyleneoctadecanoic acid (8%), whereas the cysteine *N*-terminus was blocked by different proportions: palmitic acid (65%), palmitoleic acid (11%) and *cis*-vaccenic acid (11%). The protein was found to have a highly ordered α-helical structure, which it easily regained after denaturation. Non-polar residues occurred with a frequency consistent with their presence aligned along one side of an α-helix. The structure finally proposed consisted of a non-helical C-terminal sequence of about ten residues, a helix of some 20 residues, a β-loop between residues 25 and 29, and a further α-helix extending almost to the *N*-terminal cysteine [8], (Fig. 7.8). During separation of the inner and outer membrane after lysozyme digestion it was shown that the lipoprotein (with residual labelled peptidoglycan attached to it) moved with the outer membrane. It was concluded, therefore, that in *E. coli* the peptidoglycan itself is linked to the outer membrane and is not attached to the inner one except during its biosynthesis [9]. Mutants of *E. coli* lacking the lipoprotein were exceptionally sensitive to low Mg^{2+} concentration and seemed to have leaky outer membranes [80]. It was concluded that the bound lipoprotein helps to maintain the outer envelope structure and thus stabilize it as a barrier to the environment, (see also Chapter 1.3).

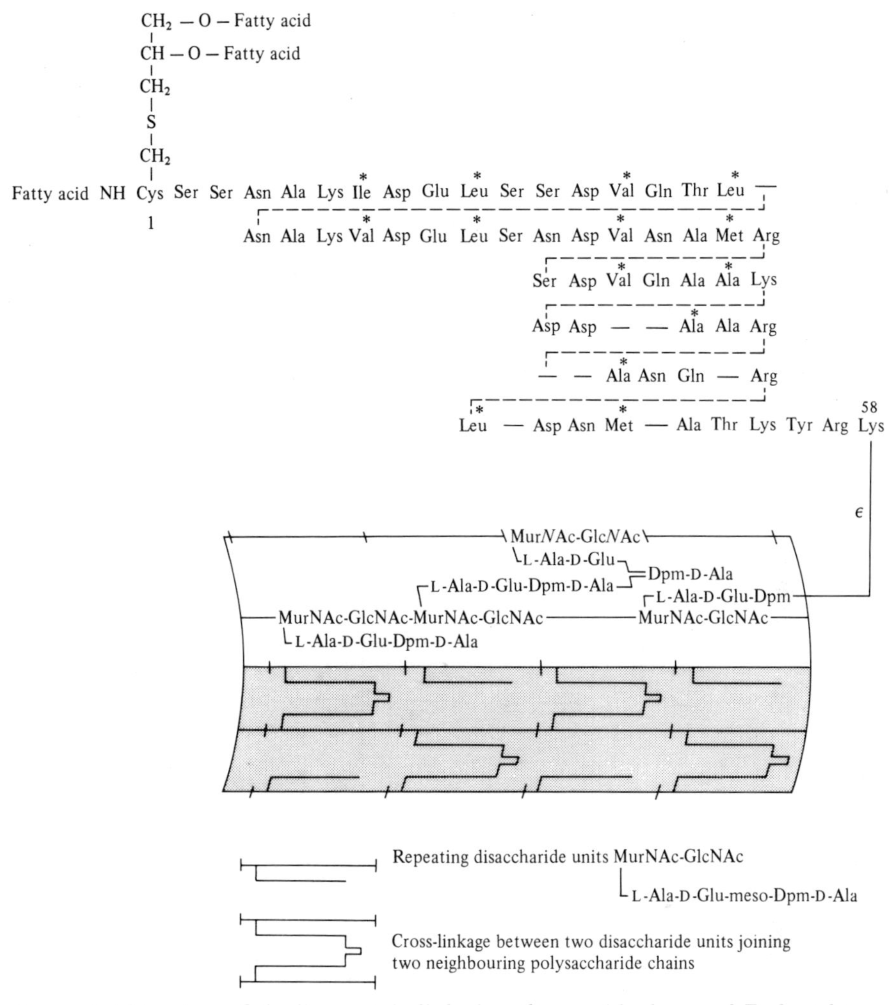

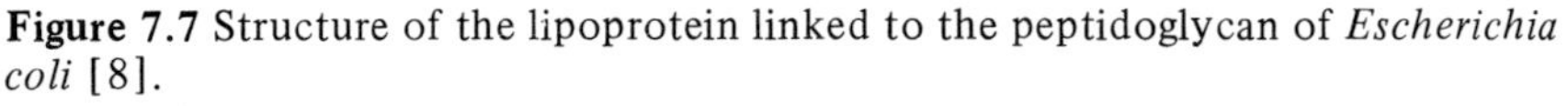

Figure 7.7 Structure of the lipoprotein linked to the peptidoglycan of *Escherichia coli* [8].

In Gram-negative bacteria other than *E. coli* lipoprotein bound to peptidoglycan has been found in Salmonella strains and in *Serratia marcescens*, but not in *Proteus mirabilis, P. vulgaris* or *Pseudomonas fluorescens* [7]. Lack of cross-reaction to anti-*E. coli* lipoprotein antiserum also showed that at least that specific structure was missing in a large number of other Gram-negative bacteria [8, 25].

Evidence has been presented for a biosynthetic precursor for the lipoprotein, namely a prolipoprotein with a peptide extension of 20 amino acid residues at the amino terminus of the lipoprotein molecule. The extension consisted of an unusual sequence of amino acids with basic but no acidic residues, three glycines (glycine is absent from the finished lipoprotein) and 60% hydrophobic residues [17, 27, 42]. This additional sequence has been proposed to be needed for binding to the cyto-

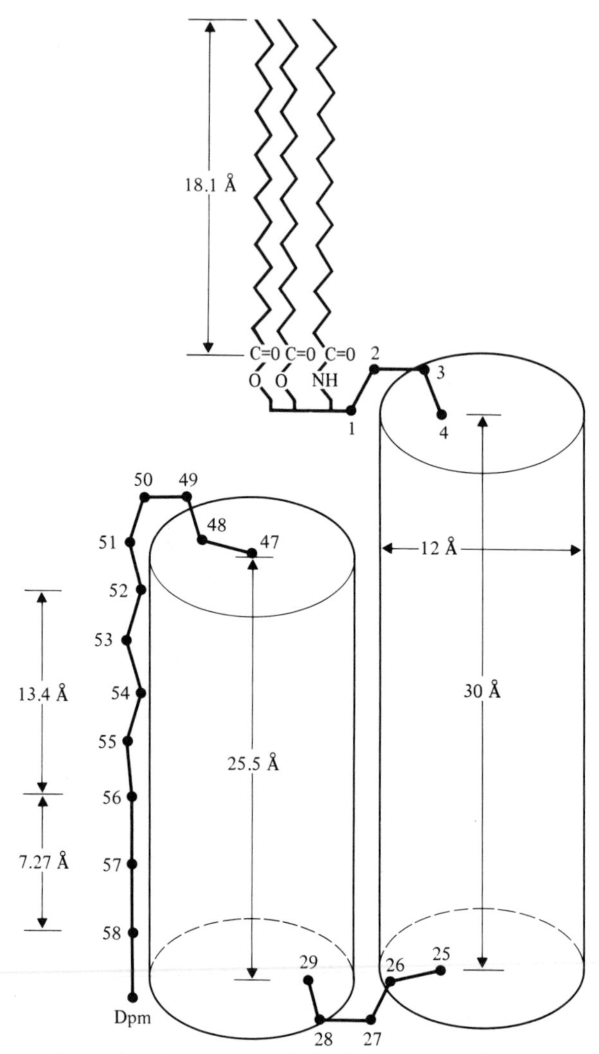

Figure 7.8 Three-dimensional structure of the lipoprotein of the outer membrane of *E. coli* [8]. The amino acid residues 1–4 form a β-loop, 5–24 an α-helix with the possible exception of 13–17, 25–29 a β-loop, 30–47 an α-helix, 48–51 a β-loop, 52–56 a β-sheet and residues 57 and 58 a coil.

plasmic membrane while protein synthesis continues and translocation across the membrane occurs. Thereafter the pro-protein is free in the periplasmic space and undergoes folding and normally, release of the extra fragment and acylation of the new *N*-terminal cysteine residue [26]. However, experiments with phenethyl alcohol treated *E. coli* [26] and with a prolipoprotein mutant [52] show that the extra sequence is not essential for incorporation into the outer membrane, nor does its continued presence prevent such incorporation.

7.2.1 *Lipopolysaccharides*

The O-somatic antigens of Gram-negative bacteria have been identified as lipopolysaccharides that form a part of the outer membrane. These molecules are also responsible for the endotoxicity of the cells. Although these compounds are part of the outer membrane component of the cell wall, they can be removed from whole cells by a number of extractants, including salt solutions, chelating agents (EDTA) and trichloroacetic acid. The most widely used method [84] involves treating dried cells or cell walls with 45% aqueous phenol at 68° C for 5 min. After cooling, two layers form and the water phase contains most O-specific lipopolysaccharides, along with some capsular antigens and nucleic acids. Some kinds of lipopolysaccharide, however, occur in the phenol phase; these are highly lipophilic, either because they have a large proportion of lipid A (as in R mutants), or because they contain much *N*-acetylamino sugar.

A method of extraction specific for R-form lipopolysaccharides and glycolipids consists of treating dry bacteria at 10–20° C with a mixture of phenol, chloroform and light petroleum [22]. Another mild method for extracting lipopolysaccharides with *n*-butanol–water has been described [61], which yields products similar in amount and composition to those obtained by phenol extraction. Lipopolysaccharide molecules are built up of two main constituent regions, the lipid A, where all the lipophilic fatty acid groups are attached, and the polysaccharide part, which is hydrophilic. The latter region has been shown to arise biosynthetically as two separate portions, the proximal core that characteristically contains ketodeoxyoctonic acid (3-deoxy-D-mannoctulosonic acid, KDO) and heptose (usually L-glycero-D-mannoheptose) as well as hexoses and amino sugars, and the distal repeating units that are specific for the particular serotype that is being considered (Fig. 7.9). The

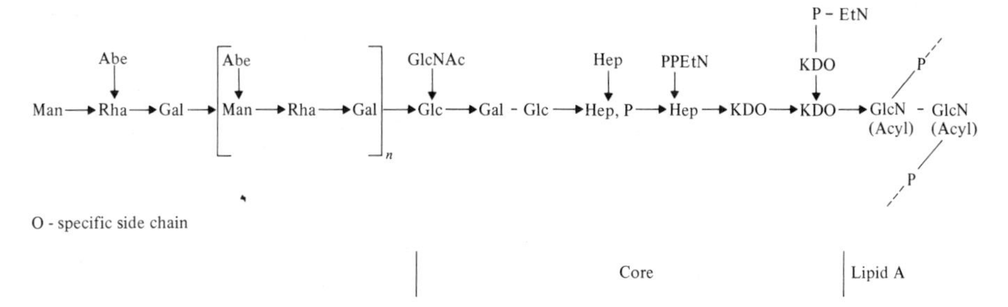

Figure 7.9 Lipopolysaccharide of *Salmonella typhimurium.*

complex relationship between immunological specificity and the chemical structure of the repeating units has been worked out over the last few decades. As is common with polysaccharide antigens, the non-reducing terminals are of greatest importance, and in the lipopolysaccharides the dideoxyhexoses abequose and colitose (respectively 3,6-dideoxy-D- and 3,6-dideoxy-L-xylohexose) paratose (3,6-dideoxy-D-ribohexose) tyvelose and ascarylose (respectively 3,6-dideoxy-D- and 3,6-dideoxy-L-arabinohexose)) occur as branch substituents on the main chain (e.g. Fig. 7.9 for

S. typhimurium) and exert a major influence. As elsewhere, the anomeric configuration of the linkages also produces immunological differences, and there are sometimes further substituents, such as *O*-acetyl groups, that provide additional specificity. Some examples of relationships between chemical structure and serological specificity are given in Table 7.3.

Table 7.3 Structure of the O-specific repeating units of *Salmonella* lipopolysaccharides [58]

Salmonella group (figures in parentheses represent the O-antigenic specificities)		*Repeating Unit*
Group A	*S. paratyphi A* (1, 2, 12)	α-Par α-Glc * |1,3 |1,4 → 2-α-Man-1,4-α-Rha-1,3-α-Gal-1 → | Ac
Group B	*S. typhimurium* (4, 5, 12)	Ac-O-2-α-Abe α-Glc * |1,3 |1,4 → 2-α-Man-1,4-β-Rha-1,3-α-Gal-1 →
	S. bredeney (wild) (1, 4, 12)	α-Abe α-Glc * |1,3 |1,6 → 2-α-Man-1,4-β-Rha-1,3-α-Gal-1 →
Group C₁	*S. thompson* (6₂, 7)	Glc † |1,3 → Man-1,2-Man-1,2-Man-1,2-Man-1 →
Group C₂	*S. newport* (6₁, 8)	α-Abe α-GlcAc † |1,3 |1,3 → 4-α-Rha-1,2-α-Man-1,2-α-Man-1,3-α-Gal → |2 Ac
Group C₃	*S. kentucky* (8, 20)	α-Abe GlcAc † |1,3 |1,4 → 4-α-Rha-1,2-Man-1,2-Man-1,3-Gal →
Group D₁	*S. typhi* (9, 12)	α-Tyv Ac-O-2-α-Glc * |1,3 | → 2-α-Man-1,4-α-Rha-1,3-α-Gal-1 →
Group D₂	*S. strasbourg* (9, 46)	α-Tyv α-Glc * |1,3 | → 6-β-Man-1,4-α-Rha-1,3-α-Gal-1 →
Group E₁	*S. muenster* (3, 10)	Glc * |1,4 → 6-β-Man-1,4-α-Rha-1,3-α-Gal →
Group E₂	*S. newington* (3, 15)	α-Glc * |1,4 → 6-β-Man-1,4-α-Rha-1,3-β-Gal →

Table 7.3 (*Cont.*)

Salmonella group		*Repeating Unit*
Group E_4	*S. senftenberg* (1, 3, 19)	α-Glc (1,6) * → 6-β-Man-1,4-Rha-1,3-α-Gal-1 →
Group G	*S.friedenau* (13, 22)	Glc-1 →? † → β-Gal-1,3-Gal*N*Ac-1,3-Gal*N*Ac-1,4-Fuc →
Group L	*S. minnesota* (21)	α-Gal α-Glc*N*Ac † → β-Gal-1,3-Gal*N*Ac-1,3-Gal*N*Ac →
Group N	*S. godesberg* (30)	Glc (1,4) † → β-Glc-1,3-Gal*N*Ac-1,4-Fuc →
Group U	*S. milwaukee* (4,3)	α-Gal (1,3) † → β-Gal-1,3-Gal*N*Ac-1,3-Glc*N*Ac-1,4-Fuc →

*Biological Unit †Chemical Unit

The structure of the core region has been elucidated particularly with the aid of mutants blocked at various stages in the biosynthesis of lipopolysaccharide. Mutants lacking O-specificity are defined as R-forms (sometimes associated with rough colonial morphology) and gradually an armoury of such mutants has been assembled, ranging from R_a, in which the whole core is present to R_e in which, of the core sugars, only KDO remains. An additional class of SR mutants exists, in which only one repeating unit of the O-specific chains is added to the core, instead of the extensively polymerized chains that occur in S forms. Apparently these mutants lack the elongation enzyme that polymerizes the repeating units once they have been synthesized on the undecaprenol carrier intermediate [92].

The most detailed information on lipopolysaccharides is available for Salmonella species, but many other Gram-negative organisms have been shown to produce similar molecules. Thus *E. coli* strains have O-antigenic specificities, which are attributable to lipopolysaccharides of the same general type as those of Salmonella, with components of lipid A, and core and side-chain regions of the polysaccharide. However, antisera even against strains with only the core sugars present often do not cross react with Salmonella strains having the same sugars, so that clearly additional structural specificity is present [55]. Thus, although both cores have the same sugars, the structures have been shown to differ in some respects (Fig. 7.10). There is also evidence that strains of *E. coli* with the same O-specificity can have different core regions in their lipopolysaccharides. Although the complete structures have not been worked out, the core designated *coli* R1 contains galactose, glucose and heptose in molar proportions of 2:3:3, whereas that called *coli* R2 has galactose, glucose, glucosamine and heptose in the proportions 2:4:1:4 [74], *coli* R3 proportions are 1:2.4:2.5:2.5, and K12, 1:3:1:4–6. These values contrast with Salmonella (galactose,

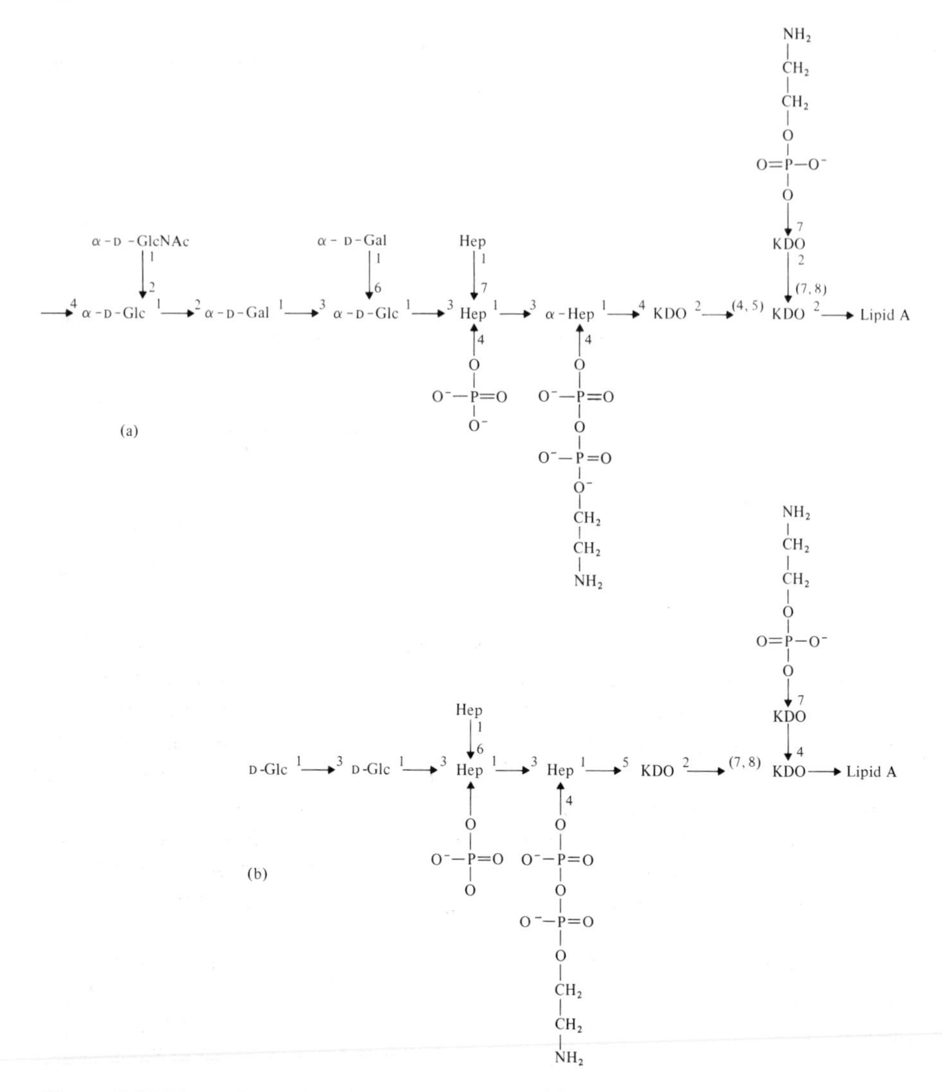

Figure 7.10 Lipopolysaccharide core structures. (a) Salmonella [56, 58]. (b) *Escherichia coli* B [69].

glucose, glucosamine, heptose; 2:2:1:3) and Shigella (galactose, glucose, glucosamine; 1:3:1 plus heptose). A structure proposed for the core of *E. coli* B is shown in Fig. 7.10.

There is considerable evidence that lipopolysaccharides form phage receptors for Gram-negative bacteria, and furthermore that the core region contains the specificity for attachment of many phages. In some cases isolated lipopolysaccharide preparations will cause phage attachment and ejection of the viral DNA [70]. Similarly *Citrobacter*, *Arizona*, *Shigella*, *Proteus*, *Yersinia*, *Chromobacterium*, *Pseudomonas* and *Xanthomonas* species all produce lipopolysaccharides with differences of detailed structure within the same general pattern.

7.2.2 *The structure of lipid A*

The structure of lipid A has been largely worked out by Lüderitz and Westphal and their colleagues. This part of the lipopolysaccharide molecule forms the backbone to which individual inner core units (i.e. KDO and heptose portions) are attached, and thence the outer core and the O-specific side chains if present. The structure is shown in Fig. 7.11 [57]. Although earlier it had been thought that lipid A molecules might be linked in a chain by pyrophosphate residues [56] this view is no longer held. The lipophilic property is provided by fatty acids that are attached to β-1,6 linked disaccharides of D-glucosamine. In *Salmonella* these fatty acids are lauric (C12), myristic (C14), palmitic (C16) and 3-D-(–)hydroxymyristic acid (β-hydroxy-myristic acid), the latter hydroxy acid being characteristic of lipid A molecules. It is this acid that substitutes the amino group of the glucosamine residues, but it also occurs as an *O*-acyl group, and often has its own hydroxy group substituted by yet another fatty acid, thus increasing the lipophilicity of the whole structure. In some lipid A molecules, probably depending on culture conditions, the terminal phosphate groups are substituted by 4-amino-L-arabinose and phosphorylethanolamine (Fig.7.11). β-Hydroxymyristic acid also occurs in *E. coli*, *Proteus mirabilis*, *Aerobacter aerogenes* and *Bordetella pertussis*. In other Gram-negative bacteria the hydroxy acid is not necessarily C14 e.g. β-hydroxylauric acid is found in the lipopolysaccharides of *Serratia*, *Neisseria*, *Pseudomonas* and *Azotobacter*, β-hydroxy tri- and penta-decanoic acids in *Veillonella* and β-hydroxyheptadecanoic acid in *Cytophaga*. Partial structures of the lipid As from a variety of Gram-negative bacteria have been proposed [57] (Fig. 7.11).

Morrison and Leive [66], showed that lipopolysaccharide extracted from *E. coli* 0111 B4 either by aqueous phenol or by aqueous butanol could be separated into two clearly defined molecular types, one of which had much longer *O*-specific side chains than the other. There is evidence that similar dual populations may occur in other bacteria, but their significance is not known.

The biological action of lipid A as the endotoxic region of lipopolysaccharide is now well established. This part of the molecule can interact with complement causing a loss of haemolytic activity and all the physiological symptoms, such as pyrogenicity, that characterize endotoxin [56].

The importance of lipopolysaccharides in the structure of the Gram-negative bacterium has been shown by the fact that it is not possible to isolate stable mutants that completely lack these entities. Mutants have been obtained, as mentioned earlier, in which the *O*-specific side chains are missing and also less and less of the oligosaccharide core is found until only KDO remains. Beyond that point blockage of synthesis appears to be lethal. However, conditional mutants of *Salmonella typhimurium* have been isolated in which synthesis or addition of KDO is temperature-sensitive [51]. At the restrictive temperature these mutants cease to grow and accumulate

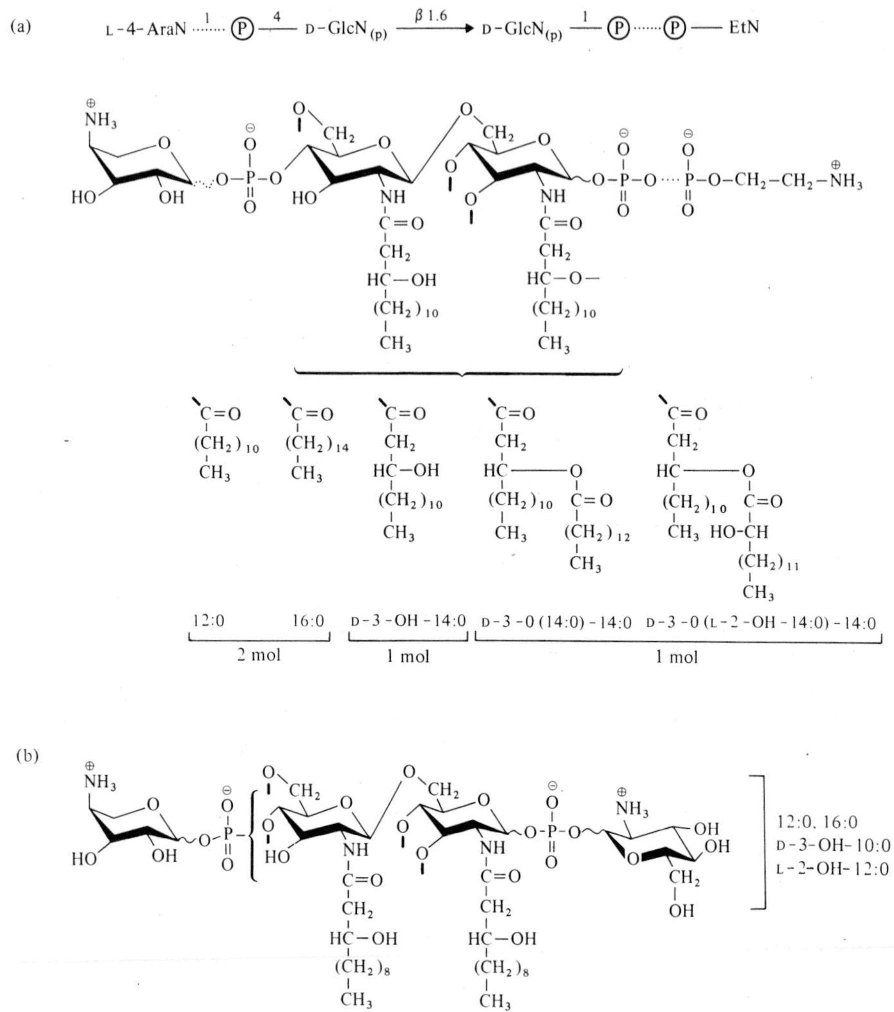

Figure 7.11 Structure proposed for lipid A moieties of lipopolysaccharides [57]. (a) *Salmonella*. The dotted lines indicate substitution on only some of the molecules. A substituent, as yet unidentified, is proposed at the OH group of the amide-linked 3-hydroxymyristic acid. The anomeric form of the linkage between L-arabinosamine has not been established. (b) *Chromobacterium violaceum*. The attachment sites of KDO and phosphate to the non-reducing glucosamine are not known. (c) *Rhodospirillum tenue*. Again attachment sites are unknown. (d) *Rhodospseudomonas viridis*.

lipid A precursors containing glucosamine, phosphate and β-hydroxymyristic acid, but lacking myristic and lauric acids. There is evidence in several members of the Enterobacteriaceae that the outer membrane contains a small (15 000 mol.wt) basic

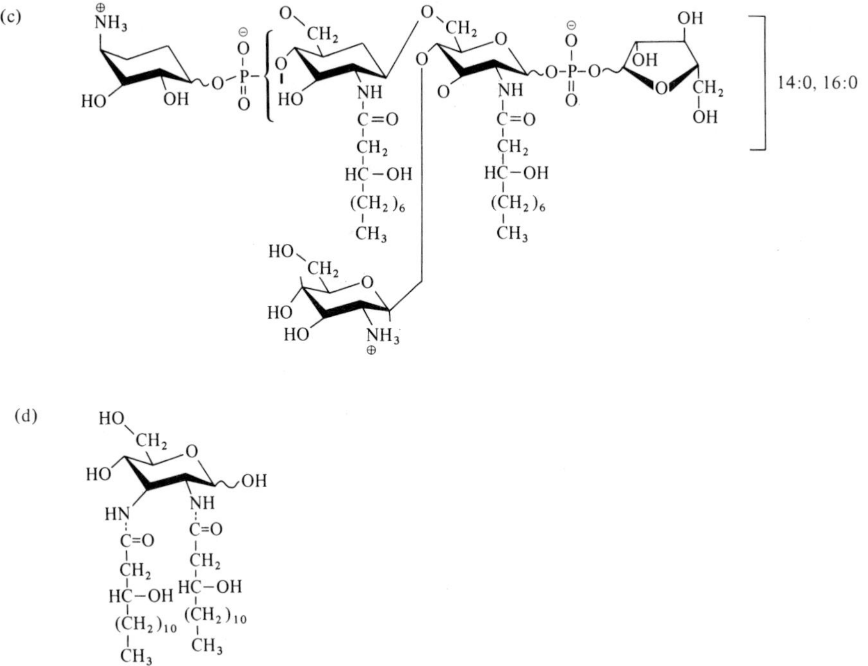

protein with a high affinity for lipopolysaccharide, mediated by the acidic groups of the lipid A [23].

7.2.3 *Role of additional polymers in bacterial cell walls*

The foregoing account gives some idea of the complexity of the cell walls of Gram-negative bacteria, also discussed in Chapter 1. There is evidence that additional polymers have specific functions (e.g. a role for teichoic acids in divalent cation control or for lipoprotein in stabilising the outer membrane) but as yet there is no clear idea of exactly how bacteria juggle with the different wall components at various stages of growth and division, or how they ensure that the right polymer is in the right place at the right time. Although some evidence is beginning to accumulate for the mechanism of assembly of outer membrane proteins in *E. coli* [17] much still remains unknown. The whole question of how directives can be transmitted from the interior of the cell to relatively remote regions of the wall is one that at present lacks an answer.

References

1. Ågren, G. and de Verdier, C. H. (1958) *Acta chem. scand.* **12**, 1927–36.
2. Archibald, A. R. and Coapes, H. E. (1972) *J.gen.Microbiol.* **73**, 581–5.

3. Archibald, A. R. and Coapes, H. E. (1976) *J.Bact.* **125**, 1195–206.
4. Armstrong, J. J., Baddiley, J., Buchanan, J. G., Carss, B. and Greenberg, G. R. (1958) *J.Chem.Soc.* 4344–5.
5. Björk, I., Petersson, B. A. and Sjöquist, J. (1972) *Eur. J. Biochem.* **29**, 579–84.
6. Braun, V. and Rehn, K. (1969) *Eur.J. Biochem*, **10**, 426–38.
7. Braun, V., Rehn, K. and Wolff, H. (1970) *Biochemistry* **9**, 5041–9.
8. Braun, V. (1975) *Biochim.biophys.Acta.* **415**, 335–77.
9. Bosch, V. and Braun, V. (1973) *FEBS Letts.* **34**, 307–310.
10. Button, D., Archibald, A. R. and Baddiley, J. (1966) *Biochem. J.* **99**, 11C-14C.
11. Cheng, K. J. and Costerton, J. W. (1977) *J. Bact.* **129**, 1506–512.
12. Cleveland, R. F., Höltje, J. V., Wicken, A. J., Tomasz, A., Daneo-Moore, L. and Shockman, G. D. (1975) *Biochem.biophys.Res.Commun.* **67**, 1128–35.
13. Coley, J., Archibald, A. R. and Baddiley, J. (1977) *FEBS Letts.*, **80**, 405–407.
14. Coyette, J. and Ghuysen, J. M. (1968) *Biochemistry*, **7**, 2385–9.
15. Coyette, J. and Ghuysen, J. M. (1970) *Biochemistry* **9**, 2935–43.
16. Diaz-Mauriño, T. and Perkins, H. R. (1974) *J.gen.Microbiol.* **80**, 533–9.
17. Di Rienzo, J. M., Nakamura, K. and Inouye, M. (1978) *A.Rev.Biochem.* **47**, 481–532.
18. Duckworth, M., Archibald, A. R. and Baddiley, J. (1975) *FEBS Letts.* **53**, 176–9.
19. Ellwood, D. C. and Tempest, D. W. (1969) *Biochem.J.* **111**, 1–5.
20. Ellwood, D. C. and Tempest, D. W. (1972) *Adv.microb.Physiol.* **7**, 83–117.
21. Fox, E. N. (1964) *Bact.Rev.* **38**, 57–86.
22. Galanos, C., Lüderitz, O. and Westphal, O. (1969) *Eur.J.Biochem.* **9**, 245–9.
23. Geyer, R., Galanos, C., Westphal, O., and Golecki, J. F. (1979) *Eur.J.Biochem.* **98**, 27–38.
24. Glaser, L., Ionesco, H. and Schaeffer, D. (1966) *Biochim.biophys.Acta.* **124**, 415–17.
25. Halegoua, S., Hirashima, A. and Inouye, M. (1974) *J.Bact.* **120**, 1204–208.
26. Halegoua, S. and Inouye, M. (1979) *J.molec.Biol.* **130**, 39–61.
27. Halegoua, S., Sekizawa, J. and Inouye, M. (1977) *J.biol.Chem.* **252**, 2324–30.
28. Hall, E. A. and Knox, K. W. (1965) *Biochem. J.* **96**, 310–318.
29. Hase, S. and Matsushima, Y. (1970) *J.Biochem.* **68**, 723–8.
30. Hase, S. and Matsushima, Y. (1971) *J.Biochem.* **69**, 559–65.
31. Hase, S. and Matsushima, Y. (1972) *J.Biochem.* **72**, 1117–28.
32. Hase, S. and Matsushima, Y. (1977) *J.Biochem.* **81**, 1181–6.
33. Hay, J. B., Davey, N. B., Archibald, A. R. and Baddiley, J. (1965) *Biochem.J.* **94**, 7c–9c.
34. Van Heijenoort, J., Menjon, D., Flouret, B., Szulmajster, J., Laporte, J. and Batelier, G. (1971) *Eur.J.Biochem.* **20**, 442–50.
35. Heptinstall, S., Archibald, A. R. and Baddiley, J. (1970) *Nature, Lond.* **225**, 519–21.
36. Hewett, M. J., Wicken, A. J., Knox, K. W. and Sharpe, M. E. (1976) *J.gen. Microbiol.* **94**, 126–30.
37. Heymann, H., Maniello, J. M. and Barkulis, S. S. (1963) *J.biol.Chem.* **238**, 502–509.
38. Höltje, J. V. and Tomasz, A. (1975) *Proc.Natn.Acad.Sci.U.S.A.* **72**, 1690–94.
39. Hughes, R. C. and Thurman, P. F. (1970) *Biochem. J.* **117**, 441–9.
40. Inouye, M. (1971) *J.biol.Chem.* **246**, 4834–8.
41. Inouye, M., Shaw, J. and Shen, C., (1972) *J.biol.Chem.* **247**, 8154–9.
42. Inouye, S., Wang, S., Sekizawa, J., Halegoua, S. and Inouye, M. (1977) *Proc. Natn.Acad.Sci.U.S.A.* **74**, 1004–1008.

43. Ivatt, R. J. and Gilvarg, C. (1977) *Biochemistry*, **16**, 2436–2440.
44. Janczura, E., Perkins, H. R. and Rogers, H. J. (1961) *Biochem.J.* **80**, 82–93.
45. Knox, K. W. and Hall, E. A. (1964) *J.gen.Microbiol.* **37**, 433–8.
46. Knox, K. W. and Hall, E. A. (1965) *Biochem.J.* **96**, 302–309.
47. Knox, K. W. and Wicken, A. J. (1973) *Bact.Rev.* **37**, 215–57.
48. Lambert, P. A., Hancock, I. C. and Baddiley, J. (1977) *Biochim.biophys.Acta* **472**, 1–12.
49. Lancefield, R. C. and Perlmann, G. (1952) *J.exp.Med.* **96**, 83–97.
50. Leduc, M., van Heijenoort, J., Kaminski, M. and Szulmajster, J. (1973) *Eur.J. Biochem.* **37**, 389–400.
51. Lehmann, V., Rupprecht, E. and Osborn, M. J. (1977) *Eur.J.Biochem.* **76**, 41–9.
52. Lim, J. J. C., Kanazawa, H., Ozols, J. and Wu, H. C., (1978) *Proc.Natn.Acad. Sci.U.S.A.* **75**, 4891–5.
53. Liu, T. Y. and Gotschlich, E. C. (1967) *J.biol.Chem.* **242**, 471–6.
54. Lüderitz, O., Staub, A. M. and Westphal, O. (1966) *Bact.Rev.* **30**, 192–255.
55. Lüderitz, O., Jann, K. and Wheat, R. (1968) *Comprehensive Biochemistry*, ed. Florkin, M. and Stotz, E. H., **26A**, 105–228. Amsterdam: Elsevier.
56. Lüderitz, O., Galanos, C., Lehmann, V., Nurminen, M., Rietschel, E. T., Rosenfelder, G., Simon, M. and Westphal, O. (1973) *J.infect.Dis.* **128**, suppl. S17–S29.
57. Lüderitz, O., Galanos, C., Lehmann, V., Mayer, H., Rietschel, E. T. and Weckesser, J. (1978) *Naturwissenschaften* **65**, 578–85.
58. Lüderitz, O., Westphal, O., Staub, A. M. and Nikaido, H. (1971) *Microbial Endotoxins* **4**, eds. Weinbaum, G., Kadis, S. and Ajl, S. J., pp. 145–233, New York and London: Academic Press.
59. McCarty, M. and Morse, S. I. (1964) *Adv.Immun.* **4**, 249–86.
60. Morse, S. I. (1962) *J.exp.Med.*, **116**, 247–51.
61. Morrison, D. C. and Leive, L. (1975) *J.biol.Chem.* **250**, 2911–19.
62. Mühlradt, P. F. (1976) *J.supramolec.Struct.* **5**, 103–108.
63. Ofek, I., Beachey, E. H., Eyal, F. and Morrison, J. C. (1977) *J.infect.Dis.* **135**, 267–74.
64. Owen, P. and Salton, M. R. J. (1975) *Biochem.biophys.Res.Commun.* **63**, 875–80.
65. Partridge, M. D., Davison, A. L. and Baddiley, J. (1973) *J.gen.Microbiol.* **74**, 169–73.
66. Perkins, H. R. (1963) *Biochem.J.* **86**, 475–83.
67. Pless, D. D., Schmit, A. S. and Lennarz, W. J. (1975) *J.biol.Chem.* **250**, 1319–27.
68. Powell, D. A., Duckworth, M. and Baddiley, J. (1975) *Biochem. J.* **151**, 387–97.
69. Prehm, P., Stirm, S., Jann, B. and Jann, K. (1975) *Eur.J.Biochem.* **56**, 41–55.
70. Rapin, A. M. C. and Kalckar, H. M. (1971) *Microbial Toxins* **4**, 267–307.
71. Robson, R. L. and Baddiley, J. (1977) *J.Bact.* **129**, 1051–1058.
72. Rohr, T. E., Levy, G. N., Stark, N. J. and Anderson, J. S. (1977) *J.biol.Chem.* **252**, 3460–65.
73. Salton, M. R. J. (1953) *Biochim.biophys.Acta* **10**, 512–23.
74. Schmidt, G., Jann, B. and Jann, K. (1969) *Eur.J.Biochem.* **10**, 501–510.
75. Shabarova, Z. A., Hughes, N. A. and Baddiley, J. (1962) *Biochem. J.* **83**, 216–19.
76. Sjödahl, J. (1977) *Eur.J.Biochem.* **73**, 343–51.
77. Sjöquist, J., Meloun, B. and Hjelm, H. (1972) *Eur.J.Biochem.* **29**, 572–8.
78. Sjöquist, J., Movitz, J., Johansson, I. B. and Hjelm, H., (1972) *Eur.J.Biochem.* **30**, 190–94.

79. Stark, N. J., Levy, G. N., Rohr, T. E. and Anderson, J. S. (1977) *J.biol.Chem.* **252**, 3466–72.
80. Suzuki, H., Nishimura, Y., Yasuda, S., Nishimura, A., Yamada, M. and Hirota, Y., (1978) *Mol.gen.Genet.* **167**, 1–9.
81. Tomasz, A.,and Waks, S. (1975) *Proc.Natn.Acad.Sci.U.S.A.* **72**, 4162–6.
82. Tomasz, A. and Höltje, J. V. (1977) *Microbiology 1977*, 209–215.
83. Tylewska, S. K., Hjertén, S. and Wadström, T. (1979) *FEMS Microbiol. Letts.* **6**, 249–53.
84. Westphal, O., Lüderitz, O. and Bister, F. (1952) *Z.Naturf*, **7b**, 148–55.
85. White, P. J. and Gilvarg, C. (1977) *Biochemistry* **16**, 2428–35.
86. Wicken, A. J. and Baddiley, J. (1963) *Biochem.J.* **87**, 54–62.
87. Wicken, A. J. and Knox, K. W. (1975) *Science* **187**, 1161–7.
88. Wu, T. C. M. and Park, J. T. (1971) *J.Bact.* **108**, 874–84.
89. Wyke, A. W. and Ward, J. B. (1977) *J.Bact.* **130**, 1055–63.
90. Yamada, M., Hirose, A. and Matsuhashi, M. (1975) *J.Bact.* **123**, 678–86.
91. Yamada, M., Matsuhashi, M. and Torii, M. (1978) *J.gen.appl.Microbiol.* **24**, 307–315.
92. Yuasa, R., Nakane, K. and Nikaido, H. (1970) *Eur.J.Biochem.* **15**, 63–71.

8
Biosynthesis of peptidoglycan

8.1 Introduction

The mechanism of biosynthesis of peptidoglycan has been elucidated in a number of organisms including *Staphylococcus aureus*, *Micrococcus luteus*, *Escherichia coli* and more recently several members of the Bacilli. Although these organisms differ widely in morphology and the chemical composition of their peptidoglycans, the process of biosynthesis in each shows sufficient common characteristics to establish the basic nature of the process. Clearly, such modifications as are made to the biosynthetic pathway in the individual organisms are characteristic of those organisms and probably represent the evolution of specific enzymes. Overall peptidoglycan synthesis can be divided into three distinct stages:

(1) The formation of the nucleotide sugar-linked precursors, UDP-*N*-acetylmuramyl-pentapeptide and UDP-*N*-acetylglucosamine;

(2) The transfer of phospho-*N*-acetylmuramyl-pentapeptide and *N*-acetylglucosamine to a lipophilic carrier, undecaprenyl phosphate, to yield a disaccharide-(pentapeptide)-pyrophosphate-undecaprenol and

(3) The transfer of this completed sub-unit to the growing peptidoglycan. At this stage cross-bridge formation occurs, together with secondary modification of the newly-synthesized peptidoglycan.

8.2 Synthesis of nucleotide sugar precursors

The initial reaction of the complex series (Fig. 8.1) leading to the synthesis of UDP-*N*-acetylmuramyl-pentapeptide is the formation of UDP-*N*-acetylglucosamine and inorganic pyrophosphate from UTP and *N*-acetylglucosamine-1-phosphate. The reaction, analogous to those involved in the synthesis of many other nucleotide-linked sugars, is catalysed by UDP-*N*-acetylglucosamine pyrophosphorylase. The next step is the formation of the first intermediate wholely characteristic of peptidoglycan, UDP-*N*-acetylmuramic acid. These reactions involve the transfer of a pyruvate enol ether from phosphoenolpyruvate to UDP-*N*-acetylglucosamine and its subsequent reducation to yield UDP-*N*-acetylmuramic acid.

The enzymes catalysing these two processes have been purified from *Enterobacter cloacae* [57, 197, 244], *E. coli* [4] and partially purified from *Staphylococcus*

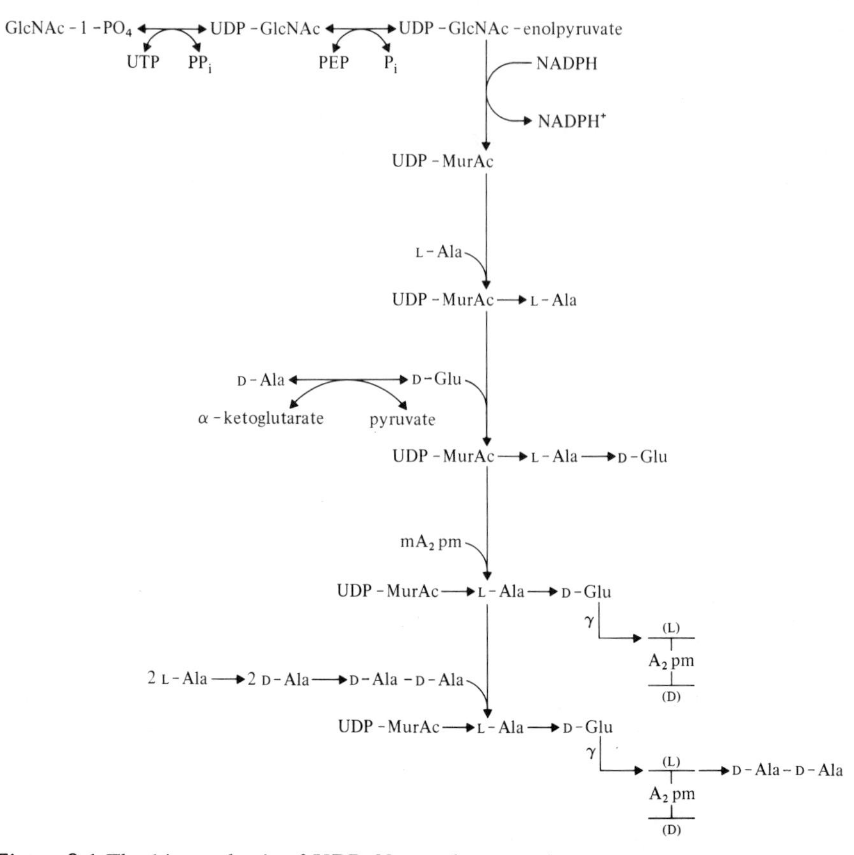

Figure 8.1 The biosynthesis of UDP-*N*-acetylmuramyl-pentapeptide. The reactions shown are those required for the biosynthesis of UDP-*N*-acetylmuramyl-pentapeptide in Bacilli and almost all Gram-negative organisms. As described in the text the ligases require either Mn^{2+} or Mg^{2+} for activity and the hydrolysis of ATP.

epidermidis [214, 232]. The formation of UDP-*N*-acetylglucosamine enolpyruvate, the first committed intermediate in the biosynthesis of peptidoglycan, is subject to feedback inhibition in each of the enzymes examined. Those from *E. cloacae* [244], *Bacillus cereus* and *E. coli* [230] were inhibited by both UDP-*N*-acetylmuramyl-pentapeptide and -tripeptide. In contrast, these precursors were without effect on the enzyme from *S. epidermidis* where inhibition was obtained only with UDP-*N*-acetylmuramic acid. However, any relationship between the precursor accumulated during inhibition of peptidoglycan synthesis by certain antibiotics such as vancomycin or cycloserine, and the inhibition of the phosphoenolpyruvate (PEP): UDP-*N*-acetylglucosamine enolpyruvyltransferase is not immediately obvious. Thus, addition of either vancomycin or cycloserine to cultures of *B. cereus* results in the accumulation of UDP-*N*-acetylmuramyl-pentapeptide and -tripeptide respectively. Should the control of precursor biosynthesis rest simply in feedback inhibition of the initial enzyme in the pathway, then marked accumulation of either of these precursors would not be expected to occur. The enolpyruvyl transferase from all the organisms investigated is irreversibly inhibited by the antibiotic phosphonomycin. The mechanism of this inhibition will be considered in Chapter 9.2.

Recent evidence [40] suggests that at this stage an additional reaction, the oxidation of the *N*-acetyl substituent of muramic acid to *N*-glycolyl, occurs in *Nocardia* and *Mycobacteria.* In the presence of benzylpenicillin *N. asteroides* accumulated UDP-*N*-glycolylmuramyl peptide and a mixture of UDP-*N*-acetyl- and -*N*-glycolylmuramic acid. Cytoplasmic extracts of the organism contained a hydroxylase which in the presence of NADPH converted the acetyl group of UDP-*N*-acetylmuramic acid to a glycolyl residue. Earlier, Petit *et al.* [154] had described the accumulation of UDP-*N*-glycolylmuramyl-L-alanyl-D-isoglutamyl-meso-diaminopimelic acid by *M. phlei* inhibited with cycloserine.

The first three amino acids of the peptide side chain L-alanine or glycine, D-glutamic acid and R_3 are then added sequentially to the carboxyl group of the muramic acid residue. The addition of each is catalysed by a specific ligase (adding enzyme) which requires for activity either Mn^{2+} or Mg^{2+} and the concomitant hydrolysis of ATP. D-Glutamic acid, one of the D-amino acids characteristic of peptidoglycan, is produced by L-glutamic acid racemase, which has been studied in *Lactobacillus fermenti* [24] or by a D-glutamate:D-alanine transaminase [118]. Unlike protein synthesis, where sequential addition of amino acids is directed by a nucleic acid template, the ordered addition of amino acids to build UDP-*N*-acetylmuramyl-pentapeptide depends upon the specificity of the ligases for their respective substrates. Consequently, the L-lysine ligase of *S. aureus* fails to add diaminopimelic acid to UDP-*N*-acetylmuramyl-L-alanyl-D-glutamic acid, while the converse result has been obtained with the meso-diaminopimelic acid ligases of *E. coli*, *Corynebacterium xerosis* and *B. cereus* [72, 74, 75, 135]. Identical results have also been obtained in *B. sphaericus* with the L-lysine ligase of the vegetative organism and the meso-diaminopimelic acid ligase required for synthesis of the peptidoglycan found in the spore cortex [204]. In contrast, the corresponding enzyme from *Bifidobacterium globosum* catalyses the addition of lysine, ornithine and even diaminobutyric acid to UDP-*N*-acetylmuramyl-L-alanyl-D-glutamic acid [64]. In *Corynebacterium poinsettiae* and *C. insidiosium* [239], the addition of the third amino acid of the peptide side chain was shown to be specific for both the amino acid and the nucleotide dipeptide involved in the peptidoglycan synthesis (Table 8.1). Thus in *C. poinsettiae*, L-homoserine was only added to UDP-*N*-acetylmuramyl-dipeptide containing glycine as the first amino acid of the peptide side chain. In contrast, the addition of D-glutamic acid was apparently unaffected by the substitution of L-alanine for glycine.

The formation of UDP-*N*-acetylmuramyl-pentapeptide is completed by addition of the preformed dipeptide D-alanyl-D-alanine. The synthesis of this dipeptide from L-alanine has been the subject of extensive investigation by Neuhaus and his colleagues [139, 141] and is described in more detail below. D-Alanine:D-alanine ligase (UDP-*N*-acetylmuramyl-tripeptide: D-Ala-D-Ala ligase (ADP)) also requires for activity the presence of either Mg^{2+} or Mn^{2+} together with the hydrolysis of ATP. The enzyme from *C. poinsettiae* [239] appears to be relatively non-specific,

Table 8.1 The synthesis of UDP–*N*-acetylmuramyl peptides by enzymes from *C. poinsettiae* (from [239])*

Position of amino acid in side chain	*1*		*2*	*3*				*4 + 5*
Substrate	*Glycine*	L-*Alanine*	D-*Glutamic*	L-*Homoserine*	L-*Diaminobutyric*	L-*lysine*	*m*A_2*pm*	D-*Alanyl*-D-*Alanine*
UDP–MurAc†	*54.5*	0.8						
UDP–MurAc-Gly			*32.7*					
UDP–MurAc-L-Ala			41.5					
UDP–MurAc-Gly-D-Glu				*38.0*	0	0	0	
UDP–MurAc-L-Ala-D-Glu				1.0	0	0	0	
UDP–MurAc-Gly-D-isoGlu-L-Hsr								*8.9*
UDP–MurAc-L-Ala-D-isoGlu-L-Lys								5.5
UDP–MurAc-L-Ala-D-isoGlu-mA_2pm								11.0

*The amounts of amino acids added are given as *n*moles/mg protein on incubation at 37°C for 4 hours and the natural components are given in italics.
†MurAc represents *N*-acetylmuramyl.

catalysing the addition of dipeptide to UDP-*N*-acetylmuramyl-tripeptides ending in L homoserine (the natural precursor) L-lysine and *meso*-diaminopimelic acid. However, this aspect of substrate specificity does not appear to have been studied in other ligases [18, 74, 143]. Earlier, Neuhaus and Struve [143] found that the ligase from *S. faecalis* had a high specificity for D-amino acids in the *C*-terminal residue of the dipeptide and a low specificity for the *N*-terminal residue. Recent experiments with *S. aureus, S. faecalis* and *B. subtilis* [29, 150] have demonstrated the ADP + P_i dependent reversibility of the D-alanine:D-alanine ligase yielding the dipeptide and UDP-*N*acetylmuramyl-L-alanyl-D-iso-glutamyl-R_3. Obviously such an activity could be significant in controlling intracellular levels of UDP-*N*-acetylmuramyl-pentapeptide and hence of peptidoglycan synthesis. However, at present the actual physiological significance of these observations remains unclear. The fact that UDP-*N*-acetylmuramyl-pentapeptide does not normally accumulate in organisms suggests an efficient link between the soluble nucleotide precursors and membrane-bound enzymes involved in the subsequent stages of synthesis. However, accumulation of UDP-*N*-acetylmuramyl-pentapeptide does occur when many organisms are treated with antibiotics, such as vancomycin and members of the *β*-lactam group, i.e. penicillins and cephalosporins. In addition, the accumulation of partially formed precursors also occurs in the presence of certain antibiotics and has been described in several series of mutants isolated from *E. coli, S. aureus* and Bacilli. The nature of these mutations and the accumulation of precursors under the influence of antibiotic action will be discussed below and in Chapter 9. The central role of Mg^{2+} in peptidoglycan synthesis is clearly illustrated by the accumulation of UDP-*N*-acetylmuramyl-pentapeptide by *B. subtilis* incubated under conditions of Mg^{2+} deprivation [41]. Meanwhile, under normal conditions the newly formed nucleotide precursors would be transferred to the lipid carrier, thus beginning the second stage of biosynthesis as described above.

8.2.1 *Synthesis of* D*-alanyl-*D*-alanine*

The conversion of L-alanine to D-alanine is catalysed by alanine racemase, an enzyme studied in detail in *S. aureus* [175], *E. coli* [95, 96, 216], *B. subtilis* [83], *S. faecalis* [112], *L. fermenti* [81] and *Pseudomonas putida* [174]. It is almost certain that all require pyridoxal phosphate as a co-factor but this has not been established unequivocally in each case. Alanine racemase is competitively and irreversibly [216] inhibited by cycloserine, *O*-carbamyl-D-serine and the haloalanines and the interaction of the enzyme with these alanine analogues is discussed in detail in Chapter 9.3.1.

D-Alanyl-D-alanine is formed by the dimerization of D-alanine according to the reaction:

$$2\ \text{D-Ala} + \text{ATP} \xrightarrow[\text{Mg}^{2+}\text{ or Mn}^{2+}]{\text{K}^+} \text{D-alanyl–D-alanine} + \text{ADP} + \text{P}_i$$

The enzyme involved, D-alanine:D-alanine ligase (ADP) (D-alanine:D-alanine synthetase) has been purified from *S. faecalis* [137, 138] and *S. aureus* [73]; the latter showing a requirement for a heat stable co-factor in addition to Mg^{2+} (or Mn^{2+}), K^+

and ATP for activity. On the basis of kinetic studies, the following reaction sequence has been proposed [138].

$$\text{E} + \text{D-Ala} \rightleftharpoons \text{E–D-Ala} \quad (1)$$

$$\text{E–D-Ala} + \text{D-Ala} \rightleftharpoons \text{E–D-Ala-D-Ala} \quad (2)$$

$$\text{E–D-Ala-D-Ala} \rightleftharpoons \text{E} + \text{D-Ala-D-Ala}. \quad (3)$$

Using enzyme purified from *S. faecalis* Neuhaus [138] determined K_m values for D-alanine of 6.6×10^{-4} M for the donor site (Equation 1) and 1×10^{-2} M for the acceptor site (Equation 2). The enzyme which is also inhibited by cycloserine ([142], Chapter 9.3.3) showed an absolute specificity for D-amino acids and glycine. However, the only analogue found binding to the donor site was D-aminobutyric acid. Incorporation of other D-amino acids only occurred in the presence of D-alanine when the product was a mixed dipeptide with a *N*-terminal D-alanyl residue. Thus D-alanine:D-alanine ligase shows the opposite substrate specificity to the UDP-*N*-acetylmuramyl-tripeptide:D-Ala–D-Ala ligase described above. That is a high specificity for D-amino acids in the *N*-terminal residue and a low specificity in the *C*-terminal residue of the dipeptide. Acting in concert, the two enzymes ensure that D-alanyl-D-alanine is preferentially synthesized and added to the UDP-*N*-acetylmuramyl-tripeptide.

Under certain conditions glycine has been shown to replace both alanine isomers in UDP-*N*-acetylmuramyl-pentapeptide. Growth of *S. aureus, B. subtilis, L. plantarum, L. cellobiosus* and certain corynebacteria in the presence of high concentrations of glycine resulted in the accumulation of modified UDP-*N*-acetylmuramyl-pentapeptides containing both glycyl-D-alanine and D-alanylglycine as the terminal dipeptide [65]. Thus, under these rather special conditions the substrate specificities of both the D-alanine:D-alanine ligase and the UDP-*N*-acetylmuramyl-tripeptide:D-Ala-D-Ala ligase had been overridden. The synthesis and subsequent incorporation into peptidoglycan of such unnatural precursors may contribute to the inhibitory effects of glycine and other D-amino acids on many bacteria. More recently Hammes [59] has described a specific role for glycine and D-amino acids in blocking the incorporation of nascent peptidoglycan into the wall of *Gaffkya homari* (see Chapter 9.7.6).

8.2.2 *Mutants in the early stages of peptidoglycan synthesis*

Conditional lethal mutants of *E. coli* [107–111, 134, 233, 234] and *S. aureus* [16, 55] have been isolated and characterized. Selection procedures were based on the ability of the mutants to grow at the non-permissive temperature, 42° C and 43° C respectively, only in the presence of an osmotic stabilizer such as sucrose or NaCl. In general, a preliminary identification of the mutation has been made on the basis of the nucleotide precursor accumulated and this identification subsequently confirmed by *in vitro* assays of the enzyme activity involved.

In this way mutants of *E. coli* having lesions in UDP-*N*-acetylglucosamine enolpyruvate reductase (proposed genetic designation: *MurB* [109]); L-alanine ligase (*MurC*); diaminopimelic acid ligase (*MurE*) and D-alanine:D-alanine ligase (*MurF*) have been isolated and studied. Additional mutations in alanine racemase [234] and and D-alanine:D-alanine synthetase [111] were also obtained. Mapping [233] revealed the *MurC, E* and *F* genes, together with the gene for D-alanine:D-alanine ligase (D-alanyl–D-alanine synthetase) (*ddl*) to be located extremely closely together between genes *leu* and *azi* (1–1.5 min) on the *E. coli* chromosome. A mutation in the first enzyme of the pathway, UDP-*N*-acetylglucosamine-2-phosphoenolpyruvate transferase (*MurA*) has been isolated as a phosphonomycin-resistant mutant of *E. coli* [238]. This gene, together with *MurB* is co-transducible with *argH* at approximately 77 min on the chromosome map. No mutation in the D-glutamate ligase (*MurD*) has so far been described. Several mutants accumulate UDP-*N*-acetyl-muramyl-pentapeptide and probably have some lesion at one of the later stages of peptidoglycan synthesis [122, 134]. These mutations also mapped in the two linkage groups described above, suggesting that genes for the synthesis of nucleotide precursor, together with genes coding for the later stages of synthesis, are located in two rather specific areas of the *E. coli* chromosome. The only exception determined to date is the gene for alanine racemase which maps separately at 83′ on the chromosome map. Wijsman [234] has suggested that the primary function of alanine racemase in *E. coli* is to convert L-alanine to pyruvate via D-alanine, rather than an involvement in peptidoglycan synthesis.

A similar, but smaller series of mutants has been isolated in *S. aureus* [16, 55]. These involve mutations in the L-alanine-, D-glutamate- and L-lysine ligases and D-alanine:D-alanine ligase. These mutations have not been mapped.

Accumulation of nucleotide precursors has also been described in L-phase variants (L-forms) of *S. aureus* [32] *S. pyogenes* [28] and several strains of *Bacilli* [219]. Transformation experiments, utilizing DNA from L-phase variants, have suggested that the L-phase state in *B. subtilis* is associated with a single genetic lesion. [240a]. Subsequent *in vitro* enzyme assays have established three groups of mutations in L-phase variants of *B. subtilis* and *B. licheniformis* [219]. These were in aspartic acid semialdehyde dehydrogenase, the enzyme which is at the branch point in the biosynthetic pathways leading to methionine and threonine and to diaminopimelic acid and lysine, the L-alanine ligase and in phospho-*N*-acetylmuramyl-pentapeptide translocase. A deficiency in phospho-*N*-acetylmuramyl-pentapeptide translocase has also been reported in L-phase variants isolated from *S. pyogenes* [164] and *S. faecium* [56].

8.3 The lipid cycle

Clearly the central role in the second stage of synthesis is held by the lipid carrier, undecaprenyl phosphate. The use of such a lipophilic molecule may enable the organism to transport hydrophilic precursors from the aqueous environment of the cytoplasm through the hydrophobic areas of the membrane for addition to the growing peptidoglycan.

8.3.1 *Biosynthesis of the carrier lipids*

Undecaprenol, originally present either as the free alcohol or as a phosphorylated derivative (Fig. 8.2), has been isolated in a pure form from *M. luteus* (*lysodeikticus*) [70], *S. aureus* [71], *S. faecalis* [210], *E. coli* [212] and *B. stearothermophilus* [179]. Where examined by either mass spectrometry or various chromatographic techniques, the isolated material is predominantly a C_{55}-isoprenoid alcohol with traces of the C_{45}-, C_{50}- and C_{60}- homologues. Whether these trace components also participate in the biosynthetic reactions described below remains unknown. Undecaprenyl-phosphate linked intermediates have been shown to function in the biosynthesis of lipopolysaccharide in *S. typhimurium* [237] of lipomannan in *M. luteus* [180] and capsular polysaccharides in *Klebsiella aerogenes* [192, 207] as well as in the biosynthesis of peptidoglycan. Isoprenyl-lipid intermediates of this type have also been implicated in the biosynthesis of exopolysaccharides [192] and other wall polymers in both Gram-positive and Gram-negative bacteria. These include teichuronic acid of *B. licheniformis* (see Chapter 10.2.3), a neuraminic acid containing polymer from the envelope of *E. coli* [208], the linkage units involved in the attachment of teichoic acids in a number of organisms (see Chapter 10.1.4) and teichuronic acid of *M. luteus* [227]. Hence, with the assumption that the various carriers are identical, a single lipid participates in the synthesis of different wall polymers of an organism. Clearly, such a situation would given the organism a single point of control over the synthesis of these polymers thus preventing an imbalance in the formation of the wall as a whole.

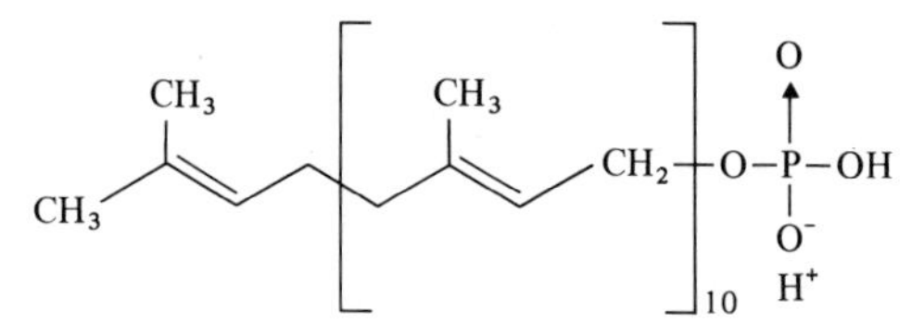

Figure 8.2 Structure of undecaprenol-phosphate.

The synthesis of undecaprenyl pyrophosphate has been studied in *S. newington, M. luteus* and *L. plantarum* [17, 88, 94]. In *S. newington* a membrane-bound enzyme catalysed the formation of polyprenyl pyrophosphates, principally C_{55}-pyrophosphate, from the pyrophosphates of farnesol and Δ^3-isopentenol. A soluble enzyme from the same organism appeared only to synthesize polyprenols of a shorter chain length. Undecaprenyl pyrophosphate synthetase from *L. plantarum*, has been studied in detail by Allen and his colleagues [1, 2, 88]. It is obtained in a soluble form from lysozyme-lysed or mechanically disrupted organisms and has a requirement for either detergent or phospholipid for activity. Thus, it seems likely that *in vivo* the enzyme is associated with the membrane, if only loosely in *L. plantarum.* In this position the water (cytosol)-soluble substrates, farnesyl pyrophosphate and Δ^3-isopentenyl pyrophosphate, would be polymerized to undecaprenyl pyrophosphate which being much more hydrophobic would enter the lipid-phase of the membrane where it would become available for use in biosynthetic reactions.

However, the carrier lipid used in the various biosynthetic cycles described above is undecaprenyl phosphate and not the pyrophosphate. Therefore it becomes necessary for cleavage of the pyrophosphate moiety to occur before newly-synthesized undecaprenol becomes available for use. This same reaction must also occur at the end of each cycle of biosynthesis (e.g. of peptidoglycan, lipopolysaccharide and teichuronic acid) where undecaprenyl pyrophosphate is produced. It was already known [88] that polyprenyl pyrophosphates were resistant to dephosphorylation by *E. coli* alkaline phosphatase when a specific enzyme undecaprenyl pyrophosphate phosphatase was isolated from *M. luteus* membranes by detergent extraction [54]. This dephosphorylation reaction is the site of action of bacitracin (see Chapter 9.4). The phosphatase had no appreciable activity on either isopentenyl pyrophosphate or nitrophenyl phosphate.

A second enzyme, catalysing the dephosphorylation of undecaprenyl phosphate to the free alcohol, is present in the membranes of *S. aureus* [235]. However, attempts to demonstrate a similar activity in the membranes of several other organisms including *M. luteus* were unsuccessful. At this time it was suggested [235] that the action of this enzyme would allow control over biosynthesis by specifically removing undecaprenyl phosphate from the system. Clearly, a control mechanism of this type would also require some means whereby the amount of undecaprenyl phosphate in the system could be increased. Such an enzyme studied in *S. aureus* [68, 69, 178], *Klebsiella aerogenes* [161, 162] and *L. plantarum* [84] is the undecaprenyl phosphate phosphokinase which catalyses the ATP-dependent phosphorylation of undecaprenol. The phosphokinase of *S. aureus*, a single polypeptide of 17 000 mol.wt. has been purified to homogeneity and its properties examined in detail by Strominger, Sandermann and their colleagues [43–45, 68, 69, 176–178]. The enzyme was initially obtained by extraction of staphylococcal membranes with acid butanol followed by differential precipitation at 0° C and −20° C. It is soluble in several organic solvents but insoluble in water and has an unusually high content (58%) of non-polar amino acids. Subsequent purification allows the separation of an apoprotein to which activity can be restored by a wide range of natural and synthetic phospholipids and detergents. Initially it was suggested that phosphatidylglycerol or diphosphatidylglycerol (cardiolipin) were specific lipid co-factors. However enzymic activity appears to be more dependent on 'lipid hydration' rather than lipid viscosity or the actual chemical structure of the polar group of the activating lipid [176, 177].

The isopentenyl residues used in the biosynthesis of undecaprenol are derived from mevalonate and radioactive mevalonic acid has been used as a method of labelling undecaprenyl intermediates in various *Lactobacilli* [201]. The forms of the lipid carrier actively involved in peptidoglycan synthesis in exponentially-growing organisms i.e. undecaprenyl phosphate, undecaprenyl pyrophosphate and the peptidoglycan lipid intermediates accounted for 57% of the total undecaprenol present (Table 8.2). In stationary-phase cultures free undecaprenol was in a large excess. Unfortunately other organisms do not appear to have the ability to utilize exogenously supplied mevalonate and amounts of lipid intermediates have to be determined chemically. This has been done with *S. aureus* [71] and *S. faecalis* [210] where free undecaprenol represented 80% and 90% of the total. As described above free prenol may be formed from the phosphorylated intermediates by the action of prenyl phosphate phosphatases although the widespread occurrence of such enzymes has not been established. The presence of these enzymes and the undecaprenol phosphokinase would allow the free prenol to be utilized as a reserve pool. The synthesis of either peptidoglycan or one of the secondary polymers could then be controlled by specific phosphorylation and dephosphorylation reactions.

8.3.2 *The reaction sequence*

The second stage of the biosynthetic process transports the newly-synthesized nucleotide precursors from the hydrophilic environment of the cytoplasm across the hydrophobic environment of the membrane to the externally situated sites of incorporation into the growing peptidoglycan (Fig. 8.3). The initial reaction involves

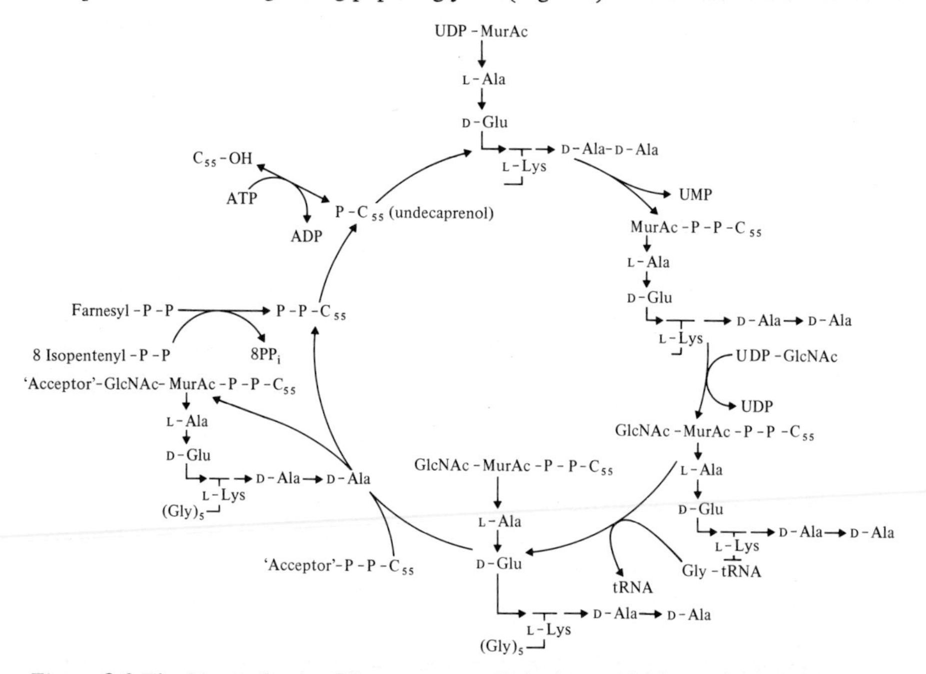

Figure 8.3 The biosynthesis of linear uncross-linked peptidoglycan in *S. aureus*. Phospho-*N*-acetylmuramyl-pentapeptide is translocated to the membrane-bound lipid carrier, undecaprenyl phosphate (C_{55}-P) followed by the transfer of *N*-acetylglucosamine to complete the disaccharide repeating unit. Addition of the five glycine residues of the cross-bridge and the amidation of the α-COOH group of D-glutamate (not shown) then occur before the disaccharide-peptide becomes polymerized. Polymerization is shown to occur while the nascent peptidoglycan remains attached to undecaprenyl-phosphate. Subsequent events include the formation of cross-links by transpeptidation and in most organisms the removal of uncross-linked D-alanine residues by carboxypeptidase activity.

Table 8.2 Effect of growth phase on the levels of prenol derivatives in *L. plantarum* (from [200]).*

Growth phase	*Lipid intermediates of peptidoglycan*	*Glyco-phosphoprenol*	*Prenol-pyrophosphate*	*Prenol-phosphate*	*Prenol*
Exponential	11.7	28.3	3.8	31.4	24.8
Late exponential	10.2	18.4	4.1	22.1	45.2
Stationary	3.3	3.0	4.8	16.3	72.6

*The figures represent the percentage of the total prenol found in each category.

the transfer of phospho-*N*-acetylmuramyl-pentapeptide from the nucleotide to undecaprenyl phosphate with the formation of undecaprenyl-P-P-*N*-acetylmuramyl-pentapeptide and UMP. This reaction, termed a translocation and the enzyme thus a translocase, involves a 'readily reversible exchange of carrier molecules with the rupture and synthesis of pyrophosphoryl bonds of equivalent reactivity' [140].

UDP-*N*-acetylmuramyl-pentapeptide + undecaprenyl phosphate $\rightleftharpoons$
undecaprenyl-P-P-*N*-acetylmuramyl-pentapeptide + UMP.

Labelling studies have shown that one of the phosphate groups of UDP-*N*-acetylmuramyl-pentapeptide is incorporated into the lipid product, undecaprenyl-pyrophospho-*N*-acetylmuramyl-pentapeptide, while the other is released as UMP [3]. As described in Chapter 9.5.2 this reaction has recently been shown to be inhibited by tunicamycin.

In a second step *N*-acetylglucosamine is glycosidically linked to the *N*-acetylmuramyl residue to form undecaprenyl-P-P-*N*-acetylmuramyl(pentapeptide)-*N*-acetylglucosamine and UDP [3]. Since these reactions involve both hydrophilic and hydrophobic substrates, i.e. of cytoplasmic and membrane origin it seems likely that both enzymes occur at the cytoplasmic surface of the membrane.

8.3.3 *Phospho-N-acetylmuramyl-pentapeptide translocase*

The translocase, which has been extensively studied by Neuhaus and his colleagues [140] can be conveniently assayed either in the forward direction, as a transferase, by following the transfer of phospho-*N*-acetylmuramyl-pentapeptide from the nucleotide precursor to the membrane-bound undecaprenyl phosphate or in the reverse direction, by an exchange reaction following the incorporation of radioactivity from UMP into UDP-*N*-acetylmuramyl-pentapeptide.

*UMP + UMPP-*N*-acetylmuramyl-pentapeptide $\rightleftharpoons$
UMP + *UMPP-*N*-acetylmuramyl-pentapeptide.

Both transfer and exchange reactions required Mg^{2+}. In addition K^+ and other monovalent cations (Rb^+, Cs^+, NH_4^+) stimulate the transfer reaction, while UMP is an effective inhibitor.

The enzyme has been obtained in a soluble form from *M. luteus* [211] and *S. aureus* [158] by treatment of membrane preparations with the non-ionic detergent Triton X-100 and from *E. coli* [42] by repeated freezing and thawing of a crude envelope fraction. The solubilized enzyme from *M. luteus* showed specific lipid requirements for activity both in the transfer and exchange reactions. The transfer reaction required the presence of undecaprenyl phosphate and was stimulated by the addition of a neutral lipid fraction, the active component of which has not been identified. In contrast, the exchange reaction showed no requirement for undecaprenyl phosphate nor was it enhanced by the addition of neutral lipid, but was stimulated by a polar lipid fraction, of which the active phosphatide has been tentatively identified as phosphatidylglycerol. However, phosphatidylserine, phosphatidylethanolamine and phosphatidylinositol were equally capable of restoring enzyme activity. The apparent discrepancy between the phospholipid requirement for exchange activity when no requirement for the transfer reaction was observed may be explained by the relative rates of the two activities. Possibly the residual activity of the exchange reaction, in the absence of added phospholipid, is sufficient to allow the maximum activity of the transfer reaction. Similar observations have been made with the enzyme from *S. aureus* [158] where solubilized enzyme was obtained in an inactive form after gel filtration. However, activity was partially restored by addition of a lipid extract from membranes of the same organism. In addition, a variety of phosphatides including phosphatidylcholine (both dioleyl and of plant origin), and phosphatidylglycerol gave partial reactivation of the enzyme in the presence of the detergent sodium lauryl sarcosinate, although none was as effective as natural lipids from *S. aureus*. Further investigation showed the increased efficiency of the bacterial extract to be due to the presence of undecaprenyl intermediates.

Thus, the reactivation of the translocase appeared to result from the provision of a lipid microenvironment rather than a requirement for a specific phospholipid. *In vivo* the microenvironment of the translocase is determined by the interaction of the enzyme with the lipids of the membrane. Certain aspects of this interaction have recently been studied in detail by Neuhaus and his colleagues using the translocase from *S. aureus* [99, 225, 226]. They have adopted two approaches. The first examined, by a variety of physical techniques, the microenvironment in membranes of the fluorescent lipid intermediate undecaprenyl-*N*-acetylmuramyl-(N^{ϵ}-dansyl)-pentapeptide. In the second, they perturbed the physical state of the membrane lipids by change in temperature and by treatment of membranes with *n*-butanol and examined the effect of these treatments on the exchange and transfer activities of the translocase. The results obtained suggest that the lipid intermediate is immobilized within a hydrophobic environment close to the membrane surface. This observation is supported by the earlier finding that spin-labelled lipid intermediate could still form complexes with vancomycin and ristocetin [82] and suggests that at least the terminal D-alanyl-D-alanine dipeptide of the pentapeptide side chain has access to the aqueous phase. Evidence was also found for strong lipid–protein interactions which may point to the lipid intermediate being organized with the translocase in a peptidoglycan synthesizing complex. The physical state of the membrane lipid was found to have marked effects on translocase activity. Since the effects are similar on both exchange and transfer activities, Weppner and Neuhaus [226] concluded that 'the physical state of the lipid matrix has a major effect on the catalytic activity of the translocase'. This presumably is an effect on the intramolecular transfer of phospho-*N*-acetylmuramyl-pentapeptide from the nucleotide precursor to undecaprenyl phosphate (Reaction 3 in the reaction sequence proposed by Pless and Neuhaus [158] and given below) since this reaction is common to both exchange and transfer activities. Whether the state of the lipid matrix also influences catalytic activity by modifying the mobility of the other lipid substrate, undecaprenyl phosphate remains unknown.

The following reaction sequence (where E is the enzyme, C_{55} is undecaprenyl, MurAc is *N*-acetylmuramyl-pentapeptide, P is phosphate, and PP is pyrophosphate) for translocase activity has been proposed:

$$\text{E} + \text{C}_{55}\text{P} \rightleftharpoons \text{E-C}_{55}\text{P} \quad (1)$$

$$\text{E-C}_{55}\text{P} + \text{UMPP-MurAc} \rightleftharpoons \text{UMPP-MurAc-E-C}_{55}\text{P} \quad (2)$$

$$\text{UMPP-MurAc-E-C}_{55}\text{P} \rightleftharpoons \text{UMP-E-C}_{55}\text{PP-MurAc} \quad (3)$$

$$\searrow\!\!\nwarrow \quad \text{UMP-}\underset{\substack{|\\ \text{P-MurAc}}}{\text{E}}\text{-C-}_{55}\text{P} \quad \nearrow\!\!\swarrow$$

$$\text{UMP-E-C}_{55}\text{PP-MurAc} \rightleftharpoons \text{UMP} + \text{E-C}_{55}\text{PP-MurAc} \quad (4)$$

$$\text{E-C}_{55}\text{PP-MurAc} \rightleftharpoons \text{E} + \text{C}_{55}\text{PP-MurAc} \quad (5)$$

In this model the transfer reaction would require the Reactions 1-5 and the presence of a pool of undecaprenyl phosphate while the exchange reaction involves Reactions 2–4 and the formation of a complex between the enzyme and undecaprenyl phosphate. The experimental evidence on which the model is based can be summarized as follows:

(1) The rate of the exchange reaction is invariably faster than that of the transfer reaction. In the model either Reaction 1, the formation of the enzyme–undecaprenyl phosphate complex or Reaction 5, the dissociation of the enzyme–undecaprenyl-P-P-*N*-acetylmuramyl-pentapeptide complex could be rate limiting for the transfer reaction without having effect on exchange activity;

(2) Under conditions where a steady-state level of undecaprenyl-P-P-*N*-acetylmuramyl-pentapeptide was formed, UMP continued to be formed. In addition, after a lag period a further product, identified as phospho-*N*-acetylmuramyl-pentapeptide, was formed in the incubation mixtures. Since this formation only occurred after a lag it suggests that the phospho-*N*-acetylmuramyl-pentapeptide residues result from the hydrolysis of an enzyme intermediate or undecaprenyl-P-P-*N*-acetylmuramyl-pentapeptide rather than the nucleotide precursor which is present throughout the experiment. Thus, the translocase has apparent pyrophosphatase activity for UDP-*N*-acetylmuramyl-pentapeptide;

(3) Certain surfactants (e.g. deoxycholate) which partially inhibit the exchange activity have a stimulating effect on the transfer reaction. These observations can be explained by the detergents influencing the ability of the enzyme to complex with the lipid substrate, undecaprenyl phosphate and the lipid product, undecaprenyl-P-P-*N*-acetylmuramyl-pentapeptide;

(4) The transfer reaction can be uncoupled by treatment of the enzyme preparation with dodecylamine resulting in the formation of phospho-*N*-acetylmuramyl-

pentapeptide. Under these conditions the enzyme behaves as though depleted of undecaprenyl-phosphate;

(5) Vancomycin and ristocetin mimic the effect of detergents described above. At low concentrations the antibiotics stimulate the transfer but not the exchange reaction. Under these conditions it is assumed that by forming an adduct with undecaprenyl-P-P-*N*-acetylmuramyl-pentapeptide, the antibiotics will prevent reassociation of the reaction product with the enzyme (i.e. the reverse of Reaction 5) and thus increase the rate of transfer. Alternatively, the interaction of vancomycin or ristocetin with the enzyme-bound lipid intermediate might stimulate dissociation of this complex and hence the rate of transfer. Inhibition at higher antibiotic concentrations of both exchange and transfer reactions probably results from the formation of an adduct with the substrate UDP-*N*-acetylmuramyl-pentapeptide. These observations are discussed in more detail in Chapter 9.

In parallel studies, the specificity requirements of the translocase have been investigated to determine the role of this enzyme in maintaining the overall fidelity of peptidoglycan structure. Some of the enzymes earlier in the process have already been shown to have the capability of incorporating alternative amino acids during the synthesis of nucleotide precursors, Hammes and Neuhaus [61, 63] have studied the effects of modified UDP-*N*-acetylmuramyl peptides on the transfer reaction of membrane-bound translocases from *S. aureus* and *G. homari* and the exchange reaction of the staphylococcal enzyme (Table 8.3). The modifications involved variations in amino acid composition at each position of the peptide side chain with the exception of the D-isoglutamic acid residue and also in the length of the peptide side chain from tripeptide to the naturally occurring pentapeptide. Substitution of glycine for both L- and D-alanine residues showed both enzymes to have a high specificity for L-alanine in position 1 and D-alanine in position 4. Substitution of the terminal D-alanine residue of the side chain (position 5) had only a marginal effect on their activity. Similar results, showing the low specificity of the translocase for the terminal position, had been obtained earlier in *S. faecalis* [158]. In this case the analogue UDP-*N*-acetylmuramyl-L-alanyl-D-isoglutamyl-L-lysyl-D-alanyl-*O*-carbamoyl-D-serine was found to be as effective a substrate in the exchange reaction as was the natural nucleotide precursor. The translocase showed only low specificity for the diamino acid in position 3, enzyme activity being similar whether the peptide side chain contained L-lysine, meso-diaminopimelic acid or L-ornithine. However, alterations in activity were noted when incomplete UDP-*N*-acetylmuramyl peptides were used as substrates. The results obtained were consistent with the above observations UDP-*N*-acetylmuramyl-tetrapeptide giving 24% of the activity found with the pentapeptide precursor, whereas with UDP-*N*-acetylmuramyl-tripeptide this value had fallen to 1.3%.

Other studies have established the importance of the uracil residue. Although 5-fluorouracil substituted nucleotide is tolerated by all the biosynthetic enzymes, leading to synthesis of the nucleotide precursor, the final product, FUDP-*N*-acetyl-

Table 8.3 The specificity of phospho-*N*-acetylmuramyl-pentapeptide translocase from *S. aureus* and *G. homari* (after [61–63]).*

	Activity (%)		
	S. aureus		*G. homari*
Substrate	*Transfer*	*Exchange*	*Transfer*
UDP–MurAc-L-Ala-D-Glu-L-Lys-D-Ala-D-Ala	100†	100†	100†
UDP–MurAc-L-Ala-D-Glu-L-Lys-D-Ala	24	40	58
UDP–MurAc-L-Ala-D-Glu-L-Lys	1.3	0.7	0.03
UDP–MurAc-L-Ala-D-Glu-L-Lys-D-Ala-Gly	67	92	64.5
UDP–MurAc-L-Ala-D-Glu-L-Lys-Gly-D-Ala	6.7	7.6	16
UDP–MurAc-Gly-D-Glu-L-Lys-D-Ala-D-Ala	11	17	24
UDP–MurAc-L-Ala-D-Glu-mA_2pm-D-Ala-D-Ala	57	66	68
UDP–MurAc-L-Ala-D-Glu-Orn-D-Ala-D-Ala	—	100	—
UDP–MurAc-Gly-D-Glu-L-Lys-D-Ala-Gly	—	9.6	—
UDP–MurAc-Gly-D-Glu-L-Lys-Gly-D-Ala	—	0.7	—

*The table shows the correlation obtained between the activity of the enzyme measured in either the transfer or exchange reactions as described in the text and the effect of modifications on the UDP–*N*-acetylmuramyl-peptide (MurAc = *N*-acetylmuramyl).

†The activity obtained by each assay using the natural substrate UDP–MurAc-L-Ala-D-Glu-L-Lys-D-Ala-D-Ala is given as 100%.

muramyl-pentapeptide, is not a substrate for the translocase [188]. In addition, UDP-*N*-acetylmuramyl-pentapeptide, in which the ϵ-amino group of the lysine residue has been substituted with L-alanine or L-serine, was an effective substrate for synthesis of lipid intermediates by membranes from *Lactobacillus viridescens* [156]. Similarly, UDP-*N*-acetylmuramyl-pentapeptide-pentaglycine was utilized by membranes of *S. aureus* and *M. lysodeikticus* at 30% and 70% of the rate obtained with the normal pentapeptide precursor [119].

The observations described above clearly establish that phospho-*N*-acetylmuramyl-pentapeptide translocase has a major role in selecting against analogues of the natural precursor of peptidoglycan. However, because of this enzyme's low specificity for residues in position 3 and 5 of the peptide chain, the organism requires additional control mechanisms to ensure that incorporation of unsuitable modifications does not occur. As described earlier (Section 8.2), these controls reside in the cytoplasmic ligases. Thus, the sequence of the peptide side chain is determined by the specificity of the D-alanine:D-alanine ligase and the dipeptide:UDP-*N*-acetylmuramyl-tripeptide ligase acting in concert and also by the ligase adding the amino acid at position 3.

8.3.4 *Synthesis of undecaprenyl-P-P-disaccharide-pentapeptide*

The formation of the second lipid intermediate involved in peptidoglycan synthesis requires the transfer of *N*-acetylglucosamine from UDP-*N*-acetylglucosamine to undecaprenyl-P-P-*N*-acetylmuramyl-pentapeptide. The products of the reaction are

N-acetylglucosaminyl-β-(1,4)-*N*-acetylmuramyl-(pentapeptide)-P-P-undecaprenol. Using membrane preparations from *M. luteus* Strominger and his colleagues [3] showed that UDP-*N*-acetylglucosamine was the only effective intermediate in peptidoglycan synthesis.

More recently the enzyme catalysing this reaction UDP-*N*-acetylglucosamine: *N*-acetylmuramyl-(pentapeptide)-P-P-undecaprenol *N*-acetylglucosamine transferase has been purified from LiCl extracts of toluene-treated *B. megaterium* [193, 196]. The isolated protein designated PG-1 stimulated the synthesis of peptidoglycan by both LiCl- and toluene-extracted bacilli and membranes prepared from these organisms. Subsequently Taku and Fan [194] showed that PG-1 catalysed the transfer of radioactive *N*-acetylglucosamine to undecaprenyl-P-P-*N*-acetylmuramyl-pentapeptide. The activity of the transferase in common with the phospho-*N*-acetylmuramyl-pentapeptide translocase (see p. 250) was markedly stimulated by the presence of a crude lipid extract in the assay mixture. The component causing increased synthesis was shown not to be undecaprenyl-P-P-*N*-acetylmuramyl-pentapeptide, but was not further identified. In view of the results of Neuhaus and his colleagues (see p. 251) in defining the importance of the lipid microenvironment for the activity of phospho-*N*-acetylmuramyl-pentapeptide translocase it seems probable that a similar environment may be necessary for the activity of the transferase. Clearly this is an area open for further investigation.

8.3.5 *Modification of the lipid intermediates*

In many organisms the lipid-linked disaccharide pentapeptide units are utilized directly for synthesis of peptidoglycan. However, in others the peptide side chains are modified either by substitution of the existing amino acids or by the addition of other amino acids which will subsequently form the cross-bridge of the polymeric peptidoglycan. Where such modifications are introduced, they invariably take place at the level of the lipid intermediate, undecaprenyl-P-P-disaccharide-pentapeptide.

Two examples of such alterations, the amidation of D-isoglutamic acid in *S. aureus* [184] and the substitution of the same α-COOH group of the D-glutamate residues by a single glycine residue in *M. luteus* [87], have been investigated in detail. Both reactions are catalysed by membrane-bound enzymes. In the former reaction the enzyme could utilize either glutamine or NH_4^+ and both undecaprenyl-P-P-*N*-acetylmuramyl-pentapeptide and -disaccharide-pentapeptide would act as substrates, whereas the nucleotide precursor UDP-*N*-acetylmuramyl-pentapeptide was completely inactive. In *M. luteus* the glycine residue is added in a reaction similar to that involved in synthesis of the peptide side chain. Again, both lipid intermediates would act as acceptor in the reaction, although addition to undecaprenyl-P-P-di saccharide-pentapeptide appeared more efficient. In contrast to the addition of cross-bridge amino acids, the reaction is independent of tRNA. The energy requirements of the reaction were supplied by the concomitant hydrolysis of ATP to ADP and P_i. It seems likely that many other minor differences found in the structure of peptidoglycan such as the *O*-acetylation of muramic acid residues in *S. aureus* [52], and the addition to the peptide side chain of single amino acid residues, not involved in cross-bridge formation, occurs at the level of the lipid intermediates.

8.4 Formation of cross-bridge peptides

In bacteria, other than those having a direct cross-link from the penultimate D-alanine residue of one peptide side chain to the free amino group of a second peptide (i.e. the peptidoglycan has a chemotype designated sub-group A1 [181]) additional amino acids, which eventually form the cross-bridge of the peptidoglycan, are added usually at the level of the lipid intermediates. Three types of such cross-bridges have been distinguished on the basis of their constituent amino acids. In the two major classes the cross-bridges consist of either a) single L-amino acids, glycine or oligomers of L-amino acids or, b) single D-amino acids which may be amidated. A third class has a mixture of both L- and D-amino acids in the same cross-bridge peptide (Table 8.4). Examples of each of these classes have now been investigated to establish the mechanism of synthesis of the peptides. This occurs by at least two distinct methods, the first of which involves the participation of tRNA and is found in those organisms having cross-bridges which contain glycine or L-amino acids.

In *S. aureus* the involvement of tRNA in the synthesis of the pentaglycine cross-bridge peptide was first suggested by the observation of Chatterjee and Park [14] that incorporation of glycine into polymeric peptidoglycan was prevented by ribonuclease. Under identical conditions the incorporation of *N*-acetylmuramyl-peptide from the nucleotide precursor was unaffected. The synthesis of the cross-bridge has now been studied in detail and establishes the unique way in which many bacteria form these peptides. Five glycine residues activated by tRNA are added; initially to the ϵ-NH_2 group of lysine and then sequentially to the amino terminus of the growing peptide chain [85, 119, 199]. This mechanism is in direct contrast to that of protein synthesis which occurs by addition at the carboxyl terminus. The residues appear to be added singly with no evidence of the involvement of peptidyl-tRNA intermediates [85, 199]. The tRNA of *S. aureus* has been found to contain at least three species of glycyl-tRNA (Gly-tRNA) all of which will act in the incorporation of glycine into peptidoglycan, whereas one of the species is inactive in protein synthesis [12]. The formation of all Gly-tRNA is catalysed by a single Gly-tRNA synthetase which has been purified [148]. In the presence of ATP and Mg^{2+} the enzyme is capable of charging tRNA from *E. coli*, *M. roseus* and yeast.

A more complex situation has been studied in *S. epidermidis* where four different pentapeptide cross-bridges have been shown to be present [155, 203]. Each contains glycine and L-serine residues in the ratio 3:2 in a characteristic sequence where glycine is always the initial substituent of the ϵ-NH_2 group of L-lysine. Addition of the amino acids occurs at the level of the lipid intermediates, the transfer of glycine

Table 8.4 Bacteria in which the biosynthesis of cross-bridge peptides has been studied

Organism	*Cross-bridge peptide*	*Intermediates involved in synthesis*	*References*
S. aureus	Gly → Gly → Gly → Gly → Gly	Glycyl-tRNA	14, 85, 119, 199
S. epidermidis	(Gly_2 L-Ser_2) → Gly	Glycyl-tRNA; Seryl-tRNA	155, 203
M. roseus	L-Ala → L-Ala → L-Ala → L-Thr →	L-Alanyl-tRNA, L-Threonyl-tRNA	171
L. viridescens	L-Ser → *L-Ala →	L-Seryl-tRNA, *L-Alanyl-tRNA	156, 157
A. crystallopoietes	L-Ala →	L-Alanyl-tRNA	170
S. faecalis (*faecium*)	D-Asp-NH_2 β └→	D-Aspartate → β-D-Aspartyl-phosphate, NH_4^+ + ATP	187
Sporosarcina ureae	D-Glu γ └→Gly	Glycyl-tRNA, D-Glutamate + ATP	105

*Alanine is added at the level of the UDP–*N*-acetylmuramyl-pentapeptide.

and serine from their respective tRNAs being catalysed by membrane-bound enzymes. In confirmation of the chemical structure, incorporation of glycine is independent of that of serine, whereas maximum serine incorporation requires the simultaneous incorporation of glycine. The ratio of glycine and serine incorporated is dependent on the ratio of the specific tRNA supplied. Fractionation of tRNA from *S. epidermidis* [155] has demonstrated the presence of four glycyl- and four seryl-tRNAs. In each case, all species were active in cross-bridge synthesis, including one species of each type which did not participate in protein synthesis. The gly–tRNA with apparent specificity for peptidoglycan synthesis, has been purified to homogeneity and shown to be made up of two distinct iso-accepting species [169]. The two sequences differ in 6 bases and the insertion of an additional base in the dihydrouridine loop of one. The species both lack the characteristic sequence GTψC and all minor bases, with the exception of a single thiouridine residue. The sequence of GTψC has been implicated in the binding of tRNA to ribosomes and the change may explain the inability of these tRNAs to function in protein synthesis. Thus, *S. epidermidis* appears to have developed a means of supplying glycine residues for cross-bridge synthesis without involving the machinery of protein synthesis.

The formation of cross-bridges has also been studied in *M. roseus* [171], *Arthrobacter crystallopoietes* [170] and *Lactobacillus viridescens* [156, 157] in each of which the incorporation of the appropriate amino acid(s) involved acyl-tRNA. In *Lactobacillus viridescens* the mechanism differed from that previously described for *S. aureus* and *S. epidermidis* in that the initial amino acid of the cross-bridge was incorporated by a soluble transferase to yield UDP-*N*-acetylmuramyl-hexapeptide [156]. The reaction was relatively non-specific in that either L-alanine or L-serine was incorporated (the cross-bridge has the structure D-alanyl-L-seryl-L-alanyl-N-$^{\epsilon}$L-lysine). The purified transferase [157] was not very specific for tRNA, since it catalysed the formation of UDP-*N*-acetylmuramyl-hexapeptide from L-seryl-tRNA, L-cysteinyl-tRNA and to a small extent glycyl-tRNA as well as from L-alanyl-tRNA^{cys}, that had been prepared by chemical modification of L-cysteinyl-tRNA. The hexapeptide was translocated to the lipid intermediate and then the second amino acid of the cross-bridge was added by a membrane bound enzyme of low specificity since *in vitro* either L-alanine or L-serine could be transferred. In contrast to the above results, in the analogous reaction in *A. crystallopoietes* [170] the transfer of the single L-alanine residue at the level of the lipid intermediates appears to be tRNA specific. Even L-alanyl-tRNA^{cys} would not participate in the reaction.

The second method of cross-bridge formation has been found predominantly in organisms having cross-bridges which consist of D-amino acids or amides. Examples of this type of cross-bridge are the single D-isoasparaginyl residue found in *S. faecalis* and *L. casei* [181]. In these organisms, activation of the amino acid does not involve the participation of tRNA. D-aspartic acid is activated by a membrane-bound enzyme to form β-D-aspartylphosphate. The enzyme has been partially purified from *S. faecalis* [187] where it can be released from the membrane by treatment with LiCl. Reattachment of the enzyme occurs on removal of the salt by dialysis, the process requiring the presence of phospholipids but not Mg^{2+}. Purified cardiolipin gave essentially similar results to those obtained with a crude lipid fraction. A second membrane-bound enzyme, D-aspartyltransferase, did not separate from the activating enzyme in these experiments. The activated D-aspartyl residue is transferred by this second enzyme to the ϵ-NH_2 group of lysine of either the nucleotide precursor, UDP-*N*-acetylmuramyl-pentapeptide or the lipid intermediate. Finally, modification of the D-aspartyl residue to form D-isoasparagine occurs by amidation of the α-carboxyl group in a reaction involving NH_4^+ and ATP.

The final type of cross-bridge found contains both L- and D-amino acids in the same cross-bridge peptide. As an example of this group, *Sporosarcina urea* contains cross-bridges of D-glutamylglycine linking D-alanine and L-lysine of adjacent peptide side chains. The synthesis of the cross-bridge peptide is based on the two mechanisms described above [105]. The addition of glycine to the ϵ-NH_2 group of lysine is mediated by Gly–tRNA, D-glutamic acid being without effect on the reaction. Fractionation of the crude tRNA of *S. ureae* showed two forms of Gly-tRNA both of which participated equally in both protein and peptidoglycan synthesis. The addition of D-glutamic acid was dependent either on the simultaneous presence of Gly–tRNA or on precharging the system with intermediates to which glycine had already been added. The D-glutamyl residue was activated in the presence of a membrane bound enzyme and ATP, presumably in a reaction analogous to that described for D-aspartic acid in *S. faecalis* [187]. Incorporation of both residues into peptidoglycan showed an absolute dependence on UDP-*N*-acetylmuramyl-pentapeptide and UDP-*N*-acetylglucosamine, suggesting that addition occurred at the level of the lipid intermediates. The finding of substituted UDP-*N*-acetylmuramyl-peptides in reaction mixtures incubated in the absence of UDP-*N*-acetylglucosamine presumably reflects the addition of glycine or of glycine and D-glutamic acid to undecaprenyl-P-P-*N*-acetylmuramyl-pentapeptide, followed by an exchange reaction of the translocase to form the substituted products.

Nothing is known of the mechanism of synthesis in organisms having other forms of cross-bridge peptides. In sub group A2 [181] the cross-bridges have the identical amino acid sequence to the peptide side chains. Ghuysen and his colleagues [48] have suggested that in *M. luteus*, a representative of this group, peptide side chains may be incorporated into the cross-bridge peptides by the sequential action of a transpeptidase and *N*-acetylmuramyl-L-alanine amidase. This mechanism is shown in Fig. 8.4, pathway A. The organism first forms a cross-link between the penultimate D-alanine residue of one peptide side chain and the ϵ-amino group of lysine in a second peptide. One of the peptides is then cleaved from the glycan by an *N*-acetylmuramyl-L-alanine amidase to release a free amino group on the L-alanine residue. A second transpeptidation then occurs to form a cross-link between the penultimate D-alanine of a third peptide side chain and the L-alanine residue. Additional cycles of this sequence will then build up a cross-bridge of multiple peptide side chains. Evidence has come from *in vivo* studies of peptidoglycan synthesis by Mirelman and Bracha [124] of a marked difference in the sensitivity of the two

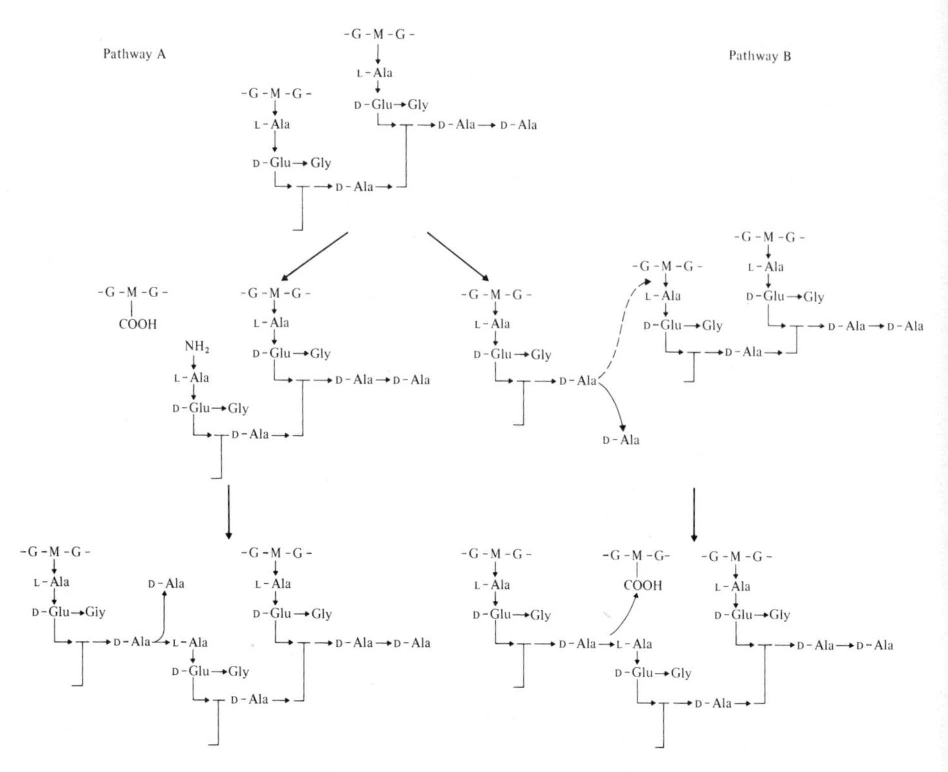

Figure 8.4 Possible mechanisms for the synthesis of cross-bridge peptides in *Micrococcus luteus.* For either mechanism the initial reaction is the formation of a direct D-alanyl-N^{ϵ}-L-lysine cross-link by transpeptidation. In pathway A this is followed by amidase action to cleave the *N*-acetylmuramyl-L-alanine bond of one peptide subunit of the cross-linked dimer and then transpeptidation to a third peptide using the NH_2-group of the released L-alanine residue as the acceptor. Alternatively (pathway B) the terminal D-alanine residue of a third peptide could be removed by D-alanine-carboxypeptidase. The resulting pentapeptide would then participate in a transamidation reaction releasing the *N*-acetylmuramic acid residue from the *N*-terminus of a cross-linked dimer and forming a D-alanyl- L-alanine bond. In either case, repetition of the latter two reactions will yield cross-bridges made up of multiple peptides in head-to-tail-linkage. In the figure] represent L-lysine.

transpeptidases to inhibition by benzylpenicillin. The formation of the D-alanyl-L-alanine bonds appeared to be some 50 times more resistant to inhibition than was the formation of D-alanyl-L-lysine bonds. Inhibition of this latter activity (50% inhibition at 0.1 μg antibiotic/ml) occurred at benzylpenicillin concentrations showing good agreement with the minimum inhibitory growth concentration.

Alternatively, the synthesis of cross-bridge peptides of the A2 sub-group may involve the *N*-acetylmuramyl-L-alanine amidase acting as a transamidase [13]. In this mechanism (Fig. 8.4, pathway B) a D-alanyl-L-lysine cross-link would first be formed as described above. A DD-carboxypeptidase then cleaves the terminal D-alanine residue from a third peptide side chain allowing the resultant tetrapeptide to participate in a transamidation reaction catalysed by the *N*-acetylmuramyl-L-alanine amidase. This would result in the formation of a D-alanyl-L-alanine bond and the release of the peptide from its linkage to the *N*-acetylmuramyl residue. However, further work is necessary to distinguish which of either of these two possible mechanisms is used by these organisms.

8.5 Polymerization of disaccharide-peptide units

The final reaction of the membrane-mediated stage of peptidoglycan synthesis is also the initial reaction in the third stage, i.e., the incorporation of the newly-synthesized disaccharide peptide unit into the growing peptidoglycan. From our knowledge of the structure of peptidoglycan it becomes obvious that in theory two possibilities exist for the mechanism of incorporation of this newly-synthesized unit into the pre-existing structure. These would be initiated by the formation of either a glycosidic or a peptide bond. In the former method, addition of the newly-synthesized unit would occur by synthesis of a β-glycosidic linkage between the disaccharide unit and the glycan chain of the pre-existing peptidoglycan. This would be followed by the formation of a second bond between the newly-incorporated peptide side-chain and another, either directly or through a cross-bridge peptide such as those described above. The alternative process would be essentially similar but with the order of reactions reversed. In fact the presence of uncross-linked side chains in all peptidoglycans examined argues strongly for polymerization to occur by extension of the glycan chain (transglycosylation) prior to peptide bond formation (transpeptidation). To yield uncross-linked peptides the reverse process would seem to require the cleavage of certain peptide bonds to form monomers from dimers synthesized during the first stage of incorporation. Moreover, work with membrane preparations from a wide variety of bacteria has demonstrated the formation of polymeric but uncross-linked peptidoglycan. On the other hand, recent work described below has established a major role for transpeptidation in the addition of newly-synthesized peptidoglycan to the pre-existing wall. However, even in these systems catalysing transpeptidation (e.g. wall-membrane preparations, toluene- and ether-treated cells) much of the newly-synthesized peptidoglycan remains uncross-linked.

Treatment of both *B. licheniformis* [209] and *M. luteus* [126] with β-lactam antibiotics also results in the synthesis of soluble uncross-linked peptidoglycan. A similar situation exists during the early stages of the reversion of both protoplasts and unstable L-forms of *B. licheniformis* [30] where the formation of uncross-linked peptidoglycan precedes the occurrence of cross-linkage. Thus, polymerization of newly-synthesized units by the formation of glycosidic bonds can occur in the absence of cross-bridge formation, whereas the reverse situation, addition of newly-synthesized units by transpeptidation in the absence of glycosidic bond formation has only been demonstrated with model peptides and not yet in any peptidoglycan synthesizing system.

The extension of growing glycan chains could, in theory, occur by the addition of disaccharide peptide units to either their reducing or non-reducing termini. In the latter case the *N*-acetylglucosaminyl residue of the newly-inserted disaccharide would be susceptible to periodate oxidation whereas this would not be so for addition at the reducing terminus. If insertion of newly-synthesized units occurred at the reducing terminus then provided the reducing group were free the *N*-acetylmuramyl residue would undergo a β-elimination reaction yielding lactyl-peptide. Using these criteria and a membrane preparation which synthesized linear uncross-linked peptidoglycan, the extension of glycan chains in *B. licheniformis* was shown to occur by addition of the newly-synthesized disaccharide-pentapeptide unit at the reducing terminus [220]. Similar results have been reported for *M. luteus* [228] and *B. megaterium* [39]. In each case the reducing terminus was blocked by a linkage labile to mild acid hydrolysis. On the basis of the conditions under which hydrolytic cleavage and the release of the muramic acid reducing terminus occured it was suggested that the muramic acid might be linked to undecaprenyl pyrophosphate [220] (Fig. 8.5). If this were the case, then synthesis would result in a completed polymer

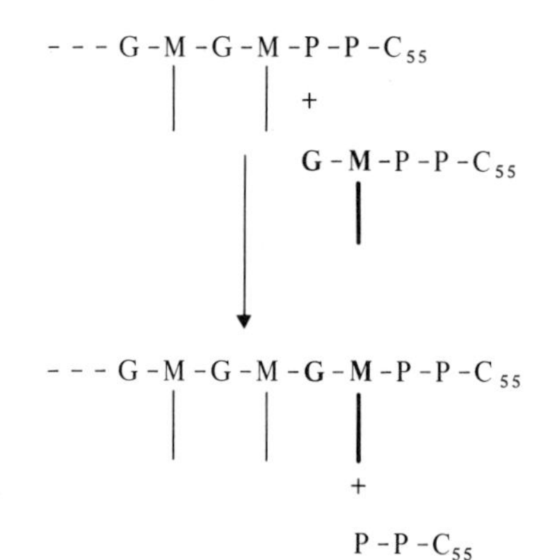

Figure 8.5 The mechanism proposed for the extension of glycan chains by polymerization of disaccharide-peptide subunits. The group blocking the reducing-terminus of the pre-existing glycan chain is shown as undecaprenyl pyrophosphate although this has not yet been unequivocally established.

still linked to the lipid carrier. The finding of free-reducing groups of muramic acid in the walls of *S. aureus* and various Bacilli [217] suggests that if the above hypothesis is correct then some mechanism must exist to terminate synthesis of a particular glycan chain by detaching the lipid carrier. At present we have no evidence as to the nature of this process.

Recently Fuchs-Cleveland and Gilvarg [39] have described the isolation from *B. megaterium* of a soluble peptidoglycan fraction. This was characterized as an un-cross-linked oligomer 11–13 disaccharide peptide units long with a blocked reducing terminus. Treatment with lysozyme gave 10–12 disaccharide peptide units and a fragment tentatively identified on the basis of its chromatographic behaviour as undecaprenyl-P-P-disaccharide peptide. The oligomer was radioactively labelled and chased with the same kinetics as both UDP-*N*-acetylmuramyl-pentapeptide and the lipid intermediates. The authors did not, however, establish a precursor-product relationship between the oligomer and cross-linked peptidoglycan. If it were an intermediate then the apparent absence of shorter oligomers, i.e. those having chain lengths of 2–10 disaccharide peptide units is surprising. Alternatively the oligomers may arise by enzymatic cleavage of nascent uncross-linked peptidoglycan. In this case the restricted size of the isolated glycan chains would be determined by the specificity of the enzyme, presumably an endo-*N*-acetylmuramidase. The release of such an oligomer suggests that *in vivo* the polymerization of uncross-linked peptidoglycan precedes cross-linking to the wall. This mechanism is supported by the earlier observation that in *B. megaterium* cross-linkage continues to increase for some 30–40 min after incorporation of the newly-synthesized peptidoglycan into the wall [33].

Detergent-soluble oligomeric intermediates have also been reported to occur during peptidoglycan biosynthesis by ether-treated *Pseudomonas aeruginosa* [128] and deoxycholate-treated *M. luteus* [202]. In the pseudomonad the oligomeric material which was a mixture of partially cross-linked peptidoglycan and lower molecular weight fragments could be labelled from the appropriate nucleotide precursors and subsequently chased into the wall (i.e. hot-detergent insoluble) fraction. However, no polyprenol-activated or lipoprotein-linked peptidoglycan fragments were detected in any analysis of either the detergent-soluble or insoluble material and the origin or role of these oligomers in biosynthesis remains unclear.

In *M. luteus* the addition of deoxycholate to wall–membrane preparations synthesizing peptidoglycan resulted in the accumulation of oligomeric material which was soluble in hot but not in cold SDS. The concentrations of detergent used blocked incorporation of nascent peptidoglycan into the pre-existing wall. The oligomeric fraction could however be isolated in very small amounts from preparations incubated in the absence of deoxycholate. Lysozyme-treatment of isolated material gave two components, the major one corresponding to disaccharide-pentapeptide and a minor fraction having the chromatographic mobility of the lipid intermediate, undecaprenyl-P-P-disaccharide-pentapeptide. The average glycan chain lengths calculated from the ratio of the radioactivity found as lipid intermediate to the total radioactivity was about 50 disaccharide units. In other experiments the muramic acid residue at the reducing terminus of the intermediate was shown to be blocked by a linkage labile to mild acid, presumably undecaprenyl pyrophosphate. Finally the oligomeric peptidoglycan could be chased into the wall after removal of the detergent and reincubation of the wall-membranes with peptidoglycan precursors. Thus it also appears that in *M. luteus* polymerization of the glycan chains can occur in the absence of transpeptidation.

The polymerization of uncross-linked peptidoglycan has been studied by Taku and Fan [195] using a reconstituted membrane system from *B. megaterium.* Membrane-preparations were solubilized by treatment with cholate and LiCl and the soluble fraction isolated by high-speed centrifugation. Membrane vesicles which synthesized peptidoglycan from UDP-*N*-acetylmuramyl-pentapeptide and UDP-*N*-

acetylglucosamine were reconstituted from this soluble material by dialysis to remove the detergent and LiCl. In addition, active vesicles were also prepared from partially purified phospho-*N*-acetylmuramyl-pentapeptide translocase, *N*-acetyl-glucosaminyl transferase (see p. 255) and a fraction of cholate solubilized membranes designated PG-II. In the absence of PG-II these vesicles synthesized only lipid intermediates suggesting that the added fraction contained either the polymerase itself or a protein whose presence was necessary for polymerization to occur. Further purification by ion-exchange and gel chromatography showed the active component to have a molecular weight of approximately 60 000. It will be of particular interest to see if this protein is indeed the polymerase and whether as described in Chapter 9.8.2 it interacts with and is inhibited by moenomycin and related antibiotics.

The other known product of the polymerization reaction is undecaprenyl-pyrophosphate. This in turn is dephosphorylated by a membrane-bound phosphatase to yield the lipid carrier, undecaprenyl-phosphate and inorganic phosphate [183]. The reaction is specifically inhibited by bacitracin as described in Chapter 9.4. The lipid carrier then becomes available to accept another phospho-*N*-acetylmuramyl-pentapeptide residue, thus beginning a second cycle of synthesis.

This mechanism of polymerization allows the synthetic enzymes to remain in the membrane, the extending glycan chain being pushed out into the pre-existing wall. Clearly, the formation of cross-links by transpeptidation, also mediated by membrane-bound enzymes, could occur. Evidence to be described below demonstrates that incorporation of newly-synthesized material into peptidoglycan involves the formation of such cross-links whereas, polymerization, as already shown, can occur in the absence of cross-linkage. In addition, synthesis by this method goes some way to explaining the finding of potential peptidoglycan synthesis in stable L-forms of *S. aureus* [15], *B. subtilis* and *B. licheniformis* [219] and in protoplast membranes of *Bacillus megaterium* [166]. Previously it had been customary to implicate an 'acceptor', assumed to be pre-existing peptidoglycan, to which newly-synthesized units were added. Obviously this situation was not applicable to the above examples, where peptidoglycan was absent. In the suggested mechanism [220], addition of the new unit could occur to any glycan, provided that it remained attached to the lipid carrier. In the simplest form this would be a second unit of undecaprenyl-P-P-disaccharide-peptide. Presumably the organism has some control mechanism which directs synthesis towards the longer glycan chains.

The method of synthesis of the glycan chains is analogous to that described for the lipopolysaccharide O-antigen of *Salmonella* [168] and in contrast to that of the poly-(glycerol phosphate) and poly-(ribitol phosphate) teichoic acids of *B. subtilis* 168 and W23 and *B. stearothermophilus* [90, 91, 240] (Chapter 10.1.1). The teichoic acids have been shown to grow by addition of units of polyol phosphate to the non-reducing terminus of the growing polymer, the reverse of the situation found in peptidoglycan and O-antigen synthesis.

8.6 Transpeptidation: the formation of cross-links

The preceding sections of this Chapter have dealt with the reactions involved in the biosynthesis of linear uncross-linked peptidoglycan and have mentioned the importance of cross-linking for the incorporation of newly synthesized material into the wall. Even when the uncross-linked polymer is many disaccharides long it remains water-soluble and only when the peptide side chains become cross-linked does peptidoglycan become insoluble and capable of fulfilling its role in maintaining the structural integrity of the organism.

By the mid 1960s detailed studies of the chemical structure of several peptidoglycans had established the presence of cross-linked peptides. Moreover, D-alanine originally present in the nucleotide precursor as the terminal D-alanyl-D-alanine dipeptide appeared to be involved in the formation of these cross-linkages. In parallel with the structural studies membrane preparations from *S. aureus* and *M. luteus* were shown to synthesize peptidoglycan *in vitro* [189]. However, in neither case was the product cross-linked and despite being inhibited by several antibiotics known to cause accumulation of peptidoglycan nucleotide precursors in whole cells, the formation of this uncross-linked peptidoglycan was not inhibited by β-lactam antibiotics even at high concentrations. Taken together these observations pointed to the cross-linking reaction as the site of inhibition of peptidoglycan synthesis by β-lactams and from this time studies of the mechanism of action both of these antibiotics and of cross-linkage have gone hand in hand.

The suggestion that cross-linkage might involve the synthesis of a peptide bond by means of a transpeptidation reaction was first made by Martin in 1964 [114–116]. An investigation of the structure of peptidoglycan synthesized by unstable L-forms of *Proteus mirabilis* grown in the presence of benzylpenicillin in which cross-linkage appeared to be markedly reduced led him to suggest that formation of such cross-links could result from a transpeptidation reaction in which the D-alanyl-D-alanine terminus of one peptide side chain was cleaved and a new peptide bond synthesized between the penultimate D-alanine and an amino group on a second peptide.

This hypothesis offered an explanation for the universal presence of a D-alanyl-D-alanine terminus in the nucleotide precursor of peptidoglycan whereas the structural studies apparently showed that cross-links involved only a single D-alanine residue. Moreover, a transpeptidation reaction should lead to little change in the total bond energy involved and consequently would allow the organism to synthesize peptide bonds to the exterior of the cytoplasmic membrane in the apparent absence of energy donors such as ATP.

Experimental support for the hypothesis came quickly from *in vivo* studies on the effects of penicillin on peptidoglycan synthesis in *S. aureus* and is discussed in Chapter 9.7. Briefly, walls radioactively labelled in both alanine and glycine residues were prepared from organisms grown in the presence of sub-lethal quantities of penicillin [236]. These walls contained an excess of alanine over those walls prepared from organisms grown in the absence of penicillin and, in addition, an increase in the free amino groups of glycine. These observations were independently confirmed by direct chemical analysis of staphylococcal walls prepared from organisms grown under similar conditions [205, 206]. After growth in the presence of low concentrations of benzylpenicillin (0.08 to 1 μg/ml) the walls radioactively labelled with ^{14}C-glycine were isolated and the peptidoglycan degraded by treatment with an *endo-N*-acetylmuramidase. The presence of the antibiotic resulted in an increased synthesis of uncross-linked peptide monomer units. These retained the complete D-alanyl-D-alanine terminus of the nucleotide precursor and were substituted with the pentaglycine cross-bridge peptide on the ϵ-amino group of lysine. During subsequent growth of the organisms in the absence of antibiotic these monomer units did not become cross-linked. To explain these observations Tipper and Strominger [206] proposed that incorporation of the newly-synthesized disaccharide-peptide occurred at the wall-membrane interface with almost immediate formation of cross-links to adjacent peptide side chains. The inability of the staphylococci to cross-link units synthesized in the presence of penicillin on the subsequent removal of the antibiotic was due to these units having been moved outside the 'reach' of the transpeptidase molecules as a consequence of the continued synthesis of linear uncross-linked peptidoglycan during the initial antibiotic treatment. While these *in vivo* studies supported the conclusion that cross-linkage in peptidoglycan occurred by means of a transpeptidation reaction inhibited by β-lactam antibiotics further investigation of this process clearly required the development of suitable *in vitro* systems.

8.6.1 *Systems catalysing transpeptidation in vitro*

The first *in vitro* systems capable of synthesizing peptidoglycan and carrying out transpeptidation were isolated as membrane preparations from the Gram-negative bacteria *E. coli* [5, 6, 77, 78] and *Salmonella newington* [78]. However, recent years have seen the isolation of an increasing number of preparations from both Gram-positive and Gram-negative organisms which catalyse transpeptidation. These systems can be divided into three types:

(1) Membrane preparations such as described above for *E. coli* and *S. newington* which are obtained by a series of differential centrifugations from mechanically disrupted bacteria;

(2) Wall-membrane preparations more commonly used with Gram-positive organisms, in which the two components of the bacterial envelope are isolated together. It is assumed that this type of preparation obtained from gently disrupted organisms, retains some of the spatial relationships of the membrane-bound enzymes and the wall which are present in the intact organism. Moreover, the products of biosynthesis can be easily fractionated on the basis of their solubility in hot detergent (sodium dodecyl sulphate is commonly used). This distinguishes between newly-synthesized material which has become covalently-linked to the wall and that which is detergent-soluble and presumably associated with the membrane;

(3) Permeabilized cells in which the integrity of the membrane is destroyed by treatment with organic solvents and the organism made permeable to exogenously supplied nucleotide precursors. Systems of this type were first developed to study nucleic acid synthesis [215] and have now been adapted for studies of wall biosynthesis in both Gram-positive and Gram-negative bacteria. It seems likely that preparations of this type which can also be fractionated by hot detergent will retain the closest association of the biosynthetic enzymes and the pre-existing wall.

Membrane preparations catalysing transpeptidation *in vitro* have been isolated from organisms having peptidoglycan of subgroup A1 with the exceptions of *M. luteus* [151] and *Sporosarcina ureae* [105] which contain peptidoglycans of subgroups A2 and A4 respectively. In subgroup A1 cross-linkage is direct to the free amino group of the diamino acid residue in position 3 of the acceptor peptide. Thus in *E. coli, S. newington, B. megaterium* [165, 229, 231] and *B. stearothermophilus* [106, 165] transpeptidation results in the formation of a D-alanyl-D-*meso*-diaminopimelic acid bond (Fig. 8.6a) with a particularly high efficiency of transpeptidation in *B. stearothermophilus* [106]. In this latter organism lysozyme degradation of peptidoglycan synthesized *in vitro* showed that 42% of the products were cross-linked dimers or higher oligomers compared with a value of 80% for the vegetative wall. In *Sporosarcina ureae* the cross-link is formed between D-alanine and the D-glutamyl residue of the D-glutamyl-glycine cross-bridge peptide [105]. Membranes from *M. luteus* were recently reported to catalyse the synthesis of both types of cross-linkage found in the native wall involving the formation of D-alanyl-L-lysine and D-alanyl-L-alanine bonds [151].

Membrane preparations of *B. megaterium* [229, 231] also catalyse the incorporation of free *meso*- or D,D-diaminopimelic acid into polymeric peptidoglycan. Direct chemical analysis has shown that approximately 15% of the total diaminopimelic acid in the peptidoglycan of *B. megaterium* is of the D,D configuration [10, 22]. The incorporation by membrane preparations occurred by a penicillin-sensitive transpeptidation reaction in which the terminal D-alanine residue of a disaccharidepentapeptide unit was replaced by D,D-diaminopimelic acid. Lysozyme degradation of products synthesized from unlabelled nucleotide precursors and radioactive diaminopimelic acid suggested that in addition some of the incorporated D,D-isomer was doubly transpeptidated, cross-linking two disaccharide-peptide units (Fig. 8.7). Studies on the inhibition by β-lactam antibiotics, of transpeptidation reactions catalysing the formation of the normal D-alanyl-D-*meso*-diaminopimelic acid cross-link and the incorporation of D,D-diaminopimelic acid suggested that they may result from the activity of separate transpeptidases [231]. Incorporation of free diaminopimelic acid was also catalysed by a membrane preparation from *B. stearothermophilus* but not by membranes from *E. coli*. There is no evidence as to the presence or absence of D,D-diaminopimelic acid in the peptidoglycan of *B. stearothermophilus*.

Although as described above, all of the earlier stages of peptidoglycan synthesis had been studied in detail in *S. aureus* and *M. luteus*, the first membrane preparations isolated from these organisms did not catalyse the final transpeptidation reaction (Fig. 8.6b) [189]. The formation of cross-links by cell-free preparations of either organism was not observed until transpeptidation was studied in wall–membrane preparations. Initially walls prepared from *S. aureus* [131] were shown to retain amounts of cytoplasmic membrane which were not removed by extensive washing. These preparations catalysed the incorporation of radioactivity from UDP-*N*-acetylmuramyl-(L-^{14}C-lysine)-pentapeptide into both peptidoglycan and lipid intermediates. The role of transpeptidation in the incorporation of newly-synthesized peptidoglycan into the wall was established in further studies [130]. The penicillin-sensitive release of D-alanine occurred when the wall–membrane preparation was incubated with the two nucleotide precursors and glycine, whereas no such release occurred in preparations incubated in the absence of UDP-*N*-acetylglucosamine. Glycine incorporation was sensitive to ribonuclease, implicating the glycyl-tRNA intermediate described previously, although addition of soluble tRNA and glycyl-tRNA synthetase was without effect. Penicillin also inhibited the incorporation of glycine although, in contrast to release of D-alanine, this was never complete even at concentrations as high as 100 μg/ml. Similar wall-membrane preparations have now been isolated from *M. luteus* [125, 228], *B. licheniformis* [218] and *Gaffkya homari* [60] in which the incorporation of newly-synthesized peptidoglycan into the wall occurs by transpeptidation with the concomitant release of D-alanine. In *M. luteus* (Fig. 8.8a) the incorporation of some 20–25% of the newly-synthesized peptidoglycan, was not inhibited by high concentrations of benzylpenicillin, whereas release of D-alanine was sensitive to as little as 1 μg/ml of the antibiotic. These findings contrast with those made with *B. licheniformis* (Fig. 8.8b), where incorporation of newly-synthesized material was totally sensitive to low concentrations (0.5–1 μg/ml) of both benzylpenicillin and cephaloridine. These results suggest that incorporation of newly-synthesized peptidoglycan into the pre-existing wall of *B. licheniformis* occurs by transpeptidation. In *M. luteus* the majority of incorporation is mediated by transpeptidation although some elongation of the glycan chains by transglycosylation appears to occur. This is reflected by the 20–25% of incorporation which is not inhibited by benzylpenicillin. Confirmation of these observations was obtained in both organisms, when the effects of benzylpenicillin on peptidoglycan synthesis by whole bacteria was studied. These experiments were based on the earlier observations that (a) synthesis of a linear uncross-linked peptidoglycan by membrane preparations was insensitive to penicillin, and (b) the incorporation of newly-synthesized peptidoglycan into the pre-existing wall occurred by penicillin-sensitive transpeptidation. It was argued that, if both reactions operated in the biosynthesis and assembly of the peptidoglycan then, in the presence of benzylpenicillin, bacteria would continue to synthesize peptidoglycan which would not be attached to the existing wall. This linear uncross-linked peptidoglycan would either remain associated with the membrane or be released into the medium, presumably by the same activity which normally terminates chain elongation, i.e. by releasing undecaprenyl-pyrophosphate, according to the hypothesis described above. Soluble uncross-linked peptidoglycan was isolated from the medium when both organisms were incubated with benzylpenicillin. In *B. licheniformis* [209] the

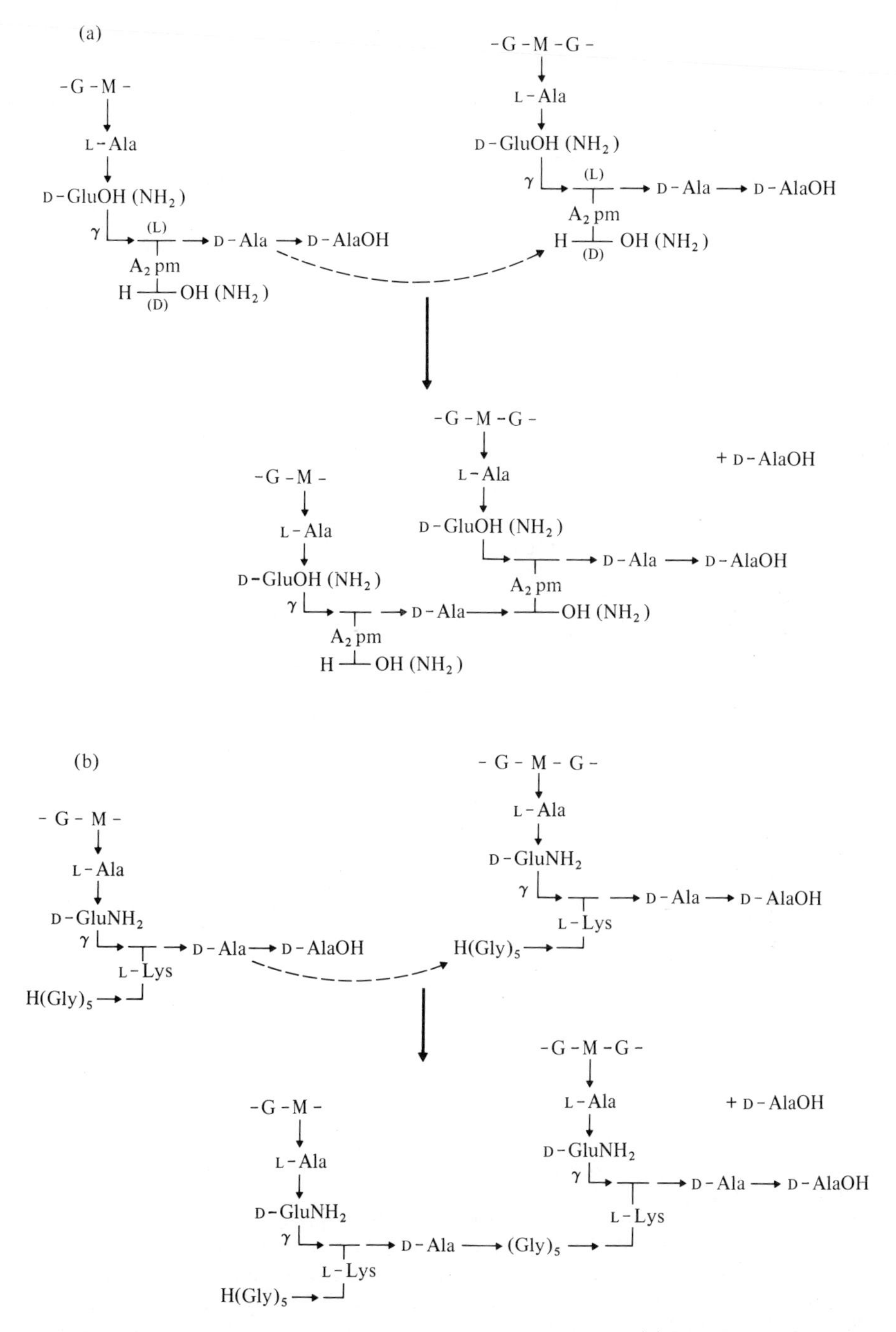

Figure 8.6 Proposed transpeptidation reactions occurring in (a) Bacilli and Gram-negative organisms and (b) *S. aureus.*

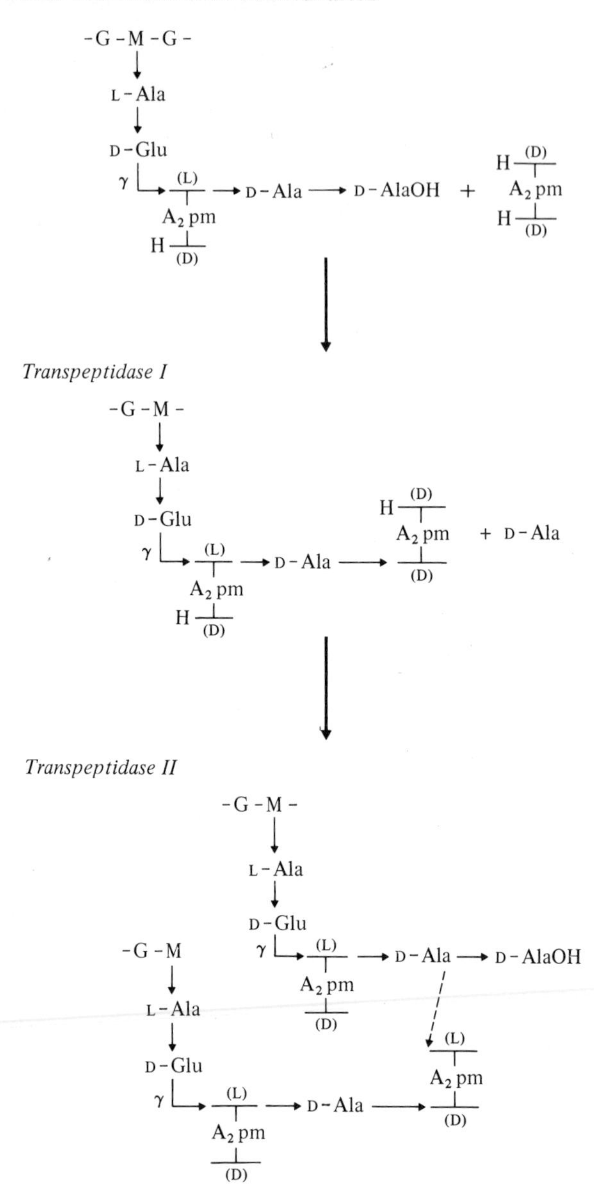

Figure 8.7 Possible mechanism for the formation of D,D-diaminopimelic acid-containing cross-links in *B. megaterium.* In the initial transpeptidation reaction D,D-diaminopimelic acid replaces the terminal D-alanine residue of a pentapeptide side chain. This is followed by a second transpeptidation which may be catalysed by the same or a different enzyme, which replaces the terminal D-alanine residue of a second pentapeptide side chain with the free D-centre of the diaminopimelic acid incorporated in the first reaction.

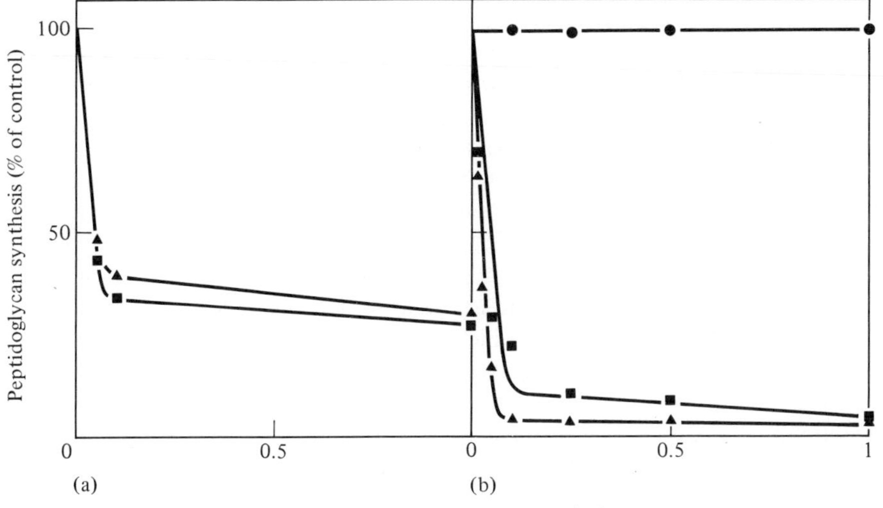

Figure 8.8 The inhibitory effect of β-lactam antibiotics on the incorporation of newly-synthesized peptidoglycan into the pre-existing walls of *M. luteus* and *B. licheniformis.* The results for *M. luteus* (a) show the incorporation of *N*-acetyl-^{14}C-glucosamine into material insoluble in hot SDS (data taken from [125]) in the presence of varying concentrations of benzylpenicillin (▲) and cephaloridine (■) The effect of the same antibiotics on the incorporation of diamino-^{3}H-pimelic acid into walls of *B. licheniformis* is given (b) together with incorporation of radioactivity into 'soluble peptidoglycan (●). Data taken from [218] and unpublished observations.

material had an average glycan chain length of 44 disaccharide units and a full complement of pentapeptide side chains, a result of the simultaneous inhibition of D-alanine carboxypeptidase, whereas the material from *M. luteus* [126], although having a similar length, was 50% deficient in peptide side chains. Subsequently, the substitution of peptides on the glycan was shown to be non-random [127, 243]. The soluble peptidoglycan contains glycan chains only poorly substituted whereas other chains appear to possess their full complement of peptides. The peptide side chains presumably released by action of an *N*-acetylmuramyl-L-alanine amidase were also secreted into the medium [126]. An enzyme of this specificity has also been shown to be active on peptidoglycan synthesized by membrane preparations of *M. luteus* (*sodonensis*) [80]. It is possible that this amidase activity is involved in the release of peptide side chains for incorporation as cross-bridge peptides in *M. luteus* (see p. 259). Soluble uncross-linked peptidoglycan has also been found secreted into the medium when both *Brevibacterium divaricatum* [89] and 'autoplasts' of *S. faecalis* [172, 173] are incubated in the presence of penicillin.

The first use of permeabilized bacteria for studies of peptidoglycan synthesis was described by Schrader and Fan [182]. Using toluene-treated *B. megaterium* they showed by density gradient centrifugation in caesium chloride that newly-synthesized peptidoglycan measured by incorporation of radioactive *N*-acetylglucosamine from the nucleotide precursor, was associated with the pre-existing wall. This did not occur when benzylpenicillin was present in the biosynthetic system although the synthesis of a soluble product was unaffected. Subsequently Taku and Fan [194, 195] have established that the reversible inactivation of toluene-treated bacilli by extraction with lithium chloride is due to solubilization of the *N*-acetylglucosaminyl transferase (see p. 255). The transpeptidase responsible for the incorporation of D,D-diaminopimelic acid is also solubilized. More recently toluene-treated cells of *B. megaterium* were used to study the direction of transpeptidation in this organism (see [53] and pp. 274–6). Similar preparations of *G. homari* and *M. luteus* also catalyse peptidoglycan biosynthesis including cross-linkage (referred to in [53] and S. J. Thorpe and H. R. Perkins unpublished observations).

In Gram-negative organisms diethyl ether rather than toluene has been used to permeabilize *E. coli* [132], *P. aeruginosa* [128, 136] and *Neisseria gonorrhoeae* [11]. In each organism newly-synthesized peptidoglycan becomes covalently-linked (i.e. SDS-insoluble) to the sacculus by a penicillin-sensitive transpeptidation reaction. A surprising observation in *N. gonorrhoeae* was that high concentrations of β-lactam antibiotics stimulated the synthesis of detergent-soluble peptidoglycan far in excess of any concomitant decrease in the amount of SDS-insoluble material. Possibly the synthesis of soluble peptidoglycan is controlled by the synthesis of the cross-linked polymer although an alternative explanation would have the β-lactams acting in some way to stimulate directly synthesis of uncross-linked polymer. Clearly this interesting observation requires further investigation to distinguish between these or other possible mechanisms of control.

Earlier Mirelman *et al.* [132] had used ether-treated bacteria to investigate peptidoglycan synthesis during division of a temperature-sensitive mutant *E. coli* PAT84. This organism grows as multinucleate nonseptate filaments at the restrictive temperature (42° C) and undergoes synchronous division when transferred to the permissive temperature (30° C). Filamentous organisms from cultures grown at the restrictive temperature incorporated more peptidoglycan with a higher degree of cross-linkage than did organisms grown at 30° C. Moreover treatment of organisms grown at 30° C with low concentrations of ampicillin (0.5 μg/ml) resulted in the partial inhibition (approximately 80%) of D-alanine carboxypeptidase activity while transpeptidation remained unaffected. These conditions also resulted in significantly more newly-synthesized peptidoglycan being incorporated again with increased cross-linkage. Treatment of *E. coli* with low concentrations of β-lactams had previously been shown to inhibit septum formation and result in continued growth of the organisms as filaments [185]. Similar results were also obtained using filaments of *E. coli* produced by treatment of cultures with either cephalexin (a β-lactam known to bind preferentially to penicillin-binding protein 3, see Chapter 9.7.2), or with nalidixic acid (an inhibitor of deoxyribonucleic acid biosynthesis) [133]. On the basis of these observations, Mirelman *et al.* concluded that the formation of septa and hence cell division was dependent upon a balance between transpeptidation, which catalysed cross-linkage and the incorporation of newly-synthesized material, and D,D-carboxypeptidase which regulated the availability of suitable pentapeptide donors in the nascent peptidoglycan. Some support for this conclusion comes from the earlier observation that certain mutants of *E. coli* selected as being temperature-sensitive and resistant to low concentrations of ampicillin, grew as filaments and showed increased cross-linkage in their peptidoglycan [86]. However, the recent results of Hirota and Matsuhashi and their colleagues described in Chapter 9.7.2 appear to conflict with this hypothesis. These groups have constructed mutants grossly deficient in or lacking D,D-carboxypeptidase activity but which grow normally, although information on the extent of cross-linkage in these organisms does not appear to be available yet.

In many respects peptidoglycan biosynthesis by permeabilized cells of *P. aeruginosa* is similar to that found in *E. coli*. The two systems differ however in that in the pseudomonad both transpeptidation and D,D-carboxypeptidase activity are sensitive to low concentrations (0.5 μg/ml) of many cephalosporins and penicillins [128, 129]. Thus, as implied from earlier investigations [191], the resistance of *P. aeruginosa* to many β-lactams *in vivo* does not reside in any intrinsic resistance of the target enzymes themselves, but rather in the inability of the antibiotics to penetrate the cell envelope and reach their targets. The pseudomonad system also provided evidence for the occurrence of a detergent-soluble precursor of cross-linked peptidoglycan [128]. As described earlier (p. 263) this material labelled with *N*-acetyl-^{14}C-glucosamine could subsequently be chased into the wall. The detergent-soluble fraction was a mixture of partially cross-linked high molecular weight material and lower molecular weight oligomers but whether there is a precursor-product relationship between these two sub-fractions remains unclear. Evidence for the presence of at least two transpeptidases in *P. aeruginosa* comes from the observation that none of the β-lactams tested completely inhibited cross-linkage [129]. It has been suggested that one β-lactam-sensitive transpeptidase is involved in 'attachment' of newly-synthesized uncross-linked peptidoglycan to the pre-existing wall, while the second transpeptidase which is relatively resistant to β-lactams, is responsible for formation of cross-links in the nascent peptidoglycan. This is the fraction which can be isolated from permeabilized cells in a detergent-soluble form. A similar situation may exist in *E. coli* where recent experiments of Schwarz and his colleagues (U. Schwarz, personal communication) have studied peptidoglycan synthesis *in vivo*. By labelling growing cells with short pulses of radioactive diaminopimelic acid and isolating sacculi, they found that peptidoglycan incorporated during these pulses was much less cross-linked than the pre-existing material. During subsequent growth the degree of cross-linkage of the labelled peptidoglycan increased to normal. In addition, treatment of the organisms with various β-lactams distinguished between incorporation of diaminopimelic acid into the sacculus and the formation of cross-links. Thus treatment with relatively high concentrations of ampicillin (100 μg/ml) or cephaloridine (50 μg/ml) completely inhibited incorporation of diaminopimelic acid and lysis of the organisms ensued. In contrast treatment with cephalexin (25 μg/ml) or piperacillin (10 μg/ml) both of which caused the organisms to grow as filaments, had no apparent effect on the incorporation of diaminopimelic acid but did have a marked effect on cross-linkage.

These observations were also interpreted as showing that 'attachment' of newly-synthesized peptidoglycan occurred via a transpeptidase which in this case was sensitive to ampicillin and cephaloridine but resistant to cephalexin and piperacillin at the concentrations tested. Secondary cross-linkage then ensued being catalysed by a transpeptidase sensitive to all the β-lactams tested. In other experiments the kinetics of cross-link formation and the attachment of lipoprotein to peptidoglycan (see Chapter 10.4.2) paralleled each other. Examination of *E. coli* mutants unable to attach lipoprotein and which lacked penicillin-binding protein 4 showed that these organisms did not catalyse the secondary transpeptidation described above. However, the degree of cross-linkage of their peptidoglycan is approximately 20% compared with the normal value of 30% suggesting that in some way the 'attachment' transpeptidase can either overcome this defect and increase the number of cross-links it forms or alternatively yet a third transpeptidase compensates for the absence of the second. A similar situation may also exist in *B. megaterium* where cross-linkage also continued for some time after the initial incorporation of diaminopimelic acid into the wall [33].

Even the above implications of complexity in the biosynthetic process may be a simplification of the situation existing *in vivo*. Bacteria probably have multiple enzyme complexes capable of catalysing peptidoglycan synthesis some of which may act continuously whereas others are only activated when the organism is subject to stress such as the presence of β-lactam antibiotics. No doubt further information on the complexity of peptidoglycan synthesis will come from studies of the effect of β-lactam antibiotics not only on transpeptidation (cross-linkage) but also on the incorporation of newly-synthesized peptidoglycan into the pre-existing wall.

8.6.2 *The direction of transpeptidation*

The formation of a cross-link in a transpeptidation reaction can be regarded as directional since one peptide side chain will provide the D-alanyl-D-alanine donor group and another the acceptor amino group. Thus, attachment of newly-synthesized peptidoglycan to the pre-existing wall may occur by cross-linkage in which the nascent polymer acts as the donor and the wall as the acceptor or *vice versa*. Nascent peptidoglycan has the capacity of fulfilling both these roles and may form cross-links with itself or with older peptidoglycan. Many organisms possessing active D,D-carboxypeptidases including Bacilli, *E. coli* and several other Gram-negative bacteria lack the terminal D-alanine residue from the peptide side chains of wall peptidoglycan. Consequently these peptides can only function as acceptors in cross-linkage.

The direction of transpeptidation was first studied in *B. licheniformis* [221] (Fig. 8.9a) where peptidoglycan was synthesized by wall-membrane preparations from UDP-*N*-acetylmuramyl-pentapeptide in which the free ε-amino group of the diaminopimelic acid residue had been blocked by acetylation. Thus, the precursor could only act as a donor peptide (Fig. 8.6a and 8.8a) in subsequent synthesis of cross-linked peptidoglycan. Incorporation of this precursor into the pre-existing wall was 23% of that observed when the normal, non-acetylated precursor was present.

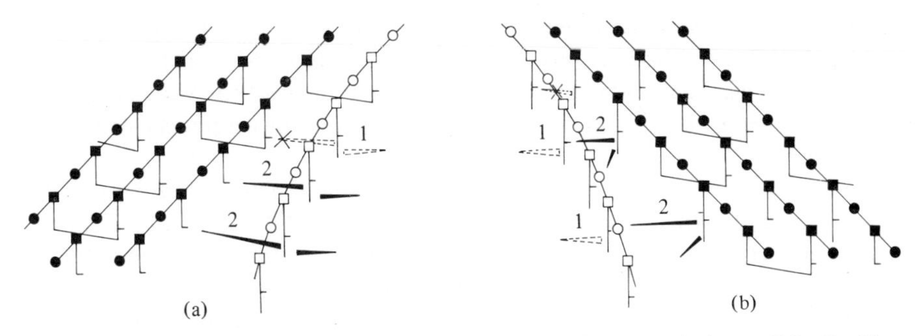

Figure 8.9 Attachment of nascent peptidoglycan to the pre-existing wall in *Bacillus licheniformis* and *Gaffkya homari.* In *B. licheniformis* (a) the nascent peptidoglycan (-□-○-) acts as the donor and the pre-existing wall (-■-●-) as the acceptor in the transpeptidation reaction. Prior D,D-carboxypeptidase activity (1) on the nascent peptidoglycan will prevent transpeptidation (2). This direction of transpeptidation has also been shown to occur in other bacilli and certain Gram-negative organisms. In *G. homari* (b) the opposite direction of transpeptidation occurs in which pentapeptide side-chains in the pre-existing wall (-■-●-) act as the donor peptides and tetra- and tripeptides of the nascent peptidoglycan (-□-○-) act as the acceptors. Prior D,D- and L,D-carboxypeptidase (1) action on the nascent peptidoglycan is essential for transpeptidation (2). However, some pentapeptide side chains must be retained to be incorporated into the wall for use as donor peptides in subsequent transpeptidation reactions.

After isolation of the wall and degradation of the radioactive peptidoglycan with an endo-*N*-acetylmuramidase, acetylated cross-linked dimers were present in about half the amount found in peptidoglycan synthesized from the natural precursor. These lower levels of incorporation could result from a decreased efficiency of the acetylated precursor as a substrate for peptidoglycan synthesis and transpeptidation or alternatively because the formation of cross-links between two newly synthesized units is prevented. The presence of the acetylated substituent precludes the formation of a cross-link in which the peptide side chain of the pre-existing wall acts as the donor and the nascent unit as the acceptor. Thus in *B. licheniformis* some if not all cross-linkage occurs from donor peptides of nascent peptidoglycan to acceptor amino groups on the pre-existing wall. More recently this direction of transpeptidation has been shown to occur in *B. megaterium* [53] and *P. aeruginosa* [128]. In the pseudomonad cross-linked peptidoglycan was synthesized from acetylated precursor although the relative efficiency of cross-linkage was not given. In *B. megaterium* the evidence of direction came from the inability of toluene-treated cells to utilize UDP-*N*-acetylmuramyl-tetrapeptide for the synthesis of cross-linked peptidoglycan although this precursor was utilized at 60% efficiency for the synthesis of uncross-linked polymer. Surprisingly, these toluene-treated cells were unable to form cross-links with acetylated precursor. This observation suggests that in *B. megaterium* the active site of the transpeptidase must differ in some way from that found in *B. licheniformis*, although the natural pentapeptide substrates are apparently identical. Evidence that cross-linkage is not a random process but is directed has been obtained in *S. faecalis* [149] and by using the soluble D,D-carboxypeptidase-transpeptidase of *Streptomyces* R61 [36]. In each case cross-linking occurred by a process of monomer addition; thus with the carboxypeptidase–transpeptidase peptide trimer was synthesized by transpeptidation of the peptide monomer acting as the donor through its D-alanyl-D-alanine terminus onto the peptide dimer acting as the acceptor.

In marked contrast to the situation in the Bacilli, *Gaffkya homari* has been shown to catalyse transpeptidation in the opposite direction [60, 58] (Fig. 8.9b). Thus the donor pentapeptide side chains come from the pre-existing wall and the acceptor amino group from the nascent peptidoglycan. Wall-membrane preparations of *G. homari* were found to synthesize cross-linked peptidoglycan from both UDP-*N*-acetylmuramyl-pentapeptide and -tetrapeptide with equal efficiency. Incorporation of *N*-acetylmuramyl-pentapeptide residues was inhibited by low concentrations of benzylpenicillin whereas even high concentrations of the antibiotic were without effect on the incorporation of *N*-acetylmuramyl-tetrapeptide. Thus in *Gaffkya* cross-linkage is resistant to penicillin. In fact the peptide side chains of nascent peptido-glycan which serve as acceptor groups in transpeptidation are tetra- and tripeptides resulting from the consecutive action of D,D- and L,D-carboxypeptidases on the pentapeptide side chains originally present [59, 66, 67]. These enzymes which act only on nascent peptidoglycan are both inhibited by a variety of β-lactam antibiotics. The evidence showing that the D,D-carboxypeptidase is the 'killing site' for β-lactams in this organism is discussed more fully in the following chapter. Some pentapeptide units must however be retained in the nascent peptidoglycan and incorporated into the wall where they serve as donor peptides in subsequent transpeptidation reactions. Thus, in nascent peptidoglycan incorporated *in vivo* the number of pentapeptide side chains present must be similar to the number converted to tri- and tetrapeptides. This ratio of modified to unmodified peptides would allow a maximum of 50% cross-linking as dimers a value which agrees closely with the degree and nature of cross-linking observed *in vivo*. It seems probable however, that the situation existing in *Gaffkya* is unusual. Clearly, in most other organisms transpeptidation is inhibited by β-lactam antibiotics and where examined attachment of nascent peptidoglycan occurs by cross-linkage in which the pre-existing wall is the acceptor.

8.7 D-Alanine carboxypeptidases

As briefly mentioned above, a second group of enzymes whose activity is expressed during the terminal stages of peptidoglycan synthesis are the D-alanine carboxy-peptidases. Chemical analysis of the peptidoglycans of several bacteria had established that D-alanine residues, not involved in the formation of cross-links, were absent from the completed peptidoglycan. This contrasted with the known biosynthetic mechanism, where the *N*-acetylmuramyl residues were substituted with a pentapeptide side chain and suggested that the excess D-alanine residues might be removed by an enzyme or enzymes having the specificity of D-alanine carboxy-peptidases. Such enzymes were first described in *E. coli* [224] and were later found to be penicillin-sensitive as was the transpeptidase from the same organism [77, 78].

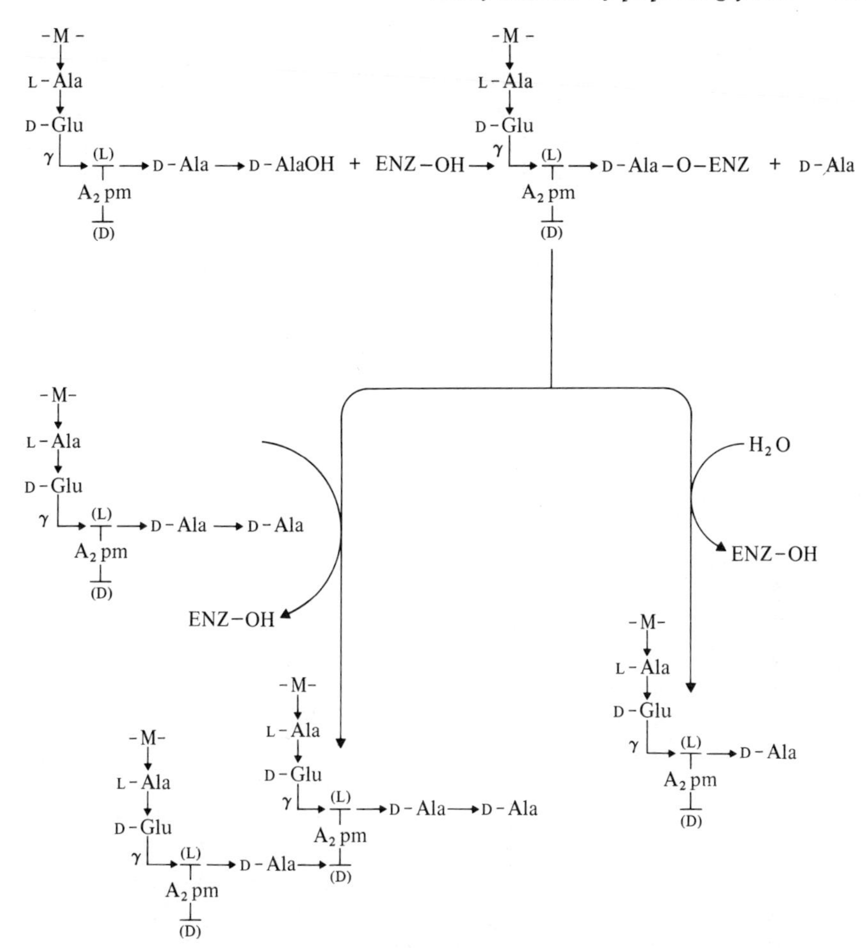

Figure 8.10 Possible mechanism of a D,D-carboxypeptidase acting as a carboxypeptidase (B). Acyl-D-alanyl-D-alanine reacts with the active site of the D,D-carboxypeptidase (a seryl residue in the carboxypeptidases of the Bacilli and certain actinomycetes) to give an acyl-D-alanyl-enzyme-intermediate. If the acceptor site then binds a peptide of appropriate configuration transpeptidation ensues with the release of enzyme. Alternatively if water becomes bound enzyme is again regenerated and acyl-D-alanine released resulting in D,D-carboxypeptidase activity.

At this time, Strominger and his colleagues postulated that the D-alanine carboxypeptidase might be an 'uncoupled' transpeptidase with water rather than the amino group of a second peptide side chain acting as the acceptor for the postulated acyl-D-alanine–enzyme intermediate (Fig. 8.10). Thus the same enzyme would be bifunctional, depending on whether it was situated in an environment from which water was excluded or not. Under the former conditions it would act as a natural transpeptidase.

D-Alanine carboxypeptidases have now been solubilized and purified from many different organisms including Bacilli, [8, 213, 241], *E. coli* [198] *Proteus mirabilis*

[117], *S. aureus* [93], *S. faecalis* [19] and several *Actinomyces* [46, 47] whereas in *Mycobacterium smegmatis* [31] and *N. gonorrhoeae* [20, 20a] they have been studied as membrane-bound enzymes. Many of the purified enzymes have activity as transpeptidases as well as D-alanine carboxypeptidases. In addition they are subject to varying inhibition by β-lactam antibiotics. This property, together with the ready availability of certain enzymes in a purified form, has led to many of the studies on the mechanism of action of penicillin being performed with these enzymes. The results of these investigations are considered in detail in the following chapter.

Recent studies of the penicillin-binding proteins of *E. coli* and the Bacilli (See Chapter 9.7.2) have established that their transpeptidases (identified by their reaction with certain β-lactam antibiotics) are quite distinct from the D-alanine carboxypeptidases. This argues strongly against the original suggestion that the two enzyme activities reflect the same protein acting in different environments. Moreover, the isolation from *E. coli* of various mutants grossly deficient in carboxypeptidase activity have virtually eliminated any requirement of this enzyme in normal growth. On the other hand, there are considerable similarities in the substrate specificities of the soluble D,D- carboxypeptidase-transpeptidases and the membrane-bound transpeptidases of several actinomycetes. In addition these specificities appear to relate to the peptidoglycan structure of the organism from which the enzymes were obtained. Although it seems doubtful that a direct relationship exists between the carboxypeptidases and transpeptidases, since both proteins are coded for by separate genes, the many similarities suggest that they may in fact have a common ancestor.

D-Alanine carboxypeptidase assayed by release of D-alanine from UDP-*N*-acetylmuramyl-pentapeptide, was first purified from sonic extracts of *E. coli* [79]. Activity was greatest when the nucleotide precursor was utilized as the substrate and the enzyme had little activity on a number of substrates more closely related to peptidoglycan. A second D-alanine carboxypeptidase (carboxypeptidase II) was also purified which released the second D-alanine residue from the pentapeptide side-chain by breaking an L–D peptide bond. In contrast to carboxypeptidase I, which was extremely sensitive to benzylpenicillin and other β-lactam antibiotics, carboxypeptidase II was resistant. Subsequently *E. coli* was shown to possess D,D-carboxypeptidases which also had activity as transpeptidases and endopeptidases [9, 190]. Ghuysen and his colleagues [145, 146, 160] utilized detergent extraction of the inner (cytoplasmic) membrane to yield one enzyme complex possessing each of the activities described above. These were measured *in vitro* as outlined in Table 8.5. Although the complex could be further fractionated on DEAE–cellulose to yield two distinct complexes differing in their affinity for β-lactam antibiotics (particularly ampicillin and cephalothin, each of the complexes still contained the four enzyme activities examined. However, the use of the glutamate-amidated tetrapeptide, L-alanyl-D-isoglutaminyl-*meso*-diaminopimelyl-D-alanine as a substrate, allowed two enzyme systems to be distinguished. These were a D,D-carboxypeptidase-transpeptidase and a D,D-carboxypeptidase-endopeptidase. The tetrapeptide functioned as a substrate for the transpeptidase (acting only as the acceptor peptide) (Fig. 8.6a) and inhibited the D,D-carboxypeptidase of this system whereas it was without effect on the D,D-carboxypeptidase-endopeptidase. On the other hand, these enzymes were neither purified nor indeed were the activities separated since as stated above each fraction of the original detergent extract contained all four activities tested. Thus it remained unclear to what extent the various reactions observed were catalysed by individual proteins. This question has now been resolved in a series of parallel investigations which studied the enzymes as both D,D-carboxypeptidases and penicillin-binding proteins. Three D,D-carboxypeptidases have been isolated and purified to homogeneity from disrupted cells of *E. coli* [198]. Enzyme 1C was present in the supernatant fraction while enzymes 1A and 1B were selectively solubilized from the membrane fraction by extraction with the non-ionic detergent Triton X-100 and LiCl respectively. Additional studies showed that enzymes 1B and 1C were extremely similar if not identical. In addition to their being D,D-carboxypeptidases both enzymes were active endopeptidases cleaving the D-alanyl-*meso*-diaminopimelic acid cross-links in peptidoglycan. On the other hand, they were relatively poor transpeptidases and neither of them bound radioactive benzylpenicillin although their enzymic activities were very sensitive to inhibition by β-lactam antibiotics including 6-amino-penicillanic acid. More recently mutants of *E. coli* have been isolated which lack both soluble carboxypeptidase activity (i.e. enzymes 1B and 1C) and penicillin-binding protein 4 (PBP 4) [76, 121]. These observations present something of a paradox since the purified enzymes do not bind the antibiotic. The independent isolation of several mutants argues against carboxypeptidase and PBP 4 being the products of independent but closely-linked genes. Possibly PBP 4 represents a membrane-bound form of the soluble enzymes, their molecular weights (49 000) are very similar if not identical. If this is the case and the two functions are directly related then solubilization must in some way prevent penicillin-binding.

In contrast carboxypeptidase 1A was an effective transpeptidase (unnatural model transpeptidase was used as the assay system), bound radioactive benzyl-penicillin but was devoid of endopeptidase activity. On gel electrophoresis in the presence of SDS purified enzyme, 1A ran as a doublet of molecular weights 40 000 and 42 000 (originally reported as 32 000 and 34 000 [198]) which correspond exactly to PBPs 5 and 6 of *E. coli* [186]. It remains unclear whether the two polypeptides differ in their enzymic activities although they can be distinguished by the half-lives of their penicillin complexes (see Chapter 9.7.2). A mutant lacking carboxypeptidase 1A activity has been isolated [120]. Surprisingly it retained both PBP 5 and 6, although earlier Tamura *et al.* [198] had shown that treatment of the purified enzyme with the sulphydryl reagent *p*-chloromercuribenzoate inhibited carboxypeptidase activity but not the binding of penicillin. These observations suggested that an SH-group was required for catalytic activity but not for antibiotic binding and raised the question whether substrate and penicillin bound to different sites in the enzyme. More recently Curtis and Strominger [20] have reported that sulphydryl reagents also inhibit the release of penicillin from the carboxypeptidase and that removal of this inhibition restores both catalytic activity and antibiotic release in parallel. Moreover, in the presence of *p*-chloromercuribenzoate an acyl-enzyme intermediate accumulated and could be isolated. Thus interaction of the

Table 8.5 Methods of measuring enzyme activities in *E. coli* (from [105]).

Enzyme activity	Substrates	Reaction products
D,D carboxypeptidase	UDP–MurAc → L-Ala → D-Glu-OH ↳ (L) → D-Ala → D-Ala DAP H (D) OH + H_2O	UDP–MurAc → L-Ala → D-Glu-OH ↳ (L) → D-Ala + D-Ala DAP H (D) OH
Unnatural model transpeptidase	UDP–MurAc → L-Ala-D-Glu-OH ↳ (L) → D-Ala → D-Ala DAP H (D) OH + gly(or D-ala*)	UDP–MurAc → L-Ala → D-Glu-OH ↳ (L) → D-Ala → gly(or D-Ala*) DAP H (D) OH + D-Ala
Natural model transpeptidase (with concomitant hydrolysis of donor)	L-Ala → D-Glu-OH ↳ (L) → D-Ala → D-Ala DAP H (D) OH + H_2O as both donor and acceptor	L-Ala → D-Glu-OH ↳ (L) → D-Ala → D-Ala DAP (D) OH L-Ala → D-Glu-OH ↳ (L) → D-Ala → DAP H (D) OH + L-Ala → D-Glu-OH ↳ (L) → D-Ala + D-Ala DAP

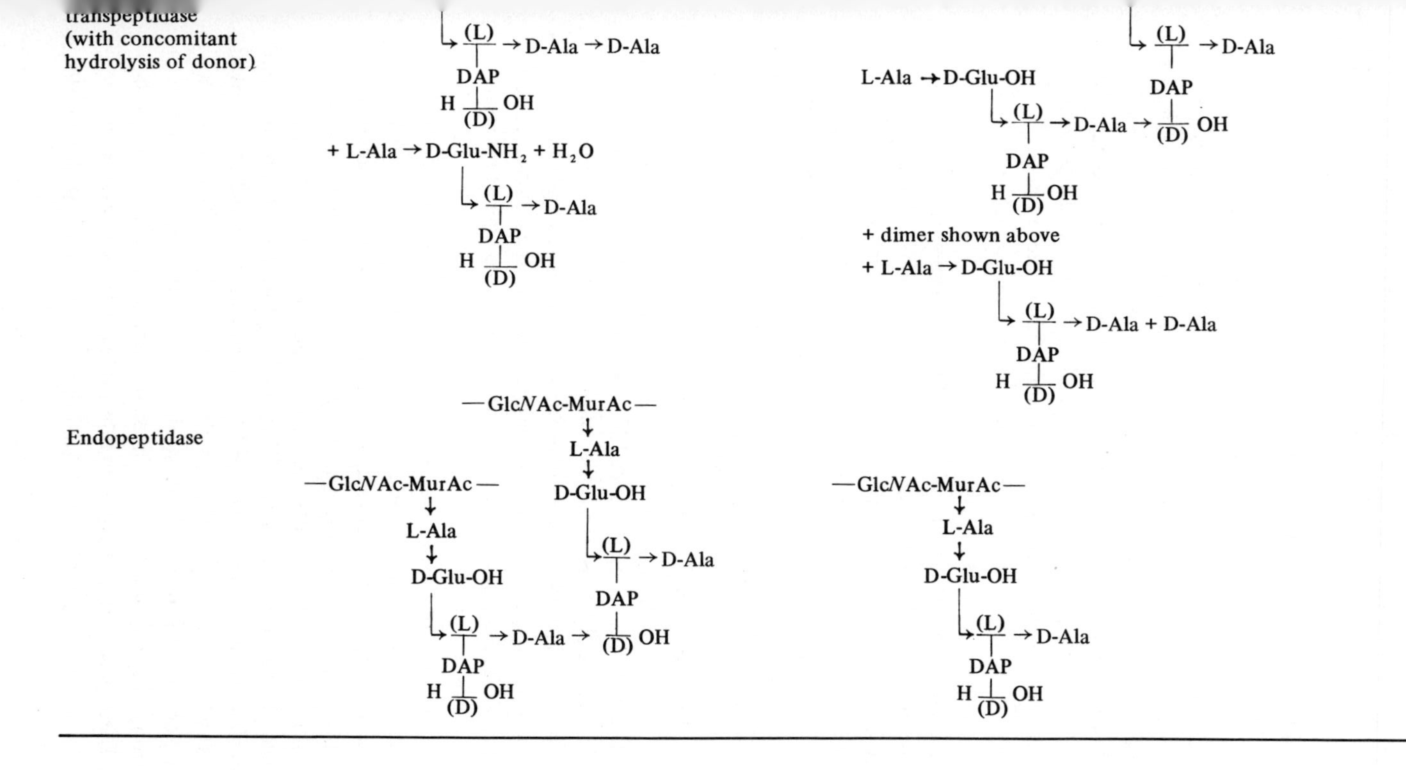

(with concomitant
hydrolysis of donor)
(L)
→ D-Ala → D-Ala
DAP
H OH
(D)
+ L-Ala → D-Glu-NH2 + H2O
(L)
→ D-Ala
DAP
H OH
(D)
(L)
→ D-Ala
L-Ala → D-Glu-OH
DAP
(L)
→ D-Ala →
(D)
OH
DAP
H OH
(D)
+ dimer shown above
+ L-Ala → D-Glu-OH
(L)
→ D-Ala + D-Ala
DAP
H OH
(D)
Endopeptidase
—GlcNAc-MurAc—
L-Ala
D-Glu-OH
(L)
→ D-Ala
DAP
—GlcNAc-MurAc—
L-Ala
D-Glu-OH
(L)
→ D-Ala →
(D)
OH
DAP
H OH
(D)
—GlcNAc-MurAc—
L-Ala
D-Glu-OH
(L)
→ D-Ala
DAP
H OH
(D)

carboxypeptidase with both substrate and penicillin appears to occur at the same active site in a two-step process involving acylation and deacylation of the enzyme. Only the second stage of the reaction requires the SH-group and is inhibited by sulphydryl reagents. Carboxypeptidases 1A and 1B/C both exhibit weak penicillinase activity with penicilloic acid as the reaction product. Clearly in the case of enzyme 1A this represents a breakdown of the enzyme–penicillin complex whereas the explanation for enzyme 1B/C which does not form a complex with the antibiotic, is not immediately obvious. These findings contrast with those reported for the purified D,D-carboxypeptidases of Bacilli and the soluble exocellular D,D-carboxypeptidases-transpeptidases of various actinomycetes where fragmentation of the penicillin molecule occurs. These results are discussed in detail in Chapter 9.7.4.

As mentioned earlier, it appears possible to dispense with both carboxypeptidases for normal growth. The only physiological response observed in the various carboxypeptidase deficient mutants was a reduced rate of ampicillin-induced lysis in organisms lacking enzyme 1B/C. Thus, *in vivo* this enzyme may in fact function as an endopeptidase rather than a carboxypeptidase.

The second group of organisms from which the D,D-carboxypeptidases have been studied in detail are the *Streptomyces* spp. and *Actinomadura.* Soluble D,D-carboxypeptidase is secreted into the culture medium and the enzymes from several organisms have been purified to homogeneity [25, 35, 38, 49, 100, 104]. With the single exception of the enzyme from *Streptomyces albus* G, these purified proteins will also catalyse transpeptidation reactions [153, 159]. The enzyme from *S. albus* G also differs in that it occurs as two forms, apparently identical other than in molecular weight; enzyme 1 (30% of the total) has an apparent molecular weight of 9000 and enzyme II (70%) of 18 500 [25]. The substrate specificity of the enzymes acting as D,D-carboxypeptidases has been studied in detail utilizing an extensive series of synthetic and naturally occurring peptides [50, 100, 102, 104]. On the basis of the model peptide acetyl-L-R_3-R_2-R_1-(OH) the following conclusions were drawn (Table 8.6): the enzymes showed a requirement for a D-amino acid or glycine at the terminal R_1 position with a preference for D-alanine. However, the requirement for D-alanine in the penultimate position was even more marked; the highest substrate specificity occurs at this residue. Finally, increased efficiency of hydrolysis was obtained if the L-amino acid at position R_3 was substituted with a long side chain. Thus, the dipeptide D-alanyl-D-alanine and its mono-acetylated derivative are not substrates, whereas the disaccharide peptide, N^{α}-*N*-acetylglucosaminyl-*N*-acetylmuramyl-L-alanyl-D-isoglutaminyl-L-lysyl-(pentaglycyl)-D-alanyl-D-alanine was a good substrate. Although the carboxypeptidases all exhibited similar substrate specificity profiles, their kinetic parameters (K_m and V_{max}) were markedly different. In *Streptomyces albus* G and *Actinomadura* R39 good substrates bound with a lower K_m than did those which were hydrolysed less efficiently; a smaller effect was noted on the V_{max}. In contrast, the enzymes from strains K11 and R61 bound all substrates at a relatively high K_m and the good substrates were distinguished by an increase in V_{max}. These observations have been interpreted to show an increased specificity of substrate binding in the carboxypeptidases from strains G and R39 in

Table 8.6 Substrate requirements of exocellular *Streptomyces* D,D- carboxypeptidase – transpeptidases acting as carboxypeptidases (release of *C*-terminal D-alanine)*

Substrate	*Encyme*											
	G			*R39*			*K11*			*R61*		
	K_m†	V_{max}†	E§	K_m	V_{max}	E	K_m	V_{max}	E	K_m	V_{max}	E
Group A: *modifications to residue 1*												
N^α, N^ϵ-diacetyl L-Lys → D-Ala → D-Ala	0.33	100	303	0.8	330	410	11	2000	180	12	890	74
N^α, N^ϵ-diacetyl L-Lys → D-Ala → D-Leu	0.33	33	100	0.7	230	320	10	160	16	10	50	5
N^α, N^ϵ-diacetyl L-Lys → D-Ala → Gly	2.5	60	24	2.5	100	40	12	220	18	36	200	6
N^α, N^ϵ-diacetyl L-Lys → D-Ala → L-Ala	no hydrolysis			no hydrolysis			no hydrolysis			no hydrolysis		
Group B: *modifications to residue 2*												
N^α, N^ϵ-diacetyl → L-Lys → D-Leu → D-Ala	no hydrolysis			no hydrolysis			13	38	2.9	10	10	1
N^α, N^ϵ-diacetyl → L-Lys → Gly → D-Ala	15	107	7	no hydrolysis			14	10	0.7	15.5	1.7	0.1
Group C: *modifications to residue 3*												
N^α, acetyl → L-Lys → D-Ala → D-Ala	6	20	3.3	0.2	600	3000	30	9	0.3	15	4	0.3
UDP–MurAc → L-Ala → D-Glu ↳ A_2pm → D-Ala → D-Ala	0.4	10	25	0.25	400	1600	17	18	1	11	8	0.3
N^α-GlcNAc-MurAc → L-Ala → D-Glu-NH_2 ↳ L-Lys → D-Ala → D-Ala; $(Gly)_5$ ↑ (on L-Lys)	0.28	9	32	0.3	420	1400	11	1050	95	14	800	57

*For details see [50, 100, 102, 104]

† K_m values are given as mM and V_{max} as μmoles D-alanine released/mg enzyme/hour.

§E is the enzyme efficiency calculated as V_{max}/K_m.

contrast to strains K11 and R61, where better substrates are postulated to bring about a conformational change in the enzyme, allowing an increased efficiency of hydrolysis. Additional information on the nature of substrate binding has come from the study of peptide analogues of the substrate N^{α}, N^{ϵ}-diacetyl-L-lysyl-D-alanyl-D-alanine. The enzymes from strains G and R61 gave similar results, suggesting that the two *C*-terminal residues were important in the initial binding of the substrate, whereas the side chain of the third residue (R_3) influences catalytic activity (Table 8.7). On the other hand, these substrate analogues did not inhibit the enzyme from strain R39.

It has already been mentioned that the purified D,D-carboxypeptidases will catalyse a transpeptidation reaction. An exception to this is the enzyme from *Streptomyces albus* G, which also differs from the purified enzymes isolated from other strains, in that it is relatively insensitive to inhibition by β-lactam antibiotics [37, 103]. Transpeptidation was demonstrated as the transfer of N^{α}, N^{ϵ}-diacetyl-L-lysyl-Dalanine to either glycine, *meso*-diaminopimelic acid or a second molecule of D-alanine as the acceptor [153, 159]. This activity has been obtained with enzymes from strains R39, R61 and K11 (Table 8.8). Under the experimental conditions used, three activities could be observed; transpeptidation as described above, and carboxypeptidase activity on either the donor peptide or the transpeptidation product. In experiments of this type, transpeptidation was favoured by raising the pH, increasing the acceptor concentration and decreasing the water content of the incubation mixture by performing the experiment in mixtures of ethylene glycol, glycerol and water. The substrate specificities of the purified enzymes as transpeptidases have been examined in some detail. Enzyme from R61 was active with a wide range of acceptors including dipeptides, antibiotics (D-cycloserine and 6-aminopenicillanic acid), 2-amino-2-deoxyglucuronic and galacturonic acids, in addition to the amino acids listed above. Greatest efficiency was observed with dipeptides having *N*-terminal glycine in agreement with the chemically determined structure of the cross-links in the peptidoglycan of strains R61 and K11 [51, 153]. In these organisms peptide side chains are linked by peptide cross-bridges of the structure D-alanyl-glycyl-L, L-diaminopimelic acid. In contrast, cross-links in strain R39 are of the direct D-alanyl-meso-diaminopimelic acid type, an observation which has led to the reclassification of this strain as *Actinomadura* [50]. Dipeptides did not serve as acceptors for the transpeptidation using enzyme from R39, whereas donors bearing analogues of the peptide side chain were active. Specificity was directed towards the sequence L-alanyl-D-isoglutamyl-*meso*-diaminopimelic acid. Complete loss in activity as an acceptor was observed when the carboxyl group of *meso*-diaminopimelic acid was amidated, whereas substitution of either the *N*-terminus with β-1,4-*N*-acetylglucosaminyl-*N*-acetylmuramic acid or the *C*-terminus with D-alanine was without significant effect. Amidation of the α-carboxyl group of glutamic acid had marked effects on both hydrolysis and transpeptidation and when amidated peptide was present in large excess it completely inhibited both carboxypeptidase and transpeptidase activities. Essentially, all the experiments described involved the use of the synthetic peptide N^{α}-N^{ϵ}-diacetyl-L-lysyl-D-alanyl-D-alanine as the donor peptide.

Table 8.7 The effect of peptide analogues as inhibitors of carboxypeptidase activity by the exocellular *Streptomyces* D,D-carboxypeptidase-transpeptidases (from [51]).

Peptide	*% Activity as substrate*	*Molecular ratio inhibitor/substrate*	*% Inhibition*	
			S. albus G	*Streptomyces R61*
N^{ϵ}-diacetyl → L-Lys → D-Ala → D-Ala	100	—	0	0
N^{ϵ}-disuccinyl → L-Lys → D-Ala → D-Ala	2–16	10–12	50	43
N^{ϵ}-disuccinyl → L-Lys → D-Ala → D-Glu	0	10	88–96	32–46
N^{ϵ}-diacetyl → L-Lys → D-Ala → D-Glu	100/10	—	0	—
N^{α}-acetyl → L-Lys → D-Ala → D-Glu	0	10	45	—
acetyl → D-Ala → D-Glu	0	10	88	25
acetyl → D-Ala → D-Asp	0	10–11	59	38
acetyl → D-Ala → D-Ala	0–1	10	10–20	9
N^{α}, N^{ϵ}-diacetyl → L-Lys → D-Glu → D-Ala	0	14	87	35
N^{α}-acetyl → L-Lys → D-Glu → D-Ala	0	11.6	67	34

Table 8.8 Transpeptidation by exocellular D,D-carboxypeptidases from *Streptomyces* R61, K11, R39, and *albus* G*. (For further details see [51, 153].)

Acceptor	*Acceptor:donor molar ratio*	*Transpeptidation (%)* *R61*	*K11*	*R39*	*S. albus* G
α-Amino acids					
^{14}C-Glycine	1:1	13.7			
	10:1	48	45	30	0
D-^{14}C-Alanine	1:1	7.5			
	10:1	50		35	0
L-^{14}C-Alanine	10:1	0		0	
meso-^{3}H-Diaminopimelic acid†	1:1	16.0	18.2	7.4	
	10:1		48	45	0
Mono-^{14}C-acetyl-L,L-diamino-pimelic acid	1:1	0			
Dipeptides					
1-^{14}C-Glycylglycine	1:1	18.4			0
	10:1			0	0
Glycyl-D-^{14}C-alanine	1:1	3.6	4.0	0	
Glycyl-L-^{14}C-alanine	1:1	25.5	24.6	0	
ε-Glycyl-α-acetyl-L-^{3}H-lysine	1:1	18.2		0	0
α-Glycyl-α′-^{14}C-acetyl-L,L-diaminopimelic acid	0.56:1	4.8		0	0
	4.1:1	2.7		0	0
D-Alanyl-^{14}C-glycine	1:1	3.6	3.4	0	
D-Alanyl-L-^{14}C-alanine	1:1	2.6	2.1	0	
D-Alanyl-D-^{14}C-alanine	1:1	0.4		0	
L-Alanyl-L-^{14}C-alanine	1:1	0.2	0.2	0	

*The donor was 1.67 mM diacetyl-L-lysyl-D-alanyl-D-alanine. Transpeptidation was determined as the percentage of the donor converted to the transpeptidation product diacetyl-L-lysyl-D-alanyl-acceptor after incubation for 1 h at 37 °C.
†Only the D-centre of *meso*-diaminopimelic acid acted as an acceptor in transpeptidation.

Whether this will influence the functioning of the various activities described is not clear and will require further investigation. At this time, transpeptidation involving naturally occurring peptides has not been described, although the formation of a dimer utilizing the pentapeptide L-alanyl-D-isoglutaminyl-*meso*-diaminopimelyl-D-alanyl-D-alanine as the donor and L-alanyl-D-isoglutaminyl-*meso*-diaminopimelyl-D-alanine as the acceptor was obtained with enzyme from R39 [51]. This dimer is identical to that found in the peptidoglycan of strain R39 with the exception of a single amide residue (Fig. 8.11).

A membrane-bound transpeptidase has been isolated from strain R61. Originally it was solubilized by treatment of membranes with 2M urea in the presence of 10mM EDTA [26, 51] although subsequently treatment of either membranes or intact organisms with the detergent cetyltrimethyl-ammonium bromide (Cetavlon)

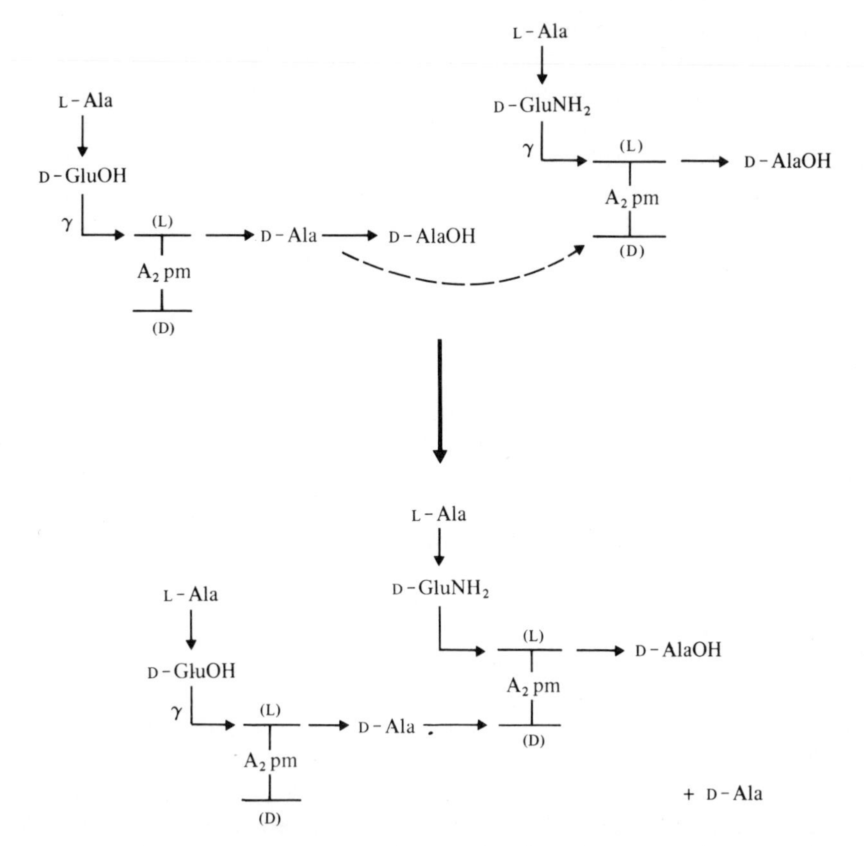

Figure 8.11 The transpeptidation reaction catalysed by the exocellular D,D-carboxypeptidase-transpeptidase from *Actinomadura* R39. The cross-linked product differs from the cross-linked peptides of the native wall by the amide residue present on the D-isoglutamyl residue of the acceptor peptide (for details see [51]).

was used [27, 101]. This procedure which has also been used with other strains, effectively solubilized the enzyme in higher yield and a more stable form. The membrane-bound enzyme functions solely as a transpeptidase, D-alanine being released from the donor N^{α}, N^{ϵ}-diacetyl-L-lysyl-D-alanyl-D-alanine only in the presence of the acceptor glycylglycine. Moreover, the transpeptidase remains active in the frozen state suggesting it exists in membranes in a lipid environment which remains fluid at very low temperatures; at least $-35°$ C is required to inhibit activity completely [27]. In contrast, the cetavlon-solubilized enzyme which requires detergent for activity is inactive at 0°C. It does however, retain the substrate specificities and β-lactam sensitivity of the membrane-bound enzyme.

More recently the relationship of soluble and membrane bound carboxypeptidases and transpeptidases has been investigated further by Ghuysen and his colleagues [101]. The formation of protoplasts from *Streptomyces* R61 and K15 by treatment of organisms with lysozyme in the presence of sucrose releases D,D-carboxypeptidase-transpeptidase into the medium. These findings suggest that the enzymes are associated either directly with the cell wall or alternatively loosely bound to the outer surface of the membrane. The lysozyme-released enzymes act mainly as D,D-carboxypeptidases possessing only weak transpeptidase activity. They are very sensitive to penicillin and have similar substrate specifities and molecular weights to their respective exocellular carboxypeptidases. The membranes of the two organisms also contain a carboxypeptidase distinct from the transpeptidase described above from strain R61. Thus *Streptomyces* R61 and K15 contain a family of carboxypeptidases which can be isolated as exocellular, lysozyme-releasable and membrane-bound activities although this latter fraction is particularly low in R61. Perhaps the membrane-bound carboxypeptidase leads via the lysozyme-released fraction to the exocellular enzyme. In contrast *Streptomyces rimosus* possesses little carboxypeptidase activity either soluble or membrane-bound and only the transpeptidase is present in any significant amount. Antiserum prepared to the exocellular carboxypeptidase of R61 inhibits all the carboxypeptidases of R61, K15 and *rimosus* [144]. However, this immunological relationship does not extend to the membrane-bound transpeptidases or the soluble carboxypeptidases of *S. albus* G and *Actinomadura* R39. Whether the antigenic cross-reactivity of the various carboxypeptidases is due to the similarity of their active sites or to other areas of the enzyme molecule is not yet known.

The extensive studies on the mechanism of inhibition by β-lactam antibiotics of both membrane-bound transpeptidases and the soluble D,D-carboxypeptidase-transpeptidases of *Streptomyces* and *Actinomadura* are discussed in Chapter 9.7.3.

The third group of organisms from which carboxypeptidases have been isolated and purified are the Bacilli. The initial studies were made on membrane-bound enzymes from *B. subtilis* [97, 98] and these have now been extended to include *B. coagulans* [123], *B. megaterium* [23, 113] and *B. stearothermophilus* [7]. Membranes from *B. subtilis* catalysed the release of both D-alanine residues from UDP-*N*-acetylmuramyl-pentapeptide and thus contained both D,D- and L,D carboxypeptidases. The D,D-carboxypeptidase which was subsequently shown to be PBP 5, the major penicillin-binding component of the membrane, was inhibited by penicillin but treatment with neutral hydroxylamine resulted in the release of penicilloyl-hydroxamate and reactivation of the enzyme [98]. Initially it was concluded that this reactivation was a chemical reaction involving a highly labile bond between the antibiotic and perhaps a thiol group in the enzyme [97]. This proposal was based on the finding that sulphydryl reagents inhibited both penicillin-binding and carboxypeptidase activity. However, it was later shown that reactivation did not occur when the enzyme was first inactivated by boiling or by treatment with SDS suggesting that the reaction leading to recovery of enzyme activity was being enzymically catalysed [92]. More recent results discussed below and in the following Chapter show that the penicilloyl moiety is released by a transpeptidase reaction in which hydroxylamine may be regarded as a more efficient nucleophilic acceptor than water. These findings provided a basis for the following method which has been used for the purification of D,D-carboxypeptidases from *B. subtilis* and *B. stearothermophilus*

[8, 241]. After treatment of membrane preparations with β-lactam antibiotics which do not bind to the carboxypeptidases (i.e. certain cephalosporins) the enzymes were solubilized by treatment with non-ionic detergents. They were purified from this extract by binding to penicillin-affinity columns from which the active enzymes were eluted with buffered hydroxylamine. D,D-Carboxypeptidase has also been purified by conventional techniques from detergent-extracts of *B. subtilis* membranes [213] and independently both D,D- and L,D-carboxypeptidases have been solubilized by *n*-butanol treatment of membranes from a second strain of *B. stearothermophilus* [7].

The enzymes purified by affinity chromatography from *B. subtilis* and *B. stearothermophilus* resemble each other in that they are the major penicillin-binding protein in the membranes, have similar molecular weights (50 000 and 46 000 respectively) and pH optima of 5 to 6. They differ in that the enzyme from *B. stearothermophilus* is thermophilic with a temperature optimum of 55° C and is not apparently activated by divalent cations. Originally both the membrane-bound and purified enzymes from *B. subtilis* were thought to have a requirement for divalent cations particularly Zn^{2+}. However the finding that Zn^{2+} was not required when the model peptide substrates, diacetyl-L-lysyl-D-alanyl-D-alanine and diacetyl-L-lysyl-D-alanyl-D-lactic acid were used shows that the cation requirement was related to the use of UDP-*N*-acetylmuramyl-pentapeptide rather than being a property of the enzyme [163].

A recent and exciting development has been the finding by Waxman and Strominger [223] that treatment of the detergent-soluble D,D-carboxypeptidase from *B. stearothermophilus* with either trypsin or chymotrypsin yields water-soluble and enzymically active fragments. This restricted proteolysis was also carried out on enzyme bound to a penicillin affinity column with apparently identical results. In each case the active fragments were derived from the native enzyme by cleavage of a hydrophobic-peptide from the COOH-terminus. The fragment had a molecular weight of 45 000 compared with 46 500 for the carboxypeptidase, increased thermostability and a loss of detergent-binding capacity. Moreover, it could not be reconstituted into bacterial lipid vesicles whereas reconstitution did occur with the native enzyme. On the other hand, the fragment did retain the full enzymic activity and penicillin-binding capacity originally present. Thus the D,D-carboxypeptidase of *B. stearothermophilus* appears to consist of a hydrophilic catalytic domain and a hydrophobic region at the COOH-terminus which is involved in anchoring the enzyme in the bacterial membrane. Similar results have been reported briefly for the *B. subtilis* carboxypeptidase [222]. Whether other carboxypeptidases and perhaps even other penicillin-binding proteins are also anchored in the membrane by hydrophobic regions at their COOH-termini remains unknown. However, it may be possible to answer this question by using the technique of limited proteolysis of covalently bound proteins as described above.

The D,D-carboxypeptidases of *B. stearothermophilus, B. subtilis* and *B. megaterium* also catalyse a model transpeptidation reaction [113, 167, 241]. This reaction is identical to the one described for *E. coli* preparations as unnatural model transpeptidation ([145] and Table 8.5) and involves the transfer of UDP-*N*-acetylmuramyltetrapeptide or diacetyl-L-lysyl-D-alanine from the respective donor peptides to either glycine, D-alanine or hydroxylamine as the acceptor with the concomitant release of D-alanine. The mechanism underlying both the carboxypeptidase and transpeptidase reactions catalysed by these enzymes has been studied in detail by Strominger and his colleagues [147, 163, 242]. Kinetic data suggested that the reactions occurred in two stages: the first was the formation of a common acyl-enzyme intermediate followed by a deacylation reaction in which the intermediate was partitioned between water (carboxypeptidase) and the nucleophile (transpeptidation). Thus the reaction could be described by the following equation:

$$E + S \underset{}{\overset{k}{\rightleftharpoons}} E\text{–}S \xrightarrow{k_2'} E\text{–}P \xrightarrow{k_3} E + P$$

where k_2 is the rate constant for the acylation reaction and k_3 for the deacylation. At this time the acyl–enzyme intermediate of *B. subtilis* carboxypeptidase was not isolated presumably due to k_2 for the substrate diacetyl-L-lysyl-D-alanyl-D-alanine being equal to or smaller than k_3. Subsequently the depsipeptide diacetyl-L-lysyl-D-alanyl-D-lactic acid was found to be a more effective substrate with enhanced acylation of the enzyme (i.e. $k_2 > k_3$) and an acyl–enzyme intermediate has now been isolated containing 0.43 mole of substrate per mole of carboxypeptidase [163]. Recently penicilloyl- and acyl-peptides have been isolated by proteolytic digestion from *B. subtilis* and *B. stearothermophilus* carboxypeptidases and both the antibiotic and substrate shown to covalently bind via an ester linkage to a serine residue in the active site of the enzyme [242]. Penicillin is also covalently bound to a serine residue in the soluble D,D-carboxypeptidase-transpeptidase purified from culture filtrates of *Streptomyces* R61 [34]. These findings are discussed in terms of the mechanism of penicillin action in Chapter 9.7.5. Finally it is worth noting that Strominger and his colleagues report considerable homology between the amino acid sequence of the two carboxypeptidases and four β-lactamases suggesting that the two groups of enzymes may be related and derived from a common ancestor.

References

1. Allen, C. M., Keenan, M. V. and Sack, J. (1976) *Archs. Biochem. Biophys.* **175**, 236–48.
2. Allen, C. M. and Muth J. D. (1977) *Biochemistry* **16**, 2908–15.
3. Anderson, J. S., Meadow, P. M., Haskin, M. A. and Strominger, J. L. (1966) *Archs. Biochem. Biophys.* **116**, 487–515.
4. Anwar, R. A. and Vlaovic, M. (1979) *Can. J. Biochem.* **57**, 188–96.
5. Araki, Y., Shimada, A. and Ito, E. (1966) *Biochem. biophys. Res. Commun.* **23**, 518–25.
6. Araki, Y., Shirai, R., Shimada, A., Ishimoto, N. and Ito, E. (1966) *Biochem. biophys. Res. Commun.* **23**, 466–72.
7. Barnett, H. J. (1973) *Biochim. biophys. Acta* **304**, 332–52.

8. Blumberg, P. M. and Strominger, J. L. (1972) *Proc. Natn. Acad. Sci. U.S.A.* **69**, 3751–5.
9. Bogdanovsky, D., Bricas, E. and Dezélée, P. (1969) *C. R. Acad. Sci. (Paris) Ser. D.* **269**, 390–93.
10. Bricas, E., Ghuysen, J-M. and Dezélée, P. (1967) *Biochemistry* **6**, 2598–607.
11. Brown, C. A. and Perkins, H. R. (1979) *Antimicrob. Agents Chemotherapy* **16**, 28–36.
12. Bumsted, R. M., Dahl, J. L., Söll, D. and Strominger, J. L. (1968) *J. biol. Chem.* **243**, 779–82.
13. Chatterjee, A. N., Doyle, R. J. and Streips, U. N. (1977) *J. theor. Biol.* **68**, 385–90.
14. Chatterjee, A. N. and Park, J. T., (1964) *Proc. Natn. Acad. Sci. U.S.A.* **51**, 9–16.
15. Chatterjee, A. N., Ward, J. B. and Perkins, H. R. (1967) *Nature, Lond.* **214**, 1311–14.
16. Chatterjee, A. N. and Young, F. E. (1972) *J. Bact.* **111**, 220–30.
17. Christenson, J. G., Gross, S. K. and Robbins, P. W. (1969) *J. biol. Chem.* **244**, 5436–9.
18. Comb, D. G. (1962) *J. biol. Chem.* **237**, 1601–608.
19. Coyette, J., Guysen, J-M. and Fontana, R. (1978) *Eur. J. Biochem.* **88**, 297–305.
20. Curtis, S. J. and Strominger, J. L. (1978) *J. biol. Chem.* **253**, 2584–8.
20a. Davis, R. H., Linder, R. and Salton, M. R. J. (1978) *Microbios.* **21**, 69–80.
21. Davis, R. H. and Salton, M. R. J. (1975) *Infect. Immun.* **12**, 1065–9.
22. Day, A. and White, P. J. (1977) *Biochem. J.* **161**, 677–85.
23. Diaz-Maurino, T., Nieto, M. and Perkins, H. R. (1974) *Biochem. J.* **143**, 391–402.
24. Diven, W. D. (1969) *Biochim. biophys. Acta* **191**, 702–706.
25. Duez, C., Frère, J-M., Geurts, F., Ghuysen, J-M., Dierickx, L. and Delcambe, L., (1978) *Biochem. J.* **175**, 793–800.
26. Dusart, J., Marquet, A., Ghuysen, J-M., Frère, J-M., Moreno, R., Leyh-Bouille, M., Johnson, K., Lucchi, Ch., Perkins, H. R. and Nieto, M. (1973) *Antimicrob. Agents Chemotherapy* **3**, 181–7.
27. Dusart, J., Marquet, A., Ghuysen, J-M. and Perkins, H. R. (1975) *Eur. J. Biochem.* **56**, 57–65.
28. Edwards, J. and Panos, C. (1962) *J. Bact.* **84**, 1202–208.
29. Egan, E., Lawrence, P. and Strominger, J. L. (1973) *J. biol. Chem.* **248**, 3122–30.
30. Elliot, T. S. J., Ward, J. B. and Rogers, H. J. (1975) *J. Bact.* **124**, 623–32.
31. Eun, H-M., Yapo, A. and Petit, J-F. (1978) *Eur. J. Biochem.* **86**, 97–103.
32. Fodor, M. and Tóth, B. (1965) *Acta microbiol. Acad. Sci. hung.* **12**, 173–9.
33. Fordham, W. D. and Gilvarg, C. (1974) *J. biol. Chem.* **249**, 2478–82.
34. Frère, J-M., Duez, C., Ghuysen, J-M. and Vanderkloee, J. (1976) *FEBS letts.* **70**, 257–60.
35. Frère, J-M., Ghuysen, J-M., Perkins, H. R. and Nieto, M. (1973) *Biochem. J.* **135**, 463–8.
36. Frère, J-M., Ghuysen, J-M., Zeiger, A. R. and Perkins, H. R. (1976) *FEBS Letts.* **63**, 112–16.
37. Frère, J-M., Guerts, F., and Ghuysen, J-M. (1978) *Biochem. J.* **175**, 801–805.
38. Frère, J-M., Moreno, R., Ghuysen, J-M., Perkins, H. R., Dierickx, L. and Delcambe, L. (1974) *Biochem. J.* **143**, 233–40.
39. Fuchs-Cleveland, E. and Gilvarg, C. (1976) *Proc. Natn. Acad. Sci. U.S.A.* **73**, 4200–204.

40. Gateau, O., Bordet, C. and Michel, G. (1976) *Biochim. biophys. Acta.* **421**, 395–405.
41. Garrett, A. J. (1969) *Biochem. J.* **115**, 419–30.
42. Geis, A. and Plapp, R. (1978) *Biochim. biophys. Acta* **527**, 414–24.
43. Gennis, R. B., Sinensky, M. and Strominger, J. L. (1976) *J. biol. Chem.* **251**, 1270–76.
44. Gennis, R. B. and Strominger, J. L. (1976) *J. biol. Chem.* **251**, 1264–9.
45. Gennis, R. B. and Strominger, J. L. (1976) *J. biol. Chem.* **251**, 1277–82.
46. Ghuysen, J-M. (1977) *J. gen. Microbiol.* **101**, 13–33.
47. Ghuysen, J-M. (1977) In *The Bacterial D,D-Carboxypeptidase-Transpeptidase Enzyme System: A New Insight into the Mode of Action of Penicillin (E. R. Squibb Lectures on Chemistry of Microbiol. Products)* ed. Brown, W. E., University of Tokyo Press.
48. Ghuysen, J-M., Bricas, E., Lache, M. and Leyh-Bouille, M. (1968) *Biochemistry* **7**, 1450–60.
49. Ghuysen, J-M., Leyh-Bouille, M., Bonaly, R., Nieto, M., Perkins, H. R., Schleifer, K. H. and Kandler, O. (1970) *Biochemistry* **9**, 2955–61.
50. Ghuysen, J-M., Leyh-Bouille, M., Campbell, J. N., Moreno, R., Frère, J-M., Duez, C., Nieto, M. and Perkins, H. R. (1973) *Biochemistry* **12**, 1243–51.
51. Ghuysen, J-M., Leyh-Bouille, M., Frère, J-M., Dusart, J. and Marquet, A. (1974) *Ann. N. Y. Acad. Sci.* **235**, 236–66.
52. Ghuysen, J-M. and Strominger, J. L. (1963) *Biochemistry* **2**, 1119–25.
53. Giles, A. F. and Reynolds, P. E. (1979) *FEBS Letts.* **101**, 244–8.
54. Goldman, R. and Strominger, J. L. (1972) *J. biol. Chem.* **247**, 5116–22.
55. Good, C. M. and Tipper, D. J. (1972) *J. Bact.* **111**, 231–41.
56. Gregory, W. W. and Gooder, H. (1978) *J. Bact.* **135**, 900–910.
57. Gunetileke, K. C. and Anwar, R. A. (1968) *J. biol. Chem.* **243**, 5570–78.
58. Hammes, W. P. (1976) *Eur. J. Biochem.* **70**, 107–113.
59. Hammes, W. P. (1978) *Eur. J. Biochem.* **91**, 501–507.
60. Hammes, W. P. and Kandler, O. (1976) *Eur. J. Biochem.* **70**, 97–106.
61. Hammes, W. P. and Neuhaus, F. C. (1974) *J. biol. Chem.* **249**, 3140–50.
62. Hammes, W. P. and Neuhaus, F. C. (1974) *Antimicrob. Agents. Chemotherapy* **6**, 722–8.
63. Hammes, W. P. and Neuhaus, F. C. (1974) *J. Bact.* **120**, 210–18.
64. Hammes, W. P., Neukam, R. and Kandler, O. (1977) *Arch. Mikrobiol.* **115**, 95–102.
65. Hammes, W. P., Schleifer, K. H. and Kandler, O. (1973) *J. Bact.* **116**, 1029–53.
66. Hammes, W. P. and Seidel, H. (1978) *Eur. J. Biochem.* **84**, 141–7.
67. Hammes, W. P. and Seidel, H. (1978) *Eur. J. Biochem.* **91**, 509–515.
68. Higashi, Y., Siewert, G. and Strominger, J. L. (1970) *J. biol. Chem.* **245**, 3683–90.
69. Higashi, Y. and Strominger, J. L. (1970) *J. biol. Chem.* **245**, 3691–6.
70. Higashi, Y., Strominger, J. L. and Sweeley, C. C. (1967) *Proc. Natn. Acad. Sci. U.S.A.* **57**, 1878–84.
71. Higashi, Y., Strominger, J. L. and Sweeley, C. C. (1970) *J. biol. Chem.* **245**, 3697–702.
72. Ito, E., Nathenson, S. G., Dietzler, D. N., Anderson, J. S. and Strominger, J. L. (1966) *Meth. Enzymol.* **8**, 324–37.
73. Ito, E. and Strominger, J. L. (1962) *J. biol. Chem.* **237**, 2696–703.
74. Ito, E. and Strominger, J. L. (1966) *J. biol. Chem.* **239**, 210–14.
75. Ito, E. and Strominger, J. L. (1973) *J. biol. Chem.* **248**, 3131–6.
76. Iwaya, M. and Strominger, J. L. (1977) *Proc. Natn. Acad. Sci. U.S.A.* **74**, 2980–84.

77. Izaki, K., Matsuhashi, M. and Strominger, J. L. (1966) *Proc. Natn. Acad. Sci. U.S.A.* **55**, 656–63.
78. Izaki, K., Matsuhashi, M. and Strominger, J. L. (1968) *J. biol. Chem.* **243**, 3180–92.
79. Izaki, K. and Strominger, J. L. (1968) *J. biol. Chem.* **243**, 3193–201.
80. Jensen, S. E. and Campbell, J. N. (1976) *J. Bact.* **127**, 319–6.
81. Johnston, M. M. and Diven, W. F. (1969) *J. biol. Chem.* **244**, 5414–20.
82. Johnston, L. S. and Neuhaus, F. C. (1975) *Biochemistry* **14**, 2754–60.
83. Johnston, R. B., Scholz, J. J., Diven, W. F. and Shepherd, S. (1968) In *Pyridoxal Catalysis: Enzymes and Model Systems,* eds. Snell, E. E., Braunstein, A. E., Severin, E. S. and Turchinsky, Y. M., pp. 537–47, New York: John Wiley & Sons Inc.
84. Kalin, J. R. and Allen, C. M. (1979) *Biochim. biophys. Acta* **574**, 112–22.
85. Kamiryo, T. and Matsuhashi, M. (1972) *J. biol. Chem.* **247**, 6306–311.
86. Kamiryo, T. and Strominger, J. L. (1974) *J. Bact.* **117**, 568–77.
87. Katz, W., Matsuhashi, M., Dietrich, C. P. and Strominger, J. L. (1967) *J. biol. Chem.* **242**, 3207–217.
88. Keenan, M. V. and Allen, C. M. (1974) *Archs. Biochem. Biophys.* **161**, 375–83.
89. Keglević, D., Ladesić, B., Hadzija, O., Tomasić, J., Valinger, Z., Pokotny, M. and Naumski, R. (1974) *Eur. J. Biochem.* **42**, 389–400.
90. Kennedy, L. D. (1974) *Biochem. J.* **138**, 525–35.
91. Kennedy, L. D. and Shaw, D. R. D. (1968) *Biochem. biophys. Res. Commun.* **32**, 861–5.
92. Kozarich, J. W., Nishimo, T., Willoughby, E. and Strominger, J. L. (1977) *J. biol. Chem.* **252**, 7525–9.
93. Kozarich, J. W. and Strominger, J. L. (1978) *J. biol. Chem.* **253**, 1272–8.
94. Kurokawa, T., Ogura, K. and Seto, S. (1971) *Biochem. biophys. Res. Commun.* **45**, 2551–6.
95. Lambert, M. P. and Neuhaus, F. C. (1972) *J. Bact.* **109**, 1156–61.
96. Lambert, M. P. and Neuhaus, F. C. (1972) *J. Bact.* **110**, 978–87.
97. Lawrence, P. J. and Strominger, J. L. (1970) *J. biol. Chem.* **245**, 3653–9.
98. Lawrence, P. J. and Strominger, J. L. (1970) *J. biol. Chem.* **245**, 3660–66.
99. Lee, P. P., Weppner, W. A. and Neuhaus, F. C. (1980) *Biochim. Biophys. Acta* **597**, 603–13.
100. Leyh-Bouille, M., Coyette, J., Ghuysen, J-M., Idczak, I., Perkins, H. R. and Nieto, M. (1971) *Biochemistry* **10**, 2163–70.
101. Leyh-Bouille, M., Dusart, J., Nguyen-Distèche, M., Ghuysen, J-M., Reynolds, P. E. and Perkins, H. R. (1977) *Eur. J. Biochem.* **81**, 19–28.
102. Leyh-Bouille, M., Ghuysen, J-M., Bonaly, R., Nieto, M., Perkins, H. R., Schleifer, K. H. and Kandler, O., (1970) *Biochemistry* **9**, 2961–70.
103. Leyh-Bouille, M., Ghuysen, J-M., Nieto, M., Perkins, H. R., Schleifer, K. H. and Kandler, O. (1970) *Biochemistry* **9**, 2971–5.
104. Leyh-Bouille, M., Nakel, M., Frère, J-M., Johnson, K., Ghuysen, J-M., Nieto, M. and Perkins, H. R. (1972) *Biochemistry* **11**, 1290–8.
105. Linnett, P. E., Roberts, R. J. and Strominger, J. L. (1974) *J. biol. Chem.* **249**, 2497–506.
106. Linnett, P. E. and Strominger, J. L. (1974) *J. biol. Chem.* **249**, 2489–96.
107. Lugtenberg, E. J. J. and de Haan, P. G. (1971) *Antoine van Leeuwenhock J. Microbiol. Serol.* **37**, 537–52.
108. Lugtenberg, E. J. J., de Haas-Menger, L. and Ruyters, W. H. M. (1972), *J. Bact.* **109**, 326–35.
109. Lugtenberg, E. J. J. and van Schijndel van Dam, A. (1972) *J. Bact.* **110**, 35–40.

110. Lugtenberg, E. J. J. and van Schijndel van Dam, A. (1972) *J. Bact.* **110**, 41–6.
111. Lugtenberg, E. J. J. and van Schijndel van Dam, A. (1973) *J. Bact.* **113**, 96–104.
112. Lynch, J. and Neuhaus, F. C. (1966) *J. Bact.* **91**, 449–60.
113. Marquet, A., Nieto, M. and Diaz-Maurino, T. (1976) *Eur. J. Biochem.* **68**, 581–9.
114. Martin, H. H. (1964) *Abst. VI Internl. Cong. Biochem. New York*, VI:70.
115. Martin, H. H. (1964) *J. gen. microbiol.* **36**, 441–50.
116. Martin, H. H. (1966) *A. Rev. Biochem.* **35**, 457–84.
117. Martin, H. H., Schilf, W. and Maskos, C. (1976) *Eur. J. Biochem.* **71**, 585–93.
118. Martinez-Carrion, M. and Jenkins, W. T. (1965) *J. biol. Chem.* **240**, 3538–46.
119. Matsuhashi, M., Dietrich, C. P. and Strominger, J. L. (1967) *J. biol. Chem.* **242**, 3191–206.
120. Matsuhashi, M., Maruyama, I. N., Takagaki, Y., Tamaki, S., Nishimura, Y. and Hirota, Y. (1977) *Proc. Natn. Acad. Sci. U.S.A.* **75**, 2631–5.
121. Matsuhashi, M., Takagaki, Y., Maruyama, I. N., Tamaki, S., Nishimura, Y., Suzuki, H., Ogino, U. and Hirota, Y. (1977) *Proc. Natn. Acad. Sci. U.S.A.* **74**, 2976-9.
122. Matsuzawa, H., Matsuhashi, M., Oka, A. and Sugino, Y. (1969) *Biochem. biophys. Res. Commun.* **36**, 682–9.
123. McArthur, H. A. I. and Reynolds, P. E. (1979) *J. gen. Microbiol.* **111**, 327–35.
124. Mirelman, D. and Bracha, R. (1974) *Antimicrob. Agents Chemotheraphy* **5**, 663–6.
125. Mirelman, D., Bracha, R. and Sharon, N. (1972) *Proc. Natn. Acad. Sci. U.S.A.* **69**, 3355–9.
126. Mirelman, D., Bracha, R. and Sharon, N. (1974) *Biochemistry* **13**, 5045–53.
127. Mirelman, D., Kleppe, G. and Jensen, H. B. (1976) *Eur. J. Biochem.* **55**, 369–73.
128. Mirelman, D. and Nuchamowitz, Y. (1979) *Eur. J. Biochem.* **94**, 541–8.
129. Mirelman, D. and Nuchamowitz, Y. (1979) *Eur. J. Biochem.* **94**, 549–56.
130. Mirelman, D. and Sharon, N. (1972) *Biochem. biophys. Res. Commun.* **46**, 1909–17.
131. Mirelman, D., Shaw, D. R. D. and Park, J. T. (1970) *J. Bact.* **107**, 239–44.
132. Mirelman, D., Yashouv-Gan, Y. and Schwarz, U. (1976) *Biochemistry* **15**, 1781–90.
133. Mirelman, D., Yashouv-Gan, Y. and Schwarz, U. (1977) *J. Bact.* **129**, 1593–1600.
134. Miyakawa, T., Matsuzawa, H., Matsuhashi, M. and Sugino, Y. (1972) *J. Bact.* **112**, 950–58.
135. Mizuno, Y. and Ito, E. (1968) *J. biol. Chem.* **243**, 2665–72.
136. Moore, B. A., Jevons, S. and Brammer, K. W. (1979) *Antimicrob. Agents Chemotheraphy* **15**, 513–17.
137. Neuhaus, F. C. (1962) *J. biol. Chem.* **237**, 778–86.
138. Neuhaus, F. C. (1962) *J. biol. Chem.* **237**, 3128–35.
139. Neuhaus, F. C. (1968) *Antimicrob. Agents Chemotheraphy* **1**, 304–13.
140. Neuhaus, F. C. (1972) *Acc. chem. Res.* **4**, 297–303.
141. Neuhaus, F. C., Carpenter, C. V., Lambert, M. P. and Wargel, R. J. (1972) In *Proc. Symp. Mol. Mechanisms of Antibiotic Action on Protein Biosynthesis and Membranes*, eds. Munoz, E., Ferrandiz, F. and Vazquez, D., pp. 339–62. Amsterdam: Elsevier Co.
142. Neuhaus, F. C. and Lynch, J. L. (1964) *Biochemistry* **3**, 471–80.
143. Neuhaus, F. C. and Struve, W. G. (1965) *Biochemistry* **4**, 120–31.

144. Nguyen-Distèche, M., Frère, J-M., Dusart, J., Leyh-Bouille, M., Ghuysen, J-M., Pollock, J. J. and Iacono, V. J. (1977) *Eur. J. Biochem.* **81**, 29–32.
145. Nguyen-Distèche, M., Ghuysen, J-M., Pollock, J. J., Reynolds, P. E., Perkins, H. R., Coyette, J. and Salton, M. R. J. (1974) *Eur. J. Biochem.* **41**, 447–55.
146. Nguyen-Distèche, M., Pollock, J. J., Ghuysen, J. M., Puig, J., Reynolds, P. E., Perkins, H. R., Coyette, J. and Salton, M. R. J. (1974) *Eur. J. Biochem.* **41**, 457–63.
147. Nishino, T., Kozarich, J. W. and Strominger, J. L. (1977) *J. biol. Chem.* **252**, 2934–9.
148. Niyomporu, B., Dahl, J. L. and Strominger, J. L. (1968) *J. biol. Chem.* **243**, 773–8.
149. Oldmixon, E. H., Dezéleé, P., Ziskin, M. C. and Shockman, G. D. (1976) *Eur. J. Biochem.* **68**, 271–80.
150. Oppenheim, B. and Patchornik, A. (1974) *FEBS Lett.* **48**, 172–5.
151. Pellon, G., Bordet, C. and Michel, G. (1976) *J. Bact.* **125**, 509–517.
153. Perkins, H. R., Nieto, M., Frère, J-M., Leyh-Bouille, M. and Ghuysen, J-M. (1973) *Biochem. J.* **131**, 707–718.
154. Petit, J. F., Adam, A. and Wietzerbin-Falzpan, J. (1969) *FEBS Letts.* **6**, 55–7.
155. Petit, J. F., Strominger, J. L. and Söll, D. (1968) *J. Biol. Chem.* **243**, 757–67.
156. Plapp, R. and Strominger, J. L. (1970) *J. biol. Chem.* **245**, 3667–74.
157. Plapp, R. and Strominger, J. L. (1970) *J. biol. Chem.* **245**, 3675–82.
158. Pless, D. D. and Neuhaus, F. C. (1973) *J. biol. Chem.* **248**, 1568–76.
159. Pollock, J. J., Ghuysen, J-M., Linder, R., Salton, M. R. J., Perkins, H. R., Nieto, M., Leyh-Bouille, M., Frère, J-M. and Johnson, K. (1972) *Proc. Natn. Acad. Sci. U.S.A.* **69**, 662–6.
160. Pollock, J. J., Nguyen-Distèche, M., Ghuysen, J-M., Coyette, J., Linder, R., Salton, M. R. J., Kim, K. S., Perkins, H. R. and Reynolds, P. E. (1974) *Eur. J. Biochem.* **41**, 439–46.
161. Poxton, I. R., Lomax, J. A. and Sutherland, I. W. (1974) *J. gen. Microbiol.* **84**, 231–3.
162. Poxton, I. R and Sutherland, I. W. (1976) *Microbios.* **15**, 93–103.
163. Rasmussen, J. R. and Strominger, J. L. (1978) *Proc. Natn. Acad. Sci. U.S.A.* **75**, 84–8.
164. Reusch, V. M. and Panos, C. (1976) *J. Bact.* **126**, 300–311.
165. Reynolds, P. E. (1971) *Biochim. biophys. Acta* **237**, 239–54.
166. Reynolds, P. E. (1971) *Biochim. biophys. Acta* **237**, 255–72.
167. Reynolds, P. E. and Barnett, H. J. (1974) *Ann. N.Y. Acad. Sci.* **235**, 269–82.
168. Robbins, P. W., Bray, D., Dankert, M. and Wright, A. (1967) *Science* **158**, 1536–42.
169. Roberts, R. J. (1972) *Nature, New Biol.* **237**, 44–5.
170. Roberts, W. S. L., Petit, J. F. and Strominger, J. L. (1968) *J. biol. Chem.* **243**, 768–72.
171. Roberts, W. S. L., Strominger, J. L. and Söll, D. (1968) *J. biol. Chem.* **243**, 749–56.
172. Rosenthal, R. S. and Shockman, G. D. (1975) *J. Bact.* **124**, 410–18.
173. Rosenthal, R. S. and Shockman, G. D. (1975) *J. Bact.* **124**, 419–23.
174. Rosso, G., Takashima, K. and Adams, E. (1969) *Biochem. biophys. Res. Commun.* **34**, 134–40.
175. Roze, U. and Strominger, J. L. (1966) *Molec. Pharmacol.* **2**, 92–4.
176. Sandermann, H. (1976) *Eur. J. Biochem.* **62**, 479–84.
177. Sandermann, H. (1976) *FEBS Letts.* **63**, 59–61.
178. Sandermann, H. and Strominger, J. L. (1972) *J. biol. Chem.* **247**, 5123–31.

179. Sasak, W. and Chojnacki, T. (1977) *Archs. Biochem. Biophys.* **181**, 402–410.
180. Scher, M., Lennarz, W. J. and Sweeley, C. C. (1968) *Proc. Natn. Acad. Sci. U.S.A.* **59**, 1313–20.
181. Schleifer, K. H. and Kandler, O. (1972) *Bact. Rev.* **36**, 407–477.
182. Schrader, W. P. and Fan, D. P. (1974) *J. biol. Chem.* **249**, 4815–18.
183. Siewert, G. and Strominger, J. L. (1967) *Proc. Natn. Acad. Sci. U.S.A.* **57**, 767–73.
184. Siewert, G. and Strominger, J. L. (1968) *J. biol. Chem.* **243**, 783–90.
185. Spratt, B. G. (1975) *Proc. Natn. Acad. Sci. U.S.A.* **72**, 2999–3003.
186. Spratt, B. G. and Strominger, J. L. (1976) *J. Bact.* **127**, 660–63.
187. Staadenbauer, W. and Strominger, J. L. (1972) *J. biol. Chem.* **247**, 5095–5102.
188. Stickgold, R. A. and Neuhaus, F. C. (1967) *J. biol. Chem.* **242**, 1331–7.
189. Strominger, J. L. (1970) *The Harvey Lectures (69) Ser.* **64**, 179–213.
190. Strominger, J. L., Willoughby, E., Kamiryo, T., Blumberg, P. M. and Yocum, R. R. (1974) *Ann. N.Y. Acad. Sci.* **235**, 210–24.
191. Suginaka, H., Ichikawa, A. and Kotani, S. (1974) *Antimicrob. Agents Chemotherapy* **6**, 672–5.
192. Sutherland, I. W. (1977) In *Surface Carbohydrates of the Prokaryotic Cell*, ed. Sutherland, I. W., pp. 27–96, London and New York: Academic Press.
193. Taku, A. and Fan, D. P. (1976) *J. biol. Chem.* **251**, 1889–95.
194. Taku, A. and Fan, D. P. (1976) *J. biol. Chem.* **251**, 6154–6.
195. Taku, A. and Fan, D. P. (1979) *J. biol. Chem.* **254**, 3991–9.
196. Taku, A., Gardner, H. L. and Fan, D. P. (1975) *J. biol. Chem.* **250**, 3375–80.
197. Taku, A., Gunetilike, K. G. and Anwar, R. A. (1970) *J. biol. Chem.* **245**, 5012–16.
198. Tamura, T., Imae, Y. and Strominger, J. L. (1976) *J. biol. Chem.* **251**, 414–23.
199. Thorndike, J. and Park, J. T. (1969) *Biochem. biophys. Res. Commun.* **35**, 642–7.
200. Thorne, K. J. I. (1973) *J. Bact.* **116**, 235–44.
201. Thorne, K. J. I. and Kodicek, E. (1966) *Biochem. J.* **99**, 123–7.
202. Thorpe, S. J. and Perkins, H. R. (1979) *FEBS Letts.* **105**, 151–4.
203. Tipper, D. J. (1969) *Biochemistry* **8**, 2192–202.
204. Tipper, D. J. and Pratt, I. (1970) *J. Bact.* **103**, 305–317.
205. Tipper, D. J. and Strominger, J. L. (1965) *Proc. Natn. Acad. Sci. U.S.A.* **54**, 1133–41.
206. Tipper, D. J. and Strominger, J. L. (1968) *J. biol. Chem.* **243**, 3169–79.
207. Troy, F. A., Frerman, F. E. and Heath, E. C. (1971) *J. biol. Chem.* **246**, 118–33.
208. Troy, F. A., Vijay, I. K. and Tesche, N. (1975) *J. biol. Chem.* **250**, 156–63.
209. Tynecka, Z. and Ward, J. B. (1975) *Biochem. J.* **146**, 253–67.
210. Umbreit, J. N., Stone, K. J. and Strominger, J. L. (1972) *J. Bact.* **112**, 1302–305.
211. Umbreit, J. N. and Strominger, J. L. (1972) *Proc. Natn. Acad. Sci. U.S.A.* **69**, 1972–4.
212. Umbreit, J. N. and Strominger, J. L. (1972) *J. Bact.* **112**, 1306–309.
213. Umbreit, J. N. and Strominger, J. L. (1973) *J. biol. Chem.* **248**, 6759–66.
214. Venkateswaran, P. S., Lugtenberg, E. J. and Wu, H. C. (1973) *Biochim. biophys. Acta* **293**, 570–74.
215. Vosberg, H. P. and Hoffmann-Berling, H. (1971) *J. molec. Biol.* **58**, 739–53.
216. Wang, E. and Wash, C. (1978) *Biochemistry* **17**, 1313–21.

217. Ward, J. B. (1973) *Biochem. J.* **133**, 395–8.
218. Ward, J. B. (1974) *Biochem. J.* **141**, 227–41.
219. Ward, J. B. (1975) *J. Bacteriol.* **124**, 669–78.
220. Ward, J. B. and Perkins, H. R. (1973) *Biochem. J.* **135**, 721–8.
221. Ward, J. B. and Perkins, H. R. (1974) *Biochem. J.* **139**, 781–4.
222. Waxman, D. J. and Strominger, J. L. (1978) *Fed. Proc.* **37**, 1393.
223. Waxman, D. J. and Strominger, J. L. (1979) *J. biol. Chem.* **254**, 4863–75.
224. Weidel, W. and Pelzer, H. (1964) *Adv. Enzymol.* **26**, 193–232.
225. Weppner, W. A. and Neuhaus, F. C. (1978) *J. biol. Chem.* **253**, 472–8.
226. Weppner, W. A. and Neuhaus, F. C. (1979) *Biochem. biophys. Acta* **552**, 418–27.
227. Weston, A. and Perkins, H. R. (1978) *FEBS Letts.* **76**, 195–8.
228. Weston, A., Ward, J. B. and Perkins, H. R. (1977) *J. gen. Microbiol.* **99**, 171–81.
229. Wickus, G. G. and Strominger, J. L. (1974) *J. biol. Chem.* **247**, 5297–306.
230. Wickus, G. G. and Strominger, J. L. (1973) *J. Bact.* **113**, 287–90.
231. Wickus, G. G. and Strominger, J. L. (1974) *J. biol. Chem.* **247**, 5307–311.
232. Wickus, G. G., Rubenstein, P. A., Warth, A. D. and Strominger, J. L. (1973) *J. Bact.* **113**, 291–5.
233. Wijsman, H. J. W. (1972) *Genet. Res.* **20**, 65–74.
234. Wijsman, H. J. W. (1972) *Genet. Res.* **20**, 269–77.
235. Willoughby, E., Higashi, Y. and Strominger, J. L. (1972) *J. biol. Chem.* **247**, 5113–5.
236. Wise, E. M. and Park, J. T. (1965) *Proc. Natn. Acad. Sci. U.S.A.* **54**, 75–81.
237. Wright, A., Dankert, M., Fennessey, P. and Robbins, P. W. (1967) *Proc. Natn. Acad. Sci. U.S.A.* **57**, 1798–803.
238. Wu, H. C. and Venkateswaran, P. S. (1974) *Ann. N.Y. Acad. Sci.* **235**, 587–92.
239. Wyke, A. W. and Perkins, H. R. (1975) *J. gen. Microbiol.* **88**, 159–68.
240. Wyke, A. W. and Ward, J. B. (1977) *J. Bact.* **130**, 1055–63.
240a. Wyrick, P. B., McConnell, M. M. and Rogers, H. J. (1973) *Nature, Lond.* **244**, 505–7.
241. Yocum, R. R., Blumberg, P. M. and Strominger, J. L. (1974) *J. biol. Chem.* **249**, 4863–71.
242. Yocum, R. R., Waxman, D. J., Rasmussen, J. R. and Strominger, J. L. (1979) *Proc. Natn. Acad. Sci. U.S.A.* **76**, 2730–34.
243. Zeiger, A. R., Eaton, S. M. and Mirelman, D. (1978) *Eur. J. Biochem.* **86**, 235–40.
244. Zemell, R. I. and Anwar, R. A. (1975) *J. biol. Chem.* **250**, 3185–92.

9
Antibiotics affecting bacterial wall synthesis

9.1 Introduction

The initial observations on the effect of benzylpenicillin on the bacterial wall came in the early 1940s shortly after its introduction as a therapeutic agent. Gardner [74] found that concentrations of penicillin, much lower than those required to kill a range of different bacteria, caused considerable morphological changes to the organisms with apparent damage to the cell envelope. In certain organisms, cell division also appeared inhibited. These observations were extended by Duguid [58] who showed that growth was a necessary requirement for penicillin action, and that killing of the bacteria was apparently associated with cell lysis. He suggested that 'penicillin at these concentrations interferes specifically with the formation of the outer cell wall, while otherwise allowing growth to proceed until the organism finally bursts its defective envelope and undergoes cell lysis'. Many Gram-positive bacteria, particularly staphylococci, are extremely sensitive to the inhibitory action of penicillin, suggesting that its action is highly specific. This observation prompted the first biochemical studies of Park and Johnson [208] who described the accumulation of a labile phosphate compound by *Staphylococcus aureus* when treated with penicillin. The accumulated material was subsequently shown to be composed of several nucleotide compounds, all of which contained an unknown amino sugar, and some, the uncommon D-isomers of glutamic acid and alanine too [201-203]; the function of these compounds was not known. However, when it was discovered the wall of *S. aureus* contained a structure analogous to part of the accumulated nucleotide-peptide, it became clear that the compounds were in fact the biosynthetic precursors of the wall [209]. At this time, considerable advances were being made in the determination of bacterial wall structure, many of which pointed to its being unique. These findings, taken in conjunction with the other studies on the mode of action of penicillin, pointed to the wall as a structure necessary for the survival of the organism and in consequence a site for the selective action of antibiotics. The biosynthesis of peptidoglycan has been discussed in detail in the previous chapter and a number of antibiotics have been mentioned where information relating to biosynthesis has resulted from investigation of their mode of action. The mode of action of these and other antibiotics will now be described in greater detail at a molecular level. This consideration will be related to their site of action in the synthesis of peptidoglycan, which may be at the level of synthesis of the nucleotide precursors, the membrane-bound stages involving the lipid carrier or the terminal stages involving transfer of the newly-synthesized unit to the growing peptidoglycan.

9.2 Phosphonomycin (fosfomycin)

This antibiotic, the isolation and characterization of which were first described in 1969 [41, 99], inhibits what might be regarded as the earliest step in peptidoglycan synthesis, that is, it inhibits the transfer of the enolpyruvate residue from phosphoenolpyruvate to UDP-*N*-acetylglucosamine.

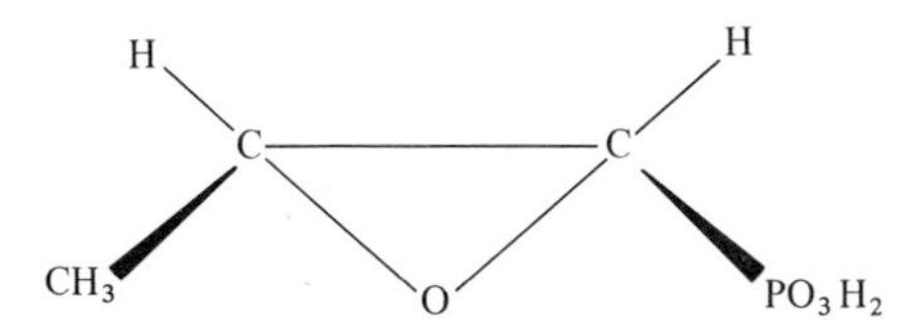

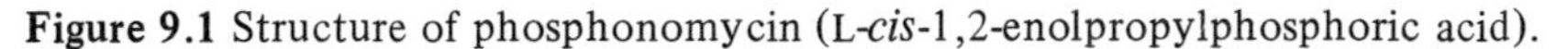
Figure 9.1 Structure of phosphonomycin (L-*cis*-1,2-enolpropylphosphoric acid).

The chemical structure (Fig. 9.1), combines the unusual features of a carbon-phosphorus bond, described for the first time among the natural products of bacteria and an epoxide ring, a structural feature also present in bacilysin [304] but not commonly found among antibiotics.

The inhibition of the enzyme catalysing the transfer reaction: phosphoenolpyruvate: UDP-*N*-acetylglucosamine-3-*O*-enolpyruvyl transferase, is competitive, the antibiotic acting as an analogue of phosphoenolpyruvate. Pre-incubation of the partially purified enzyme with phosphonomycin, in the absence of phosphoenolpyruvate, leads to a complete loss of transfer activity at concentrations of the antibiotic comparable to those giving inhibition of growth. This inhibition is irreversible. In addition, UDP-*N*-acetylglucosamine is an obligatory co-factor for inhibition although, unlike phosphonomycin, radioactive UDP-*N*-acetylglucosamine does not become acid-precipitable with the enzyme during the course of inactivation.

Earlier studies [87] showing that transferase activity was rapidly inactivated by sulphydryl reagents led to the suggestion that inactivation by phosphonomycin was through covalent binding of the antibiotic to a cysteine residue in the active site of the enzyme. This has now been confirmed by the isolation of 2-*S*-L-cysteinyl-1-hydroxypropyl-phosphonate [129]. Further information on the phosphoenolpyruvate: enolpyruvate transferase complex has come from the work of Cassidy and Kahan [37]. The stable complex can be formed either by combination of the enzyme with phosphoenolpyruvate, a reaction in which UDP-*N*-acetylglucosamine is an obligatory co-factor as in the case of inactivation with phosphonomycin, or by phosphorolysis of UDP-*N*-acetylglucosamine-3-*O*-enolpyruvate. Exposure of the complex, prepared by either method, to UDP-*N*-acetylglucosamine results in the release of phosphoenolpyruvate in addition to the reaction products, UDP-*N*-acetylglucosamine-enolpyruvate and inorganic phosphate. On the other hand, pre-inactivation by incubation of the enzyme with antibiotic and UDP-*N*-acetylglucosamine, as described above, prevents the subsequent formation of the stable complex. The covalent attachment of phosphoenol-pyruvate to a cysteine residue of the enzyme is also supported by the facts that the complex, once formed, is resistant to the sulphydryl reagent *N*-ethylmaleimide and that the release of inorganic phosphorus from the complex is 50 times more sensitive to mild acid hydrolysis than is the release from phosphoenolpyruvate. In addition, the incorporation of a proton from water on to the enol-carbon of the product UDP-*N*-acetylglucosamine-enolpyruvate has been established by mass spectrometry [129]. These observations, together with specificity of phosphonomycin for the UDP-*N*-acetylglucosamine-enolpyruvate transferase, have led to the suggested reaction (Fig. 9.2) of the antibiotic with the

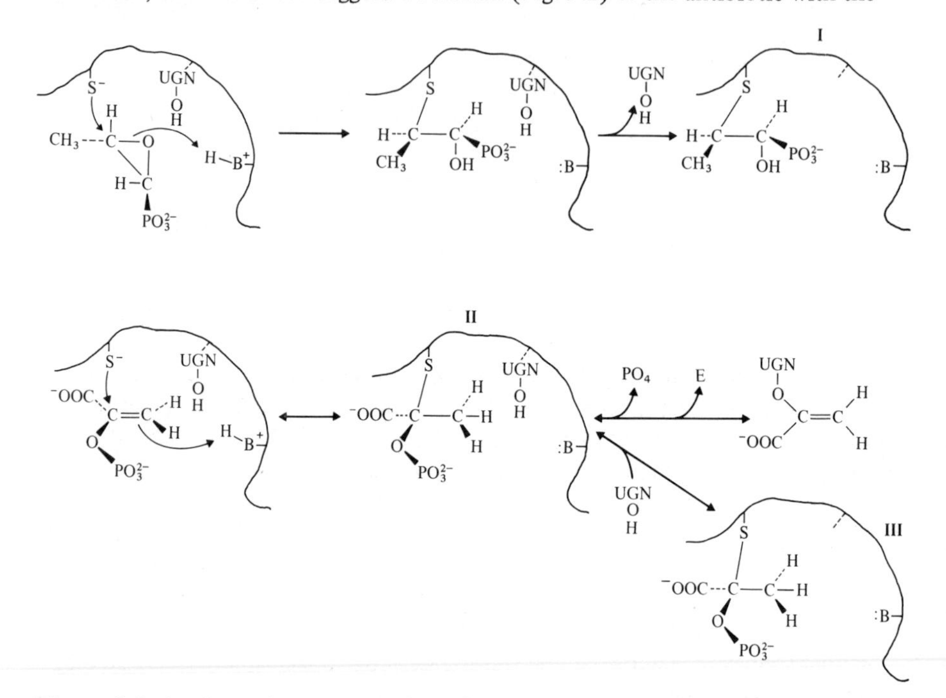

Figure 9.2 A schematic representation of a possible relationship between the reactions of phosphonomycin and phosphoenolpyruvate at the active site of UDP-*N*-acetylglucosamine-3-*O*-enolpyruvyl transferase. In the initial reaction of both substrate and inhibitor with the enzyme, UDP-*N*-acetyl-glucosamine (UGN–OH; the 3-hydroxyl is shown) is an obligatory cofactor which does not react with either the enzyme or the other substrate. In the case of phosphonomycin the proton donor site H^+–B is thought to activate the epoxide which is then subject to attack by a cysteinyl-sulphur at C_2. This leads to formation of an inactive enzyme-phosphonomycin adduct (I). Similar steps occur with phosphoenolpyruvate and a substrate enzyme complex (II) is formed which can become stabilized by removal of UDP-*N*-acetylglucosamine (III). In the final stages the enolpyruvyl residue is transferred with release of phosphate and enzyme. All these reactions are reversible in contrast to the reactions with phosphonomycin. After [129].

active site of the enzyme. For other enzymes that utilize phosphoenolpyruvate, phosphonomycin is either a weak competitive inhibitor or has no detectable activity. These include pyruvate kinase, phosphoenolpyruvate: carboxykinase enolase and phosphoenolpyruvate: shikimate-5-phosphate-enolpyruvyl transferase.

9.2.1 *Uptake of phosphonomycin by sensitive organisms*

Early investigations of the mode of action of phosphonomycin encountered a wide variation (from 1-200 μM) in the *in vivo* susceptibility of organisms. However, subsequent studies of the activity of various enolpyruvate transferases, including enzyme from a resistant strain, failed to reveal significant differences, either in the amount of enzyme activity present or in their sensitivity to the antibiotic. In addition, no evidence for the metabolic inactivation of the antibiotic was obtained. Taken together, these findings suggest that permeability of the organism to the antibiotic is the primary reason for variation in the *in vivo* susceptibility. This supposition has now been established by the finding that sensitive bacteria can accumulate phosphonomycin by the L-α-glycerophosphate transport system (genetic designation *glpT*) [100, 145]. The evidence for use of this particular pathway is as follows. Sensitivity to phosphonomycin in 24 strains of bacteria from several species is without exception associated with the ability to metabolize L-α-glycerophosphate. The uptake of radioactive phosphonomycin by sensitive bacteria can be effectively blocked by inclusion of L-α-glycerophosphate in the medium. Inorganic phosphorus, which inhibits the transport of L-α-glycerophosphate by the *glpT* system, antagonizes the action of phosphonomycin. Finally, mutations in the *glpT* system confer the expected properties on the action of phosphonomycin, i.e. those constitutive for *glpT* were some three times more sensitive to phosphonomycin, whereas transport-negative mutants *glpT*$^-$, were at least 30 times more resistant than were the parent strains.

An alternative pathway for uptake of the antibiotic is the inducible hexose phosphate uptake system (genetic designation *uhp*) [317]. The activation of this system explains the increased sensitivity of several strains, including those known to be *glpT*$^-$, when blood is present in the growth medium. Kadner and Winkler [128] have established that phosphonomycin resistance and the inability to transport hexosephosphate (*uhp*$^-$) are located at the same genetic locus in a *glpT*$^-$ strain.

The mechanism of uptake in *Klebsiella* remains unknown but a third alternative uptake mechanism has been suggested by the observation that strains of *Escherichia coli* constitutive for alkaline phosphatase show an enhanced sensitivity to phosphonomycin. This enzyme has recently been implicated in the transport of inorganic phosphate and the possibility remains that phosphonomycin could also enter the cell by this mechanism.

9.2.2 *Resistance to phosphonomycin*

The majority of strains which have been isolated as resistant to the action of phosphonomycin have been found to be defective in one of the transport systems outlined above. However, a temperature-sensitive mutant of *E. coli* has been isolated [301] in which resistance appears to be associated with a decreased affinity of the enolpyruvate transferase for the antibiotic and also the natural substrate, phosphoenolpyruvate.

9.3 Antibiotics inhibiting D-alanine metabolism in peptidoglycan biosynthesis: cycloserine, *O*-carbamoyl-D-serine, alaphosphin (L-alanyl-L-1-aminoethyl phosphonic acid) and the haloalanines

The inhibitory effect of D-cycloserine on bacterial growth can be reversed by the addition of D-alanine to the growth medium [23, 250]. Similarly the presence of D-alanine prevents the conversion to spheroplasts of several bacteria incubated with cycloserine in osmotically stabilized media [42, 139]. These observations led Park [204] to suggest that cycloserine might act by preventing the 'normal incorporation of D-alanine into the wall'. Confirmation of this supposition was quickly provided by Park and others [42, 205, 272] who described the accumulation of UDP-*N*-acetylmuramyl-L-alanyl-D-isoglutamyl-L-lysine in cycloserine-treated *S. aureus*. The addition of D-alanine to the medium specifically and competitively reversed the accumulation of the UDP-*N*-acetylmuramyl-tripeptide. These findings were subsequently extended to investigate the effects of cycloserine on the enzymes involved in the incorporation of D-alanine into UDP-*N*-acetylmuramyl-pentapeptide [270]. The pathway of alanine in the biosynthesis of this precursor is shown in Fig. 9.3.

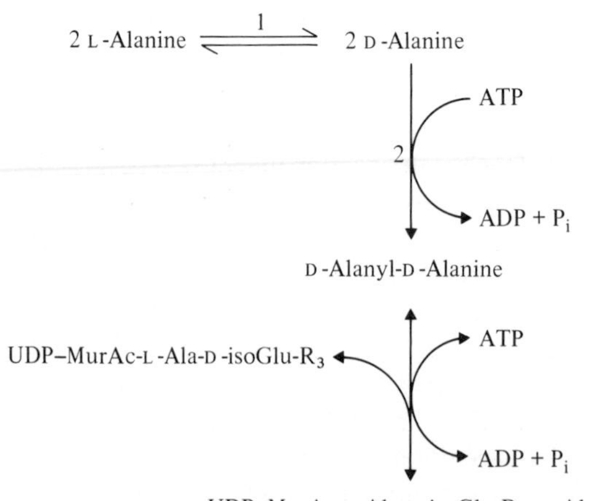

Figure 9.3 Pathway of alanine in the biosynthesis of UDP-*N*-acetylmuramyl-pentapeptide. Alanine racemase (Reaction 1) is inhibited by D-cycloserine, *O*-carbamoyl-D-serine, the haloalanines and alaphosphin while D-cycloserine alone appears to inhibit D-alanine:D-alanine ligase (Reaction 2).

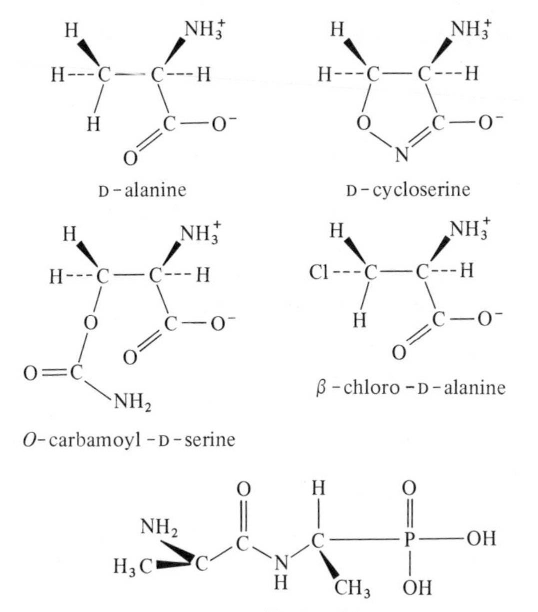

Figure 9.4 Structure of D-alanine and of certain antibiotics interfering with D-alanine metabolism.

At this time only D-cycloserine and *O*-carbamoyl-D-serine were known to interfere with these stages of peptidoglycan biosynthesis. In recent years alaphosphin and the halogenated derivatives of D-alanine have been added to this group (Fig. 9.4). All these antimicrobial agents inhibit the alanine racemase and D-cycloserine also inhibits D-alanine: D-alanine ligase (D-alanyl-D-alanine synthetase). The third enzyme, the D-alanyl-D-alanine adding enzyme (UDP-*N*-acetylmuramyl-L-Ala-D-isoglu-L-lys: D-Ala-D-Ala ligase (ADP)) is not affected by any of these antibiotics.

The early observations of Neuhaus and his colleagues [184–186] established the efficiency of D-cycloserine as a competitor of the substrate for both alanine racemase and D-alanine:D-alanine ligase. Their detailed studies describe both the kinetic properties of the enzymes and the structural requirements necessary for D-cycloserine to have antibiotic action. In brief, the effects of structural modifications on the D-cycloserine molecule are as follows. Cleavage of the ring between positions 4 and 5 and either acetylation or alkylation of the amino group abolished activity. Cleavage of the ring between positions 2 and 3 yielding β-aminoxy-D-alanine results in 75% loss of activity, while substitution of the heterocyclic ring or replacement of the hydrogens on position 5 by bulky substituents also causes large decreases in inhibitory activity.

9.3.1 *Effects on alanine racemase*

D-cycloserine, O-carbamoyl-D-serine, haloalanines

Roze and Strominger [238] investigated the inhibition of the alanine racemase of *S. aureus* in some detail in an effort to determine the underlying cause of the high affinity of this enzyme for D-cycloserine. L-cycloserine is not an inhibitor. On the basis of molecular models, they proposed that both D- and L-alanine could adopt the same conformation with respect to the functional NH_3^+ and COO^- groups and that this conformation is found in D-cycloserine. Since the enzyme catalyses the racemization of alanine in both directions, one might expect that L-cycloserine would also be an inhibitor of the enzyme. However, L-cycloserine cannot attain the conformation found in the alanine isomers and D-cycloserine. To explain the high affinity of enzyme and antibiotic, it was suggested that D-cycloserine is fixed in the conformation required by the enzyme, whereas D- and L-alanine have to adopt this conformation to allow enzyme activity [271]. However, studies with alanine racemases from *E. coli* [138] and *B. subtilis* [124] revealed that the above hypothesis is not generally applicable. In these organisms, L-cycloserine is an effective inhibitor of racemization (Table 9.1). On the basis of a single binding site

Table 9.1 Summary of inhibition constants for alanine racemase

	K_1(M)		
Organism	*D-cycloserine*	*L-cycloserine*	*Reference*
S. aureus	5.0×10^{-5}	>0.01	238
S. faecalis	2.4×10^{-4}	>0.01	155
B. subtilis	1.0×10^{-3}	5.0×10^{-4}	124
E. coli	6.5×10^{-4}	2.1×10^{-3}	138

for D- and L-alanine, as suggested by Roze and Strominger [238], Neuhaus *et al.* [186] postulated sub-binding sites for the methyl group in both D- and L-configurations, together with inhibition of the racemase by α-amino-isobutyric acid. Such inhibition was not detected using *E. coli* racemase. This result, together with the inhibition by L-cycloserine, were taken as indicative of two distinct binding sites, one for D-alanine and one for L-alanine. Certain kinetic data, particularly substrate inhibition found in the D- to L-alanine assay have also been interpreted in favour of the two site model [186].

Alanine racemase is also competitively inhibited by *O*-carbamoyl-D-serine [155], whereas *O*-carbamoyl-L-serine is without inhibitory activity. These findings were also taken as indicative of either two substrate binding sites [185] or of steric hinderance preventing access of the L-isomer to the alanine binding site. Evidence favouring the latter possibility came from the observation that both D- and L-β-aminoxy-alanine, a molecule with a less bulky side chain (resulting from cleavage of the cycloserine ring between positions 2 and 3) were effective inhibitors of both growth and alanine racemase of *S. faecalis*.

In addition to the two possible models of alanine racemase activity described above (the one [238] and two [186] site models) a third, in which two forms of the enzyme bind L-alanine and D-alanine respectively, has been postulated by Johnston and Diven [123]. However, the evidence with the racemase of *E. coli* is not consistent with this hypothesis [138].

Although as described above, D-cycloserine and *O*-carbamoyl-D-serine were known to be competitive inhibitors of alanine racemase from several organisms, the mechanism underlying this inhibition remained unclear. In 1968 Neuhaus [185] reported that pre-incubation of the racemase from *S. faecalis* with either D- or L-cycloserine, *O*-carbamoyl-D-serine or either isomer of β-aminoxy-alanine resulted in an irreversible inhibition of the enzyme. On the other hand, in similar pre-incubation studies with D-alanine:D-alanine ligase the inhibition obtained was reversible. More recently β-chloroalanine was reported to cause time-dependent irreversible inactivation of alanine racemase from *B. subtilis* and *E. coli* [127, 156]. At this time Rando [219, 220] postulated that D-cycloserine and *O*-carbamoyl-D-serine were in fact 'K_{cat}' inhibitors of alanine racemase; that is, inhibitors requiring catalytic conversion by the target enzyme to their active form. This has now been confirmed experimentally in a series of elegant experiments by Wang and Walsh [306] who use the term 'suicide substrate' to describe this kind of inhibitor. They found that both D- and L-isomers of β-chloroalanine and β-fluoroalanine undergo an enzyme-catalysed modification leading either to inactivated enzyme or an α,β-elimination reaction yielding pyruvate, ammonia and the respective halide ion. Racemization of the isomers was not observed but the possibility remains that it occurs in the enzyme–substrate complex. Their finding that all the haloalanines, together with *O*-carbamoyl-D-serine and *O*-acetyl-D-serine, showed a similar partition ratio (of 790–920:1) between the elimination reaction and the irreversible inactivation of the enzyme, led them to the conclusion that a common intermediate, an *ene*aminoacid-pyridoxalphosphate complex is responsible for the inactivation. This mechanism is shown in Fig. 9.5. The observed partition ratio of approximately 830:1 reflects the ratio of the two constants k_3 and k_4 for the release of dehydroalanine and for an enzyme-directed nucleophilic attack leading to inactivation. Wang and Walsh [306] also reported that incubation with D-cycloserine resulted in the time-dependent inactivation of alanine racemase thus confirming the earlier studies of Neuhaus [185] but did not investigate the mechanism of inhibition further. Rando [219, 220] has postulated for D-cycloserine a similar mechanism to that found for the haloalanines and *O*-carbamoyl-D-serine but this remains unproved.

Alaphosphin

This recently developed antibacterial agent is the L-alanyl derivative of L-1-aminoethylphosphonic acid (Fig. 9.4) [2, 3, 10]. L-1-Aminoethylphosphonic acid (L-Ala-Ⓟ) was known from the studies of Lambert and Neuhaus [138] to be an inhibitor of alanine racemase from *E. coli* and *S. faecalis*, although addition of the amino acid analogue to the culture medium did not inhibit growth. The key to its use as an antibacterial agent was the finding that incorporation of the analogue into a suitable dipeptide facilitates active transport and hence accumulation in the target organism.

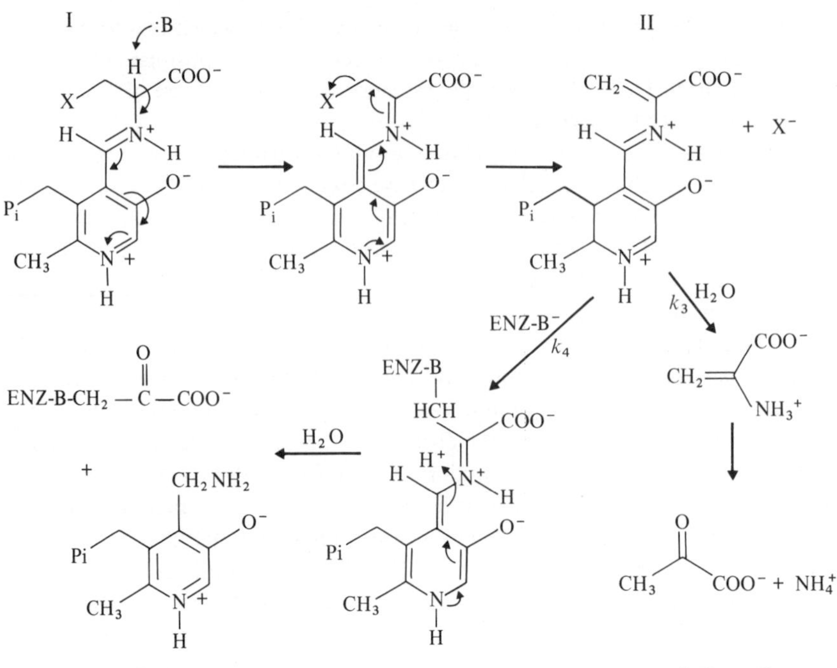

Figure 9.5 Proposed mechanism for the irreversible inactivation of alanine racemase. In the scheme proposed by Wang and Walsh [306] the initial reaction of the substrate (shown here as one of the haloalanines; X = halide), is with the pyridoxal phosphate moeity of the enzyme and requires the participation of a proton donor in the active site of the enzyme. The complex formed (I) is then converted into an ene-amino aldime complex (II) with the elimination of X^-. As described in the text intermediate II is partitioned between pathways A and B in the ratio 1:~830. Pathway B leads to the formation of pyruvate and NH_4^+ via dehydroalanine (aminoacrylate) whereas pathway A requires the further addition of some enzyme-nucleophile (ENZ-B⁻) and leads to enzyme inactivation. After [306].

For this reason sensitive organisms are those having an active peptide permease system.

Confirmation that L-l-aminoethylphosphonic acid acts as an analogue of L- rather than D-alanine in inhibiting alanine racemase comes from the following observations [2]. *Escherichia coli* treated with alaphosphin radioactively labelled with ^{14}C-D-Ala-Ⓟ (where Ⓟ indicates the phosphonic acid) did not accumulate any ^{14}C-D-Ala-Ⓟ but only the L-isomer showing that racemization did not occur. The major peptidoglycan precursor accumulated by alaphosphin-treated *S. aureus* was UDP-*N*-acetylmuramyl-L-alanyl-D-*iso*glutamyl-L-lysine. More recently the alanine racemase from *E. coli* and *P. aeruginosa* were reported to be reversibly and competitively inhibited by alaphosphin whereas the inhibition of the racemase from *S. aureus* and *S. faecalis* became irreversible with time [10].

A second, but minor site of inhibition, was demonstrated by the accumulation of

UDP-*N*-acetylmuramyl-^{14}C-L-Ala-Ⓟ in *E. coli.* Thus L-Ala-Ⓟ appears to compete with L-alanine as a substrate for the UDP-*N*-acetylmuramyl:L-alanine ligase but the resulting phosphonic acid substituted precursor is not a substrate for the UDP-*N*-acetylmuramyl-L-Ala:D-glutamic acid ligase. However, there seems little doubt that alanine racemase is the primary site of inhibition.

9.3.2 D-*alanine:*D-*alanine ligase*

The second enzyme of the D-alanine branch of peptidoglycan synthesis which is competitively inhibited by D-cycloserine is D-alanine:D-alanine ligase. This enzyme is of particular interest, since it has two distinct binding sites for D-alanine both of which are sensitive to D-cycloserine [187]. In the normal uninhibited state, the enzymic reactions leading to formation of the D-alanyl-D-alanine dipeptide are as follows:

$$\text{Enzyme} + \text{D-Ala} \xrightleftharpoons{K_A} \text{Enzyme–D-Ala} \tag{1}$$

$$\text{Enzyme–D-Ala} + \text{D-Ala} \xrightleftharpoons{K_{AA}} \text{Enzyme–D-Ala–D-Ala} \tag{2}$$

$$\text{Enzyme–D-Ala–D-Ala} \longrightarrow \text{Enzyme} + \text{D-Ala–D-Ala.} \tag{3}$$

The kinetics of the ligase are consistent with Reactions 1 and 2 being first order with respect to D-alanine with binding constants of 6×10^{-4} M and 10^{-2} M for K_A and K_{AA} respectively [183]. The following reactions can occur in the presence of both substrate and antibiotic [187].

$$\text{Enzyme} + \text{D-cycloserine (1)} \xrightleftharpoons{K_I} \text{Enzyme–I} \tag{4}$$

$$\text{Enzyme–D-Ala} + \text{I} \xrightleftharpoons{K_{AI}} \text{Enzyme–D-Ala–I} \tag{5}$$

$$\text{Enzyme–I} + \text{I} \xrightleftharpoons{K_{II}} \text{Enzyme–I–I} \tag{6}$$

$$\text{Enzyme–I} + \text{D-Ala} \xrightleftharpoons{K_{IA}} \text{Enzyme–I–D-Ala} \tag{7}$$

The K_I for Reaction 4 is 2.2×10^{-5} M and that of K_{AI} for Reaction 5 is 1.4×10^{-4} M. In both cases the inhibition is competitive with respect to D-Alanine, and as described above, completely reversible.

Neuhaus and Lynch [187] have also studied structural modifications of D-cycloserine to determine which parts of the molecule are required for inhibition of both growth and the ligase. Certain of these modifications have been mentioned briefly above. Cleavage of the cycloserine ring at either positions 1–2 or 2–3 results in a marked decrease in inhibitory action; β-aminoxy-D-alanine resulting from cleavage of positions 2–3 is not an inhibitor. The effects of *cis*- and *trans*-5-methyl substitution has also been studied with the results shown in Table 9.2 [185]. Thus *cis*-5-methyl substitution affects binding of the antibiotic primarily at the donor site, shown by the marked increase in the binding constant K_I, whereas *trans*-5-methyl

Table 9.2 Effect of 5-methyl substitution of D-cycloserine on D-alanine:D-alanine ligase, alanine racemase and the minimum inhibitory growth concentration of *Streptococcus faecalis* (data from [185]).

		D-alanine:D-alanine ligase (M)		*Alanine racemase*	*MIGC*
cis	*trans*	K_I	K_{AI}	*(M)*	*(M)*
H	H	2.2×10^{-5}	1.4×10^{-4}	2.4×10^{-4}	$1\text{–}2 \times 10^{-4}$
CH_3	H	1.2×10^{-4}	1.9×10^{-4}	2×10^{-3}	$5\text{–}6 \times 10^{-4}$
H	CH_3	5.4×10^{-4}	5.6×10^{-4}	$>5 \times 10^{-4}$	$3\text{–}4 \times 10^{-3}$

substitution affects both donor (K_I) and acceptor sites (K_{AI}). Substitution at C_4 by proton removal and removal of the amino group abolished inhibitory activity.

Any correlation between those concentrations inhibiting growth (MIGC) on the one hand and the enzymic activities on the other, is relatively poor for both alanine racemase and the binding of alanine to the acceptor site of the ligase. In each case the concentration of antibiotic required to cause 50% inhibition of the racemase is higher than that required to cause lysis, so that low antibiotic concentrations will cause lysis while the racemase remains uninhibited and therefore does not appear to be the primary site of action. Moreover, if the primary site were the acceptor site of the synthetase, for which D-cycloserine and its *cis*-5 methyl substituted analogue, D-cyclothreonine, are equally efficient inhibitors then each substance would have the same minimum inhibitory growth concentration provided each were taken up equally well by *S. faecalis*. In fact the MIGC of D-cycloserine is some 3-5 fold lower. Thus, at low antibiotic concentrations inhibition of the donor site rather than the acceptor site seems likely to be the primary site of action of D-cycloserine [187]. This does not preclude alanine racemase from being a potential 'target enzyme' at higher antibiotic concentrations.

9.3.3 *Uptake of D-cycloserine and O-carbamoyl-D-serine and resistance to these antibiotics*

Analysis of D-cycloserine and *O*-carbamoyl-D-serine-resistant mutants of *S. faecalis* [222] has revealed three types of resistance to these antibiotics. Mutants resistant to *O*-carbamoyl-D-serine had elevated activities of alanine racemase, whereas the other enzymes of the alanine branch of peptidoglycan synthesis were unaffected. Mutants resistant to D-cycloserine fell into two classes, the first of which showed increased activities of both alanine racemase and D-alanine:D-alanine ligase. In contrast, the second class of resistant mutants had normal activities of the antibiotic-sensitive enzymes but were unable to concentrate alanine. D-cycloserine prevented the accumulation of both D- and L-alanine by the parent organism. Thus, it appeared that resistance in these organisms was associated with an inability to accumulate D-cycloserine. Clearly, this observation gives an integral role to the transport system (or systems) for alanine in the action of D-cycloserine and the acquisition of resistance to the antibiotic.

The transport of alanine in *E. coli* has been resolved into two systems. One is involved in the transport of both D-alanine and glycine and the other in the transport of L-alanine [309]. D-cycloserine was an effective inhibitor of the D-alanine–glycine system, but not of L-alanine transport. On the other hand, inhibition of L-alanine accumulation was obtained with both L-cycloserine and *O*-carbamoyl-D-serine. The kinetics of uptake of D-cycloserine were similar to those of D-alanine and could be differentiated from those found for the uptake of L-alanine. These observations allowed Wargel *et al.* [309] to establish that glycine was a more efficient antagonist of D-cycloserine than was L-alanine. This antagonism occurred at the level of the transport system and did not affect the sensitivity of both alanine racemase and D-alanine:D-alanine ligase to the antibiotic. The importance of the role of the D-alanine–glycine transport system in the antibiotic activity of D-cycloserine has been clearly established in a series of D-cycloserine resistant mutants of *E. coli* [310]. These mutants were isolated in a stepwise manner against increasing concentrations of the antibiotic. In all cases, the loss of a D-cycloserine transport system was accompanied by a concomitant loss of D-alanine–glycine transport. Thus a first-step mutant isolated as resistant to D-cycloserine had lost the 'high affinity segment' of the D-alanine–glycine transport system. A second-step mutation involved the additional loss of the 'low affinity segment'. A multistep mutant of *E. coli* W, approximately 80-fold more resistant than the parent organism, retained 75% of the L-alanine transport activity but only 6–8% of the D-alanine–glycine transport activity. *E. coli* W could utilize either D- or L-alanine as a carbon source for growth, whereas the multistep mutant could only use L-alanine. Clearly, a functional transport system for D-alanine–glycine is necessary for both D-cycloserine action and for growth on D-alanine [310].

In contrast Clark and Young [45, 46] have recently reported the occurrence in *B. subtilis* of inducible resistance to D-cycloserine. Treatment with low concentrations of the antibiotic led to the acquisition of resistance (measured as the ability to survive incubation with 0.5 mM D-cycloserine for 2 h) but this was rapidly lost on subsequent growth of the bacilli in antibiotic-free medium. Surprisingly, resistant organisms remained fully sensitive to the inhibitory action of D-cycloserine on peptidoglycan synthesis and appeared to owe their survival to a decreased accumulation of the antibiotic. This was not due however, to a decrease in their ability to transport D-cycloserine but rather to an increased efflux of the antibiotic coupled with an enhanced uptake of both glycine and D-alanine.

9.4 Bacitracin

The bacitracins are a mixture of cyclic peptide antibiotics isolated from *B. licheniformis* [121]. The major component is bacitracin A (Fig. 9.6). The molecule contains a thiazoline ring formed between the L-cysteine and L-isoleucine residues at the *N*-terminal end of the acyclic peptide side chain. The free NH_2-group of L-isoleucine, adjacent to the thiazoline ring is essential for antimicrobial activity. Deamination of this residue, as in bacitracin F, leads to a loss of antimicrobial activity

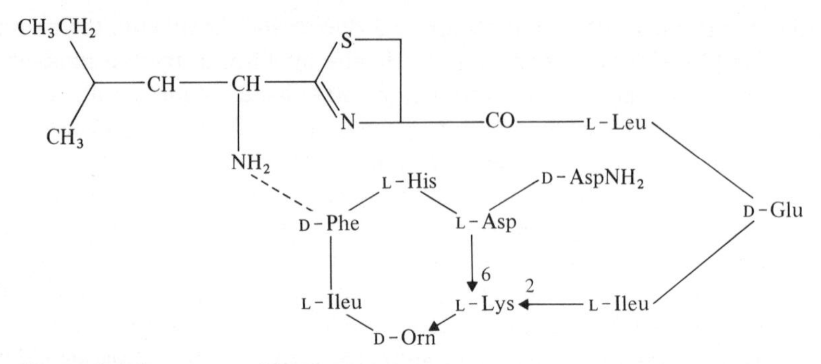

Figure 9.6 Structure of Bacitracin A.

[268]. In addition, the presence of a divalent metal ion is essential for antimicrobial activity, which can be inhibited by metal-chelating agents such as EDTA [267].

In common with many of the other antibiotics discussed here, the initial observations on the mode of action of bacitracin showed that it inhibited the incorporation of lysine into the peptidoglycan of *S. aureus* and caused the accumulation of UDP-*N*-acetylmuramyl peptides [205]. Under identical conditions the synthesis of protein was unaffected. In cell-free systems bacitracin was shown to inhibit peptidoglycan synthesis [4], although the inhibition obtained was variable, even at concentrations greatly in excess of those required to inhibit growth. However, when the individual reactions involved in peptidoglycan synthesis were examined none was found to be sensitive to bacitracin [131]. These observations led Siewert and Strominger [252] to investigate directly the lipid cycle of peptidoglycan synthesis and to the discovery that bacitracin specifically inhibited the dephosphorylation of undecaprenyl-pyrophosphate (see Fig. 8.3). By following the synthesis of peptidoglycan from *N*-acetylglucosaminyl-*N*-acetylmuramyl-(^{14}C-pentapeptide)-^{32}P-P-undecaprenol, they showed that incorporation of ^{14}C-pentapeptide into peptidoglycan was unaffected by bacitracin. However, with increasing concentrations of the antibiotic, the release of ^{32}P-inorganic phosphate was progressively inhibited and a simultaneous increase in undecaprenyl-P-^{32}P was observed. Undecaprenyl-P-^{32}P, isolated from large-scale incubation mixtures, was dephosphorylated by membrane (particulate) preparations from both *M. lysodeikticus* (*luteus*) and *S. aureus* in addition to *E. coli* alkaline phosphomonoesterase and alkaline phosphatase from calf intestinal mucosa. In all cases, dephosphorylation was inhibited by bacitracin, whereas dephosphorylation of glucose-1-phosphate or *p*-nitrophenyl phosphate by the *E. coli* enzyme was unaffected. Thus the reaction inhibited by bacitracin is as follows:

$$C_{55}\text{-isoprenyl-pyrophosphate} \rightleftharpoons C_{55}\text{-isoprenyl-phosphate} + P_i.$$

The results described above, particularly those obtained with the *E. coli* and calf intestinal enzymes, provide the first clear indication that bacitracin action is associated with an interaction between the antibiotic and the substrate (undecaprenyl pyrophosphate), rather than the antibiotic and the dephosphorylating enzyme.

Stone and Strominger [267] extended these findings and demonstrated the formation of a complex between undecaprenyl-pyrophosphate and bacitracin. Complex formation was dependent on the presence of a divalent metal ion, since the presence of a metal-chelating agent abolished the inhibition. On the other hand when EDTA was added after bacitracin had been pre-incubated with the lipid substrate and metal ion in then the chelating agent had only a minimal effect establishing that complex formation did not require participation of the enzyme. From a study of molecular models, it was suggested that the lipid pyrophosphate fitted into a pocket in the bacitracin molecule with the metal ion forming a bridge between the antibiotic and the polar end of the substrate. However, the hydrophobic side chain also played a part in complex formation, since neither inorganic pyrophosphate nor isopentenyl-pyrophosphate formed complexes, whereas C_{15}-farnesyl-pyrophosphate was bound almost as efficiently as C_{15}-isoprenyl-pyrophosphate [267]. Further studies on the mechanism of bacitracin action have been described in detail by Storm and Strominger [268, 269]. Binding constants for the interaction of bacitracin, Mg^{2+} and either undecaprenyl pyrophosphate or certain analogues of the lipid pyrophosphate, were measured. Columns of Sephadex G25 were equilibrated with buffer containing radioactive undecaprenyl pyrophosphate and Mg^{2+}. The subsequent addition of a known concentration of bacitracin A resulted in complex formation which was demonstrated by the appearance of a peak of radioactivity followed by a trough of equal area in the column profile (Fig. 9.7). From these curves the ratio of bound to unbound lipid and the equilibrium constant between *n* molecules of bacitracin (B) and lipid (L) were calculated according to the following equation:

$$nB + L \rightleftharpoons nBL$$

$$\log \frac{[nBL]}{[L]} = n \log B + \log K$$

Thus a plot of log $[nBL]/[L]$ against the -log of the bacitracin concentration gave both the stoicheometry of the complex and the association constant [268]. The values obtained for the association constant for undecaprenyl-pyrophosphate and various analogues are given in Table 9.3. Storm and Strominger [268] concluded that effective complex formation had the following requirements:

(1) A specific peptide (bacitracin) structure which included an intact thiazoline ring and also the imidazole ring of the histidine residue. This latter part of the molecule had been implicated in an earlier investigation as necessary for the binding of metal ions to bacitracin [73];
(2) A divalent cation of which Zn^{2+} was the most effective of those tested and finally;
(3) A lipid moiety which had both a pyrophosphate residue and also a hydrocarbon chain of some minimum length between C_5 and C_{15}.

Complex formation was shown to result principally from polar interactions with some contribution to overall stability of the complex from secondary hydrophobic interactions of the hydrocarbon chain with the aliphatic portions of the bacitracin peptide.

Table 9.3 Association constants for interaction of lipid analogues with bacitracin A (from [268]).

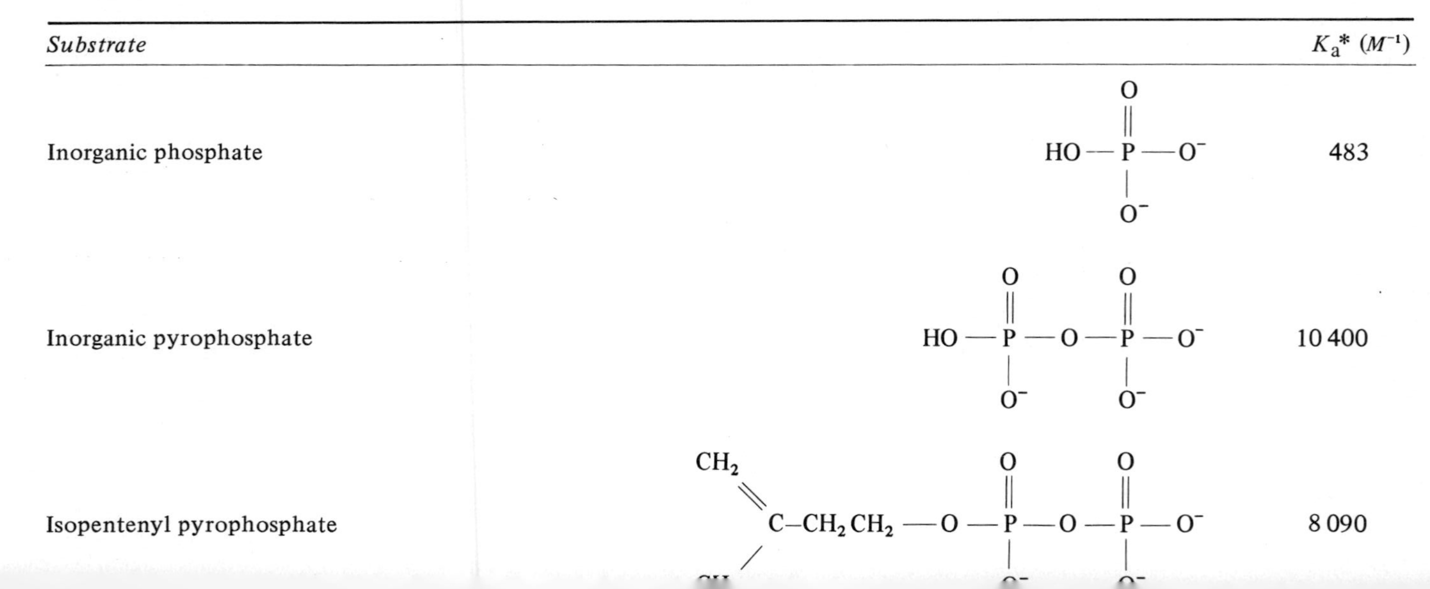

Substrate		K_a* (M^{-1})
Inorganic phosphate	$HO-P(=O)(O^-)-O^-$	483
Inorganic pyrophosphate	$HO-P(=O)(O^-)-O-P(=O)(O^-)-O^-$	10 400
Isopentenyl pyrophosphate	$CH_2{=}C(CH_3)-CH_2CH_2-O-P(=O)(O^-)-O-P(=O)(O^-)-O^-$	8 090

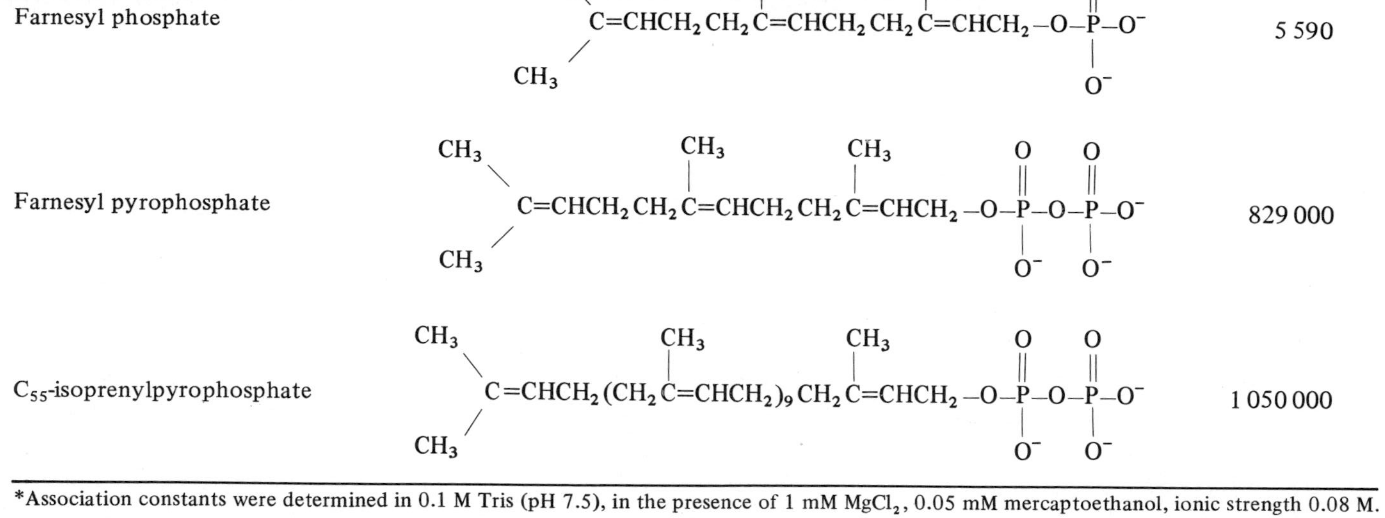

Compound	Structure	Association constant*
Farnesyl phosphate	$(CH_3)_2C{=}CHCH_2CH_2C(CH_3){=}CHCH_2CH_2C(CH_3){=}CHCH_2{-}O{-}P(O)(O^-){-}O^-$	5 590
Farnesyl pyrophosphate	$(CH_3)_2C{=}CHCH_2CH_2C(CH_3){=}CHCH_2CH_2C(CH_3){=}CHCH_2{-}O{-}P(O)(O^-){-}O{-}P(O)(O^-){-}O^-$	829 000
C_{55}-isoprenylpyrophosphate	$(CH_3)_2C{=}CHCH_2(CH_2C(CH_3){=}CHCH_2)_9CH_2C(CH_3){=}CHCH_2{-}O{-}P(O)(O^-){-}O{-}P(O)(O^-){-}O^-$	1 050 000

*Association constants were determined in 0.1 M Tris (pH 7.5), in the presence of 1 mM $MgCl_2$, 0.05 mM mercaptoethanol, ionic strength 0.08 M.

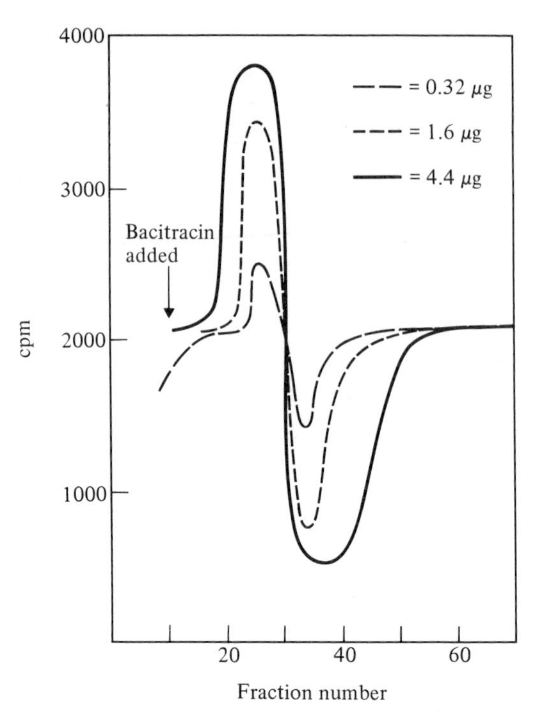

Figure 9.7 Binding of bacitracin A with C_{55}-isoprenylpyrophosphate. Sephadex G25 (1.5 ml) was equilibrated with buffer containing radioactive C_{55}-isoprenylpyrophosphate. Bacitracin A added at the concentrations shown, resulted in complex formation with the lipid intermediate. This was observed as deviation of radioactivity from the baseline. From binding curves such as the ones shown it is possible to calculate the ratio of lipid bound to a known concentration of antibiotic and from this the association constant for lipid and peptide. (From [268]).

In view of the clearly delineated role of bacitracin as an inhibitor of peptidoglycan synthesis, an unexpected observation was the finding that protoplasts of *B. megaterium* [97] and *S. faecalis* [251] were as sensitive to bacitracin as were the intact bacteria. Incubation of *B. megaterium* protoplasts with the antibiotic resulted in an efflux of potassium ions [97], which suggested that part of the action of bacitracin was directed against the cytoplasmic membrane and resulted in permeability changes. In contrast, Reynolds [226] showed that if the addition of bacitracin to a suspension of *B. megaterium* protoplasts was delayed until such time as they became metabolically active then the antibiotic had little effect on the subsequent increase in turbidity of the suspension. In an attempt to clarify this situation, Storm and Strominger [269] investigated the binding of ^{3}H-bacitracin by protoplasts of *M. luteus*. They found that both intact bacteria and protoplasts bound similar amounts of bacitracin ($\sim 2 \times 10^5$ molecules per organism) and that the binding was dependent on the presence of a divalent cation. The number of C_{55}-isoprenyl-pyrophosphate molecules present in an organism was estimated at approximately 5×10^5. In addition, the K_m for binding bacitracin to cells

(3×10^{-6} M) was similar to the *in vitro* association constant for bacitracin and undecaprenyl-pyrophosphate (1×10^{-6} M). From these findings Storm and Strominger [269] concluded that the reported effects on protoplasts [97, 251] could be explained by a disruption of the protoplast membrane resulting from complex formation between the antibiotic and undecaprenyl-pyrophosphate in the membrane. Whether permeability changes induced in this way also play a part, together with the inhibition of peptidoglycan synthesis, in the bactericidal action of bacitracin, remains open to question. Similarly, whether the apparent conflict between the observations on the sensitivity of *B. megaterium* protoplasts [97, 226] reflects a decrease in the concentration of undecaprenyl-pyrophosphate during the period of incubation as the protoplasts become metabolically active, is also unknown.

9.5 Tunicamycin

Tunicamycin was originally isolated from cultures of *Streptomyces lysosuperificus* as an antiviral agent although subsequent work showed that it was also active against many Gram-positive bacteria, yeasts and fungi [281]. The antibiotic was found to contain glucosamine but not amino acids. This observation apparently distinguished it from a second closely similar antiviral antibiotic, mycospocidin, which was reported to contain glycine but not glucosamine [182]. However Tkacz and Wong [292] have now established that hydrolysates of both tunicamycin and mycospocidin contain glucosamine and the two antibiotics appear identical. More recently, cultures of *Streptomyces clavuligerus* were found to contain an antibiotic complex (MM 19290) which is closely related to tunicamycin [133].

9.5.1 *Structure*

The chemical structure of tunicamycin has recently been established (Fig. 9.8) [113, 282]. Earlier analysis showed that it was a nucleoside antibiotic containing uracil, *N*-acetylglucosamine, unsaturated fatty acids and a novel amino sugar tunicamine, a C_{11}-aminodeoxy-dialdose. More detailed studies have allowed the structure in Fig. 9.8 to be assigned and revealed tunicamycin to be a family of homologous compounds differing only in the chain length of the unsaturated fatty acid esterified to the amino group of the tunicamine moiety.

9.5.2 *Mode of action*

The initial observations of Takatsuki and Tamura [284] showed that tunicamycin selectively inhibited glycoprotein synthesis in Newcastle disease virus and that it was without effect on nucleic acid and protein synthesis. These findings were extended to glycoprotein synthesis in yeast by Kuo and Lampen [137] where it was shown that tunicamycin inhibited synthesis of wall mannan and invertase by preventing the incorporation of glucosamine. On the other hand the synthesis of

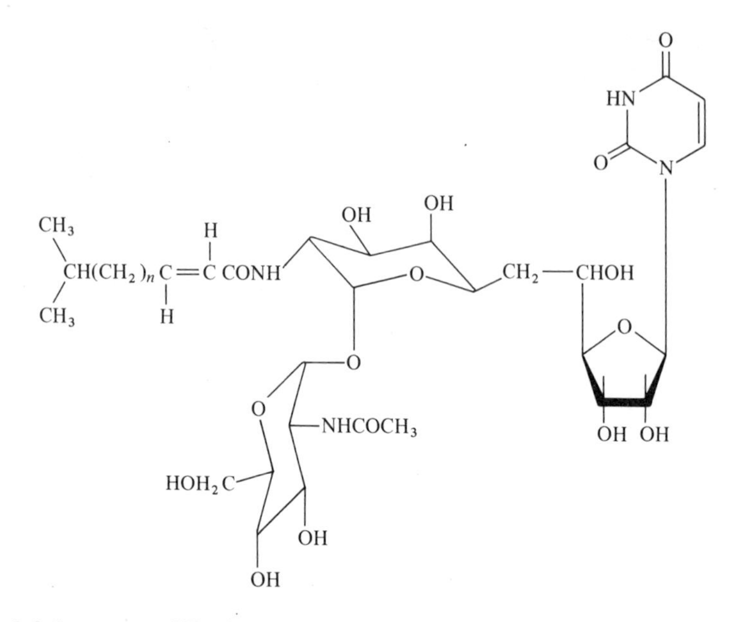

Figure 9.8 Structure of Tunicamycin. Tunicamycin is in fact a family of homologous compounds differing in the chain length of the fatty acid. The series includes tunicamycin C, $n = 8$; A, $n = 9$; B, $n = 10$ and D, $n = 11$. Whether these homologues differ in activity is not yet known.

chitin and the incorporation of mannose into mannan were unaffected. In contrast, the chitin synthases from *Neurospora crassa* and *Mucor rouxii* have been reported to be competitively inhibited by the antibiotic [246] with a K_i of approximately 480 μM (~ 400 μg/ml) for the *Neurospora* enzyme. Tunicamycin is known to inhibit UDP-*N*-acetylglucosamine:dolichyl-phosphate-*N*-acetylglucosamine-1-phosphate transferase and thus prevents the formation of dolichyl-pyrophosphoryl-*N*-acetyl-glucosamine [144]. This lipid intermediate participates in the first step of the pathway leading to glycoprotein synthesis. Studies on the transferase from porcine aorta suggest that tunicamycin may act as a substrate-product transition state analogue binding irreversibly to the enzyme and thus inactivating it [98].

In *B. subtilis*, tunicamycin was shown to inhibit the incorporation of glucosamine into macromolecular material, presumably the wall of the organism [283]. It is known to inhibit phospho-*N*-acetylmuramyl-pentapeptide translocase, that is the formation of the first lipid intermediate in peptidoglycan synthesis [287, 308]. Conflicting reports have been published as to whether the transferase catalysing the formation of the second lipid intermediate undecaprenyl-P-P-*N*-acetylmuramyl-(pentapeptide)-*N*-acetylglucosamine is also inhibited [13, 287, 308]. Inhibition of the translocase alone would be consistent with the observations in *Saccharomyces cerevisiae* where tunicamycin inhibits the synthesis of dolichyl-P-P-*N*-acetylglucosamine but not the subsequent transfer of a second residue of *N*-acetylglucosamine to this precursor [144].

N-acetylglucosamine-containing lipid intermediates have been described in *S. aureus*, and several bacilli participating in the biosynthesis of a specific oligomer linking the teichoic acids of these organisms to peptidoglycan [25, 169, 308, 320]. A similar situation exists in *M. luteus* where a *N*-acetylglucosamine-containing lipid is involved in the linkage of the teichuronic acid to peptidoglycan [313]. The biosynthesis of these linkage units is discussed in detail in Chapters 10.1.4 and 10.2.1. In each case the formation of the first lipid intermediate, polyprenyl-P-P-*N*-acetylglucosamine is inhibited by tunicamycin. Whether the inhibition of the biosynthesis of this intermediate relative to the inhibition of peptidoglycan synthesis is of importance in the antimicrobial activity of tunicamycin remains unclear. Recently tunicamycin has been found to be without effect on the formation of a lipid intermediate, presumed to be undecaprenyl-P-P-*N*-acetylgalactosamine, involved in the biosynthesis of a teichuronic acid in wall–membrane preparations from *B. licheniformis* [308a]. Thus at least some of the specificity of tunicamycin to inhibit a particular reaction appears to involve the phospho-*N*-acetylglucosamine residue. In this context *N*-acetylmuramyl-pentapeptide can be regarded as a derivative (i.e. a peptide substituted 3-*O*-lactyl-ether) of *N*-acetylglucosamine.

Mutants resistant to tunicamycin have been isolated from *B. subtilis* and mapped in two site on the chromosome [197]. Whether resistance in either case is directly related to the prevention of the inhibition of phospho-*N*-acetyl muramyl-pentapeptide translocase is not yet known.

9.6 The vancomycin group of antibiotics: vancomycin, ristocetins, ristomycins, actinoidin

The initial observations that vancomycin and other members of this group of antibiotics acted by inhibition of peptidoglycan synthesis came from the studies of Reynolds [223] and Jordan [125], both of whom showed that vancomycin inhibition was accompanied by the accumulation of UDP-*N*-acetylmuramyl peptides. Similar results were obtained for ristocetin [305]. In the case of vancomycin, inhibition occurred extremely rapidly and binding of the antibiotic to the bacteria took place within 20 seconds of treatment [126, 224]. This binding appeared irreversible, since the antibiotic was not removed by simple washing procedures [126]. A clear involvement of the bacterial wall in the binding process came with the demonstration that isolated walls of *B. subtilis* [12] and *S. aureus* [126] would effectively bind the antibiotic.

In earlier reports, vancomycin has been implicated in the inhibition of growth and the induction of permeability changes in the membranes of protoplasts from *B. megaterium* [97, 324] and *S. faecalis* [251]. However, none of these effects occurred as quickly as did the inhibition of peptidoglycan synthesis and for this reason they may be considered as secondary effects. These secondary effects appear a likely explanation for the inability of Williams [316] to isolate L-forms from *S. aureus* by growth in the presence of vancomycin, although it should be noted that L-forms of this organism isolated by other means, were subsequently found to be resistant to vancomycin [178, 316]. The reasons underlying these apparently anomalous observations may be in some modification of the membrane occurring on isolation of the L-form but the actual change involved remains unknown.

9.6.1 *Structure*

In contrast to most of the other antibiotics described in this chapter, the molecular structure of this group of antibiotics, particularly vancomycin itself, has only recently been fully established ([248] and references cited therein). Earlier chemical analysis of vancomycin showed the presence of glucose, aspartic acid, *N*-methyl-leucine, phenolic and chlorophenolic residues [160] whereas the ristocetins contained mannose, arabinose and rhamnose in addition to glucose, phenols and an amino sugar [215]. The amino sugar present in vancomycin, which may be a common component of this group of antibiotics, has been identified as vancosamine, 2,3,6-trideoxy-3-*C*-methyl-L-lyxo-hexopyranose [120, 312]. The removal of the sugars to form the aglycone, does not abolish the antibiotic activity of either vancomycin, ristomycin or the risocetins. Thus, the carbohydrate moiety does not appear to participate in the binding of the specific peptide structures such as those described below. Sheldrick *et al.*, [248] have investigated a derivative of vancomycin by X-ray analysis and the structure is as shown (Fig. 9.9a). It contains a disaccharide of glucose and vancosamine linked to three aromatic rings, two of which are substituted with

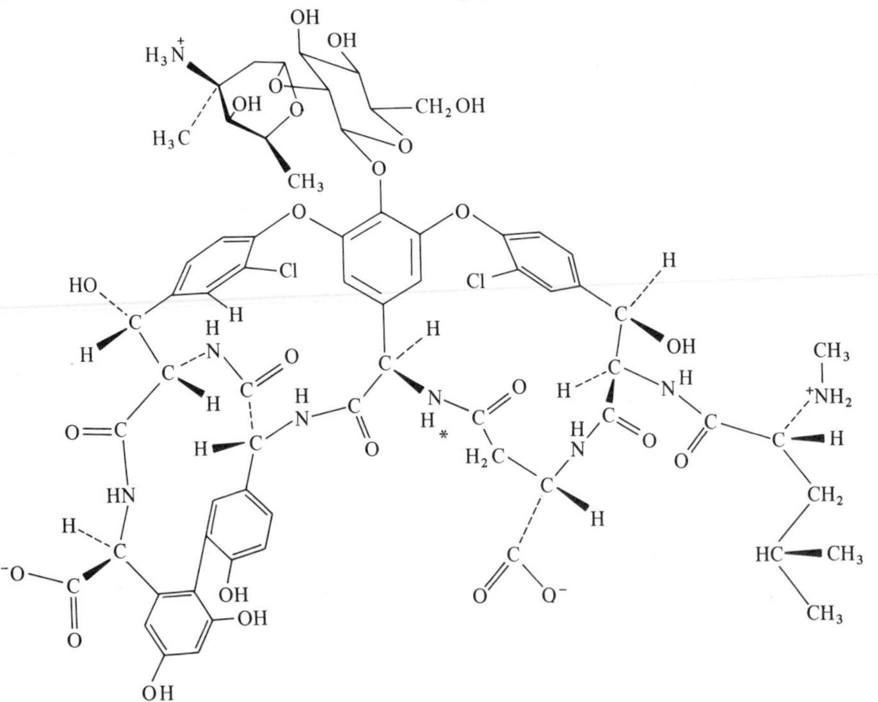

(a)

chlorine; a biphenyl system, aspartic acid and *N*-methylleucine. These units are linked by secondary amide bonds to form a tricyclic molecule containing two free carboxyl groups and *N*-terminal-*N*-methylleucine. Earlier Lomakina *et al.*, [150] investigated the structure of the aglycone of ristomycin A and proposed the structure shown in Fig. 9.9b. This contains a cyclic dipeptide composed of the closely related diaminodicarboxylic acids, ristomycinic and actinoidinic acids, both in phenolic ether linkage. Similar structures have been isolated from ristocetin A [64].

9.6.2 *Interaction with bacteria, walls and peptides*

Bacteria and isolated walls

Some of the earlier studies showing the rapid and apparently irreversible binding of vancomycin to bacteria have already been described. Best and Durham [12] in their investigation of the binding of vancomycin to the isolated walls of *B. subtilis*, found the walls to bind almost their own weight of the antibiotic. Clearly, this amount is in vast excess of the growth inhibitory concentration. The amount of antibiotic bound

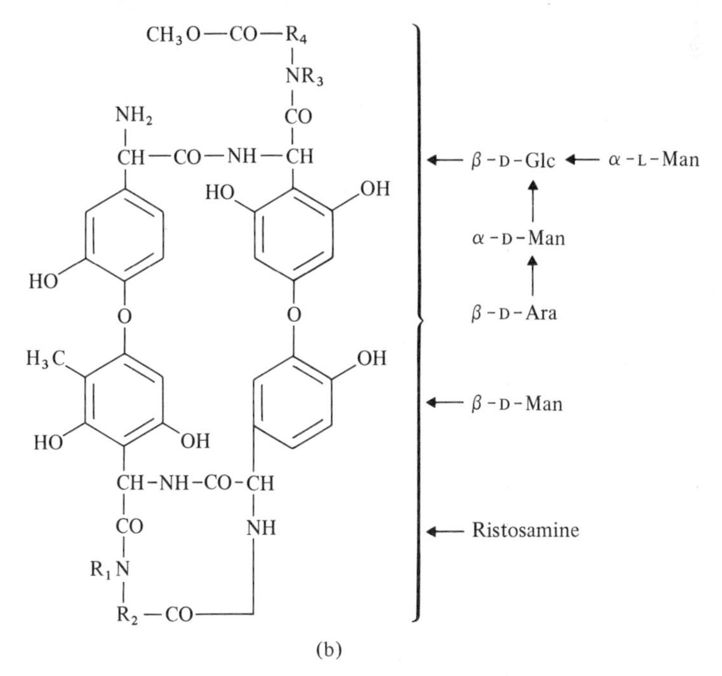

Figure 9.9 Partial structures of (a) vancomycin and (b) the aglycone of ristomycin A. The structure of vancomycin shown is that of a degradation product obtained with loss of ammonia. The main components are a disaccharide made up of glucose and vancosamine, linked to three phenolic rings, *N*-methylleucine, aspartic acid and a biphenyl system. The units are linked by secondary amide bonds to form a tricyclic molecule with two free carboxyl groups and *N*-terminal *N*-methylleucine. From [248]. The structure of the aglycone of ristomycin A is that proposed by [150]. It contains a cyclic dipeptide composed of the diaminodicarboxylic acids, ristomycinic and actinoidinic acids in phenolic either linkage.

was suppressed by the presence of divalent cations and up to 85% of the antibiotic could be removed by washing the walls with a solution of the cation. Thus, much of the observed binding appeared to be of a relatively non-specific nature. Confirmation of the results with *B. subtilis* was reported by Perkins and Nieto [211], whereas, in contrast, 10mM Mg^{2+} had little effect on the binding of iodinated vancomycin to the walls of *M. luteus.* Subsequently, Perkins and Nieto [213] suggested that the ability of vancomycin to form aggregates might well provide an explanation for the large amounts of antibiotic bound, particularly when this binding was apparently reversed by treatment with divalent cations. Reynolds [224] has calculated that the maximum binding obtained with *S. aureus* represents some 10^7 molecules of antibiotic bound per organism. For the above reasons, it seems unlikely that this represents the actual number of binding sites, although the extent of aggregate formation under these circumstances remains unknown.

The affinity of walls for vancomycin was further demonstrated by Sinha and Neuhaus [253], who showed that the presence of walls abolished the inhibition by vancomycin of peptidoglycan synthesis by a particulate enzyme preparation from *M. luteus*. Similarly, the addition of walls to an enzyme preparation previously treated with the antibiotic, would reverse the inhibition.

The initial observation that the mode of action of these antibiotics involved the formation of specific complexes between the antibiotics and certain peptide structures specific to the bacterial wall came from the finding [40] that trichloroacetic acid extracts of certain bacteria, inhibited with vancomycin and ristocetin, contained, in addition to the UDP-*N*-acetylmuramyl peptides previously observed [125, 223], complexes of the nucleotide precursors and the antibiotics. These complexes were subsequently shown to be formed by simply mixing the peptidoglycan precursor and antibiotic [210]. Degradation of the precursor molecule showed that the acyl-D-alanyl-D-alanine terminus was essential for complex formation to occur. A possible mechanism of interaction between the peptide and vancomycin is shown in Fig. 9.10. Addition of suitable peptides to solutions of either vancomycin or risto-

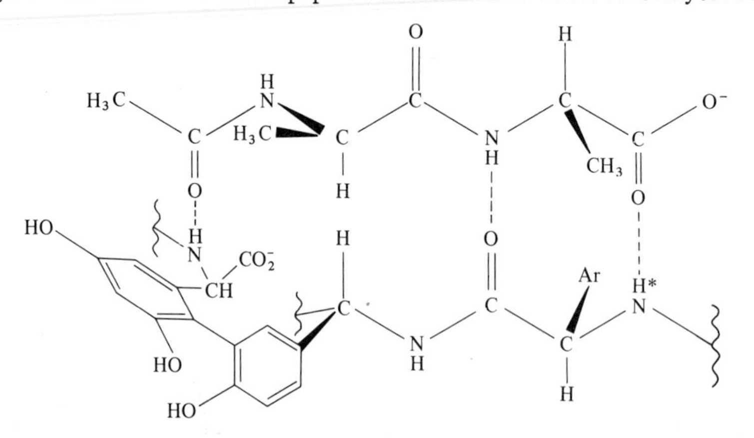

Figure 9.10 A portion of the vancomycin molecule schematically drawn to show a possible interaction with Acetyl-D-alanyl-D-alanine. H* is also shown in Fig. 9.9a to allow the structures to be aligned. (From [248].)

cetin resulted in a change in the ultraviolet spectrum of the solution. The use of difference spectra allowed the formation of the complex to be titrated. In the same investigation, Perkins [210] showed that complex formation required that both alanine residues were of the D-configuration and also that the carboxyl group of the terminal residue was free. Thus vancomycin and ristocetin both combined with peptides having an identical terminal structure to the cytoplasmic and membrane bound intermediates of peptidoglycan synthesis.

Further insight into the binding of vancomycin to these potential sites of inhibition came with the finding that ^{125}I-iodinated vancomycin retained almost all the antibiotic activity of the parent molecule [211]. Hence, the radioactively-labelled antibiotic could be used as a marker to determine the distribution of bound antibiotic in the whole organism. Fractionation of *M. luteus* treated with radioactively-labelled vancomycin revealed that the major portion of the radioactivity was associated with the walls and only some 8% was bound to the membrane fraction [211] (Table 9.4). However, on prolonged incubation a greater proportion of the antibiotic

Table 9.4 Distribution of bound ^{125}I-vancomycin in growing cultures of *M. luteus* and *B. subtilis* when labelled with radioactive antibiotic (see [211] of details).

Cellular fraction	*% of bound radioactivity*			
	M. luteus			*B. subtilis*
	Labelling time (min)			*Labelling time (min)*
	5	12	235	12
Solubilized walls (lysozyme digest)	66.3	56.3	22.2	58.7–82.2
Protoplast membranes	8.2	16.6	52.3	10.8–25.3
Cytoplasmic extract	6.8	9.8	11.4	5.6–13.0
Total recovery %	81.3	82.7	85.9	94.2–99.0

was found associated with the membranes. The mechanism of, and reason for, this change in distribution is not yet understood. On the other hand, protoplasts of *M. luteus*, prepared prior to treatment with the antibiotic, bound considerably less radioactivity than was found associated with the membrane fraction of whole bacteria treated under identical conditions. In addition, radioactivity was essentially absent from the cytoplasmic fraction in all cases. This latter finding effectively eliminates UDP-*N*-acetylmuramyl-pentapeptide as the site of action, whereas the small amount of antibiotic found associated with the membrane might be of particular importance. The interaction of these antibiotics with systems synthesizing peptidoglycan *in vitro* is discussed below (p. 324).

Peptides

The formation of complexes between two of the antibiotics of this group, vancomycin and ristocetin B, and a large series of synthetic and naturally occurring peptides, has been studied in detail by Perkins and Nieto [190–192]. Complex formation, between the various peptides and a fixed concentration of antibiotic, was measured as UV difference spectra and from Scatchard plots the association

Table 9.5 Association constants and free energy changes for the combination of vancomycin and ristocetin B with peptides (from [212, 213]).

	Vancomycin		*Ristocetin B*	
	K_a (l/mol)	ΔG (cal/mol)	K_a (l/mol)	ΔG (cal/mol)
Changes in Residue 1				
Ac_2-L-Lys-D-Ala-D-Ala	1.5×10^6	−8400	5.9×10^5	−7850
Ac_2-L-Lys-D-Ala-Gly	1.3×10^5	−6950	2.2×10^4	−5900
Ac_2-L-Lys-D-Ala-D-Leu	9.2×10^3	−5390	6.1×10^5	−7860
Ac_2-L-Lys-D-Ala-D-Lys	1.4×10^4	−5620	1.0×10^5	−6800
Ac_2-L-Lys-D-Ala-L-Ala	no combination		no combination	
Ac-D-Ala-D-Ala	2.0×10^4	−5840	7.2×10^4	−6600
Ac-D-Ala-Gly	5.4×10^3	−5070	1.9×10^3	−4470
Changes in Residue 2				
Ac_2-L-Lys-Gly-D-Ala	9.4×10^4	−6760	1.6×10^5	−7070
Ac_2-L-Lys-D-Leu-D-Ala	2.9×10^5	−7420	5.8×10^4	−6470
Ac_2-L-Lys-L-Ala-D-Ala	no combination		no combination	
Ac_2-L-Lys-Aib-Gly†	no combination		no combination	
Ac-Gly-D-Ala	1.1×10^4	−5500	4.9×10^4	−6390
Changes in Residue 3				
Ac-Gly-D-Ala-D-Ala	9.4×10^4	−6760	1.6×10^5	−7070
Ac-L-Ala-D-Ala-D-Ala	3.1×10^5	−7450	2.2×10^5	−7270
N-Ac-L-Tyr-D-Ala-D-Ala	1.9×10^5	−7180	2.9×10^5	−7430
Ac-D-Ala-D-Ala-D-Ala	5.0×10^4	−6380	1.3×10^5	−6960
Ac_2-L-Dbu-D-Ala-D-Ala	7.7×10^5	−8000		
Ac_2-L-Orn-D-Ala-D-Ala	1.3×10^4	−8320		
Ac_2-L-Lys-D-Ala-D-Ala‡	1.1×10^6	−8220		
Myristoyl-D-Ala-D-Ala‡	5.0×10^4	−6500		
Influence of Free Amino Groups				
L-Lys-D-Ala-D-Ala	1.2×10^4	−5510	8.2×10^3	−5320
α-Ac-L-Lys-D-Ala-D-Ala	4.7×10^5	−7700	1.9×10^5	−7200
Other Peptides				
Ac-L-Ala-Gly-Gly	4.9×10^3	−5200	2.5×10^3	−4620
Ac-Gly-Gly-Gly-Gly	1.5×10^3	−4300	8.0×10^2	−3950
Ac-L-Ala-D-Glu-Gly	4.8×10^5	−7720	6.8×10^2	−3850
C. poinsettiae dimer §	9.4×10^4	−6760	9.8×10^5	−8150
Ac_2-L-Lys-D-Leu-D-Leu	5.0×10^3	−5000		
Ac_2-L-Lys-L-Ala-L-Ala	no combination			
Influence of Peptide Chain Length				
D-Glu-γ-L-Lys-D-Ala-D-Ala	7.6×10^5	−8000		
D-Glu(α-Bzl)-γ-L-Lys-D-Ala-D-Ala	3.2×10^5	−7670		
L-Ala-D-Glu-γ-L-Lys-D-Ala-D-Ala	6.3×10^5	−7870		
L-Tyr-D-Glu-γ-L-Lys-D-Ala-D-Ala	8.3×10^5	−8050		
UDP-MurAc-L-Ala-D-Glu-γ-L-A_2pm-D-Ala-D-Ala D-CO_2H	7.2×10^5	−7960		

†Abbreviations: A_2pm, α,α′-diaminopimelic acid; Hsr, homoserine; Mur, muramic acid; Dbu, 2, 4-diaminobutyric acid; Aib, α-aminoisobutyric acid; Bzl, benzyl.
‡Experiment performed in 50% ethanol.
§Structure of dimer:

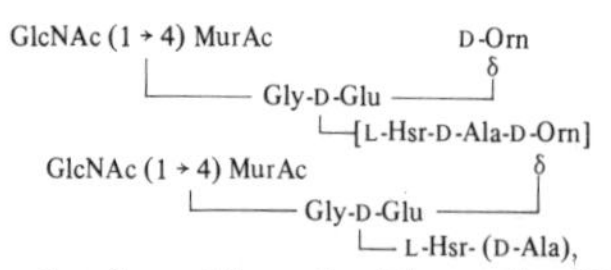

which represents a mixture of molecules with and without the (D-Ala) residue. The region that should form complexes with the antibiotics is enclosed within square brackets.

constants and free energy changes were calculated. The peptides were based on the structure of the pentapeptide residue found in the UDP-*N*-acetylmuramyl pentapeptide precursor of *S. aureus* peptidoglycan (Fig. 9.11). The results obtained (Table 9.5) showed that complex formation, with either vancomycin or ristocetin B, occurred only when both residues 1 and 2 were either D-amino acids or glycine. However, when considered in detail contrasting results were obtained for the two antibiotics. D-Alanine was the preferred residue in both positions for interaction with vancomycin, this being particularly pronounced with residue 1. Apparently, the binding site for this residue on the vancomycin molecule has a high degree of specificity, whereas a larger degree of difference was tolerated in residue 2. In contrast, ristocetin B showed a greater tolerance to an increase in size of residue 1, although the presence of an amino group in this position hindered complex formation with this antibiotic, but not with vancomycin. On the other hand, the change from glycine to D-leucine in residue 2 decreased the affinity of the peptide for ristocetin while increasing the affinity for vancomycin. This result suggests a greater steric restriction for residue 2 in ristocetin than is present in vancomycin. The peptide containing α-aminoisobutyric acid as residue 2 did not react with either antibiotic, showing that even in the presence of an appropriate D-substituent, an L-side chain prevented combination.

```
              5              4
H —— L-Ala ——→ D-Glu —— OH
                  |
                γ |          3              2              1
                  └——→ L-Lys ——→ D-Ala ——→ D-Ala —— OH
```

Figure 9.11 Pentapeptide residue found in the UDP-*N*-acetylmuramyl-pentapeptide precursor of *S. aureus* peptidoglycan.

However, although the various side chains had striking effects on the affinity of both antibiotics for the peptides, measurable complex formation was observed with acetyltetraglycine which represents the basic peptide backbone, thus showing that this structure accounts for some binding of the antibiotic molecule. Additional binding was concerned specifically with the side chains in the D-configuration at residues 1 and 2 and the L-configuration at residue 3. Modification of the peptide beyond residue 3 appeared to have little effect on complex formation with either antibiotic. On the other hand, Hammes and Neuhaus [94] found that the substitution of glycine for L-alanine in the peptidoglycan precursor, UDP-*N*-acetylmuramyl-L-alanyl-D-isoglutamyl-L-lysyl-D-alanyl-D-alanine, gave a two-fold stimulation of the inhibitory effect of vancomycin on *in vitro* peptidoglycan synthesis by *Gaffkya homari*, thus implicating in some way that this residue also participates in the interaction of the antibiotic and the peptide side chain.

The above peptides have all been considered, as described at the outset, as analogues of the uncross-linked precursor, terminating in a D-alanyl-D-alanine dipeptide, whereas acetyl-L-alanyl-D-glutamyl-glycine, is analogous to part of the peptide side chains of *M. luteus*, whether or not they are cross-linked. In contrast to ristocetin, vancomycin complexes effectively with this peptide, an observation which may account for the high affinity of *M. luteus* walls for vancomycin. In addition, this observation is in agreement with the findings of Sinha and Neuhaus [253], that while *M. luteus* walls readily reversed the inhibition by vancomycin of *in vitro* synthesis of peptidoglycan, a similar inhibition by ristocetin was only reversed by some 30%. In contrast, the cross-bridge peptide of many Bacilli, in which the sequence *meso*-diaminopimelic acid(L-centre)-D-alanyl-*meso*-diaminopimelic acid (D-centre) occurs, should form complexes with both antibiotics. However, since the structure might be regarded as having a long side-chain of the D-configuration at residue 1, the affinity for ristocetin should be higher than that for vancomycin. Complex formation with this type of structure may well account for the bulk of the antibiotic bound to the bacterial walls, although binding to the D-alanyl-D-alanine termini must be responsible for the inhibition of peptidoglycan synthesis.

9.6.3 *Interaction with systems synthesizing peptidoglycan in vitro*

The inhibition of peptidoglycan synthesis by vancomycin has been studied with particulate (membrane) preparations from *S. aureus* [4, 5, 122], *M. luteus* [6], *B. megaterium* [225, 226], *B. licheniformis* [307] and *G. homari* [94]. In all cases the addition of antibiotic to the biosynthetic system resulted in decreased synthesis, usually measured as the material remaining on the origin of a paper chromatogram, accompanied by an increase in the amount of radioactivity present as the lipid intermediates. The biosynthetic system from *G. homari* differs from the others described, in that synthesis of polymeric peptidoglycan occurs from the modified precursors, UDP-*N*-acetylmuramyl-tetrapeptide and -tripeptide, (in which either the terminal or both of the D-alanine residues have been removed) in addition to the normal precursor UDP-*N*-acetylmuramyl-pentapeptide. As expected from the detailed studies of complex formation between the antibiotic and specific peptides described above, vancomycin only inhibits synthesis from the complete precursor. Clearly, the antibiotic has no effect on the enzyme polymerizing the newly synthesized disaccharide peptide units. Thus, complex formation with either the lipid intermediates or the growing glycan chain appears to underlie inhibition of peptidoglycan synthesis by these antibiotics. In membrane preparations of *B. licheniformis* [307] the inhibition of peptidoglycan synthesis and the accumulation of lipid intermediates occurred at much lower vancomycin concentrations, measured in terms of the ratio of antibiotic to UDP-*N*-acetylmuramyl-pentapeptide present, than did the inhibition of D-alanine carboxypeptidase. As discussed in the previous chapter, this enzyme releases the terminal D-alanine residue from the pentapeptide side chain. Direct evidence for the formation of a complex between the lipid intermediates and both vancomycin and ristocetin has come through the use of the spin-labelled intermediate UDP-*N*-acetylmuramyl-L-Ala-D-isoGlu-L-Lys-(N^{ϵ}-tempyo)pentapeptide [122]. Membrane preparations of *S. aureus* utilized this modified precursor to synthesize undecaprenyl-P-P-*N*-acetylmuramyl-N^{ϵ}-tempyo-pentapeptide and complex formation between this lipid intermediate and both antibiotics was observed.

Thus, the acyl-D-alanyl-D-alanine termini of this membrane-bound intermediate are accessible to these antibiotics. However, in both cases the association constants for the spin-labelled lipid intermediate undecaprenyl-P-P-*N*-acetylmuramyl-(N^{ϵ}-tempyo)pentapeptide were lower (20 fold with vancomycin and 3-fold with ristocetin) than were those observed for the spin-labelled nucleotide. These results argue strongly against complex formation with undecaprenyl-P-P-*N*-acetylmuramyl-pentapeptide being the principal site of inhibition by these antibiotics. Unfortunately, evidence for the affinity of vancomycin for the disaccharide-pentapeptide-lipid intermediate and the equivalent residue present at the reducing terminus of the glycan chain is not yet available. Earlier, reactions leading to modification of the lipid intermediates, such as those occurring in the amidation and substitution of the α-COOH group of the D-isoglutamate residue of *S. aureus* [132] and *M. luteus* [163] respectively, were shown to be resistant to both vancomycin and ristocetin at concentrations where the synthesis of peptidoglycan was inhibited by 80–90%. Similarly, the addition of amino acids to form cross-bridge peptides was not affected.

However, the absence of radioactive vancomycin from the cytoplasm of treated organisms [211] argues against inhibition of peptidoglycan synthesis *in vivo* being the result of complex formation with the nucleotide precursor. The primary site of antibiotic action is probably located at the wall-membrane interface thus preventing the transfer of newly-synthesized units into the pre-existing wall.

9.6.4 *Vancomycin, ristocetin and phospho-N-acetylmuramyl-pentapeptide translocase*

As described in the previous chapter, vancomycin and ristocetin have been shown to have variable effects on phospho-*N*-acetylmuramyl-pentapeptide translocase, dependent on the concentration of the antibiotic used [216, 253, 274, 275]. Thus, at low concentrations they stimulate the transfer reaction (Reactions 1–5), while inhibiting the exchange reaction (Reactions 2–4). In these reactions E is the enzyme, C_{55} is undecaprenyl, MurAc is *N*-acetylmuramyl, P is phosphate, and PP is pyrophosphate.

$$\text{E} + \text{C}_{55}\text{P} \rightleftharpoons \text{E-C}_{55}\text{P} \qquad (1)$$

$$\text{E-C}_{55}\text{P} + \text{UMPP-MurAc} \rightleftharpoons \text{UMPP-MurAc-E-C}_{55}\text{P} \qquad (2)$$

$$\text{UMPP-MurAc-E-C}_{55}\text{P} \rightleftharpoons \text{UMP-E-C}_{55}\text{PP-MurAc} \qquad (3)$$

$$\text{UMPP-MurAc-E-C}_{55}\text{P} \rightleftharpoons \underset{\text{P-MurAc}}{\text{UMP-E-C}_{55}\text{P}} \rightleftharpoons \text{UMP-E-C}_{55}\text{PP-MurAc}$$

$$\text{UMP-E-C}_{55}\text{PP-MurAc} \rightleftharpoons \text{UMP} + \text{E-C}_{55}\text{PP-MurAc} \qquad (4)$$

$$\text{E-C}_{55}\text{PP-MurAc} \rightleftharpoons \text{E} + \text{C}_{55}\text{PP-MurAc} \qquad (5)$$

At higher concentrations the transfer reaction is also inhibited. Inhibition of the exchange reaction is thought to occur as a result of complex formation between the antibiotic and the wall peptide present, either as the nucleotide precursor, UDP-*N*-acetylmuramyl-pentapeptide or as the proposed enzyme–phospho-*N*-acetylmuramyl-pentapeptide intermediate. Under conditions where the exchange reaction with the formation of the enzyme–phospho-*N*-acetylmuramyl-pentapeptide complex, was proceeding at a faster rate than the overall transfer reaction, then higher concentrations of the antibiotic would be required to inhibit the transfer reaction. Vancomycin was also postulated to act as a surfactant and stimulate the transfer reaction by influencing the ability of the enzyme protein to form complexes with both the lipid substrate, undecaprenyl-phosphate and the lipid product, undecaprenyl-P-P-*N*-acetylmuramyl-pentapeptide. However, Hammes and Neuhaus [94] have found that the stimulation of the transfer reaction by low concentrations of vancomycin was obtained only with UDP-*N*-acetylmuramyl-pentapeptide and not -tetrapeptide as the substrate. This absence of stimulation with the modified substrate eliminated the detergent effect of vancomycin previously postulated. They concluded that complex formation, resulting in the formation of an ineffective intermediate, increased the rate of transfer by preventing the re-association of undecaprenyl-P-P-*N*-acetylmuramyl-pentapeptide with the enzyme, i.e. the reverse of Reaction 5. Alternatively, the reverse situation, increased dissociation resulting from complex formation with the enzyme bound lipid intermediate, would give the same stimulation.

9.6.5 *Resistance to vancomycin and ristocetin*

Gram-negative bacteria are relatively insensitive to this group of antibiotics. Since the peptidoglycan of the majority of these organisms has a similar structure to that found in many of the Bacilli, this insensitivity presumably reflects the inability of the antibiotics to penetrate the outer membrane of these organisms. On the other hand, the lack of resistance encountered among Gram-positive bacteria reflects the ubiquitous presence of the acyl-D-alanyl-D-alanine terminus. Clearly, the development of resistance, other than by preventing the penetration of the antibiotic, will require dramatic changes in the transpeptidation reaction and consequently in the mechanism of peptidoglycan assembly.

9.7 β-Lactam antibiotics: the penicillins and cephalosporins

Much of the earlier investigation which led to the discovery of the nucleotide-linked precursors of peptidoglycan [201–203] (the so-called Park nucleotides) and their implication in peptidoglycan synthesis has been described in the introduction to this chapter. Subsequent work showed that peptidoglycan synthesis, measured as the incorporation of a radioactive amino acid into material isolated as peptidoglycan by the Park–Hancock technique and other methods, was inhibited by benzylpenicillin and other semisynthetic penicillins at concentrations similar to those causing a cessation of growth. Park and Strominger [209] had earlier suggested that the site of penicillin action was the transglycosylation reaction, thus preventing the polymerization of newly-synthesized disaccharide-pentapeptide subunits into peptidoglycan. However, the first cell-free system [170] showed that this was not the case, since polymeric product was synthesized from the appropriate nucleotide precursors, even in the presence of high concentrations of penicillin. On the other hand, these particulate enzyme systems from *S. aureus* did not catalyse the postulated final reaction namely the formation of bonds between the free terminal amino group of the pentaglycine cross-bridge peptide and the carboxyl terminal of the penultimate D-alanine residue of a neighbouring peptide side chain. Since this reaction did not take place in the particulate enzyme preparation it could not be inhibited by penicillin. Wise and Park [318] subsequently examined the inhibition of cross-linking by growing *S. aureus* in the presence of low concentrations of penicillin under conditions which allowed the continued incorporation of ^{3}H-alanine or -lysine and ^{14}C-glycine into peptidoglycan. After isolation of the radioactive peptidoglycan, again by the Park–Hancock technique, the ratios of ^{3}H:^{14}C were determined. The results showed that peptidoglycan synthesized in the presence of penicillin contained an increased amount of alanine relative to the amount of lysine incorporated and this was accompanied by an increase in the number of free amino groups of glycine. A more direct examination of the effects of penicillin on cross-linking was made by Tipper and Strominger [290, 291] also in *S. aureus.* They reasoned that if penicillin inhibited cross-linking, then peptidoglycan synthesized in the presence of the antibiotic, whose glycan chains were subsequently hydrolysed by specific enzymes, should yield material of lower molecular weight than similar hydrolytic products obtained from control cells where normal cross-linking had occurred. In brief, the molecular size of the fragments obtained on enzymic hydrolysis would be dependent on the extent of cross-linking between the peptide side chains. *Staphylococcus aureus* was grown in the presence of various low concentrations of benzylpenicillin (0.09–1.0 μg/ml) and the peptidoglycan synthesized under these conditions labelled with ^{14}C-glycine. The bacteria were harvested when the cultures ceased exponential growth and the walls were isolated. After digestion with *Chalaropsis* muramidase the teichoic acid–peptidoglycan complex was separated on Ecteola-cellulose and the remaining soluble peptidoglycan fractionated on a column of Sephadex G25. In control cultures the majority of the incorporated radioactivity was found in oligomeric material (Table 9.6), whereas peptidoglycan synthesized in the presence of benzylpenicillin contained an increased proportion of low molecular weight material. The amount of this material was relatively constant, irrespective of the antibiotic concentration, although the total radioactivity incorporated decreased with increasing penicillin concentrations. Chemical analysis of the 'monomer' fraction, isolated from both control and penicillin treated cultures, showed

Table 9.6 Effect of benzylpenicillin on cross-linkage in *Staphylococcus aureus* (from [291]).*

Concentration of benzylpenicillin	*Radioactivity incorporated in wall (%)*			
(μg/ml)	*Total*	*Oligomer*	*Dimer*	*Monomer*
None	100	76	14	10
0.086	50	15	13	21.2
0.15	35	7.6	8.5	18.3
1.00	18	4.5	2.9	12.1

*Cultures of *S.aureus* Copenhagen were grown in the presence of sub-lethal concentrations of benzylpenicillin in medium containing ^{14}C-glycine. After incubation at 37° C for 30 min, walls were isolated and the peptidoglycan treated with *Chaloropsis* muramidase. The various fractions were then separated by gel filtration chromatography and the relative amounts of radioactivity incorporated into each was determined.

the presence of 2 molecules of D-alanine to 1 of L-alanine. Similar results were obtained with cultures treated with ampicillin, methicillin and cephalothin. In addition, pulse-labelling experiments showed that the uncross-linked material synthesized in the presence of penicillin was a direct precursor of cross-linked peptidoglycan. These experiments provided the first direct evidence that the action of penicillin is to inhibit the formation of cross-links.

Similar investigations were also carried out with the Gram-negative organisms *E. coli* and *Proteus mirabilis*. Martin, studying the effects of penicillin on the peptidoglycan of *P. mirabilis* and an L-form, which had lost the rod morphology, concluded that penicillin was inhibiting cross-linking [161, 162]. However, later chemical analysis of the isolated peptidoglycan failed to substantiate the original conclusions [132]. Similarly, no significant difference was found in the amount of dimer found in walls of *E. coli* grown in the presence of various concentrations of penicillin [244].

In contrast, the cross-linking reactions carried out by particulate enzyme from *E. coli* and *Salmonella newington*, the first *in vitro* demonstration of the transpeptidation reaction, were very sensitive to inhibition by penicillins [8, 9, 116, 117]. Thus cross-linking, which in these organisms is of the direct type (sub-group A1) [242], was inhibited by 50% using as little as 3 μg/ml of benzylpenicillin. Peptidoglycan was synthesized from UDP-*N*-acetylmuramyl-pentapeptide radioactively labelled in both the diaminopimelic acid and terminal D-alanyl-D-alanine residues. Lysozyme digestion and chromatography of the newly-synthesized material showed that while control preparations contained both dimers (i.e. cross-linked material) and monomers, only monomers were present when penicillin was present in the incubation mixture. In addition, a second penicillin sensitive enzyme was also present. In control preparations, this acted as a D-alanine carboxypeptidase, releasing the terminal D-alanine residue from both UDP-*N*-acetylmuramyl-pentapeptide and also from those peptide side chains not involved in cross-linking whereas the monomeric material synthesized in the presence of penicillin contained twice the

D-^{14}C-alanine content of the identical fraction purified from control preparations. In addition transpeptidation remained inhibited in enzyme preparations in which pre-incubation with penicillin was followed by extensive washing or treatment with penicillinase to destroy the antibiotic, a result which suggested that the inhibition of transpeptidation was irreversible [117].

9.7.1 *Models for the mechanism of action of penicillin*

A number of models to account for the mechanism of action of penicillin have been based on the various unique structural features of peptidoglycan. In the first, the structural similarities of *N*-acetylmuramic acid and penicillin were suggested as the basis of the mechanism [49]. However, this model was proposed prior to the inhibition of transpeptidation being demonstrated and was abandoned once this fact was established. Wise and Park [318] postulated that the β-lactam ring of penicillin was an analogue of the L-alanyl-D-glutamyl portion of the peptide side chain. They suggested that the free α-COOH groups of D-glutamic acid and penicillin had a similar spatial relationship to the peptide bond between the D- and L-amino acids in each molecule. The finding that the β-COOH group of D-glutamic acid was amidated and not free in the peptidoglycan of penicillin-sensitive *S. aureus* and that these substituents were incorporated at the level of the lipid intermediate and before cross-linking, rendered this hypothesis unlikely.

At approximately the same time, Tipper and Strominger [290] proposed that the similarity between peptidoglycan structure and penicillin lay in the terminal D-alanyl-D-alanine dipeptide of the peptidoglycan precursor. More specifically, they suggested that penicillin was an analogue of acyl-D-alanyl-D-alanine. Dreiding stereomodels (Fig. 9.12) of benzylpenicillin and one of the possible conformations of acyl-D-alanyl-D-alanine showed a strong similarity in structure, although it should be noted that D-alanyl-L-alanine has even greater analogy. Probably of greater importance, was the observation that the -CO-N-bond in the highly reactive β-lactam ring of penicillin was held in a closely similar position to the peptide bond cleaved during transpeptidation. From these observations, Tipper and Strominger [290] postulated the mechanism of transpeptidation and its inhibition by penicillin shown in Fig. 9.13. Under normal (i.e. non-inhibited) conditions, transpeptidation would proceed by the formation of an acyl-D-alanyl-enzyme intermediate with the release of the terminal D-alanine residue. Subsequently, the acylated enzyme would react with a suitable free amino group of a neighbouring peptide side chain to complete the transpeptidation reaction with the formation of a cross-link and the concomitant release of the enzyme. In the presence of penicillin, acting as an active site inhibitor, the initial reaction would result in the binding of the antibiotic with the formation of an inactive penicilloylated enzyme. Thus, treatment with penicillin would result in the irreversible inhibition of enzyme activity and penicilloylation of the enzyme protein. Evidence in support of this hypothesis came from the findings described previously that both transpeptidation in *E. coli* [116, 117] and D-alanine carboxy-

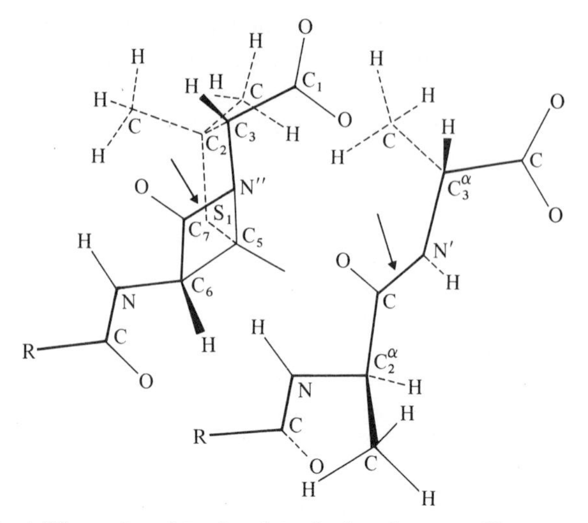

Figure 9.12 Penicillin and acyl-D-alanyl-D-alanine drawn to illustrate their structural analogy. The conformation of the acyl dipeptide given is that which most closely resembles that of the penicillin nucleus. The bonds broken in penicilloylation of penicillin-sensitive enzymes and by carboxypeptidases and transpeptidases are arrowed.

peptidase activity of *B. subtilis* were apparently [141, 142] irreversibly inhibited by penicillin. Reversal of inhibition, in this latter case, could be accomplished by treatment of the inhibited enzyme complex with neutral hydroxylamine and certain thiol reagents, an observation which led the authors to conclude that inhibition resulted from the formation of a protein-linked thiol ester derivative of penicilloic acid.

However, a more critical study of the stereomodels revealed differences other than the one described above. In particular certain of the bond angles associated with the β-lactam ring are markedly different from those found in acyl-D-alanyl-D-alanine. Thus the angle of the carboxyl group of the β-lactam ring is 90.5° compared with 117° for the equivalent angle in the dipeptide. Secondly the dihedral angle around the peptide bond of the β-lactam (135.7°) is different from that found normally in peptide bonds (approximately 180°). Moreover, the peptide bond linking the two D-alanine residues is some 25% longer than the equivalent bond in the antibiotic [143, 290]. These changes all serve to make the β-lactam nitrogen pyramidal rather than planar in character. The 'substrate analogue' hypothesis requires that the same active site of the transpeptidase both recognizes and binds either the natural substrate, acyl-D-alanyl-D-alanine or the antibiotic. In view of the differences found in the stereomodels such a situation seems unlikely unless one or other of the two 'substrates' becomes modified by the enzyme in a manner which will allow both to interact with the active site. Thus, either the acyl-dipeptide can be modified to a configuration more closely resembling that of the antibiotic or alternatively the converse can occur. In fact Tipper and Strominger [290] and subsequently Lee [143] noting that the carboxyl–nitrogen bond of the β-lactam ring has more single-

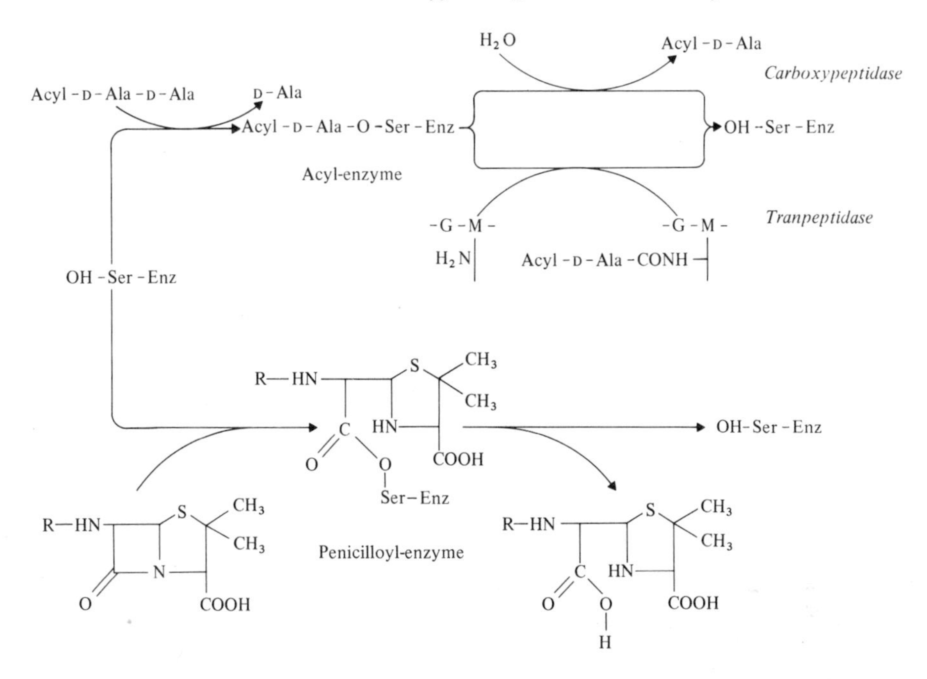

Figure 9.13 Formation of acyl and penicilloyl intermediates in D,D-carboxypeptidase and transpeptidase reactions. A seryl residue in the donor site of the enzyme reacts either with acyl-D-alanyl-D-alanine or penicillin to yield either acyl-D-alanyl–enzyme and D-alanine or penicilloyl–enzyme intermediates. If water binds to the acceptor site the enzyme functions as a D,D-carboxypeptidase; acyl-D-alanine is released and the active enzyme regenerated. Alternatively if a suitable amino acceptor binds to the acceptor site then transpeptidation ensues. Although both functions are shown for the same enzyme individual enzymes tend to favour one reaction, i.e. they are D,D-carboxypeptidases with inefficient transpeptidase activity or the reverse may be the case. In the case of penicilloyl-enzyme intermediates these may also react with water to regenerate active enzyme and yield penicilloic acid or alternatively as shown in Fig. 9.18, the penicillin nucleus may undergo further degradation to *N*-formylpenicillamine and in the case of benzylpenicillin, phenylacetylglycine.

than double-bond character, postulated that the antibiotics are analogues of an enzyme-bound transition state formed during cleavage of the D-alanyl-D-alanine peptide bond. In this case the dipeptide is being enzymically modified so as to resemble the antibiotic. Supportive evidence for this view has recently been provided by the computer-simulation studies of Boyd [24] in which the nucleophilic attack of a hydroxyl ion on glycylglycine (used as an analogue of D-alanyl-D-alanine) was studied. Here attack on the α-face of the carbonyl carbon of the dipeptide yielded a transition intermediate, similar in structure to the bicyclic nucleus of the penicillins and cephalosporins.

Of potentially greater significance as evidence against the hypothesis of Tipper and Strominger was the finding that 6-methylpenicillins and 7-methylcephalosporins were almost inactive as antibiotics both *in vitro* [315] and *in vivo* [21, 102].

Since these molecules appeared to resemble more closely acyl-D-alanyl-D-alanine in structure, it had been postulated [290] that they might be even better antibiotics than the unsubstituted molecules. In contrast the naturally occurring 7-methoxy-cephalosporins (cephamycins) proved potent antibiotics [172, 173] (Cefoxitin; Fig. 9.14). This apparent conflict has recently been resolved by Virudachalam and Rao [302] who used stereochemical criteria to examine the effect of changes in both substituents and configuration at the C_6 and C_7 positions of the penicillin and cephalosporin nuclei respectively. These changes were then related to the conformation of acyl-D-alanyl-D-alanine. They found that the configuration of the stereo-model (Fig. 9.12) used by both Tipper and Strominger [290] and Lee [143] which was based on the conformation of crystalline benzylpenicillin could not be attained

Penicillins

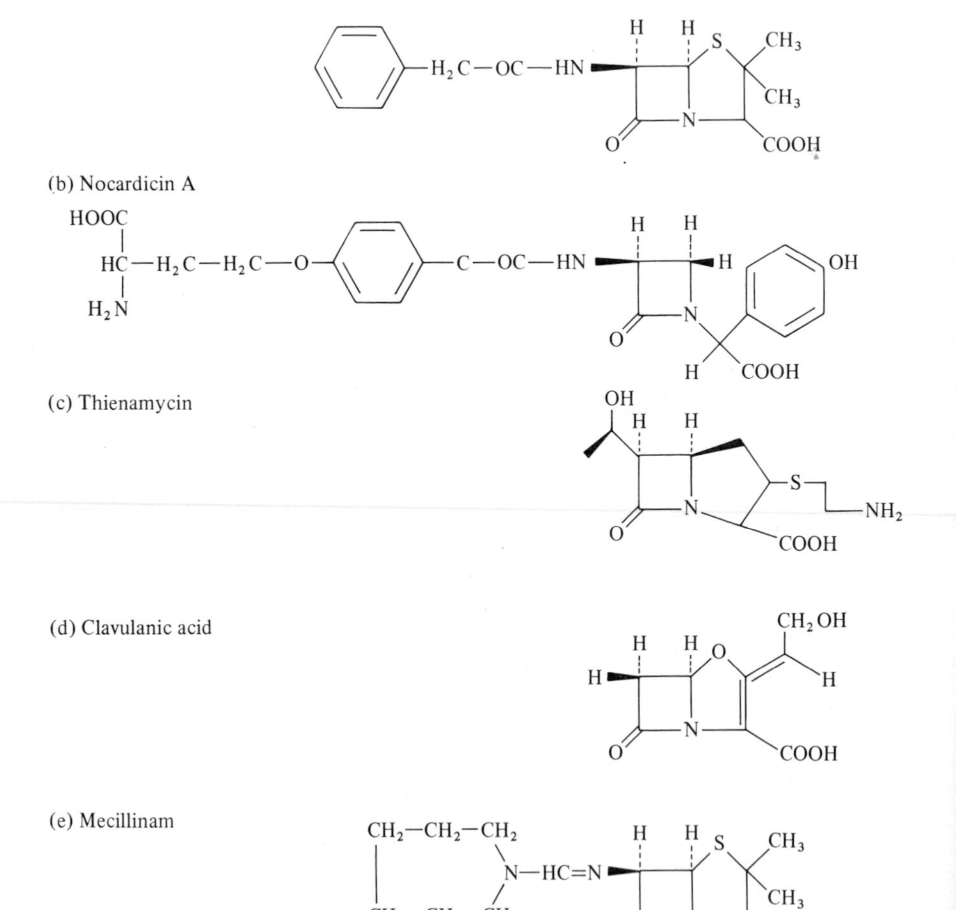

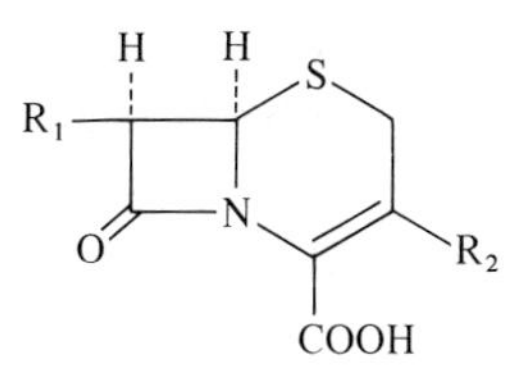

(a) Cephaloridine R_1

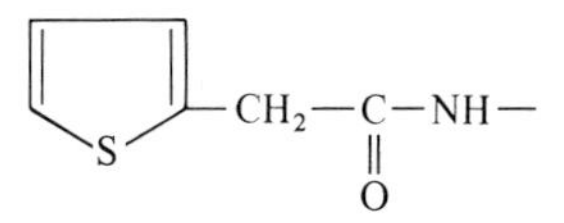

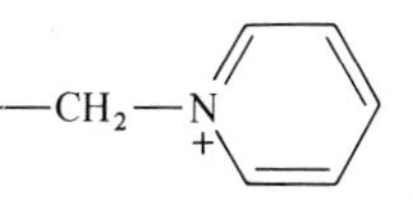

(b) Cefsulodin

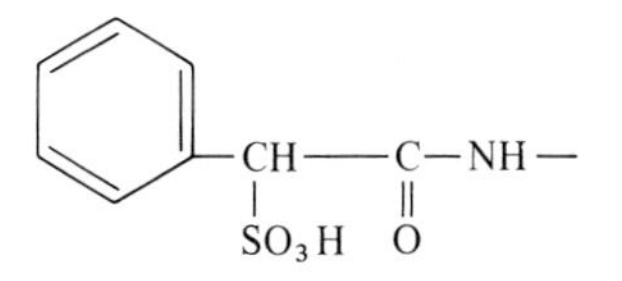

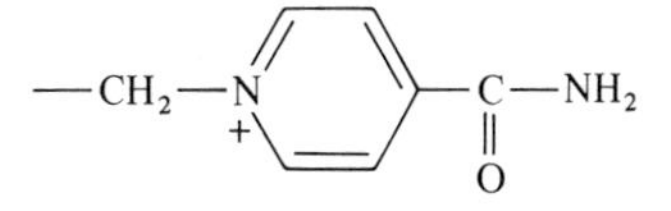

(c) Cephalothin

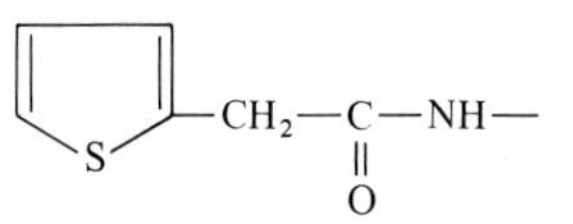

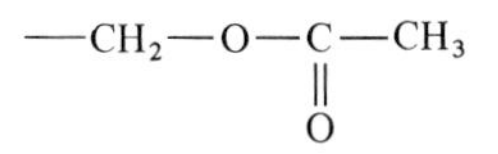

(d) Cephalexin

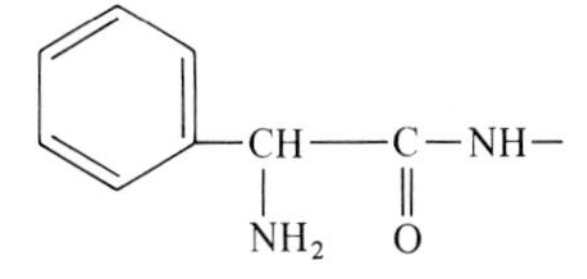

(e) Cefoxitin

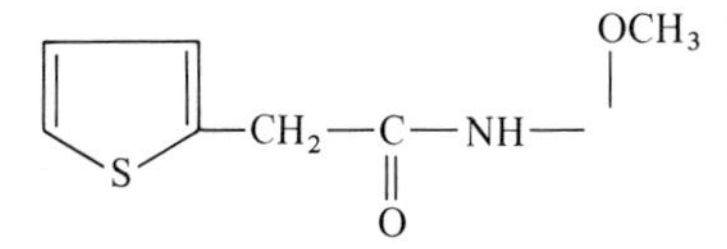

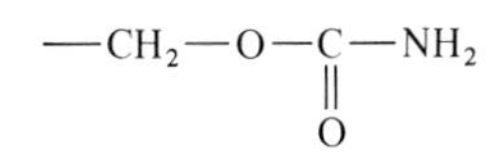

(f) Cephaloglycine

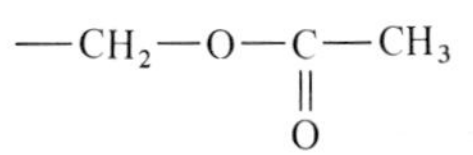

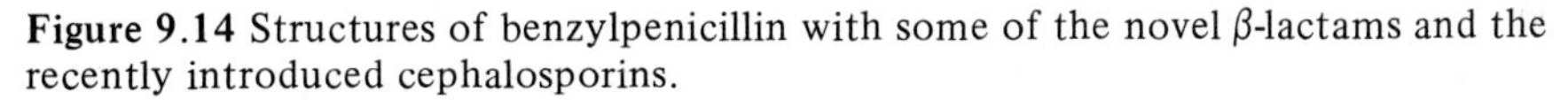

Figure 9.14 Structures of benzylpenicillin with some of the novel β-lactams and the recently introduced cephalosporins.

by acyl-D-alanyl-D-alanine because of constraints on the amino acyl group of the dipeptide. However, other evidence suggested that, in solution, some conformational flexibility of the antibiotics was possible, thus allowing them to attain a greater similarity with the acylated dipeptide. This could not occur where either the C_6 or C_7 residues of the respective β-lactam rings were in the D-configuration or were substituted with α-methyl residues although α-methoxy substituents were allowed. Moreover, an intact β-lactam ring was shown to be required to hold the L-cysteinyl residue of the antibiotic nucleus in conformational similarity to the acyl-D-alanine moiety of the acylated dipeptide. Thus Virudachalam and Rao argue that the β-lactam ring is essential for activity for conformational reasons, any particular reactivity it confers on the antibiotic molecule probably being of secondary importance. These findings clarify many previous observations regarding the structure and function of β-lactam antibiotics and also give an explanation of the inactivity of synthetic derivatives based on D-cysteinyl-D-valine [86] which again had been postulated to resemble D-alanyl-D-alanine.

More recently Rando [219, 220] has proposed an alternative mechanism based on the theories of 'K_{cat}' inhibitors, which in effect is the direct opposite of the mechanism described above. The designation 'K_{cat}' inhibitor has been applied to those situations where the inhibitory compound requires catalytic conversion by the target enzyme to the active form. In the model proposed by Rando, the antibiotic is being made to attain a conformation more closely related to that of acyl-D-alanyl-D-alanine rather than the reverse. He argues that the active site of the transpeptidase binds the antibiotic in such a way that the β-lactam nitrogen is forced to approach planarity. In this way, the strain energy applied by the enzyme would be transferred to the already reactive carbonyl-nitrogen bond in the β-lactam ring and in comparison with the free antibiotic the enzyme-bound form would be in a higher energy state. Cleavage of the carbonyl–nitrogen bond would release the strain. This however would be manifest as an increase in the chemical reactivity of the molecule. The net result of these changes would be to increase the probability of a reaction between the antibiotic and an active site residue.

In this model the β-lactam ring is essential for the chemical reactivity it confers upon the molecule although it remains unclear what conformational role it also has. The model also appears to clarify an earlier observation related to the isomers of cephalosporins. The Δ^3-cephalosporins (cephaloridine, cephaloglycine, etc., Fig. 9.14) are active antibiotics whereas the Δ^2-cephalosporins, although more closely resembling the stereomodels of penicillin, are inactive. However, the chemical reactivity of Δ^2-cephalosporins is relatively low [279]. These observations appear to support the view that chemical reactivity is more important than isosterism for antibiotic activity.

In addition, the 'K_{cat}' mechanism was thought to provide an explanation for a requirement in the β-lactam antibiotics for both a fused ring system and the presence or absence of certain substituents particularly in relation to the free amino and carboxylic acid moieties. Thus for antibiotic activity, the presence of a free carboxylic acid function adjacent to the β-lactam ring and a substituted free amino group were thought to be required. These groupings were postulated by Rando to react with binding sites in the transpeptidases and in this way function as 'handles' through which the enzyme could apply the forces necessary to strain the antibiotic further as described above. Such a role would provide an explanation for the dramatic effect of certain substituents on the antimicrobial activity of penicillins and cephalosporins particularly when it was not immediately apparent how the isosterism with acyl-D-alanyl-D-alanine had been changed. Whether the recent observations of Virudachalam and Rao [302] influence this conclusion remains unclear. Stronger evidence against the 'handle' hypothesis has come from the recent isolation of new β-lactam antibiotics such as thienamycin [85] and clavulanic acid [31] (Fig. 9.14). These antibiotics resemble penicillins in having a five-membered ring fused to the β-lactam, although of a different structure but they do not have the substituted amino group attached to C_6. In fact clavulanic acid has no side chain at this point although one of a different structure is present in thienamycin. Clavulanic acid has only weak antibacterial activity although its action against *Neisseria gonorrhoeae* appears to be an exception [32], but is a potent inhibitor of β-lactamases and used in conjunction with other β-lactam antibiotics may potentiate their activity against β-lactamase-producing organisms. Thienamycin, on the other hand, is a broad-spectrum antibiotic active against many Gram-negative as well as Gram-positive organisms.

The fused ring system appeared necessary to induce a certain basal level of strain and hence reactivity into the β-lactam ring [279]. This, it was argued, explained the virtual absence of activity in penicillin derivatives where the thiazoidine ring had been opened thus destroying the fused ring character. However, the recent isolation of the nocardicins [7] (Fig. 9.14), a family of antibiotics which contain the β-lactam ring alone has clearly rendered this part of the hypothesis untenable. These antibiotics do, however, contain the free carboxylic acid and substituted amino group described above.

Currently it remains unclear whether these various contradictions represent only minor discrepancies to the hypothesis as put forward by Rando. Perhaps only when the various antibiotics have been examined as inhibitors of individual enzymes will the truth become apparent.

Finally in this discussion of the 'substrate analogue' hypothesis, it is necessary to describe one piece of evidence which suggests unequivocally that β-lactam antibiotics are not simply analogues of acyl-D-alanyl-D-alanine. Membranes of *S. faecium* (*faecalis*) exhibit in adddtion to a D,D-carboxypeptidase (carboxypeptidase I) an L,D-transpeptidatase activity [53]. This latter enzyme catalyses the model transpeptidation reaction:

$$(\text{acyl})_2\text{-L-lys-D-ala} + {}^{14}\text{C-D-ala} \rightarrow (\text{acyl})_2\text{-L-lys-}{}^{14}\text{C-D-ala} + \text{D-ala}$$

but appears not to possess L,D-carboxypeptidase activity. The enzymes are both inhibited to varying extents by a range of β-lactams. Thus antibiotics were found which were more effective inhibitors of either the D,D-carboxypeptidase, the L,D-transpeptidase or were equally effective against both enzymes. To date no physiological significance has been found for the L,D-transpeptidase in this organism but its inhibition by the β-lactam antibiotics appears strong evidence against their being simply analogues of acyl-D-alanyl-D-alanine.

An alternative hypothesis for the action of penicillin, based largely on studies of the inhibition of the *Streptomyces* D,D-carboxypeptidase-transpeptidases, was postulated by Ghuysen and his colleagues. In this case, penicillin was supposed to bring about a 'conformational response' in the enzyme protein by binding to a site on the enzyme, other than the active site. This mechanism of inhibition could be likened to an allosteric effect. The experimental observations leading to this hypothesis have been reviewed in detail [79–81]. In brief, they showed that inhibition of the *Streptomyces* R39 enzyme by penicillin is not truly competitive as would be expected for an active-site inhibitor. Secondly, physicochemical studies of the binding of benzylpenicillin to the *Streptomyces* R61 enzyme revealed that the addition of neither the acceptor or donor peptides could successfully compete with the binding process. In addition, penicillin could still bind to enzyme that had been catalytically inactivated by treatment with guanidinium hydrochloride. Earlier studies had shown that inhibition of the R61 enzyme by benzylpenicillin was readily reversible by either dialysis or by treatment with β-lactamase; neither were sulphydryl groups implicated in either the catalytic or inhibitory process. This postulated mechanism would also account for those situations where an enzyme, particularly that from *Streptomyces albus* G was resistant to the inhibitory action of penicillins but which was subject to inhibition by certain substrate analogues.

Support for this hypothesis came from studies on the D-alanine carboxypeptidase of *B. stearothermophilus* [11]. However, the same enzyme from a second strain of *B. stearothermophilus*, in this case solubilized and purified to homogeneity rather than studied as the membrane-bound activity, as in the example described above, showed no evidence for independent catalytic and penicillin-binding sites [322].

Blumberg and Strominger [18], in their comprehensive review of penicillin-binding proteins and penicillin-sensitive enzymes of bacteria, have suggested alternative explanations for each of the experimental observations cited as evidence in support of the 'conformational response' or 'allosteric site' hypothesis. Some of the detailed evidence on the interaction of both penicillins and cephalosporins with specific enzymes is considered in the following sections dealing with the inhibition by these antibiotics of both transpeptidases and D-alanine carboxypeptidases.

9.7.2 *Penicillin-binding components*

Among the earlier studies of the effects of penicillin on bacteria, was the observation that sensitive organisms bound small amounts of the antibiotic (reviewed by Cooper [50]). This irreversible fixation occurred with both growing and non-growing cells and appeared unaffected by the temperature of incubation. Such an irreversible fixation of the antibiotic would agree with the hypothesis of Tipper and Strominger [290] that inhibition of transpeptidation resulted from penicilloylation of the enzyme and hence a covalent binding of penicillin to the membrane-bound protein.

This led to a considerable experimental effort being put into a study of the penicillin binding components of bacteria in the hope that this approach would lead to the isolation and characterization of the transpeptidase and in particular to demonstrate the mechanism of penicillin inhibition. The earlier literature covering these studies to 1974 has been comprehensively reviewed by Blumberg and Strominger [18].

Binding to whole organisms

Much of the earlier work was carried out using *S. aureus* (*Micrococcus pyogenes*) as the penicillin-sensitive organism. In brief, it established that the binding of radioactive penicillin was relatively specific in that saturation of the binding sites occurred using levels of the antibiotic, similar to those required to inhibit growth. Subsequent treatment of the bacteria with a large excess of non-radioactive penicillin did not displace the bound radioactive antibiotic, nor could it be removed by washing in a variety of buffers or by treatment with β-lactamase. However, treatment with alkali did remove the radioactivity as penicilloic acid [241]. The amount of penicillin bound to *S. aureus* appeared remarkably constant according to a number of investigators, ranging from 4.0–6.0 nmol/g dry weight [50, 235, 276], whereas the calculations of the actual number of molecules of antibiotic bound by a single cell shows no such similarities, ranging from 170 [235] to 4000 [276] depending on the method used for estimating the number of individual organisms present. *S. aureus*, growing as it does in clumps, presents a particular problem in this context.

The suggestion that the bound penicillin was important for the inhibitory action of the antibiotic, stemmed from the apparent correlation between the sensitivity of the organism and the amount of fixed antibiotic. One major drawback was the observation implied from the earliest studies and examined in more detail by Eagle [61] that cells pretreated with penicillin and placed in fresh growth medium continued to grow. This difficulty was overcome by the supposition that the organisms rapidly synthesized additional sites for peptidoglycan synthesis, while those previously present remained blocked by the fixed antibiotic [50]. This point was specifically investigated by Rogers [235], who studied the synthesis of peptidoglycan by non-growing cultures of *S. aureus*. Suspensions of the bacteria were pretreated with benzylpenicillin, washed to remove the excess antibiotic and then transferred to a wall amino acids–glucose medium containing chloramphenicol. Under these conditions, the synthesis of peptidoglycan resumed after a short lag period at a rate identical to that found in the control preparation, which showed no such lag. The possibility that such synthesis, which almost certainly involves cross-link formation in view of the findings of Mirelman and Sharon [177], resulted from the synthesis of new enzyme molecules, was precluded by the inclusion of chloramphenicol as an inhibitor of protein synthesis, in the incubation medium. Under these conditions, more than 90% of the fixed benzylpenicillin remained bound to the bacteria, suggesting that this amount at least was unimportant for the inhibition of peptidoglycan synthesis. Moreover, continued incubation of the penicillin-treated organisms resulted in the appearance of a small number of additional penicillin-binding sites which were 'exceedingly sensitive' to the addition of further small concentrations of penicillin. In fact, subsequent inhibition of peptidoglycan synthesis was directly related to the degree of saturation of these newly-appeared sites. One possible explanation for these observations is that the inhibition of transpeptidation is readily reversible by simple washing procedures, a conclusion apparently in conflict with the 'substrate analogue' hypothesis [290]. On the other hand, it should not be forgotten that these incubations took place under conditions designed to prevent protein synthesis and continued growth of the penicillin-treated organisms. To draw conclusions one must therefore assume that peptidoglycan synthesized under such conditions is strictly comparable with that synthesized during normal growth. Chemical analysis of peptidoglycan from *B. subtilis* [111] and *B. licheniformis* [297] have shown that the overall composition remains the same, although the wall formed is thickened and irregular in section. The findings of Rogers have been confirmed by Tynecka and Ward [297], using an autolysin and penicillinase-deficient mutant of *B. licheniformis*. In these experiments the reversal of the inhibition of transpeptidation was directly measured as the synthesis of cross-linked dimers. Further consideration of these results will be left for inclusion in the following section.

The question 'is the bound penicillin responsible for its biological activity' was investigated by Edwards and Park [62]. They studied the binding of benzylpenicillin together with a series of penicillins and cephalosporins to *S. aureus* and related the results obtained to the biological activity of the antibiotic measured as the minimum growth inhibitory concentration (MGIC). In all cases (Table 9.7) the values obtained in the binding experiments, either by direct competition between radioactive benzylpenicillin and non-radioactive analogue or by determining the concentration of the analogue required for prebinding which would prevent the

Table 9.7 Correlation of the MGIC values of various penicillins and cephalosporins with their rate of binding to *S. aureus* H (from [207]).

Antibiotic	*Effective concentration* (μg/ml)		
	*By growth inhibition**	*By pre-binding†*	*By direct competition‡*
Benzylpenicillin	0.06	0.05	0.07
Cephaloridine	0.05	0.06	0.05
Ampicillin	0.05	0.12	—
Cephalothin	0.1	0.03	0.06
Oxacillin	0.2	0.4	0.9
Cloxacillin	0.4	0.4	0.7
Methicillin	1	0.8	2.0
6-Aminopenicillanic acid	20	12	14
Cephalosporin C	30	6	25
7-Aminocephalosporanic acid	150	40	90

*Minimum concentration that inhibits growth of *S. aureus* H
†Concentration at which the effectiveness of the antibiotic in preventing subsequent binding of ^{14}C-benzylpenicillin is equal to that of 0.05 μg/ml of benzylpenicillin.
‡Concentration that will reduce the amount of radioactivity specifically bound to *S. aureus* by 50% when mixed with 0.06 μg/ml of ^{14}C-benzylpenicillin.

subsequent binding of radioactive benzylpenicillin, gave similar results. These values agreed closely with the MGIC and with theoretical values calculated on the assumption that both benzylpenicillin and the other antibiotics examined bind to the same site(s).

Penicillin-binding proteins

The development in recent years of techniques for the study of bacterial membrane proteins, particularly their separation by sodium dodecylsulphate-polyacrylamide slab gel electrophoresis (SDS–PAGE) has stimulated the investigation of penicillin-binding proteins (PBPs) in an increasing number of bacteria. The basic premise underlying these studies is that penicillin-binding proteins are equivalent to penicillin-sensitive enzymes. In each of the organisms examined fractionation of the membrane proteins to which radioactive penicillin has been bound has revealed the presence of multiple penicillin-binding proteins (PBPs). Thus as shown in Fig. 9.15 membranes of *E. coli*, *Enterobacter cloacae*, *Klebsiella aerogenes*, *Proteus rettgeri* and *Pseudomonas aeruginosa* contain from 7 to 9 PBPs [55, 195a, 196, 199, 258, 262]; *S. faecalis*, 6 [52]; *S. typhimurium*, 5 [249]; *B. subtilis* from 5 to 8 [16, 134]; *B. stearothermophilus*, *B. licheniformis*, *B. megaterium* and *B. cereus* from 3 to 5 [18, 38, 39]; *S. aureus* H, 4 [136] and multiple components are also present in the several strains of *Streptomyces* examined (J. M. Ghuysen, personal communication).

The proteins observed are those which retain the bound antibiotic in the presence of an excess of non-radioactive penicillin and during the denaturing conditions of detergent-treatment in preparation for electrophoresis. Clearly such procedures will tend to exclude proteins which, although reacting with penicillin, do not bind the antibiotic covalently or, do so only transiently. Enzymes of this type are the D,D-carboxypeptidase 1B of *E. coli* [286] and the 46 000 molecular weight carboxypeptidase-transpeptidase of *S. aureus* [136]. However, in each of the organisms studied, one or more of the PBPs have properties consistent with their being the 'killing site'. Thus, a requirement for interaction of the antibiotic with other unidentified sites does not seem to be necessary.

Initially two complementary approaches were adopted but, with continued investigation and in the light of the significant advances made, these have now more or less combined. However, particularly in *E. coli*, initial studies were designed to establish the relationship of the PBPs to the various morphological changes induced in this organism by treatment with β-lactam antibiotics. It had been known for several years that the presence of low concentrations of benzylpenicillin induced filament formation in cultures of *E. coli* [58, 74]. More recently, a variety of other morphological responses including for example cell lysis, bulge formation and growth of osmotically-stable ovoid cells were observed on treatment with other penicillins and cephalosporins [258].

The alternative approach adopted, particularly in the studies with *Bacilli*, was the purification and characterization of the individual PBPs. However, as stated above, the distinction between these approaches has now become blurred and in the follow-

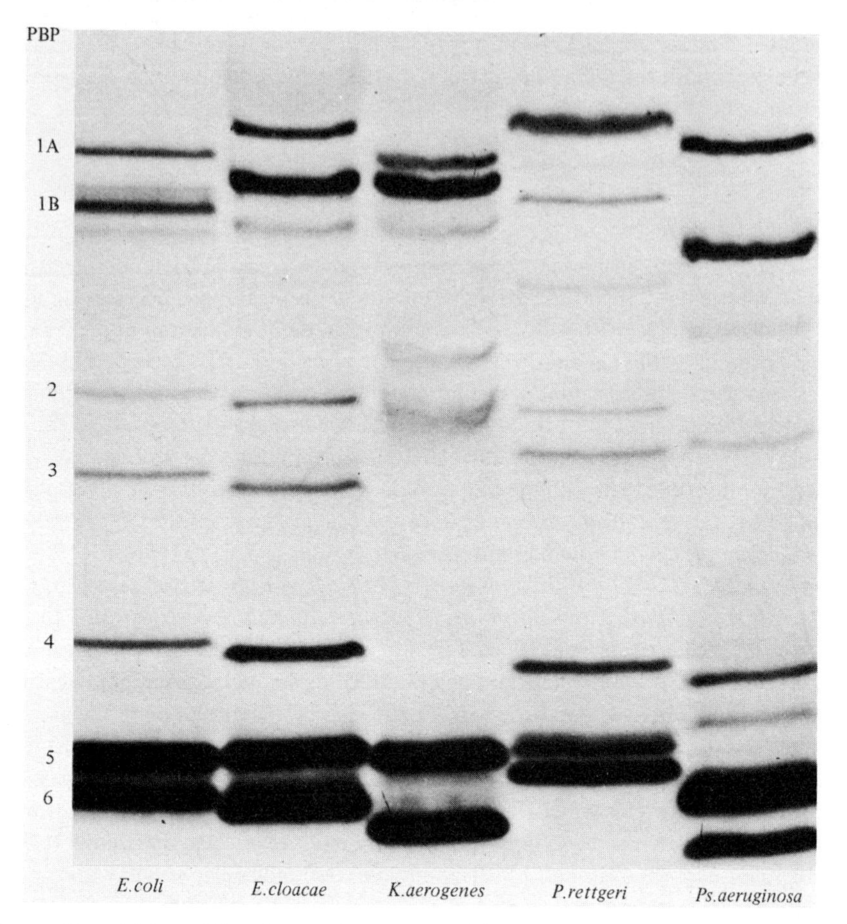

ing sections PBPs will be discussed in terms of the organism, or group of organisms in which they occur.

(*i*) *E. coli*

Investigation of the PBPs of *E. coli* (Table 9.8) has largely been the work of Spratt and his colleagues and has recently been reviewed by him [258, 260-262]. In these studies they have used the dual approach of either examining the interaction of various penicillins and cephalosporins with PBPs or selecting mutants having growth characteristics similar to those described above, i.e. filament formation or growth as ovoid cells, and examining such mutants for concomitant changes in their PBPs.

The cytoplasmic membrane of *E. coli* K_{12}, solubilized with sarkosyl (sodium lauroyl sarcosinate, 1% w/v) has consistently been shown to contain a minimum of six PBPs. More recently a modification of the SDS-PAGE system previously used, separated PBP-1 into two or possibly more components [263] and *E. coli* B/r appears to have additional PBPs [260]. The treatment with sarkosyl served not only to prevent loss of the bound antibiotic by denaturing the binding proteins but also

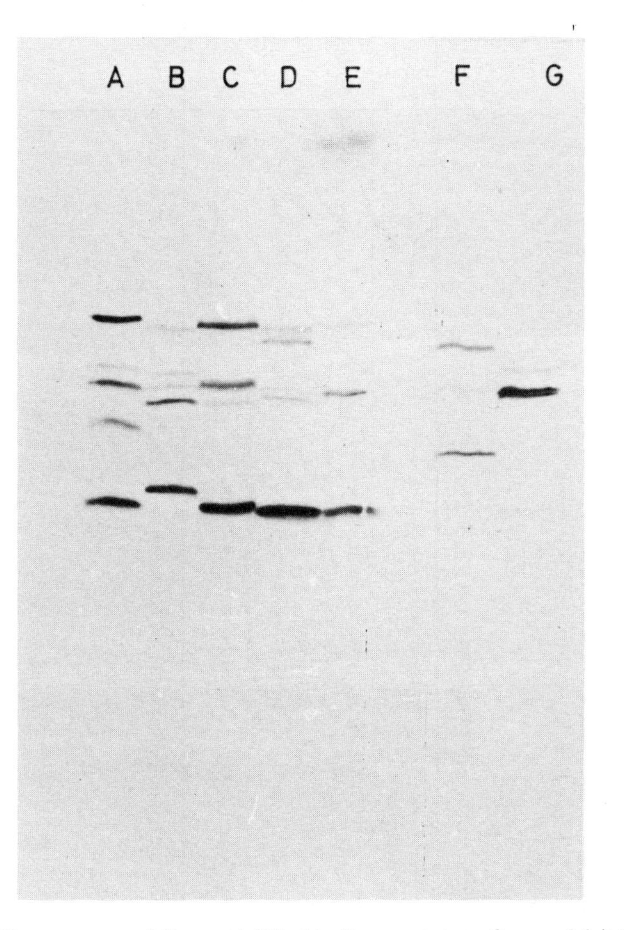

Figure 9.15 Fluorograms of the penicillin-binding proteins of several (a) Gram-negative bacteria; PBPs of *E. coli* have been numbered as described in the text and (b) Gram-positive bacteria, showing A, *B. megaterium*; B, *B. subtilis*; C, *B. licheniformis* 94; D, *B. stearothermophilus*; E, *B. cereus*; F, *M. luteus*; G, *Staph. aureus.* Kindly provided by N. A. C. Curtis (a) and H. Chase (b).

selectively solubilize the cytoplasmic membrane [66, 258]. Binding of radioactive β-lactams to either cytoplasmic proteins or to those of the outer membrane has not been observed.

Initially studies of the binding of various β-lactams with individual PBPs was dependent on their competition with ^{14}C-benzylpenicillin. More recently other radioactive antibiotics have become available allowing a more direct examination in these cases. Additional PBPs have been found particularly after treatment of membrane preparations with ^{125}I-ampicillin in which the amino group of the β-lactam has been iodinated (J. T. Park, personal communication). However the significance of these observations remains unclear. Other radioactive penicillins and cephalosporins appear to bind either to all or only to some of those proteins which bind benzylpenicillin.

Table 9.8 Penicillin-binding proteins of *E. coli* K12 (data from [55a, 181, 258, 260, 262, 289]).*

PBP	*Symbol*	*Map site (min)*	*Molecular weight*	*Number of molecules/cell*	*Proposed or known function*	*Antibiotics showing marked affinity*
1A	*ponA*	73.5	91	230	Peripheral wall extension	Cephaloridine/Cefsulodin
1B	*ponB/mrc*	3.3	86.5–81.5		Transpeptidase/Transglycosylase	
2	*rodA*	14.4	66	20	Rod shape	Mecillinam
3	*fts1*	1.8	60	50	Septum formation	Cephalexin
4	*dacB*	68.0	49	110	Endopeptidase/D,D-carboxypeptidase 1B/1C	
5	*dacA*	13.7	42	1800	D,D-carboxypeptidase 1A	Cefoxitin
6			40	570	?	

*Molecular weights given are based on relative mobilities on SDS–polyacrylamide gel electrophoresis. Map position is the locus of mutants affecting the amount of the particular PBP.

Correlations between PBPs and penicillin-sensitive enzymes have been made: PBP 4, 5 and 6 are all equivalent to D,D-carboxypeptidases and recently PBP 1B has been shown to have both transglycosylase and transpeptidase activity [115, 166, 167, 181, 262].

The carboxypeptidases have been purified to electrophoretic homogeneity by conventional techniques after extraction from the cytoplasmic membrane of *E. coli* [286]. Thus, D,D-carboxypeptidase 1B was solubilized by treatment with 0.5 M LiCl and carboxypeptidase 1A by subsequent extraction of the membranes with Triton-X100 (2%). The enzymic activities associated with these proteins and a soluble form (carboxypeptidase 1C), which appears identical with the membrane-bound carboxypeptidase 1B, are discussed in detail below. On SDS-PAGE purified carboxypeptidase 1A ran as a doublet of molecular weights 40 000 and 42 000 (originally reported as 32 000 and 34 000 [282]) which corresponds exactly to PBP 5 and 6 [266]. Initially, it was unclear whether the two polypeptides differed in their enzymic activities or represented modifications of the same polypeptide. They could, however, be distinguished by the rate of release of bound benzylpenicillin. The half-lives of the enzyme-penicillin complex at 30° C being 5 min and 19 min for PBP 5 and 6 respectively. More recently mutants have been found lacking one or other of the two proteins and the fact that they are different polypeptides coded for by separate genes has been established [167].

Penicillin-binding protein 4 has been shown to correspond to D,D-carboxypeptidase 1B. Mutants (*dac*B, mapping near *arg*G at 68 min) isolated independently by two groups [115, 166] lack carboxypeptidase 1B, the soluble form 1C and PBP 4. A second mutant (*dac*A mapping near *leu*5 13.7 min) lacks carboxypeptidase 1A activity although, surprisingly, it retains both PBP 5 and 6 [167]. On further investigation, evidence was obtained that the mutation results in a defect in the release of bound penicillin from PBP 5 whereas the release of antibiotic from PBP 6 is unchanged [167]. This defect appears analogous to the inhibition by sulphydryl reagents of the deacylation reaction catalysed by D,D-carboxypeptidase 1A [56] (p. 279, Chapter 8.7). Since the release of antibiotic from PBP 6 was not accompanied by restoration of enzyme activity it appears that PBP 5 is the major D,D-carboxypeptidase 1A of wild-type *E. coli*; whether PBP 6 is in fact a penicillin-sensitive enzyme is not yet known. A double mutant *dac*A *dac*B constructed to contain the two lesions has less than 10% of the D,D-carboxypeptidase activity of the wild-type [278].

The only physiological response observed in the absence of these enzymes was in the *dac*B mutants, where a reduced rate of lysis in the presence of ampicillin was found [115]. Thus, the role of this enzyme *in vivo*, may be as an endopeptidase rather than a carboxypeptidase. As discussed earlier (Chapter 8.7), D,D-carboxypeptidases have been implicated in the control of transpeptidation in *E. coli*. Although the evidence from the mutants described above does not appear to support this conclusion, it remains possible that the residual carboxypeptidase activity present in these organisms is sufficient to maintain normal growth.

Treatment of *E. coli* with penicillins and cephalosporins leads to characteristic morphological responses ranging from the inhibition of septation where the organism continues to grow, but as non-septate filaments, to cell lysis, a response that has been ascribed to inhibition of cell elongation. Filamentous growth occurs when *E. coli* is treated with low concentrations of many β-lactam antibiotics although in the majority of cases the use of slightly higher concentrations results in cell lysis. However, there are exceptions such as cephalexin which inhibits division over a wide range of concentrations. As might be expected, the majority of β-lactam antibiotics fall between these two extremes in the morphological responses they induce.

At the lowest concentration inhibiting division cephalexin binds preferentially to PBP 3. It does not bind to PBP 2, 5 and 6 even at much higher concentrations. The relationship between cell division and PBPs has been established by the independent isolation of mutants (designated the *fts* phenotype) which are resistant to low levels of cephalexin and thermo-sensitive in division [257, 278]. At the restrictive temperature, the inhibition of division characterized as the onset of filamentous growth, is accompanied by a loss of the ability to bind benzylpenicillin to PBP 3. In both cases, it is thought that the mutation which is co-transducible with *leu* and maps at 1.8 min, is in the structural gene for PBP 3. In view of its involvement in cell division, there are several possibilities for the role of this protein. Thus, it may itself be a transpeptidase directly responsible for initiating cross-wall synthesis, presumably coupled in some way to the termination of a round of DNA replication. Alternatively, it may act to alter the direction of insertion of new peptidoglycan chains such that the formation of a septum occurs rather than the continued synthesis of peripheral wall. The former suggestion is supported by the recent report of Ishimo & Matsuhashi [112a] that membrane preparations from a strain of *E. coli* lacking PBP 1Bs but with a ten-fold over-production of PBP 3, catalyse cross-linkage which is particularly sensitive to both cephalexin and benzylpenicillin. Cross-linkage catalysed by PBP 1Bs alone is relatively resistant to cephalexin.

PBP 2 specifically binds the 6-amidinopenicillanic acid, mecillinam (Fig. 9.14a) and through the use of this antibiotic as a probe, has been shown to be involved in the maintenance of rod morphology in *E. coli*. Thus, treatment with mecillinam produces large osmotically stable ovoid cells. Only with relatively high concentrations of antibiotic and prolonged incubation is lysis observed [119, 154, 171]. At this time, the only other antibiotic found to induce a similar response was 6-aminopenicillanic acid [265]. More recently, a second amidinopenicillanic acid, thienamycin and clavulanic acid have all been shown to bind with highest affinity to PBP 2 at concentrations similar to, or slightly higher than, their minimal morphological change concentration (the MMCC) [259, 264]. Under these conditions incubation of *E. coli* with thienamycin produced large osmotically stable round cells, whereas the use of higher concentrations resulted in rapid lysis. In contrast to mecillinam, thienamycin and clavulanic acid were bound to all other PBPs but with only low affinity to PBP 3. However, it is interesting in terms of the structure-function relationships of β-lactam antibiotics to note as pointed out by Spratt and his colleagues [264] that all antibiotics known to bind preferentially to PBP 2 are either unsubstituted or have unusual substitution of the β-lactam ring.

Again the relationship between PBP 2, the antimicrobial activity of mecillinam and the control of cell shape has been established by the isolation of two classes of mutants. These were obtained by selection either for resistance to mecillinam [114, 257] or as temperature-sensitive mutants which grew normally as rods at 30° C but which were converted to osmotically-stable round cells at 42° C [260]. The former mutants, which also had round morphology, failed to bind either mecillinam or benzylpenicillin to PBP 2 as did the latter class (designated *rod*A) at the restrictive temperature. The majority of these mutants map at 14 min [101, 104, 168, 260] although others have been reported to map near *envA* at 70 min [114].

The nature of the activity catalysed by PBP 2 remains unknown. Initial investigations, made soon after mecillinam became available, had found that lethal concentrations did not inhibit cross-linking or the other penicillin-sensitive enzymes of *E. coli* [164, 206]. Since it appeared that the lethal target would be some process involved in the synthesis of Gram-negative rather than Gram-positive wall, Park and Burman [206] suggested this was the enzyme linking lipoprotein [26] to the peptidoglycan, but Braun and Wolff [28] disproved this suggestion. In fact growth in the presence of mecillinam increased rather than decreased the proportion of attached lipoprotein. Moreover, peptidoglycan synthesis in living cells was inhibited by 50% at antibiotic concentrations (1 μg/ml) close to the MMCC [28]. This finding suggests that PBP 2 is involved directly in peptidoglycan synthesis and may be a specific transpeptidase required for the synthesis of peripheral wall.

Thus mutants of *E. coli* K_{12} have been isolated that lack one or more of the PBPs 2–6 yet continue to grow, although in the cases of mutants in PBP 2 and 3 growth is morphologically disturbed. These results suggest that none of these PBPs is the 'killing target' for benzylpenicillin and thus implicate the PBP 1 complex. Perhaps more compelling evidence for this conclusion is the finding that treatment of *E. coli* with the minimum inhibitory concentration of cephaloridine and cefsulodin alone among the β-lactams examined, resulted in lysis of the organisms without the earlier appearance of filamentous growth or other changes in morphology [55a, 257]. Subsequently cephaloridine and cefsulodin were shown to have strongest affinity for PBP 1, all of which led [257] to the conclusion that this protein was the transpeptidase involved in cell elongation.

Mutants in PBP 1 had not then been described. More recently Spratt *et al.* [263] have demonstrated that PBP 1 is in fact a complex of two and possibly more binding proteins and have simultaneously described a mutant lacking PBP 1A. This organism was obtained as a rod-shaped revertant from a spherical mutant lacking both PBP 1A and 2. The authors were unable to establish any relationship which would account for the loss of the two binding proteins. Similar mutants have been studied by Hirota and his colleagues [278]. The absence of PBP 1A did not appear to have any effect on growth of *E. coli*. The mutation (*ponA*: *P*enicillin-binding protein *one A*) has been mapped at 73.5 min.

A temperature-sensitive mutant in PBP 1B, hypersensitive to cephalosporins and which lyses at the restrictive temperature, has been studied by Matsuhashi and his colleagues [285]. The mutant also contained a second lesion in PBP 4 but this was unrelated to thermo-sensitivity and was absent in thermo-resistant revertants. Similar ts-mutants lacking PBP 1B (*ponB*, mapping at 3.3 min) were isolated by Suzuki *et al.* [278] but in this case the mutation in PBP 1B was separated from that causing thermo-sensitive growth, and growth in the absence of PBP 1B appeared normal. The fact that these mutants were also hypersensitive to cephalosporins was attributable to the high affinity of PBP 1A for these antibiotics. The authors concluded that *in vivo* peptidoglycan biosynthesis necessary for cell elongation, requires the gene product of either *ponA* or *ponB*. In the absence of PBP 1B cell elongation is maintained by PBP 1A which is exquisitely sensitive to cephalosporins and hence growth shows similar characteristics. This hypothesis suggests that double mutants for both PBP 1A and 1B would be non-viable and this has proved to be the case [278].

Peptidoglycan biosynthesis by membrane preparations from mutants lacking either PBP 1A or 1B has been examined. In both investigations, mutants defective in PBP 1B were found to be capable of synthesizing lipid intermediates but not peptidoglycan whereas only a slight decrease in overall synthesis was observed in the absence of PBP 1A [278, 285]. Such a result would not be expected if the mutation was only in a transpeptidase and many examples are known (discussed in the previous chapter) where membrane preparations catalyse the synthesis of polymeric but un-cross-linked peptidoglycan. One possible explanation is that the transpeptidase and polymerase are in some way coupled and that the mutation in the PBP-1B affects both. Evidence in agreement with this supposition has come from the recent findings that purified PBP 1B will catalyse the synthesis of peptidoglycan from the lipid intermediate, undecaprenyl-P-P-disaccharide-pentapeptide [181, 278a]. Lysozyme-treatment of the newly synthesized peptidoglycan revealed a low degree of cross-linkage ([278a]; M. Matsuhashi, personal communication). This was markedly increased giving a dimer to monomer ratio of approximately 0.55 when purified PBP 1B was incubated with a particulate preparation from a mutant lacking PBP 1B and both D,D-carboxypeptidases (i.e. *ponB*$^-$, *dacA*$^-$, *dacB*$^-$) [278a]. As expected penicillin inhibited cross-linkage and the release of D-alanine while the polymerization reaction was inhibited by moenomycin. It therefore appears that PBP 1B catalyses both peptidoglycan transglycosylase and transpeptidation, the relationship of those two activities clearly deserves additional investigation.

Thermo-resistant revertants isolated from the mutant by Tamaki *et al.* [285] simultaneously regained the ability to synthesize cross-linked peptidoglycan *in vitro*, penicillin-binding to PBP 1B, increased resistance to cephalosporins and resistance to lysis, observations which suggest that all these characters result from a single mutation. However, it is also possible that thermo-sensitivity does not reside in the same but in a closely-linked mutation such as found in the mutants isolated by Suzuki *et al.* [278a]. Other revertants were found with partial recovery of normal β-lactam sensitivity but which did not synthesize peptidoglycan *in vitro* nor appear to contain PBP 1B. They did, however, contain increased amounts of PBP 1A, 2 and occasionally 4 suggesting that under certain conditions, these PBPs may act to compensate for a lack of activity in PBP 1B [185].

(ii) S. typhimurium

The profile of PBPs in *S. typhimurium* resembles that of *E. coli* K_{12} with the exception that the PBP of lowest molecular weight (PBP 5) whilst still the major penicillin-binding component, is not a doublet (PBP 5 and 6 of *E. coli*) [249]; whether the PBP 1 observed is a single polypeptide or a complex as in *E. coli* remains unknown.

PBP 1, 4 and 5 have been isolated from membrane preparations by treatment with the non-ionic detergent Genapol X-100 and lithium chloride. The extract prepared in this way catalysed D,D-carboxypeptidase, natural model transpeptidase and endopeptidase activities utilising UDP-*N*-acetylmuramyl-pentapeptide and the pentapeptide L-Ala-D-isoGlu-m-A_2pm-D-Ala-D-Ala as substrates. Subsequent purification of the individual proteins by covalent affinity chromatography revealed PBP 4 and 5 and probably also PBP 1, to catalyse carboxypeptidase and transpeptidase activities with varying efficiency. Examination of the binding capacity of membrane preparations showed PBP 4 to have the highest affinity for benzylpenicillin and it was suggested that in *S. typhimurium*, this protein rather than PBP 1 may be the 'killing target'. However, in this context it should be noted that in *E. coli* cefoxitin has the highest affinity for PBPs 4, 5 and 6, but evidence described in the previous section argues strongly against inhibition of these proteins being lethal [55a]. PBP 4 and 5 of *S. typhimurium* release bound benzylpenicillin with half-lives of 90 min and 5 min at 37° C. The nature of the products released has not been reported. Whether the other PBPs have roles similar to those described for PBP 2 and 3 in *E. coli* remains unknown. It would be of particular interest if they too were shown to have specific functions in the maintenance of cell shape and division.

(iii) Bacilli

The initial studies of PBPs of *B. subtilis* and *B. cereus* were made on membrane preparations pre-treated with radioactive benzylpenicillin, which were then solubilized by detergent and fractionated by ion-exchange chromatography and isoelectric focusing on polyacrylamide gels [276]. Although these studies revealed multiple penicillin-binding components, their further interpretation was complicated by the inability of the authors to eliminate different states of aggregation or of protein charge as being responsible for multiple bands of the same binding protein. The situation was resolved by separation of the binding components from the solubilized membrane by covalent affinity chromatography on columns of 6-amino-penicillanic acid–Sepharose, followed by SDS–PAGE [14, 17]. Using these conditions, *B. subtilis* was shown to contain five distinct PBPs and *B. cereus* three. More recently the number of different bacilli examined has been extended with the demonstration that *B. megaterium* also contains five PBPs; *B. licheniformis* four or possibly five and *B. stearothermophilus* three [38, 39, 322].

In *B. subtilis* pre-treatment of the membranes with cephalothin prevented binding of radioactive benzylpenicillin to PBP 1, 2 and 4. PBP 5, again the major binding protein representing at least 60% of the radioactivity bound to membranes, was identified as the D,D-carboxypeptidase. This enzyme was known to be cephalothin-resistant [16, 17]. Subsequently this protein and the analogous ones from *B. stearothermophilus* and *B. megaterium* have been purified from cephalothin-treated membranes by covalent affinity chromatography on penicillin substituted Sepharose. As described below, the interaction of these proteins with penicillins has been intensively studied. The recovery of PBP 3 of *B. subtilis* was poor and this protein has not been studied further. However the lack of cephalothin-binding to these proteins, effectively eliminates PBP 3 and 5 as the 'killing target' in *B. subtilis*. The selective binding of cephalosporins to PBP 1, 2 and 4 has been used by Kleppe and Strominger [134] to investigate these proteins further. Using cephalosporin-affinity chromatography they have isolated a mixture of PBP 1AB, 2B and 4 uncontaminated with PBP 5 the D,D-carboxypeptidase. PBP 2B which has a low affinity for penicillins and a high affinity for cephalosporins had not been detected in earlier studies. The authors were however, unable to demonstrate any enzyme activity in these proteins other than their interaction with β-lactam antibiotics.

The question of which if any of PBP 1, 2 and 4 can be recognized as the lethal site has been investigated by Buchanan and Strominger [34, 35], in a series of mutants of *B. subtilis* showing stepwise increased resistance to cloxacillin. They found among the most resistant mutants (those showing a 180-fold increase in the LD_{50} for the antibiotic) an organism showing normal growth and viability but which lacked PBP 1. Only with PBP 2 was decreased affinity to cloxacillin associated with increased resistance of the mutants, suggesting that this protein is the probable target for killing by penicillins. Surprisingly both the mutants and the individual binding proteins failed to show a corresponding resistance to benzylpenicillin. Thus, it appears that penicillin-sensitive proteins can be modified in their sensitivity to one β-lactam antibiotic, but not others, even though the antibiotics are closely related.

These results contrast with those of Reynolds and his colleagues who have shown that in *B. megaterium* and *B. licheniformis* benzylpenicillin at the minimum inhibitory growth concentration, effectively binds only to PBP 1 ([38, 228], and unpublished results). Additional evidence that this protein is the killing target in these bacilli comes from the finding that characteristics of the inhibition by benzylpenicillin and of the breakdown of the protein-penicillin complex are both consistent with PBP 1 being the major transpeptidase. These proteins have been purified to greater than 95% protein homogeneity. In *B. megaterium* this was done by detergent extraction of membranes followed by ion-exchange chromatography of the extract and covalent affinity chromatography of appropriate fractions on ampicillin-affinose. In *B. licheniformis* advantage was taken of the finding that only the penicillin complex with PBP 1 undergoes degradation on subsequent incubation (Fig. 9.16). Membranes of *B. licheniformis* were treated with saturating concentrations of benzylpenicillin and after removal of excess antibiotic were incubated to allow complex breakdown to occur. After solubilization of the membranes PBP 1 alone was bound onto a column of ampicillin–affinose and eluted in pure form with neutral hydroxylamine. Purified protein 1 bound 0.7 equivalents of benzylpenicillin and the complex had a half-life of 16 min at 35° C. This value is identical with that measured for

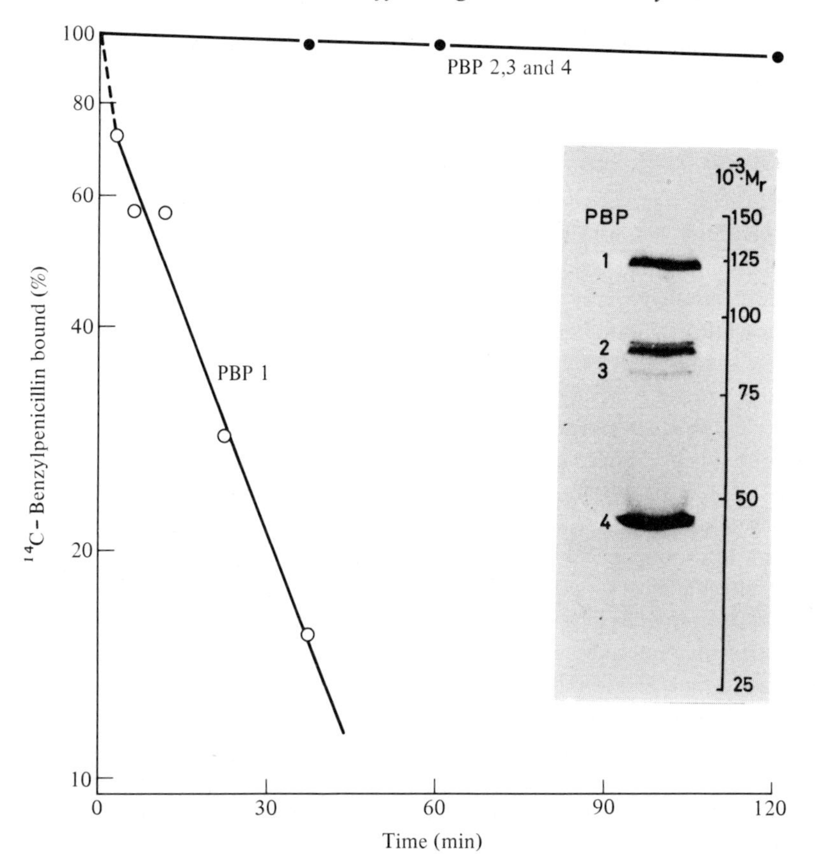

Figure 9.16 The breakdown of benzylpenicillin–protein complexes in membranes of *Bacillus licheniformis*. Membranes were incubated with radioactive benzylpenicillin and then treated with penicillinase. During subsequent incubation at 37° C samples were taken, the PBPs separated by SDS–polyacrylamide gel electrophoresis and located by fluorography. The proportion of bound penicillin was then determined by microdensitometry of the fluorogram. (From [38].)

the release of antibiotic from membrane-bound PBP 1. However, it has not yet proved possible to demonstrate any enzymic activity associated with either of the purified PBP 1s. In both *B. megaterium* and *B. licheniformis* only the PBPs of lowest molecular weight have been shown to have activity as D,D-carboxypeptidases.

Giles and Reynolds [82] have also investigated the PBPs of a series of cloxacillin-resistant mutants from *B. megaterium* with different results from those of Buchanan and Strominger. In *B. megaterium* acquisition of resistance was accompanied by an increase in the ratio of PBP 3 to PBP 1, a decrease in the growth rate and filamentation of the organism. A partial revertant selected on the basis of a faster growth rate showed a decrease in the ratio of the PBPs and increased septation suggesting that all three changes were the consequence of a single mutation. In these mutants development of resistance to the β-lactam appears to result from a decrease in the lethal target, this loss being in some way compensated by an increase in a less-sensitive binding protein. Whether the mutants resistant to cloxacillin also acquired resistance to other β-lactams remains unknown.

(iv) S. aureus H.

S. aureus was originally reported to contain two PBPs although modification of the techniques used has now allowed Kozarich and Strominger [136] to recognize four. These have molecular weights of 115 000, a doublet of 100 000 and the 46 000 protein mentioned briefly in the introduction to this section. Penicillin-binding protein 4 releases bound benzylpenicillin extremely rapidly and this is almost certainly the reason why it was originally undetected. The rapid breakdown of the antibiotic-protein complex has allowed PBP 4 to be purified by a method similar to that described above for PBP 1 of *B. licheniformis*. The purified protein catalyses D,D-carboxypeptidase, an unnatural model transpeptidation reaction using diacetyl-L-Lys-D-Ala-D-Ala as the donor substrate, and penicillinase activities. The preferred acceptor in the transpeptidase is glycine which at 1 mM completely inhibits concomitant hydrolysis of the donor peptide.

The physiological function of PBP 4 remains unknown. D,D-carboxypeptidase or penicillinase activities have not previously been detected in membranes of *S. aureus H*. One possible role suggested by Kozarich and Strominger is that of a penicillin-resistant transpeptidase. The occurrence of such an enzyme was deduced from studies of peptidoglycan synthesis, using wall-membrane preparations in which inhibition by benzylpenicillin of the addition of newly-synthesized material to the pre-existing wall, appeared biphasic. Thus, approximately 60% of the addition i.e. the transpeptidase activity, was inhibited by 0.5 μg/ml of penicillin whereas inhibition of the remaining 40% required much higher concentrations. In the case of PBP 4 penicillin-resistance would not reside in any intrinsic resistance of the protein itself but rather result from the rapid breakdown of the antibiotic-protein complex with the accompanying degradation of the penicillin molecule. Moreover, even if this postulated role were correct, it does not point to PBP 4 being the lethal target for penicillin. This presumably resides in one of the higher molecular weight PBPs and examination of cloxacillin-resistance mutants has provided evidence in support of this conclusion [135].

9.7.3 *β-Lactam antibiotics and sensitive enzymes*

Simultaneously with, and as a corollary to the investigations of the penicillin-binding components, other studies have been made on the individual penicillin-sensitive enzyme activities associated with peptidoglycan biosynthesis. Three basic enzymes have been described which are subject to penicillin-inhibition. These are transpeptidases, D-alanine carboxypeptidases and endopeptidases, that is, enzymes which will cleave the peptides of cross-linked dimers [20]. Unfortunately, the situation is not as clear as this statement implies, since as described above a number of examples exist where a purified enzyme has been shown to catalyse more than one of the above activities. In the following sections various examples are described, with particular reference to their interaction with this group of antibiotics. For further information the reader is directed towards the preceding chapter where these enzymes are discussed with reference to peptidoglycan synthesis; to the recent reviews of Ghuysen [77, 78] and Tipper [289] and to the sections of this Chapter dealing with penicillin-binding proteins (p. 339) and the degradation of benzylpenicillin by penicillin-sensitive enzymes.

Transpeptidases

The first cell-free preparations catalysing transpeptidation and the formation of cross-links were isolated from the Gram-negative organisms, *E. coli* and *Salmonella newington* [8, 9, 116, 117] (see Section 9.7). In both cases, the radioactive products of peptidoglycan synthesis were characterized after degradation with lysozyme to yield monomers and cross-linked dimers. Under these circumstances the absence of cross-linked material (i.e. dimers) from peptidoglycan synthesized in the presence of penicillin was clearly shown. Moreover, attempts to reverse the inhibition of transpeptidation by washing or treatment with penicillinase were unsuccessful More recently, transpeptidation with the formation of cross-linked dimers has been described in *B. megaterium* [225, 314], *B. stearothermophilus* [149, 225] and *Sporosarcina ureae* [147]. In these cases particulate enzyme preparations were isolated to minimize contamination with the bacterial wall and may be regarded as membrane preparations. The effects of β-lactam antibiotics on the preparations isolated from *B. megaterium* [227, 315] and *B. stearothermophilus* [227] have been studied in some detail. The membrane preparations from *B. megaterium* catalysed two types of transpeptidation, both of which were inhibited by penicillins. The reactions catalysed are discussed in detail in Chapter 8.6.1. The two transpeptidases were both inhibited by penicillins; concentrations of antibiotic giving a 50% inhibition of growth also caused a 50% inhibition of transpeptidase activity. However, a similar correlation was not observed with cephalosporins where concentrations effective in the inhibition of growth, had only slight inhibitory action on the incorporation of ^{14}C-diaminopimelic acid. The inhibition of both types of transpeptidase by cloxacillin (the membrane preparations contained an active penicillinase) was reversible: that involved in the formation of cross-linked dimer almost completely, whereas the second, catalysing the incorporation of diaminopimelic acid, was only partially reversible. Similar results obtained by Reynolds and Barnett [227], using a second strain of *B. megaterium* and *B. stearothermophilus*, led these authors to conclude that the membrane preparations contained two enzymes catalysing transpeptidation reactions which could be distinguished on the basis of their sensitivities to β-lactam antibiotics.

Membrane preparations from *S. aureus* [6], *M. luteus* [6] and *B. licheniformis* [307] do not catalyse the formation of cross-linked peptidoglycan. However, in each of these organisms the use of membrane-wall preparations has shown that a penicillin-sensitive transpeptidation reaction is responsible for the incorporation of newly-synthesized peptidoglycan into the pre-existing wall [175, 177, 307]. In all cases, this incorporation was inhibited by low concentrations of penicillin, similar to those inhibiting growth. Reversibility of the inhibition was only examined in *B. licheniformis* [307]. After pre-treatment of the transpeptidase with either benzyl-penicillin or cephaloridine, treatment with β-lactamase and washing with buffer resulted in a recovery of some 50% of the initial activity. At this time it was not clear whether this represented a rather inefficient reversal or the presence of two enzyme activities, one completely reversible, and the second not.

However, recent studies of penicillin-binding proteins of *B. licheniformis* provide evidence for the former explanation [38]. The half-life of the PBP 1 complex was found to be 16 min at 35° C whereas the time allowed for the reversal of penicillin-inhibition of transpeptidation was 5 min at 24° C followed by washing at 4° C, conditions which would not allow complete breakdown of the antibiotic–enzyme complex.

Confirmation of the results obtained with membrane–wall preparations from both *B. licheniformis* and *M. luteus* has come from studies of the effects of penicillin on wall synthesis *in vivo* [176, 297] where inhibition of transpeptidation by penicillin resulted in the synthesis of uncross-linked peptidoglycan which was excreted into the medium. An extension of these investigations on the effects of penicillin *in vivo* resulted in Mirelman and Bracha [174] suggesting that *M. luteus* contains two penicillin-sensitive transpeptidases catalysing the formation of D-alanyl → L-alanine, and D-alanyl → ϵ'-NH_2-L-lysine cross-links respectively (p. 259). The two activities were distinguished by their sensitivity to penicillin, that involved in the formation of D-alanyl → L-alanine cross-links being some 50-fold more resistant to the action of the antibiotic. The sensitivity of the second activity, forming D-alanyl → L-lysine cross-links, more closely resembled the minimum inhibitory growth concentration for penicillin, suggesting that in *M. luteus* the 'killing site' of action is more likely to be this enzyme.

D-Alanine carboxypeptidases

The second major group of penicillin-sensitive enzymes are the D-alanine carboxy-peptidases. With three exceptions, *S. aureus, S. faecalis* and *M. luteus* [53, 136, 146, 200] they have only been reported in organisms containing diaminopimelic acid in the peptidoglycan and have generally been assayed by following the release of D-alanine from UDP-MurAc-pentapeptide although recently model peptide substrates (i.e. acyl-D-alanyl-D-alanine) have been widely used. As mentioned briefly above, certain of these enzymes will function as transpeptidases, particularly in model transpeptidation reactions, i.e. the incorporation of single amino acid residues rather than the formation of the natural cross-links of the parent organism. Also included in this section are the D,D-carboxypeptidase–transpeptidases of various Actinomycetes studied in detail by Ghuysen and his colleagues [79–81]. Since carboxypeptidases were until relatively recently the only penicillin-sensitive enzymes which have retained activity on purification their interaction with β-lactams has been studied in some detail.

(*i*) *E. coli*

The initial observation of D-alanine-carboxypeptidase activity was made in *E. coli* [116, 117]. The enzyme was partially purified [118] and shown to be extremely sensitive to a number of β-lactam antibiotics. Surprisingly, inhibition was also observed with certain penicillin derivatives, particularly benzylpenicilloic acid, which had no antimicrobial activity. In contrast to the transpeptidase activity from the same particulate enzyme preparations, the inhibition of D-alanine-carboxypeptidase by these antibiotics appeared freely reversible. At this time, the suggestion was made that D-alanine-carboxypeptidase might be an uncoupled transpeptidase, where water rather than an amino group was involved in enzyme activity. Thus, carboxypeptidase activity would result in hydrolytic cleavage of the D-alanyl-D-alanine bond rather than transpeptidation.

Strominger and his colleagues [273, 286] subsequently re-investigated the penicillin-sensitive enzymes of *E. coli* by a combination of detergent and LiCl-extraction of particulate enzyme preparations, followed by ion-exchange and penicillin-affinity chromatography of both these extracts and the initial cytoplasmic supernatant fraction. Initially the application of these techniques led to the separation of numerous penicillin-sensitive enzyme activities, the physiological significance of which remained unclear [273]. The complexity of the situation is demonstrated by the combination of activities observed: carboxypeptidase with no endopeptidase activity; endopeptidase with no carboxypeptidase activity and enzymes catalysing both reactions. Technical difficulties prevented the demonstration of transpeptidase activity so that it was not known if any of these enzymes would also catalyse this reaction. A similar complex situation has been reported for the D,D-carboxypeptidase–transpeptidase system of *E. coli* 44 [188, 189, 217] where enzymes having the activities described above, in addition to natural and model transpeptidase activity, were found.

More recently the purification to homogeneity of three D-alanine carboxypeptidases from *E. coli* has been described [286]. There is strong evidence however, that two of these enzymes, although they were purified separately from the membrane and cytoplasmic fractions respectively, are in fact the same protein. Thus *E. coli* contains two major D-alanine carboxypeptidases. These differ markedly in both their interaction with benzylpenicillin and other β-lactam antibiotics and also in the additional enzymic reactions that they catalyse. Enzyme 1A (now shown to be PBP 5, p. 343) isolated by detergent extraction of membranes, catalyses transpeptidation (determined as unnatural model transpeptidation), effectively binds radioactive benzylpenicillin but does not possess any endopeptidase activity. In contrast, the second enzyme, isolated from the membrane by LiCl-extraction (IB; PBP 4) and as the soluble form from the cytoplasm (1C), is an effective endopeptidase but not transpeptidase, nor does it bind benzylpenicillin. The benzylpenicillin bound to the carboxypeptidase–transpeptidase (1A) is released on continued incubation in the absence of antibiotic with a half-time for release of 5 min at 37° C and 60 min at 4° C [286]. The released radioactivity was chromatographically indistinguishable from benzylpenicilloic acid. This observation contrasts with the findings obtained with the isolated carboxypeptidases from Bacilli and *Streptomyces*, described in detail below, and suggests the purified enzymes will also act as weak penicillinases. Moreover, the second carboxypeptidase (1B and 1C) which did not bind benzylpenicillin also catalysed a similar penicillinase reation although at a much slower rate. Additional studies utilising the enzyme inhibitor *p*-chloromercuribenzoate eliminated the possibility that the degradation of the antibiotic was being catalysed by a small amount of contaminating penicillinase, thus confirming that the penicillinase activity observed was an intrinsic property of the carboxypeptidases. Further studies on the *E. coli* carboxypeptidases as enzymes involved in the late stages of peptidoglycan synthesis and as PBPs are described in this and Chapter 8.7 (pp. 276 and 343).

(*ii*) Bacilli

The D-alanine carboxypeptidases of bacilli, particularly *B. megaterium*, *B. subtilis* and *B. stearothermophilus* have also been studied in detail. They are membrane-bound enzymes which may be solubilized with detergent and subsequently purified to homogeneity by conventional techniques [11, 57, 159, 227, 298, 299, 322] and also by covalent affinity chromatography on columns of 6-aminopenicillanic acid–Sepharose [17]. This latter procedure was based on the initial observations of Lawrence and Strominger [141, 142], that inhibition of the D-alanine carboxypeptidase of *B. subtilis* could be reversed by treatment with neutral hydroxylamine, whereas washing with buffer and treatment with penicillinase were without effect. This apparently irreversible inhibition of the enzyme was used by Blumberg and Strominger [15] to investigate whether inhibition of the carboxypeptidase was lethal to the organism. On the basis of the inhibitory effects of 6-aminopenicillanic acid and cephalothin on both growth and carboxypeptidase activity they concluded that inhibition of carboxypeptidase had no significant effects on growth of the organism under the conditions they examined and that, in addition to the carboxypeptidase, *B. subtilis* contained a cephalothin-sensitive 'killing' site. These investigations led to further studies on penicillin-binding components, described in the previous section.

However the 'irreversible inhibition' initially described can no longer be regarded as such. Purified D-alanine carboxypeptidase of *B. stearothermophilus*, pre-treated and consequently inhibited by ^{14}C-benzylpenicillin, was shown to release the bound antibiotic with a concomitant restoration of enzyme activity [19]. At 55° C, the enzyme is thermostable, the half-time for antibiotic release was approximately 10 min and complete recovery of carboxypeptidase activity occurred in 60 min. The kinetics of antibiotic release and recovery of enzyme activity were identical (Fig. 9.17). Re-examination of the *B. subtilis* enzyme showed that it also underwent exactly the same reactions but with a half-time of 200 min at 37° C. Since denaturation of the enzyme, either by boiling or by treatment with sodium dodecyl sulphate, prevented the release of the antibiotic, it was concluded that the release must involve enzyme action. The fragmentation of the benzylpenicillin molecule which occurs as breakdown of the antibiotic-enzyme complex and the isolation of penicilloyl- and acyl-peptides from these enzymes are discussed in detail below (pp. 358 and 360).

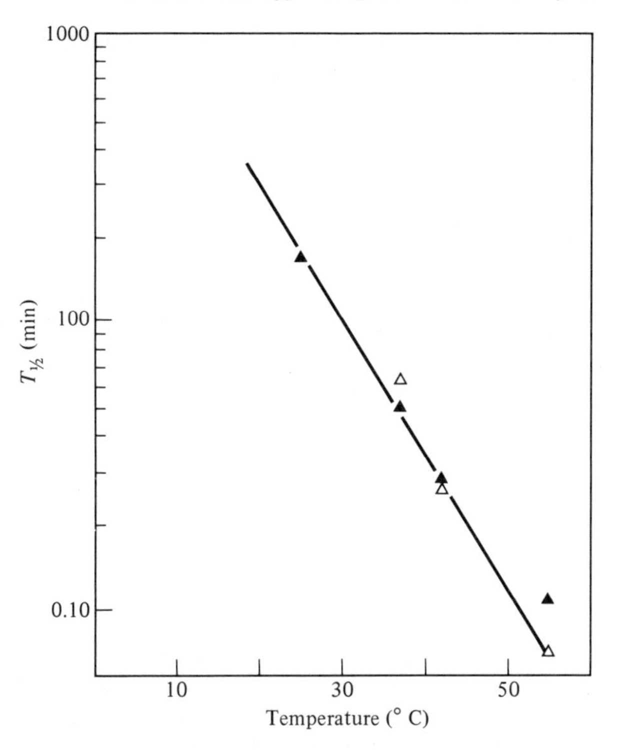

Figure 9.17 Effect of temperature on the half-life of the benzylpenicillin–D,D-carboxypeptidase complex in the thermophile *Bacillus stearothermophilus*. The release of radioactive benzylpenicillin (▲) from membranes pretreated with cephalothin and the recovery of enzymic activity (△) after inhibition of the purified D,D-carboxypeptidase. After [19] where details of the experimental procedures may be found.

(*iii*) Actinomycetes

The isolation and characterization of the carboxypeptidases from several Actinomycetes and the subsequent discovery that the purified enzymes would catalyse transpeptidation reactions analogous to those found in the parent organism, has been described in the previous chapter. Similarly, the observations on the inhibition of these enzymes by β-lactam antibiotics, which led Ghuysen and his colleagues to postulate the 'conformational response' model for the mode of action of β-lactams, have been described briefly in the section dealing with possible models for penicillin action. A more detailed account of the interaction of β-lactam antibiotics and these carboxypeptidase-transpeptidases is given below.

The exocellular enzymes from *Streptomyces* strain R61 and *Actinomadura* R39 have been purified to homogeneity and studied in some detail [79, 81]. They are inhibited by a variety of penicillins and cephalosporins with the formation of equimolar enzyme–antibiotic complexes. On the basis of steady-state kinetics, the inhibition of the exocellular enzyme from R61 was competitive, with respect to the donor peptide (usually ^{14}C-Ac_2-L-lysyl-D-alanyl-D-alanine) and non-competitive to the acceptor substrate (*meso*-diaminopimelic acid). Previously, physiochemical studies, particularly fluorescence and circular dichroism, had suggested that binding of benzylpenicillin occurred at a site other than that occupied by the acceptor substrate [193]. However, these studies did not allow such a distinction to be made with reference to the donor peptide. In this case, although the kinetics of inhibition were competitive, saturating concentrations of donor substrate had no appreciable effect on the binding of penicillin. Similarly, binding was observed under conditions where the donor peptide was not a substrate and the rate of binding of the two compounds, antibiotic and donor peptide, showed considerable differences. These observations led to the conclusion that the donor and antibiotic binding sites were distinct, if not totally, then at least partially. However, they could not be totally independent, since the binding of the antibiotic appeared to affect the binding of the donor substrate.

The interaction of the antibiotics with the R61 and R39 enzymes has been suggested to occur by a mechanism shown in the following model [79]

$$E + I \overset{K}{\rightleftharpoons} EI \xrightarrow{k_3} EI^* \xrightarrow{k_4} E + X$$

where E is the active enzyme; I is the inhibitor (the antibiotic); EI an intermediate inactive enzyme inhibitor complex; EI* the inactive enzyme inhibitor complex, where both constituents have been modified; X is released as a chemically altered antibiotic; K is the association constant and k_3 and k_4 are rate constants. At low concentrations of antibiotic the ratio of the rate constant k_3 to the association constant K is of importance in defining the inhibitory activity of the particular antibiotic. Thus, the higher this ratio k_3/K the more active is the antibiotic. Clearly, this high ratio may result from either a high value for k_3 or a low value for K. The interaction of R61 enzyme with benzylpenicillin and carbenicillin are examples of the two types (Table 9.9). At very low concentrations of antibiotic the rate constant k_4 is also of importance in defining activity. At equal K and k_3 values, antibiotics for which k_4 is low are the most effective inhibitors. Table 9.9 also shows the half-life of the complex E1*, formed between the two enzymes and a range of β-lactam antibiotics. The breakdown of the complexes, determined under conditions chosen to maintain enzyme activity, is always relatively slow, whereas the formation of the complex may occur quite quickly. The complex formation has also been studied between both enzymes and the chromogenic cephalosporin, nitrocefin, (3-[2,4-dinitrostyryl] -[6R-7R] -7-[2-thienylacetamido] -ceph-3-em-4-carboxylic acid, E isomer) [198]. In this antibiotic, hydrolysis of the β-lactam bond by penicillinase results in a shift of the absorption maximum, from 386 to 482 nm. Using the R39 enzyme, a similar change was observed, even when the addition of the antibiotic to the enzyme was made at 0° C, suggesting that formation of E1* occurred with the concomitant hydrolysis of the β-lactam bond. In contrast, with the R61 enzyme a similar spectral change was only observed on breakdown of the enzyme-antibiotic complex (i.e. E1* → E + X). Apparently, this process resulted in the hydrolysis of

Table 9.9 Rate constants for the breakdown of the complexes formed between the D,D-carboxypeptidases of *Streptomyces* R61 and *Actinomadura* R39 and various β-lactam antibiotics (data from [63, 70, 72]).

Antibiotic	*R39 enzyme*		*R61 enzyme*	
	k_4 (S^{-1})*	*Half-life of the EI* complex (min)*	k_4 (S^{-1})*	*Half-life of the EI* complex (min)*
Benzylpenicillin	2.8×10^{-6}	4100	1.4×10^{-4}	80
Ampicillin	4.4×10^{-6}	2600	1.4×10^{-4}	80
Carbenicillin	5.4×10^{-6}	2125	1.4×10^{-4}	80
Cephalosporin C	0.28×10^{-6}	40 000	1.0×10^{-6}	11 200
Cephaloglycine	0.8×10^{-6}	14 000	3.0×10^{-6}	3700

*The rate constant k_4 is for the reaction $EI^* \rightarrow E + products$.

the β-lactam bond with the release of the degraded antibiotic and the restoration of enzyme activity. In both cases, the enzyme recovered after breakdown of the complex was as sensitive to fresh antibiotic as was the untreated enzyme.

The observations made on the soluble carboxypeptidase-transpeptidase of R61 have now been extended to a membrane-bound transpeptidase from the same organism [59, 60, 157]. The specificity profile of the membrane-bound activity closely resembled that of the soluble enzyme with regard to both donor and acceptor substrates although D,D-carboxypeptidase activity was not present in the isolated membranes. Membrane preparations also bound ^{14}C-benzylpenicillin and the relationship of this bound penicillin to the membrane-bound transpeptidase was demonstrated with the finding that the concentrations of several other β-lactam antibiotics, which decreased the binding of radioactive benzylpenicillin by 50%, also inhibited transpeptidation by the same extent. Other experiments showed that as found previously with the soluble enzyme [72], incubation of the enzyme–antibiotic complex in buffer resulted in its dissociation together with the release of a chemically modified antibiotic and the concomitant restoration of enzyme activity. Thus, the membrane-bound enzyme behaved in many ways in a similar manner to the soluble carboxypeptidase-transpeptidase except that it lacked detectable D,D-carboxypeptidase activity.

9.7.4 *Degradation of benzylpenicillin by penicillin-sensitive enzymes*

As described in the previous sections the study of the interaction of β-lactam antibiotics with specific enzymes has proceeded furthest with the D,D-carboxypeptidases. Antibiotic–enzyme complexes have been shown to break down with concomitant restoration of enzymic activity and fragmentation of the β-lactam molecule. Moreover, the structural changes made in the antibiotic were in many cases incompatible with the D-alanine carboxypeptidases of both Bacilli and *Actinomycetes* having a simple β-lactamase activity such as that described (Section 9.7.3) for carboxypeptidases from *E. coli* (carboxypeptidase 1A) and *Proteus mirabilis* [243].

The radioactive molecule released from the isolated carboxypeptidase–transpeptidase of *Streptomyces* strains R61 and *rimosus* and *Actinomadura* R39, pretreated with 8-^{14}C-benzylpenicillin has been identified as ^{14}C-phenylacetylglycine [68]. Identical results have been obtained for the product of the reversible binding of benzylpenicillin to the D-alanine carboxypeptidases of *B. stearothermophilus* [89] and *S. faecalis* [51]. These investigations have been extended by Ghuysen and his colleagues [71] to identify the second product of the reaction, derived initially from the thiazolidine nucleus of the penicillin molecule. Using ^{3}H-benzylpenicillin labelled in the β-methyl group and 5-^{14}C-benzylpenicillin the radioactive degradation product was characterised as di-(*N*-formyl-D-penicillamine)-disulphide. Additional experiments carried out over shorter time periods and under N_2 provided evidence that the primary degradation product released was *N*-formyl-D-penicillamine and that during the prolonged incubation periods initially used this was oxidized to the dimer previously found. Somewhat surprisingly, this oxidation reaction did not occur to any great extent when the enzyme was denatured suggesting that enzymic activity was involved in the reaction. In contrast, Hammarstrom and Strominger [90] claimed that 5,5-dimethyl-Δ^2-thiazoline-4-carboxylic acid and not *N*-formyl-D-penicillamine was the fragment arising from the thiazolidine moiety of benzylpenicillin. This conclusion has been disputed on the basis of the relative instability of the proposed intermediate [1]. However, a short-lived intermediate (half-life 10 min in 3mM-phosphate buffer pH 7.5 at 37° C) which gives rise to *N*-formyl-D-penicillamine does occur in the fragmentation of phenoxymethylpenicillin [69]. The nature of this intermediate (Z) which does not contain any free sulphydryl residue remains unknown although it is very unlikely to be 5,5-dimethyl-Δ^2-thiazoline-4-carboxylic acid which has a half-life of 45 min under the conditions described.

At this point the influence of acyl substituents and environment in the degradation of β-lactams should be mentioned. Although the acyl substituent of the 6-aminopenicillanic acid nucleus does not necessarily affect the degradative process (phenoxymethylpenicillin is fragmented to phenoxyacetylglycine and *N*-formyl-D-penicillamine [71], as described above) it clearly influences both the rate of formation and the stability of the enzyme–antibiotic complex. Secondly, the fragmentation products released from isolated membranes of *Streptomyces* R61 and *rimosus*, pretreated with 8-^{14}C-benzylpenicillin, is benzylpenicilloate [59, 157]. A similar result has been obtained with the isolated enzyme from *Actinomadura* R39 incubated under conditions of low rather than high ionic strength [70]. In this case however, reactivation of the enzyme was not observed. Thus the products obtained from the antibiotic on breakdown of the E1* complex appear dependent upon the environmental conditions under which the complex is incubated. The two pathways of E1* complex breakdown using benzylpenicillin as the example are shown in Fig. 9.18 [158]. In pathway I, the benzylpenicilloyl-enzyme is subject to an enzyme-catalysed nucleophilic attack on C_7 which results in the release of benzylpenicilloate and reactivated enzyme. Alternatively (pathway II) the penicilloyl-residue is further degraded by cleavage of the C_5–C_6 bond followed by protonation of C_6 to yield enzyme-linked phenylacetylglycine and intermediate Z. Again nucleophilic attack on the original C_7 results in release or transfer of the phenylacetylglycyl residue and reactivation of the enzyme. Whether the nucleophilic attack on C_7 occurring either before or after the cleavage of C_5–C_6, is catalysed by the same active site of the enzyme remains unknown. This site(s) may also be the same or independent of the site catalysing nucleophilic attack on the carbonyl carbon of the penultimate D-alanyl residue in the natural substrate. The particular, degradative pathway followed by an enzyme–antibiotic complex may either be a characteristic of the enzyme itself or of the environmental conditions in which the complex is incubated. Thus breakdown of the *E. coli* (1A) and *P. mirabilis* D,D-carboxypeptidase E1* complexes in water yields benzylpenicilloate whereas the isolated enzymes from *B. stearothermophilus* and the Actinomycetes follow pathway II. The fragmentation products from *Streptomyces* R61 E1* complex are dependent upon either the environment of the enzyme (i.e. membranes or isolated enzyme) and of the nucleophiles present.

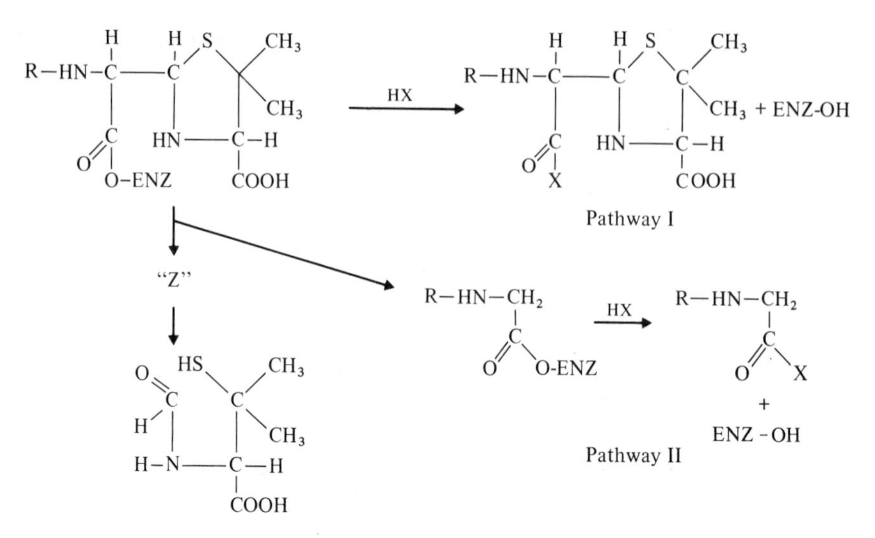

Figure 9.18 Pathways for the breakdown of benzylpenicilloyl–enzyme complexes. Penicilloylation of the enzyme occurs by the formation of an ester linkage between the penicilloyl-moiety and a serine residue of the enzyme. This complex can breakdown either by interaction with a nucleophilic agent (HX) to yield the penicilloyl derivative of the nucleophile (e.g. penicilloyl hydroxamate or in the presence of water, penicilloic acid) (Pathway I). Alternatively the penicilloyl nucleus undergoes further degradation by an enzymically catalysed cleavage of C_5–C_6 bond to yield R-glycyl–enzyme and an unknown intermediate Z. Nucleophilic attack on R-glycyl–enzyme regenerates active enzyme and the R-glycyl derivative of the nucleophile (in the case of benzylpenicillin with water as the nucleophile the product is phenylacetylglycine). 'Z' undergoes further breakdown to eventually yield *N*-formyl-D-penicillamine.

9.7.5 *Penicilloyl- and acyl- intermediates of penicillin-sensitive enzymes*

The previous section has dealt with fragmentation of the penicillin molecule after the initial formation of a penicilloylated enzyme. Clearly Tipper and Strominger's [290] substrate analogue hypothesis for the mechanism of penicillin action would receive strong support if, following interaction of a penicillin-sensitive enzyme with either penicillin or the substrate (e.g. acyl-D-alanyl-D-alanine) identical penicilloyl- or acyl-D-alanyl-peptides could be isolated.

Penicilloyl-peptides have been isolated and characterized from the D,D-carboxypeptidases of *Streptomyces* R61 [67] *B. subtilis* and *B. stearothermophilus* [76, 323] after denaturation and proteolytic digestion of the enzyme–antibiotic complex. In each case the penicilloyl residue is linked to L-serine, the sequence of the R61 peptide being Val-Gly-Ser and in the bacilli Ala-Ser-Met.

A prerequisite for the isolation of an acyl-D-alanyl-peptide requires that the enzyme should form a relatively stable intermediate with a natural substrate. Complexes of this type have now been obtained with the D,D-carboxypeptidases of *B. subtilis* and *B. stearothermophilus* and in smaller amounts from the carboxypepti-

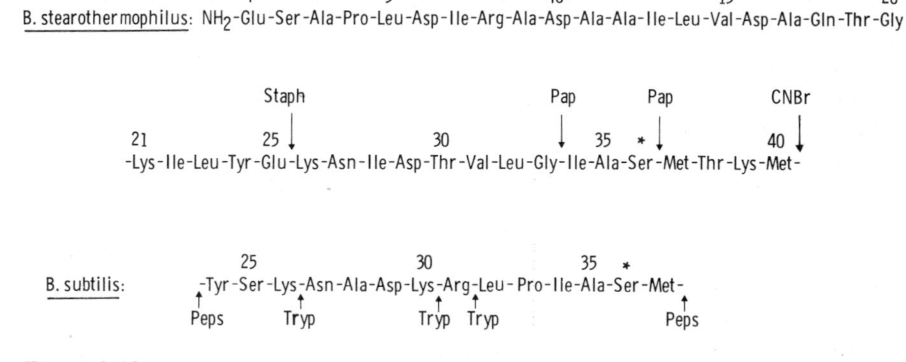

Figure 9.19 Amino acid sequences of the active sites of the D,D-carboxypeptidases of *Bacillus stearothermophilus* and *B. subtilis*. Enzymes labelled with either radioactive benzylpenicillin or acyl-D-alanine were cleaved with cyanogen bromide (CNBr), staphylococcal protease (Staph); papain (Pap); pepsin (Peps) and trypsin (Tryp). The labelled peptides were purified and sequenced allowing the serine (Residue 36) to be identified as the site of both acylation and penicilloylation. After [323].

dases of *E. coli* (1A) and *S. aureus* [56, 221, 323]. In the case of the bacilli rapid denaturation of incubation mixtures containing the depsipeptide substrate ^{14}C-diacetyl-L-lysyl-D-alanyl-D-lactate and purified enzyme, allowed the isolation of covalently-linked ^{14}C-acyl-D-alanyl-enzyme complexes [323]. These contained 0.6 to 0.85 mol of substrate bound per mol of enzyme. Active site peptides were then isolated by chemical and enzymic cleavage of the carboxypeptidases and their amino acid sequences of penicillin- and substrate-labelled peptides shown to be identical. In both cases the acyl-substituent was bound by ester-linkage to a serine residue at position 36 in the *B. stearothermophilus* carboxypeptidase and the corresponding residue in the *B. subtilis* enzyme. The structure of these peptides is shown in Fig. 9.19. Tipper and Strominger [290] also postulate that β-lactamases and penicillin-sensitive enzymes involved in peptidoglycan biosynthesis may be related and have been derived from a common ancestor. In this context it is interesting to note that Strominger and his colleagues [323] report a considerable degree of homology between the two carboxypeptidases and four β-lactamases of known sequence. Moreover, seryl residues also appear to participate as active site residues in the β-lactamases.

9.7.6 *L,D- and D,D-carboxypeptidases of Gaffkya homari and Streptococcus faecalis*

The detailed studies of Hammes and Kandler [91, 93] on the biosynthesis of peptidoglycan in *G. homari* have been described in Chapters 8.6.1 and 8.6.2. In brief they found that membrane-wall preparations synthesized cross-linked peptidoglycan from both UDP-*N*-acetylmuramyl-pentapeptide and UDP-*N*-acetylmuramyl-tetrapeptide. Incorporation of the *N*-acetylmuramyl-pentapeptide residues was sensitive to benzylpenicillin (ID_{50} 0.25 μg/ml) whereas incorporation of *N*-acetylmuramyl-tetrapeptide was not (ID_{50} 630 μg/ml). However if both penta- and tetrapeptide substrates were present simultaneously then incorporation of *N*-acetylmuramyl-pentapeptide residues occurred even in the presence of relatively high concentrations of benzylpenicillin (5 μg/ml) although the terminal D-alanine residue was not released. Thus, the incorporated pentapeptide residues did not participate in any transpeptidation reaction. These observations led the authors to conclude that in *G. homari* incorporation of newly-synthesized peptidoglycan occurs by a transpeptidation reaction, insensitive to inhibition by β-lactams, in which the D-alanyl-D-alanine sequences are donated by the pre-existing wall and the nascent peptidoglycan functions as the acceptor. The 'killing site' of β-lactams was considered to be the D,D-carboxypeptidase known to be present in the wall-membrane preparations. A similar situation may exist in *S. faecalis* where D,D-carboxypeptidase activity appears 'important if not essential' for growth [52, 71]. In both organisms ability of a particular β-lactam to inhibit growth appeared related to its ability to inhibit the D,D-carboxypeptidase. These observations suggest that in both organisms carboxypeptidase activity is vital for normal peptidoglycan synthesis and that in *G. homari* at least, pentapeptide residues are poor substrates for the 'acceptor site' of the transpeptidase.

More recently Hammes and Seidel [92, 95, 96] have described an L,D-carboxypeptidase of *G. homari* whose natural substrate, like that of the D,D-carboxypeptidase, appears to be nascent peptidoglycan. This enzyme catalyses both carboxypeptidase activity and an exchange reaction where D-amino acids or glycine are incorporated into position 4 of the tetrapeptide subunit L-Ala^1-D-Glu^2-L-Lys^3-D-Ala^4. The presence of these D-amino acids in incubation mixtures inhibits the synthesis of cross-linked peptidoglycan from UDP-MurAc-tetrapeptide. The exchange reaction is also responsible for the inhibition of peptidoglycan synthesis in *G. homari* by nocardicin A, and is partially responsible for inhibition by cephalosporin C. In each case D-amino acid substituents on the antibiotics (D-homoserine on nocardicin A; D-α-aminoadipic acid on cephalosporin C) are thought to be incorporated into the tetrapeptide. This appears to be the only mechanism whereby nocardicin A inhibits peptidoglycan synthesis in *G. homari* and recalls the observation that D-cycloserine and 6-aminopenicillanic acid can act as acceptors for *in vitro* transpeptidation by the soluble enzymes from Actinomyces [214]. Cephalosporin C in common with other cephem antibiotics also inhibits the D,D-carboxypeptidase although its inhibitory action on synthesis of cross-linked peptidoglycan from UDP-MurAc tetrapeptide (i.e. inhibition of L,D-carboxypeptidase) appears unique among this group of compounds. In contrast, both benzylpenicillin and thienamycin inhibit the L,D-carboxypeptidase directly. However, both these antibiotics and cephalosporin C show their greatest activity against D,D-carboxypeptidase. These findings and the knowledge that UDP-MurAc-pentapeptide is the *in vivo* precursor of peptidoglycan serve to reinforce the conclusion that in *G. homari* the 'killing target' for the majority of β-lactams is the D,D-carboxypeptidase. The L,D-carboxypeptidase does, however, appear to have an important role in peptidoglycan synthesis and tripeptide substituents on the nascent peptidoglycan appear to be a required substrate of the transpeptidase.

A membrane bound L,D-transpeptidase catalysing an exchange reaction between the terminal D-alanyl residue of a range of peptides of general composition X-L-R_3-R_2 (OH) and a variety of acceptor compounds, especially D-amino acids and glycine, has been described in *S. faecalis* [53]. L,D-carboxypeptidase activity was not observed and the physiological role of the enzyme remains unknown. The exchange activity is inhibited by β-lactams, both penicillins and cephalosporins, although the results obtained do not point to its being a 'killing target'. Thus, similar β-lactam sensitive enzymes have been found in both *G. homari* and *S. faecalis.* Whether this also means that both organisms have similar mechanisms of peptidoglycan synthesis will only become known with further study of *S. faecalis*.

9.7.7 *The role of autolysins in the bactericidal activity of β-lactam antibiotics*

One of the earliest observations on the antibacterial activity of penicillin was that it appeared to be lethal only to growing organisms. Since penicillin had no inhibitory activity on protein synthesis it was argued that synthesis of cellular components, including autolysins, would continue in its presence [236, 294]. Earlier Weidel and Pelzer [311] had proposed that autolysins played an essential role in the incorporation of newly-synthesized peptidoglycan into the pre-existing wall. Thus, the hydrolysis of appropriate bonds would release acceptor sites for the incorporation of the newly-synthesized material. Clearly, if such conditions prevailed in the growing organism, then inhibition of the synthesis of cross-linked peptidoglycan would eventually upset the balance between synthesis and hydrolysis, resulting in the peptidoglycan becoming weak and unable to withstand the high internal osmotic pressure. As a result cell lysis would occur. Alternatively, even in the absence of obvious lysis the organism could suffer sufficient membrane damage to prove lethal. Evidence to support this general mechanism came from the observation that chloramphenicol antagonized the lytic and bactericidal action of methicillin on *S. aureus* [236].

More direct evidence for the involvement of autolysins in the bactericidal rather than the bacteriostatic action of β-lactams and indeed other antibiotics came initially from the studies of Rogers and Tomasz and their colleagues [236, 237, 293, 295]. Two mutants of *Streptococcus (Diplococcus) pneumoniae* were used by Tomasz *et al.*: one having a defect in the autolysin *N*-acetylmuramyl-L-alanine amidase the second having a modified wall in which ethanolamine replaced choline in pneumococcal C substance a teichoic acid-like secondary polymer covalently attached to peptidoglycan. This change rendered the walls resistant to autolysin and recent work [103, 104] has suggested that teichoic acid containing choline is necessary for activation of the amidase. The parent organism lyses readily and is extremely sensitive to treatment with antibiotics inhibiting wall synthesis (i.e. β-lactams, vancomycin and cycloserine) and detergents such as deoxycholate. In contrast the mutants did not lyse in the presence of these agents although growth was inhibited. Similar results have been obtained with autolysin-deficient mutants of *B. licheniformis* [237] and *B. subtilis* [65]. These organisms were some 10–20-fold more resistant to lysis in the presence of wall-antibiotics than were the parent bacilli and remained viable during prolonged antibiotic-treatment. In all cases peptidoglycan synthesis remained sensitive to inhibition by the various antibiotics. This phenomenon resulting from some modification of the bacterial autolytic system, was termed 'tolerance' [294] to distinguish it from the well characterized forms of antibiotic resistance.

In contrast to the activation of pneumococcal amidase by choline-containing C-polysaccharide, Holtje and Tomasz [103, 104] found that Forssman antigen from *S. pneumoniae* inhibited the autolysin. This polymer has been reported also to contain choline and as a membrane teichoic acid is probably analogous to the lipoteichoic acids of other Gram-positive bacteria [29]. Treatment of autolysin-defective pneumococci, streptococci and other organisms with penicillin or other wall antibiotics resulted in the rapid release of the autolysin-inhibitor and lipid-containing substances in the absence of cell-lysis [88, 105, 107, 296, 303]. However, the Forssman antigen inhibited only the pneumococcal amidase and was without effect on the other autolysins tested. In these cases lipoteichoic acids and a number of other amphipathic substances including cardiolipin and diphosphatidylglycerol, have proved more or less effective inhibitors of *N*-acetylmuramidases in *S. faecalis* and *Lactobacillus acidophilus* [47, 48] and of *N*-acetylglucosaminidases in *S. aureus* [277] and *B. subtilis* (C. Taylor, personal communication). Lipoteichoic acid was also reported to inhibit the *N*-acetylmuramyl-L-alanine amidase of *B. subtilis* [47, 48] but this finding is now open to doubt (C. Taylor, personal communication). In each case where inhibition was observed the esterified fatty acids were essential for the various polymers to show inhibitory activity. As a result of their observations Tomasz and his colleagues [295, 296] proposed that autolytic enzymes are under negative control and that the bactericidal effects of wall antibiotics are mediated by the loss of an autolysin inhibitor. This 'indirect' mechanism postulates that wall antibiotics interact with a specific target in the bacterium resulting in the inhibition of peptidoglycan synthesis. As a consequence the autolysin inhibitor is released and autolytic activity is triggered. Thus in the 'indirect' model the irreversibility of β-lactam inhibition results, totally or partially, from secondary factors which do not themselves have any interaction with the β-lactam. This mechanism contrasts with the alternative 'direct' model where the interaction of the antibiotic with a target enzyme is alone responsible for loss of viability. The indirect model is supported by the observation that during treatment of both *B. subtilis* and *Neisseria gonorrhoeae* with benzylpenicillin inhibition of growth can be separated from loss of viability and lysis by manipulating the pH and Mg^{2+} concentration of the medium [84, 151, 295]. Moreover, *Streptococcus sanguis* appears to be completely 'tolerant' to wall antibiotics [106]. Although the organism shows a characteristic response to treatment with β-lactams and other wall antibiotics (i.e. inhibition of peptidoglycan synthesis, morphological changes and secretion of uncross-linked peptidoglycan, lipid and lipoteichoic acid) these are not accompanied by a loss of viability (lysis). *S. sanguis* has no detectable autolytic enzymes. Tolerant strains of *S. aureus* have recently been described among clinical isolates [239]. Growth of these organisms is inhibited by low (i.e. MGIC) concentrations of β-lactamase-resistant penicillins and cephalosporins and by vancomycin but even in the presence of high concentrations the organisms remain viable. Phenotypically the tolerant strains were autolysin-deficient, this deficiency being due to the production of an autolysin-inhibitor in large excess. Whether this inhibitor is lipoteichoic acid is not yet established.

Although the indirect model provides a working hypothesis for the mechanisms underlying the bactericidal effects of wall-antibiotics in Gram-positive cocci its general applicability remains unknown. The secretion of polymeric glycerol (wall and lipoteichoic acid) was not observed in penicillin-treated *B. licheniformis* (Tynecka and Ward, unpublished observations) and lipoteichoic acids are absent from *M. luteus* [218] and have not been positively identified in Gram-negative organisms. A recent study has investigated the response of *N. gonorrhoeae*, an organism highly-sensitive to β-lactams and with an active autolysin(s), to treatment with benzylpenicillin [84]. Addition of the antibiotic to growing cultures resulted in the inhibition of peptidoglycan synthesis, a marked increase in the turnover of the total glucosamine of the cell but not of diaminopimelic acid, the 'shedding' of lipid and lipopolysaccharide and the loss of viability. All these changes were observed while 'growth' (measured as the increase in turbidity) continued. On further incubation the organisms lysed although as described above lysis could be separated from inhibition of growth. Whether an alternative autolysin-inhibitor, analogous to lipoteichoic acid, is present in the released macromolecular material requires further investigation.

9.7.8 *Resistance to β-lactam antibiotics*

One of the methods by which normally sensitive organisms can develop resistance to the inhibitory action of both penicillins and other antibiotics is 'tolerance' as described in the preceding section. Probably a more obvious mechanism would be the modification of the target enzyme to prevent the interaction of the target and the antibiotic while retaining the catalytic activity of the enzyme. The third mechanism of resistance is provided by a modification of outer surfaces of the organism which will prevent the access of the antibiotic molecules to the target enzymes. Since these enzymes are, for reasons explained in the previous chapter, likely to be situated at the membrane-wall interface, it follows that to prevent the antibiotic molecules from reaching such a site the modification must alter the porosity (with respect to the antibiotic) of the outer layers of the wall. Alternatively the modification may be in the membrane, such that the topography of the membrane surface is altered, preventing access of the antibiotic while the enzyme molecules retain their catalytic activity.

Many bacteria, both Gram-positive and Gram-negative, are resistant to β-lactam antibiotics by virtue of their ability to modify the antibiotic itself. This is accomplished either by hydrolysis of the β-lactam ring (penicillinases and cephalosporinases) [43, 44, 232] or by cleavage of the amide-linkage between the penicillin nucleus, 6-aminopenicillanic acid in the case of penicillins, and the acyl side chain (penicillin amidases) [44]. The former enzymes are by far the most common and are of considerable clinical importance. Consequently they have been the subject of intensive investigation. As a mechanism of resistance, they will not be considered further but much of the information on these enzymes from both Gram-positive and Gram-negative organisms can be found in the following reviews [43, 44, 232, 280].

Mutants resistant to penicillins have been isolated in both Gram-positive and Gram-negative organisms [18]. However, in both cases, none of the particular mutations have been attributed to a modification of one of the penicillin-sensitive enzymes. In the case of *S. aureus* [61, 207] much of this early work was directed towards the determination of the amount of penicillin bound by a resistant strain in comparison with the amount bound by the sensitive parent organism. As discussed in the section dealing with 'the binding of penicillin', these investigations proved relatively inconclusive.

A more direct approach for the isolation of mutants, with altered penicillin-sensitive enzymes, of both *B. subtilis* [247] and *E. coli* [130] has been attempted by Strominger and his colleagues. In the former case, mutants of *B. subtilis* resistant to 100 and 200 μg/ml of 6-aminopenicillanic acid (the normal growth inhibitory concentration is 30 μg/ml), have been isolated. The resistant strains, grown at the high concentrations of antibiotic, showed no alteration in the amount of cross-linking (measured by dinitrophenylation of the uncross-linked diaminopimelic acid residues), whereas the D-alanine carboxypeptidase retained only 5% of the activity present in the control cells. Neither were morphological changes detectable by either electron or phase-contrast microscopy. The nature of the underlying reasons for the observed resistance is not yet known. Clearly, it could be either a modification of the target enzyme, the transpeptidase, or some alteration in the accessibility of the antibiotic to the target site. If this latter alternative is the case, then the modification must be highly specific, since the D-alanine carboxypeptidase, also a membrane-bound enzyme, retains full sensitivity to 6-aminopenicillanic acid.

The attempt to isolate mutants of *E. coli* [130] with defective penicillin-sensitive enzymes was based on a selection technique designed to eliminate mutations showing resistance to penicillin by virtue of changes in permeability or in some penicillinase activity. The mutants selected were ampicillin-resistant at the permissive temperature (32° C) and, in the absence of ampicillin, lysed or were unable to grow at the restrictive temperature (42° C). The rationale for the isolation procedure was that both penicillinase producers and permeability mutants would be resistant at the permissive temperature, but in addition they would not lyse or cease to grow at the restrictive temperature. Thus, mutations involving resistance to ampicillin at 32° C, but which were thermolabile at 42° C, would be selected. Among the mutants examined were two groups which at the restrictive temperature synthesized hyper- and hypo-cross-linked peptidoglycan respectively. The nature of the mutation underlying these changes is not yet known. However, analysis of revertants from both groups suggested that resistance to ampicillin, temperature sensitivity, cross-linking, growth characteristics and morphological changes were all related to a single mutation.

The resistance of a wide range of Gram-negative bacteria to β-lactam and other antibiotics resulting from the selective permeability of the outer membrane has been discussed earlier in Chapter 4.10 and reviewed several times in recent years [22, 194, 230]. Briefly, the outer membrane has been shown to present a barrier to hydrophilic molecules above a certain size. The elegant experiments of Nikaido and his colleagues and other recent work on the penetration of oligopeptides and oligosaccharides have established that in the *Enterobacteriaceae* passage of these substances is markedly reduced at molecular weights of approximately 600 to 700 and those having molecular weights of 1000 and above are essentially excluded from the cell. It should be noted however that the exclusion limit determined for the *Enterobacteriaceae* appears to be an exception perhaps related to their natural habitat. The outer membranes of other Gram-negative organisms allow the passage of substances of much higher molecular weights (H. Nikaido, personal communication). In all cases penetration of the outer membrane by β-lactams and many other antibiotics is thought to occur through aqueous pores formed by certain of the major outer membrane proteins which have been called porins [179, 180, 194a; see also Chapters 3.4.2 and 4.10]. In *P. aeruginosa* the major porin has an exclusion limit of approximately 6000 daltons. [97a]

Although, almost all β-lactam antibiotics have molecular weights below the exclusion limit even of the enterobacteria there is considerable evidence linking the hydrophobicity of the various penicillins and cephalosporins and their rates of penetration through the outer membrane. An increase in the hydrophobicity of the β-lactam tends to decrease the rate of diffusion (Fig. 9.20). At this time it is worth remembering that in the few cases so far examined cross-linkage in β-lactam resistant Gram-negative and β-lactam sensitive Gram-positive bacteria has proved equally sensitive to inhibition by these antibiotics. Thus, the ability of a particular β-lactam to penetrate the outer membrane together with its resistance to hydrolysis by β-lactamases is of considerable importance in determining whether it will be of any therapeutic use. Subsequently, there have been in recent years several investigations attempting to define the nature and characteristics of the barrier functions of the outer membrane. In general the methods used fall into two categories. The first of these is based on the findings of Zimmerman and Rosselet [325] who showed that the rates of hydrolysis of β-lactam antibiotics by *E. coli* containing a periplasmically-located β-lactamase, could be explained by assuming that diffusion across the outer membrane was balanced by the hydrolysis of the antibiotic in the periplasm. Under the conditions used little difference was observed in the diffusion rates of five of the six cephalosporins examined; the exception was cephalothin which resembled ampicillin and benzylpenicillin. These three antibiotics were the most hydrophobic examples tested; a finding which led to the conclusion stated above that the more lipophilic β-lactams diffuse slower. The method of Zimmermann and

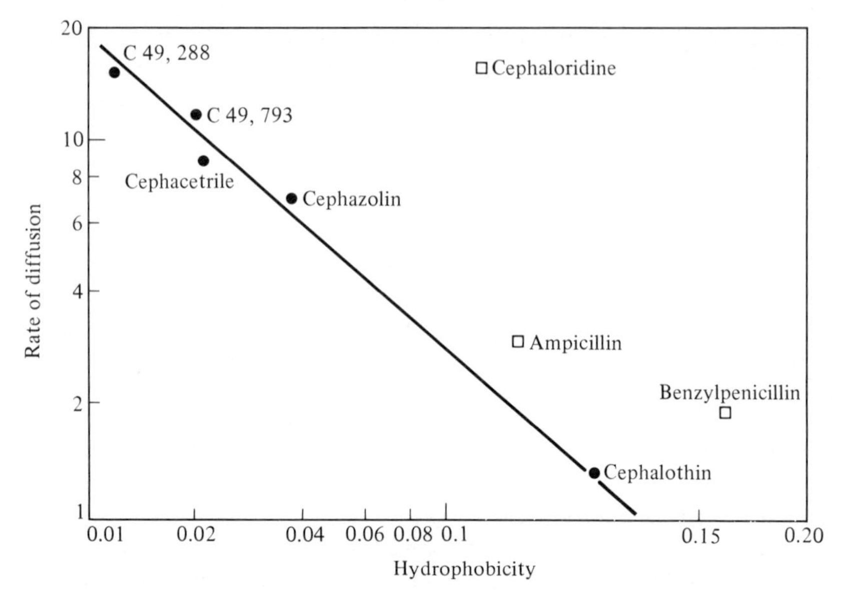

Figure 9.20 The relationship of hydrophobicity to the rate of diffusion of various cephalosporins and penicillins through the outer membrane of *E. coli*. From the data of [325]. Hydrophobicity was determined as the partition coefficient of the antibiotics in isobutanol-0.02 M phosphate buffer pH 7.4–0.9% (w/v) NaCl at 37°C. The calculation for the rate of diffusion is given by [325].

Rosselet [325] has also been used to study the role of individual porins in the diffusion of cephaloridine across the outer membrane of *S. typhimurium* [195]. The permeability coefficient for the antibiotic (calculated from the diffusion constant multiplied by the surface area of the organism) was reduced in mutants lacking one or other of the two major outer membrane proteins of molecular weights of 34 000 and 36 000. In mutants lacking both proteins the permeability coefficient was reduced 10-fold. These proteins had earlier been shown to function as porins in the reconstitution system [180].

The outer membrane of *E. coli* contains three porins designated 1a, 1b and E. Using a series of mutants constructed to contain only one of these proteins Nikaido and his colleagues examined the effect of hydrophobicity and alteration of change on the permeability coefficient of a series of cephalosporins (H. Nikaido, personal communication). Comparing cephaloridine with cefsulodin (negative charge) and cephaloglycine (positive charge) they found that the presence of a negative charge resulted in a marked reduction in the rate of diffusion through outer membranes containing any one of the three porins whereas the presence of a positive charge enhanced the rate of diffusion through outer membranes containing porins 1a and 1b but had less effect with E. In all the mutants an increase in the hydrophobicity of the cephalosporins resulted in a decreased rate of diffusion.

One of the major disadvantages of the method of Zimmermann and Rosselet is that it can only be used to evaluate the penetration of β-lactams which are sensitive to the periplasmic β-lactamases. Since resistance to hydrolysis is clearly important in any β-lactam of therapeutic use alternative methods of measuring the barrier function of the outer membrane have been developed. These rely either on the use of mutants selected as showing increased sensitivity to an antibiotic to which the parent organism is normally resistant [229] or on using bacteria in which the outer membrane has been specifically disrupted by brief treatment with EDTA [245]. Increased sensitivity to one antibiotic is often accompanied by increased sensitivity to a wide range of other antibacterial agents and these 'permeability' mutants appear to have some defect in the integrity of their outer membranes. Using these mutants or EDTA-treated bacteria the penetration of a particular β-lactam through the outer membrane is measured indirectly by determining the difference in antibacterial activity (measured either as the MGIC [54, 231] or turbidometrically over short incubation periods [245]) against the parent organisms and these 'permeabilized' cells. The results obtained by these two methods appear to be in general agreement with those based on the method of Zimmermann and Rosselet.

The activity of any therapeutically useful β-lactam antibiotic against Gram-negative organisms is dependent upon three interacting factors. These are the ease with which the particular antibiotic penetrates the outer membrane, its stability to periplasmically located β-lactamases and finally its affinity for the target enzymes, the transpeptidases and carboxypeptidases located at the cytoplasmic membrane-wall interface. Clearly, the search for new and useful semisynthetic β-lactam antibiotics must take into account each of these factors all of which can now be measured in a wide range of organisms.

9.8 Antibiotics inhibiting biosynthesis of wall polymers but whose site of action is not yet established

A number of other antibiotics have been described as inhibiting wall polymer biosynthesis as judged by their ability to cause accumulation of the nucleotide precursors of peptidoglycan. However, an increasing number of these have now been shown to inhibit peptidoglycan synthesis by particulate enzyme preparations. Among this group of antibiotics are novobiocin [319], moenomycin [108, 300], janiemycin [33], the diumycin group [33, 152], enduracidin [165], gardimycin [256], amphomycin [288], the prasinomycins and antibiotic 11837 RP [140, 153, 300].

9.8.1 *Novobiocin*

Earlier work [319] had shown that treatment of *S. aureus* with novobiocin resulted in the accumulation of nucleotide precursors of both peptidoglycan and teichoic acid but that there was a lack of specificity of antibiotic action in that both protein and nucleic acid synthesis were also inhibited. Moreover, high concentrations of novobiocin did not appear to inhibit peptidoglycan synthesis by membrane preparations from *S. aureus* and *M. luteus* [44] although inhibition of teichoic acid biosynthesis was reported by several authors [36, 83, 112]. Smith and Davis [255], using various strains of *E. coli*, found that the earliest and most pronounced effect of novobiocin treatment was on the synthesis of DNA. These observations led Brock [30] to conclude that novobiocin exerted its inhibitory effect by complexing with Mg^{2+} in the cell and thus removing an essential cofactor from many enzymes. This rather unselective mode of action was questioned by the demonstration of a selective inhibition of teichoic acid biosynthesis in *B. licheniformis* [109]. Membrane preparations from this organism were shown to catalyse the synthesis of both polyglycerol phosphate and polyglucosylglycerol phosphate. The synthesis of the polyglycerol phosphate teichoic acid was at least five times more sensitive to novobiocin than was the synthesis of the glucosylated derivative although both systems showed similar requirements for Mg^{2+}. On the basis of these results, the authors concluded that the general inhibition of enzymes requiring magnesium ions was a secondary effect resulting from the inhibition of membrane teichoic acid biosynthesis (i.e. polyglycerol phosphate). The major consequence of this primary effect would be an interference with the accumulation of Mg^{2+} by the membrane. Clearly, their hypothesis is at variance with the findings of Smith and Davis [255] which have themselves been supported by the recent observation that novobiocin is a specific inhibitor of DNA gyrase, an enzyme involved in DNA replication [75]. Moreover, subsequent work discussed in Chapter 10.1.7, has revealed that the precursor of membrane teichoic acid is phosphatidylglycerol and not CDP-glycerol as used in the experiments described above. Thus, the inhibition observed in *B. licheniformis* was of wall rather than membrane teichoic acid biosynthesis. Whether both forms of teichoic acid have identical roles in the maintenance of high Mg^{2+} concentrations at the membrane remains unclear. More recently novobiocin-resistant mutants of *B. licheniformis* have been isolated and their walls found, perhaps surprisingly in view of the results described above, to be deficient not in teichoic but in teichuronic acid [234]. Earlier Hughes [110] had reported that novobiocin at high concentrations inhibited teichuronic acid biosynthesis by membrane preparations of another strain of *B. licheniformis* and more recently Yamada *et al.* [321] have reported a similar inhibition of teichuronic acid biosynthesis in *M. luteus*. Thus novobiocin has been shown to inhibit the biosynthesis of the main wall polymers of Gram-positive organisms whereas little similar work appears to have been done in Gram-negative bacteria. However, on balance it appears that inhibition of DNA biosynthesis is the primary site of novobiocin action.

9.8.2 *Phosphorus-containing glycolipid antibiotics*

The diumycins, prasinomycins, moenomycin and antibiotic 11837RP are all members of a group of phosphorus-containing glycolipid antibiotics [254]. The chemistry of the individual antibiotics is complex and not known with certainty and as with tunicamycin all may be families of closely related substances. A second group, of a different structure, is probably made up of janiemycin and enduracidin [33]. The two groups of antibiotics have similar antimicrobial spectra, being active *in vivo* only against Gram-positive bacteria. The resistance of Gram-negative organisms to the phosphoglycolipid group is related, at least in part, to the inability of the antibiotic to reach the target site [33, 254]. Particulate enzyme preparations from *E. coli* and *B. stearothermophilus* were equally sensitive to inhibition by these antibiotics [148].

The use of these particulate enzyme preparations [148, 153] has shown that all the antibiotics of both groups inhibit the *in vitro* synthesis of peptidoglycan but not the incorporation of radioactivity from the nucleotide precursors into one or both of the lipid intermediates. Thus, the results obtained are similar to those described previously for the action of vancomycin and ristocetin on similar preparations. This similarity, however appears to be superficial, since the concentrations of the antibiotic (diumycins, moenomycin and prasinomycins) required to completely inhibit peptidoglycan synthesis, is considerably less than the amount that would be necessary for a mechanism involving stoichiometric binding between the antibiotic and the D-alanyl-D-alanine terminus, as shown for vancomycin and ristocetin [210]. Moreover, moenomycin was shown to inhibit the incorporation into peptidoglycan of *G. homari* of both *N*-acetylmuramyl-pentapeptide and -tetrapeptide, whereas vancomycin, as expected, only inhibited the incorporation of *N*-acetylmuramyl-pentapeptide [94]. A specific site of action for moenomycin, diumycin, prasinomycin, 11837RP and 8036RP has been suggested by the recent finding (Table 9.10)

Table 9.10 Inhibition of peptidoglycan transglycosylase (polymerase) by various antibiotics (data provided by J. Van Heijenoort. Details of the assay of enzymic activity are given in [300]).

Antibiotic added	*Final concentration (μg/ml)*	*% Inhibition of transglycoslylase*
Ristocetin	100	87
	10	8
	1	5
Vancomycin	100	100
	10	38
	1	3
11837 R.P.	1	100
	0.1	95
	0.01	5
8036 R.P.	1	100
	0.1	100
	0.01	56
Moenomycin	1	100
	0.1	100
	0.01	39
Diumycin	1	100
	0.1	100
	0.01	51
Prasinomycin	1	100
	0.1	100
	0.01	60

that in *E. coli* the transglycosylation reaction synthesizing uncross-linked peptidoglycan from the lipid-linked precursor undecaprenyl-pyrophosphoryl-*N*-acetylmuramyl-(pentapeptide)-*N*-acetylglucosamine is inhibited by low concentrations of these antibiotics [300, 300a].

On the basis of structural similarities between the lipid component of the diumycins and related antibiotics, and C_{25}-isoprenyl alcohols, Linnett and Strominger [148] suggested that these antibiotics may function as analogues of the C_{55}-isoprenyl phosphates involved in peptidoglycan synthesis. In this context it is worth noting that moenomycin inhibits C_{55}-isoprenyl alcohol kinase of *S. aureus* [240]. However, as the author points out inhibition of this enzyme does not seem to be the 'killing site' for this antibiotic. If the Linnett and Strominger hypothesis were correct, then it might be expected that the synthesis of other wall polymers in which the undecaprenyl phosphate carriers participated would also be sensitive to the dumycins. Further experiments will be required to investigate this possibility.

References

1. Adriaens, P., Meesschaert, B., Frère, J-M., Vanderhaeghe, H., Degelaen, J., Ghuysen, J-M. and Eyssen, H. (1978) *J. biol. Chem.* **253**, 3660–65.
2. Allen, J. G., Atherton, F. R., Hall, M. J., Hassall, C. H., Holmes, S. W., Lambert, R. W., Nisbet, L. J. and Ringrose, P. S. (1978) *Nature, Lond.* **272**, 56–8.
3. Allen, J. G., Atherton, F. R., Hall, M. J., Hassall, C. H., Holmes, S. W., Lambert, R. W., Nisbet, L. J. and Ringrose, P. S. (1979) *Antimicrob. Agents Chemotherapy* **15**, 684–95.
4. Anderson, J. S., Matsuhashi, M., Haskin, M. A. and Strominger, J. L. (1965) *Proc. Natn. Acad. Sci. U.S.A.* **53**, 881–9.
5. Anderson, J. S., Matsuhashi, M., Haskin, M. A. and Strominger, J. L. (1967) *J. biol. Chem.* **242**, 3180–90.
6. Anderson, J. S., Meadow, P. M., Haskin, M. A. and Strominger, J. L. (1966) *Archs. Biochem. Biophys.* **116**, 487–515.
7. Aoki, H., Sakai, H., Kohsaka, M., Konomi, T., Hosoda, J., Kubochi, Y., Iguchi, E. and Imanaka, H. (1976) *J. Antibiot.* **29**, 492–500.
8. Araki, Y., Shimada, A. and Ito, E. (1966) *Biochem. biophys. Res. Commun.* **23**, 518–25.
9. Araki, Y., Shirai, R., Shimada, A., Ishimoto, N. and Ito, E. (1966) *Biochem. biophys. Res. Commun.* **23**, 466–72.
10. Atherton, F. R., Hall, M. J., Hassall, C. H., Lambert, R. W., Lloyd, W. J. and Ringrose, P. S. (1979) *Antimicrob. Agents Chemotherapy*. **15**, 696–705.
11. Barnett, H. J. (1973) *Biochim. biophys. Acta* **304**, 332–52.
12. Best, G. K. and Durham, N. N. (1965) *Archs. Biochem. Biophys.* **111**, 685–91.
13. Bettinger, G. E. and Young, F. E. (1975) *Biochem. biophys. Res. Commun.* **67**, 16–21.
14. Blumberg, P. M. (1974) *Ann. N. Y. Acad. Sci.* **235**, 310–25.
15. Blumberg, P. M. and Strominger, J. L. (1971) *Proc. Natn. Acad. Sci. U.S.A.* **68**, 2814–17.
16. Blumberg, P. M. and Strominger, J. L. (1972) *J. biol. Chem.* **247**, 8107–113.

17. Blumberg, P. M. and Strominger, J. L. (1972) *Proc. Natn. Acad. Sci. U.S.A.* **69**, 3751–5.
18. Blumberg, P. M. and Strominger, J. L. (1974) *Bact. Rev.* **38**, 291–335.
19. Blumberg, P. M., Yocum, R. R., Willoughby, E. and Strominger, J. L. (1974) *J. biol. Chem.* **249**, 6828–35.
20. Bogdanovsky, D., Bricas, E. and Dezélée, P. (1969) *C. R. Acad. Sci. Paris Ser. D.* **269**, 390–93.
21. Böhme, E. H. W., Applegate, H. E., Toeplitz, B., Dolfini, J. F. and Gongontas, J. Z. (1971) *J. Am. chem. Soc.* **93**, 4324–6.
22. Boman, H. G., Nordstrom, K. and Normark, S. (1974) *Ann. N. Y. Acad. Sci.* **235**, 569–85.
23. Bondi, A., Kornblum, J. and Forte, C. (1957) *Proc. Soc. exp. Biol. Med.* **96**, 270–72.
24. Boyd, D. B. (1977) *Proc. Natn. Acad. Sci. U.S.A.* **74**, 5239–43.
25. Bracha, R. and Glaser, L. (1976) *Biochem. biophys. Res. Commun.* **72**, 1091–8.
26. Braun, V., Bosch, V., Hantke, K. and Schaller, K. (1974) *Ann. N. Y. Acad. Sci.* **235**, 66–82.
28. Braun, V. and Wolff, H. (1975) *J. Bact.* **123**, 888–97.
29. Briles, E. B. and Tomasz, A. (1973) *J. biol. Chem.* **248**, 6394–7.
30. Brock, T. D. (1967) In *Antibiotics Vol. 1, Mechanism of Action,* eds. Gottleib, D. and Shaw, P. D., pp. 651–65, Berlin: Springer-Verlag.
31. Brown, A. G., Butterworth, D., Cole, M., Hanscomb, G., Hood, J. D., Reading, C. and Rolinson, G. N. (1976) *J. Antibiot.* **29**, 668–9.
32. Brown, C. A. and Perkins, H. R. (1979) *Antimicrob. Agents Chemotherapy.* **16** 28–36.
33. Brown, W. E., Seinerova, V., Chan, W. M., Laskin, A. I., Linnett, P. and Strominger, J. L. (1974) *Ann. N. Y. Acad. Sci.* **235**, 399–405.
34. Buchanan, C. E. (1977) In *Microbiology 1977*, ed. Schlessinger, D. pp. 191–4, Washington, D.C: American Soc. for Microbiol.
35. Buchanan, C. E. and Strominger, J. L. (1976) *Proc. Natn. Acad. Sci. U.S.A.* **73**, 1816–20.
36. Burger, M. M. and Glaser, L. (1964) *J. biol. Chem.* **239**, 3168–77.
37. Cassidy, P. J. and Kahan, F. M. (1973) *Biochemistry* **12**, 1364–74.
38. Chase, H. A., Reynolds, P. E. and Ward, J. B. (1978) *Eur. J. Biochem.* **88**, 275–85.
39. Chase, H. A., Shepherd, S. T. and Reynolds, P. E. (1977) *FEBS Letts.* **76**, 199–203.
40. Chatterjee, A. N. and Perkins, H. R. (1966) *Biochem. biophys. Res. Commun.* **24**, 489–94.
41. Christensen, B. G., Leanza, W. J., Beattie, T. R., Patchett, A. A., Arison, B. H., Ormond, R. E., Kuehl, F. A., Albers-Schonberg, G. and Jardetzky, O. (1969) *Science* **166**, 123–4.
42. Ciak, J. and Hahn, F. E. (1959) *Antibiotics and Chemotherapy* **9**, 47–54.
43. Citri, N. (1971) In *The Enzymes Vol. 4* (*3rd edn.*), ed. Boyer, P. D., pp. 23–46, London and New York: Academic Press.
44. Citri, N. and Pollock, M. R. (1966) *Adv. Enzymol.* **28**, 237–323.
45. Clark, V. L. and Young, F. E. (1977) *Antimicrob. Agents Chemotherapy.* **11**, 871–6.
46. Clark, V. L. and Young, F. E. (1977) *Antimicrob. Agents Chemotherapy.* **11**, 877–80.
47. Cleveland, R. F., Holtje, J. V., Wicken, A. J., Tomasz, A., Daneo-Moore, L. and Shockman, G. D. (1975) *Biochem. biophys. Res. Commun.* **67**, 1128–35.

48. Cleveland, R. F., Wicken, A. J., Danoe-Moore, L. and Shockman, G. D. (1976) *J. Bact.* **126**, 192–7.
49. Collins, J. F. and Richmond, M. H. (1962) *Nature, Lond.* **195**, 142–3.
50. Cooper, P. D. (1956) *Bact. Rev.* **20**, 28–48.
51. Coyette, J., Ghuysen, J-M., Binot, F., Adriaens, P., Meesschaert, B. and Vanderhaeghe, H. (1977) *Eur. J. Biochem.* **75**, 231–9.
52. Coyette, J., Ghuysen, J-M. and Fontana, R. (1978) *Eur. J. Biochem.* **88**, 297–305.
53. Coyette, J., Perkins, H. R., Polacheck, I., Shockman, G. D. and Ghuysen, J-M. (1974) *Eur. J. Biochem.* **44**, 459–68.
54. Curtis, N. A. C., Brown, C., Boxall, M. and Boulton, M. G. (1979) *Antimicrob. Agents Chemotherapy.* **15**, 332–6.
55. Curtis, N. A. C., Orr, D., Ross, G. W. and Boulton, M. G. (1979) *Antimicrob. Agents Chemotherapy.* **16**, 325–8.
55a. Curtis, N. A. C., Orr, D., Ross, G. W. and Boulton, M. G. (1979) *Antimicrob. Agents Chemotherapy.* **16**, 533–9.
56. Curtis, S. J. and Strominger, J. L. (1978) *J. biol. Chem.* **253**, 2584–8.
57. Diaz-Maurino, T., Nieto, M. and Perkins, H. R. (1974) *Biochem. J.* **143**, 391–402.
58. Duguid, J. P. (1946) *Edinburgh med. J.* **53**, 401–412.
59. Dusart, J., Leyh-Bouille, M. and Ghuysen, J-M. (1977) *Eur. J. Biochem.* **81**, 33–44.
60. Dusart, J., Marquet, A., Ghuysen, J-M. and Perkins, H. R. (1975) *Eur. J. Biochem.* **56**, 57–65.
61. Eagle, H. (1954) *J. exp. Med.* **100**, 103–115.
62. Edwards, J. R. and Park, J. T. (1969) *J. Bact.* **99**, 459–62.
63. Faud, N., Frère, J.-M., Ghuysen, J.-M. and Duez, C. (1976) *Biochem. J.* **155**, 623–9.
64. Fehlner, J. R., Hutchinson, R. E. J., Tarbell, D. S. and Schenck, J. R. (1972) *Proc. Natn. Acad. Sci. U.S.A.* **69**, 2420–21.
65. Fein, J. E. and Rogers, H. J. (1976) *J. Bact.* **127**, 1427–42.
66. Filip, C., Fletcher, G., Wulff, J. L. and Earhart, C. F. (1973) *J. Bact.* **115**, 717–22.
67. Frère, J-M., Duez, C., Ghuysen, J-M. and Vandekerkhove, J. (1976) *FEBS Letts.* **70**, 257–60.
68. Frère, J-M., Ghuysen, J-M., Degelaen, J., Loffet, A. and Perkins, H. R. (1975) *Nature, Lond.* **258**, 168–170.
69. Frère, J-M., Ghuysen, J-M. and de Graeve, J. (1978) *FEBS Letts.* **88**, 147–50.
70. Frère, J-M., Ghuysen, J-M., Reynolds, P. E., Moreno, R. and Perkins, H. R. (1974) *Biochem. J.* **143**, 241–9.
71. Frère, J-M., Ghuysen, J-M., Vanderhaeghe, H., Adriaens, P., Degelaen, J. and de Graeve, J. (1976) *Nature, Lond.* **260**, 451–4.
72. Frère, J-M., Leyh-Bouille, M., Ghuysen, J-M. and Perkins, H. R. (1974) *Eur. J. Biochem.* **50**, 203–214.
73. Garbutt, J. T., Morehouse, A. L. and Hanson, A. M. (1961) *J. agr. Fd. Chem.* **9**, 285–9.
74. Gardner, A. D. (1940) *Nature, Lond.* **146**, 837–8.
75. Gellert, M., O'Dea, M. H., Itoh, T. and Tomizawa, J. (1976) *Proc. Natn. Acad. Sci. U.S.A.* **73**, 4474–8.
76. Georgopapadakou, N., Hammarström, S. and Strominger, J. L. (1977) *Proc. Natn. Acad. Sci. U.S.A.* **74**, 1009–1012.

77. Ghuysen, J.-M. (1977) *The Bacterial* D,D*-carboxypeptidase-transpeptidase Enzyme System*, University of Tokyo Press.
78. Ghuysen, J.-M. (1977) *J. gen. Microbiol.* **101**, 13–33.
79. Ghuysen, J-M., Frère, J-M., Leyh-Bouille, M., Dusart, J., Nguyen-Disteche, M., Coyette, J., Marquet, A., Perkins, H. R. and Nieto, M. (1975) *Bull. Inst. Pasteur* **73**, 101–140.
80. Ghuysen, J-M., Leyh-Bouille, M., Frère, J-M., Dusart, J., Johnson, K., Nakel, M., Coyette, J., Perkins, H. R. and Nieto, M. (1972) In *Molecular Mechanisms of Antibiotic Action in Protein Biosynthesis and Membranes*, eds. Munoz, E., Garcia-Ferrandiz, F. and Vazquez, D., pp. 406–426, London: Elsevier.
81. Ghuysen, J-M., Leyh-Bouille, M., Frère, J-M., Dusart, J. and Marquet, A., Perkins, H. R. and Nieto, M. (1974) *Ann. N. Y. Acad. Sci.* **235**, 236–66.
82. Giles, A. F. and Reynolds, P. E. (1979) *Nature, Lond.* **280**, 167–8.
83. Glaser, L. (1964) *J. biol. Chem.* **239**, 3178–86.
84. Goodell, E. W., Fazio, M. and Tomasz, A. (1978) *Antimicrob. Agents Chemotherapy.* **13**, 514–26.
85. Gorman, M. and Huber, F. (1977) *Annual Reports on fermentation processes Vol. 1.*, eds. Perlman, D. and Tsao, G. T., pp. 327–46, London and New York: Academic Press.
86. Gorman, M. and Ryan, C. W. (1972) In *Cephalosporins and Penicillins: Chemistry and Biology* ed. Flynn, E. H., pp. 533–79, London and New York: Academic Press.
87. Gunetileke, K. G. and Anwar, R. A. (1968) *J. biol. Chem.* **243**, 5770–78.
88. Hakenbeck, R., Waks, S. and Tomasz, A. (1978) *Antimicrob. Agents Chemotherapy.* **13**, 302–311.
89. Hammarstrom, S. and Strominger, J. L. (1975) *Proc. Natn. Acad. Sci. U.S.A.* **72**, 3463–7.
90. Hammarstrom, S. and Strominger, J. L. (1976) *J. biol. Chem.* **251**, 7947–9.
91. Hammes, W. P. (1976) *Eur. J. Biochem.* **70**, 107–113.
92. Hammes, W. P. (1978) *Eur. J. Biochem.* **91**, 501–507.
93. Hammes, W. P. and Kandler, O. (1976) *Eur. J. Biochem.* **70**, 97–106.
94. Hammes, W. P. and Neuhaus, F. C. (1974) *Antimicrob. Agents Chemotherapy.* **6**, 722–8.
95. Hammes, W. P. and Seidel, H. (1978) *Eur. J. Biochem.* **84**, 141–7.
96. Hammes, W. P. and Seidel, H. (1978) *Eur. J. Biochem.* **91**, 509–515.
97. Hancock, R. and Fitz-James, P. C. (1964) *J. Bact.* **87**, 1044–50.
97a. Hancock, R. E. W., Decad, G. M. and Nikaido, H. (1979) *Biochim. Biophys. Acta* **554**, 323–31.
98. Heifetz, A., Keenan, R. W. and Elbein, A. D. (1979) *Biochemistry* **18**, 2186–92.
99. Hendlin, D., Stapley, E. O., Jackson, M., Wallick, H., Miller, A. K., Wolf, F. J., Miller, T. W., Chaiet, L., Kahan, F. M., Foltz, E. L., Woodruff, H. B., Mata, J. M., Hernandez, S. and Mochales, S. (1969) *Science* **166**, 122–3.
100. Hayashi, S., Koch, J. P. and Lin, E. C. C. (1964) *J. biol. Chem.* **239**, 3098–3105.
101. Henning, U., Rehn, K., Braun, V., Höhn, B. and Schwarz, U. (1972) *Eur. J. Biochem.* **26**, 570–86.
102. Ho, P. P. K., Towner, R. K., Indelicato, J. M., Spitzer, W. A. and Koppel, G. A. (1972) *J. Antibiot.* **25**, 627–8.
103. Holtje, J. V. and Tomasz, A. (1975) *Proc. Natn. Acad. Sci. U.S.A.* **72**, 1690–94.
104. Holtje, J. V. and Tomasz, A. (1975) *J. biol. Chem.* **250**, 6072–6.

105. Horne, D., Hakenbeck, R. and Tomasz, A. (1977) *J. Bact.* **132**, 704–717.
106. Horne, D. and Tomasz, A. (1977) *Antimicrob. Agents Chemotherapy.* **11**, 888–96.
107. Horne, D. and Tomasz, A. (1979) *J. Bact.* **137**, 1180–84.
108. Huber, G. and Nesemann, G. (1968) *Biochem. biophys. Res. Commun.* **30**. 7–13.
109. Hughes, A. H., Stow, M., Hancock, I. C. and Baddiley, J. (1971) *Nature, New Biol.* **229**, 53–5.
110. Hughes, R. C. (1966) *Biochem. J.* **101**, 692–97.
111. Hughes, R. C., Tanner, P. J. and Stokes, E. (1970) *Biochem. J.* **120**, 159–70.
112. Ishimoto, N. and Strominger, J. L. (1966) *J. biol. Chem.* **241**, 638–50.
112a. Ishino, F. and Matsuhashi, M. (1979) *Agric. biol. Chem.* **43**, 2641–2.
113. Ito, T., Kodama, Y., Kawamura, K., Suzuki, K., Takatsuki, A. and Tamura, G. (1977) *Agric. biol. Chem.* **41**, 2303–305.
114. Iwaya, M., Jones, C. W., Korana, J. and Strominger, J. L. (1978) *J. Bact.* **133**, 196–202.
115. Iwaya, M. and Strominger, J. L. (1977) *Proc. Natn. Acad. Sci. U.S.A.* **74**, 2980–84.
116. Izaki, K., Matsuhashi, M. and Strominger, J. L. (1966) *Proc. Natn. Acad. Sci. U.S.A.* **55**, 656–63.
117. Izaki, K., Matsuhashi, M. and Strominger, J. L. (1968) *J. biol. Chem.* **243**, 3180–92.
118. Izaki, K. and Strominger, J. L. (1968) *J. biol. Chem.* **243**, 3193–201.
119. James, R., Haga, R. Y. and Pardee, A. B. (1975) *J. Bact.* **122**, 1283–92.
120. Johnson, A. W., Smith, R. M. and Guthrie, R. D. (1972) *J. Chem. Soc. Perkin*, **I**, 2153–9.
121. Johnson, B. A., Anker, H. and Meleney, F. L. (1945) *Science* **102**, 376–7.
122. Johnston, L. S. and Neuhaus, F. C. (1975) *Biochemistry* **14**, 2754–60.
123. Johnston, M. M. and Diven, W. F. (1969) *J. biol. Chem.* **244**, 5414–20.
124. Johnston, R. B., Scholz, J. J., Diven, W. F. and Shepherd, S. (1968) In *Pyridoxal Catalysis: Enzymes and Model Systems*, eds. Snell, E., Braunstein, A. E., Severin, E. S. and Torchnisky, Y. M. pp. 537–47, New York: John Wiley & Sons Inc.
125. Jordan, D. C. (1961) *Biochem. biophys. Res. Commun.* **6**, 167–70.
126. Jordan, D. C. (1965) *Can. J. Microbiol.* **11**, 390–93.
127. Kaczorowski, G., Shaw, L., Laura, R. and Walsh, C. T. (1975) *J. biol. Chem.* **250**, 8921–30.
128 Kadner, R. J. and Winkler, H. H. (1973) *J. Bact.* **113**, 895–900.
129. Kahan, F. M., Kahan, J. S., Cassidy, P. J. and Kropp, H. (1974) *Ann. N. Y. Acad. Sci.* **235**, 364–86.
130. Kamiryo, T. and Strominger, J. L. (1974) *J. Bact.* **117**, 568–77.
131. Katz, W., Matsuhashi, M., Dietrich, C. P. and Strominger, J. L. (1967) *J. biol. Chem.* **242**, 3207–217.
132. Katz, W. and Martin, H. H. (1970) *Biochem. biophys. Res. Commun.* **39**, 744–9.
133. Kenig, M. and Reading, C. (1979) *J. Antibiotics* **32**, 549–54.
134. Kleppe, G. and Strominger, J. L. (1979) *J. biol. Chem.* **254**, 4856–62.
135. Kozarich, J. W. (1977) In *Microbiology 1977* ed. Schlessinger, D., pp. 203–208, Washington, D.C: American Soc. Microbiol.
136. Kozarich, J. W. and Strominger, J. L. (1978) *J. biol. Chem.* **253**, 1272–8.
137. Kuo, S-C. and Lampen, J. O. (1976) *Archs. Biochem. Biophys.* **172**, 574–81.
138. Lambert, M. P. and Neuhaus, F. C. (1972) *J. Bact.* **110**, 978–87.

139. Lark, C. and Schichtel, R. (1962) *J. Bact.* **84**, 1241–4.
140. Laskin, A. I., Chan, W. M., Smith, D. A. and Mayers, E. (1968) *Antimicrob. Agents Chemotherapy*. pp. 251–6.
141. Lawrence, P. J. and Strominger, J. L. (1970) *J. biol. Chem.* **245**, 3653–9.
142. Lawrence, P. J. and Strominger, J. L. (1970) *J. biol. Chem.* **245**, 3660–66.
143. Lee, B. (1971) *J. molec. Biol.* **61**, 463–9.
144. Lehle, L. and Tanner, W. (1976) *FEBS Letts.* **71**, 167–70.
145. Lin, E. C. C., Koch, J. P., Chused, T. M. and Jorgensen, S. E. (1962) *Proc. Natn. Acad. Sci. U.S.A.* **48**, 2145–50.
146. Linder, R. and Salton, M. R. J. (1975) *Eur. J. Biochem.* **55**, 291–7.
147. Linnett, P. E., Roberts, R. J. and Strominger, J. L. (1974) *J. biol. Chem.* **249**, 2497–506.
148. Linnett, P. E. and Strominger, J. L. (1973) *Antimicrob. Agents Chemotherapy*. **4**, 231–6.
149. Linnett, P. E. and Strominger, J. L. (1974) *J. biol. Chem.* **249**, 2489–96.
150. Lomakina, N. N., Bognar, R., Brazhnikova, M. G., Sztaricskai, F. and Muravyeva, L. I. (1970) *7th Intl. Symp. Chem. Nat. Products.* pp. 625–6, Riga.
151. Lopez, R., Lain, C. R., Tapia, A., Waks, S. B. and Tomasz, A. (1976) *Antimicrob. Agents Chemotherapy*. **10**, 697–706.
152. Lugtenberg, E. J. J., Hellings, J. A. and van de Berg, G. J. (1972) *Antimicrob. Agents Chemotherapy*. **2**, 485–91.
153. Lugtenberg, E. J. J., Schijndel van Dam, A. and van Bellegem, T. H. M. (1971) *J. Bact.* **108**, 20–29.
154. Lund, F. and Tybring, L. (1972) *Nature, New Biol.* **236**, 135–7.
155. Lynch, J. L. and Neuhaus, F. C. (1966) *J. Bact.* **91**, 449–60.
156. Manning, J. M., Merrifield, N. E., Jones, W. H. and Gotschlich, E. C. (1974) *Proc. Natn. Acad. Sci. U.S.A.* **71**, 417–21.
157. Marquet, A., Dusart, J., Ghuysen, J-M. and Perkins, H. R. (1974) *Eur. J. Biochem.* **46**, 515–23.
158. Marquet, A., Frère, J-M., Ghuysen, J-M. and Loffet, A. (1979) *Biochem. J.* **177**, 909–916.
159. Marquet, A., Nieto, M. and Diaz-Mauriño, T. (1976) *Eur. J. Biochem.* **68**, 581–9.
160. Marshall, F. J. (1965) *J. med. Chem.* **8**, 18–22.
161. Martin, H. H. (1964) *J. gen. Microbiol.* **36**, 441–50.
162. Martin, H. H. (1966) *A. Rev. Biochem.* **35**, 457–84.
163. Matsuhashi, M., Dietrich, C. P. and Strominger, J. L. (1967) *J. biol. Chem.* **242**, 3191–206.
164. Matsuhashi, S., Kamiryo, T., Blumberg, P. M., Linnett, P., Willoughby, E. and Strominger, J. L. (1974) *J. Bact.* **117**, 578–87.
165. Matsuhashi, M., Ohara, I. and Yoshiyama, Y. (1970) *Prog. Antimicrob. Anticancer Chemoth. Proc. Int. Congr. Chemoth. 6th 1969, Vol. 1,* pp. 226–9.
166. Matsuhashi, M., Takagaki, Y., Maruyama, I. N., Tamaki, S., Nishimura, Y., Suzuki, H., Ogino, U. and Hirota, Y. (1977) *Proc. Natn. Acad. Sci. U.S.A.* **74**, 2976–9.
167. Matsuhashi, M., Tamaki, S., Curtis, S. J. and Strominger, J. L. (1979) *J. Bact.* **137**, 644–7.
168. Matzuzawa, H., Hayakawa, K., Sata, T. and Imahori, K. (1973) *J. Bact.* **115**, 436–42.
169. McArthur, H. A. I., Roberts, F. M., Hancock, I. C. and Baddiley, J. (1978) *FEBS Letts.* **86**, 193–200.
170. Meadow, P. M., Anderson, J. S. and Strominger, J. L. (1964) *Biochem. biophys. Res. Commun.* **14**, 382–7.

171. Melchior, N. H., Blom, J., Tybring, L. and Birch-Anderson, A. (1973) *Acta path. microbiol. scand. Ser. B.* **81**, 393–407.
172. Miller, A. K., Celozzi, E., Pelak, B. A., Stapley, E. O. and Hendlin, D. (1972) *Antimicrob. Agent. Chemotherapy.* **2**, 281–6.
173. Miller, T. W., Goegelman, R. T., Weston, R. G., Putker, I. and Wolf, F. J. (1972) *Antimicrob. Agent. Chemotherapy.* **2**, 132–5.
174. Mirelman, D. and Bracha, R. (1974) *Antimicrob. Agents Chemotherapy.* **5**, 663–6.
175. Mirelman, D., Bracha, R. and Sharon, N. (1972) *Proc. Natn. Acad. Sci. U.S.A.* **69**, 3355–9.
176. Mirelman, D., Bracha, R. and Sharon, N. (1974) *Biochemistry* **13**, 5045–53.
177. Mirelman, D. and Sharon, N. (1972) *Biochem. biophys. Res. Commun.* **46**, 1909–917.
178. Molander, C. W., Kagan, B. M., Weinberger, H. J., Heimlich, E. M. and Busser, R. J. (1964) *J. Bact.* **88**, 591–4.
179. Nakae, T. (1976) *Biochem. biophys. Res. Commun.* **64**, 1224–30.
180. Nakae, T. (1976) *J. biol. Chem.* **251**, 2176–8.
181. Nakagawa, J., Tamaki, S. and Matsuhashi, M. (1970) *Agric. biol. Chem.* **43**, 1379–80.
182. Nakamura, S., Arai, M., Karasawa, K. and Yonehara, H. (1957) *J. Antibiotics Ser. A.* **10**, 248–53.
183. Neuhaus, F. C. (1962) *J. biol. Chem.* **237**, 3128–35.
184. Neuhaus, F. C. (1967) In *Antibiotics Vol. 1 – Mechanism of Action,* eds. Gottlieb, D. and Shaw, P. D., pp. 40–83, Berlin: Springer-Verlag.
185. Neuhaus, F. C. (1968) *Antimicrob. Agents Chemotherapy* **1**, 304–13.
186. Neuhaus, F. C., Carpenter, C. V., Lambert, M. P. and Wargel, R. J. (1972) In *Molecular Mechanisms of Antibiotic Action in Protein Biosynthesis and Membranes*, eds. Munoz, E., Garcia-Ferrandiz, F. and Vazquez, D., pp. 339–62, London: Elsevier.
187. Neuhaus, F. C. and Lynch, J. L. (1964) *Biochemistry* **3**, 471–80.
188. Nguyen-Distèche, M., Ghuysen, J-M., Pollock, J. J., Reynolds, P. E., Perkins, H. R., Coyette, J. and Salton, M. R. J. (1974) *Eur. J. Biochem.* **41**, 447–55.
189. Nguyen-Distèche, M., Pollock, J. J., Ghuysen, J-M., Puig, J., Reynolds, P. E., Perkins, H. R., Coyette, J. and Salton, M. R. J. (1974) *Eur. J. Biochem.* **41**, 457–63.
190. Nieto, M. and Perkins, H. R. (1971) *Biochem. J.* **123**, 773–87.
191. Nieto, M. and Perkins, H. R. (1971) *Biochem. J.* **123**, 789–803.
192. Nieto, M. and Perkins, H. R. (1971) *Biochem. J.* **124**, 845–52.
193. Nieto, M., Perkins, H. R., Frère, J. M. and Ghuysen, J-M. (1973) *Biochem. J.* **135**, 493–505.
194. Nikaido, H. (1979) *Angewante Chem.* **91**, 394–407.
194a. Nikaido, H. and Nakae, T. (1979) *Adv. microbial Physiol.* **20**, 163–250.
195. Nikaido, H., Song, S. A., Shaltiel, L. and Nurminen, M. (1977) *Biochem. biophys. Res. Commun.* **76**, 324–330.
195a. Noguchi, H., Matsuhashi, M. and Mitsuhashi, S. (1979) *Eur. J. Biochem.* **100**, 41–9.
196. Noguchi, H., Matsuhashi, M., Nikaido, T., Itoh, J., Matsubara, N., Takaoka, M. and Mitsuhashi, S. (1978) In *Microbial Drug Resistance, vol. 2*, ed. Mitsuhashi, S., pp. 361–387, Baltimore: University Park Press.
197. Nomura, S., Yamane, K., Sasaki, T., Yamasaki, M., Tamura, G. and Maruo, B. (1978) *J. Bact.* **136**, 818–21.
198. O'Callaghan, C., Morris, A., Kirby, S. A. and Shingler, A. H. (1972) *Antimicrob. Agent. Chemotherapy.* **1**, 283–8.

199. Ohya, S., Yamasaki, M., Sugawara, S. and Matsuhashi, M. (1979) *J. Bact.* **137**, 474–9.
200. Oppenheim, B., Koren, R. and Patchornik, A. (1974) *Biochem. biophys. Res. Commun.* **57**, 562–71.
201. Park, J. T. (1952) *J. biol. Chem.* **194**, 877–84.
202. Park, J. T. (1952) *J. biol. Chem.* **194**, 885–95.
203. Park, J. T. (1952) *J. biol. Chem.* **194**, 897–904.
204. Park, J. T. (1958) *Symp. Soc. gen. Microbiol.* **8**, 49–61.
205. Park, J. T. (1958) *Biochem. J.* **70**, 2P.
206. Park, J. T. and Burman, L. (1973) *Biochem. biophys. Res. Commun.* **51**, 863–8.
207. Park, J. T., Edwards, J. R. and Wise, E. M. (1974) *Ann. N. Y. Acad. Sci.* **235**, 300–309.
208. Park, J. T. and Johnson, M. J. (1949) *J. biol. Chem.* **179**, 585–92.
209. Park, J. T. and Strominger, J. L. (1957) *Science* **125**, 99–101.
210. Perkins, H. R. (1969) *Biochem. J.* **111**, 195–205.
211. Perkins, H. R. and Nieto, M. (1970) *Biochem. J.* **116**, 83–92.
212. Perkins, H. R. and Nieto, M. (1972) In *Molecular Mechanisms of Antibiotic Action in Protein Biosynthesis and Membranes*, eds. Munoz, E., Garcia-Ferrandiz, F. and Vazquez, D., pp. 363–87. London and New York: Elsevier.
213. Perkins, H. R. and Nieto, M. (1974) *Ann. N. Y. Acad. Sci.* **235**, 348–63.
214. Perkins, H. R., Nieto, M., Frère, J.-M., Leyh-Bouille, M. and Ghuysen, J.-M. (1973) *Biochem. J.* **131**, 707–718.
215. Philip, J. E., Schenck, J. R., Hargie, M. P., Holper, J. C. and Grundy, W. E. (1960) *Antimicrob. Agent Ann.* 10–16.
216. Pless, D. D. and Neuhaus, F. C. (1973) *J. biol. Chem.* **248**, 1568–76.
217. Pollock, J. J., Nguyen-Distèche, M., Ghuysen, J-M., Coyette, J., Linder, R., Salton, M. R. J., Kim, K. S., Perkins, H. R. and Reynolds, P. E. (1974) *Eur. J. Biochem.* **41**, 439–46.
218. Powell, D. A., Duckworth, M. and Baddiley, J. (1974) *FEBS Letts.* **41**, 259–63.
219. Rando, R. R. (1975) *Acc. chem. Res.* **8**, 281–88.
220. Rando, R. R. (1975) *Biochem. Pharmacol.* **24**, 1153–60.
221. Rasmussen, J. R. and Strominger, J. L. (1978) *Proc. Natn. Acad. Sci. U.S.A.* **75**, 84–8.
222. Reitz, R. H., Slade, H. D. and Neuhaus, F. C. (1967) *Biochemistry* **6**, 2561–70.
223. Reynolds, P. E. (1961) *Biochim. biophys. Acta* **52**, 403–405.
224. Reynolds, P. E. (1966) *Symp. Soc. gen. Microbiol.* **16**, 47–69.
225. Reynolds, P. E. (1971) *Biochim. biophys. Acta* **237**, 239–54.
226. Reynolds, P. E. (1971) *Biochim. biophys. Acta* **237**, 255–72.
227. Reynolds, P. E. and Barnett, H. J. (1974) *Ann. N.Y. Acad. Sci.* **235**, 269–82.
228. Reynolds, P. E., Shepherd, S. T. and Chase, H. A. (1978) *Nature, (Lond.)* **271**, 568–70.
229. Richmond, M. H., Clark, D. C. and Wotton, S. (1976) *Antimicrob. Agents Chemotherapy.* **10**, 215–18.
230. Richmond, M. H. and Curtis, N. A. C. (1974) *Ann. N. Y. Acad. Sci.* **235**, 553–67.
231. Richmond, M. H. and Wotton, S. (1976) *Antimicrob. Agents Chemotherapy.* **10**, 219–22.

232. Richmond, M. H. and Sykes, R. B. (1973) In *Adv. Microbiol. Physiol. Vol. 9.* eds. Rose, A. H. and Tempest, D. W., pp. 31–88, London and New York: Academic Press.
234. Robson, R. L. and Baddiley, J. (1977) *J. Bact.* **129**, 1051–8.
235. Rogers, H. J. (1967) *Biochem. J.* **103**, 90–102.
236. Rogers, H. J. (1967) *Nature, Lond.* **213**, 31–3.
237. Rogers, H. J. and Forsberg, C. W. (1971) *J. Bact.* **108**, 1235–43.
238. Roze, U. and Strominger, J. L. (1966) *Molec. Pharmacol.* **2**, 92–4.
239. Sabath, L. D., Wheeler, N., Laverdiere, M., Blazevic, D. and Wilkinson, B. J. (1977) *Lancet* 443–7.
240. Sandermann, H. (1976) *Biochim. biophys. Acta* **444**, 783–8.
241. Schepartz, S. A. and Johnson, M. J. (1956) *J. Bact.* **71**, 84–90.
242. Schleifer, K. H. and Kandler, O. (1972) *Bact. Rev.* **36**, 407–477.
243. Schilf, W., Frère, P., Frère, J-M., Martin, H. H., Ghuysen, J-M., Adriaens, P. and Meesschaert, B. (1978) *Eur. J. Biochem.* **85**, 325–30.
244. Schwarz, U., Asmus, A. and Frank, H. (1969) *J. molec. Biol.* **41**, 419–29.
245. Scudamore, R. A., Beveridge, T. J. and Goldner, M. (1979) *Antimicrob. Agents. Chemotherapy.* **15**, 182–9.
246. Selitrennikoff, C. P. (1979) *Archs. Biochem. Biophys.* **195**, 243–4.
247. Sharpe, A., Blumberg, P. M. and Strominger, J. L. (1974) *J. Bact.* **117**, 926–7.
248. Sheldrick, G. M., Jones, P. G., Kennard, O., Williams, D. H. and Smith, G. A. (1978) *Nature, Lond.* **271**, 223–5.
249. Shepherd, S. T., Chase, H. A. and Reynolds, P. E. (1977) *Eur. J. Biochem.* **78**, 521–32.
250. Shockman, G. D. (1959) *Proc. Soc. Exp. Biol. Med.* **101**, 693–5.
251. Shockman, G. D. and Lampen, J. O. (1962) *J. Bact.* **84**, 508–512.
252. Siewert, G. and Strominger, J. L. (1967) *Proc. Natn. Acad. Sci. U.S.A.* **57**, 767–73.
253. Sinha, R. K. and Neuhaus, F. C. (1968) *J. Bact.* **96**, 374–82.
254. Slusarchyk, W. A. (1971) *Biotech. Bioeng.* **13**, 399–407.
255. Smith, D. H. and Davis, B. D. (1967) *J. Bact.* **93**, 71–9.
256. Somma, S., Merati, W. and Parenti, F. (1977) *Antimicrob. Agents Chemotherapy.* **11**, 396–401.
257. Spratt, B. G. (1975) *Proc. Natn. Acad. Sci. U.S.A.* **72**, 2999–3003.
258. Spratt, B. G. (1977) *Eur. J. Biochem.* **72**, 341–52.
259. Spratt, B. G. (1977) *Antimicrob. Agents Chemotherapy.* **11**, 161–6.
260. Spratt, B. G. (1977) In *Microbiology 1977*, ed. Schlessinger, D. pp. 182–190, Washington, D.C: American Soc. for Microbiol.
261. Spratt, B. G. (1978) *Sci. Prog. Oxf.* **65**, 101–128.
262. Spratt, B. G. (1979) In *Microbial Drug Resistance, Vol. 2*, ed. Mitsuhashi, S., pp. 349–60 Baltimore: University Park Press.
263. Spratt, B. G., Jobanputra, V. and Schwarz, U. (1977). *FEBS Letts.* **79**, 374–8.
264. Spratt, B. G., Jobanputra, V. and Zimmermann, W. (1977) *Antimicrob. Agents Chemotherapy.* **12**, 406–409.
265. Spratt, B. G. and Pardee, A. B. (1975) *Nature, Lond.* 516–17.
266. Spratt, B. G. and Strominger, J. L. (1976) *J. Bact.* **127**, 660–63.
267. Stone, K. J. and Strominger, J. L. (1971) *Proc. Natn. Acad. Sci. U.S.A.* **68**, 3223–7.
268. Storm, D. R. and Strominger, J. L. (1973) *J. biol. Chem.* **248**, 3940–45.
269. Storm, D. R. and Strominger, J. L. (1974) *J. biol. Chem.* **249**, 1823–927.

270. Strominger, J. L., Ito, E. and Threnn, R. H. (1960) *J. Am. Chem. Soc.* **82**, 998–9.
271. Strominger, J. L., Izaki, K., Matsuhashi, M. and Tipper, D. J. (1967) *Fedn. Proc. Fedn. Am. Socs. exp. Biol.* **26**, 9–22.
272. Strominger, J. L., Threnn, R. H. and Scott, S. S. (1959) *J. Am. Chem. Soc.* **81**, 3803–804.
273. Strominger, J. L., Willoughby, E., Kamiryo, T., Blumberg, P. M. and Yocum, R. R. (1974) *Ann. N. Y. Acad. Sci.* **235**, 210–24.
274. Struve, W. G. and Neuhaus, F. C. (1965) *Biochem. biophys. Res. Commun.* **18**, 6–12.
275. Struve, W. G., Sinha, R. K. and Neuhaus, F. C. (1966) *Biochemistry* **5**, 82–93.
276. Suginaka, H., Blumberg, P. M. and Strominger, J. L. (1972) *J. biol. Chem.* **247**, 5279–88.
277. Suginaka, H., Shimatini, M., Ogawa, M. and Kotani, S. (1979) *J. Antibiotics* **32**, 73–7.
278. Suzuki, H., Nishimura, Y. and Hirota, Y. (1978) *Proc. Natn. Acad. Sci. U.S.A.* **75**, 664–8.
278a. Suzuki, H., van Heijenoort, Y., Tamura, T., Mizoguchi, J., Hirota, Y. and van Heijnoort, J. (1980) *FEBS Letts.* **110**, 245–9.
279. Sweet, R. M. and Dahl, L. F. (1970) *J. Am. Chem. Soc.* **92**, 5489–507.
280. Sykes, R. B. and Matthew, M. (1976) *J. Antimicrob. Chemotherapy.* **2**, 115–57.
281. Takatsuki, A., Arima, K. and Tamura, G. (1971) *J. Antibiotics* **24**, 215–23.
282. Takatsuki, A., Kawamura, K., Okina, M., Kodama, Y., Ito, T. and Tamura, G. (1977) *Agric. biol. Chem.* **41**, 2307–309.
283. Takatsuki, A., Shimizu, K. and Tamura, G. (1972) *J. Antibiotics* **25**, 75–85.
284. Takatsuki, A. and Tamura, G. (1971) *J. Antibiotics* **24**, 785–94.
285. Tamaki, S., Nakajima, S. and Matsuhashi, M. (1977) *Proc. Natn. Acad. Sci. U.S.A.* **74**, 5472–6.
286. Tamura, T., Imae, Y. and Strominger, J. L. (1976) *J. biol. Chem.* **251**, 414–23.
287. Tamura, G., Sasaki, T., Matsuhashi, M., Takatsuki, A. and Yamasaki, M. (1976) *Agric. biol. Chem.* **40**, 447–9.
288. Tanaka, H., Iwai, Y., Oiwa, R., Shinohara, S., Shimizu, S., Oka, T. and Omura, S. (1977) *Biochim. biophys. Acta* **497**, 633–40.
289. Tipper, D. J. (1979) *Rev. Infect. Dis.* **1**, 39–53.
290. Tipper, D. J. and Strominger, J. L. (1965) *Proc. Natn. Acad. Sci. U.S.A.* **54**, 1133–41.
291. Tipper, D. J. and Strominger, J. L. (1968) *J. biol. Chem.* **243**, 3169–79.
292. Tkacz, J. and Wong, A. (1978) *Fedn. Proc. Fedn. Am. Socs. exp. Biol.* **37**, 1766.
293. Tomasz, A. (1974) *Ann. N. Y. Acad. Sci.* **235**, 439–48.
294. Tomasz, A., Albino, A. and Zaneti, E. (1970) *Nature, Lond.* **227**, 138–40.
295. Tomasz, A. and Holtje, J. V. (1977) In *Microbiology 1977*, ed. Schlessinger, D., pp. 209–215, Washington, D.C: American Soc. Microbiol.
296. Tomasz, A. and Waks, S. (1975) *Proc. Natn. Acad. Sci. U.S.A.* **72**, 4162–6.
297. Tynecka, Z. and Ward, J. B. (1975) *Biochem. J.* **146**, 253–67.
298. Umbreit, J. N. and Strominger, J. L. (1973) *J. biol. Chem.* **248**, 6759–66.
299. Umbreit, J. N. and Strominger, J. L. (1973) *J. biol. Chem.* **248**, 6767–71.
300. Van Heijenoort, Y., Derrien, M. and van Heijenoort, J. (1978) *FEBS Letts.* **89**, 141–4.
300a. Van Heijenoort, Y. and Van Heijenoort, J. (1980) *FEBS Letts.* **110**, 241–4.

301. Venkateswaran, P. S. and Wu, H. C. (1972) *J. Bact.* **110**, 935–44.
302. Virudachalam, R. and Rao, V. S. R. (1977) *Int. J. Peptide Protein Res.* **10**, 51–9.
303. Waks, S. and Tomasz, A. (1978) *Antimicrob. Agents Chemotherapy.* **13**, 293–301.
304. Walker, J. E. and Abraham, E. P., (1970) *Biochem. J.* **118**, 563–70.
305. Wallas, C. H. and Strominger, J. L. (1963) *J. biol. Chem.* **238**, 2264–6.
306. Wang, E. and Walsh, C. T. (1978) *Biochemistry* **17**, 1313–21.
307. Ward, J. B. (1974) *Biochem. J.* **141**, 227–41.
308. Ward, J. B. (1977) *FEBS Letts.* **78**, 151–4.
308a. Ward, J. B., Wyke, A. W. and Curtis, C. A. M. (1980) *Biochem. Transact.* **8**, 164–6.
309. Wargel, R. J., Shadur, C. A. and Neuhaus, F. C. (1970) *J. Bact.* **103**, 778–88.
310. Wargel, R. J., Shadur, C. A. and Neuhaus, F. C. (1971) *J. Bact.* **105**, 1028–35.
311. Weidel, W. and Pelzer, H. (1964) *Adv. Enzymol.* **26**, 193–232.
312. Weringa, W. D., Williams, D. H., Feeney, J., Brown, J. P. and King, R. W. (1972) *J. Chem. Soc. Perkin* **1**, 443–6.
313. Weston, A. and Perkins, H. R. (1977) *FEBS Letts.* **76**, 195–8.
314. Wickus, G. G. and Strominger, J. L. (1972) *J. biol. Chem.* **247**, 5297–306.
315. Wickus, G. G. and Strominger, J. L. (1972) *J. biol. Chem.* **247**, 5307–311.
316. Williams, R. E. O. (1963) *J. gen. Microbiol.* **33**, 325–34.
317. Winkler, H. H. (1966) *Biochim. biophys. Acta* **117**, 231–40.
318. Wise, E. M. and Park, J. T. (1965) *Proc. Natn. Acad. Sci. U.S.A.* **54**, 75–81.
319. Wishnow, R. M., Strominger, J. L., Birge, C. H. and Threnn, R. H. (1965) *J. Bact.* **89**, 1117–23.
320. Wyke, A. W. and Ward, J. B. (1977) *J. Bact.* **130**, 1055–63.
321. Yamada, M., Matsuhashi, M. and Torii, M. (1978) *J. Gen. Appl. Microbiol.* **24**, 307–315.
322. Yocum, R. R., Blumberg, P. M. and Strominger, J. L. (1974) *J. biol. Chem.* **249**, 4863–71.
323. Yocum, R. R., Waxman, D. J., Rassmussen, J. R. and Strominger, J. L. (1979) *Proc. Natn. Acad. Sci. U.S.A.* **76**, 2730–34.
324. Yudkin, M. D. (1963) *Biochem. J.* **89**, 290–96.
325. Zimmermann, W. and Rosselet, A. (1977) *Antimicrob. Agents Chemotherapy* **12**, 368–72.

10
Biosynthesis of other bacterial wall components

10.1 Biosynthesis of teichoic acids

The isolation and structure of many teichoic acids has been discussed in detail in Chapter 7.1.1. As described there the teichoic acids were initially thought to be polymers of either polyglycerol or polyribitol phosphate, together with associated sugars and ester-linked D-alanine [7]. However, in recent years the term teichoic acid, in the case of wall-associated material, has come to mean any one of a range of phosphate-containing polymers found covalently linked to peptidoglycan [4, 8]. In contrast, the membrane teichoic acids so far examined have all been found to be polyglycerol phosphates [4, 8]. In this Chapter we will consider the biosynthesis of both wall and membrane teichoic acids and, in the case of the wall polymers, the formation of the linkage to peptidoglycan.

The biosynthetic enzymes for all teichoic acids appear to be located in the cytoplasmic membrane. Earlier investigations had used particulate enzyme preparations, obtained by differential centrifugation, from sonically disrupted organisms or from osmotically lysed protoplasts that had been prepared by lysozyme digestion of the peptidoglycan. More recently, both teichoic acid and peptidoglycan synthesis have been studied by using toluenized cells or membrane and wall–membrane preparations isolated from bacteria disrupted by grinding with alumina (see Chapter 8.6.1).

The teichoic acids differ from all other bacterial wall polymers in that the discovery of their nucleotide-linked precursors preceded the discovery of the polymers themselves. Thus in 1956 Baddiley *et al.* [10] isolated both CDP-ribitol and CDP-glycerol from *Lactobacillus arabinosus*, an observation which led to the discovery of poly (ribitol phosphate) in this and other organisms [6]. Subsequently, Shaw [193] demonstrated the presence of the enzymes, CDP–glycerol pyrophosphorylase and CDP–ribitol pyrophosphorylase in cell-free extracts from a number of Gram-positive bacteria. These enzymes catalyse the synthesis of the nucleotide precursors from cytidine-triphosphate and polyol-phosphate according to the following equation.

$$\text{CTP} + \text{polyolphosphate} \rightleftharpoons \text{CDP-polyol} + \text{PP}_i.$$

D-Ribitol-5-phosphate is formed by an NADH-dependent reduction of D-ribulose-1-phosphate [65]; glycolysis is generally assumed to be the source of D-glycerol-1-phosphate.

10.1.1 *Biosynthesis of poly(glycerol-phosphate) and poly(ribitol-phosphate) teichoic acids*

The first experiments utilizing these cytidine-linked precursors and membrane preparations for the synthesis of teichoic acid were those of Burger and Glaser [29]. These studies, using particulate enzyme preparations from *Bacillus licheniformis* and *B. subtilis*, established that CDP-glycerol was the only substrate required for poly(glycerol-phosphate) polymerase and the reaction which required the presence of either Ca^{2+} (10 mM) or Mg^{2+} (40 mM) for optimum activity, proceeded according to the following equation

$$\text{CDP–glycerol} + (\text{glycero-phosphate})_n \rightarrow \text{CMP} + (\text{glycero-phosphate})_{n+1}.$$

Degradation studies and end-group analysis were both consistent with the product of the reaction being poly-1, 3-glycero-phosphate. Burger and Glaser [29] also suggested that newly-synthesized material was being linked to teichoic acid already present in the enzyme preparation. Similar conclusions were drawn by Kennedy and Shaw [104] who showed that newly-synthesized glycero-phosphate units were added to pre-existing teichoic acid chains in membrane preparations of *B. subtilis*. Degradation of the glycerol-terminus by periodate oxidation released 45% of the incorporated radioactivity as formaldehyde. Newly-synthesized units added at the phosphate terminus would not be susceptible to periodate oxidation and consequently radioactive formaldehyde would not be formed. Thus the direction of synthesis is similar to that reported for the addition of glucosyl units in glycogen synthesis, and differed from the direction of synthesis observed for both the O-antigen of *Salmonella anatum* [172] and peptidoglycan [208], where the newly-synthesized units are added at the reducing termini of the growing chains.

Glucosylation of poly(glycerol-phosphate) involving the transfer of glucosyl residues from UDP–glucose to the free hydroxyl groups of the polymer has also been reported [30]. The particulate enzyme from *B. subtilis* utilized for the synthesis of poly(glycerol-phosphate) [29] also catalysed this glucosylation reaction. Either poly(glycerol-phosphate) extracted from the walls of the organism, or polymer synthesized *in vitro* were substrates for the reaction; the simultaneous synthesis of poly(glycerol-phosphate) and its glucosylation were not described. Non-glucosylated teichoic acid extracted from the wall of a phage-resistant mutant of *B. subtilis* was used as the substrate to study the glucosyl transferase enzyme in *B. subtilis* 168 [28]. In this case, soluble enzyme was obtained from the cytoplasm of exponentially growing cells, although the majority of the enzyme was associated with the membrane fraction. In stationary phase organisms, all the transferase was membrane-bound. Evidence for the participation of a lipid intermediate in the glucosylation reaction was not obtained with either form of the enzyme.

Similar particulate enzyme preparations from *L. plantarum* [66] and *Staphylococcus aureus* Copenhagen [99] have been used to study the synthesis of poly(ribitol-phosphate) from CDP–ribitol. The reaction was analogous to that described above for the synthesis of poly(glycerol-phosphate) with CMP being the reaction product. Of particular interest was the finding that the poly(ribitol-phosphate) synthetase (polymerase) of *L. plantarum* remained closely associated with the wall of the organism. The particulate enzyme used was essentially equivalent to the wall-membrane preparations described in previous chapters; preparations substantially free from wall were much less active. In contrast, the polymerase from *S. aureus* was present in a particulate fraction - presumably membranes, which sedimented at between 30 000 and 100 000 g. Optimum polymer synthesis required the presence of relatively high concentrations (10-30 mM) of divalent cations, whereas the reaction was inhibited by the addition of CMP. The addition of potential acceptors such as isolated walls, teichoic acid–peptidoglycan complexes or poly(ribitol-phosphate) itself were without effect on the reaction. The polymer synthesized by preparations from *L. plantarum* had an average chain length of 7.9 units [66], although in both cases the newly synthesized material appeared linked to some unidentified acceptor. More recently, Fiedler and Glaser [56, 57] have purified poly(ribitol-phosphate) polymerase by conventional techniques from Triton X-100 extracts of *S. aureus* H membranes. The purified enzyme, which showed a requirement for phospholipid and divalent cation or spermidine, was totally dependent for activity on the presence of an acceptor identified as lipoteichoic acid carrier. The use of this carrier and the nature of the biosynthetic reaction *in vivo* are discussed in detail below.

Particulate enzyme preparations from *S. aureus* also catalysed the transfer of *N*-acetylglucosamine from the uridine nucleotide to poly(ribitol-phosphate) according to the following equation [99, 141].

$$\text{(ribitol–phosphate)}_n + n\text{UDPGlc}N\text{Ac} \longrightarrow \underset{\underset{\text{Glc}N\text{Ac}}{\uparrow}}{\text{(ribitol–phosphate)}_n} + n\text{UDP.}$$

In the earlier investigation [141] glycosylation occurred using pre-formed poly(ribitol phosphate) as the acceptor. This was prepared by removal of β-linked *N*-acetylglucosamine from the teichoic acid of *S. aureus* Copenhagen using β-*N*-acetylglucosaminidase. The native teichoic acid contains both α- and β-linked substituents. Later investigations showed, however, that a substantially greater rate of glycosylation was obtained if CDP-ribitol and UDP-*N*-acetylglucosamine were added simultaneously to the enzyme preparation [99]. Moreover, delayed addition of UDP-*N*-acetylglucosamine to the reaction mixture resulted in a marked decrease in the amount of *N*-acetylglucosamine incorporated. Thus glycosylation of poly(ribitol-phosphate) synthesized concomitantly appears to occur more readily than does glycosylation of both preformed and exogenously added polymer. These observations suggest the existence in the membrane of a close spatial relationship between the synthetase and the glycosyl transferase. In an extension of their earlier observations, Nathenson *et al.* [140] showed that various strains of *S. aureus* contain different proportions of α- and β-linked *N*-acetylglucosamine in their teichoic acids. Using as the acceptor teichoic acid, from which β-linked substituents were removed as described above, they found that enzyme preparations from the various organisms synthesized α- and β-linkages in a ratio close to that found in their teichoic acids. Thus *S. aureus* strains H and Duncan synthesized linkages predominantly of the β- configuration, whereas linkages of the α-configuration were obtained with strain 3528. In accordance with its known composition, *S. aureus* Copenhagen gave results between these two extremes.

Glucosylation of poly(ribitol-phosphate) has also been studied with enzyme preparations from *B. subtilis* W23 [37]. The walls of this organism contain both poly(ribitol-phosphate) and poly(glucosylribitol-phosphate). The former polymer was used as acceptor for the glucosyl transferase and on incubation with UDP-glucose became fully glucosylated. The linkages synthesized were of the β-configuration. In addition, teichoic acid present in isolated walls of the organism would also act as acceptor, some 20% of the unsubstituted ribitol residues becoming glucosylated. The enzyme preparation used in these experiments was 'solubilized' from highly fragmented membrane preparations. Experiments in which poly(ribitol-phosphate) synthesis and glucosylation occurred simultaneously were not reported so that the relative rates of glucosylation are not known. The observation that poly(ribitol-phosphate) present in the walls of the organism could act as a substrate for the glucosyl transferase strongly suggests the presence of two teichoic acid synthesizing systems, one for each polymer. Moreover, poly(ribitol-phosphate) synthesized by one of these systems is unavailable to the other (ie the glucosylating system). Further evidence to support this hypothesis came from the finding that germinating spores of *B. subtilis* W23 contained the enzymes necessary for synthesis of poly(glucosylribitol-phosphate). On the other hand, synthesis of unsubstituted poly(ribitol-phosphate) required the prior synthesis of the necessary enzyme [38].

10.1.2 *Biosynthesis of teichoic acids in which glycosyl residues form part of the polymer chain*

As described above, the teichoic acids with a basic polyol-phosphate backbone required only the presence of the appropriate nucleotide-linked precursor for synthesis to occur. The formation of glycosylated derivatives of these polymers, while enhanced by the simultaneous presence of both nucleotide precursors (ie that of the polyol-phosphate and the glycosyl substituent) can be studied as the glycosylation of pre-formed polymer. In contrast, the synthesis of teichoic acids in which the glycosyl residues form part of the polymer chain show an absolute requirement for the presence of both precursors. The biosynthesis of such polymers was again first investigated by Burger and Glaser [31] using particulate enzyme from *B. licheniformis* 9945A. The walls of this organism contain, in addition to poly(glycerol-phosphate), poly(glucosyl-glycerol-phosphate) and poly(galactosyl-glycerol-phosphate). Synthesis of these polymers occurred when CDP-glycerol and, respectively, UDP-glucose and UDP-galactose were present. The simultaneous synthesis of poly(glycerol-phosphate) made interpretation of the results difficult but it appeared that synthesis occurred by the sequential addition of monomer units (ie hexose and glycerol-phosphate) to some unknown acceptor, according to the reaction

Glucosyl-X + CDP-glycerol → CMP + glycerol-P-glucosyl-X

Glycerol-P-glucosyl-X + UDP-glucose → UDP + Glucosyl-glycerol-P-glucosyl-X.

An analogous reaction was postulated for the galactose-containing polymer. Moreover, preincubation with UDP-glucose did not prevent the subsequent incorporation of galactosyl residues from UDP-galactose, suggesting the presence of independent acceptors for the two glycosyl substituents. More recently, Baddiley and his colleagues [3, 73] have confirmed these results showing unequivocally that the phosphate residue incorporated has its origins in CDP-glycerol rather than UDP-glucose. On the basis of the evidence outlined below, they further concluded that the synthesis of both poly(glycerol-phosphate) and poly(glucosyl-glycerol-phosphate) teichoic acids involved the participation of undecaprenyl-phosphate intermediates [3, 73]. A similar deduction had been made earlier for the synthesis of the wall teichoic acid of *Staphylococcus lactis* I3 (now *Micrococcus* sp. I3) [9, 45, 90, 209] where particulate enzyme preparations catalysed the transfer of glycerol phosphate and *N*-acetylglucosamine-1-phosphate from CDP-glycerol and UDP-*N*-acetylglucosamine respectively. Synthesis of the teichoic acid required the simultaneous presence of both nucleotide precursors and occurred by addition of newly-synthesized units to the non-reducing terminal of the growing polymer chain [91].

More recently the biosynthesis of poly(glucosyl-*N*-acetylgalactosamine-phosphate) the minor teichoic acid present in the walls of *B. subtilis* 168 has been investigated using a phosphoglucomutase-negative mutant of this strain [78]. As a consequence of this enzyme deficiency the mutant cannot synthesize UDP-glucose and this teichoic acid is not present in the walls. However, membrane and wall-membrane preparations supplied with the appropriate precursors can still synthesize the polymer *in vitro*. Formation of the teichoic acid occurs by the addition of alternating residues of *N*-acetylgalactosamine-phosphate and glucose. Thus the biosynthetic repeating unit glucosyl-*N*-acetylgalactosamine-phosphate differs from that obtained on mild acid hydrolysis of the polymer. The acid-lability of the *N*-acetylgalactosamine-1-phosphate linkage results in the formation of *N*-acetylgalactosaminylglucose-6-phosphate [194]. No evidence was obtained for the participation of lipid intermediates in biosynthesis and although wall-membrane preparations will catalyse the attachment of newly-synthesized teichoic acid to peptidoglycan the nature of the linkage and the process involved in this remain unknown.

10.1.3 *Nature of the lipophilic acceptor*

As mentioned briefly above, the possible participation of polyprenol phosphate-linked intermediates in the biosynthesis of teichoic acids was first postulated by

Baddiley and his colleagues [4, 8, 27, 45]. Examination of the particulate enzyme preparations catalysing the synthesis of poly(*N*-acetylglucosamine-1-phosphate) in *S. lactis* (now *M. varians*) 2102 [27] and poly(glycerol-phosphate-*N*-acetyl-glucosamine-1-phosphate) in *S. lactis* I3 [45, 90] revealed the presence of labelled material which could be extracted from the incubation mixtures with butan-l-ol and other organic solvents. In *S. lactis* 2102 this material was identified as *N*-acetylglucosamine linked through a pyrophosphate bridge to a lipid acceptor. Pulse-chase experiments showed the transfer of the *N*-acetylglucosamine-1-phosphate residue from the nucleotide via the lipid intermediate to the polymer. Essentially, similar observations of the occurrence of butan-1-ol soluble intermediates were also made for the synthesis of poly(glycerol phosphate) in *B. subtilis* [3] and poly(glucosyl-glycerol-phosphate) in *B. licheniformis* [73]. It should be noted, however, that unusually long times (20 to 30 min) were required to chase the radioactivity from the lipid-linked intermediates into the teichoic acid. This observation is surprising since it suggests both a slow turnover of the lipid intermediates and the presence of a large pool of polyprenol-phosphate.

As described below and in Chapter 8.3, the lipid intermediate participating in the synthesis of peptidoglycan, lipopolysaccharide and various other bacterial polymers has been shown to be undecaprenol-phosphate. In all these cases identification of the carrier was achieved by mass spectroscopy of purified material. In the case of the 'lipid intermediates' involved in teichoic acid biosynthesis, this has not yet been achieved. Instead, identification of the lipid carrier as undecaprenol-phosphate was first attempted by indirect methods. Underlying these studies was the hypothesis that if the lipid carrier participating in the synthesis of both peptidoglycan and teichoic acid was the same, and that both biosynthetic pathways were supplied from a common pool of undecaprenol-phosphate, then by using a cell-free system capable of synthesizing both peptidoglycan and teichoic acid it should be possible to show an interdependence of the two biosynthetic pathways. Using membranes from *B. licheniformis* and *B. subtilis* [3], which catalysed the synthesis of both peptidoglycan and teichoic acid, Baddiley and his colleagues showed that the presence of the nucleotide precursors of teichoic acid partially inhibited the synthesis of peptidoglycan. Similar results were obtained with the reverse experiment when the presence of the nucleotide precursors of peptidoglycan caused inhibition of teichoic acid synthesis. Maximum levels of inhibition were obtained when the experiments were carried out under conditions optimum for the competing synthesis. A second line of evidence came from a study of the effects of bacitracin, a known inhibitor of peptidoglycan but not of teichoic acid synthesis, on the synthesizing system. As described in Chapter 9.4, bacitracin is a specific inhibitor of the dephosphorylation of undecaprenol-pyrophosphate. Thus in the presence of the antibiotic, the lipid carrier becomes blocked in this form with a consequent depletion of the levels of undecaprenol-phosphate in the membrane and a cessation of peptidoglycan synthesis. Bacitracin is not inhibitory to teichoic acid synthesis *in vitro*, an observation explained by the non-participation of undecaprenol-pyrophosphate in the biosynthetic cycle [3]. However, when bacitracin and the nucleotide precursors of peptidoglycan were incubated together with the membrane preparation, marked inhibition of teichoic acid synthesis was observed. Surprisingly, the level of inhibition obtained with both antibiotic and precursors (76.5%) showed only a 10–16% increase over that obtained on incubation of the system with the nucleotide precursors alone. Similar results were also obtained for the inhibition, by bacitracin and the nucleotide precursors of peptidoglycan, of teichoic acid synthesis by a membrane preparation of *S. lactis* I3 (*Micrococcus* sp I3) [209]. In this case, the cumulative inhibition observed in the presence of both bacitracin and UDP-*N*-acetylmuramyl-pentapeptide was more marked than that obtained with the nucleotide precursor alone. The results obtained in these experiments were interpreted as showing the interdependence of the two biosynthetic systems due to the participation of undecaprenol-phosphate, as the common lipid carrier.

However, more recent work discussed in detail below (p. 393) has established both the presence of an oligomeric linkage unit interposed between the main teichoic acid chain and peptidoglycan and the participation of polyprenyl-lipid intermediates in the biosynthesis of these units (Fig. 10.1). Moreover, the teichoic acids of a number of bacteria are now known to be polymerized on a linkage unit before attachment to the peptidoglycan occurs. When Baddiley and his colleagues made the observations described above these facts were not known. With hindsight it seems apparent that most if not all their findings about lipid intermediates can now be explained in terms of the involvement of linkage units.

At the time, the participation of polyprenol-linked intermediates in the biosynthesis of teichoic acid was thrown into doubt by the work of Glaser and his colleagues. In the purification of poly(glycerol-phosphate) polymerase from

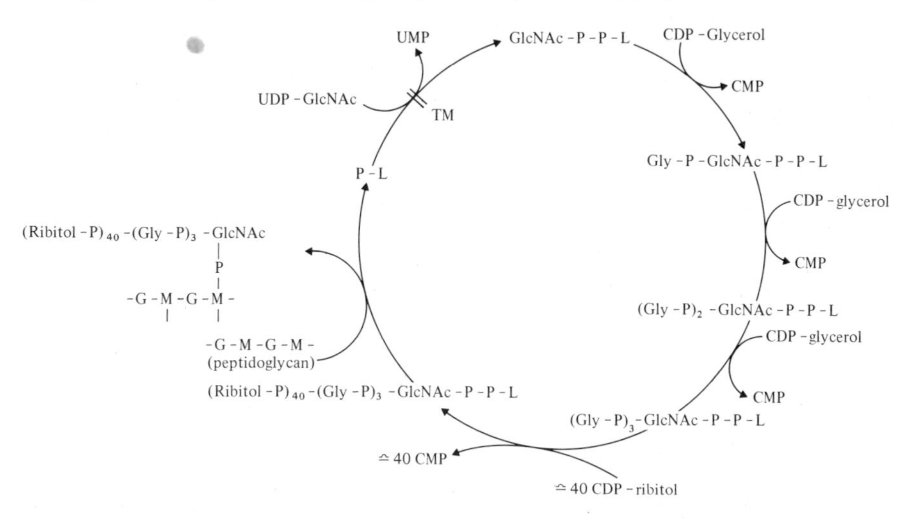

Figure 10.1 Proposed pathway for the synthesis of poly(ribitol-phosphate) teichoic acid and its linkage to peptidoglycan in *S. aureus.* The site of inhibition by tunicamycin (TM) is shown and L is the lipid carrier, presumably undecaprenol.

membranes of *B. subtilis*, Mauck and Glaser [122] found the activity of the enzyme to be dependent on the presence of a heat-stable acceptor. Both the enzyme and the acceptor were solubilized by extraction of isolated membranes with Triton X-100 and separation of the two components was achieved by either ion-exchange chromatography or sucrose density-gradient centrifugation. Glycerol-phosphate units were transferred from the nucleotide precursor CDP–glycerol to the acceptor and a continuous increase in the average chain length of the poly(glycerol-phosphate) was obtained. Preliminary analysis of the acceptor showed it to contain equimolar amounts of glycerol and phosphorus, together with 0.2 molar equivalents of both glucosamine and fatty acids [122]. Glucose was also present [59]. Thus, in composition the acceptor resembled the lipoteichoic acids obtained by phenol extraction from the membranes of several Gram-positive bacteria [108]. However, attempts to use phenol-extracted material as acceptor in assays of polymerase activity gave highly variable results. In addition, undecaprenol-phosphate did not substitute for the acceptor in poly(glycerol-phosphate) synthesis and no exchange reactions were observed between CMP and CDP–glycerol. Such an exchange reaction, which occurs readily between UMP and UDP-*N*-acetyl-muramyl-pentapeptide in peptidoglycan synthesis ([142], and Chapter 8.3.3) would be expected if undecaprenol phosphate were involved as an intermediate in the system. At this time the possibility remained that the difference in the nature of the lipid carrier (acceptor) found in these experiments and those described earlier [3, 73] was that Mauck and Glaser were studying membrane teichoic acid synthesis, whereas the systems of the Baddiley group were involved in the synthesis of the wall teichoic acids.

This possibility was eliminated by a series of elegant experiments mentioned briefly above, which studied the synthesis of poly(ribitol-phosphate), the wall teichoic acid of *S. aureus* H [56, 57, 59]. The poly(ribitol-phosphate) polymerase purified by Fiedler and Glaser showed an absolute requirement for an acceptor designated lipoteichoic acid carrier (LTC) which was also purified to homogeneity from Triton extracts of the membranes and shown to contain glycerol, phosphate, glucose and fatty acids in the ratio 1:1:0.1:0.1. Trace amounts of glucosamine and ribitol were also present. Thus, the LTC isolated from *S. aureus* was similar in composition to that described above for *B. subtilis*. In fact, the two carriers were shown to be interchangeable and could be used for the synthesis of either poly(ribitol-phosphate) or poly(glycerol-phosphate). LTC prepared from a number of other Gram-positive organisms would also function in synthesis of poly(ribitol-phosphate) with the polymerase from *S. aureus*.

LTC and LTC loaded with poly(ribitol-phosphate) were separable by poly-acrylamide gel electrophoresis. Surprisingly, when formation of the polymeric material was followed as a function of time, material of an intermediate size was not found. This observation suggests that synthesis of poly(ribitol-phosphate) is an ordered process involving the addition of units to a single chain rather than the random addition of units to several chains simultaneously. Moreover, the chain length of the newly synthesized material must be limited. Confirmation of this came from the finding that pre-loaded LTC could no longer function as the acceptor of additional ribitol phosphate units. The average chain length of the poly(ribitol-phosphate)linked to LTC was some 30 units, a value in good agreement with that found for the teichoic acid isolated from walls of *S. aureus.* The mechanism, by which the enzyme appears to regulate the length of the polymer synthesized, is not understood. One possibility put forward by Fiedler and Glaser [57] is the formation of a hydrophobic complex of polymerase, LTC and phospholipid. They then envisage the more hydrophilic poly(ribitol-phosphate) linked to the LTC becoming dissociated from the complex on completion of a chain of some 30 ribitol phosphate units. In contrast, the poly(glycerol-phosphate) polymerase of *B. subtilis* did not synthesize a chain of finite length when LTC from either *B. subtilis* or *S. aureus* was used as the acceptor. Thus, the size differences observed between LTC-linked poly(ribitol-phosphate) and poly(glycerol-phosphate) are a function of the specific polymerases and not of the LTC.

The linkage between LTC and poly(ribitol-phosphate) was studied by chemical and enzymic degradation of the isolated complex [46, 58]. The specific point of attachment was shown to be a phosphodiester bond formed between ribitol-phosphate and a glycerol residue in the LTC, the phosphate residue being derived from CDP–ribitol. This linkage has now been studied in detail by Fischer *et al.* [60,61] in their investigation of the structural requirements of LTC for recognition by the poly(ribitol-phosphate) polymerase. The ribitol phosphate units are polymerized on the terminal glycerol residue of the carrier. This conclusion was reached after the authors found that conversion of the terminal residue to either ethylene glycol or simple phosphomonoester abolished all LTC-acceptor activity and that in the latter case activity could be restored by treatment of the polymer with phosphomonoesterase (Table 10.1). On the other hand, both derivatives proved to be competitive inhibitors of the polymerase as was deacylated LTC. These observations suggest that other parts of the poly(glycerol-phosphate) chain and not the hydrophobic terminus of the molecule are important for recognition by the enzyme. To examine further the structural requirements for enzyme recog-

Table 10.1 Poly(ribitol-phosphate) acceptor activity of lipoteichoic acid and various derivatives. (Data from [60, 61] where details of the assay of acceptor activity can be found.)

Lipoteichoic acid or derivative	*Acceptor activity (pmol ribitol polymerized)*	
LTA (LTC)	925	580
deacylated	134	–
oxidised (ethylenediol terminus)	147	–
phosphomonoester terminus	227	–
phosphomonoester after dephosphorylation	531	–
native (contains alanine ester)	–	90
alanine-free derivative	–	510
no additions	169	88

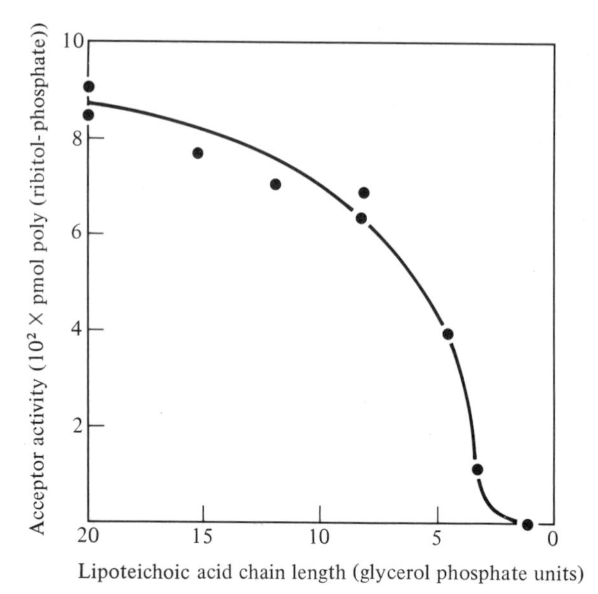

Figure 10.2 Effect of lipoteichoic acid chain length on poly(ribitol-phosphate) acceptor activity. Lipoteichoic acid from *Leuconostoc mesenteroides* was enzymically shortened as described in the text and samples varying in average chain length from 20 to 1.2 glycerol-phosphate residues were assayed for poly(ribitol-phosphate) acceptor activity. (Data from [60] where further details may be found.)

nition Fischer *et al.* utilized a partially purified enzyme preparation from *Aspergillus niger* which contains both phosphodiesterase and phosphomonoesterase to shorten systematically the poly(glycerol-phosphate) chain of LTC. In this way they isolated a series of analogues of LTC which differ only in the number of glycero-phosphate units in the hydrophilic chain. Using these as acceptors for poly(ribitol-phosphate) they established a length of 20 units for optimum acceptor activity with a minimum of four unsubstituted glycero-phosphate residues being required for enzyme recognition (Fig. 10.2). From the evidence described above it appears that the terminal glycero-phosphate is not involved in this process since LTC in which this residue has been chemically modified can still interact with and inhibit the enzyme. Nor does the hydrophobic terminus of the LTC appear to be important in enzyme recognition, whereas it is clearly required for acceptor activity [56, 57, 61]. On the basis of their results, Fischer *et al.* have suggested that *in vitro* in the presence of phospholipid and detergent the polymerase exists in mixed micelles and that the hydrophobic region functions in maintaining an appropriate orientation of LTC therein.

It was also found that glycosyl-substitution of the poly(glycero-phosphate) chain reduced acceptor activity. As a result of these observations a method was developed to isolate LTC from which the naturally occurring ester alanine substituents had not been removed. These preparations did not function as acceptors for poly(ribitol-phosphate) [60]. The implication of these observations for the biosynthesis of teichoic acid *in vivo* are discussed in detail below.

10.1.4 *Biosynthesis of teichoic acid linkage units*

Chemical studies reviewed recently by Baddiley and his colleagues [41] have shown that the wall teichoic acids of *S. aureus* H, *Micrococcus* strains 2102 and I3 and *B. subtilis* W23, although differing in structure, are all covalently linked to peptidoglycan by a linkage unit composed of *N*-acetylglucosamine and three residues of glycerophosphate. This unit is interposed between the teichoic acid chain and the 6-position of a muramic acid residue in the glycan (see Chapter 7.1.1).

More or less simultaneously with the first of these chemical studies [80] Bracha and Glaser [18] reported that the biosynthesis and linkage of poly(ribitol-phosphate) teichoic acid to peptidoglycan in wall–membrane preparations of *S. aureus* H was greatly stimulated by the presence of CDP–glycerol and UDP–*N*-acetylglucosamine. At the same time Hancock and Baddiley [74] described the formation by membrane preparations of *S. aureus* H, *B. subtilis* W23 and *Micrococcus* sp. 2102 of water-soluble polymers containing glycerol. The biosynthesis of this polymeric material required the presence of the nucleotide precursors of the particular wall teichoic acid, CDP–glycerol and UDP–*N*-acetylglucosamine. Treatment of the incubation mixtures with butan-1-ol extracted lipids containing both radioactive glycerol and *N*-acetylglucosamine. Subsequently the biosynthesis and linkage of poly(ribitol-phosphate) to peptidoglycan in wall–membrane preparations of *B. subtilis* W23 was shown to occur by utilizing the nucleotide precursors in the order UDP–*N*-acetylglucosamine, CDP–glycerol and CDP–ribitol [226]. Butanol-soluble lipids containing *N*-acetylglucosamine and glycerol were also present in this system. In each of these investigations it was assumed that the teichoic acid chain was synthesized on LTC and transferred to the linkage unit before the assembled polymer became covalently-linked to peptidoglycan.

The biosynthesis of the linkage unit, for this is what the lipid extracts proved to be, was first studied using membrane preparations of *S. aureus* H [17]. Lipids extracted with 70% ethanol which contained both radioactive *N*-acetylglucosamine and glycerol were shown to be converted into water-soluble polymer on incubation with CDP–ribitol and fresh membranes. The polymeric material synthesized under these conditions had the electrophoretic characteristics of poly(ribitol-phosphate). In independent experiments [75] membranes of *S. aureus* H and *B. subtilis* W23 were found to synthesize two compounds containing poly(ribitol-phosphate). One of these, synthesized in the absence of CDP–glycerol and UDP–*N*-acetyl-glucosamine, was characterized as teichoic acid linked to LTC. The second product, which was formed only in the presence of the two additional nucleotide precursors, contained radioactive glycerol and thus appeared to be teichoic acid attached to the linkage unit. The formation of this second compound and of the lipid extracted into 70% ethanol were both inhibited by tunicamycin as was the covalent linkage of teichoic acid to peptidoglycan in wall–membrane preparations of *B. subtilis* W23

[17, 75, 226]. As discussed in Chapter 9.5.2 this antibiotic inhibits the transfer of *N*-acetylglucosamine-1-phosphate to polyprenol-phosphate. These findings of tunicamycin sensitivity and preliminary chemical studies on the lipid suggested that the lipid intermediate contained glycerol-phosphate residues attached to *N*-acetylglucosamine-P-P-polyprenol. This has now been established by the isolation and detailed characterization of three lipid intermediates isolated from *S. aureus* H and *M. varians* (formerly *Micrococcus* sp 2102) [124, 177]. These were polyprenol-P-P-*N*-acetylglucosamine and polyprenol-P-P-*N*-acetylglucosamine with one or two residues of glycerol phosphate attached (Fig. 10.1). The complete linkage unit, i.e. lipid-linked *N*-acetylglucosamine with three glycerol phosphate units, was not isolated although in *M. varians* the addition of CDP–glycerol greatly stimulated the formation of polymeric material from the lipid intermediate containing two glycerol phosphate units and UDP–*N*-acetylglucosamine [177]. The teichoic acid of *M. varians* is a polymer of *N*-acetylglucosamine-1-phosphate.

At present little detailed information is available concerning the enzymes catalysing the individual reactions in linkage unit biosynthesis, although methods to follow the process as a whole have recently been described [15]. For example, it has not yet been established whether the addition of the glycerol-phosphate residues to *N*-acetylglucosamine-P-P-polyprenol is catalysed by a single enzyme or three separate enzymes. The inhibition by tunicamycin of the *N*-acetylglucosamine-1-phosphate transferase has been mentioned earlier (Chapter 9.5.2). An inability to catalyse this reaction is responsible for the absence of ribitol teichoic acid from the walls of *S. aureus* strain 52A5. This organism was orginally isolated [32] as being resistant to bacteriophage 52A part of the receptor of which is known to be ribitol teichoic acid [5]. Inactivation of this same enzyme has been shown to occur when *B. subtilis* W23 is placed in phosphate-free medium [69]. Under these conditions the organism ceases to synthesize teichoic acid and begins instead the synthesis of teichuronic acid (see p. 396).

The possibility that LTC is an analogue of the natural acceptor, ie the linkage unit, was first suggested by Bracha and Glaser [17]. However, at that time they had difficulty in separating poly(ribitol phosphate) attached to either LTC or the linkage unit. Consequently, they were not able to eliminate the possibility that polymerization of ribitol phosphate occurred on LTC before the completed teichoic acid chain was transferred to the linkage unit. Only then would the completed polymer become linked to peptidoglycan.

Also at this time in their independent investigations Baddiley and his colleagues achieved the separation by ion-exchange chromatography of poly(ribitol-phosphate)–LTC and poly(ribitol-phosphate)–linkage unit [75]. Since the two compounds were formed simultaneously by membrane preparations from *S. aureus* H they concluded that both acceptors were required and that teichoic acid was synthesized by the mechanism described above.

However, the recent studies of Fischer *et al.* [60, 61] clearly support the analogue hypothesis. As described above they [61] found that poly(ribitol-phosphate) cannot become polymerized on LTC that retains the ester-linked

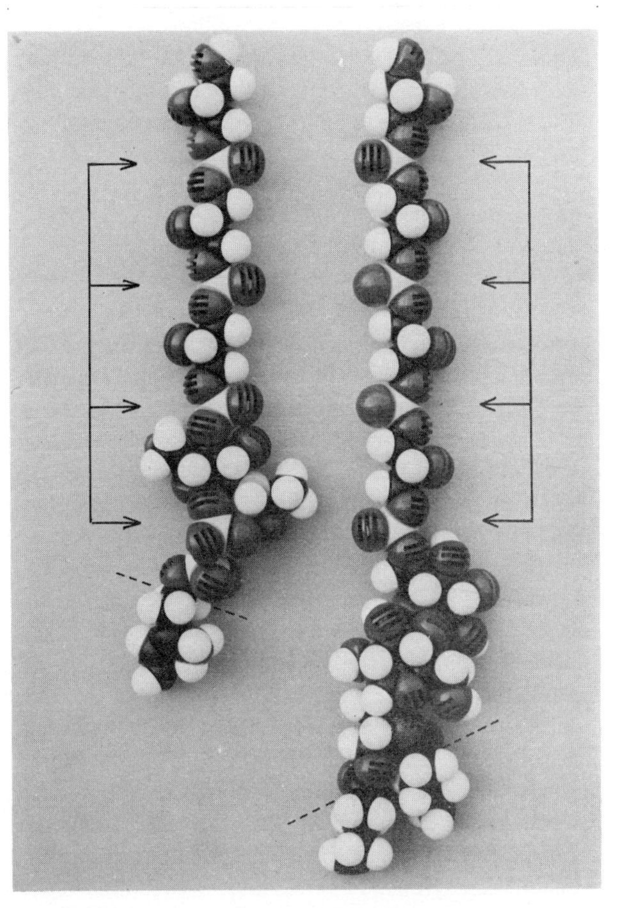

Figure 10.3 Space-filling models of the lipid-bound linkage unit of *S. aureus* (left) and the tetra(glycero-phosphate) derivative of lipoteichoic acid. The *N*-acetyl-glucosaminyl residue is the α-anomer and substituted by glycerophosphate at C_{-4}. Only part of the hydrophobic chains are shown with the hydrophobic/hydrophilic boundary given by dashed lines. The four phosphate residues required for recognition are marked by arrows. (Photograph kindly provided by Professor W. Fischer, and taken with permission from [60].)

D-alanine substituents. Since this is the form in which LTC occurs *in vivo* it follows that under these conditions polymerization must occur on an alternative acceptor, presumably the linkage units. The fact that LTC acts as an acceptor *in vitro* is a consequence of the extreme lability to alkaline hydrolysis of the ester-linkage. The preparation of membranes or LTC in the commonly used weakly alkaline buffers (pH 7.5 to 8.5) results in a loss of alanine substituents. This enables the LTC whether membrane-bound or not, to be recognized by poly(ribitol-phosphate) polymerase. The remaining ambiguity that the polymerase requires four unsubstituted glycerol-phosphate residues for recognition whereas the linkage unit contains only three has also been resolved. Space-filling models of both compounds revealed that the presence of the *N*-acetylglucosaminyl residue in the linkage unit makes both structures of the same length with their phosphate residues in more or less identical positions (Fig. 10.3). Taken together these observations justify the conclusion that *in vivo* LTC does not function in teichoic acid biosynthesis and that polymerization of poly(ribitol-phosphate) occurs directly onto the terminal glycerol of the linkage units.

10.1.5 *Attachment of teichoic acid to peptidoglycan*

The first evidence on the interrelationship between the synthesis of both teichoic acid and peptidoglycan came from the observations of Mauck and Glaser [123]. Their experiments were based on the earlier studies of Ellwood and Tempest [50] who showed that the walls of *B. subtilis* grown in the presence of excess phosphate contained teichoic acid but not teichuronic acid. However, on transfer to growth in a low-phosphate medium, the synthesis of teichoic acid ceased and it was replaced in the wall by teichuronic acid. This switch from one polymer to the other, which was readily reversible, occurred without a demonstrable lag. Using these conditions, both teichoic acid and teichuronic acid were shown to be linked only to peptidoglycan synthesized concomitantly and not to pre-existing material [123]. Further evidence has come from examination of the effects of penicillin on wall synthesis by *M. luteus* [126] and *B. licheniformis* [203]. Incubation of both organisms in the presence of benzylpenicillin resulted in the continued synthesis of uncross-linked peptidoglycan. However, in neither case was evidence obtained for the attachment of the secondary polymers, teichuronic acid and teichoic acid respectively, to the uncross-linked material. Thus synthesis of the linkages between peptidoglycan and either teichoic acid or teichuronic acid appears to require the concomitant synthesis of cross-links between the newly-synthesized peptidoglycan and the pre-existing wall. Evidence to support this conclusion has come from experiments carried out *in vitro* with wall–membrane preparations from *B. licheniformis* [224]. These preparations catalyse the synthesis of covalently-linked (as judged by insolubility in hot sodium dodecyl sulphate and aqueous 80% phenol) peptidoglycan and poly-1, 3(glycerol-phosphate) teichoic acid. In the presence of benzylpenicillin, overall synthesis of the two polymers was unaffected, whereas the incorporation of radioactive glycerol into the detergent-insoluble fraction was inhibited by 80%. Earlier studies [205] had shown that this concentration of penicillin completely inhibited the incorporation of newly-synthesized peptidoglycan into the pre-existing wall. Therefore, it seems that while these results lend general support to the ideas of Mauck and Glaser [123], wall–membrane preparations were also capable of incorporating a small amount of poly-(glycerol-phosphate) into the pre-existing wall.

In contrast wall–membrane preparation of *B. subtilis* W23 linked newly-synthesized poly(ribitol-phosphate) exclusively to pre-existing wall [226] whereas in similar preparations from *S. aureus* 52A2 linkage of teichoic acid occurred to both pre-existing and newly-synthesized peptidoglycan [15, 16]. The reasons underlying these various observations remains unclear but may reflect differences in the specificity of the enzyme catalysing the actual linkage of the two polymers. In the *S. aureus* preparations the two fractions, ie teichoic acid linked to pre-existing as opposed to newly-synthesized peptidoglycan, can be further distinguished by their lysozyme sensitivity [15, 18]. The peptidoglycan present in the wall of *S. aureus* is largely resistant to lysozyme digestion because of the presence of *O*-acetyl groups on muramic acid whereas that synthesized *in vitro* lacks these substituents and is lysozyme-sensitive. Chromatography of lysozyme-released material on Sephadex-G100 revealed size differences in the teichoic acids synthesized under the various conditions [18]. Material attached to the pre-existing wall in the absence of peptidoglycan synthesis chromatographed mainly as a single peak of high molecular weight. In the presence of UDP-*N*-acetylmuramyl-pentapeptide when peptidoglycan biosynthesis was also occurring, chromatography of the teichoic acid revealed an additional peak of lower molecular weight. The structural difference between the two polymers is not known.

In the absence of ribitol-phosphate polymerization, the linkage unit itself can become attached to peptidoglycan. However, in *B. subtilis* W23 at least, there is no evidence that polymerization of poly(ribitol-phosphate) can occur on these transferred units [226].

Linkage of both teichoic and teichuronic acids to peptidoglycan occurs through a phosphodiester bond to a 6-hydroxyl group of a muramic acid residue in the glycan chain (see Chapter 7.1.1). Newly-synthesized muramic acid phosphate has been isolated from wall–membrane preparations of *B. licheniformis* in which *in vitro* synthesis of poly(glycerol-phosphate) teichoic acid and peptidoglycan had occurred [225]. Formation of muramic acid phosphate required the simultaneous synthesis of both polymers and the phosphate moiety was derived from UDP-*N*-acetylglucosamine. This observation provided the first evidence for the participation of linkage units in organisms with glycerol-containing teichoic acids. Subsequently additional evidence has come from chemical studies on the linkage of glucosylated poly(glycerol-phosphate) teichoic acid in *B. subtilis* [41] and the finding that in *B. licheniformis* linkage of teichoic acid to peptidoglycan is sensitive to inhibition by tunicamycin [206].

10.1.6 *Control of teichoic acid synthesis*

As described above, transfer of either Staphylococci or Bacilli to growth in phosphate-limited conditions leads to the rapid cessation of teichoic acid synthesis and its replacement in the wall by teichuronic acid. This changeover occurs at rates much faster than can be accounted for by a simple dilution of pre-existing wall and turnover of the wall is now known to play an important part in this process (see Chapter 11.5.3).

Initially Ellwood and Tempest [51] suggested that phosphate-limitation simply prevented the formation of phosphate-containing intermediates required for teichoic acid biosynthesis. The teichoic acid precursors were postulated to repress enzymes of teichuronic acid biosynthesis and only when their intracellular concentration fell would this pathway become operative. Subsequently Rosenberger [181] demonstrated in *B. subtilis* the repression of enzyme synthesis in both teichoic and teichuronic acid biosynthesis. Thus the amount of CDP-glycerol pyrophosphorylase present was low under conditions of phosphate-limitation and similarly UDP-glucose dehydrogenase was decreased in phosphate-rich medium. These enzymes were studied as examples of enzymes involved in the biosynthesis of teichoic and teichuronic acid respectively. There was, however, no evidence supporting the suggestion that the nucleotide precursors of one polymer inhibited enzymes functioning in the biosynthesis of the other. Earlier Baddiley and his colleagues [2] had shown that CDP-glycerol pyrophosphorylase was inhibited by UDP-*N*-acetylmuramyl-pentapeptide and stimulated by UDP-*N*-acetylglucosamine. Unfortunately, the effect of teichuronic acid precursors did not appear to be investigated.

Rosenberger [181] also showed that if *B. subtilis* was transferred from phosphate-rich to phosphate-limited conditions and protein synthesis was inhibited then the organism continued to synthesize teichoic acid. In the alternative experiment ie, transfer from phosphate-limited to phosphate-rich conditions, teichuronic acid synthesis continued. Thus in either case phosphate levels as such did not influence formation of the alternative polymer. Rather it appeared that any change in the polymer being synthesized required the formation of new protein(s). When synthesis of the alternative polymer began this appeared to inhibit synthesis of the original polymer presumably by an effect on one or more of the enzymes involved. More recently, changes in specific enzymes particularly CDP-glycerol pyrophosphorylase have been studied in *B. licheniformis* [92]. Under conditions where the organisms were changing to growth under conditions of phosphate-limitation a dramatic decrease in the activity of several enzymes involved in teichoic acid biosynthesis was observed. However, only when the activity of CDP-glycerol pyrophosphorylase reached very low levels did teichoic acid synthesis cease. Loss of the pyrophosphorylase activity occurred at rates much faster than the theoretical dilution rate which would be observed if cessation of protein synthesis alone were responsible. A similar decrease was found if protein synthesis was inhibited even in the presence of excess phosphate. If, however, phosphate-limited cells were transferred to phosphate-rich conditions then there was an immediate and dramatic increase in pyrophosphorylase activity. Again this required *de novo* protein synthesis. Thus, a potential control mechanism over teichoic acid biosynthesis could reside in the relative instability of CDP-glycerol pyrophosphorylase and intracellular phosphate concentrations may in some way influence this control. Clearly this whole area, although presenting technical difficulties in controlling cultural conditions and measuring intracellular metabolite concentrations, deserves further investigation.

10.1.7 *Biosynthesis of lipoteichoic acid*

The biosynthesis of lipoteichoic acid and consequently of lipoteichoic acid carrier

(LTC) was first studied in pulse-chase experiments using *S. aureus* H [68] and *Streptococcus sanguis* [52]. Organisms in the exponential phase of growth were pulsed for 2 min and 10 min respectively with radioactive glycerol and then the radioactivity was replaced with an excess of unlabelled glycerol. Under these conditions, the major fraction of radioactivity was initially located in phosphatidylglycerol. During the chase period which lasted 45 min [68] and 60 min [52] a turnover of some 68% to 87% of the radioactivity present in phosphatidylglycerol was observed. This was accompanied by an increase in radioactive material having the properties of lipoteichoic acid. In each case this material was identified by polyacrylamide gel electrophoresis and in *S. aureus* by the finding that 50% of the extracted LTC was able to act as the acceptor for the poly(ribitol-phosphate) polymerase [56, 57]. The suggestion originally put forward that the remaining 50% represents LTC partially loaded with poly(ribitol-phosphate) is invalidated by the observations of Fischer *et al.* [60, 61]. It now seems probable that the acceptor activity observed by Glaser and Lindsay [68] represents the extent to which ester-alanine substituents were lost during the extraction and purification of the lipoteichoic acid. Emdur and Chiu [53] subsequently extended their observations with *S. sanguis* to demonstrate the synthesis *in vitro* of both lipoteichoic acid and poly(glycerol-phosphate) from phosphatidyl-glycerol. Synthesis of lipoteichoic acid required the presence of UDP–glucose in addition to phosphatidylglycerol; in the absence of UDP-glucose only poly(glycerol-phosphate) was synthesized. In addition, radioactivity from phosphatidylglycerol labelled from ^{14}C-acetate was also found in the lipoteichoic acid, suggesting that the phospholipid may also be the precursor of certain of the fatty acid residues in addition to the glycerol phosphate units. The relationship of the poly(glycerol-phosphate) synthesized to the lipoteichoic acid is not yet clear. One possible explanation is that much of this material represents lipoteichoic acid from which the lipophilic moieties have been removed. In this context Kessler and Shockman [106] have recently reported the enzymic deacylation of lipoteichoic acid by protoplasts of *S. faecium*. It was concluded that the deacylase is a membrane-bound enzyme since no activity could be detected in the protoplast medium. Clearly the activity of an enzyme of this specificity would explain the results obtained with *S. sanguis*.

The precursors of lipoteichoic acid synthesized by membrane preparations from *S. faecalis* have been shown to be phosphatidyl-kojibiosyl-diacylglycerol and phosphatidylglycerol [64]. When incorporation of the glycolipid was studied as a function of time the product was initially soluble in chloroform–methanol–water. However, with addition of further glycerolphosphate residues the product became soluble in water. The overall synthesis of lipoteichoic acid was stimulated by Triton X-100 and mercaptoethanol, whereas Mg^{2+} was without effect.

10.1.8 *Addition of ester-linked alanine residues*

Ester-linked D-alanine residues are found as components of both wall- and membrane-associated teichoic acids in a wide range of bacteria. However, until recently attempts to study the biosynthesis of such linkages had been unsuccessful. The earlier observations of Baddiley and Neuhaus [11] had shown the existence in *Lactobacillus casei, L. arabinosus* and other Gram-positive bacteria of a soluble enzyme which in the presence of ATP catalysed the formation of D-alanine–AMP–enzyme complex according to the following equation.

$$\text{D-alanine} + \text{enzyme} + \text{ATP} \rightleftharpoons \text{D-alanine–enzyme–AMP} + \text{PP}_i$$

This complex reacts spontaneously with hydroxylamine to yield D-alanine hydroxamate. Although they postulated that the enzyme might have a role in teichoic acid synthesis, attempts to use the preparation to catalyse the addition of ester-linked alanine to teichoic acids were unsuccessful [8].

More recently, a membrane-bound enzyme has been demonstrated in the same strain of *L. casei*, which catalyses the transfer of D-alanine into the membrane fraction [117, 143, 167]. This enzyme which requires the presence of Mg^{2+}, ATP and a supernatant fraction for activity, was termed D-alanine: membrane acceptor ligase. The acceptor, with ester-linked D-alanine attached, has been isolated from the membrane fraction and shown by normal chemical procedures and precipitation with specific antisera to be poly-1,3(glycerol-phosphate). There is no chemical evidence, although it seems highly probable, that the acceptor is in fact membrane bound lipoteichoic acid. D-alanine residues were not linked to exogenously added teichoic acid.

The factor required for ligase activity which was present in the supernatant fraction, could not be separated by conventional techniques from D-alanine-activating enzyme. Moreover, the stimulating factor and the activating enzyme had identical profiles to heat inactivation and inhibition with either *p*-hydroxymercuri-benzoate or pyrophosphate. These observations led to the conclusion that the addition of ester-linked D-alanine is a two step process. This involves the activation of the amino acid as described above, followed by a transfer reaction catalysed by the ligase, according to the following equation:

$$\begin{aligned}&\textbf{D-alanyl—activating enzyme—AMP + acceptor} \xrightarrow{\textbf{ligase}}\\&\qquad\textbf{D-alanyl—acceptor + AMP + activating enzyme}\end{aligned}$$

Of particular interest in this context are the studies of Neuhaus, Panos and their colleagues on the formation of ester-alanine linkages in *Streptococcus pyogenes* and an L-phase variant derived from it. The L-phase variant synthesizes a membrane associated poly(glycerol-phosphate) lacking ester-linked alanine [195]. Further studies [36] have shown that the L-phase variant is deficient in ligase activity, the amount of D-alanine activating enzyme present being identical to that found in the parent coccus. Further examination of this system may show whether the lesion is in the ligase itself or whether some other membrane factor is involved in transfer of the D-alanine residues. Although the systems described catalyse the addition of ester-linked D-alanine residues to the membrane teichoic acids, it seems likely that a similar, if not identical process will function in the synthesis of these ester-linkages in wall teichoic acids.

10.2 Biosynthesis of other components of the Gram-positive bacterial wall

In contrast to the detailed information available on the biosynthesis of teichoic acids by cell-free systems, relatively little work has yet been reported on the biosynthesis of the numerous polysaccharides and other polymers known to be covalently linked to the peptidoglycan. Among the exceptions are the *N*-acetyl-mannosaminuronic acid-containing teichuronic acid of *Micrococcus luteus (lysodeikticus)*, various reports on the synthesis of rhamnose-containing poly-saccharides in Streptococci and the teichuronic acid of *B. licheniformis.*

10.2.1 *The teichuronic acid of M. luteus (lysodeikticus)*

The teichuronic acid present in the walls of *M. luteus* first shown by Perkins [164] to contain equimolar amounts of D-glucose and *N*-acetylmannosaminuronic acid (Man*N*AcUA) has now been investigated in detail and shown [76, 77] to have the following structure (see Chapter 7.1.2).

$$\left[\text{D-Man}N\text{AcUA} \xrightarrow{\beta-1,6} \text{D-Glc} \xrightarrow{\alpha-1,4} \right]_{10-40} .$$

In the wall the teichuronic acid has been reported to be linked to peptidoglycan through a reducing *N*-acetylglucosamine residue and a phosphodiester bond to the 6-position of a muramic acid residue in the glycan chain [77].

In vitro synthesis of the teichuronic acid was originally studied using membrane preparations from *M. luteus* [1, 162, 178, 196] although more recently both wall-membrane preparations [214] and toluene-treated cells [165] have been used to investigate linkage to peptidoglycan. In the earlier studies [1, 162] the formation of polymeric material was shown to require the presence of the nucleotide precursors UDP-glucose, UDP-*N*-acetylmannosaminuronic acid and UDP-*N*-acetylglucosamine. Biosynthesis was also stimulated by a heat stable cofactor which was not further identified. The newly-synthesized polymer contained equal amounts of glucose and Man*N*AcUA together with one residue of *N*-acetylglucosamine for each 15 to 20 residues of glucose. At this time the significance of this observation remained unknown and evidence for the participation of lipid intermediates was not obtained. However, recent studies described below have established that *N*-acetylglucosamine is involved in the linkage of the teichuronic acid to peptidoglycan. UDP-*N*-acetylglucosaminuronic acid could not replace UDP-Man*N*AcUA although the pool of UDP-*N*-acetylhexosaminuronic acid extracted from *M. luteus* inhibited with chloramphenicol and benzylpenicillin appeared to contain both isomers. The synthesis of UDP-Man*N*AcUA from UDP-*N*-acetylglucosamine has been studied in *E. coli* [93, 94, 103]. In this organism two enzymes which have been extensively purified, catalyse the epimerization of the *N*-acetylhexosamine moiety followed by oxidation at C-6 as follows:

$$\text{UDP–Glc}N\text{Ac} \rightleftharpoons \text{UDP–Man}N\text{Ac}$$
$$\text{UDP–Man}N\text{Ac} \xrightarrow{\text{NAD}^+} \text{UDP–Man}N\text{AcUA}.$$

The dehydrogenase activity catalysing the second reaction was also found in *M. luteus* suggesting that the above mechanism also occurs in the micrococcus.

More recently Anderson and Perkins and their colleagues have established that in common with the teichoic acids a linkage unit containing *N*-acetylglucosamine participates in the attachment of the teichuronic acid to peptidoglycan [165, 178, 196, 214]. Thus the biosynthesis of the polymer proceeds in two stages. In the first a polyprenol lipid carrier presumed to be undecaprenol phosphate accepts one residue of *N*-acetylglucosamine and two of Man*N*AcUA and in the second stage the main polysaccharide is built up by transferring alternating residues of glucose and Man*N*AcUA (Fig. 10.4). Indirect evidence for the lipid carrier being undecaprenol comes from the finding that concomitant synthesis of peptidoglycan and lipomannan (biosynthetic processes known to use this lipid carrier) inhibit the synthesis of the teichuronic acid [178]. Unlike the initial reaction in teichoic acid biosynthesis described above, that in the teichuronic acid appears to be the transfer of *N*-acetylglucosamine rather than *N*-acetylglucosamine phosphate. The product of the reaction is UDP and it is this nucleotide that was found to participate in an exchange reaction and not UMP. However, in separate studies Perkins and his colleagues [165, 214] have found that incorporation of *N*-acetylglucosamine into teichuronic acid is sensitive to tunicamycin, complete inhibition being obtained at an antibiotic concentration of 10 μg/ml. Together these observations represent something of a paradox since, as described in Chapter 9.5.2, tunicamycin at low concentrations is thought to inhibit only those reactions involving the transfer of *N*-acetylglucosamine-l-phosphate and which lead to the formation of polyprenol-P-P-*N*-acetylglucosamine. A possible explanation for the results obtained would be requirement in biosynthesis for a reaction involving a phosphotransferase of the above type but there is no evidence to support this.

In wall-membrane preparations linkage of newly-synthesized teichuronic acid occurred predominantly to pre-existing wall and was not significantly stimulated by further peptidoglycan synthesis [214]. Attempts to determine the biosynthetic origin of the phosphate attached to the 6-position of muramic acid have not been successful. In this context it is worth noting that the kinetics of the release of free reducing groups of *N*-acetylglucosamine and the removal of phosphate presumably from muramic acid-6-phosphate in the glycan did not agree (see Fig. 1 in [77]). Whether these results point to some additional residue interposed between the two components or simply reflect the inability of alkaline phosphatase to interact with and hydrolyse the phosphomonoester residues on the glycan chains is unknown.

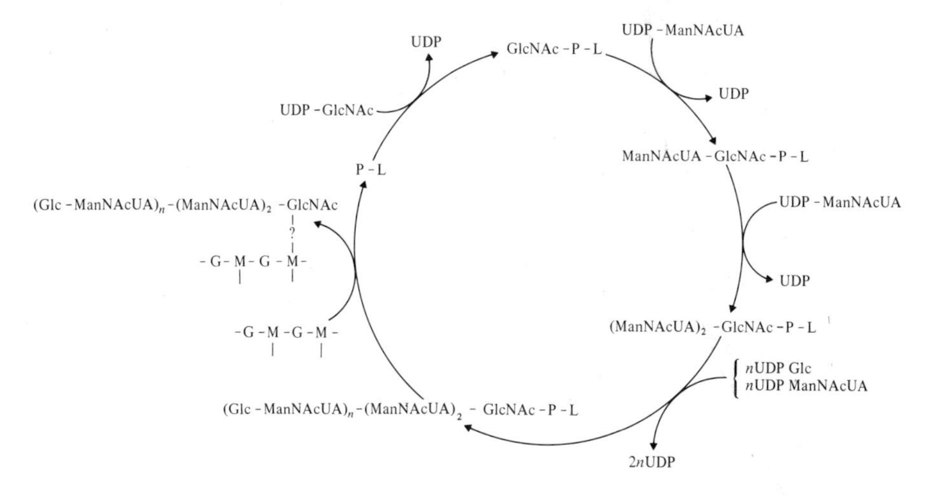

Figure 10.4 Proposed pathway for the biosynthesis and linkage of the teichuronic acid to peptidoglycan in *M. luteus.* As described in the text the formation of this polymer is sensitive to tunicamycin although the site of inhibition has not yet been established. Similarly the nature of the linkage to peptidoglycan is not known. L is the lipid carrier, presumably undecaprenol.

The two transferases catalysing the addition of alternating residues of glucose and Man*N*AcUA to the carrier lipid have been solubilized from membranes of *M. luteus* by treatment with Triton X-100 [196]. The glucosyltransferase was selectively released by detergent treatment in the presence of 5 to 10 mM Mg^{2+} whereas at a Mg^{2+} concentration of 1 mM or less the *N*-acetylmannosaminuronyl-transferase was also solubilized.

Thus, the biosynthesis of the teichuronic acid of *M. luteus* has been established in detail although the mechanism of linkage of the polymer to peptidoglycan requires further investigation.

10.2.2 *Rhamnose-containing polysaccharides of Streptococci*

The walls of *Streptococcus pyogenes, S. faecalis* and *S. sanguis* all contain polysaccharides with rhamnose and *N*-acetylglucosamine as major components. In *S. pyogenes* this polymer is the group specific polysaccharide [125, 135]. The composition of these polysaccharides is generally known but not their structure.

The direct incorporation of rhamnose from TDP–rhamnose into polymeric material was first shown using membrane preparations from both *S. faecalis* [163] and *S. pyogenes* [228]. In *S. pyogenes* small amounts of *N*-acetylglucosamine were also incorporated. Later Cohen and Panos [40] demonstrated the synthesis of polymeric material from TDP–rhamnose using membrane preparations from *S. pyogenes* and a stable L-phase variant derived from this organism. In the streptococcal membranes the polymeric rhamnose was found, after detergent-extraction or mild acid treatment, to be associated with both muramic acid and glucosamine. In contrast these peptidoglycan components were absent from the small amount of polymer synthesized by the L-phase variant membranes. On further investigation Reusch and Panos [168] isolated a glucosamine-containing lipid from the streptococcal membranes and showed that incorporation of rhamnose was partially dependent upon the presence of UDP-*N*-acetylglucosamine and *vice versa.* On the basis of their results they also suggested that a lipid intermediate might be involved in biosynthesis although incorporation of rhamnose into lipid was not observed. However, the recent detailed studies of Chiu and Saralkar [39] on the biosynthesis of an oligosaccharide-lipid by membranes of *S. sanguis* established that at least in this organism and probably in *S. pyogenes* lipid intermediates are involved. In *S. sanguis* the oligosaccharide was synthesized by the sequential transfer of *N*-acetylglucosamine, rhamnose and glucose to a lipid, presumably phosphorylated undecaprenol (Fig. 10.5). The formation of the *N*-acetylglucosamine-lipid was stimulated 2-fold by ATP and was strongly inhibited by UDP but only slightly by UMP. Using radioactive ATP it was shown that the stimulation observed was due to phosphorylation of the lipid carrier. The inhibition

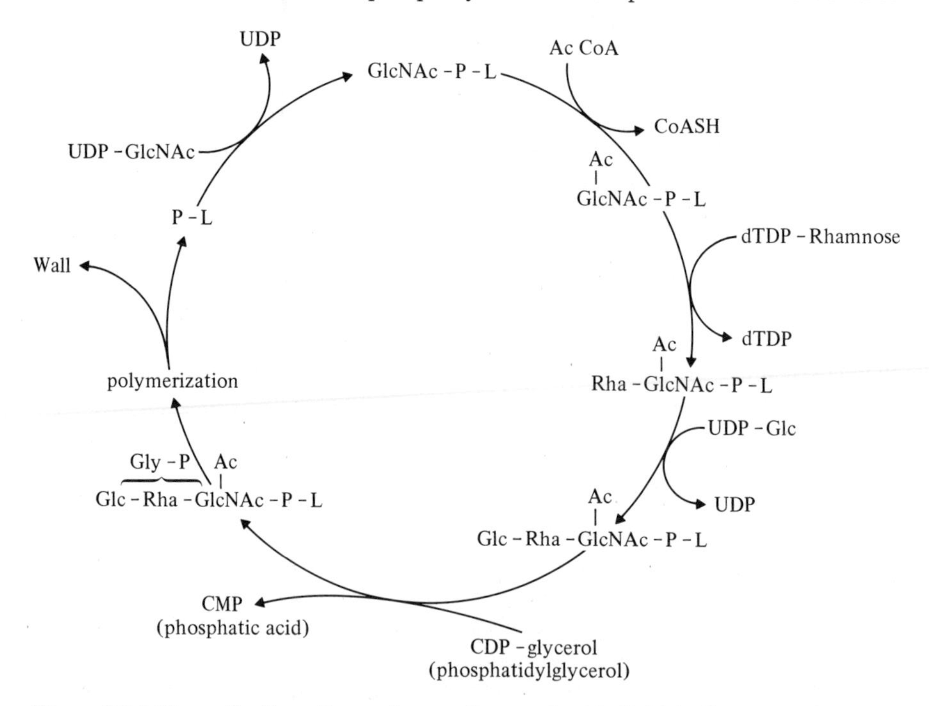

Figure 10.5 Biosynthetic pathway for an oligosaccharide-lipid in *S. sanguis.* The wall polysaccharide is known to contain both acyl- and glycerophosphate substituents [54]. The incorporation of these is shown although this has not been demonstrated *in vitro* for glycerophosphate. The nature of the linkage to peptidoglycan which presumably occurs after polymerization is not known. L is the lipid carrier, presumably undecaprenol.

by UDP suggests that the initial reaction of the biosynthetic cycle is the transfer of *N*-acetylglucosamine rather than the translocation of *N*-acetylglucosamine-phosphate as found in the biosynthesis of teichoic acid linkage units. This was confirmed when chemical analysis of the lipid intermediate gave equimolar amounts of glucosamine and phosphate. This reaction is therefore equivalent to that proposed by Anderson and his colleagues [178] for the first step in the biosynthesis of teichuronic acid in *M. luteus* (see p. 402). Unfortunately the sensitivity of the streptococcal system to antibiotics particularly tunicamycin and bacitracin, does not appear to have been investigated. It therefore remains unknown whether a similar paradox to that described earlier (p. 402) also exists in *S. sanguis.* Synthesis of polysaccharide by the streptococcal membranes was relatively inefficent and nothing is known about linkage to peptidoglycan. However, it has been established that the native polysaccharide contains glycerol phosphate substituents [54] and additional residues may be involved in the formation of a linkage unit. Whether the presence of CDP–glycerol and/or phosphatidylglycerol as potential precursors of glycerol phosphate would stimulate polysaccharide biosynthesis and linkage requires further investigation.

10.2.3 *The teichuronic acid of Bacillus licheniformis*

The teichuronic acid present in the walls of *B. licheniformis* grown in batch culture [88] and in *B. subtilis* W23 grown under conditions of phosphate-limitation [221] is a linear polysaccharide composed of alternating residues of D-glucuronic acid and *N*-acetylgalactosamine as follows:

$$\left[\text{D-GlcUA} \xrightarrow{1,3} \text{Gal}N\text{Ac} \xrightarrow{1,4} \right]_{23-25} .$$

Chemical analysis ([87] and Chapter 7.1.2) and biosynthetic studies described below have established that the polysaccharide is linked to peptidoglycan via a phosphodiester bond from the reducing terminal *N*-acetylgalactosamine of the polymer to the 6-position of a muramic acid residue in the glycan.

Using a particulate enzyme preparation from *B. licheniformis* 6346, Hughes [86] demonstrated the synthesis of polymeric material from UDP–glucuronic acid and UDP-*N*-acetylgalactosamine. The biosynthesized material behaved like isolated teichuronic acid on paper electrophoresis and ion-exchange chromatography.

More recently, the mechanism of biosynthesis of teichuronic acid has been studied in detail in a phosphoglucomutase-deficient mutant of *B. licheniformis* [207]. This organism, which is deficient in phosphoglucomutase, cannot synthesize UDP–glucose and hence UDP–glucuronic acid. The absence of this enzyme prevents the formation of teichuronic acid *in vivo* although cell-free preparations supplied with the appropriate precursors will synthesize the polymer *in vitro.* With wall–membrane preparations of *B. licheniformis* the majority of newly-synthesized teichuronic acid was found linked to peptidoglycan synthesized

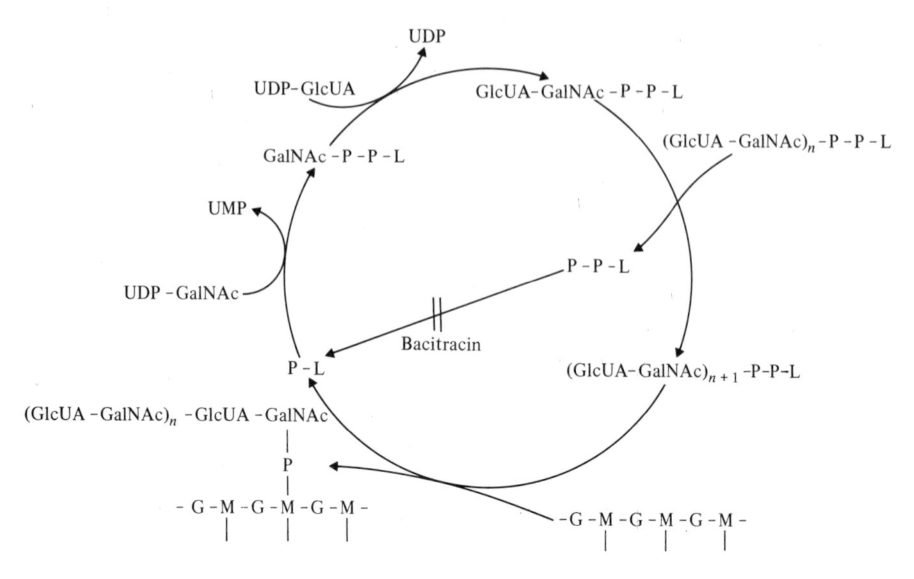

Figure 10.6 The pathway for biosynthesis of teichuronic acid and its linkage to peptidoglycan in *B. licheniformis.* The site of inhibition of bacitracin is shown. L is the lipid carrier, presumably undecaprenol.

concomitantly although as observed with the linkage of teichoic acid in these preparations (p. 396) a minor fraction becomes covalently-bound in the absence of peptidoglycan synthesis [207]. Again this observation is in good agreement with the findings of Mauck and Glaser [123] discussed earlier.

The initial reaction of teichuronic acid biosynthesis is the formation of a lipid intermediate by the translocation of *N*-acetylgalactosamine-1-phosphate from its uridine nucleotide to polyprenol phosphate (Fig. 10.6). The translocase catalysing the reaction is not inhibited by tunicamycin in contrast to the analogous reaction involving *N*-acetylglucosamine-1-phosphate in the formation of teichoic acid linkage unit (p. 393). The disaccharide repeating unit of teichuronic acid is then formed by the transfer of glucuronic acid from the nucleotide precursor with the release of UDP. Polymerization of these repeating units then occurs by the incorporation of new disaccharide units at the reducing terminus of the growing polysaccharide chain while this remains attached to the polyprenol pyrophosphate. This direction of chain extension is the same as that found in peptidoglycan and O-antigen biosynthesis. The common product of all these reactions is lipid pyrophosphate which requires conversion to lipid monophosphate before it can be re-utilized in biosynthesis. It is this dephosphorylation reaction which is specifically inhibited by bacitracin (Chapter 9.4). Thus teichuronic acid biosynthesis is also inhibited by this antibiotic. Once the newly-synthesized polymer has reached the correct length it is linked to peptidoglycan by a phosphate residue derived biosynthetically from UDP-*N*-acetylgalactosamine. This forms a phosphodiester bond between the reducing terminus of the teichuronic acid and the 6-position of muramic acid in the glycan, polyprenol phosphate being the other product of the reaction. Thus

B. licheniformis has evolved quite distinct mechanisms for the biosynthesis of the two major secondary polymers present in its walls. However, the involvement of polyprenol phosphates, presumably all derivatives of undecaprenol, as carriers in the biosynthesis of both these secondary polymers and of peptidoglycan gives the organism a point of control over wall biosynthesis in general.

In recent years there have been several reports that the growth of certain organisms including *S. aureus* and other Bacilli under conditions of phosphate-limitation results in the replacement of the normally occurring teichoic acids with alternative polymers identical with or closely resembling the teichuronic acid of *B. licheniformis* [51]. However, to date there have been no *in vitro* studies on either the biosynthesis of these teichuronic acids or their linkage to peptidoglycan.

10.3 Biosynthesis of the lipopolysaccharides

The lipopolysaccharides of the Enterobacteriaceae are probably the most thoroughly investigated components of the Gram-negative bacterial wall. These heat-stable antigens contain the somatic or O-antigenic determinants of the enteric and probably many other Gram-negative bacteria. They are also responsible for the endotoxic activity of the parent cell and represent the receptor sites for numerous bacteriophages. They are thus of considerable importance, not only for their involvement in the pathogenicity of many of these bacteria, but also in the enterobacteria from both the immunological and taxonomic points of view. Consequently they have been studied in detail and a series of reviews dealing with their chemistry [119. 120, 147]; biosynthesis [147, 153, 154, 159, 175, 220], the genetic control of their synthesis [147, 197] and their biological activity [101] as endotoxins has been published. As described in detail in Chapter 7.2.1 they are composed of a complex polysaccharide (the O-polysaccharide, O-antigen, O-side chain) linked covalently through an oligosaccharide (the R-core) to a glucosamine-containing lipid, lipid A. In addition, the oligosaccharide core region can be further subdivided into the outer core and backbone regions. The outer core has a structure more like that of a conventional polysaccharide, whereas, the backbone region in the enterobacteria is characterised by the presence of the specific sugars 3-deoxy-D manno-octulosonate, (2-keto-3-deoxyoctonate) (KDO) and heptose in addition to ethanolamine and phosphate (Fig. 7.9). Since the majority of the biosynthetic and structural studies have been carried out on the lipopolysaccharides of various *Salmonella sp.*, the discussion of the biosynthetic process will be confined to these except where significant advances have been made in other organisms. At this point it is also worth emphasizing that the isolation of mutants blocked at specific points in the synthesis of LPS have proved of immense importance in furthering both the biosynthetic and structural studies. However, since the mechanisms of biosynthesis for at least two parts of the molecule (the O-polysaccharide and the outer region of the oligosaccharide core) are quite distinct, there are valid reasons for considering the biosynthesis of each of the components separately before turning to the mechanism of assembly of the complete macromolecule. Relatively little is known about the biosynthesis of lipid A and the inner core regions of lipopolysaccharide and these will be considered together. Finally, the translocation of the completed lipopolysaccharide from its site of synthesis in the cytoplasmic membrane to its ultimate position in the outer membrane of the bacterial wall, will be discussed.

10.3.1 *Lipid A and the inner core region*

Until recently relatively little was known concerning the biosynthesis of this part of the lipopolysaccharide. In the case of lipid A this was restricted to an isolated observation describing the incorporation by an envelope preparation of *Pseudomonas aeruginosa* of 3-hydroxydodecanoic acid into material identified as lipopolysaccharide [89]. In this organism [58, 89] 3-hydroxydodecanoic acid is linked to lipid A replacing the 3-hydroxytetradecanoic acid (β-hydroxymyristic acid) residues found in amide-linkage to the glucosamine residues of lipid A in *S. typhimurium.*

The addition of KDO residues to lipid A has been studied in both *E. coli* and more recently in *S. typhimurium.* In the initial investigation in *E. coli* [79] the transfer of KDO from the nucleotide precursor CMP–KDO onto a partially degraded lipid A acceptor was catalysed by a soluble enzyme. Mild alkaline hydrolysis of lipid A to remove ester-linked but not amide-linked fatty acids gave material with the highest acceptor activity. These results gave the first indication that KDO is incorporated at an intermediate stage in lipid A biosynthesis prior to complete O-acylation of the molecule. However, the detailed structure of lipid A was not then known and characterization of the KDO transferase system and the nature of the product was therefore extremely difficult.

More recently mutants of *S. typhimurium* conditionally defective in KDO biosynthesis and growth have been described [114, 157, 170]. The biosynthesis of KDO from D-ribulose-5-phosphate involves three sequential reactions (Fig. 10.7) before the free KDO is converted into the nucleotide sugar CMP–KDO for use in lipopolysaccharide biosynthesis. The mutants studied in greatest detail are all defective in KDO-8-phosphate synthetase. When transferred to the restrictive temperature growth ceases after one generation and the organisms accumulate precursors of lipid A (acidic precursor) which lack KDO and certain O-acyl fatty acid substituents [110, 169]. Chemical analysis of the purified precursors showed them to be composed of a glucosamine disaccharide carrying two phosphate groups as phosphomonoesters. The disaccharide was further substituted by two amide and one or two ester-linked residues of β-hydroxymyristic acid. Pulse-chase experiments showed that precursor synthesized at the restrictive temperature was rapidly converted into lipopolysaccharide when the organisms were transferred to the permissive temperature and the ability to synthesize KDO restored [110, 139].

The lipid A precursor has also been used to study the CMP–KDO:lipid A–KDO transferase system of *S. typhimurium* [136, 137]. Enzyme activity which is

$$\text{D-ribulose-5-phosphate} \rightleftharpoons \text{D-arabinose-5-phosphate}$$

$$\text{D-arabinose-5-phosphate} + \text{phosphoenolpyruvate} \longrightarrow \text{KDO-8-phosphate} + P_i$$

$$\text{KDO-8-phosphate} \longrightarrow \text{KDO} + P_i$$

$$\text{CTP} + \text{KDO} \rightleftharpoons \text{CMP-KDO} + PP_i$$

Figure 10.7 Reactions leading to the biosynthesis of KDO, 2-keto-3-deoxyoctonate (3-deoxy-D-manno-octulosonate). The reactions are catalysed by D-ribulose-5-phosphate isomerase, KDO-8-phosphate synthetase, KDO-8-phosphate phosphatase and cytidine monophosphate KDO pyrophosphorylase.

membrane-bound was partially purified from the soluble fraction obtained by treatment of envelope preparations with Triton X-100 at alkaline pH. Using purified precursors as the acceptor with either partially purified enzyme or the membrane fraction a single reaction product containing two residues of KDO was obtained. Efforts to obtain a product containing a single residue of KDO or to distinguish between two separate KDO transferases were unsuccessful. Analysis of the product suggested that the transferred KDO residues formed the branch of the KDO trimer found *in vivo.* Why the transferase system fails to add the third main chain residue of KDO remains unclear. One possibility is that additional substituents such as phosphorylethanolamine must be added prior to incorporation of the third KDO. Evidence to support this has come from structural studies on a deep-rough mutant of *E. coli* which makes only the disaccharide branch of the KDO trimer and lacks the phosphorylethanolamine substituent in addition to the third KDO [166].

Again, using one of the KDO-8-phosphate synthetase mutants, Lehmann and Rupprecht [113] found that transfer of the organism to a 'semi permissive' temperature (30°–36° C) resulted in the biosynthesis of a small amount of lipopolysaccharide and the accumulation of a second neutral lipid A precursor. In the earlier experiments described above cultures grown at the permissive temperature (25° C) were transferred directly to 37°–42° C where only the acidic precursor accumulates. The second neutral precursor differed from the acidic precursor in that it also contained 4-aminoarabinose, phosphorylethanolamine and polyamines (Fig. 10.8). Substitution of the ester-linked phosphate by 4-amino-arabinose had already been found in *Chromobacterium violaceum* [63] and suggested in Salmonella lipopolysaccharide. The phosphorylethanolamine appeared to be linked by pyrophosphate bonds to the reducing end of the glucosamine disaccharide. Pulse-chase experiments showed that the neutral precursor was also converted into lipopolysaccharide when organisms were returned to the permissive temperature and suggested that it represented an intermediate stage between the acidic precursor and lipid A. Subsequently Lehmann *et al.* [112] demonstrated that the acidic precursor would in fact act as the direct acceptor for 4-aminoarabinose, phosphorylethanolamine and KDO. From these findings and detailed analysis of pulse-chase experiments they proposed the sequence of reactions shown in Fig. 10.9 for the biosynthesis of lipid A.

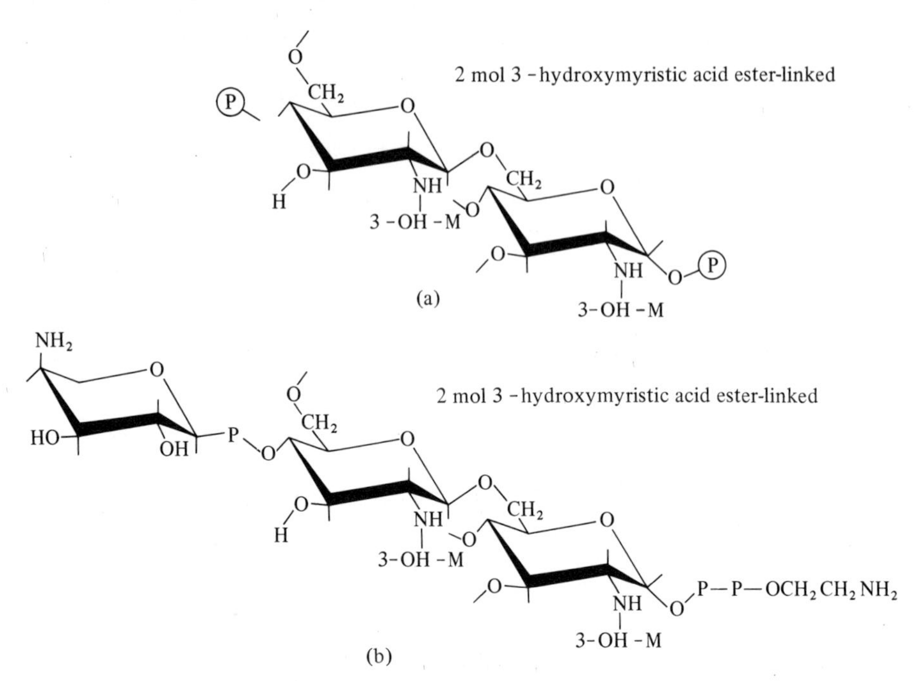

Figure 10.8 Acidic (a) and Neutral (b) intermediates in the biosynthesis of lipid A in *S. typhimurium.* Both intermediates contain ester-linked 3-hydroxymyristic acid residues as shown. In the neutral precursor 4-amino arabinose is linked to the ester-bound phosphate of the glucosamine disaccharide while the phosphate at the reducing terminus is substituted with ethanolamine phosphate.

However, nothing is known as yet about the addition of the third KDO residue or of the heptose of the backbone region. Indeed, to date there is no direct evidence as to the nature of the nucleotide-precursor of these heptose residues. Genetic and biochemical evidence [48, 49, 111] suggests that sedoheptulose-7-phosphate is the precursor, synthesis of the nucleotide sugar probably occurring as shown in the following equations:

$$\text{Sedoheptulose-7-phosphate} \xrightleftharpoons{\text{isomerase}} \text{D-glycero-D-manno-heptose-7-phosphate}$$

$$\text{D-glycero-D-manno-heptose-7-phosphate} \xrightleftharpoons{\text{mutase}} \text{D-glycero-D-manno-heptose-1-phosphate}$$

$$\text{D-glycero-D-manno-heptose-1-phosphate} + \text{NDP} \xrightleftharpoons{\text{NDP heptose pyrophosphorylase}} \text{NDP-D-glycero-D-manno-heptose} + \text{PP}_i$$

$$\text{NDP-D-glycero-D-manno-heptose} \xrightleftharpoons{\text{epimerase}} \text{NDP-L-glycero-D-manno-heptose}$$

where NDP = nucleotide diphosphate.

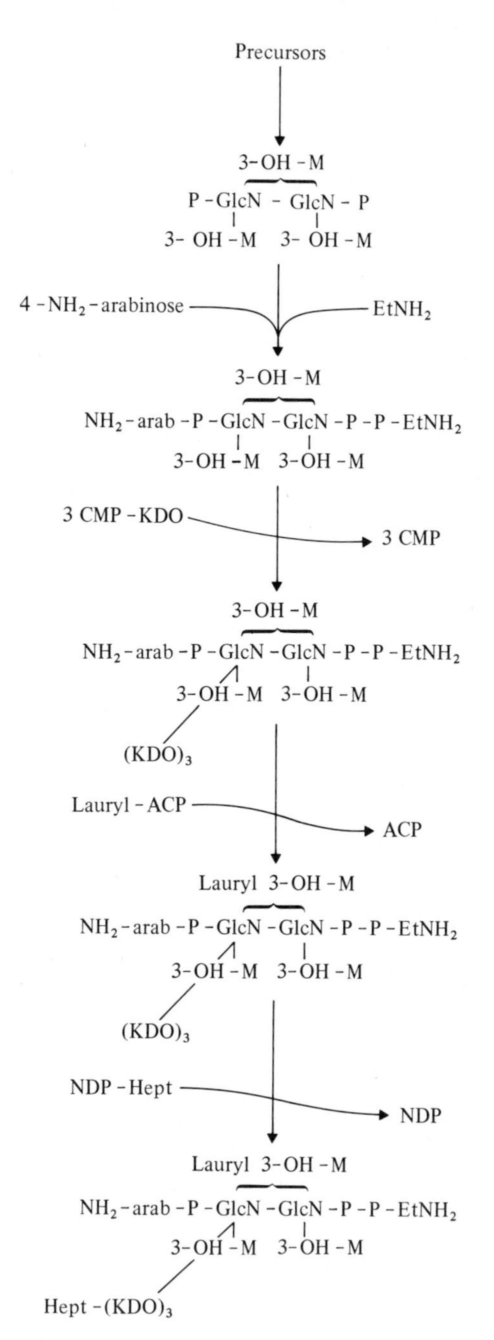

Figure 10.9 Proposed pathway for the biosynthesis of lipid A and part of the inner core of lipopolysaccharide in *S. typhimurium.* The neutral precursor shown in Fig. 10.8 is further substituted with 3 mol of KDO prior to incorporation of additional fatty acid residues. These are shown as being derived from lauryl-acyl carrier protein (lauryl-ACP) although this may not be the precursor and other fatty acids, eg palmitate, are present. Finally heptose is linked to one of the KDO residues. NDP is nucleotide diphosphate.

Mutants of *S. typhimurium* in which both D-glycero-D-manno-heptose and the more common L-glycero-D-manno-heptose are incorporated into the lipopolysaccharide have been described. However, the mechanism by which the heptose residues are added can only be inferred. Mutants lacking heptose but containing the normal amount of KDO have been isolated. In view of the mechanism of synthesis of the outer core region described below, it seems likely that the heptose units are also added in a stepwise manner.

The phosphorylation of the inner core region has been studied by Mühlradt [130, 131]. In a mutant of *S. minnesota* lacking heptose-linked phosphate, he showed the transfer of phosphate residues from ATP to the heptose residues of the lipopolysaccharide. The phosphotransferase, isolated as a soluble fraction by repeated washing of EDTA–lysozyme treated cells, was present in organisms containing phosphorus but not in the phosphate-deficient mutants. Lipopolysaccharide lacking the initial glucose residue of the outer-core oligosaccharide acted less efficiently as acceptor than did material containing the glucosyl residue. In addition, lipopolysaccharide from a strain of *S. minnesota* defective in the incorporation of galactose from UDP-galactose was used. In this case phosphorylation of the endogenous lipopolysaccharide significantly increased the capacity of the particulate fraction to incorporate galactose. These observations have led to the conclusion that phosphorylation occurs, at least in part, between the transfer of the first glucosyl and galactosyl residues to the growing core oligosaccharide.

10.3.2 *The outer core oligosaccharide*

The usefulness of mutants defective in the synthesis of nucleotide sugars was first recognized when the studies of Nikaido and his colleagues on mutants of *E. coli, S. typhimurium* and *S. enteritidis,* deficient in UDP–galactose-4-epimerase, showed that these organisms synthesized incomplete lipopolysaccharides. The latter contained only glucose, heptose and KDO as sugar components in addition to lipid A. Moreover, growth of the mutants in medium containing galactose, which enabled the enzyme defect to be bypassed, resulted in the synthesis of lipopolysaccharide of normal structure [144]. Nikaido [145] then utilized both intact cells and cell-free extracts of the mutant strains, to demonstrate the incorporation of galactose into the galactose-deficient lipopolysaccharide. Other authors [62, 158, 171] both confirmed and extended these observations by using other mutants lacking either UDP–glucose pyrophosphorylase or phosphoglucoisomerase activity and therefore unable to synthesize UDP–glucose. Mutants of this type contained lipopolysaccharide with only the lipid A and inner core region components. From these investigations it became possible to define partially the mechanism of synthesis of the outer-core oligosaccharide. Thus the glycosylation of the inner core region proceeds by the stepwise addition of single glycosyl residues to the non-reducing terminal of the pre-existing oligosaccharide (Fig. 10.10). Moreover, there was no indication that any material, other than the incomplete lipopolysaccharide, would act as the acceptor of these residues. For example, Osborn and her colleagues [158] found that incorporation of ^{14}C-glucose from UDP-glucose into the endogenous lipopolysaccharide of a galactose-deficient mutant of *S. typhimurium*, would occur only if UDP-galactose was also present in the incubation mixture. Similarly, the incorporation of *N*-acetylglucosamine from UDP-*N*-acetylglucosamine could only be demonstrated in the presence of both UDP-glucose

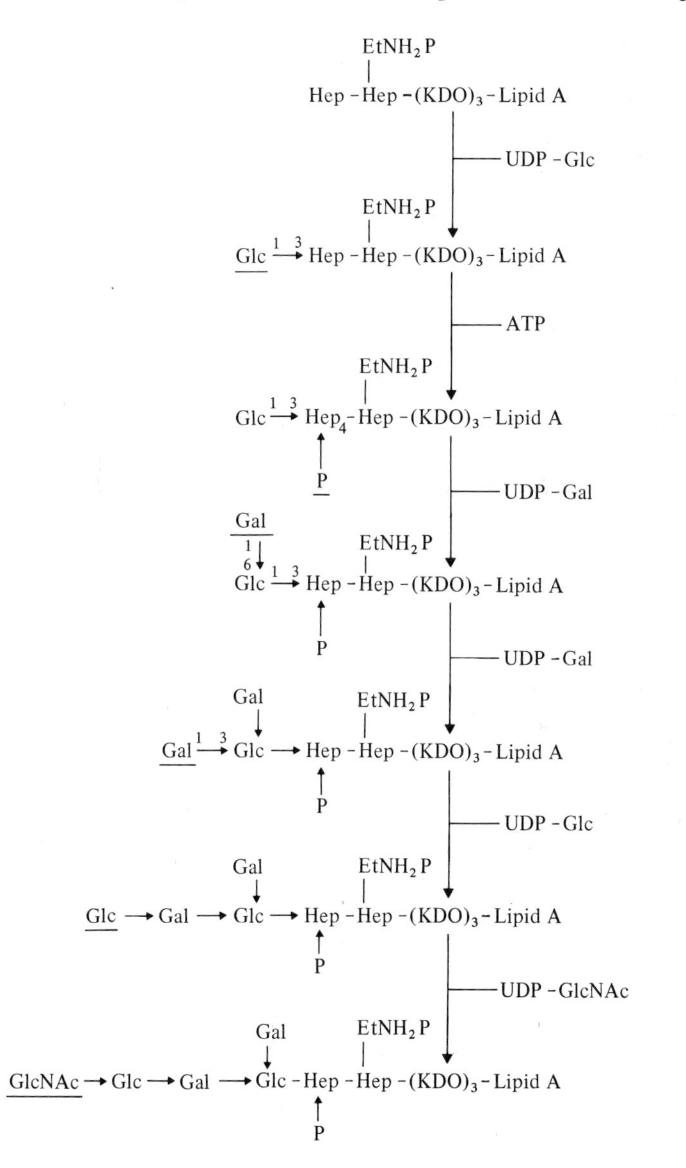

Figure 10.10 Pathway proposed for the biosynthesis of the core of lipopolysaccharide in *S. typhimurium*. Material isolated *in vivo* contains an additional heptose residue linked to the outer heptose shown. When this is incorporated into the core is not known.

and UDP-galactose. It should be emphasized, however, that simultaneous incorporation was not a prerequisite of the system. Prior incubation of the enzyme preparation with the unlabelled nucleotide sugars, followed by their removal by washing and replacement with the radioactively labelled nucleotide sugar, resulted in efficient incorporation of the radioactivity into the lipopolysaccharide. Missing from the lipopolysaccharide synthesized *in vitro* was the second galactosyl residue (α-1, 6-linked galactose), although chemical studies [198] had characterized it as a component of the core oligosaccharide. Furthermore, mutants specifically lacking galactosyl transferase 1 incorporated equimolar amounts of glucose and galactose into their lipopolysaccharide [152, 216]. The nature of the linkage of the two sugars was established by the isolation in high yield of α-6-galactosylglucose [152] (melibiose) from partial acid hydrolysates of the polysaccharides. More recently, Mühlradt [131] has demonstrated the addition of both galactosyl residues in a cell-free system. From the phosphorylation results described earlier, together with those of Osborn [152], he concluded that synthesis of the core oligosaccharide proceeded by the mechanism shown in Fig. 10.10. Sequential synthesis of the core oligosaccharide from the appropriate nucleotide precursors has also been reported in *E. coli* [47] and *Pseudomonas aeruginosa* [7a].

10.3.3 *Phospholipids as cofactors of the glycosyl transferases*

Among the earlier studies of the glycosyl transferases, was the finding that prolonged sonication solubilized a significant portion of the enzymic activity [183]. Using such preparations of both glucosyl transferase 1 and galactosyl transferase 1, the glycosylation of lipopolysaccharide present in heat-inactivated envelope preparations was demonstrated. If, on the other hand, the lipopolysaccharide was first extracted and purified, then it lost acceptor activity completely. Thus, transferase activity appeared to require some additional component of the envelope preparation. The situation was clarified by the findings of Rothfield and Horecker [182] that phospholipids played an important role in the enzyme activity. These conclusions were based on the observations that in the presence of phospholipids acceptor activity was restored to both lipid-depleted envelope fractions and to the purified lipopolysaccharide. Moreover, this restoration was virtually complete when the mixed phospholipid and lipopolysaccharide components were first heated and then allowed to cool slowly to room temperature. Transferase activity also required the presence of either unsaturated or cyclopropane fatty acid residues as acyl substituents of the phospholipids [184]. More recently, the effect of changes in fatty acid composition on the activity of galactosyl transferase had been studied using an unsaturated fatty acid auxotroph of *E. coli* [13]. In such mutants it is possible to modify the fatty acids incorporated into the lipids by alteration in the supply of such acids in the growth medium. Beacham and Silbert [13] found that transferase activity was sensitive to quite minor changes in the fatty acid composition. In general, however, they found that activity was enhanced by changes which would restrict the packing of the acyl chains of the phospholipids.

Galactose - deficient LPS + Phospholipid ⟶ LPS - PL complex
(LPS) (PL)

LPS - PL complex + galactose transferase I $\xrightarrow{Mg^{2+}}$ LPS - PL - transferase complex

LPS - PL - transferase + UDP - Gal ⟶ Galactosyl - LPS + UDP + transferase + PL

Figure 10.11 Reactions involved in the activity of purified glycosyl-transferases.

These results are in good agreement with the earlier findings described above. Moreover, they suggest, perhaps not unreasonably in view of the nature of the enzymic process involved and the situation in which it occurs, that fluidity of the membrane plays a major role in determining the efficiency of transferase activity. Earlier, a specificity towards the polar region of the phospholipid had also been demonstrated [184]. Although some phospholipids could be substituted with varying success for phosphatidylethanolamine, no enzyme activity was obtained with phosphatidylcholine.

In a continuing series of elegant experiments, Rothfield and his colleagues [186, 187] investigated in depth the mechanism of both the galactosyl and glucosyl transferases. On the basis of their findings, the reaction sequence shown in Fig. 10.11 appears to be involved in these and perhaps all the transferase reactions. Using enzymes purified to homogeneity, together with purified phospholipids and lipopolysaccharide, this reaction sequence has been studied in both aqueous suspensions and in artificial monolayers. Thus the binary complex formed on mixing lipopolysaccharide and phosphatidylethanolamine and the ternary complex obtained on the subsequent addition of the purified galactosyl transferase have been isolated and purified from the individual components by density gradient centrifugation [213]. The presence of Mg^{2+} is necessary for the formation of the ternary complex. Glycosylation was initiated by addition of the appropriate nucleotide sugar and did not require the further addition of divalent cation. Essentially similar results were obtained in the monolayer system except that the yield of glycosylated product was some 50-fold lower [179, 180]. In the intact membrane or aqueous suspension it appears that each enzyme molecule can catalyse the transfer of many glycosyl residues, suggesting a certain fluidity in terms of movement between enzyme and lipopolysaccharide in this system. This is in marked contrast to the monolayer system where both components appear to be more rigidly held.

More recently, the isolation by density gradient centrifugation of a quaternary complex of phosphatidylethanolamine, lipopolysaccharide and both glucosyl-1 and galactosyl-1 transferases has been reported [81]. Addition of the appropriate nucleotide sugars to this system resulted in the coupled transfer of both glucose and galactose. The high initial rates of galactose transfer obtained suggest strongly that the two enzymes are inserted into a common lipopolysaccharide–phospholipid matrix, although this has not been proved unequivocally. Moreover, one can conclude that insertion of a particular enzyme into such a matrix is independent of the presence of lipopolysaccharide molecules with the appropriate deficiency to act as acceptor of the glycosyl residue. These experiments have also been repeated using the monolayer system [188]. In neither system was evidence obtained for the release of the transferases after completion of the reaction. On the basis of these results it seems highly probable that the introduction into the system of the other transferases, not yet available in a purified form, would under the appropriate conditions allow synthesis of the complete oligosaccharide of the outer core region.

Levy and Leive [115] have studied a ternary complex of phospholipid, lipopolysaccharide and galactosyl transferase released by treatment with EDTA from a mutant of *E. coli* lacking UDP–galactose-4-epimerase. This complex was active *in vitro* in the transfer of galactose to the complexed lipopolysaccharide; other transferases that may have been present were not tested. Evidence that such enzymes were present in the complex would clearly strengthen the suggestion that this complex represents the biosynthetic unit *in vivo* which has been released intact by the EDTA treatment.

10.3.4 *The O-polysaccharide*

In much the same way that the biosynthetic pathway leading to the synthesis of the core oligosaccharide was elucidated through the use of mutants lacking the ability to synthesize one of the nucleotide sugar precursors, so the pathway leading to biosynthesis of the O-polysaccharide was established. Mutants of *S. typhimurium* blocked in the synthesis of dTDP–rhamnose [151] and GDP–mannose [158, 229] were isolated and shown to lack all the other specific components of the O-polysaccharide, abequose and mannose or rhamnose respectively. Using these mutants [148, 158, 229] and others from *S. anatum* [176], several groups demonstrated the incorporation of appropriate radioactive sugar residues into cell envelope fractions. In all cases there was a marked stimulation of the incorporation if all the component sugars of the polysaccharide were provided as their nucleotide sugar derivatives. At this point it is worth emphasizing that there was already strong evidence in existence that the major portion, if not all the O-polysaccharide, was composed of repeating oligosaccharide units which in *S. typhimurium* had the structure shown in Fig. 10.12 (see also Chapter 7.2.1). Chemical analysis of the biosynthesized material showed the sugars to be linked in the correct sequence. However, the chemical and the biosynthetic repeating units were different, due to differences in acid-lability of the linkages in the repeating unit. Whereas acid hydrolysis yielded a galactosyl–mannosyl–rhamnosyl sequence, the biosynthetic sequence was established as mannosyl–rhamnosyl–galactose [148]. Thus, the incorporation of rhamnose was dependent on the presence of galactose and similarly that of mannose on both rhamnose and galactose.

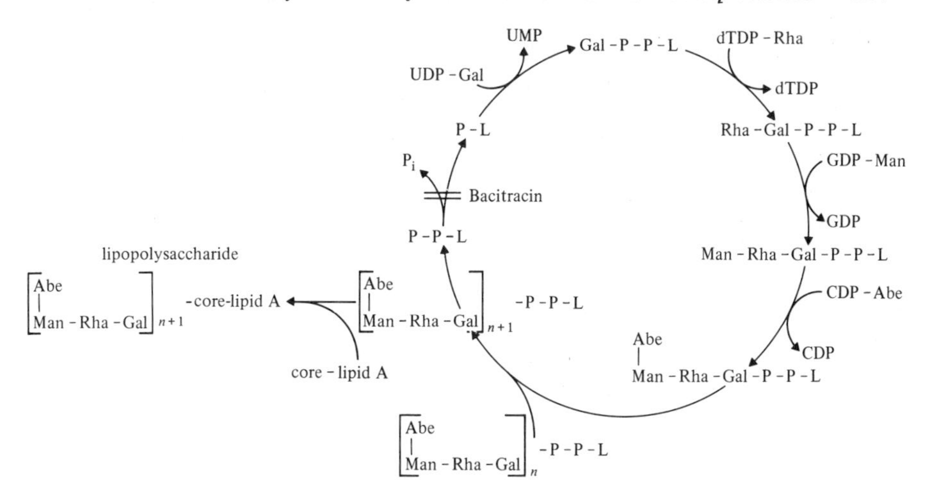

Figure 10.12 Pathway for the biosynthesis of O-side chains of lipopolysaccharide in *S. typhimurium.* The tetrasaccharide abequosyl-mannosyl-rhamnosyl-galactosyl-phosphate is synthesized on the lipid carrier undecaprenyl phosphate. Polymerization which occurs at the reducing terminus leads to the formation of oligosaccharide pyrophosphate undecaprenol and the release of undecaprenyl pyrophosphate. The dephosphorylation of this pyrophosphate is the site of inhibition by bacitracin.

10.3.5 *Assembly of the repeating unit*

The first indication that a lipid carrier was involved in the synthesis of O-polysaccharides came from the studies of Wright *et al.* [219]. Using an envelope fraction from *S. newington* they found that O-polysaccharide synthesis occurred on addition of the nucleotide precursors UDP-galactose, dTDP-rhamnose and GDP-mannose. If, however, GDP-mannose was omitted from the system then an intermediate soluble in organic solvents and containing the disaccharide rhamnosyl-galactose accumulated. Moreover, readdition of GDP-mannose to an envelope fraction in which this intermediate had accumulated, but from which the residual UDP-galactose and dTDP-rhamnose had been removed by washing, allowed the synthesis of O-polysaccharide to occur. The observations were extended [43, 219] to show that the first reaction in the synthesis of the repeating unit was the transfer of galactose-1-phosphate from UDP-galactose to the lipid carrier which was termed antigen carrier lipid (ACL). A rhamnose residue was then added to yield the intermediate isolated previously. The addition of mannose to complete the trisaccharide intermediate was not observed but rather, as described above, O-polysaccharide synthesis occurred. Osborn and her colleagues [160, 161, 211, 212] have described a similar lipid cycle in *S. typhimurium* in which the initial reaction sequence was identical. However, by incubating at low temperatures the synthesis of the trisaccharide intermediate was established. In addition abequose was transferred from CDP-abequose to yield a tetrasaccharide-lipid intermediate. These reactions are shown in Fig. 10.12. Using endogenous intermediates generated by incubation of the system at low temperature, Osborn and Weiner [161] showed that the tetrasaccharide intermediate became polymerized more rapidly than the trisaccharide intermediate when the incubation temperature was raised. This suggestive evidence that the tetrasaccharide–lipid was in fact the natural intermediate has now been established unequivocally by the finding that a mutant of *S. typhimurium* blocked in the synthesis of CDP–abequose lacked the O-polysaccharide in its wall [227].

Subsequent studies [189, 218] have shown that antigen carrier lipid is in fact undecaprenol phosphate and thus the identical carrier to that participating in peptidoglycan synthesis (see Chapter 8.3.1). Hence, the initial reactions in the synthesis of both lipid intermediates are analogous and result in the formation of a pyrophosphate bridge between the undecaprenol residue and the primary unit of the intermediate. On the one hand this is *N*-acetylmuramyl-(pentapeptide)-1-phosphate and on the other galactose-l-phosphate. UMP is the second product of the reactions and both are freely reversible.

Polymerization of the lipid-linked intermediates has been studied *in vitro* by Kanegasaki and Wright [102]. The isolated lipid intermediate was added to the envelope preparation and after several cycles of freezing and thawing, polymerization at high efficiency was observed. The authors concluded that the freezing and thawing of the membrane–intermediate mixture resulted in some change in the membrane either by phase transition or mechanical alteration which allowed enzyme–substrate interaction. The polymerase catalyses the addition of repeating units to the reducing terminal of the growing polysaccharide chain [24, 172].

The studies of Kanegasaki and Wright [102] also suggest that the lipid intermediates are free to move within the hydrophobic environment of the membrane. In this way intermediates would not be restricted to involvement with a particular polymerase complex, but rather would be free to move from one enzyme molecule to another. This type of movement may provide a mechanism whereby intermediates could be translocated through the membrane, from the cytoplasm, where the nucleotide precursors exist and synthesis presumably occurs, to the outer surface for polymerization and synthesis of the O-polysaccharide. In cell-free systems some specificity of the polymerase enzyme is clearly lost, as demonstrated by the polymerization of the trisaccharide, rather than the tetrasaccharide–lipid intermediate in *S. typhimurium* [161, 212].

As expected from the biosynthetic cycle, bacitracin inhibits O-polysaccharide synthesis in both *S. anatum* and *S. typhimurium* with the concomitant accumulation of undecaprenol pyrophosphate. However, the *in vitro* polymerase reaction was not inhibited [102]. In this respect the antibiotic inhibits an idential reaction in both O-polysaccharide and peptidoglycan synthesis, that is the dephosphorylation of undecaprenol pyrophosphate, by a specific pyrophosphatase, to yield undecaprenol phosphate and inorganic phosphate.

10.3.6 *Modification of the O-polysaccharide*

Again, in reactions analogous to the modification of the lipid-linked intermediates in peptidoglycan synthesis, described in Chapter 8.3.5, the corresponding intermediates in the synthesis of the O-polysaccharide undergo certain modifications. Two of these modifications, glucosylation and O-acetylation, have been investigated biosynthetically, and will be discussed below.

10.3.7 *Glucosylation*

Glucosylation of the O-polysaccharide chains in lipopolysaccharide has been studied in salmonellae of both Group E (*S. anatum*) and Group B (*S. typhimurium* and *S. enteritidis*). In the former case, glucosylation of the chains results from lysogenisation of the organism by the phage Σ34. Incubation of envelope preparations from these organisms with UDP–glucose resulted in the glucosylation of the endogenous lipopolysaccharide without concomitant O-polysaccharide synthesis [204].

In salmonellae of Groups A, B and D, glucosylation of the lipopolysaccharide determines the expression of the somatic antigen 12_2. The glucosyl residues are present as single non-reducing units α (1,4)-linked to the D-galactosyl residues of the polysaccharide chains. The extent of glucosylation appears to be highly variable. Thus, overnight incubation of the progeny from a single colony results in the appearance of both 12_2-positive and -negative forms. This indication that populations of these organisms are made up of cells containing glucosylated lipopolysaccharide (12_2-positive) and others from which glucosylation is absent (12_2-negative) is reminiscent of the situation in *B. subtilis* W23, where both glucosylated and non-glucosylated teichoic acids are found [37]. Whether, in the salmonellae, the two forms of the lipopolysaccharide are present in the same organism is not known. What is clear, however, is the apparent indifference of the O-polysaccharide polymerase to the glucosyl substituents, in contrast to the requirement for the abequosyl residues demonstrated in other strains of *S. typhimurium* [227].

Detailed studies of the mechanism of the glucosylation reaction have now been made in both *S. anatum* (Group E) [191, 217] and organisms of Group B [149, 150]. In both cases the reaction sequence has been established as follows where ACL is the antigen carrier lipid (undecaprenol phosphate)

$$\text{UDP-glucose} + \text{P-C}_{55}\ (\text{ACL}) \rightleftharpoons \text{glucosyl-P-C}_{55} + \text{UDP}$$

$$\text{Acceptor} + \text{glucosyl-P-C}_{55} \rightarrow \text{glucosyl-acceptor} + \text{P-C}_{55}.$$

Thus the glucose residue is transferred from the uridine nucleotide to undecaprenol phosphate and from there to the acceptor polysaccharide. During this process it has also been established that the glucose residue undergoes a double inversion of the anomeric configuration. Thus the α-linked glucose of the nucleotide precursor becomes a β-glucosyl derivative of undecaprenol phosphate. The second inversion occurs when the lipid-linked glucose becomes incorporated into the O-polysaccharide as an α-glucoside. Evidence for the glucosylation of lipid-linked O-polysaccharide intermediates comes from the observations of Takeshita and

Mäkelä [200]. Mutants of *S. typhimurium* defective in synthesis of the core oligosaccharide, continued to synthesize O-polysaccharide which remained linked to the lipid carrier. In organisms having the determinant for glucosylation, these O-polysaccharides were fully glucosylated with the exception of the galactose-residue at the reducing terminal. Since this residue represents the most recently added trisaccharide unit of the polysaccharide chain, it suggests that glucosylation occurs not at the level of the lipid-linked repeating units but after polymerization of the repeating unit has occurred. As described by Takeshita and Mäkelä [200], such a mechanism would result in both glucosylated and non-glucosylated polysaccharide chains having identical units at the reducing terminus. These observations have subsequently been confirmed by Sasaki *et al.* [191] using a mutant of *S. anatum.* Hence, a single polymerase should be capable of the addition of newly-synthesized units to either type of chain and the presence or absence of glucosylation would not interfere with such a reaction. Furthermore, if as described previously the synthesis of the repeating units occurs at the inner surface of the cytoplasmic membrane and the polymerisation reaction occurs at the outer surface, then the use of undecaprenol phosphate as a lipophilic carrier of the glucose residue may have evolved as a result of the need to transport glucose across the membrane.

10.3.8 O-*acetylation*

O-acetylation of the O-polysaccharides of *S. anatum* has been investigated by Robbins and his colleagues, particularly with respect to the modifications resulting from infection of the organism with bacteriophage ϵ15. The structure of the repeating unit obtained on acid hydrolysis of the lipopolysaccharide of *S. anatum* is *O*-acetyl-α-D-galactosyl-mannosyl-rhamnose which, on infection with ϵ15, undergoes modification to β-D-galactosyl-mannosyl-rhamnose [174]. Particulate enzyme from the uninfected organism catalyses the acetylation of both purified oligosaccharides with the correct sequence and also of endogenous lipopoly-saccharide associated with the enzyme preparation. Acetyl-coenzyme A is the donor of the acetyl group [173]. Other studies of Keller (referred to by Robbins and Wright [175] have shown the acetylation of lipid-linked oligosaccharides synthesized by mutants of *S. anatum* defective in synthesis of the core region. Thus it appears that acetylation normally precedes the linking of O-polysaccharides and the core region of the lipopolysaccharide.

Lysogenization of *S. anatum* with ϵ15 results in the repression of the trans-acetylase and the absence of *O*-acetyl substituents from the lipopolysaccharide synthesized [173, 174]. Further experiments have shown that infection also results in the production of a new polymerase synthesizing β-1,6-galactosylmannose bonds, replacing the α-1,6-galactosylmannose linkages found in the uninfected bacteria. The synthesis of this enzyme is coupled with inhibition of the α-polymerase [25, 118]. Mutants of the bacteriophage defective in their ability either to repress the transacetylase or to initiate production of the β-polymerase have been described. Whereas infection with mutants of the former type results in the synthesis of lipopolysaccharide with the repeating sequence *O*-acetyl-β-D-galactosyl-mannosyl-rhamnose, infection with the latter type of mutant results in the complete loss of O-polysaccharide. Thus the phage-induced inhibition of the α-polymerase can clearly be separated from the production of the β-polymerase.

10.3.9 *O-antigen-lipopolysaccharide ligase: linkage of the O-polysaccharides to the lipopolysaccharide core region*

As described in the preceding sections, the O-polysaccharides and the core region of the lipopolysaccharides are synthesized independently and by completely different mechanisms. The final stage of the assembly process is the linkage of these two macromolecules to form the complete lipopolysaccharide. The enzyme catalysing this process, O-antigen-lipopolysaccharide ligase, has been studied by Osborn and her colleagues both *in vitro* [42] and *in vivo* [105]. In the cell-free system, Cynkin and Osborn [42] utilized as acceptor the lipopolysaccharide core region isolated from a mutant of *S. typhimurium* defective in the synthesis of the O-polysaccharide side chains. Lipid-linked intermediates of the O-polysaccharide and the ligase were isolated from a second mutant of *S. typhimurium*, defective in the synthesis of the lipopolysaccharide core region. Incubation of this particulate preparation, together with the acceptor in the presence of nonionic detergent, resulted in the transfer of the O-polysaccharide to the core. Preincubation of the isolated core region with the detergent resulted in a stimulation of the ligase reaction.

The extent of polymerization of the polysaccharide chains normally transferred to the core and the mechanism by which this polymerization is controlled remain unknown. Whereas the O-polysaccharides usually contain several repeating sequences, mutants defective in polymerase activity have been shown to contain lipopolysaccharide with single repeating sequences [138]. In addition, Nikaido [146] has demonstrated the linkage of the incomplete unit rhamnosylgalactose to the core region in a mutant of *S. typhimurium*. Thus the control over the extent of polymerization of the polysaccharide side chains does not appear to reside in the ligase itself.

In a study complementary to the *in vitro* experiments described above, Kent and Osborn [105], using a mutant of *S. typhimurium* defective in phosphomannoisomerase, have demonstrated *in vivo* a precursor-product relationship between the lipid-linked O-polysaccharides and the O-polysaccharide linked to the lipopolysaccharide core.

The reactions leading to both the polymerization of the repeating sequences and also the final ligase reaction result in the release of the lipid carrier as undecaprenol pyrophosphate. Consequently, the biosynthesis of lipopolysaccharide, and in particular the O-polysaccharide, is subject to inhibition by bacitracin (Chapter 9.4). In the analogous situation in peptidoglycan synthesis, bacitracin has been shown to exert its antimicrobial activity through inhibition of a specific pyrophosphatase. This enzyme cleaves the released undecaprenol pyrophosphate to yield inorganic phosphate and undecaprenol phosphate, which is then available to participate in further cycles of biosynthesis. Whether the same pyrophosphatase functions in both peptidoglycan and lipopolysaccharide biosynthesis is not known. Clearly the use of a common pool of undecaprenol phosphate in both reaction sequences would provide an excellent point of control over the synthesis of both peptidoglycan and lipopolysaccharide as components of the bacterial envelope. However, recent results of Rundell and Shuster [190] have suggested that the initial enzymes of the two biosynthetic pathways leading to O-polysaccharide and peptidoglycan synthesis; galactose-diphosphoglycosyl carrier lipid (Gal-P-P-GCL) synthetase and phospho-*N*-acetylmuramyl-pentapeptide translocase respectively, react with different pools of undecaprenol phosphate at any given time. These conclusions were based on the observations that while the synthetase was markedly inhibited by the presence of either glucose (Glc-P-GCL) or rhamnosyl-galactose (Rha-Gal-P-P-GCL) linked to the lipid intermediate present in the membrane preparation, the translocase activity of the same membrane preparation remained unaffected. On the other hand, the inhibition of the synthetase by lipid-linked glucose (Glc-P-GCL) suggests that both the activities involved in O-polysaccharide synthesis do react with the same pool of the lipid carrier. Whether the pools of undecaprenol phosphate remain separate or equilibrate over a period of time remains unknown. Clearly these interesting observations deserve further investigation.

10.3.10 *Incorporation of lipopolysaccharide into the outer membrane*

Before entering into a consideration of the mechanism by which lipopolysaccharide becomes incorporated into the outer membrane of the bacterial envelope, it is worth pointing out that the majority of the biosynthetic studies described in the previous sections had not established whether the enzymes involved were located in the inner (cytoplasmic) or outer membranes of the various bacteria examined. This question has now been clearly answered by Osborn and her colleagues [155, 156], who have separated the two membrane components of the *S. typhimurium* envelope. Similar preparations have also been obtained from *E. coli* [127, 128, 192]. The enzymes of O-polysaccharide synthesis were located almost entirely in the cytoplasmic membrane. The results obtained for the distribution of the enzymes involved in synthesis of the core region were less clear. However, it seems certain that their location is also in the cytoplasmic membrane and that the activity found elsewhere resulted from a certain amount of redistribution of the enzymes during preparation of the various envelope fractions. In addition, Osborn *et al.* [155] were able to show by pulse-chase experiments that both O-polysaccharide and the core region were synthesized initially in the cytoplasmic membrane, but that in a matter of 1-2 min radioactivity was also present in the outer membrane. Further experiments with mutants conditionally defective in the synthesis of the O-polysaccharide revealed that, although these mutants incorporated incomplete lipopolysaccharide into the outer membrane during growth under non-permissive conditions, on return to the permissive conditions, this incomplete lipopoly-saccharide could not act as acceptor for the newly synthesized O-polysaccharide.

Thus the translocation process is not reversible. Moreover, the mechanism by which it is controlled is not readily apparent, for although as determined in the experiments described above it is unidirectional, there is no obvious control over the size of the lipopolysaccharide molecule exported. In mutants of various bacteria, lipopolysaccharide containing from 1 to 2 glycosyl residues of the core region to the complete O-polysaccharide chain are all located in the outer membrane.

In addition, the site of synthesis of phospholipids in both *S. typhimurium* and *E. coli* is the cytoplasmic membrane [14, 215]. Again the cell is faced with the necessity of transporting or translocating certain of these molecules from the cytoplasmic to the outer membrane. The possibility exists that some form of unit of the outer membrane, containing lipopolysaccharide, phospholipid and protein, is formed in association with the cytoplasmic membrane and translocated intact for insertion into the outer membrane. In this context, Wu and Heath [222] have isolated a lipopolysaccharide-protein complex by detergent extraction of the envelope of *E. coli*. The detergent treatment would probably solubilize and cause the removal of any associated phospholipids. On the other hand, cessation of protein synthesis, whether by amino acid starvation or by inhibition with chloramphenicol, does not prevent the continued synthesis of both lipopoly-saccharide and phospholipids [107, 185]. Under such conditions a complex of lipopolysaccharide, phospholipid and a small amount of protein is excreted into the medium. Moreover, the complex excreted at the end of the period of inhibition contained much less protein than did material excreted at the commencement of inhibition [185]. Clearly at least during unbalanced growth the process of translocation from the cytoplasmic membrane has no absolute requirement for protein synthesis. However, this does not preclude the possibility that under normal growth conditions the translocated unit contains protein as suggested above.

The insertion of newly-synthesized lipopolysaccharide into the outer membrane has been studied by Mühlradt and his colleagues [132, 134] and by Kulpa and Leive [109]; who used mutants lacking UDP-galactose-4-epimerase from *S. typhimurium* and *E. coli* respectively. As a result of this mutation, these organisms require the presence of galactose in the growth medium to enable them to synthesize the complete lipopolysaccharide. By separating membrane particles containing newly-synthesized lipopolysaccharide from the pre-existing membrane by density gradient centrifugation, Kulpa and Leive [158] were able to show that insertion of newly-synthesized material occurred at discrete areas of the membrane rather than by diffuse intercalation over the whole surface.

Mühlradt *el al.* [133, 134] visualized complete lipopolysaccharide in the outer membrane by the use of a specific ferritin-conjugated antibody directed against the complete lipopolysaccharide molecule. Some 30 s after addition of galactose to the growing cells freeze-etching revealed the presence of an average of 220 discrete patches of lipopolysaccharide per cell. Moreover, in thin sections of plasmolysed cells the majority (some 86%) of the ferritin patches were located over the adhesion sites between cytoplasm and outer membrane. Similar adhesion sites had previously been demonstrated in *E. coli* and shown to be connected with several bacteriophage receptors [12]. A further 2-3 min incubation resulted in newly synthesized lipopolysaccharide covering the entire bacterial surface. However, this redistribution of the lipopolysaccharide occurred in the absence of both growth and a supply of energy but not when the cells were incubated at 0° C rather than 37° C. Further examination of the cells held at 0° C revealed that the lipopolysaccharide was confined to the outer surface of the outer membrane. Removal of the peptidoglycan by incubation with lysozyme at 0° C followed by incubation of the membrane preparation for as little as 1 min at 37° C resulted in a redistribution of the lipopolysaccharide to both sides of the membrane. Again this redistribution did not occur if the membranes were maintained at 0° C. Prefixation of the membrane proteins with glutaraldehyde did not entirely prevent the transmembrane movement, arguing that the majority of the lipopolysaccharide could not be covalently-linked in lipopolysaccharide-protein complexes, such as those isolated by Wu and Heath [222]. Taken together, these observations have been interpreted to show a considerable potential for transmembrane movement of lipopolysaccharide, particularly when the peptidoglycan layer has been disorganized. The presence of an intact peptidoglycan appears to be necessary to maintain an asymmetric distribution of lipopolysaccharide. In this context it is interesting to speculate that the adhesion sites between cytoplasmic and outer membranes may be in fact associated with areas where the peptidoglycan layer is discontinuous.

10.4 Lipoprotein from the outer membrane of Gram-negative bacteria

One of the major proteins of the cell envelope of *E. coli* and other related Gram-negative bacteria is the lipoprotein found covalently linked to the peptidoglycan (murein) of these organisms (see Chapter 7.2.3). In recent years this protein has become, largely through the work of Braun, Inouye and their associates, one of the most thoroughly characterized proteins of the outer membrane of *E. coli.* This work has recently been discussed in detail in six excellent reviews [19, 20, 44, 95-97]. Consequently in this section we will consider only those points of particular interest in relation to the biosynthesis of the protein itself and the formation of the linkage between the protein and peptidoglycan.

The lipoprotein, (the structure and linkage to peptidoglycan is shown in Fig. 7.7) can be isolated from envelope preparations of *E. coli* in two forms; approximately one third is covalently linked to the peptidoglycan layer, the remaining two thirds exists as a free-form located in the outer membrane of the organism [21, 97]. Estimates of the total number of lipoprotein molecules present in a single bacterium range from 225 000 to 750 000, depending on the growth conditions and size of the organism [22].

It should be noted, however, that the presence of lipoprotein is not essential for growth and division. An *E. coli* mutant lacking both free and bound forms of lipoprotein has been isolated [85]. This organism appears to grow and divide normally although it is extremely sensitive to EDTA and periplasmic enzymes are released, suggesting that outer membrane structure is in some way disturbed. The mutation (*lpo*) which maps at 36.5 min appears to be a deletion in the structural gene for lipoprotein since no active mRNA is produced and chromosomal DNA lacks a restriction fragment which will hybridize with purified mRNA [85, 139]. Other mutations in the structural gene for lipoprotein and in control functions have been isolated in *E. coli* [199, 202, 223] and *S. typhimurium* [210]. Some of the observations made with these organisms have suggested roles for lipoprotein particularly in division which are difficult to reconcile with the findings described above.

Mutants of *S. typhimurium* originally selected as leaking periplasmic enzymes (*lkyD*), were found to contain decreased amounts of bound lipoprotein with an associated increase in the amount of free lipoprotein present [210]. During septum formation the outer membrane of these mutants did not invaginate although apparently normal invagination of the cytoplasmic membrane and peptidoglycan did occur. Consequently the outer membrane formed large 'blebs' over the septal region. Clearly these results suggest that bound lipoprotein is required for effective invagination of the outer membrane during division. Torti and Park [202] have described a temperature-sensitive mutant of *E. coli* which forms filaments at the restrictive temperature (42° C). This change was accompanied by a marked decrease in the amount of both free and bound lipoprotein formed at 42° C. Since revertants of the mutant synthesized normal amounts of lipoprotein and appeared to grow normally at 42° C it was again concluded that lipoprotein was important if not essential for proper division to occur. The mutant maps at 74 min (reported in [44]) and thus appears to involve some regulatory function and not the structural gene for lipoprotein.

Mutants which synthesize structurally altered lipoprotein have also been isolated [199, 223]. One of these which is altered in the signal sequence for lipoprotein [116] has been used to study the biosynthesis of the covalently-linked diglyceride moiety. The results obtained are discussed in detail below.

10.4.1 *Biosynthesis of the lipoprotein*

The biosynthesis of the lipoprotein polypeptide chain was initially investigated in two ways. These were based on the continued synthesis of the lipoprotein when the synthesis of both cytoplasmic and inner membrane proteins was inhibited. From a study of the effects of various antibiotics inhibiting both protein and RNA synthesis it was found that the lipoprotein and indeed other outer membrane proteins continued to be synthesized in the presence of relatively high concentrations of puromycin, kasugamycin and rifampicin, whereas all protein synthesis was sensitive to chloramphenicol and tetracycline [82]. Formation of lipoprotein was particularly resistant to puromycin even in cell-free synthesizing systems. However, the reasons underlying this observation remain unknown, although it has been suggested that puromycin resistance may result from the mode of binding to the cytoplasmic membrane of the polyribosomes involved in lipoprotein synthesis [44].

The alternative approach made use of the fact that histidine, proline, tryptophan, glycine and phenylalanine are not present in the lipoprotein. Using auxotrophic strains of *E. coli* deprived of histidine for example, Hirashima and Inouye [83] demonstrated the continued synthesis of lipoprotein although some 95% of total protein synthesis was suppressed. During the period of amino-acid starvation the proportion of 'bound' lipoprotein decreased: an observation which the authors concluded resulted from inhibition of the synthesis of the enzyme catalysing the formation of lipoprotein-peptidoglycan bonds. Examination of the effects of various inhibitors of protein synthesis on the formation of lipoprotein under conditions of amino acid starvation again revealed biosynthesis to be inhibited by tetracycline and chloramphenicol but not by rifampicin. Thus it appeared that, while synthesis was occurring *de novo* on ribosomes, it also involved the participation of a particularly stable messenger RNA. This observation was subsequently investigated and the messenger RNA for lipoprotein shown to have a half-life of 11.5 min, twice that of the average of other major envelope proteins and some five times longer than the average for cytoplasmic proteins [82]. Furthermore, based on the figures quoted above, the lipoprotein appeared to be one of the most, if not the most, abundant polypeptide in *E. coli* with up to 750 000 molecules being synthesized in each generation. These factors taken together with the assumption that because of the small size of the polypeptide the messenger RNA was likely to be considerably smaller than other messenger RNAs present, were the basis of a purification of the messenger RNA involved in lipoprotein biosynthesis. This was achieved by Inouye and his colleagues [84] and the product of synthesis in a cell-free system directed by the messenger RNA has been shown to be the lipoprotein polypeptide by immunological precipitation and peptide mapping. Although the carboxyl-terminal of the synthesized and native polypeptides were the same, no evidence was available as to possible modification of the amino-terminal end.

At this time Inouye and his colleagues [70] developed a protein-synthesizing system using toluenized *E. coli*. Similar systems had been used previously to study DNA, RNA and peptidoglycan synthesis (see Chapter 8.6.1). Protein synthesis was totally dependent upon added ATP and was sensitive to the antibiotics chloramphenicol and puromycin. Moreover, examination of the products of synthesis revealed them to consist only of membrane proteins. Among these were two polypeptides which reacted with antilipoprotein serum. One of these co-migrated on SDS-polyacrylamide gel electrophoresis with authentic lipoprotein whereas the apparent molecular weight of the second was almost double (15 000) [70]. This increase resulted from the presence of a hydrophobic peptide of 20 amino acids, 'the signal sequence', linked to the amino terminus of the lipoprotein [72, 98]. This new form of the molecule designated prolipoprotein was thought to be a precursor of lipoprotein and accumulated in toluenized cells because of some unknown blockage of the processing reactions. The complete amino acid sequence of prolipoprotein has been determined on material synthesized in a cell-free system using purified mRNA.

Other outer membrane and periplasmic proteins have been shown to be synthesized from precursors with additional peptide sequences at their *N*-terminus [44]. In all cases cleavage of these polypeptides to yield the native protein must occur. A model which suggests that translocation across the cytoplasmic membrane and processing of lipoprotein are tightly coupled has been proposed. This is described in detail in the recent reviews of DiRienzo *et al.* [44], Inouye *et al.* [96] and by Halegoua and Inouye [71].

On the other hand, prolipoprotein was found in both the inner and outer membranes of the envelope of an *E. coli* mutant with an amino acid alteration in the signal sequence of prolipoprotein [116]. In the parent organism lipoprotein is located exclusively in the outer membrane. Thus, the alteration in the signal sequence prevents subsequent processing. However, it appears that proteolytic cleavage is not essential for translocation and assembly of at least some lipoprotein into the outer membrane.

Treatment of *E. coli* with phenethyl alcohol leads to the accumulation of other intermediates in the biosynthesis and assembly of outer membrane proteins [71]. Accumulation of promatrix protein and protol G protein occurred in the presence of the alcohol. Although no accumulation of prolipoprotein was observed, partially processed (trypsin-sensitive) intermediate was detected. Lipoprotein containing all three fatty acid substituents (Fig. 7.7) is trypsin-resistant and it was suggested that the accumulated intermediate lacked one or more of these substituents although this has not yet been established. The data, do, however, suggest that the signal sequence may not be essential for insertion of the lipoprotein into the outer membrane.

Synthesis of the complete lipoprotein requires modification of the polypeptide by addition of glycerol in thioester-linkage to the amino-terminal cysteine residue, together with the incorporation of both amide- and ester-linked fatty acids (Fig. 7.7). These fatty acids are typical of those esterified to cellular phospholipids, an observation which led to the conclusion that the diglyceride moiety is derived from normal phospholipid metabolism [20]. Subsequently pulse-chase experiments showed that the diglyceride attached to lipoprotein contained glycerol derived from a pool of relatively long half-life and pointed to phosphatidylglycerol as the biosynthetic precursor [35]. This has recently been confirmed in *S. typhimurium* [34]. Using the technique devised by Jones and Osborn [100] of fusing exogenous phospholipid vesicles with intact *S. typhimurium* it was shown that phosphatidylglycerol was an excellent donor of the glycerol residue, whereas, cardiolipin and phosphatidylethanolamine were not. These findings were further strengthened by studies using the antibiotic cerulenin, which inhibits fatty acid synthesis, and 3,4-dihydroxybutyl-1-phosphonate (a four carbon analogue of glycerol-3-phosphate) [33, 35]. On the basis of these results Wu and his colleagues concluded that the reaction sequence is

cyteine–lipoprotein + phosphatidylglycerol → glycerylcysteine–lipoprotein + phosphatidic acid.

Processing of the lipoprotein must then be completed by the incorporation of the ester- and amide-linked fatty acid substituents. Whether this occurs in the inner or outer membranes remains unclear. If the suggestion of Halegoua and Inouye [71] based on their observations with phenethyl alcohol treated *E. coli* is confirmed, then it would appear that some may be attached in the inner membrane whereas completion of processing finally occurs in the outer membrane.

10.4.2 *Formation of the lipoprotein-peptidoglycan linkage*

Pulse-chase experiments designed to study lipoprotein synthesis *in vivo* clearly demonstrated that the free-form of the lipoprotein was a precursor of the lipoprotein linked to peptidoglycan [95, 97]. Thus only the free-form of lipoprotein was synthesized during a four minute pulse of radioactive arginine. If, however, the pulse was subsequently chased with non-radioactive arginine, then after 1 generation time some 40% of the incorporated radioactivity (ie newly-synthesized lipoprotein) was now linked to peptidoglycan. Surprisingly, no significant change in the amount of bound-lipoprotein was observed if the chase was continued for a further two generation times. Two possible explanations for these findings were suggested by Inouye *et al.* [95]. In the first they suggested that there is a reversible interconversion of the free and bound forms of the lipoprotein with the newly synthesized material becoming equilibrated with a large pre-existing pool of the free-form. The results obtained require that the ratio of free to bound lipoprotein is maintained at two. Alternatively, the formation of the lipoprotein–peptidoglycan complex is irreversible. Thus, during the first generation time, some 40% of the newly synthesized material is incorporated as the bound form, while the remaining 60% remains in the outer membrane as the free-form. This free-form would have to be either modified or segregated in such a way as to be unavailable for subsequent linkage to the peptidoglycan. Currently no evidence is available to distinguish between these possibilities.

More recently Movva *et al.* [129] have examined the effect of gene dosage of the structural gene (*lpp*) for lipoprotein using a F-prime factor containing the *lpp* locus. A merodiploid strain containing this factor approximately twice the amount of free-lipoprotein found in the haploid strain whereas the amount of bound form was not significantly different. This observation suggests that the *lpp* gene is expressed constitutively. Moreover, some form of control clearly exists over the conversion of free to bound lipoprotein. This does not appear to be the lack of appropriate sites on peptidoglycan since as described below mecillinam-treatment of *E. coli* can double the ratio of bound to free form.

The antibiotic bicyclomycin inhibits both RNA and protein synthesis in *E. coli*. In organisms deprived of histidine it inhibits the biosynthesis of lipoprotein having a more profound effect on the bound, rather than the free-form [201]. On the basis of this result it was concluded that the primary site of action of this antibiotic was the conversion of free- to bound-lipoprotein.

The synthesis of the lipoprotein–peptidoglycan complex has also been studied with regard to the peptidoglycan component. Using *E. coli* auxotrophic for either diaminopimelic acid or *N*-acetylglucosamine, Braun and Wolff [23] have specifically labelled the newly-synthesized peptidoglycan. After isolation of material insoluble in boiling SDS the peptidoglycan was hydrolysed with lysozyme and the radioactive products separated by paper chromatography. In this way approximately 30% of the peptidoglycan which was linked to lipoprotein could be separated from the other products of lysozyme digestion. Molecular weight determinations showed these to be single lipoprotein molecules with attached peptidoglycan, rather than longer units of peptidoglycan, with several lipoprotein molecules attached. Pulse-label experiments of 1 min duration failed to reveal the presence of newly-synthesized peptidoglycan associated with the lipoprotein. Thus, unlike the teichoic acids, lipoprotein does not appear to be predominantly associated with peptidoglycan which is being synthesized concomitantly. Moreover, newly-synthesized peptidoglycan was only slowly incorporated into the lipoprotein–peptidoglycan complex released by lysozyme digestion. Even after some 2–5 generations of growth less than 50% of the peptidoglycan linked to lipoprotein was radioactively-labelled, whereas random attachment of lipoprotein would require that some 65% of the peptidoglycan isolated at this time should be labelled. This result may reflect either a particular stability inherent in the peptidoglycan associated with lipoprotein, or a relatively slow linkage of lipoprotein to the newly-synthesized peptidoglycan when this rate is compared with the rate of peptidoglycan synthesis [23]. On the other hand, an outer membrane location for the enzyme synthesizing the linkage, together with a layering of peptidoglycan in terms of age such as that described in Bacilli (Chapter 15.3) may give the same result. The newly-synthesized peptidoglycan would be located close to its sites of synthesis in the cytoplasmic (inner) membrane, whereas the older pre-existing peptidoglycan would be associated with the outer membrane. Under such conditions the attachment of the lipoprotein to the pre-existing peptidoglycan would be favoured. Clearly the feasibility of this hypothesis could be tested by establishing the localization of the linking enzyme.

The lipoprotein replaces D-alanine at the L-centre of meso-diaminopimelic acid in approximately 10% of the peptide side chains of the peptidoglycan. The bond is formed between the ϵ-amino group of the carboxyl-terminal lysine of the lipoprotein and the carboxy-group of diaminopimelic acid. The nature of this linkage has led to the suggestion that the linking enzyme is a transpeptidase using the energy of the diaminopimelyl-D-alanine bond to effect synthesis of the linkage. D-Alanine would be a product of the reaction. Conversion of the free lipoprotein to the bound form occurs in the presence of the energy uncoupler carboxyl cyanide *m*-chlorophenylhydrazone, suggesting that ATP is not required for the process [97]. The proposal that the linking enzyme was the site of action of mecillinam (FL 1060) has now proved incorrect. In fact, the ratio of bound lipoprotein to peptidoglycan doubled when the organism was incubated in the presence of the antibiotic. The possibility that this result arises because of the selective removal of peptidoglycan not associated with the lipoprotein attachment sites has been eliminated [23].

Clearly the ratio of lipoprotein molecules to peptidoglycan found in normally growing cells does not represent the maximum substitution of the peptide side chains. It will be of particular interest to see if other means of increasing the extent of substitution by lipoprotein can be found.

References

1. Anderson, J. S., Page, R. L. and Salo, W. L. (1972) *J. biol. Chem.* **247**, 2480–85.
2. Anderson, R. G., Douglas, L. J., Hussey, H. and Baddiley, J. (1973) *Biochem. J.* **136**, 871–6.
3. Anderson, R. G., Hussey, H. and Baddiley, J. (1972) *Biochem. J.* **127**, 11–25.
4. Archibald, A. R. (1974) *Adv. Microbiol. Physiol.* **11**, 53–95.
5. Archibald, A. R. (1981) In *Virus Receptors (Receptors and Recognition, Series B. Volume 7)*, eds. Philipson, L., and Randall, L. London: Chapman and Hall.
6. Armstrong, J. J., Baddiley, J., Buchanan, J. G., Carss, B. and Greenberg, G. R. (1958) *J. Chem. Soc.* 4344–5.
7. Armstrong, J. J., Baddiley, J., Buchanan, J. G., Davison, A. L., Kelemen, M. V. and Neuhaus, F. C. (1959) *Nature, Lond.* **184**, 247–8.

7a. Asonganyi, T. M. and Meadow, P. M. (1980) *J. Gen. Microbiol.* **117**, 1–7.

8. Baddiley, J. (1972) *Essays in Biochemistry* **8**, 35–77.
9. Baddiley, J., Blumson, N. L. and Douglas, L. J. (1968) *Biochem. J.* **110**, 567–71.
10. Baddiley, J., Buchanan, J. G., Carss, B., Mathias, A. P. and Sanderson, A. R. (1956) *Biochem. J.* **64**, 599–603.
11. Baddiley, J. and Neuhaus, F. C. (1960) *Biochem. J.* **95**, 579–87.
12. Bayer, M. E. (1974) *Ann. N.Y. Acad. Sci.* **235**, 6–28.
13. Beacham, I. R. and Silbert, D. F. (1973) *J. biol. Chem.* **248**, 5310–18.
14. Bell, R. M., Mavis, R. D., Osborn, M. J. and Vagelos, P. R. (1971) *Biochim. Biophys. Acta* **249**, 628–35.
15. Bracha, R., Chang, M., Fiedler, F. and Glaser, L. (1978) *Meth. Enzymol.* **50**, 387–402.
16. Bracha, R., Davidson, R. and Mirelman, D. (1978) *J. Bact.* **134**, 412–17.
17. Bracha, R. and Glaser, L. (1976) *Biochem. biophys. Res. Commun.* **72**, 1091–8.
18. Bracha, R. and Glaser, L. (1976) *J. Bact.* **125**, 872–9.
19. Braun, V. (1975) *Biochim. Biophys. Acta* **415**, 335–77.
20. Braun, V., Bosch, V., Hantke, K. and Schaller, K. (1974) *Ann. N.Y. Acad. Sci.* **235**, 66–82.
21. Braun, V., Hantke, K. and Henning, U. (1975) *FEBS Letts.* **60**, 26–8.
22. Braun, V., Rehn, K. and Wolff, H. (1970) *Biochemistry* **9**, 5041–9.
23. Braun, V. and Wolff, H. (1975) *J. Bact.* **123**, 888–97.
24. Bray, D. and Robbins, P. W. (1967) *Biochem. biophys Res. Commun.* **28**, 334–9.
25. Bray, D. and Robbins, P. W. (1967) *J. molec. Biol.* **30**, 457–90.

27. Brooks, D. and Baddiley, J. (1969) *Biochem. J.* **115**, 307–314.
28. Brooks, D., Mays, L. L., Hatefi, Y. and Young, F. E. (1971) *J. Bact.* **107**, 223–9.
29. Burger, M. M. and Glaser, L. (1964) *J. biol. Chem.* **239**, 3168–77.
30. Burger, M. M. and Glaser, L. (1964) *J. biol. Chem.* **239**, 3187–91.
31. Burger, M. M. and Glaser, L. (1966) *J. biol. Chem.* **241**, 494–506.
32. Chatterjee, A. N. (1969) *J. Bact.* **98**, 519–27.
33. Chattophadhyay, P. K., Engel, R., Tropp, B. E. and Wu, H. C. (1979) *J. Bact.* **138**, 944–8.

34. Chattopadhyay, P. K., Lai, J.-S. and Wu, H. C. (1979) *J. Bact.* **137**, 309–12.
35. Chattopadhyay, P. K., and Wu, H. C. (1977) *Proc. Natn. Acad. Sci. U.S.A.* **74**, 5318–22.
36. Chevion, M., Panos, C., Linzer, R. and Neuhaus, F. C. (1974) *J. Bact.* **120**, 1026–32.
37. Chin, T., Burger, M. M. and Glaser, L. (1966) *Archs. Biochem. Biophys.* **116**, 358–67.
38. Chin, T., Younger, J. and Glaser, L. (1968) *J. Bact.* **95**, 2044–50.
39. Chui, T. H. and Saralkar, C. (1978) *J. Bact.* **133**, 185–95.
40. Cohen, M. and Panos, C. (1971) *J. Bact.* **106**, 347–55.
41. Coley, J., Tarelli, E., Archibald, A. R. and Baddiley, J. (1978) *FEBS Letts.* **88**, 1–9.
42. Cynkin, M. A. and Osborn, M. J. (1968) *Fedn. Proc. Am. Socs. exp. Biol.* **27**, 293.
43. Dankert, M., Wright, A., Kelley, W. S. and Robbins, P. W. (1966) *Archs. Biochem. Biophys.* **116**, 425–35.
44. DiRienzo, J., Nakamura, K. and Inouye, M. (1978) *A. Rev. Biochem.* **47**, 481–532.
45. Douglas, L. J. and Baddiley, J. (1968) *FEBS Letts.* **1**, 114–16.
46. Duckworth, M., Archibald, A. R. and Baddiley, J. (1975) *FEBS Letts.* **53**, 176–9.
47. Edstrom, R. D. and Heath, E. C. (1967) *J. biol. Chem.* **242**, 3581–8.
48. Eidels, L. and Osborn, M. J. (1971) *Proc. Natn. Acad. Sci. U.S.A.* **68**, 1673–7.
49. Eidels, L. and Osborn, M. J. (1974) *J. biol. Chem.* **249**, 5642–8.
50. Ellwood, D. C. and Tempest, D. W. (1969) *Biochem. J.* **111**, 1–5.
51. Ellwood, D. C. and Tempest, D. W. (1972) In *Advances in Microbiol. Physiol., Vol 7.,* eds. Rose, A. H. and Tempest, D. W. pp. 83–117. London and New York: Academic Press.
52. Emdur, L. I. and Chui, T. H. (1974) *Biochem. biophys. Res. Commun.* **59**, 1137–44.
53. Emdur, L. I. and Chui, T. H. (1975) *FEBS Letts.* **55**, 216–19.
54. Emdur, L. I., Saralkar, C., McHugh, J. G. and Chui, T. H. (1974) *J. Bact.* **120**, 724–32.
55. Fensom, A. H. and Gray, G. W. (1969) *Biochem J.* **114**, 185–96.
56. Fiedler, F. and Glaser, L. (1974) *J. biol. Chem.* **249**, 2684–9.
57. Fiedler, F. and Glaser, L. (1974) *J. biol. Chem.* **249**, 2690–95.
58. Fiedler, F. and Glaser, L. (1974) *Carbohydrate Res.* **37**, 37–46.
59. Fiedler, F., Mauck, J. and Glaser, L. (1974) *Ann N.Y. Acad. Sci.* **235**, 198–209.
60. Fischer, W., Kock, H. U., Rösel, P. and Fiedler, F. (1980) *J. biol. Chem.* **255**, 4550–6.
61. Fischer, W., Koch, H. U., Rösel, P., Fiedler, F. and Schmuk, L. (1980) *J. biol. Chem.* **255**, 4557–62.
62. Fukasawa, T., Jokura, K. and Kurahashi, K. (1962) *Biochem. biophys. Res. Comm.* **7**, 121–5.
63. Galanos, Ch., Lüderitz, O. and Westphal, O. (1969) *Eur. J. Biochem.* **9**, 245–9.
64. Ganfield, M.-C. and Pieringer, R. (1978) *Abst. A30 XII Int. Cong. Microbiol. Munich.* p. 73.
65. Glaser, L. (1963) *Biochim. biophys. Acta* **67**, 525–30.
66. Glaser, L. (1964) *J. biol. Chem.* **239**, 3178–86.
68. Glaser, L. and Lindsay, B. (1974) *Biochem. biophys Res. Commun.* **59**, 1131–6.
69. Glaser, L. and Loewy, A. (1979) *J. biol. Chem.* **254**, 2184–6.
70. Halegoua, S., Hirashima, A., Sekizawa, J. and Inouye, M. (1976) *Eur. J. Biochem.* **69**, 163–7.

71. Halegoua, S. and Inouye, M. (1979) *J. molec. Biol.* **130**, 39–61.
72. Halegoua, S., Sekizawa, J. and Inouye, M. (1977) *J. biol. Chem.* **252**, 2324–30.
73. Hancock, I. C. and Baddiley, J. (1972) *Biochem. J.* **127**, 27–37.
74. Hancock, I. C. and Baddiley, J. (1976) *J. Bact.* **125**, 880–86.
75. Hancock, I. C., Wiseman, G. and Baddiley, J. (1976) *FEBS Letts.* **69**, 75–80.
76. Hase, S. and Matsushima, Y. (1972) *J. Biochem.* **72**, 1117–28.
77. Hase, S. and Matsushima, Y. (1977) *J. Biochem.* **81**, 1181–6.
78. Hayes, M. V., Ward, J. B., and Rogers, H. J. (1977) *Proc. Soc. gen. Microbiol.* **4**, 85–6.
79. Heath, E. C., Mayer, R. M., Edstrom, R. D. and Beaudreau, C. A. (1966) *Ann. N.Y. Acad. Sci.* **133**, 315–33.
80. Heckels, J. E., Archibald, A. R. and Baddiley, J. (1975) *Biochem. J.* **149**, 637–47.
81. Hinckley, A., Müller, E. and Rothfield, L. (1972) *J. biol. Chem.* **247**, 2623–8.
82. Hirashima, A., Childs, G. and Inouye, M. (1973) *J. molec. Biol.* **79**, 373–89.
83. Hirashima, A. and Inouye, M. (1973) *Nature, Lond.* **242**, 405–7.
84. Hirashima, A., Wang, S. and Inouye, M. (1974) *Proc. Natn. Acad. Sci. U.S.A.* **71**, 4149–53.
85. Hirota, Y., Suzuki, H., Nishimura, Y. and Yasuda, S. (1977) *Proc. Natn. Acad. Sci. U.S.A.* **74**, 1417–20.
86. Hughes, R. C. (1966) *Biochem. J.* **101**, 692–7.
87. Hughes, R. C. (1970) *Biochem. J.* **117**, 431–9.
88. Hughes, R. C. and Thurman, P. F. (1970) *Biochem J.* **117**, 441–9.
89. Humphreys, G. O., Hancock, I. C. and Meadow, P. M. (1972) *J. gen. Microbiol.* **71**, 221–30.
90. Hussey, H. and Baddiley, J. (1972) *Biochem. J.* **127**, 39–50.
91. Hussey, H., Brooks, D. and Baddiley, J. (1969) *Nature, Lond.* **221**, 665–6.
92. Hussey, H., Sueda, S., Cheah, S.-C. and Baddiley, J. (1978) *Eur. J. Biochem.* **82**, 169–74.
93. Ichihara, N., Ishimoto, N. and Ito, E. (1974) *FEBS Letts.* **39**, 46–8.
94. Ichihara, N., Ishimoto, N. and Ito, E. (1974) *FEBS Letts.* **40**, 309–11.
95. Inouye, M., Hirashima, A. and Lee, N. (1974) *Ann. N.Y. Acad. Sci.* **235**, 83–90.
96. Inouye, M., Pirtle, R., Pirtle, I., Seikizawa, K., Nakamure, K., DiRienzo, J., Inouye, S., Wang, S. and Halegona, S. (1979) *Microbiology 1979*, ed. Schessinger, D., pp. 34–7, American Society of Microbiology.
97. Inouye, M., Shaw, J. and Sheu, C. (1972) *J. biol. Chem.* **247**, 8154–9.
98. Inouye, S., Wang, S., Sekizawa, J., Halegoua, S. and Inouye, M. (1977) *Proc. Natn. Acad. Sci. U.S.A.* **74**, 1004–8.
99. Ishimoto, N. and Strominger, J.L. (1966) *J. biol. Chem.* **241**, 639–50.
100. Jones, N. C. and Osborn, M. J. (1977) *J. biol. Chem.* **252**, 7405–412.
101. Kadis, S., Weinbaum, G. and Ajl, S. J. (1971) In *Microbial Toxins Vol. 5.*, New York and London: Academic Press.
102. Kanegasaki, S. and Wright, A. (1970) *Proc. Natn. Acad. Sci. U.S.A.* **67**, 951–8.
103. Kawamura, T., Ishimoto, N. and Ito, E. (1979) *J. biol. Chem.* **254**, 8457–65.
104. Kennedy, L. D. and Shaw, D. R. D. (1968) *Biochem. biophys. Res. Commun.* **32**, 861–5.
105. Kent, J. L. and Osborn, M. J. (1968) *Biochemistry* **7**, 4419–22.
106. Kessler, R. E. and Shockman, G. D. (1979) *J. Bact.* **137**, 1176–9.
107. Knox, K. W., Cullen, J. and Work, E. (1967) *Biochem. J.* **103**, 192–201.
108. Knox, K. W. and Wicken, A. J. (1975) *Science* **187**, 1161–7.

109. Kulpa, C. F. and Leive, L. (1972) In *Membrane Research: First ICN-UCLA Symposium on Molecular Biology,* ed. Fox, C. F., pp. 155–60. New York and London: Academic Press.
110. Lehmann, V. (1977) *Eur. J. Biochem.* **75**, 257–66.
111. Lehmann, V., Hämmerling, G., Nurminen, M., Minner, I., Ruschmann, E., Lüderitz, O., Kuo, T.–T. and Stocker, B. A. D. (1973) *Eur. J. Biochem.* **32**, 268–75.
112. Lehmann, V., Redmond, J., Egan, A. and Minner, I. (1978) *Eur. J. Biochem.* **86**, 487–96.
113. Lehmann, V. and Rupprecht, E. (1977) *Eur. J. Biochem.* **81**, 443–52.
114. Lehmann, V., Rupprecht, E. and Osborn, M. J. (1977) *Eur. J. Biochem.* **76**, 41–9.
115. Levy, S. B. and Leive, L. (1970) *J. biol. Chem.* **245**, 585–94.
116. Liu, J. J. C., Kanazawa, H., Ozols, J. and Wu, H. C. (1978) *Proc. Natn. Acad. Sci. U.S.A.* **75**, 4891–5.
117. Linzer, R. and Neuhaus, F. C. (1973) *J. biol. Chem.* **248**, 3196–201.
118. Losick, R. and Robbins, P. W. (1967) *J. molec. Biol.* **30**, 445–55.
119. Luderitz, O., Galanos, Ch., Lehmann, V. and Rietschel, E. Th. (1974) *J. Hyg. Epidemiol. Microbiol. Immunol.* **18**, 381–90.
120. Luderitz, O., Westphal, O., Staub, A. M. and Nikaido, H. (1971) In *Microbial Toxins, Vol. 4,* eds. Weinbaum, G., Kadis, S. and Ajl, S. J., pp. 145–233, New York and London: Academic Press.
122. Mauck, J. and Glaser, L. (1972) *Proc. Natn. Acad. Sci. U.S.A.* **69**, 2386–90.
123. Mauck, J. and Glaser, L. (1972) *J. biol. Chem.* **247**, 1180–7.
124. McArthur, H. A. I., Roberts, F. M., Hancock, I. C. and Baddiley, J. (1978) *FEBS Letts.* **86**, 193–200.
125. McCarty, M. (1956) *J. exp. Med.* **104**, 629–43.
126. Mirelman, D., Bracha, R. and Sharon, N. (1974) *Biochemistry* **13**, 5045–53.
127. Miura, T. and Mizushima, S. (1968) *Biochim. Biophys. Acta* **150**, 159–61.
128. Miura, T. and Mizushima, S. (1969) *Biochim. Biophys. Acta* **193**, 268–76.
129. Movva, N. R., Katz, E., Asdourian, P. L., Hirota, Y. and Inouye, M. (1978) *J. Bact.* **133**, 81–4.
130. Mühlradt, P. F. (1969) *Eur. J. Biochem.* **11**, 241–8.
131. Mühlradt, P. F. (1971) *Eur. J. Biochem.* **18**, 20–27.
132. Mühlradt, P. F. and Golecki, J. R. (1975) *Eur. J. Biochem.* **51**, 343–52.
133. Mühlradt, P. F., Menzel, J., Golecki, J. R. and Speth, V. (1973) *Eur. J. Biochem.* **35**, 471–81.
134. Mühlradt, P. F., Menzer, J., Golecki, J. R. and Speth, V. (1974) *Eur. J. Biochem.* **43**, 533–9.
135. Munoz, E., Ghuysen, J.-M. and Heyman, H. (1967) *Biochemistry* **6**, 3659–70.
136. Munsen, R. S. and Osborn, M. J. (1975) *Fedn. Proc. Am. Socs. exp. Biol.* **34**, 669.
137. Munsen, R. S., Rasmussen, N. S. and Osborn, M. J. (1978) *J. biol. Chem.* **253**, 1503–11.
138. Naide, Y., Nikaido, H., Mäkelä, P. H., Wilkinson, R. G. and Stocker, B. A. D. (1965) *Proc. Natn. Acad. Sci. U.S.A.* **53**, 147–53.
139. Nakamura, K., Katz-Wurtzel, E., Pirtle, R. M. and Inouye, M. (1979) *J. Bact.* **138**, 715–20.
140. Nathenson, S. G., Ishimoto, N., Anderson, J. S. and Strominger, J. L. (1966) *J. biol. Chem.* **241**, 651–8.
141. Nathenson, S. G. and Strominger, J. L. (1963) *J. biol. Chem.* **238**, 3161–9.
142. Neuhaus, F. C. (1971) *Acc. Chem. Res.* **4**, 297–303.

143. Neuhaus, F. C., Linzer, R. and Reusch, V. M. (1974) *Ann N.Y. Acad. Sci.* **235**, 502–18.
144. Nikaido, H. (1962) *Proc. Natn. Acad. Sci. U.S.A.* **48**, 1337–41.
145. Nikaido, H. (1962) *Proc. Natn. Acad. Sci. U.S.A.* **48**, 1542–8.
146. Nikaido, H. (1965) *Biochemisty* **4**, 1550–61.
147. Nikaido, H. (1973) In *Bacterial Membranes and Walls,* ed. Leive, L., pp. 131–208, New York: Marcel Dekker Inc.
148. Nikaido, H. and Nikaido, K. (1965) *Biochem. biophys. Res. Commun.* **19**, 322–7.
149. Nikaido, K. and Nikaido, H. (1971) *J. biol. Chem.* **246**, 3912–19.
150. Nikaido, H., Nikaido, K., Nakae, T. and Mäkelä, P. H. (1971) *J. biol. Chem.* **246**, 3902–11.
151. Nikaido, H., Nikaido, K., Subbaiah, T. V. and Stocker, B. A. D. (1964) *Nature, Lond.* **201**, 1301–2.
152. Osborn, M. J. (1968) *Nature, Lond.* **217**, 957–60.
153. Osborn, M. J. (1969) *A. Rev. Biochem.* **38**, 501–38.
154. Osborn, M. J. (1971) In *Structure and Function of Biological Membranes*, ed. Rothfield, L. I., pp. 343–400, New York and London: Academic Press.
155. Osborn, M. J., Gander, J. E. and Parisi, E. (1972) *J. biol. Chem.* **247**, 3973–86.
156. Osborn, M. J., Gander, J. E., Parisi, E. and Carson, J. (1972) *J. biol. Chem.* **247**, 3962–72.
157. Osborn, M. J., Rick, P. D., Lehmann, V., Rupprecht, E. and Singh, M. (1974) *Ann N. Y. Acad. Sci.* **235**, 52–65.
158. Osborn, M. J., Rosen, S. M., Rothfield, L., Zeleznick, L. D. and Horecker, B. L. (1964) *Science* **145**, 783–9.
159. Osborn, M. J. and Rothfield, L. I. (1971) In *Microbial Toxins Vol. 4.,* eds Weibaum, G., Kadis, S. and Ajl, S. J. pp. 331–50, New York and London: Academic Press.
160. Osborn, M. J. and Tze-Yuen, R. Y. (1968) *J. biol. Chem.* **243**, 5145–52.
161. Osborn, M. J. and Weiner, I. M. (1968) *J. biol. Chem.* **243**, 2631–9.
162. Page, R. L. and Anderson, J. S. (1972) *J. biol. Chem.* **247**, 2471–9.
163. Pazur, J. H. and Anderson, J. S. (1963) *Biochim biophys. Acta* **74**, 788–90.
164. Perkins, H. R. (1963) *Biochem. J.* **86**, 475–83.
165. Perkins, H. R., Thorpe, S. J. and Brown, C. A. (1980) *Biochem. Transact.* **8**, 163–4.
166. Prehm, P., Stirm, S., Jann, B. and Jann, K. (1975) *Eur. J. Biochem.* **56**, 41–55.
167. Reusch, V. M. and Neuhaus, F. C. (1971) *J. biol. Chem.* **246**, 6136–43.
168. Reusch, V. M. and Panos, C. (1977) *J. Bact.* **129**, 1407–414.
169. Rick, P. D., Fung, L. M–W., Ho, C. and Osborn, M. J. (1977) *J. biol. Chem.* **252**, 4904–12.
170. Rick, P. D. and Osborn, M. J. (1977) *J. biol. Chem.* **252**, 4895–903.
171. Risse, H. J., Luderitz, O. and Westphal, O. (1967) *Eur. J. Biochem.* **1**, 233–42.
172. Robbins, P. W., Bray, D., Dankert, M. and Wright, A. (1967) *Science* **158**, 1536–42.
173. Robbins, P. W., Keller, J. M., Wright, A. and Bernstein, R. L. (1965) *J. biol. Chem.* **240**, 384–90.
174. Robbins, P. W. and Uchida, T. (1965) *J. biol. Chem.* **240**, 375–83.
175. Robbins, P. W. and Wright, A. (1971) In *Microbial Toxins, Vol. 4,* eds. Weinbaum, G., Kadis, S. and Ajl, S. J. pp. 351–69. New York and London: Academic Press.
176. Robbins, P. W., Wright, A. and Bellons, J. L. (1964) *Proc. Natn. Acad. Sci. U.S.A.* **52**, 1302–308.

177. Roberts, F. M., McArthur, H. A. I., Hancock, I. C. and Baddiley, J. (1979) *FEBS Letts.* **97**, 211–16.
178. Rohr, T. E., Levy, G. N., Stark, N. J. and Anderson, J. S. (1977) *J. biol. Chem.* **252**, 3460–5.
179. Romeo, D., Girard, A. and Rothfield, L. (1970) *J. molec. Biol.* **53**, 475–90.
180. Romeo, D., Hinckley, A. and Rothfield, L. (1970) *J. molec. Biol.* **53**, 491–501.
181. Rosenberger, R. F. (1976) *Biochim. biophys. Acta* **428**, 516–24.
182. Rothfield, L. and Horecker, B. L. (1964) *Proc. Natn. Acad. Sci. U.S.A.* **52**, 939–46.
183. Rothfield, L., Osborn, M. J. and Horecker, B. L. (1964) *J. biol. Chem.* **239**, 2788–95.
184. Rothfield, L. and Pearlman, M. (1966) *J. biol. Chem.* **241**, 1386–92.
185. Rothfield, L. and Pearlman-Kothencz, M. (1969) *J. molec. Biol.* **44**, 477–92.
186. Rothfield, L. and Romeo, D. (1971) *Bact. Rev.* **35**, 14–38.
187. Rothfield, L. and Romeo, D. (1971) In *Structure and Function of Biological Membranes,* ed. Rothfield, L. I., pp. 251–82. New York and London: Academic Press.
188. Rothfield, L., Romeo, D. and Hinckley, A. (1972) *Fedn. Proc. Am. Socs. exp. Biol.* **31**, 12–17.
189. Rundell, K. and Shuster, C. W. (1973) *J. biol. Chem.* **248**, 5436–42.
190. Rundell, K. and Shuster, C. W. (1975) *J. Bact.* **123**, 928–36.
191. Sasaki, T., Uchida, T. and Kurahashi, K. (1974) *J. biol. Chem.* **249**, 761–72.
192. Schnaitman, C. A. (1970) *J. Bact.* **104**, 890–901.
193. Shaw, D. R. D. (1962) *Biochem. J.* **82**, 297–312.
194. Shibaev, V. N., Duckworth, M., Archibald, A. R. and Baddiley, J. (1973) *Biochem. J.* **135**, 383–4.
195. Slabyj, B. M. and Panos, C. (1973) *J. Bact.* **114**, 934–42.
196. Stark, N. J., Levy, G. N., Rohr, T. E. and Anderson, J. S. (1977) *J. biol. Chem.* **252**, 3466–72.
197. Stocker, B. A. D. and Mäkelä, P. H. (1971) In *Microbial Toxins Vol. 4,* eds. Weinbaum, G., Kadis, S. and Ajl, S. J., pp. 369–438, New York and London: Academic Press.
198. Sutherland, I. W., Lüderitz, O. and Westphal, O. (1965) *Biochem. J.* **96**, 439–48.
199. Suzuki, H., Nishimura, Y., Iketani, H., Campsi, J., Hirashima, A., Inouye, M. and Hirtoa, Y. (1976) *J. Bact.* **127**, 1494–501.
200. Takeshita, M. and Mäkelä, P. H. (1971) *J. biol. Chem.* **246**, 3920–27.
201. Tanaka, N., Ishei, M., Myoshi, T., Aoki, H. and Imanaka, H. (1976) *J. Antibiotics* **29**, 155–68.
202. Torti, S. V. and Park, J. T. (1976) *Nature, Lond.* **263**, 323–6.
203. Tynecka, Z. and Ward, J. B. (1975) *Biochem. J.* **146**, 253–67.
204. Uchida, T., Makind, T., Kurahashi, K. and Uetake, H. (1965) *Biochem. biophys. Res. Commun.* **21**, 354–60.
205. Ward, J. B. (1974) *Biochem. J.* **141**, 227–41.
206. Ward, J. B. (1977) *FEBS Letts.* **78**, 151–4.
207. Ward, J. B. and Curtis, C. A. M. (1980) in preparation.
208. Ward, J. B. and Perkins, H. R. (1974) *Biochem. J.* **135**, 721–8.
209. Watkinson, R. J., Hussey, H. and Baddiley, J. (1971) *Nature, Lond.* **229**, 57–9.
210. Weigand, R. A., Vinci, K. D. and Rothfield, L. I. (1976) *Proc. Natn. Acad. Sci. U.S.A.* **73**, 1882–6.
211. Weiner, I. M., Higuchi, T., Osborn, M. J. and Horecker, B. L. (1966) *Ann N.Y. Acad. Sci.* **166**, 391–404.

212. Weiner, I. M., Higuchi, T., Rothfield, L., Saltmarsh-Andrew, M., Osborn, M. J. and Horecker, B. L. (1965) *Proc. Natn. Acad. Sci. U.S.A.* **54**, 228–35.
213. Weiner, I. M. and Rothfield, L. (1968) *J. biol. Chem.* **243**, 1320–8.
214. Weston, A. and Perkins, H. R. (1977) *FEBS Letts.* **76**, 195–8.
215. White, D. A., Albright, F. R., Lennarz, W. J. and Schnaitman, C. A. (1971) *Biochim. biophys. Acta* **249**, 636–42.
216. Wilkinson, R. G. and Stocker, B. A. D. (1968) *Nature, Lond.* **217**, 955–7.
217. Wright, A. (1971) *J. Bact.* **105**, 927–36.
218. Wright, A., Dankert, M., Fennessey, P. and Robbins, P. W. (1967) *Proc. Natn. Acad. Sci. U.S.A.* **57**, 1798–803.
219. Wright, A., Dankert, M. and Robbins, P. W. (1965) *Proc. Natn. Acad. Sci. U.S.A.* **54**, 235–41.
220. Wright, A. and Kanegasaki, S. (1971) *Physiol. Rev.* **51**, 749–84.
221. Wright, J. and Heckels, J. E. (1975) *Biochem. J.* **147**, 187–9.
222. Wu, M–C. and Heath, E. C. (1973) *Proc. Natn. Acad. Sci. U.S.A.* **70**, 2572–6.
223. Wu, H. C. and Liu, J. J. C. (1976) *J. Bact.* **126**, 147–56.
224. Wyke, A. W. and Ward, J. B. (1975) *Biochem. biophys. Res. Commun.* **65**, 877–885.
225. Wyke, A. W. and Ward, J. B. (1977) *FEBS Letts.* **73**, 159–63.
226. Wyke, A. W. and Ward, J. B. (1977) *J. Bact.* **130**, 1055–63.
227. Yuasa, R., Levinthal, M. and Nikaido, H. (1969) *J. Bact.* **100**, 433–44.
228. Zeleznick, L. D., Boltralik, J. J., Barkulis, S. S., Smith, C. and Heyman, H. (1963) *Science* **140**, 400–401.
229. Zeleznick, L. D., Rosen, S. M., Saltmarsh-Andrew, M., Osborn, M. J. and Horecker, B. L. (1965) *Proc. Natn. Acad. Sci. U.S.A.* **53**, 207–14.

11
The bacterial autolysins

11.1 Introduction

Among the earliest knowledge about microbes is the fact that cells, suspended and incubated under unfavourable conditions for growth and anabolism, disintegrate. Indeed autolysates of yeasts and other cells were some of the earliest sources of soluble enzymes, and did trojan work in the heroic days of the investigations of intermediary carbohydrate metabolism. With the development of the electron microscope and consequent ultrastructural awareness, cell walls were recognized, isolated and their chemistry studied. Autolysis could then be seen to be due to dissolution of the walls. As the chemical structure of the polymers making up the walls was elucidated, it became possible to designate the bonds specifically hydrolysed by various autolytic enzymes and thus to account for the solubilization of the walls and the disintegration of the cells. Since the supportive polymers in microbial cell walls are often either complex, such as, for example, are the peptidoglycans of bacteria or, consist of a number of polymers interacting to provide mechanical support for the cell as in some microfungi, autolysins with a variety of bond specificities are often to be found in the same cell. It is not always possible to distinguish which autolysin is responsible for any particular effect, or to exclude the necessity for the combined action of more than one enzyme.

At the outset, the purely degradative aspect of autolysins was emphasized and one of their functions is presumably to remove what might otherwise be an embarrassing accumulation of insoluble microbial matter. Recently, however, the possibility of more positive functions has been widely discussed and tested. One of the paradoxes inherent in microbes with cell walls is that, although they have a surface completely covered with insoluble polymers, they must be able to expand during growth, and modify the surface so that division into two new individuals can take place. Obvious candidates that might modify surfaces are the autolysins but unambiguous evidence that they are necessarily involved during growth is difficult to obtain. Work with micro-fungi and with bacteria has been actively proceeding for some ten years to test such hypotheses and with the organisms such as *Neurospora crassa* hopeful correlative evidence has been obtained. In bacteria, evidence has so far been rather negative but certainly not unambiguously so.

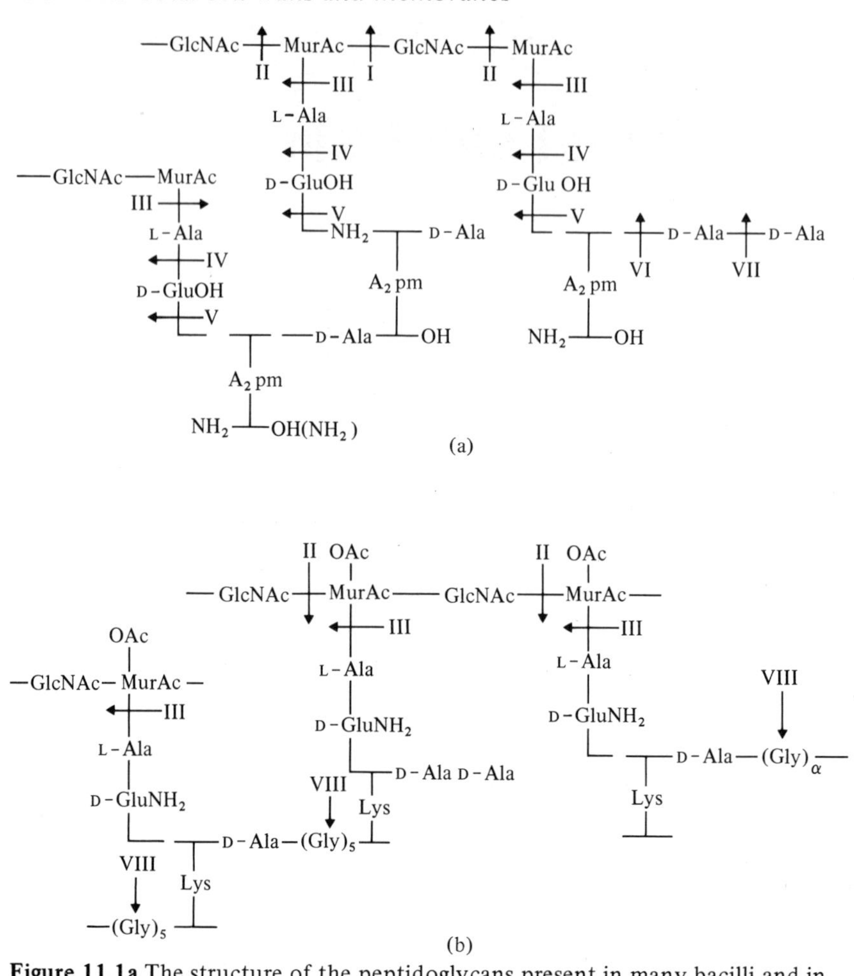

Figure 11.1a The structure of the peptidoglycans present in many bacilli and in all the Gram-negative species examined. The arrows indicate the bonds hydrolysed by the known autolytic enzymes. These enzymes are (i) *endo*-muramidase, (ii) *endo*-β-*N*-acetylglucosaminidase, (iii) *N*-acetylmuramyl-L-alanine amidase (amidase), (iv) and (v) peptidases associated with spores or sporulating cultures, (vi) and (vii) D-alanine carboxypeptidases present in many organisms.
Figure 11.1b The structure of a peptidoglycan with a 'bridge' peptide. The one illustrated is present in *Staphylococcus aureus.* The notation for autolytic enzymes is as above together with (viii) which is a peptidase present in lysostaphin and produced by other staphylococci.

11.2 Bond specificity and distribution of bacterial autolysins

The bond specificities of the four common types of autolytic enzyme are shown in Fig. 11.1. The autolytic enzymes are:
(1) A lysozyme-like enzyme (ie a muramidase) that hydrolyses the *N*-acetylmuramyl, 1, 4-β-*N*-acetylglucosamine bonds in the glycans to liberate free-reducing groups of *N*-acetylmuramic acid.

(2) A *β-N*-acetylglucosaminidase that liberates the free-reducing groups of *N*-acetyl-glucosamine.
(3) An *N*-acetylmuramyl-L-alanine amidase (amidase) that hydrolyses the bond between the glycan chains and the peptide.
(4) Peptidases that can hydrolyse some of the main peptides and the bridge peptides when they occur between the D-alanyl terminus and the amino group of a contiguous peptide chain.
Both (1) and (2) are *endo*-enzymes and the corresponding *exo*-enzymes, which are also known [33,88] do not seem to act as autolysins being unable to hydrolyse bonds in whole peptidoglycans [7]. Also, commonly occurring in many organisms are D-alanine carboxypeptidases which can cleave off any carboxy terminal D-alanyl residues of the peptidolgycan (see Chapters 6 and 9). These enzymes however, are not known to act as autolysins and therefore will not be considered. A further most interesting autolytic enzyme is formed by *E. coli.* This not only hydrolyses the 1, 4-*β* bond between *N*-acetylmuramic acid and *N*-acetylglucosamine but also carries out a dehydration reaction so that the disaccharide formed from peptidoglycan is non-reducing because a 1,6-anhydro-*N*-acetylmuramic acid residue is present. This enzyme was first unambiguously identified as the result of the action of phage λ and VII endolysins on *E. coli* [130]. Subsequently it was isolated and purified from disintegrated cells of *E. coli* strain W7. It completely degraded peptidoglycan sacculi from the organism into the non-reducing disaccharide [65]. The distribution of this enzyme is not known.

Table 11.1 shows the distribution of the four autolytic enzymes. The commonest ones are the amidase and the endo-*β-N*-acetylglucosaminidase which most frequently occur together in the same organisms. The muramidases appear, from existing evidence, to exist more often as the sole or predominant autolytic enzyme of the bacteria concerned, but relatively few organisms have been examined in sufficient detail to make this certain. Among those that have been studied are *Streptococcus faecalis* [118] and *Lactobacillus acidophilus* [30]. Such enzymes as those formed by *Streptomyces,* have been used extensively to investigate the structures of bacterial peptidoglycans but it is not entirely clear whether all of them are involved in autolysis of the organisms that produce them. They have at present, therefore, to be regarded as bacteriolytic enzymes rather than as autolytic ones.

11.3 Purification and properties of the autolytic enzymes

Rather few autolysins have been purified to yield homogeneous proteins when examined by techniques such as polyacrylamide gel electrophoresis. Undoubtedly one of the reasons for this is the difficulty that has frequently been met in isolating them from wall autolysates where the enzymes are often firmly bound to wall constituents, such as the teichoic acids [12, 14]. Success has been more readily met in purifying autolysins that are exported from the cells into the culture fluid. For example, each of the three types of staphylococcal enzyme has been obtained

Table 11.1 The distribution of common bacterial autolysins

Micro-organism	*N-acetylmuramyl-L-alanine amidase (amidase)*	*β-N-acetylglucosaminidase*	*Muramidase*	*Pepidase*	*Reference*
Arthrobacter crystallopoietes	0	0	+	0	76
Staphylococcus aureus	+ (E) §	+ (E)	0	+	134
Bacillus subtilis 168	+	+	0	0	157
B. subtilis W23	+	+	0	0	12
B. subtilis YT25	?	?	+ (E)	?	129
B. subtilis K77	?	?	+ (E)	?	87
B. cereus	+	+	0	0	70
B. licheniformis	+	+	0	0	*, 69
B. thuringiensis var thuringiensis (sporulating)	+	0	+	+	74
B. stearothermophilus	+	0	0	0	53
Clostridium welchii	0	+	+	0	82, 154
Cl. botulinum type A	+	+	0	0	128
Streptococcus faecalis	0	0	+	0	118
Micrococcus luteus	+	+	0	0	122, 121
Strep. pyogenes†	0	+	0	0	6
Diplococcus pneumoniae	+				85, 67
Lactobacillus acidophilus (AM Gasser)	0	0	+	0	30
Mycobacterium smegmatis	+	0	0	+	73
Escherichia coli	+	+	0	+	89
Proteus vulgaris	+	?	?	?	142
Brucella abortus	+	?	?	?	142
Listerella monocytogenes	+	(+)‡	(+)‡	0	133
Streptomyces griseus	?	0	+ (E)	?	150
Streptomyces sp.	+	+	+ (E)	?	86
Acromonas hydrophila				+ E	28
Myxobacter sp.	+			+ E	34, 71

* J. S. Thompson and H. J. Rogers (1969), unpublished observation
† Bacteriophage infected
‡ Glycanase present but bond specificity not determined
§ Extracellular

as a homogeneous protein [145, 147] as have the *endo*-muramidases of the strains of *Bacillus subtilis* [87, 129]. Among the autolysins of Gram-negative bacteria the amidase of *E. coli* strain K12 has been purified from the supernatant fluid of spheroplasts [142]. The bacteriolytic amidase, peptidase and two glycanases have been purified from the supernatants from cultures of Streptomyces species [46, 150].

Nevertheless, some successes have been obtained by direct fractionation of cell wall hydrolysates. For example, at an early stage a streptococcal *endo-β-N*-acetyl-glucosaminidase was purified by some 5300-fold from phage lysates with a yield of 36% [6] and the autolytic amidase was rather simply purified from autolysins walls of *B. megaterium* by centrifugation [23]. In many other examples, however, only partial purification was likely to have been achieved. The difficulties of separating the autolytic enzymes from wall polymers have been overcome in recent times by applying [10, 36] the observation first made by Pooley *et al.* [93]. These workers found that the autolytic muramidase could be removed from cell walls of *S. faecalis* by extraction with strong salt solutions. Solutions of CsCl, NH_4Cl and LiCl at concentrations of up to 10 M, were used and both the latent (see p. 443) and active forms of the enzyme were removed by solutions of 4.0 M and 10.0 M respectively. LiCl was chosen for further study because the relatively low density of strong solutions facilitated deposition of wall preparations during centrifuging. Application of this method has allowed the purification of the amidase and the *β N*-acetyl-glucosamindase from the walls of *B. subtilis* [59, 140a].

11.3.1 *Properties of the isolated autolysins*

The molecular weights of the isolated enzymes show considerable differences, ranging from 20 000 for the amidase of *B. megaterium* [23] to 100 000 for the *endo-β-N*-acetylglucosaminidase of streptococci [6] and 90 000 for the muramidase from one strain of *B. subtilis* [87]. There are differences in molecular weight for enzymes of the same specificity isolated even from different strains of an organism or from closely related species. For example, the staphylococcal *endo-β-N*-acetyl-glucosaminidase isolated from culture filtrates of strain M18 had a molecular weight of 70 000 whereas that in lysostaphin was 55 000. Even more impressive, *endo*-muramidases isolated from two strains of *B. subtilis* had molecular weights of 13 000 [129] and 90 000 [87].

The glucosaminidases at least from some sources appear to depend upon -SH groups for their activity. The inactivation of the enzyme from staphylococci during storage at 4° C was prevented by the presence of cysteine, or dithioerythritol, and enzyme activity was reduced to 50% by 1 mM *N*-ethylmaleimide, 0.1 mM *p*-hydroxy-mercuribenzoate or 0.1 mM iodoacetate [143]. Likewise, the enzyme from haemo-litic streptococci was inhibited by iodoacetamide [6]. On the other hand, the activity of the muramidase of *S. faecalis* in partially purified preparations did not depend on reduced -SH groups [118] and neither did that of this enzyme from one strain of *B. subtilis* [129]. The amidase from *B. subtilis* [59], the streptococcal *β-N*-acetylglucosaminidase [6] and the muramidase of *B. subtilis* [129] are critically influenced in activity by their ionic environment at low concentrations of salt, whereas the *β-N*-acetylglucosaminidase from staphylococci is not so, except at high concentrations (0.2–0.4 M) when it is inactivated [143]. The latter enzyme, which like lysozyme has a high iso-electric point of pH 9.5 [146] is stable to heat at pH 3–4, resisting 100° C for 30 min but becomes inactivated at pH 9.5 in 5 min at the same temperature. Despite the heat stability at low pH it is 50% inactivated by keeping for 2 h at 4° C in a pH of 3.2 and all activity had disappeared in 24 h [143]. The muramidase of *B. subtilis* on the contrary was of intermediate heat stability being inactivated by 1 h at 90° C at pH 6.2. Thus, the autolysins seem to be a rather widely assorted set of enzyme proteins which irrespective of bond specificity, show a variety of characteristics which differ not only according to the species of micro-organism from which they come but even, in some examples, according to the strain of the same species.

11.3.2 *Specificity of the autolysins*

Each of the four groups of enzymes seems quite specific for a particular bond in the peptidoglycan, except possibly for some of the peptidases which have not been investigated in detail. The ability of autolysins to act on walls from different species is probably as much related to their interaction with the polymers attached to the peptidoglycan as upon the structure of the polymer itself. For example, the presence and type of teichoic acid attached to peptidoglycan in strains of *B. subtilis* was shown to be important to the action of the isolated amidase and even more to its avidity for the walls [60]. In general, the strength with which the autolytic amidases bind to walls, however, does not seem a reliable guide to their enzymic effectiveness. For example, the amidase from *B. megaterium* walls does not bind significantly to homologous wall preparations [23] whereas that from *B. subtilis* binds strongly with a specificity dependent upon the teichoic acid present [60]; both enzymes rapidly lysed their homologous walls.

The staphylococcal *endo-β-N*-acetylglucosaminidase [145] is rather catholic in its taste for walls of different species of bacteria, lysing cells of 11 species at a significant rate – a wider choice than egg-white lysozyme. It could also hydrolyse chito-dextrins at a slow rate [144] with an optimum pH of 4.1 compared with a value of about 6.0 when acting on bacterial cell walls. The product was chitobiose and not free *N*-acetylglucosamine. Some muramidases are highly specific, for example, the enzyme from *L. acidophilus* is able to hydrolyse walls from only one other species of bacteria out of nine tested [30]. Unlike egg-white lysozyme, the bacterial muramidases usually appear unable to hydrolyse chitin and its dextrins.

The amidases may be rather more specific than the β-glucosaminidases but less so than the muramidases. For example, the amidase from *B. subtilis* will hydrolyse only walls of *B. megaterium* as rapidly as homologous walls, but those from *B. licheniformis* are hydrolysed at about 30% of the rate, and surprisingly those from one strain but not another of *S. lactis* are hydrolysed slightly faster (50%)

than those from *B. licheniformis* [23]. Neither of the isolated enzymes from bacilli would break down peptidoglycan in which its glycoside bonds had been extensively hydrolysed [23, 59]. The enzyme from staphylococci appeared able to do this [147] but very slowly. In contrast one of the best substrates for the amidase from *E. coli* was the *N*-acetylmuramyl-pentapeptide [142]. Peptidoglycans, freed from their associated polymers such as the teichoic or teichuronic acids, are frequently not such good substrates for autolytic enzymes as are whole cell walls. The combined result of this observation and the acute specificity demands for the correct teichoic acid to be attached to the peptidoglycan, already exemplified, has led to suggestions of a necessary specific interaction between the teichoic acids and the autolysins. Alternatively, it has been suggested [105] that teichuronic and not teichoic acid is essential for rapid autolysis of walls of *B. licheniformis* strain ATCC 9945. A chain-forming, novobiocin-resistant mutant was almost completely deficient in cell wall teichuronic acid. Selective removal of teichoic acid from walls of the wild-type had little effect on sensitivity to added lysin whereas removal of teichuronic acid made them as resistant as those of the mutant strain.

11.4 Location of autolytic enzymes

In order to be able to lyse the walls around bacteria without prior damage to the underlying protoplast membrane, a proportion at least of one or more of the autolytic enzymes produced must be associated with the walls, even if the steady state level during growth is low. Intact functioning protoplasts, for example, have been produced from *S. faecalis* by incubating suspensions of cells in buffered sucrose and allowing the wall associated autolysin to remove the walls – they were referred to as autoplasts [113, 114]. It is technically very difficult, nevertheless, to design methods which unambiguously define the level during growth because, as soon as the cells are disrupted, redistribution of the enzymes can take place. Alternatively, attempts to estimate enzyme activity associated with walls in growing cells from the known or suspected functions of autolysins, such as wall turnover, are bedevilled by insufficient knowledge about the processes themselves, and about the importance of the chemical topology of the surface substrate which may regulate autolytic activity. In *S. faecalis* a rather high proportion of the total cellular enzyme is in a latent form that can be activated by trypsin treatment of the walls. Both this latent form and the active enzyme have very high affinities for wall preparations with the result that most of the total autolytic activity appears as wall-associated in a disintegrated cell preparation [118]. Other work [92] suggested that a high proportion of the latent enzyme might be associated with the cytoplasm in the living organism. In L-forms of *B. licheniformis* strain 6346, on the other hand, 93% of the autolytic enzyme was associated with the membrane in disrupted cell preparations, from which it could be removed by incubation with isolated walls [43]. In the vegetative cells, significant amounts of the enzyme were also found in the membranes. In *L. acidophilus* about 25% of the muramidase remained associated with the cytoplasm in disrupted preparations and this was thought to be due to the limited number of binding sites in the walls [31]. In *E. coli* [89] and in some strains of *S. aureus*, [123] very high proportions of the autolytic activities remained in the soluble cytoplasmic fractions. More recently [155] a rather high proportion of the amidase in the former organism was found in the outer membrane of the bacteria. Presumably these observations may reflect the relative affinities of the enzymes for the different cell fractions rather than their true *in vivo* locations. At first sight, the most surprising feature of autolysins is that, in a number of strains of Gram-positive species such as *S. aureus* and some bacilli [68, 87, 98, 129], a large proportion of the activity is found soluble in the supernatant fluid of the cultures. Examination of the distribution of lytic factors in cultures of *S. aureus* strain Oeding [8] for example, showed the presence of four times as much activity, soluble in the supernatant fluid as there was associated with the cells. Subsequent to the demonstration [134], that three autolysins were present in staphylococci, all three were shown to be present in the supernatant fluids of another strain of *S. aureus* M18 [170]. Lysostaphin indeed is a soluble bacteriolytic preparation produced from cultures of a colony of micro-cocci occurring on staphylococcal plates and causing bacteriolysis of the surrounding staphylococci [115]. It contains all three autolysins [9, 148]. A soluble extra-cellular glycanase attacking the walls of *Micrococcus luteus* was found to be produced by a strain of *B. subtilis* [103, 104] and extra-cellular *endo*-muramidases were recognized in two other strains of the organism [87, 129]. These latter observations point to the great strain variation in this respect since other common laboratory strains of *B. subtilis* such as 168 and W23, do not form detectable soluble autolysins. Likewise, a strain of a closely related species, *B. licheniformis* with extremely active autolysins, as compared with some strains of *B. subtilis,* also did not liberate the enzymes into the supernatant fluid. A so-called 'superlytic' mutant of the *B. licheniformis* strain, however, did so to some extent during the exponential phase of growth. Examination of cultures of this mutant strain for a known cytoplasmic enzyme, α-glucosidase, showed that an exactly equivalent proportion was present in the supernatant fluids. Thus, it seems probable that, in this instance, some of the cells were lysing even during the exponential phase of growth and leakage, rather than export of both enzymes, was occurring [44]. It, nevertheless, seems unlikely that such an explanation applies to other organisms where very high proportions of the autolytic enzymes are extra-cellular during exponential growth. Despite the apparent absence of soluble autolytic enzymes in some strains of bacilli, they can be transferred between cells in cultures of *B. subtilis.* If wild-type cells are grown in the same culture as an autolytic-deficient mutant, aspects of the phenotype such as cell separation and wall turnover of the latter, are restored towards those of the wild-type [41, 111]. The two organisms must be grown together in mixed culture, no soluble factor is present and no permanent change of the mutant cell is involved. Thus, the autolytic enzymes are apparently so superficially located in the wild-type, that they can be transferred to the surfaces of the mutant cells. Indeed two quite different approaches [41, 63] have suggested that effective autolytic enzymes might always be exported and then act from the outside of the cell, rather than acting within the wall or on its inner face as might be expected.

Another approach, attempting to answer questions about the location of autolytic enzymes, is to examine sections of cells that are undergoing autolysis. In this way it was shown that in an *S. faecalis* cell the leading edge of the forming septum that will eventually divide the cells, is the first site of action of the autolytic muramidase [62]. Only after the septum has been extensively damaged is the peripheral wall in the vicinity removed. This technique certainly gives a clear impression of localization of enzyme action in cells placed under conditions that are suitable for autolysis (ie non-growing suspensions incubated in buffer), but it cannot, of course, be extrapolated with certainty to the situation in rapidly growing bacteria. The results obtained may also depend on the local susceptibility of wall polymers as well as on localization of the enzyme. Other evidence [117] supports the presence of active autolytic enzyme in the region of the septum. When exponentially growing cultures are first labelled with ^{14}C-L-lysine for 0.1–0.4 of a generation, the walls isolated from disrupted cells and allowed to autolyse, radioactivity due to the L-lysine in the peptidoglycan is solubilized more rapidly than the walls as a whole. In other words, the most recently synthesized wall is most rapidly released as might be expected since quite different evidence (see Chapter 15.1 and [112, 119] for summaries) shows that the walls of streptococci grow from the septal region, near to where autolysin is seen to be acting most vigorously in non-growing cells. Thus, despite the known relocalization of latent, but apparently not active autolysin [92] during cell disruption, autolysis still occurs near the same place as it does in whole cells. Further evidence about differences in local susceptibility to autolytic enzymes in walls is needed since it, clearly, raises questions about the interpretation of the localization of autolysin as seen in whole cells by the sectioning technique (see also [119]).

11.5 Function of autolysins

11.5.1 *The autolysins and the growth of microbial cells*

If indeed walls consist of polymers covalently linked together into 'huge bag-shaped' molecules [153], or even if they contain polymers like cellulose covalently linked together in one plane but with the fibrils held together by secondary valency bonds in two other planes, enzymic nicking would be an expeditious method of allowing the introduction of new units to expand the network during growth. If newly-biosynthesized units are to be added to the nicked polymers, the appropriate acceptor groups must be made available by the hydrolases present. This is particularly relevant of course, when complex wall polymers such as the peptidoglycans in bacteria are involved. Either the reducing group of the *N*-acetylmuramyl residues, a suitable amino group of a diamino acid, or a –COOH group of D-alanine in the peptide, must be available for the biosynthetic enzymes to function. In fungal cell walls, the requirements are less stringent since most of the autolytic hydrolysases break bonds in the main backbone of homo-polysaccharides, even when branching is present. They thus, inevitably liberate suitable functional groups for the addition of new material.

As far as bacteria are concerned, the first organism to be seriously considered in the context of a possible essential role for autolytic enzymes in growth, was *S. faecalis* [118, 137]. In some ways this was fortunate since it has a muramidase as the sole autolytic enzyme (see Table 11.1) and the specificity of this enzyme is theoretically suitable for revealing potentially effective functional *N*-acetyl-muramyl reducing groups as acceptors (see Section 8.5). Secondly, evidence already mentioned showed that autolytic activity is concentrated in just the region of the cells in which new wall material is being deposited during growth and division. Thus, it seemed, and still seems, a wholly reasonable hypothesis to be applied to this organism. Unfortunately, as can be seen from Table 11.1 many other bacteria have autolytic enzymes of unsuitable specificity which could not easily liberate suitable functional groups, and the location of autolytic activity has not been so easy to demonstrate in other organisms as it was with *S. faecalis.* Where some localization has been demonstrated in the septal regions, as for example in *L. acidophilus* [63], it has appeared to be acting from the outside of the cells and to be part of cell separation events, rather than of cell growth. Suggestive negative evidence against an essential role for autolysins in growth has also now accumulated from the study of defective mutants. It should immediately be said, however, that this evidence does not yet exclude unambiguously, a role for a very small proportion of the wild-type autolytic activity, strategically positioned in the cells. Successful isolation of autolytic-deficient mutants has been made from strains of *S. faecalis* [29, 95] *Streptococcus pneumoniae* [77] *S. aureus* [25, 75], *B. licheniformis* [42, 44], and *B. subtilis* [5, 37, 41, 156]. Conditions have also been found [139] whereby, due to a change in the wall teichoic acid of *S. pneumoniae*, the autolytic activity is grossly reduced. The autolytic activities of the strains in these various experiments were reduced by 50–95% but only in two instances [37, 120] was a significant effect on either growth rate or individual cell morphology recorded; this experiment will be discussed later. In the work of Fein and Rogers [41] the mutants of *B. subtilis* were 90–95% deficient in both of the autolytic enzyme activities present (ie an *endo*-β-*N*-acetylglucosaminidase and an amidase). Despite an effect on both activities, all the genetic evidence supports the presence of only a single lesion. The growth rates of these mutants in five different media in which the doubling times varied from 19.3–65.7 min, were identical within 15% to those of the isogenic wild-type. The morphology of the individual cells of the mutants appeared normal under the light microscope but detailed ultrastructural work was not done. It is against this evidence that the observation [37] of a positive role for autolysins in the growth of *B. subtilis* has to be assessed. A mutant strain was shown to grow better at 51° C in a broth medium when either lysozyme, or an extract containing autolysin from the wild-type, was added and the distored morphology of the individual cells was corrected. Although very interesting, the meaning of this work has to be queried (see [41]).

Mutants of *Streptococcus faecium* with reduced autolytic activity correlated with reduced growth rates have also been isolated [120]. The mutants were obtained after treatment of cultures with *N*-methyl-*N*-nitro-*N*-nitrosoguanidine and selected either by repeated challenge with penicillin G (10 μg/ml) and cycloserine (360 μg/ml) or by being 'metabolically poisoned' with 5×10^{-5} M iodoacetate and then selected for resistance to the non-ionic detergent Triton X-100 (0.6 mg/ml). No genetic manipulation could be done to test for multiplicity of damage. The mutants were highly pleiotrophic: for example, the cell size of two out of three mutants was different from the wild-type; two out of the three strains grew more slowly in broth cultures; all three mutants in the defined medium failed to reach the same optical density as the wild-type indicating a different nutritional requirement; two of the mutants grew as long chains of cells in exponential cultures one of which formed clumps in the stationary phase (this strain was the only one grossly deficient in autolysins after activation by trypsin treatment), the third apparently grew in the usual way as pairs of cells. The total autolysin activities in wall preparations from the three mutants were 78%, 57% and 45% of that of the wild-type despite the fact that the rates of autolysis of whole cells, even after trypsin activation, were respectively only 53%, 24% and 1.3% of the wild-type. Further work is clearly needed on these and other [29] fascinating mutants before weight can be put on the correlation between slower growth rates and reduced autolysin content.

One final piece of evidence shows that the bulk of the expressed autolytic activity of bacteria is not necessary for morphological changes. Conditional mutants of several species of bacteria have been described which can grow either as rods or cocci (see Chapter 15.5) the so-called *rod* mutants. Examination of one such mutant of *B. subtilis* has shown that the coccal form, like *S. faecalis*, forms its peripheral wall from the septal region, unlike the rod form which extends within the cylindrical region of the cell [16]. One might expect that the coccus would show some disability in growth if deprived of autolysin since present thinking [61] would demand the presence of autolytic activity at the base of the septum to allow the 'pealing' apart of the peripheral walls as they are formed. However, insertion of a *lyt* gene [41] into strains already carrying *rod* genes had little or no effect on either the morphological change or the growth of the coccal forms [109, 110].

A clear weakness, on the other hand, of the negative evidence is that none of the mutants so far described, is entirely deficient in autolysins and there is no yardstick to measure how much is enough. An indication of our ignorance in this respect is illustrated by the finding (J. B. Ward and C. Taylor, unpublished work) that, in one of the mutants isolated by Fein and Rogers [41], the glycan chains in the peptidoglycan had been hydrolysed to yield free *N*-acetylglucosamine-reducing groups to nearly the same extent as those in the walls of the isogenic wild-type

Table 11.2 Average chain lengths of glycans in peptidoglycan from an autolytic-deficient mutant (*lyt*) and its isogenic wild-type of *B. subtilis* (J. B. Ward and C. Taylor, unpublished work)

	Average chain lengths					
	*Biosynthetic**		*In vivo**		*ratio of biosynthetic: in vivo*	
Mutant	487.6	313.4	192.9	158.3	2.53	1.98
Wild-type	604.1		209.3		2.89	

*These values for two separate batches of cell walls are calculated from results by the method of Ward [149]. The biosynthetic value indicates the proportion of free-reducing groups of *N*-acetylmuramyl residues to the total whilst the *in vivo* length also includes free-reducing groups of *N*-acetylglucosamine.

(Table 11.2), yet this mutant appeared to have only 3–4% of the wild-type *endo-N*-acetylglucosaminidase activity in LiCl extracts made from it. The exact interpretation of the results for 'biosynthetic' and *in vivo* chain lengths [149] is possibly open to some debate. Nevertheless, the appearance of an undiminished amount of free-reducing groups of *N*-acetylglucosamine in the walls, strongly suggests continued enzyme action despite the low enzymic activity as measured in extracts acting on exogenous substrates.

11.5.2 *Cell separation and motility*

Study of the deficient mutants has demonstrated clearly that depriving bacterial cells of a large proportion of their autolytic activity leads to their failure to separate from each other. Mutants of rods or streptococci form long chains of cells [41, 42, 44, 95], as wild-type strains of pneumococci do when phenotypic wall changes lead to gross reduction in autolytic activity [139]. With mutants of organisms such as staphylococci which usually grow as small irregular groups of cells, large regular packets are formed [75] from some strains though possibly not others [25], as would be expected if cells of these mutants dividing successively in planes at right angles to each other cannot separate. When normal amounts of autolysin are formed, it must be supposed the cells can move relative to one another after septation and division. Another mutant possibly not forming regular packets was described as growing in large clumps of cells [25].

Most of the mutants so far examined for the specificity of their remaining enzyme complement have less of all of those present in the wild-type and they must be supposed to have lesions in regulatory rather than structural genes. For this reason, it is not possible to know which enzyme is responsible for separation or whether all must be present. A recently examined mutant of *B. subtilis* isolated by Tilby [132] however, has only 2% of the *endo*-β-acetylglucosaminidase activity but 60% of the amidase and separates normally to grow as single motile cells (H. J. Rogers and C. Taylor, unpublished work). This may suggest that the amidase alone can effect cell separation. No mutant depressed only in the amidase has been found.

Another characteristic of the autolysin-deficient mutants of *B. subtilis* and *B. licheniformis* is that the bacteria have lost their flagella and are therefore non-motile [5, 40, 41, 156]. Some of these [156] have been shown to have a pool of flagellin in the cytoplasm, suggesting that the reduced autolytic activity in some way prevents the extrusion and/or organization of the flagella from the protein subunits. Motile revertants are all fully autolytic. Again the single mutant of *B. subtilis* so far available with a grossly depressed β-*N*-acetylglucosaminidase but a reasonably high amidase activity is motile (H. J. Rogers and C. Taylor, unpublished work).

Since the lytic-deficient mutants of *B. subtilis* have complex phenotypes and are most likely to have a lesion in some regulatory gene, it is possible to argue that the lack of flagella is one more aspect of the phenotype not directly connected in a causative manner with the absence of active autolytic enzymes. Examination of autolytic-deficient phosphoglucomutase-deficient mutants of *B. licheniformis* and *B. subtilis* [42, 44] showed that these too did not separate and were non-flagellated [40]. This observation alone would not help the argument since their phenotype is equally complicated, but it is possible to circumvent the effect of the phosphoglucomutase lesion in a biochemical way. When grown in media containing galactose together with an assimilable carbon source such as glycerol, the phenotype of these mutants is partially reversed. Polymers dependent upon the formation of UDP–glucose appear in the walls and the autolytic enzyme activity is partially restored [45]. More important in the present context, flagella are formed and the bacteria are once more motile, like the wild-type [40].

Evidence has also been obtained [155] that cell separation and the splitting of the peptidoglycan septum [17, 18] may be related to the autolytic amidase activity in Gram-negative species such as *E. coli.* The specific activity of amidase in the *envA* mutant, which forms chains of unseparated cells is some five times lower than in *envA*$^+$ strain at high growth rates but only 30% lower at slow rates. Correspondingly the chains were longer at high growth rates and the specific activity of the amidase varied as would be expected if it were related to separation. Addition of *N*-acetyl-muramyl-L-Ala-D-Glu-*meso*-DAP stopped cell separation altogether suggesting competitive inhibition of the amidase by a good enzyme substrate [142].

11.5.3 *Turnover of bacterial wall polymers*

Examination [8, 21, 22, 25, 84, 90, 91] of the fate of pulses of radioactive compounds labelling peptidoglycans, or of cells already so labelled but growing in non-radioactive medium, has shown that the walls of many organisms such as bacilli, lactobacilli, or staphylococci are in a state of flux. In other species such as *S. faecalis* this is not so [8] and the walls do not turnover a term usually applied to this process of loss and renewal of wall polymers. It is of course, not strictly turnover, as applied to small molecules, or even to proteins, since it probably represents the continued partial breakdown of macromolecules and either the further growth of their remaining parts or the initiation of new chains. The negative side of such a process involves the formation of soluble material from the insoluble wall and it is, therefore, reasonable to suppose that autolytic enzymes are involved. One report [47] very disturbing to this hypothesis, however, is that of a strain of *B. subtilis* W23 with a normal complement of autolysins, and normal walls but with a wall turnover reduced by 90%. It has also been stated [8] that a strain of *L. acidophilus* autolysing at $<2\%$ of the wild-type rate, showed a similar rate of wall turnover. Although little other exacting work has been done to test the idea, such results as are available suggest a role for autolysins in turnover. An autolytic-deficient strain of *Staph. aureus* failed to show wall turnover although the wild-type rate was rather slow [25]. Examination [111] of an autolysin-deficient, phosphoglucomutase-negative strain of *B. licheniformis* and the chain forming strain Ni15 of *B. subtilis* [90] by first labelling the walls with ^{14}C-*N*-acetylglucosamine and then growing the bacteria in medium containing the non-radioactively labelled compound, showed that both had greatly reduced turnover compared with the wild-types from which they were derived. When small volumes of a crude autolysate from the wild-type *B. subtilis* were added to cultures of Ni15, turnover was increased in relation to the volume of autolysate added [90]. Examination of the turnover of the walls of the autolytic-deficient mutant of *B. subtilis* FJ6 isolated by Fein and Rogers [41], showed that it was greatly depressed (R. S. Buxton and J. E. Fein, unpublished work). This was again done by the complete labelling technique. Work with another of the latter mutants (FJ3) by treatment with short pulses of ^{14}C-*N*-acetylglucosamine followed by a chase, appears, however, (J. Mandelstam, personal communication) to demonstrate that turnover can occur under some circumstances. On the other hand, similar pulse experiments with Ni5 did not allow a different conclusion from the experiment with completely labelled walls already described (H. M. Pooley, personal communication) since greatly reduced turnover was found. An autolytic-deficient strain of *Staphylococcus aureus* also failed to show wall turnover [25]. Again, it is not possible from these experiments to say whether one or both of the autolytic enzymes in the organisms are likely to be necessary for the process. It could be argued that some activity of both would be expected to be necessary since both glycan and peptide bonds hold the peptidoglycan together. However, only one enzyme, a muramidase, is present in *L. acidophilus* [30] yet the peptidoglycan of this organism actively turns over [8, 32]. This raises interesting problems about the geometry of the process. Two possibilities can be envisaged:

(1) Two contiguous glycan strands may be hydrolysed that are cross-linked by peptides but are not thus joined to other glycans.

(2) One strand may be hydrolysed but only in regions where peptide cross-linking is incomplete.

It is particularly interesting that whereas the peptidoglycan of *S. faecalis* does not turnover, that of *L. acidophilus* does, despite the presence in both organisms of only muramidase as an autolytic enzyme. It may be relevant to note that a minimum fraction, equivalent to 10–20% of the peptidoglycan in the latter organism, was immune from the process [32]. This proportion corresponds well with the figure of 13% for the conserved radioactivity in the poles of bacilli [94], and with 15% for the measured area of the pole formed as a proportion of the whole surface [19]. These facts might with profit be related to those suggesting that autolysins act from the outside of the cell both in promoting turnover [90, 91, 111] and in cell separation [62]. The poles of bacilli, but not the cylindrical region, are partially resistant to the action of autolysin, at least under some conditions [38] and exponentially growing whole cells of *S. faecalis* are resistant to high conentrations of lysozyme [138]. Supernatant fluids from cultures of another strain of this latter organism, however, dechain the streptococci [79] suggesting the presence therein of autolysin. The strain used by Toennies *et al.* [138] was not dechained by the culture fluid but was by a cytoplasmic preparation. One might see in these various observations a reason for the failure of the walls of *S. faecalis* to turnover whereas bacilli, for example, do so; namely, that the walls of growing streptococci cannot be enzymically attacked readily from the outside of the bacteria and that this is necessary for turnover to occur. Exponentially growing lactobacilli on the other hand are highly sensitive to lytic enzyme, along the cylinders of the rod-shaped organisms. Nevertheless, despite these suggestive observations, the possibility must still be kept in mind that the difference between *L. acidophilus* and *S. faecalis* may lie in the organization of the wall in such a way that muramidase can liberate soluble material from the former but not the latter microbe. On grounds of resistance to external lytic enzyme, one might suppose that the fraction of the wall of *L. acidophilus* resistant to turnover is the polar region of the cells. However, a claim has been made that the polar regions of *B. subtilis* turnover at nearly the same rate as the wall of the cylindrical part [39] despite the resistance of its poles to externally applied lytic enzyme [38].

11.5.4 *Autolysins in transformation*

Another attractive role for the autolysins of bacteria is to allow entry to the cell through the cell wall, of macromolecules such as DNA. Such a role in transformation has been repeatedly suggested over a period of many years in a number of different systems [1-3, 35, 99, 101, 116, 132, 160]. A variety of evidence supporting the idea has been obtained. For example, correlations have been found between competence and the rates of autolysis of cells or walls isolated from them in *B. subtilis* [158] and in Group H streptococci [99, 101]. No doubt, the extremely complicated nature of the competent state is responsible for difficulties met in providing unambiguous evidence about the role of autolysis of walls in transformation. When competence factor (CF) is added to streptococci, the rate at which either Group A streptococci [99] or pneumococci [116] lyse, increases. In the former, this also corresponds with increasing appearance of extra-cellular lytic activity. Competence reaches a peak at the maximum of autolytic rate and extra-cellular lytic activity. Seto and Tomasz [116] propose that the CF causes some form of membrane change allowing exit of autolytic enzymes and access to the walls. That autolytic activity has a role to play in the transformation of streptococci is further suggested by the finding [100] that reagents such as *N*-ethylmaleimide or mercuric chloride inhibit both autolysis and the development of competence. The inhibition of both is reversed by 2-mercaptoethanol. When placed in hypertonic media, pneumococci form spheroplasts rather than protoplasts since they retain about 13–14% of the wall muramic acid. Competence and the rate of spheroplast formation are usually correlated [78]. Once again, as in studies with other organisms [159], this correlation, although undoubtedly suggesting involvement of autolytic activity in some way in transformation, is not complete. For example, the mutation *ntr* causes pneumococci to be genetically incompetent and whilst *ntr-2* mutants form spheroplasts very slowly indicating reduced autolytic activity, other *ntr* mutations such as *ntr-9* have no such effect. The autolytic-deficient mutant cwl-1 is fully transformable [77] as the *lyt* mutants of *B. subtilis* appeared to be [41]. Complete removal of the wall renders some Gram-positive species unable to take up DNA. However, its partial removal in *B. subtilis* has been claimed [131] to increase DNA penetration and hence competence. Fractionation of water extracts from competent cells of *B. subtilis* yields a material showing both amidase and competence increasing activities [1] and it also contains a nuclease, forming single-stranded DNA [4]. Thus, the evidence for a role of autolysins in transformation, is teasingly suggestive without being, as yet, compelling.

11.5.5 *Autolysins in spore formation and germination*

One might expect enzymes attacking peptidoglycans to be involved at two points during the spore formation and germination. The first point is when the mature spore is liberated from the vegetative or 'mother cell' and the second when the cortex of the spore which has a high content of modified sparsely cross-linked peptidoglycan [135, 152] is broken down during germination. Apart from these two obvious needs, however, specific enzymes are formed during the process of sporulation, with so far unknown functions. Examinations of sporulating cultures of *B. thuringiensis* [74], *B. subtilis* and *B. sphaericus* [54–57] have shown the presence of endopeptidases as well as the better recognized amidases and carboxy-peptidases. These endopeptidases hydrolyse bonds in the main peptide chains of the peptidoglycan. In *B. thuringiensis* the bond between L-alanine and D-glutamate is hydrolysed whereas in *B. subtilis* and *B. sphaericus* it is that between the γ-D-glutamyl and the mesodiaminopimelic acid residues. In the latter examples, it seems highly probable that the endopeptidase is specific to the developmental process. It is not present in vegetative cells, increases during sporulation and the activity is depressed if sporulation is inhibited by using carefully controlled concentrations of the drug netropsin that inhibits RNA synthesis [55, 56]. Moreover, it is principally located in the forespore integument, other fractions from the bacteria having only 20% of the specific activity of this region [57]. It will hydrolyse the spore cortical peptidoglycan but not that in the wall of the vegetative cell. A dipeptidase is also specifically formed late in sporulation of *B. sphaericus* and probably of *B. subtilis*, hydrolysing the bond in the dipeptides, L-Lys-D-Ala and *meso*-A_2pm-D-Ala [141]. This presumably, however, does not attack peptidoglycans but breaks down the products resulting from the action of the endopeptidase. A role for autolytic enzymes in the process of sporulation and germination of *B. cereus* was proposed at an earlier time when soluble fragments containing diaminopimelic acid, glutamate, alanine and hexosamines were discovered to increase in sporulating cultures [95]. The hexosamines included an unknown compound subsequently identified as muramic acid [124]. It was deduced that these soluble fractions arose by activity of an enzyme liberating soluble material from the insoluble fraction of the vegetative cell [96]. In cultures of the bacilli in an advanced stage of sporulation, two enzymic activities V and S, were identified [126] with a pH optima of 4.5 and 8 respectively. One of the enzymes, S, was also found in disintegrated spore suspensions [127]. The authors considered that both enzymes were formed during sporulation and that either or both were concerned with release of spores from the vegetative cell, although they clearly favoured the idea that enzyme V was concerned with spore release and S with spore germination. Lysis of spore integuments by enzyme extracted from spores was shown to liberate reducing sugars but not amino groups [50, 51] and was presumably a glycanase acting on the cortical peptidoglycan. Examination of extracts from intact spores of *B. cereus* made with 7.2 M urea containing 10% 2-mercaptoethanol took us a step forward by showing the presence of enzymes capable of solubilizing fragments of spore integument [13]. This experiment allowed the deduction that the enzyme or enzymes were superficially located on the spores. Meanwhile Warth [151] claimed the presence of a β-*N*-acetylglucosaminidase and an amidase in spore integuments of *B. subtilis*, and the soluble autolysins from *B. cereus* spores appeared to have the same specificites. Preliminary examination of the specificity and function of the superficial enzyme from *B. cereus* spores suggested that it was specific for cortical spore fragments and would not act on sporangia but it did not promote germination changes occur early in spore germination. Lysozyme, among a range of other substances, also triggers similar changes [49]. The superficial spore enzyme has since been purified to homogeneity and shown to be an *endo*-β-*N*-acetylhexosaminidase dark and suspensions to show a reduction in optical density [52]. Both of these changes occur early in spore germination. Lysozyme, among a range of other substances, also triggers similar changes [49]. The superficial spore enzyme has since been purified to homogeneity and shown to be *endo*-β-*N*-acetylhexosaminidase with a monomer weight of 43 000 [15]. Thus, enzymes both with unique bond specificity, as well as autolysins with the usual preferences, may be involved in some way with sporulation and germination of bacillary spores. Where enzymes with common specificity are involved, it is not yet possible to say dogmatically either whether they are coded for by the spore chromosome, or whether they are distinct in characteristics other than bond specificity from those of the vegetative cell. The preliminary result that the superficial enzyme in *B. cereus* spores is unable to attack vegetative walls suggests that this β-*N*-acetylglucosaminidase is in some way different from the vegetative enzyme. It may be pointed out that mutants of *B. subtilis* lacking 95% of the complement of vegetative cell, autolysins were still able to sporulate normally and the spores to germinate well [41].

11.5.6 *Autolysins and the action of antibiotics*

As we have seen, during growth of many species, wall synthesis and degradation are balanced and so-called turnover of the wall ensues. If wall synthesis is inhibited in such situations, without a corresponding inhibition of either the formation or action of the autolysins, cell lysis might be expected and indeed occurs. Wall synthesis may be inhibited either by temperature-sensitive lesions affecting it [20, 24, 48, 80, 83], by removal of specifically essential amino acids [102, 136], or by adding antibiotics or other substances that are more or less specifically inhibitory [97, 108].

Antibiotics that inhibit wall synthesis are bactericidal, unlike most of those inhibiting protein synthesis which are bacteriostatic. The question posed many years ago [106] but not frequently mentioned, is the following: is simple inhibition of wall synthesis sufficient alone to kill micro-organisms, and when the cells lyse is this because their walls can no longer expand so that the cell bursts [81], or is autolytic action essential? One possible way to answer this question would seem to be to stop the formation of autolysins by blocking protein synthesis. Combinations of inhibitors of protein synthesis, such as chloramphenicol, with the penicillins are not bactericidal for *E. coli* [72, 97] or for staphylococci [106]. Lysis of staphylococci could also be stopped [106] as could that of *B. subtilis* [108] and *S. faecalis* [117] by the inhibition of protein synthesis.

Another more specific approach to the relationship between autolysins of bacteria and the bactericidal action of wall inhibitors is to study mutants or circumstances making the bacteria deficient in autolytic activity. Mutants of *B. licheniformis* [108] and wild-type *Diplococcus pneumoniae* with modified walls [140], died very much less rapidly then the unmodified wild-type strains, when treated with a variety of antibiotics inhibiting wall synthesis. There are, however, certain points to be noted about both of these studies. Both organisms had walls that were modified so that they were resistant to the action of the cells' own autolytic enzymes as well as being deficient in the autolytic enzymes themselves. The *B. licheniformis* mutants were lacking phosphoglucomutase [45] and therefore could not glycosylate the teichoic acid in their walls, and nor could they form teichuronic acid. The non-autolytic *D. pneumoniae* cells were produced by growing the wild-type strain in medium containing ethanoline instead of choline. Nevertheless, in this latter study a mutant of the organism, apparently with normal walls but with low autolytic ability, was also resistant to the killing action of several wall inhibitors. The above work is all consistent with the hypothesis that autolysis is necessary for cell death when wall synthesis is inhibited but a note of caution is introduced by a study of autolytic-deficient mutants of *B. subtilis* [41]. These organisms had walls of normal composition and susceptibility to autolysins. Their autolysis in the presence of three different wall-inhibiting antibiotics was undoubtedly very much slower but only when rather low concentrations were used. With larger amounts (5 μg/ml) only small, barely significant differences between wild-type and mutant could be shown. A study of *S. aureus* H Str [25] superficially led to still more disturbing conclusions. The addition of sufficient benzylpenicillin to the wild-type of this strain to stop growth only began to lead to slow lysis four or five generations later. This might be explained however, by the fact that the penicillin was added to the cultures towards the end of the exponential phase of growth. Lysis might be expected to be, and indeed is, more rapid during the true exponential growth phase when rapid protein synthesis is occurring; despite the negative conclusions drawn by the authors, clear differences were still apparent between the autolysin-deficient mutant and the wild-type, even under their unfavourable conditions for autolysis of the latter strain [25].

A problem, possibly raised by the results with the *lyt* strains of *B. subtilis* mentioned above, is that of just how far the presence of penicillin itself modifies autolytic enzyme formation and action. Clearly it is possible that penicillins may induce enzyme activity, or that some results could be explained by inhibition of the formation of autolytic enzymes by higher concentrations of some penicillins [107]. No evidence on these matters has yet appeared so that they must remain speculations but a different explanation for increased activity of autolysins during inhibition of wall synthesis has been suggested [66]. It is known from other work [26, 27, 64] that amphipathic substances such as lipoteichoic acids, cardiolipin, dipalmitoyl phosphatidylglycerol and the Forssman antigen from *D. pneumoniae* can strongly inhibit the action of some autolysins. It has now been found [58, 66, 148] that during inhibition of peptidoglycan synthesis by growing cultures of several species of bacteria, lipid-containing materials are released from the bacteria without lag. This was shown by first incorporating ^{3}H-acetate into the bacteria and measuring the proportion of soluble radioactivity appearing in the cultures. In streptococci 80–90% of the radioactivity in the cells had been incorporated into lipid and representatives of all the membrane phospholipids could be recognized in the culture supernatants after the addition of wall inhibiting antibiotics. Clearly, dilution of substances such as lipoteichoic acids in the culture supernatants might lead to greater activities of cellular autolysins. Suginaka *et al.* [127a] have studied the converse and have shown that the addition of homologous lipoteichoic acid to cultures of *Staph. aureus* treated with penicillin prevents lysis and death of the bacteria. Forssman antigen, which is probably the lipoteichoic acid formed by pneumococci, can also prevent penicillin induced lysis of these organisms [64]. Further work is required to show that lipoteichoic acids or lipids certainly regulate the action of autolysins *in vivo.* Other potential methods for regulation of autolytic activity are also possibly used by bacteria, such as the protease activation of muramidase in *S. faecalis* that has already been referred to and the production of protein modifiers as observed for the amidase of *B. subtilis* [59]. The total picture for the regulation of autolysins in bacteria is as yet unclear.

References

1. Akrigg, A. and Ayad, S. R. (1970) *Biochem. J.* **117**, 397–403.
2. Akrigg, A., Ayad, S. R. and Barker, G. R. (1967) *Biochem. biophys. Res. Commun.* **28**, 1062–7.
3. Akrigg, A., Ayad, S. R. and Blamire, J. (1969) *J. Theor. Biol.* **24**, 266–72.
4. Ayad, S. R. and Shimmin, E. R. A. (1974) *Biochem. Genet.* **11**, 455–74.
5. Ayusawa, D., Yoneda, Y., Yamane, K. and Maruo, B. (1975) *J. Bact.* **124**, 459–69.
6. Barkulis, S. S., Smith, C., Boltralik, J. J. and Heymann, H. (1964) *J. biol. Chem.* **239**, 4027–33.
7. Berkeley, R. C. W., Brewer, S. J., Ortiz, J. M. and Gillespie, J. B. (1973) *Biochim. biophys. Acta* **309**, 157–68.
8. Boothby, D. L., Daneo-Moore, L., Higgins, M. L., Coyette, J. and Shockman, G. D. (1973) *J. biol. Chem.* **248**, 2161–9.
9. Browder, H. P., Zygmunt, W. A., Young, J. R. and Tavormina, P. A. (1965) *Biochem. biophys. Res. Commun.* **19**, 383–9.
10. Brown, W. C. (1972) *Biochem. biophys. Res. Commun.* **47**, 993–6.
11. Brown, W. C. (1977) In *Microbiology 1977* ed. Schlessinger, D., pp. 75–84, Washington, D.C: Am. Soc. Microbiol.
12. Brown, W. C. and Young, F. E. (1970) *Biochem. biophys. Res. Commun.* **38**, 564–8.
13. Brown, W. C. and Cuhel, R. L. (1975) *J. gen. Microbiol.* **91**, 429–32.
14. Brown, W. C., Fraser, D. K. and Young, F. E. (1970) *Biochim. biophys. Acta* **198**, 308–15.
15. Brown, W. C., Vellom, D., Schnepf, E. and Greer, C. (1978) *FEBS Letts.* **3**, 247–9.
16. Burdett, I. D. J. (1979) *J. Bact.* **137**, 1395–405.
17. Burdett, I. D. J. and Murray, R. G. E. (1974a) *J. Bact.* **119**, 303–24.
18. Burdett, I. D. J. and Murray, R. G. E. (1974b) *J. Bact.* **119**, 1039–53.
19. Burdett, I. D. J. and Higgins, M. L. (1978) *J. Bact.* **133**, 959–71.
20. Buxton, R. S. (1978) *J. gen. Microbiol.* **105**, 175–85.
21. Chaloupka, J. (1967) *Folia microbiol.* **12**, 264–73.
22. Chaloupka, J., Rihova, L. and Krekova, P. (1962) *Experentia, Basel* **18**, 362–3.
23. Chan, L. and Glaser, L. (1972) *J. biol. Chem.* **247**, 5391–7.
24. Chatterjee, A. N. and Young, F. E. (1972) *J. Bact.* **111**, 220–30.
25. Chatterjee, A. N., Wong, W., Young, F. E. and Gilpin, R. W. (1976) *J. Bact.* **125**, 961–7.
26. Cleveland, R. F., Holtje, J–V., Wicken, A. J., Tomasz, A., Daneo-Moore, L. and Shockman, G. D. (1975) *Biochem. biophys. Res. Commun.* **67**, 1128–35.
27. Cleveland, R. F., Wicken, A. J., Daneo-Moore, L. and Shockman, G. D. (1976) *J. Bact.* **126**, 192–7.
28. Coles, N. W., Gilbo, C. M. and Broad, A. J. (1969) *Biochem. J.* **111**, 7–15.
29. Cornett, J. B., Redman, B. E. and Shockman, G. D. (1978) *J. Bact.* **133**, 631–40.
30. Coyette, J. and Ghuysen, J–M. (1970) *Biochemistry* **9**, 2952–6.
31. Coyette, J. and Shockman, G. D. (1973) *J. Bact.* **114**, 34–41.
32. Daneo-Moore, L., Coyette, J., Sayare, M., Boothby, D. and Shockman, G. D. (1975) *J. biol. Chem.* **250**, 1348–53.
33. Del Rio, L. A. and Berkeley, R. C. W. (1976) *Eur. J. Biochem.* **65**, 3–12.
34. Ensign, J. C. and Wolfe, R. S. (1965) *J. Bact.* **90**, 395–402.
35. Ephrussi-Taylor, H. and Freed, B. A. (1964) *J. Bact.* **87**, 1211–15.

36. Fan, D. P. (1970) *J. Bact.* **103**, 488–93.
37. Fan, D. P. and Beckman, M. M. (1971) *J. Bact.* **105**, 629–36.
38. Fan, D. P., Pelvit, M. C. and Cunningham, W. P. (1972) *J. Bact.* **109**, 1266–72.
39. Fan, D. P., Beckman, B. E. and Beckman, M. M. (1974) *J. Bact.* **117**, 1330–4.
40. Fein, J. E. (1979) *J. Bact.* **137**, 933–46.
41. Fein, J. E. and Rogers, H. J. (1976) *J. Bact.* **127**, 1427–42.
42. Forsberg, C. W. and Rogers, H. J. (1971) *Nature, Lond.* **229**, 272–3.
43. Forsberg, C. W. and Ward, J. B. (1972) *J. Bact.* **110**, 878–88.
44. Forsberg, C. W. and Rogers, H. J. (1974) *J. Bact.* **118**, 358–68.
45. Forsberg, C. W., Ward, J. B., Wyrick, P. D. and Rogers, H. J. (1973) *J. Bact.* **113**, 969–84.
46. Ghuysen, J–M. (1968) *Bact. Rev.* **32**, 425–64.
47. Glaser, L. (1973) *A. Rev. Biochem.* **42**, 91–112.
48. Good, C. M. and Tipper, D. J. (1972) *J. Bact.* **111**, 231–41.
49. Gould, G. W. and Hitchins, A. D. (1963) *J. gen. Microbiol.* **33**, 413–23.
50. Gould, G. W. and Hitchins, A. D. (1965) In *Spores III*, eds. Campbell, L. L. and Halvorson, H. O., pp. 213–21, Washington, D.C: Am. Soc. Microbiol.
51. Gould, G. W. and King, W. L. (1969) In *Spores IV*, ed. Campbell, L. L., pp. 276–86, Washington, D.C: Am. Soc. Microbiol.
52. Gould, G. W., Hitchins, A. D. and King, W. L. (1966) *J. gen. Microbiol.* **44**, 293–302.
53. Grant, W. D. and Wicken, A. J. (1970) *Biochem. J.* **118**, 859–68.
54. Guinand, M., Michel, G. and Tipper, D. J. (1974) *J. Bact.* **120**, 173–84.
55. Guinand, M., Michel, G. and Balassa, G. (1976) *Biochem. biophys. Res. Commun.* **68**, 1287–93.
56. Guinand, M., Vacheron, M. J. and Michel, G. (1978) *Biochem. biophys. Res. Commun.* **80**, 429–34.
57. Guinand, M., Vacheron, M. J., Michel, G. and Tipper, D. J. (1979) *J. Bact.* **138**, 126–32.
58. Hakenbeck, R., Waks, S. and Tomasz, A. (1978) *Antimicrob. Agents Chemotherapy* **13**, 302–11.
59. Herbold, D. R. and Glaser, L. (1975a) *J. biol. Chem.* **250**, 1676–82.
60. Herbold, D. R. and Glaser, L. (1975b) *J. biol. Chem.* **250**, 7231–8.
61. Higgins, M. L. and Shockman, G. D. (1976) *J. Bact.* **127**, 1346–58.
62. Higgins, M. L., Pooley, H. M. and Shockman, G. D. (1970) *J. Bact.* **103**, 504–12.
63. Higgins, M. L., Coyette, J. and Shockman, G. D. (1973) *J. Bact.* **116**, 1375–82.
64. Holtje, J–V. and Tomasz, A. (1975) *Proc. Natn. Acad. Sci. U.S.A.* **72**, 1690–94.
65. Holtje, J–V., Mirelman, D., Sharon, N. and Schwarz, U. (1975) *J. Bact.* **124**, 1067–76.
66. Horne, D., Hakenbeck, R. and Tomasz, A. (1977) *J. Bact.* **132**, 704–17.
67. Howard, L. V. and Gooder, H. (1974) *J. Bact.* **117**, 796–804.
68. Huff, E. and Silverman, C. S. (1968) *J. Bact.* **95**, 99–106.
69. Hughes, R. C. (1970) *Biochem. J.* **119**, 849–60.
70. Hughes, R. C. (1971) *Biochem. J.* **121**, 791–802.
71. Hungerer, K. D., Fleck, J. and Tipper, D. J. (1969) *Biochemistry* **8**, 3567–73.
72. Jarretz, E., Gunnison, J. B., Speck, R. C. and Coleman, V. R. (1951) *Arch. intern. Med.* **57**, 349–59.
73. Kilburn, J. O. and Best, G. K. (1977) *J. Bact.* **129**, 750–55.
74. Kingau, S. L. and Ensign, J. C. (1968) *J. Bact.* **96**, 629–38.
75. Koyama, T., Yamada, M. and Matsuhashi, M. (1977) *J. Bact.* **129**, 1518–23.
76. Krulwich, T. A. and Ensign, J. C. (1968) *J. Bact.* **96**, 857–9.

77. Lacks, S. (1970) *J. Bact.* **101**, 373–83.
78. Lacks, S. and Neuberger, M. (1975) *J. Bact.* **124**, 1321-9.
79. Lominski, I., Cameron, J. and Wylie, G. (1958) *Nature, Lond.* **181**, 1477.
80. Lugtenberg, E. J. J., De Haas-Menger, J. and Ruyters, W. H. M. (1971) *J. Bact.* **109**, 326–35.
81. McQuillen, K. (1958) *J. gen. Microbiol.* **18**, 498–512.
82. Martin, H. H. and Kemper, S. (1970) *J. Bact.* **102**, 347–52.
83. Matsuzawa, H., Matsuhashi, M., Oka, A. and Sugino, Y. (1969) *Biochem. biophys. Res. Commun.* **36**, 682–9.
84. Mauck, J., Chan, L. and Glaser, L. (1971) *J. biol. Chem.* **246**, 1820–21.
85. Mosser, J. L. and Tomasz, A. (1970) *J. biol. Chem.* **245**, 287–98.
86. Munoz, E., Ghuysen, J–M., Leyh-Bouille, M., Petit, J–F. and Tinelli, R. (1966) *Biochemistry* **5**, 3091–8.
87. Okada, S. and Itahata, S. (1973) *J. ferment. Technol.* **51**, 705–12.
88. Ortiz, J. M., Gillespie, J. B. and Berkeley, R. C. W. (1972) *Biochim. biophys. Acta* **289**, 174–86.
89. Pelzer, von H. (1963) *Z. Naturf.* **18b**, 950–6.
90. Pooley, H. M. (1976a) *J. Bact.* **125**, 1127–38.
91. Pooley, H. M. (1976b) *J. Bact.* **125**, 1139–47.
92. Pooley, H. M. and Shockman, G. D. (1969) *J. Bact.* **100**, 617–24.
93. Pooley, H. M., Porres-Juan, J. M. and Shockman, G. D. (1970) *Biochem. biophys. Res. Commun.* **38**, 1134–40.
94. Pooley, H. M., Schlaeppi, J–M. and Karamata, D. (1978) *Nature, Lond.* **274**, 264–6.
95. Pooley, H. M., Shockman, G. D., Higgins, M. L. and Porres-Juan, J. (1972) *J. Bact.* **109**, 423–31.
96. Powell, J. F. and Strange, R. E. (1956) *Biochem. J.* **63**, 661–8.
97. Prestidge, L. S. and Pardee, B. (1957) *J. Bact.* **74**, 48–59.
98. Priest, F. G. (1977) *Bact. Rev.* **41**, 711–53.
99. Ranhand, J. M. (1973) *J. Bact.* **115**, 607–14.
100. Ranhand, J. M. (1974) *J. Bact.* **118**, 1041–50.
101. Ranhand, J. M., Leonard, C. G. and Cole, R. M. (1971) *J. Bact.* **106**, 257–68.
102. Rhuland, L. E. (1957) *J. Bact.* **73**, 778–93.
103. Richmond, M. H. (1959a) *Biochim. biophys. Acta* **33**, 78–91.
104. Richmond, M. H. (1959b) *Biochim. biophys. Acta* **33**, 92–101.
105. Robson, R. L. and Baddiley, J. (1977) *J. Bact.* **129**, 1051–8.
106. Rogers, H. J. (1964) In *Experimental Chemotherapy,* eds. Schnitzer, R. J. and Hawking, F., pp. 37–76, London and New York: Academic Press.
107. Rogers, H. J. (1967) *Nature, Lond.* **213**, 31–3.
108. Rogers, H. J. and Forsberg, C. W. (1971) *J. Bact.* **108**, 1235–43.
109. Rogers, H. J. and Thurman, P. F. (1978) *J. Bact.* **133**, 298–305.
110. Rogers, H. J. and Taylor, C. (1978) *J. Bact.* **135**, 1032–42.
111. Rogers, H. J., Pooley, H. M., Thurman, P. F. and Taylor, C. (1974) *A. Microbiol. (Inst. Past.)* **125B**, 135–47.
112. Rogers, H. J., Ward, J. B., Burdett, I. D. J. (1978) *Soc. gen. Microbiol. Symp.* **28**, 139–75.
113. Rosenthal, R. S. and Shockman, G. D. (1975) *J. Bact.* **124**, 419–23.
114. Rosenthal, R. S., Jungkind, D., Daneo-Moore, L., Shockman, G. D. (1975) *J. Bact.* **124**, 398–409.
115. Schindler, C. A. and Schuhardt, V. T. (1964) *Proc. Natn. Acad. Sci. U.S.A.* **51**, 414–21.
116. Seto, H. and Tomasz, A. (1975) *J. Bact.* **121**, 344–53.

117. Shockman, G. D., Pooley, H. M. and Thompson, J. S. (1967) *J. Bact.* **94**, 1525–30.
118. Shockman, G. D., Thompson, J. S. and Conover, M. J. (1967) *Biochemistry* **6**, 1054–65.
119. Shockman, G. D., Daneo-Moore, L. and Higgins, M. L. (1974) *Ann. N.Y. Acad. Sci.* **235**, 161–97.
120. Shungu, D. L., Cornett, J. B. and Shockman, G. D. (1979) *J. Bact.* **138**, 598–608.
121. Silcock, R. (1979) *Ph.D. Thesis, Univ. Liverpool.*
122. Silcock, R., Weston, A. and Perkins, H. R. (1978) *Proc. Soc. gen. Microbiol.* **5**, 63–4.
123. Singer, H. J., Wise, E. M. and Park, J. T. (1972) *J. Bact.* **112**, 932–9.
124. Strange, R. E. (1956) *Biochem. J.* **64**, 23P.
125. Strange, R. E. and Powell, J. F. (1954) *Biochem. J.* **58**, 80–5.
126. Strange, R. E. and Dark, F. A. (1957) *J. gen. Microbiol.* **17**, 525–37.
127. Strange, R. E. and Dark, F. A. (1957) *J. gen. Microbiol.* **16**, 236–49.
127a. Suginaka, H., Shimatini, M., Ogawa, M. and Kotani, S. (1979) *J. Antibiot.* **32**, 73–7.
128. Takumi, K., Kawata, T. and Hisatsune, K. (1971) *Jap. J. Microbiol.* **15**, 131–41.
129. Takahara, Y., Machigaki, E. and Maruo, S. (1974) *Agr. biol. Chem.* **38**, 2357–65.
130. Taylor, A., Das, B. C. and van Heijenoort, J. (1975) *Eur. J. Biochem.* **53**, 47–54.
131. Tichy, P. and Landman, O. E. (1969) *J. Bact.* **97**, 42–51.
132. Tilby, M. J. (1978) *J. Bact.* **136**, 10–18.
133. Tinelli, R. (1968) *Bull Soc. Chim. Biol.* **51**, 283–97.
134. Tipper, D. J. (1969) *J. Bact.* **97**, 837–47.
135. Tipper, D. J. and Gauthier, J. J. (1972) In *Spores V,* eds. Halverson, H. O., Hanson, R. and Campbell, L. L., pp. 3–12, Washington, D.C: Am. Soc. Microbiol.
136. Toennies, G. and Gallant, D. L. (1949) *J. biol. Chem.* **177**, 831–9.
137. Toennies, G. and Shockman, G. D. (1958) *Proc. 4th Internat. Congress Biochem.* **13**, 365–94.
138. Toennies, G., Izard, L., Rogers, N. B. and Shockman, G. D. (1961) *J. Bact.* **82**, 1054–65.
139. Tomasz, A. (1968) *Proc. Natn. Acad. Sci. U.S.A.* **59**, 86–93.
140. Tomasz, A., Albino, A. and Zaneti, E. (1970) *Nature, Lond.* **227**, 138–40.
140a. Taylor, C., Rogers, H. J. and Ward, J. B. (1980) *Society Gen. Microbiol. Quarterly.* **7**, 73–4.
141. Vacheron, M. J., Guinand, M. and Michel, G. (1978) *FEMS Letts.* **3**, 71–5.
142. Van Heijenoort, J., Parquet, C., Flouret, B. and van Heijenoort, Y. (1975) *Eur. J. Biochem.* **58**, 611–19.
143. Wadström, T. (1970) *Biochem. J.* **120**, 745–52.
144. Wadström, T. (1971) *Acta. chem. scand.* **25**, 1807–12.
145. Wadström, T. and Hisatune, K. (1970a) *Biochem. J.* **120**, 725–34.
146. Wadström, T. and Hisatune, K. (1970b) *Biochem. J.* **120**, 735–44.
147. Wadström, T. and Vesterberg, O. (1971) *Acta path. microbiol. scand., Sect. B* **79**, 248–64.
148. Waks, S. and Tomasz, A. (1978) *Antimicrob. Ag. Chemotherapy* **13**, 293–301.
149. Ward, J. B. (1973) *Biochem. J.* **133**, 395–8.
150. Ward, J. B. and Perkins, H. R. (1968) *Biochem. J.* **106**, 69–76.

151. Warth, A. D. (1972) In *Spores V,* eds. Halverson, H. O., Hanson, R. and Campbell, L. L., pp. 28–34, Washington D.C.: Am. Soc. Microbiol.
152. Warth, A. D. and Strominger, J. L. (1971) *Biochemistry* **10**, 4349.
153. Weidel, W. and Pelzer, H. (1964) *Adv. Enzymol.* **26**, 193.
154. Williamson, R. and Ward, J. B. (1979) *J. gen. Microbiol.* **114**, 349–54.
155. Wolf-Watz, H. and Normark, S. (1976) *J. Bact.* **128**, 580–6.
156. Yoneda, Y. and Maruo, B. (1975) *J. Bact.* **124**, 48–54.
157. Young, F. E. (1966) *J. biol. Chem.* **241**, 3462–7.
158. Young, F. E. and Spizizen, J. (1963) *J. biol. Chem.* **238**, 3126–30.
159. Young, F. E. and Wilson, G. A. (1972) In *Spores V,* eds. Halverson, H. O., Hanson, R. and Campbell, L. L., pp. 77–106, Washington D.C: Am. Soc. Microbiol.
160. Young, F. E., Tipper, D. J. and Strominger, J. L. (1964) *J. biol. Chem.* **239**, 2600–2.

12
Cell walls of Mycobacteria

12.1 Wall composition

The walls of Mycobacteria, like those of most other prokaryotes, contain peptidoglycan but characteristically they also contain glycolipid. Much glycolipid is covalently bound to the structural wall components, but there is also an unbound lipid fraction that is extractable into organic solvents representing in *Mycobacterium bovis* BCG, for example, about 34–40% of the wall [6, 18]. The walls of Mycobacteria have formed the subject of a valuable review [23] and a more general account of mycobacterial physiology is given by Ratledge [30]. In its amino acid composition the peptidoglycan resembles that of Bacilli, but as mentioned in Chapter 6.1, the most noticeable difference is that the acylation of the amino group of the muramic acid residue is by a glycolyl rather than the more common acetyl group [3, 7] a substitution also found in Nocardia [17]. There is evidence that oxidation of acetyl to glycolyl occurs at the level of UDP-*N*-acetylmuramic acid, since it has been shown that extracts of *Nocardia asteroides* will oxidize the latter substrate to UDP-*N*-glycolylmuramic acid, whence the precursor nucleotide pentapeptide presumably is synthesized in the usual way. Another major difference from most other peptidoglycans is in the type of cross-linking. Some links are made directly from the sub-terminal D-alanine of one pentapeptide side-chain to the D-centre of meso-diaminopimelic acid in another chain, just as in Bacilli or in Gram-negative species, but other links involve a direct link between the meso-diaminopimelic acid residue of one chain and that of another, or even the presence of tripeptides of diaminopimelic acid [34]. The exact linkages between the diaminopimelic acid residues (ie whether the L-centre carboxyl group of one is linked to the D-centre amino group of the next and so on) have not been established, and attempts to demonstrate a specific LD-transpeptidase that would catalyse the formation of that type of linkage have not so far succeeded [28].

The peptidoglycan of Mycobacteria is attached to a glycolipid that contains mycolic acids, which are in turn esterified to arabinogalactan. The mycolic acids are long chain fatty acids, branched at the α- and hydroxylated at the β- position. Their structures from various sources are shown in Table 12.1, from which it will be seen that unsaturation, cyclopropane rings or additional methyl side chains are also common features. The general method of synthesis seems to be for the primary chain to be built up to the required length by condensation of the pre-existing

Table 12.1 Mycobacterial mycolic acids

Strain	*Principal mycolic acid*	*Structure*
M. tuberculosis var. *hominis*	β-Mycolic acid	$CH_3-(CH_2)_{17}-\underset{CH_3}{CH}-\overset{O}{\overset{\|}{C}}-(CH_2)_{17}-\underbrace{CH-CH}_{CH_2}-(CH_2)_{19}-\overset{OH}{CH}-\underset{C_{24}H_{49}}{CH}-COOH$
	Methoxylated mycolic acid	$CH_3-(CH_2)_{17}-\underset{CH_3}{CH}-\overset{OCH_3}{CH}-(CH_2)_{16}-\underbrace{CH-CH}_{CH_2}-(CH_2)_{17}-\overset{OH}{CH}-\underset{C_{24}H_{49}}{CH}-COOH$
M. kansasii	α-Kansa mycolic acid	$CH_3-(CH_2)_{17}-\underbrace{CH-CH}_{CH_2}-(CH_2)_{14}-\underbrace{CH-CH}_{CH_2}-(CH_2)_{17}-\overset{OH}{CH}-\underset{C_{22}H_{45}}{CH}-COOH$
M. smegmatis	α-Smegmamycolic acid	$CH_3-(CH_2)_{17}-CH=CH-(CH_2)_{13}-CH=CH-\underset{CH_3}{CH}-(CH_2)_{17}-\overset{OH}{CH}-\underset{C_{22}H_{45}}{CH}-COOH$
M. phlei	Dicarboxylic mycolic acid	$HOOC-(CH_2)_{14}-\underset{CH_3}{CH}-CH=CH-(CH_2)_{16}-\overset{OH}{CH}-\underset{C_{22}H_{45}}{CH}-COOH$

fatty acid residues to yield the appropriate meromycolic acid, to which is then added an additional chain to provide the branch on the α- carbon atom. Many of these structures have been reviewed by Goren [15].

As mentioned above, the mycolic acids are linked to the chief antigenic polysaccharide of Mycobacteria, the D-arabino-D-galactan. This is a branched polysaccharide with rather large repeating units that contain some α-1,5- and some α-1,2-D-arabinofuranoside residues, as well as β-1,5-D-galactopyrano- (or furano-) side units. There are side chains with non-reducing arabinofuranoside terminals and they are attached to the main chain both at C_3 of arabinose and probably at C_6 of galactose [26, 31]. A possible formulation of the arabinogalactan is shown in Fig. 12.1. This polysaccharide is esterified by residues of mycolic acid, each of which is linked to the 5-OH group of an arabinofuranose residue [8, 9]. A similar linkage to nocardic acids of *Nocardia* has also been found [21].

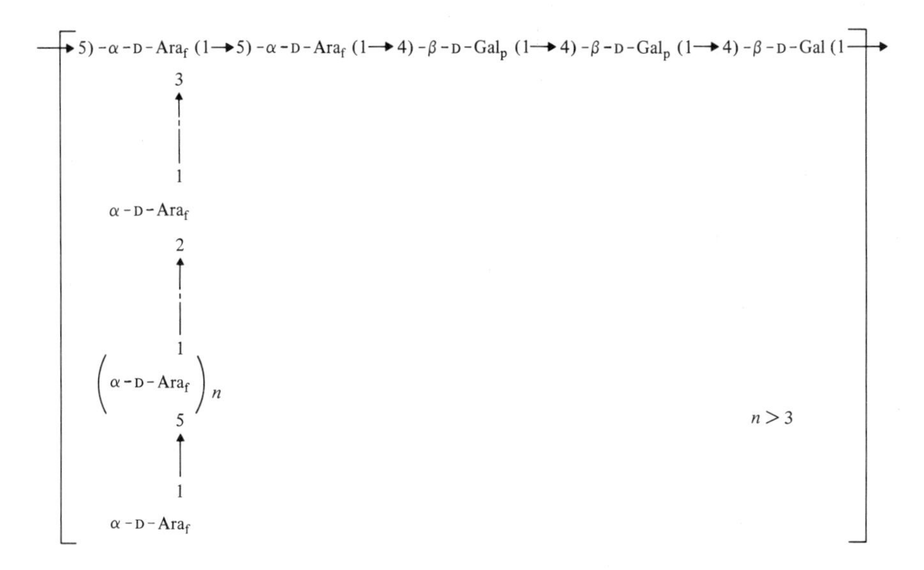

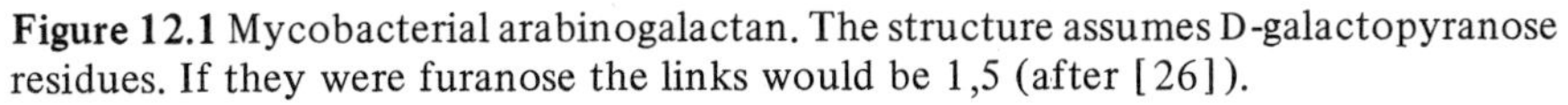

Figure 12.1 Mycobacterial arabinogalactan. The structure assumes D-galactopyranose residues. If they were furanose the links would be 1,5 (after [26]).

As with other secondary polymers of bacterial walls, linkage to peptidoglycan occurs and may be mediated by way of a phosphate residue at C_6 of muramic acid. Certainly, as in other bacteria, muramic-6-phosphate can be shown in Mycobacterial cell walls. This and other evidence led Lederer [22] to propose the structure shown in Fig. 12.2, but it must be emphasized that there is no direct evidence for a linkage between arabinose and muramic acid by the sole intervention of a phosphate bridge. In view of the evidence in some other systems that specialized linkage regions join accessory polymers to peptidoglycan, similar results may well transpire for Mycobacteria (see Chapter 7).

As mentioned above, apart from glycolipids bound to peptidoglycan there are

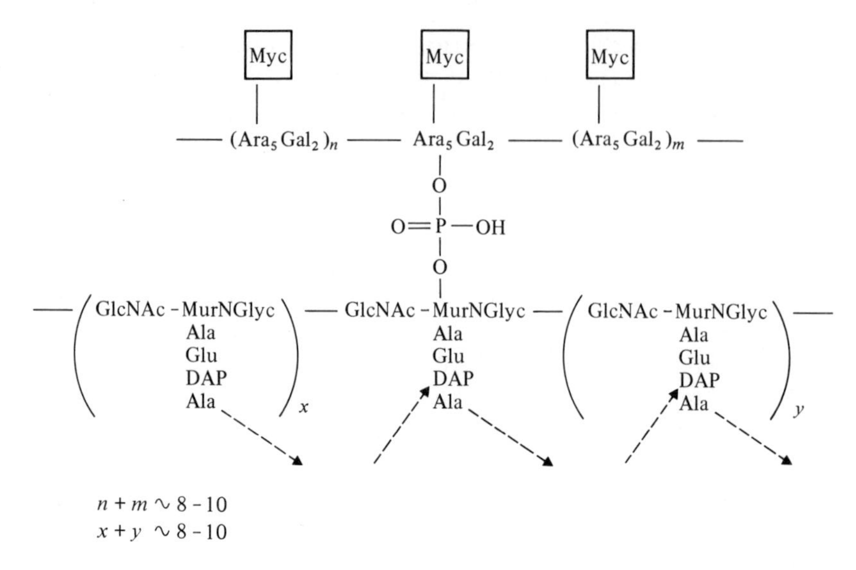

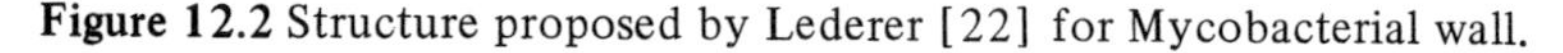
Figure 12.2 Structure proposed by Lederer [22] for Mycobacterial wall.

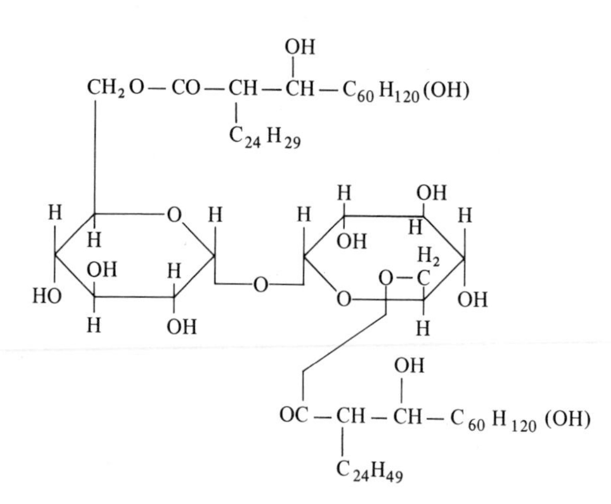

Figure 12.3 Trehalose-6,6′-dimycolate, 'cord factor'.

other important lipid substances associated with the wall of Mycobacteria. The so-called wax D is now recognized as being a degradation product of the mycobacterial wall that develops in old cultures undergoing partial autolysis [10]. The 'cord factor' originally associated with the ability of virulent strains to form intertwining cords when growing *in vitro*, has been identified as trehalose 6,6′-dimycolate [27] (Fig. 12.3), and similar compounds have been isolated from Corynebacteria and Nocardiae. There are also type-specific glycolipids known as mycosides. Mycosides consist of methylated sugar residues linked glycosidically to a phenol *para* substituted by a long chain branched alkyl group, itself substituted again via fatty acylated OH-groups [14]. In mycoside B the sugar is 2-*O*-methyl-rhamnose (ie 2-*O*-methyl-6-deoxy-6-methylmannose) and in mycoside A it consists of a trisaccharide made up of one residue each of 2-*O*-methylfucose, 2-*O*-methyl-rhamnose and 2,4-di-*O*-methylrhamnose (Fig. 12.4). Mycosides C, found in *M. avium, M. butyricum* and *M. marianum* are complex lipoglycopeptide substances, an example of which is shown in Fig. 12.5.

The third chief type of unbound lipid associated with the mycobacterial wall comprises the sulphatides, which consist of multiacylated trehalose sulphates. These occur with structural variations, one of which is shown in Fig. 12.6 [16]. The walls of *Mycobacteria* may also contain a glucan [47] and a poly-α-L-glutamic acid has been found in human and bovine strains, [33] that may reach a molecular weight of 30 000, and account for as much as 8% of the cell wall of *M. tuberculosis* Brevannes [35]. The amount of poly-L-glutamic acid found in the cell walls of various species of Mycobacteria and in various strains of *M. bovis* BCG varies from 0 to 8%, and its molecular weight from 3500 to 35 000, but neither of these parameters could be correlated with the biological properties of the strains [29].

12.2 Adjuvant and other immunostimulant properties

Freund's complete adjuvant, consisting of mycobacterial cells in a water-in-oil emulsion with a desired antigen in the aqueous phase, has long been used to enhance the production of antibodies. Omission of the mycobacteria gives greatly inferior results. It was shown in 1958 that wax D could effectively replace the whole mycobacteria in the adjuvant [32] and since that time the portion of the degraded wall (wax D) responsible for the stimulant action has been progressively refined. The first water soluble adjuvant represented a portion of arabinogalactan linked to peptidoglycan [2] and eventually it became clear that the key structure

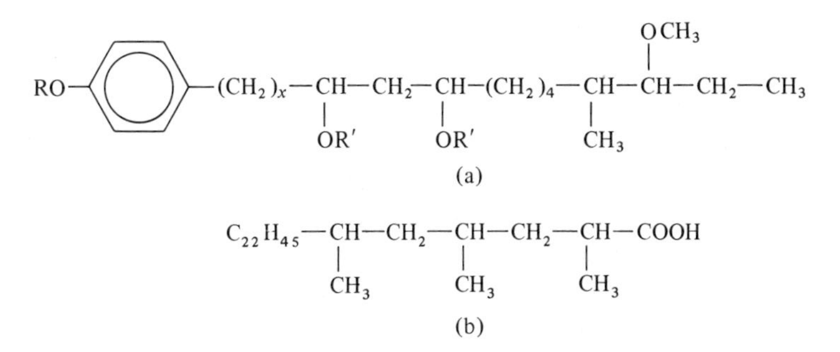

Figure 12.4 (a) Mycosides A and B (after [15]). In Mycoside A, R = trisaccharide 2-O-methylfucose; 2-O-methylrhamnose; 2,4-di-O-methylrhamnose; $x = 16, 17, 18, 19, 20$. In Mycoside B, R = 2-O-methylrhamnose; $x = 14, 15, 16, 17, 18$. For both R′ = acyl functions such as palmitic acid and mycocerosic acids (b).

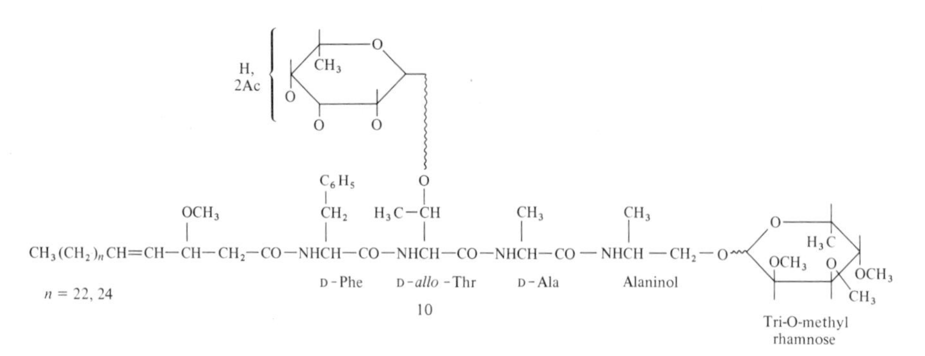

Figure 12.5 Mycoside C of *M. butyricum* (after [23]).

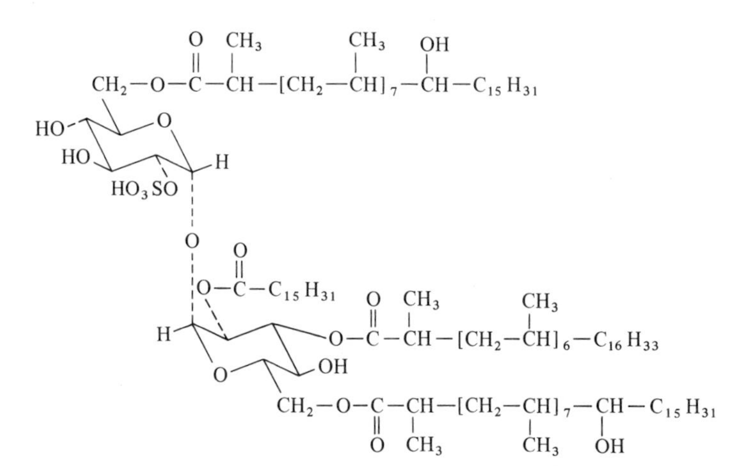

Figure 12.6 The principle sulphatide of *M. tuberculosis* (after [16]).

required was the peptidoglycan fragment *N*-acetylmuramyl-L-alanyl-D-isoglutamyl-meso-diaminopimelic acid [1, 13]. Replacement of diaminopimelic acid by lysine allowed full activity [1] and in fact even synthetic *N*-acetyl-muramyl-L-alanyl-D-isoglutamine exerted adjuvant stimulation when added to Freund's incomplete adjuvant in place of the whole mycobacteria [24]. However, omission of the *N*-acetylmuramyl residue, or of the peptide portion, gave inactive compounds. Evidence has also been obtained that inactivity results if the α-carboxyl group of the D-glutamyl residue is not free, or simply amidated, but substituted by another amino acid, amino acid amide or amino acid peptide as in Type B cross-linking (see Chapter 6.1) [19, 20].

Apart from adjuvant activity, *N*-acetylmuramyl-L-alanyl-D-isoglutamine and certain analogues enhanced the non-specific immunity of mice to challenge by *Klebsiella pneumoniae* [12]. A list of synthetic peptides active in protecting mice and also as adjuvants is shown in Table 12.2.

Table 12.2 Synthetic glycopeptides as adjuvants and anti-infectious agents (after [12]).

Glycopeptide	*Anti-infectious activity*	*Adjuvant activity*		
		In saline	*In water-oil emulsion*	
			Antibody response	*Delayed hypersensitivity*
MurNAc-L-Ala-D-isoGln	+	+	+	+
MurNAc-L-Ala-D-Glu	+	+	±	+
MurNAc-L-Ser-D-isoGln	—	+	+	+
MurNAc-L-Ala-D-isoGln-L-Lys	+	+	+	+
MurNAc-L-Ala-D-isoGln-L-Lys-D-Ala	—	+	+	+
MurNAc-L-Ala-D-Glu-diamide	—	+	+	+
MurNAc-L-Ala-D-isoGln-γ-Me-ester	+	+	±	±
MurNAc-L-Ala-D-Glu-di Me ester	+	+	+	+
MurNAc-L-Ala-D-Glu-α-methylamide	—	+	+	+

12.3 Antitumour activity

There is a considerable literature on the ability of certain intracellular bacterial parasites, including Mycobacteria, to stimulate their host's immunity to tumours. This effect could be extended to isolated cell walls; thus, for instance, those of *M. kansasii* were shown to protect mice against Ehrlich ascitic carcinoma or a syngeneic lymphoid leukemia more effectively than whole killed cells of *M. bovis* BCG [11]. Furthermore, these wall preparations lacked some of the undesirable side effects of whole mycobacterial cells eg they did not sensitize to endotoxins, produce hypertrophy of the liver or spleen, or sensitize to histamine. If the cell walls of *M. bovis* BCG are treated with proteolytic enzymes, acid, alkali, or organic solvents they lose some of their tumour-suppressive properties, but these can be restored by the addition of 'cord factor' (trehalose-6,6′-dimycolate) [25]. However, 'cord factor' is not an absolute requirement for the biological activities of living BCG, since it could not be found in BCG Montreal [5].

References

1. Adam, A., Ciorbaru, R., Ellouz, F., Petit, J–F. and Lederer, E. (1974) *Biochem. biophys. Res. Commun.* **56**, 561–7.
2. Adam, A., Ciorbaru, R., Petit, J–F. and Lederer, E. (1972) *Proc. Natn. Acad. Sci. U.S.A.* **69**, 851–4.
3. Adam, A., Petit, J–F., Wietzerbin-Falszpan, J., Sinaÿ, P., Thomas, D. W. and Lederer, E. (1969) *FEBS Letts.* **4**, 87–92.
4. Amar-Nacasch, C. and Vilkas, E. (1970) *Bull. Soc. Chim. Biol.* **52**, 145–51.
5. Asselineau, J. and Portelance, V. (1974) *Recent Res. Cancer Res.* **47**, 214–20.

6. Azuma, I., Ribi, E. E., Meyer, T. J. and Zbar, B. (1974) *J. Natn. Cancer Inst.* **52**, 95–101.
7. Azuma, I., Thomas, D. W., Adam, A., Ghuysen, J–M., Bonaly, R., Petit, J–F. and Lederer, E. (1970) *Biochim. biophys. Acta* **208**, 444–51.
8. Azuma, I. and Yamamura, Y. (1962) *J. Biochem.* **52**, 200–6.
9. Azuma, I., Ribi, E. E., Meyer, T. J. and Zbar, B. (1963) *J. Natn. Cancer Inst.* **53**, 275–81.
10. Brennan, P. J., Rooney, S. A. and Winder, F. G. (1970) *Ir. J. med. Sci.* **3**, 371–90.
11. Chedid, L., Lamensans, A., Parant, F., Parant, M., Adam, A., Petit, J–F. and Lederer, E. (1973) *Cancer Res.* **33**, 2187–95.
12. Chedid, L., Parant, M., Parant, F., Lefrancier, P., Choay, J. and Lederer, E. (1977) *Proc. Natn. Acad. Sci. U.S.A.* **74**, 2089–93.
13. Fleck, J., Mock, M., Tytgat, F., Nauciel, C. and Minck, R. (1974) *Nature, Lond.* **250**, 517–18.
14. Gastambide-Odier, M. (1973) *Eur. J. Biochem.* **33**, 81–6.
15. Goren, M. B. (1972) *Bact. Rev.* **36**, 33–64.
16. Goren, M. B., Brokl, O., Roller, P., Fales, H. M. and Das, B. C. (1976) *Biochemistry* **15**, 2728–35.
17. Guinand, M., Vacheron, M. J. and Michel, G. (1970) *FEBS Letts.* **6**, 37–9.
18. Kotani, S., Kitaura, T., Hirano, T. and Tanaka, A. (1959) *Biken J.* **2**, 129–41.
19. Kotani, S., Narita, T., Stewart-Tull, D. E. S., Watanabe, Y., Kato, K. and Iwata, S. (1975) *Biken J.* **18**, 77–92.
20. Kotani, S., Watanabe, Y., Kinoshita, F., Kato, K., Schleifer, K. H. and Perkins, H. R. (1977) *Biken J.* **20**, 1–4.
21. Lanéelle, M. A. and Asselineau, J. (1970) *FEBS Letts.* **7**, 64–7.
22. Lederer, E. (1971) *Pure Appl. Chem.* **25**, 135–65.
23. Lederer, E., Adam, A., Ciorbaru, R., Petit, J–F. and Wietzerbin, J. (1975) *Molec. Cell Biochem.* **7**, 87–104.
24. Merser, C., Sinaÿ, P. and Adam, A. (1975) *Biochem. biophys. Res. Commun.* **66**, 1316–22.
25. Meyer, T. J., Ribi, E. E., Azuma, I. and Zbar, B. (1974) *J. Natn. Cancer Inst.* **52**, 103–11.
26. Misaki, A., Seto, N. and Azuma, I. (1974) *J. Biochem.* **76**, 15–27.
27. Noll, H., Bloch, H., Asselineau, J. and Lederer, E. (1956) *Biochim. biophys. Acta* **20**, 299–309.
28. Petit, J–F. and Lederer, E. (1978) *Symp. Soc. Gen. Microbiol.* **28**, 177–99.
29. Phiet, P. H., Wietzerbin, J., Zissman, E., Petit, J–F. and Lederer, E. (1976) *Infect. Immun.* **13**, 677–81.
30. Ratledge, C. (1976) *Adv. microbial Physiol.* **13**, 115–244.
31. Vilkas, E., Amar, C., Markovits, J., Vliegenhart, J. F. G. and Kamerling, J. P. (1973) *Biochim. biophys. Acta* **297**, 423–35.
32. White, R. G., Bernstock, L., Johns, R. G. S. and Lederer, E. (1958) *Immunology* **1**, 54–66.
33. Wietzerbin-Falszpan, J., Das, B. C., Gros, C., Petit, J–F. and Lederer, E. (1973) *Eur. J. Biochem.* **32**, 525–32.
34. Wietzerbin, J., Das, B. C., Petit, J–F., Lederer, E., Leyh-Bouille, M. and Ghuysen, J–M. (1974) *Biochemistry* **13**, 3471–6.
35. Wietzerbin, J., Lederer, F. and Petit, J–F. (1975) *Biochem. biophys. Res. Commun.* **62**, 246–52.

13
Cell walls of filamentous fungi

13.1 Introduction

Fungal walls, like those of plants and bacteria, consist of a rigid layer outside the protoplast, which they protect from osmotic and other changes in the environment. In addition they are responsible for the characteristic shape of the cell and have to be modified when the cell changes, as for instance during the growth of a hyphal tip, the initiation of a branching hypha, the change to a conidiospore, or from mycelial to yeast-like growth or *vice versa.* The wall is composed largely of polysaccharides, with some protein and lipid, although the latter represents only a small proportion. It has been known for a long time that one of the chief polysaccharides is chitin, the homopolymer of β-1,4-*N*-acetylglucosamine that also occurs in the integument of arthropods, and another β-1, 4-glucose, ie cellulose. Where present, both chitin and cellulose have been identified by X-ray powder crystallography and shown to be the same as authentic samples from other sources [3, 37]. These two polymers form the fibrils that make the rigid component of most fungal walls, in contrast to the yeasts where other glucose polymers, along with a small amount of chitin, take over this function [4, 7]. Before looking in detail at the other polysaccharides present, we will examine the evidence for the minor components, protein and lipid, as integral parts of the wall. Walls that have been isolated and washed as thoroughly as possible [43] still contain about 5-10% of lipid, but so far this has not been well characterized [40]. It is noteworthy, however, that mycelial walls of *Penicillium charlesii* contained as much as 37.5% lipid [15] and the wall of a strain of *Aspergillus nidulans* has been reported to have half its 10% of galactose in the form of a glycolipid [47]. The residual protein, on the other hand, can sometimes be extracted by detergents as in *Aspergillus nidulans* [14], whereas in other instances such as *Aspergillus niger* and *Chaetomium globosum* it resists even protracted extraction with 8 M urea [33]. Analysis of the latter firmly-attached protein and its peptic digests suggested that many acidic peptide sequences were present and showed high serine and threonine contents, as often observed in the protein of yeast walls. The importance of protein in the wall was emphasized by Hunsley and Burnett [24], who used a succession of glycanases and the proteolytic enzyme pronase to 'dissect' fungal walls. They concluded that in the wall of *Neurospora crassa*, for instance, there was a glycoprotein reticulum embedded in easily-removable protein, beneath which lay a separate protein layer,

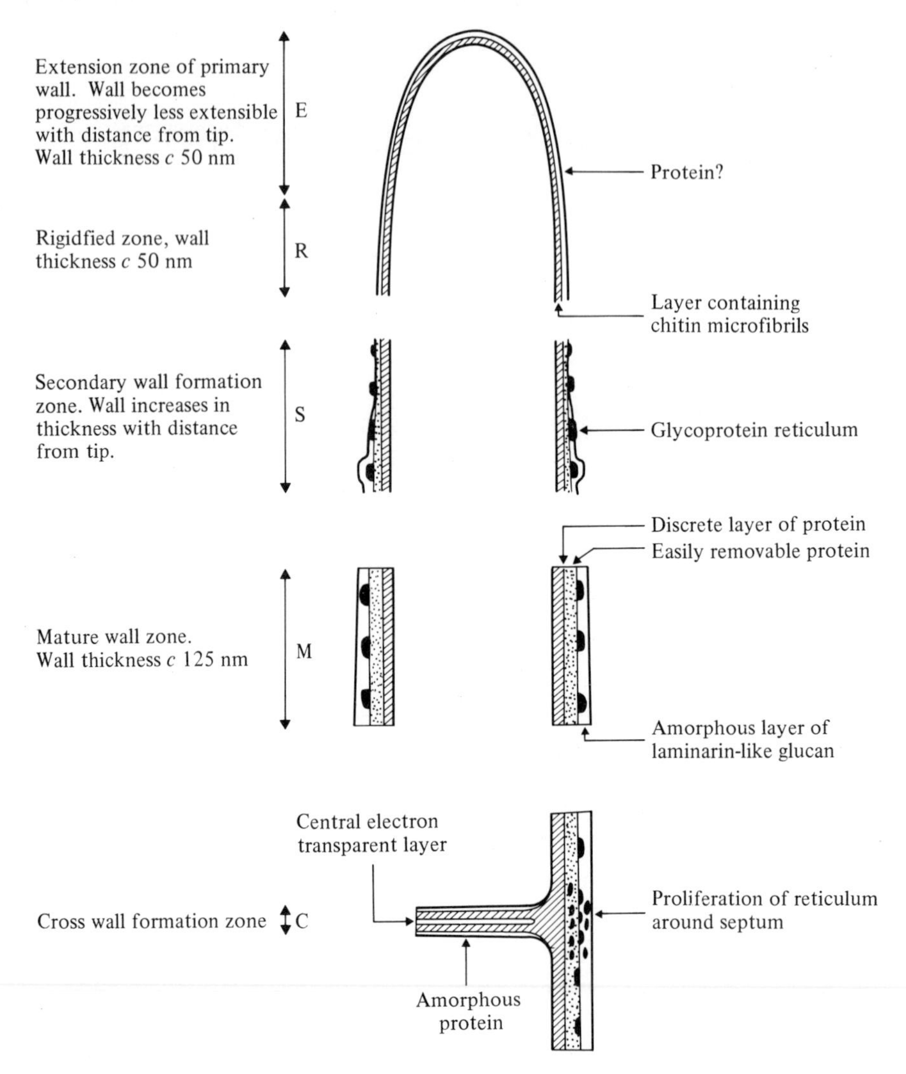

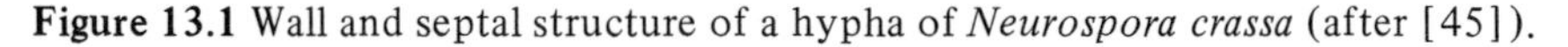

Figure 13.1 Wall and septal structure of a hypha of *Neurospora crassa* (after [45]).

which was in turn outside and possibly intermingled with the innermost layer of chitin microfibrils. It is now clear that, in some although not all fungi [46], as hyphae extend two separate kinds of wall are laid down, a primary wall consisting of chitin microfibrils covered with protein which is laid down as tip extension proceeds, and a layer of secondary wall that is later deposited on the outside. All layers may thicken as maturation occurs, except that the inner chitin layer of *N. crassa* appears not to change [45] (Fig. 13.1). Another process that intervenes, but is at present little understood, is rigidification, whereby the primary wall that is first deposited in a relatively plastic state becomes more rigid as it thickens and adopts the shape and diameter of the final hyphal tube. It has been suggested that rigidification may correspond either to a diminution in the activity of autolysins or to an increasing resistance to these enzymes as the wall matures [45], though no mechanism for either process is known at present.

Experiments showing that, in the walls of non-growing mycelium, enzymic removal of the constituent polymers one by one [24] does not destroy the shape of the wall are not necessarily valid for growing wall. Thus a mutant of *Aspergillus nidulans* blocked in chitin synthesis produced normally-shaped hyphae when grown

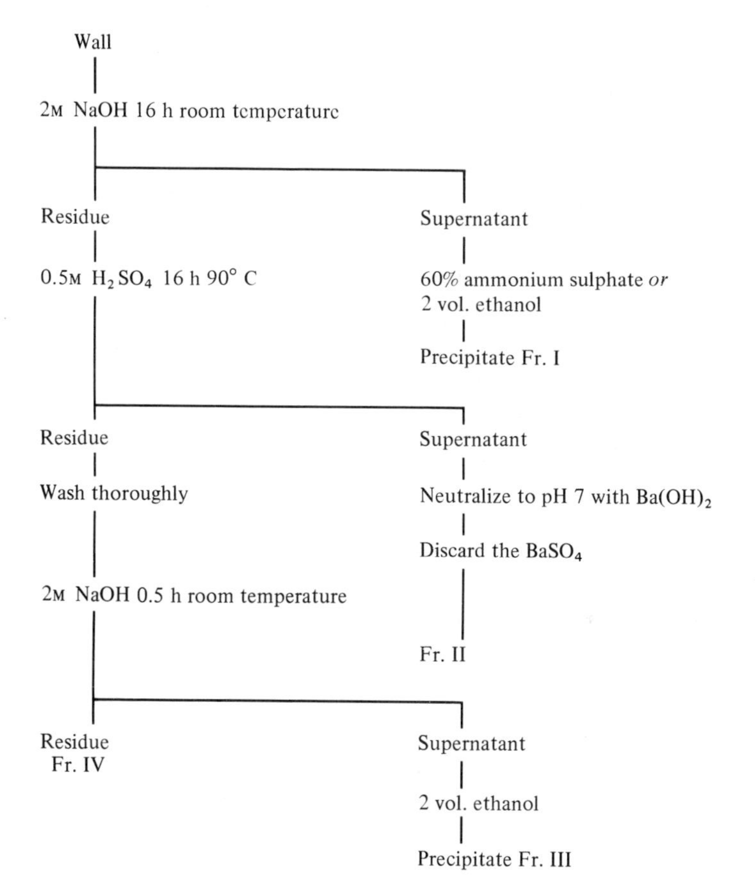

Figure 13.2 Extraction of fungal walls according to [32]. Some examples are as follows:
(a) In *Neurospora crassa* [32] Fr I contained α-1,3-glucan, protein and polygalactosamine; Fr II, galactose, glucose, mannose, glucosamine, protein; Fr III, glucan, predominantly β-1,3; Fr IV chitin.
(b) In *Helminthosporium spiciferum* [12] Fr I, 19%, contained hexoses, mainly glucose with a little galactose and mannose, some protein and phosphate and a trace of glucosamine; Fr II, 40%, hexoses, glucose and galactose in a 10:1 ratio, some protein and more glucosamine; Fr III, traces only; Fr IV, 38%, glucosamine and protein.

in osmotically-stabilized medium. These hyphae were not in themselves osmotically sensitive, but transfer to growth medium without stabilizer led to swelling and lysis [26]. Thus at least in the early stages of wall extension the presence of all normal components appears to be necessary to provide stability [40].

13.2 Carbohydrates in the wall

As indicated already, fungal walls contain a number of different polymers and various extraction procedures have been employed to give a primary fractionation scheme. Ethylenediamine has been used as a gentle procedure [27] and so has snail digestive juice [13], but selective extraction by cold alkali, hot acid and cold alkali again has been most widely used (Fig. 13.2) [32].

Bartnicki-Garcia [7] pointed out the close correlation between the wall carbohydrates and fungal taxonomy. He proposed a general classification based on dual combinations of the chief wall polysaccharides (Table 13.1). This system shows

Table 13.1 Cell wall taxonomy of fungi (from [7]).

Chemical category	*Taxonomic group**	*Distinctive features*
I. Cellulose-glycogen**	Acrasiales	pseudoplasmodia
II. Cellulose-glucan††	Oomycetes	biflagellate zoospores
III. Cellulose-chitin	Hyphochytridiomycetes	anteriorly uniflagellate zoospores
IV. Chitosan-chitin	Zygomycetes	zygospores
V. Chitin-glucan††	Chytridiomycetes	posteriorly uniflagellate zoospores
	Ascomycetes†	septate hyphae, ascospores
	Basidiomycetes‡	septate hyphae, basidiospores
	Deuteromycetes§	septate hyphae
VI. Mannan-glucan††	Saccharomycetacceae	yeast cells, ascospores
	Cryptococcaceae	yeast cells
VII. Mannan-chitin	Sporobolomycetaceae	yeasts (carotenoid pigment) ballistospores
	Rhodotorulaceae	yeasts (carotenoid pigment)
VIII. Polygalactosamine-galactan	Trichomycetes	heterogeneous group, arthropod parasites

* Not all orders or families within each group have been examined for wall composition. Further segregation is possible.
† Except Saccharomycetaceae.
‡ Except Sporobolomycetaceae.
§ Except Cryptococcaceae (and Rhodotorulaceae).
** Incompletely characterized.
†† Incompletely characterized; probably β1,3- and β1,6-linked.

that the great majority of fungi, including all those forms with typical septate mycelium, fall in category V with walls composed of chitin and glucan. Many, but not all, of the lower fungi have cellulose as a characteristic component. On the other hand, where the yeast growth habit supervenes there is a considerable increase in the amount of mannan in the wall. Although the Vaucheria algae and the Saprolegnia fungi are related phylogenetically, there are notable wall differences such as the amount and degree of crystallinity of the cellulose [37]. Cellulose represented only about 15% of the weight of these fungal walls, the other 85% consisting of glucans and mannans. The principle wall component in a number of members of the Oomycetes (*Phytophthora, Pythium, Achlya, Saprolegnia, Atkinsella, Brevilegnia, Dictyuchus*) is a glucan insoluble in both alkali and cuprammonium salts and containing β-1,3 and β-1,6 linkages [2, 6, 8, 17]. As remarked earlier, the removal of a single component, in this case cellulose by cuprammonium, did not affect the appearance of *Phytophthora walls*, but the same arguments as before may have to be applied in assessing the need for particular polymers to establish the morphology in the first place.

The Oomycetes were considered to lack chitin [7] but evidence has accumulated that this *N*-acetylglucosamine polymer occurs in some species. Thus in *Apodachlya* chitin occurs along with cellulose [30, 31] and it has also been shown that polyoxin D, a specific inhibitor of chitin synthetase [22] causes an inhibition of *N*-acetylglucosamine uptake into the walls of *Achlya radiosa* [18]. In the latter species the effect of polyoxin D on morphology was relatively slight, being confined to a 25% increase in hyphal diameter, compared with its effect in fungi having chitin as a major wall polymer, where it leads to tip-bursting [22].

Another feature of the Oomycetes is the presence in the wall proteins of hydroxyproline, an amino acid characteristic of the cellulosic walls of fungi, algae or higher plants [7]. It may have a role in providing an anchorage for a glycosidic link between the polysaccharides and the proteins [29].

Apart from the chitin that occurs so widely in fungal walls, there is good evidence for the presence of its non-acetylated analogue, chitosan, in the walls of the Mucorales. Thus in *Mucor rouxii* chitosan occurs in all growth forms, and is accompanied by glucuronic acid polymers not generally found in other fungi and also by phosphate that may serve to neutralize the free amino groups of the chitosan (Table 13.2) [7]. The glucuronic acid is present in two forms of polymer, one a homopolymer resistant to acid hydrolysis and containing at least 90% glucuronic acid (mucoric acid) and the other an easily hydrolysed heteropolymer containing D-glucuronic acid, L-fucose, D-mannose and D-galactose (mucoran) [10]. Chitosan occurs in *Phycomyces* too [28] and has been reported in the spore walls of *Agaricus bisporus* and *A. campestris,* where the chitin/chitosan ratio is 0.38:1 and 2.8:1 respectively [21]. In other Mucorales acidic polysaccharides have been isolated from the walls and culture filtrates [34, 35]. An alkaline extract of the wall of *Absidia cylindrospora* contained a polysaccharide composed of L-fucose, D-galactose and D-glucuronic acid [34] reminiscent of the mucoran mentioned above.

A possible role for the matrix acidic polymers of the wall has been suggested by

Table 13.2 Chemical differentiation of the cell wall in the life cycle of *Mucor rouxii* (from [7])*.

Wall Component	*Yeasts*	*Hyphae*	*Sporangiophores*	*Spores*
Chitin	8.4	9.4	18.0	2.1†
Chitosan	27.9	32.7	20.6	9.5†
Mannose	8.9	1.6	0.9	4.8
Fucose	3.2	3.8	2.1	0.0
Galactose	1.1	1.6	0.8	0.0
Glucuronic Acid	12.2	11.8	25.0	1.9
Glucose	0.0	0.0	0.1	42.6
Protein	10.3	6.3	9.2	16.1
Lipid	5.7	7.8	4.8	9.8
Phosphate	22.1	23.3	0.8	2.6
Melanin	0.0	0.0	0.0	10.3

* The values are given as the percentage of dry weight of the cell wall.
† Not confirmed by X-ray. Value of spore chitin represents *N*-acetylated glucosamine; chitosan is non-acetylated glucosamine.

Dow and Rubery [20], who found that the bursting of hyphal tips of *Mucor rouxii* at pH 3.4 to 5.5 or in the presence of chelating agents [9] was moderated by the opposing action of Ca^{2+} and H^{+} ions. It seemed possible that the physical state of the acidic polymers found in the hyphal tips may exercise some control over the plasticity of the extension region, as required in growth.

Another hexosamine, galactosamine, has been identified with certainty in only a few species. *Aspergillus* species had the highest mycelial galactosaminoglycan content among many species examined [19] and a non-acetylated aminogalactan accounted for the polyphosphate binding properties of *Neurospora crassa* [23]. The walls of a strain of *Aspergillus nidulans* contained 10.8% of their dry weight as galactosamine compared with about 15% as *N*-acetylglucosamine, much of which was attributable to chitin [14]. Since the galactosamine could be extracted by alkali, to which the chitin was resistant, it was clearly present as a separate polymer. Similar proportions of galactosamine were found in the walls of *Helminthosporium sativum* [1]. However, other strains of *A. nidulans* contained a much lower proportion of galactosamine in their walls (about 1-2%) [47], comparable to the amount reported in *N. crassa* [32].

The non-cellulosic glucans form a large proportion of many fungal walls. The alkali soluble or S-glucan appears in Fr I of the extraction procedure (Fig. 13.2) and is an α-1,3-linked straight chain polysaccharide that forms a major part of the walls of Ascomycetes and Basidiomycetes [40]. Although usually a biologically stable polymer [48], in *A. nidulans* it appears to be degraded during fruiting-body formation, when it acts as a reserve polysaccharide [50, 51]. The alkali-resistant R-glucan, which appears in Fr III (Fig. 13.2) is a major component of a wide range of fungal walls, although it does not appear in Zygomycetes or *Agaricus* [7, 40]. Although the R-glucan contains many β-1,3-linkages it is not a simple straight chain polymer, but also has β-1,6 branch points. Rosenberger [40] has suggested that such a branched polymer might be an ideal candidate to provide bridgeheads for cross-linking in fungal walls, but as yet there is no evidence for this suggestion, beyond the existence of unattributed phenomena like the 'rigidification' of growing hyphal tips.

A relatively new and rapid approach to characterizing the surface carbohydrates of fungi has been provided by the use of lectins [5, 16, 44]. By means of fluorescein-labelled soybean agglutinin, specific for D-galactose and *N*-acetyl-D-galactosamine, peanut agglutinin (D-galactose), wheat germ agglutinin (*N*-acetyl-D-glucosamine) and concanavalin A (α-D-mannose, α-D-glucose) it was possible to test for the presence of particular glycans in a variety of species [5]. In addition, the lectins were shown to inhibit various stages of fungal development. An even wider range of lectin specificities would give this technique enormous potential.

13.3 Wall composition and dimorphism

Differences between the walls of the various forms of *Mucor rouxii* have already been mentioned (Table 13.2). Many other dimorphic fungi are known to have altered wall compositions and structures in their various phases, and indeed such changes have been invoked to provide a mechanism for morphological differentiation [36]. Thus, for instance, the human pathogen *Paracoccidioides brasiliensis*, which exists in a yeast-form at 37° C and a mycelial form at room temperature, has different wall glucans in the two forms [25]. There is some evidence that the presence of the characteristic yeast-form α-1,3-glycan is related to pathogenicity, and that during continuous sub-culture of the yeast form *in vitro* the amount of this polysaccharide becomes very small, its place being taken by a galactomannan [42]. In another human pathogen, *Sporothrix schenckii*, it was found that the mycelial and conidial walls were qualitatively and quantitatively similar, whereas the yeast-form had more water and alkali-soluble polysaccharides and less protein and lipids in its walls. Among the lipids of the latter was a large proportion of a $C_{18:3}$ acid that was absent from mycelial or conidial walls [39]. Further examination of the various glucans present showed that, in all forms, the soluble glucans had β-1,3-,1,6- and 1,4-linkages in the proportions 44:28:28 whereas in the insoluble glucans the proportions were 66: 29: 5 respectively [38]. In spite of the mixture of linkages these were linear polymers.

The linear β-D-1,3-glucans of fungi have a protective effect against animal tumours [49] and the insoluble glucan of *S. schenckii* was active in the same way [38]. These results recall the anti-tumour effects of the walls from the Mycobacteria (Chapter 12.3) but if there is any underlying similarity of structure involved it is certainly not evident.

13.4 Melanins and depsipeptides

Other components of fungal walls not mentioned above are the melanins. These pigments, characteristic of many fungi particularly during ageing, are thought to inhibit autogenous or exogenous lytic enzymes, as for instance in *Helminthosporium spiciferum* [12].

The spores of certain *fungi imperfecti* are covered with crystalline spicules of depsipeptide [11]. These cyclic compounds of both amino acids and hydroxy acids can be removed from the surface by the action of benzene, or will dissolve in chloroform, leaving the residual spore coat behind. The hyphal walls of one of the fungi in question, *Pithomyces chartarum*, contain some lipid (8.5%), most amino acids found in protein, glucosamine, large quantities of glucose and less galactose and mannose [41].

13.5 Conclusion

It is clear that the structure of fungal walls is a field of most active research, in which a confusing array of components is somehow assembled and modified during the processes of growth and differentiation. Some of the biosynthetic evidence is discussed in Chapter 14, but evidently much further endeavour will be required before a comprehensive overall picture can emerge.

References

1. Applegarth, D. A. and Bozoian, G. (1969) *Archs. Biochem. Biophys.* **134**, 285–9.
2. Aronson, J. M., Cooper, B. A. and Fuller, M. S. (1967) *Science* **155**, 332–5.
3. Aronson, J. M. and Preston, R. D. (1960) *Proc. R. Soc. B.* **152**, 346–52.
4. Ballou, C. E. (1976) *Adv. microbial Physiol.* **14**, 93–158.
5. Barkai-Golan, R., Mirelman, D. and Sharon, N. (1978) *Arch. Microbiol.* **116**, 119–24.
6. Bartnicki-Garcia, S. (1966) *J. gen. Microbiol.* **42**, 57–69.
7. Bartnicki-Garcia, S. (1968) *A. Rev. Microbiol.* **22**, 87–105.
8. Bartnicki-Garcia, S. and Lippman, E. (1967) *Biochim. biophys. Acta.* **136**, 533–43.
9. Bartnicki-Garcia, S. and Lippman, E. (1972) *J. gen. Microbiol.* **73**, 487–500.
10. Bartnicki-Garcia, S. and Reyes, E. (1968) *Biochim. biophys. Acta.* **170**, 54–62.
11. Bertaud, W. S., Morice, I. M., Russell, D. W. and Taylor, A. (1963) *J. gen. Microbiol.* **32**, 385–95.
12. Berthe, M. C., Bonaly, R. and Reisinger, O. (1976) *Can. J. Microbiol.* **22**, 929–36.
13. Bonaly, R. (1972) *Carbohydrate Res.* **24**, 355–64.
14. Bull, A. T. (1970) *J. gen. Microbiol.* **63**, 75–94.
15. Bulman, R. A. and Chittenden, G. J. F. (1976) *Biochim. biophys. Acta,* **444**, 202–11.
16. Cassone, A., Mattia, E. and Boldrini, L. (1978) *J. gen. Microbiol.* **105**, 263–73.
17. Cooper, B. A. and Aronson, J. M. (1967) *Mycologia.* **59**, 658–70.
18. Dietrich, S. M. C. and Campos, G. M. A. (1978) *J. gen. Microbiol.* **105**, 161–4.
19. Distler, J. J. and Roseman, S. (1960) *J. biol. Chem.* **235**, 2538–41.

20. Dow, J. M. and Rubery, P. H. (1975) *J. gen. Microbiol.* **91**, 425–8.
21. Garcia Mendoza, C., Leal, J. A. and Novaes-Ledieu, M. (1979) *Can. J. Microbiol.* **25**, 32–9.
22. Gooday, G. W. (1978) In *The Filamentous Fungi, Vol. 3,* eds. Smith, J. E. and Berry, D. R., pp. 51–77, London: Edward Arnold.
23. Harold, F. M. (1962) *Biochim. biophys. Acta,* **57**, 59–66.
24. Hunsley, D. and Burnett, J. H. (1970) *J. gen. Microbiol.* **62**, 203–18.
25. Kanetsuna, F., Carbonell, L. M., Azuma, I. and Yamamura, Y. (1972) *J. Bact.* **110**, 208–18.
26. Katz, D. and Rosenberger, R. F. (1971) *Arch. Mikrobiol.* **30**, 284–92.
27. Korn, F. C. and Northcote, D. H. (1960) *Biochem. J.* **75**, 12–17.
28. Kreger, D. R. (1954) *Biochim. biophys. Acta.* **13**, 1–9.
29. Lamport, D. T. A. (1970) *A. Rev. Pl. Physiol.* **21**, 235–70.
30. Lin, C. C. Y. and Aronson, J. M. (1970) *Arch. Mikrobiol.* **72**, 111–14.
31. Lin, C. C., Sicher, jr., R. C. and Aronson, J. M. (1976) *Arch. Microbiol.* **108**, 85–91.
32. Mahadevan, P. R. and Tatum, E. L. (1965) *J. Bact.* **90**, 1073–81.
33. Mitchell, A. and Taylor, I. F. (1969) *J. gen. Microbiol.* **59**, 103–9.
34. Miyazaki, T. and Irino, T. (1970) *Chem. Pharm. Bull.* **18**, 1930–1.
35. Miyazaki, T. and Irino, T. (1972) *Chem. Pharm. Bull.* **20**, 330–5.
36. Nickerson, W. J. and Bartnicki-Garcia, S. (1964) *A. Rev. Pl. Physiol.* **15**, 327–44.
37. Parker, B. C., Preston, R. D. and Fogg, G. E. (1963) *Proc. R. Soc. B.,* **158**, 435–45.
38. Previato, J. O., Gorin, P. A. J., Haskins, R. H. and Travassos, L. R. (1979) *Exp. Mycol.* **3**, 92–105.
39. Previato, J. O., Gorin, P. A. J. and Travassos, L. R. (1979) *Exp. Mycol.* **3**, 83–91.
40. Rosenberger, R. F. (1976) In *The Filamentous Fungi, Vol. 2,* eds. Smith, J. E. and Berry, D. R., pp. 328–44, London: Edward Arnold.
41. Russell, D. W., Sturgeon, R. J. and Ward, V. (1964) *J. gen. Microbiol.* **36**, 289–96.
42. San-Blas, F., San-Blas, G. and Cova, L. J. (1976) *J. gen. Microbiol.* **93**, 209–18.
43. Taylor, I. F. P. and Cameron, D. S. (1973) *A. Rev. Microbiol.* **27**, 243–60.
44. Travassos, L. R., Souza, W., Mendonca-Previato, L. and Lloyd, K. O. (1977) *Exp. Mycol.* **1**, 293–305.
45. Trinci, A. P. J. (1978) *Sci. Prog. Oxf.* **65**, 75–99.
46. Trinci, A. P. J. and Collinge, A. J. (1975) *J. gen. Microbiol.* **91**, 355–61.
47. Valentine, B. P. and Bainbridge, B. W. (1978) *J. gen. Microbiol.* **109**, 155–68.
48. Wessels, J. G. H., Kreger, D. R., Marchant, R., Regensburg, B. A. and De Vries, O. M. H. (1972) *Biochim. biophys. Acta.* **273**, 346–58.
49. Whistler, R. L., Bushway, A. A., Singh, P. P., Nakahara, W. and Tokuzen, R. (1976) *Adv. Carbohyd. Chem.* **32**, 235–75.
50. Zonneveld, B. J. M. (1972) *Biochim. biophys. Acta.* **273**, 174–87.
51. Zonneveld, B. J. M. (1974) *J. gen. Microbiol.* **81**, 445–51.

14
Biosynthesis of wall components in yeast and filamentous fungi

14.1 Introduction

In much the same way as the information available describing the biosynthesis of bacterial wall polymers has increased dramatically during the last decade, so a corresponding increase has occurred for the wall polymers of yeasts and other fungi. In view of its ready availability perhaps it is not surprising that much of this information concerns the yeast *Saccharomyces cerevisiae*. Less immediately obvious are the reasons underlying the relatively little information available concerning the biosynthesis of the glucans, undoubtedly the major structural polymers of the yeast wall. Thus, the detailed evidence available is concentrated on the biosynthesis of chitin, found in *S. cerevisiae* specifically as a component of the bud scar but as a major wall polymer in many other fungi, and on mannan, the immunodeterminant of the yeast wall. In the following sections, the biosynthesis of each of these polymers and of the glucans is considered, taking together evidence from both yeast and other fungi in an attempt to show both the similarities and differences in the various systems. Further details of the biosynthesis of fungal wall polymers are to be found in the reviews of Ballou [6, 7], Cabib [22] and Farkas [37].

14.2 Biosynthesis of chitin

The initial observations on the biosynthesis of chitin were made in 1957 [40] using a particulate enzyme preparation isolated from the disrupted mycelium of *Neurospora crassa.* Since that time particulate preparations of chitin synthase (UDP-2-acetamido-2-deoxy-D-glucose : chitin 4-β-acetamidodeoxyglucosyl transferase) have been obtained from a wide range of fungi including yeasts [2, 17, 28, 40, 42–4, 49, 57, 72, 82, 84]. The enzyme has been solubilized from several of these preparations by treatment with butanol [40] or digitonin [31, 33, 42]. These conditions remove lipid and destroy the integrity of the membrane. In contrast apparently soluble enzyme was obtained by incubating particulate enzyme from *Mucor rouxii* with substrate (UDP-*N*-acetylglucosamine) and activator (*N*-acetyl-glucosamine) [101]. However, it was subsequently shown by electron microscopy that this 'soluble' enzyme was in fact still particulate and these chitin synthase-containing particles of approximately 35 to 100 nm diameter, were named chitosomes [102, 104]. Similar structures have now been isolated from several different fungi including yeast [10]. Whether the chitosomes are artifacts or have a definite role in chitin synthesis *in vivo* remains unclear; the dimensions of the particles preclude their being components of the cytoplasmic membrane. Chitin synthase obtained by digitonin treatment from *Coprinus cinereus* was an aggregate with a molecular weight of several million. In the presence of a high salt concentration this dissociated into active units with molecular weights of approximately 150 000 [42].

All the particulate enzymes appear to have similar properties [42] and since those from *M. rouxii, C. cinereus* and yeast, particularly *Saccharomyces cerevisiae* and *S. carlsbergensis* have been investigated in greatest detail the following discussion will concentrate on biosynthesis in these organisms.

The reaction catalysed by all preparations is the transfer of *N*-acetylglucosamine from the nucleotide precursor UDP-*N*-acetylglucosamine to an endogenous receptor with the release of UDP as follows:

$$\text{UDP-Glc}N\text{Ac} + (\text{Glc}N\text{Ac})_n \xrightarrow{Mg^{2+}} (\text{Glc}N\text{Ac})_{n+1} + \text{UDP}.$$

Divalent cations particularly Mg^{2+} are normally required for enzyme activity. Clearly during extension of the chain the acceptor will be chitin itself; whether an endogenous acceptor is required for the initiation of a chitin chain remains unknown. The recent demonstrations of the formation of microfibrillar chitin by solubilized synthase from *S. cerevisiae* and chitosomes from *M. rouxii* suggests that the presence of an acceptor is not essential for the effective synthesis of polymeric material. On the other hand, butanol-solubilized enzyme from *N. crassa* showed a requirement for primer in the form of *N*-acetylchitodextrins [40]. Similarly a particulate enzyme preparation from *M. rouxii,* although not showing an absolute requirement for acetylchitodextrins, was stimulated by their presence [82]. In this case, the stimulation observed with di-*N*-acetylchitobiose and tri-*N*-acetylchitotriose the dimer and trimer of the series, was not as great as that obtained with higher oligomers. This effect, which could be clearly distinguished from the allosteric activation obtained with *N*-acetylglucosamine [42], led to the conclusion that the *N*-acetylchitodextrins were in fact acting as primers and that the enzyme preparations were in some way deficient in acceptor activity. In contrast, the synthase from other sources, including *S. carlsbergensis* [57] and *Mortierella vinacea* [96] was not stimulated by *N*-acetylchitodextrins. Whether these various observations simply reflect some unrecognized difference in the amount of endogenous acceptor, presumably pre-existing chitin which remains associated with the enzyme preparation, remains unclear.

As mentioned briefly above, all the particulate chitin synthases examined from fungal sources appear to be allosteric enzymes and are activated by *N*-acetylglucosamine. The nature of this activation has not yet been clearly defined. *N*-acetylglucosamine is not incorporated to any great extent into the newly-synthesized chitin. However, it has been proposed that the hexosamine has an additional role in the initiation of new chains of chitin during biosynthesis by a particulate preparation from *Blastocladiella emersonii* [28]. This does not appear to be the case in *M. rouxii* [82]. Initiation of new chains by *N*-acetyl-^{14}C-glucosamine would result in the formation of radioactive diacetylchitobiose upon the addition of a second *N*-acetylglucosamine residue from UDP-*N*-acetylglucosamine. However, radioactive diacetylchitobiose was not detected among the products of chitinase digestion of the newly-synthesized chitin. Moreover, addition of non-radioactive diacetylchitobiose to the biosynthetic system appeared to enchance, rather than as would be expected from simple isotopic dilution, to reduce the incorporation of *N*-acetyl-^{14}C-glucosamine into polymeric material. On the other hand, should the newly initiated chain remain in close association with the enzyme, or be blocked by some other means, then perhaps the exchange with the unlabelled diacetyl-chitobiose would be prevented. There appears to be no direct evidence to support the alternative suggestion that incorporation of *N*-acetyl-^{14}C-glucosamine results as a consequence of the synthesis of UDP-*N*-acetyl-^{14}C-glucosamine by the particulate enzyme preparations used.

Finally, it is worth emphasizing that no evidence has been obtained for the participation of lipid intermediates in chitin synthesis.

14.2.1 *Activation of chitin synthase zymogen*

The occurrence of an inactive form of the enzyme, chitin synthase zymogen, was first described in *S. cerevisiae* [24]. Mild sonic oscillation followed by centrifugation and washing served to fractionate particulate enzyme preparations isolated from osmotically lysed spheroplasts into particulate and soluble components. Separately, each of these two fractions was inactive but synthase activity was recovered on incubation of the two fractions together. Moreover, the activating component present in the solubilized fraction could by replaced by trypsin. Thus it appeared that the chitin synthase activity was normally present in an inactive form, termed the chitin synthase zymogen, which could be activated by limited proteolysis mediated either by the activating factor present in the particulate enzyme preparation or by trypsin. This conclusion has been further substantiated by the

Table 14.1 The effect of sonication and trypsin on chitin synthase (from [24]).*

		N-*acetylglucosamine incoporated (nmol)*		
Organism	*Treatment*	*No addition*	*+ Inhibitor*	*+ Trypsin*
S. cerevisiae	None	1.1	0.65	10.5
	Sonication	0.76	0.76	12.7
S. carlsbergensis	None	6.8	3.1	14.9
	Sonication	0.91	0.44	11.7

*Chitin synthase was prepared as the particulate fraction from organisms in the exponential phase of growth. The inhibitor protein, the inhibitor of activating factor present in association with the particulate enzyme preparation was isolated and purified from *S. carlsbergensis* [25].

Table 14.2 The separation of chitin synthase and activating factor in *S. cerevisiae* (from [27]).*

	% Recovery			
	Chitin synthase (after trypsin activation)		*Activating factor*	
Experiment	*1*	*2*	*1*	*2*
Fraction				
Particulate fraction	100	100	100	100
Vacuole fraction	1.9	0.53	39.1	40.4
Pellet	50.8	–	not detectable	–
Membrane band 1	25.6		not detectable	–

*Particulate material prepared from osmotically lysed protoplasts of *S. cerevisiae* was fractionated by centrifugation in discontinuous gradients of Ficoll, ranging in concentration from 6% to 10%. The material at the top of the gradient (vacuole fraction) and the pellet were recovered and the pellet was further fractionated on Ficoll gradients to yield among others membrane band 1 which was located at the interphase between 10% and 20% Ficoll.

recent finding that chitin synthase solubilized from *S. cerevisiae* membranes by treatment with digitonin is in the zymogen form [33]. A comparison of the total chitin synthase activity present (ie after trypsin treatment) and the activity present before proteolysis suggested that only some 10% was orginally in the active form (Table 14.1). Essentially similar results have been obtained by Hasilik [46] who found that 5-10% of the enzyme was in the active form in log phase *S. cerevisiae* but that in the stationary phase of growth all the chitin synthase was present as the zymogen.

The relative ease with which the activating factor could be separated from the particulate fraction led to the conclusion that it was located in some easily disrupted organelle such as vesicles, rather than being incorporated within the membrane itself. These have now been isolated from *S. cerevisiae* as a fraction enriched in activating factor but without chitin synthase activity [27] (Table 14.2). Yeast protoplasts were subject to metabolic lysis, thus allowing extremely gentle disruption, and the lysate was then fractionated by centrifugation in a discontinuous gradient of Ficoll. In this way the vesicles were separated from a second membrane-rich fraction which contained the chitin synthase almost entirely as the zymogen.

Earlier Cabib and Keller [25] had isolated from the cytoplasm of *S. carlsbergensis* a heat-stable protein which appeared to inhibit chitin synthase. Subsequent work, however, showed it to be without effect on enzyme which had previously been activated, while it was extremely efficient at preventing the activation of synthase zymogen by the activating factor but not by trypsin. Thus the heat-stable protein inhibited the activating factor rather than the chitin synthase itself. This protein with a molecular weight of 8500, has now been purified to homogeneity from *S. cerevisiae* [121]. It had earlier been shown that the inhibitor protein isolated from either *S. carlsbergensis* or *S. cerevisiae* effectively bound to and inhibited activating factor from either source [24]. In addition, it was also found to inhibit a neutral protease isolated from *S. cerevisiae*, an observation which led to the conclusion that this enzyme and the activating factor were one and the same [26]. Their identification as proteinase B was based on the similarities in proteolytic activity and sensitivity to inhibitors reported for this enzyme and those obtained with the activating factor. Moreover, proteinase B had previously been isolated as an inactive complex with a heat-stable protein inhibitor [71]. These preliminary conclusions on the identity of the two enzyme activities were subsequently confirmed and extended by Hasilik and Holzer [47] to show that the activating factor for chitin synthase zymogen, proteinase B and a protease isolated and subsequently partially purified as tryptophan synthase inactivase II [107] were in fact one and the same enzyme. Other yeast proteinases do not appear to be capable of activating chitin synthase zymogen. However, the recent finding [126] that mutants of *S. cerevisiae* deficient in proteinase B are able to synthesize chitin suggests that this enzyme is not essential for activation of the zymogen *in vivo.* The results do not rule out the possibility that in the absence of proteinase B an alternative activation mechanism may be used. At this point it is perhaps worth noting that a protease-inhibitor system similar to that described for yeast has been found in *N. crassa* [115, 125].

Activators do not appear to have been isolated from filamentous fungi although there is abundant evidence that the chitin synthases of *M. rouxii* [101, 104] *Aspergillus nidulans* [105, 106], *A. fumigatus* [2] and *Phycomyces blakesleeanus* [122] exist in the form of zymogens which can be activated by proteolysis. In contrast the synthase from *C. cinereus* did not require proteolytic activation and appeared to be always present in an active form [42]. Inhibitors of chitin synthase have been reported to occur in *M. vincea* [96] and *M. rouxii* [74, 83]. The inhibitor from *M. rouxii* is a protein of 17 500 molecular weight which directly inhibits the activity of chitin synthase in contrast to the situation in *S. cerevisiae* described above. Thus, unlike proteinase B, it appears specific for chitin synthesis.

Taken together these observations suggest that in yeast and filamentous fungi no common mechanism for the activation and control of chitin synthase exists. However, in each organism some form of compartmentalization must occur, which in the case of *S. cerevisiae* will localize the inhibitor, the activating protease and the chitin synthase zymogen and thus serve to control the specific site of chitin synthesis.

14.2.2 *Localization of chitin synthesis*

Although much of the work described in the previous sections has been carried out using particulate preparations as a source of chitin synthase activity, some evidence has been obtained as to cellular localization. In the earlier studies the enzyme was said to be located primarily in the microsomal fraction [28, 40, 43, 97] whereas in the *Mucorales* [82, 84, 86, 96], the highest activities were found associated with cell wall material. However, in all these latter organisms, a small amount of chitin

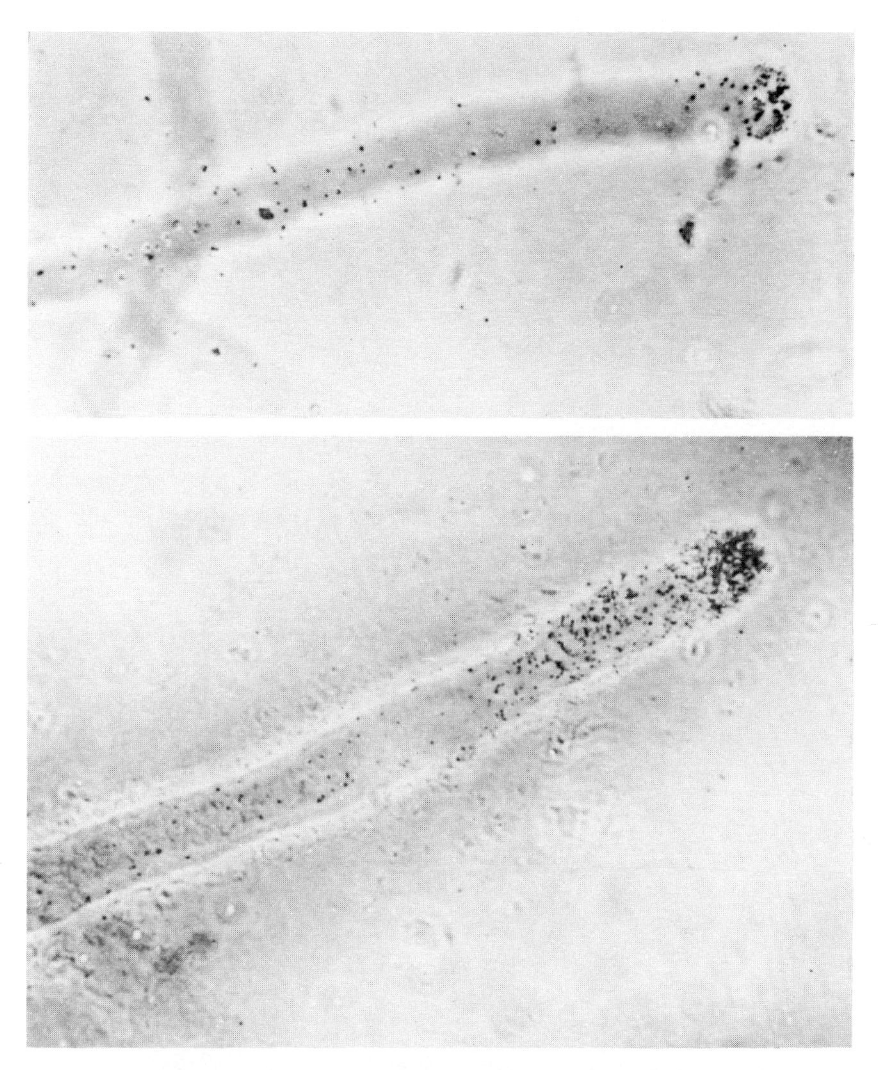

Figure 14.1 Autoradiographs of the apical region of hyphal walls of *Neurospora crassa*. Cultures were incubated with either ^{3}H-glucosamine or ^{3}H-glucose for 1 min. The walls were then prepared for autoradiography as described by Gooday [41]. Photograph kindly provided by Dr G. W. Gooday.

synthase was found in a microsomal fraction, an observation which led to the suggestion that the 'microsomes' were the site of chitin synthase formation before the enzyme migrated to its alternative site of action in the wall of the organism [5, 6]. In this context it was also proposed that the cytoplasmic inhibitor, described in the previous section, would prevent premature activation of the enzyme, before it reached its site of action. However, it now appears that these observations were artifactual and more recent studies with *S. cerevisiae* [27, 32,

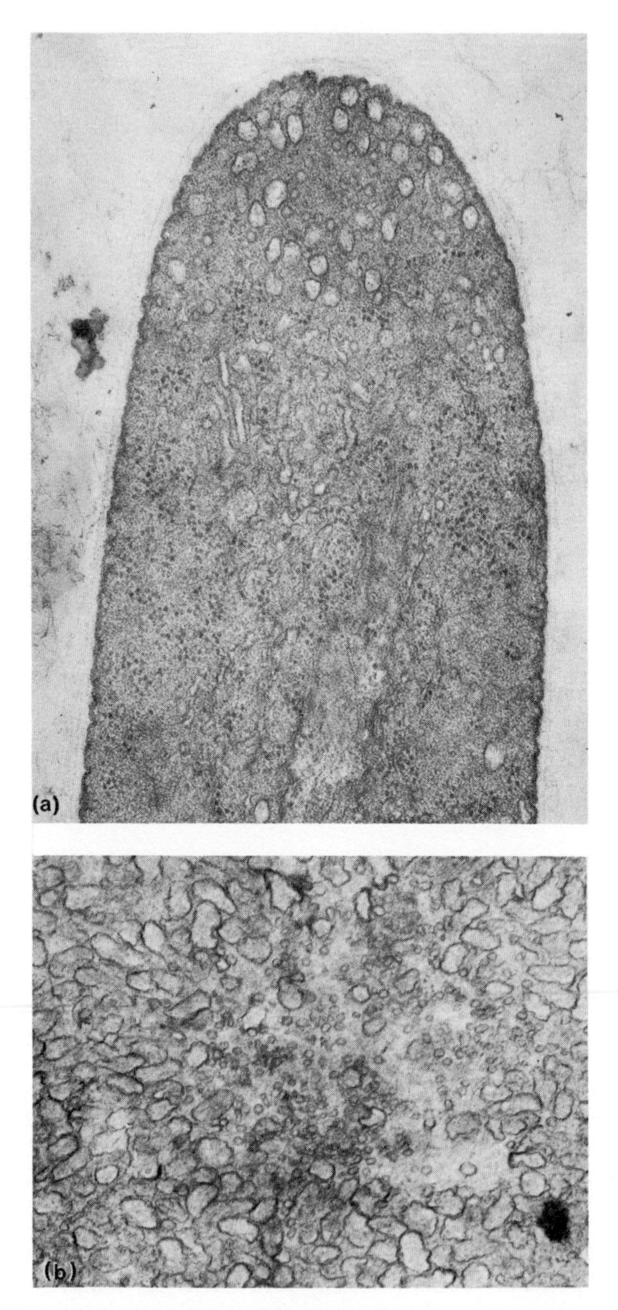

Figure 14.2 (a) Longitudinal section of a hyphal tip of *Penicillium chrysogenum* showing the concentration of vesicles at the tip. (b) Transverse section of a hyphal tip of *Neurospora crassa* showing the central region in which micro-vesicles are surrounded by apical vesicles. Photograph kindly provided by Dr A. J. P. Trinci and taken with permission from [44a].

34], *Candida albicans* [17], *P. blakesleeanus* [49] and *M. rouxii* [101, 104] have established that the enzyme is located in the cytoplasmic membrane. Earlier Cabib and Farkas [24] had suggested that in *S. cerevisiae* chitin synthase zymogen was uniformly distributed throughout the membrane. Evidence to support this has come recently with the finding that isolated yeast membranes are capable of synthesizing chitin at many sites. Cabib and Farkas had based their hypothesis on the results of experiments showing that the total amount of chitin synthase activity appeared to be unaffected by either the stage of growth or the growth medium. To explain the observed localization of chitin, which in *S. cerevisiae* is found only in the septum separating mother cell and bud [23], they argued that only chitin synthase molecules formed in the septal area would be active. Thus, in agreement with the results obtained, only a small fraction of the enzyme would be active at any one time. Moreover, the finding that the activating factor, although soluble, was contained within vesicles, provided a mechanism whereby chitin synthesis would be initiated at selected sites. Clearly the presence of zymogen throughout the cytoplasmic membrane would allow synthesis to occur over the whole surface. It was suggested, however, that activation occurred only when and where vacuoles coalesced with the cytoplasmic membrane, thus allowing the interaction of protease and zymogen. Such a mechanism may also be applicable to the filamentous fungi. Microscopic observation [100], radioautography [41, 84] and studies using fluorescent antibodies directed specifically against wall components [77] have all shown that hyphae grow by synthesizing new wall at their apex. As the hyphae continue to extend, secondary subapical sites of chitin deposition occur which ultimately give rise to lateral branches in which wall synthesis is again confined to the apical region (Fig. 14.1). Alternatively, these subapical sites function in septum formation. Evidence has also been obtained for wall thickening occurring in subapical regions and a much lower level of wall synthesis along the length of the hyphae as a whole. In addition, as shown in Fig. 14.2, electron microscopy has revealed vesicles to be concentrated at the apices of the growing hyphae together with fusion of the vesicles with the cytoplasmic membrane in these regions [45, 79]. Thus, there exists in both yeasts and filamentous fungi the possibility that activation of the chitin synthase zymogen occurs only when vesicles loaded with the activating protease fuse with the cytoplasmic membrane allowing proteolysis of the zymogen to occur. Evidence for the uniform distribution of the zymogen throughout the cytoplasmic membrane of *Aspergillus nidulans* has come from the work of Katz and Rosenberger [56]. In this organism it was found that treatment of the hyphae either by osmotic shock or by inhibition of protein synthesis resulted in a generalized incorporation of *N*-acetylglucosamine into chitin rather than the localized apical growth observed under normal growth conditions (Fig. 14.3). These findings can be interpreted to show non-specific activation of chitin synthase zymogen resulting either from non-specific proteolysis or perhaps from some disturbance in the directional flow of vesicles to the hyphal tips.

This latter point emphasizes one of the areas of uncertainty in the whole case for the localized activation of chitin synthesis. How does the organism direct the

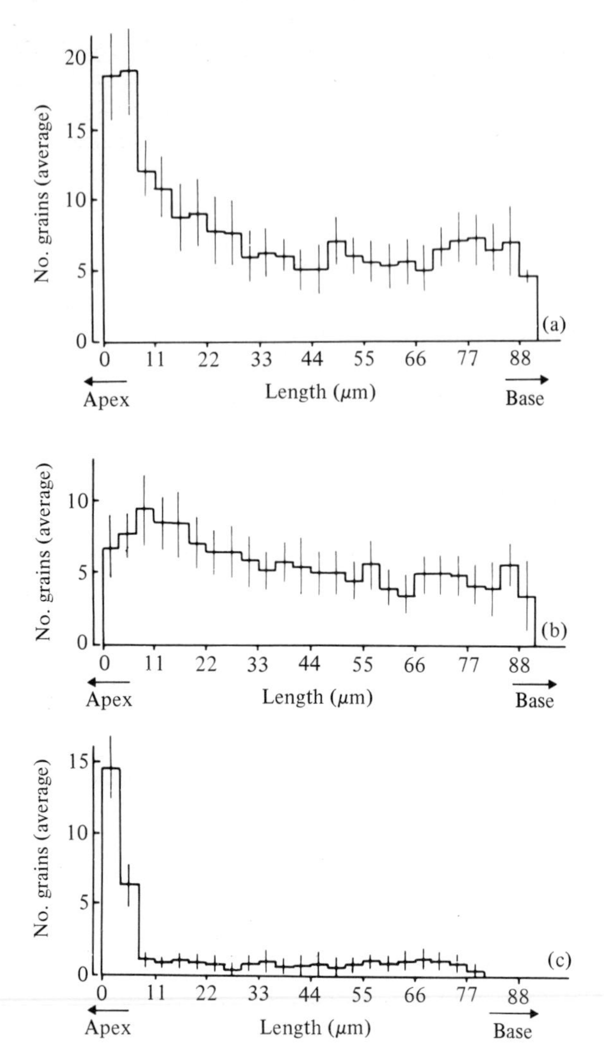

Figure 14.3 The effect of cycloheximide on chitin synthesis in *Aspergillus nidulans*. Hyphae were labelled with *N*-acetyl-^{3}H-glucosamine and the distribution of radioactivity along the hyphae determined. (a) Labelled during the first 10 min of cycloheximide treatment. (b) Labelled during 10–20 min after addition of cycloheximide. (c) Hyphae washed free from cycloheximide and labelled from 30–40 min after removal of the inhibitor. Figure kindly supplied by Dr R. F. Rosenberger and reproduced with permission from [56].

vesicles to the appropriate site? For although the fusion of vesicles with the cytoplasmic membrane provides a mechanism whereby chitin synthase can be specifically activated, the means by which the filamentous fungi and yeast respectively direct their vesicles to the apices of growing hyphae and to the site of bud formation, remain unknown. Among possible agents providing such a directional flow would be cytoplasmic streaming or gradients within the cytoplasm either of some metabolite or alternatively of osmotic pressure, particularly if the wall at the hyphal apex or at the site of bud formation were weakened. However, in the absence of any definite evidence these suggestions must be regarded as purely speculative.

The second major area of uncertainty is the mechanism by which chitin synthase activity is terminated. Evidence has come from the studies of Hasilik [46] that proteinase B is not only specific among the yeast proteinases in activiting chitin synthase zymogen as described earlier, but also prolonged incubation of the two enzymes results in inactivation of the synthase. How the cell would control this inactivation process remains unclear. Presumably the proteinase and synthase would have to remain in reasonably close association in the cytoplasmic membrane otherwise it is difficult to envisage how inhibition of the proteinase by the cytoplasmic inhibitor protein could be avoided.

14.2.3 *Inhibition of chitin synthase by the polyoxins*

The polyoxins are a series of at least twelve closely related nucleotide antibiotics

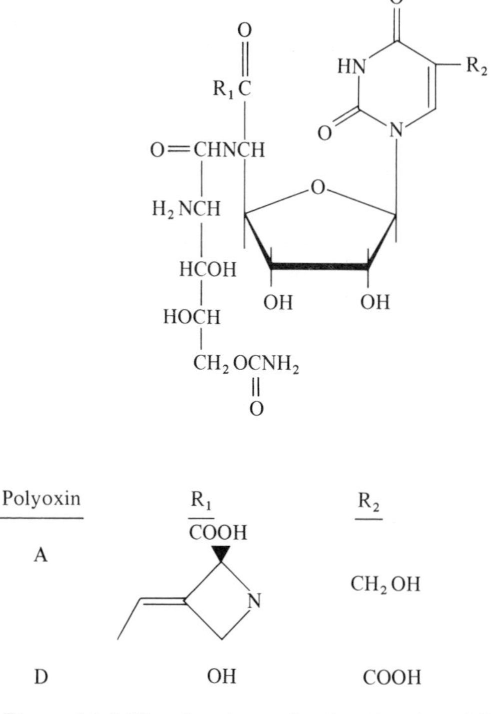

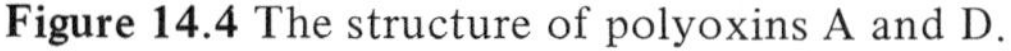

Figure 14.4 The structure of polyoxins A and D.

produced by *Streptomyces cacaoi* var. *asoensis* [48]. The structures of the individual members are known, those of polyoxins A and D being given in Fig. 14.4. Early investigations established their antifungal activity whereas they were apparently without effect on the growth of bacteria and yeast. Polyoxin D inhibited the incorporation of ^{14}C-glucosamine into wall chitin of *N. crassa* and also induced the accumulation of UDP-*N*-acetylglucosamine [36]. Using a particulate preparation as a source of chitin synthase, polyoxin D was shown to be a competitive inhibitor of synthesis with a K_i for polyoxin D of 1.4×10^{-6} M in contrast to the K_m of 1.43×10^{-3} M of UDP-*N*-acetylglucosamine. Similar results have now been described for a number of fungi [42] in addition to *M. rouxii* [11] and *Phycomyces blakesleeanus* [49].

On the other hand, although the polyoxins apparently act by the inhibition of chitin synthesis, differences have been observed in the morphological changes induced by addition of the antibiotic to the growth medium. Thus in *N. crassa* [36] spore germination proceeded normally, but the germ tubes were subject to considerable distortion although lysis did not ensue. In *M. rouxii* germination appeared inhibited [11]. Polyoxin D also induced the lysis of growing hyphal tips in *M. rouxii* and the formation of osmotically fragile protoplast-like structures from the germ-tubes of *Cochliobolus miyabeanus* [36] and *P. blakesleeanus* [49]. The formation of protoplast-like structures under appropriate conditions of osmotic stability and the lysis of growing hyphae all presumably represent some imbalance in wall synthesis relative to cellular growth and thus are analogous to the similar effects observed as a result of the inhibition of bacterial peptidoglycan synthesis by a number of antibiotics.

The apparent anomaly that the polyoxins were without inhibitory effect on yeast has recently been explained. Peptides present in growth media were shown earlier to be antagonists of polyoxin activity; thus the use of a synthetic medium has allowed a re-investigation of the effects of polyoxin D on the growth and division of *S. cerevisiae* [16]. Earlier experiments [57] with *S. carlsbergensis* had shown that the chitin synthase of this organism showed a similar sensitivity to polyoxin A as did the synthases of the filamentous fungi described above. Thus the inclusion of polyoxin D at relatively high concentrations in the synthetic growth medium resulted in the inhibition of primary septum synthesis and as a consequence the formation of abnormal secondary septa. As described earlier the primary septum has been shown to consist largely of chitin [23]. Additional biochemical studies demonstrated that the inhibition of chitin synthesis was specific, the incorporation of ^{3}H-glucose into the other wall polysaccharides, the glucans and mannan being relatively unaffected.

14.3 The biosynthesis of mannan

Although mannans appear to be present in the walls of a range of filamentous fungi and yeasts, investigations into both the structure and the biosynthesis of these polymers have been carried out principally in yeast. Almost certainly the initial impetus was provided by the ready availability of *S. cerevisiae* from commercial sources for the isolation of mannan. More recently additional emphasis has been provided by the recognition that the mannan is the immunodominant component of the yeast wall [6, 7].

Yeast mannan is a glycopeptide composed of a mannose polysaccharide linked to a peptide. Among the numerous strains of yeast that have been investigated, variable amounts of both *N*-acetylglucosamine and phosphate have been reported as additional components. The detailed structures of several mannans have now been elucidated largely through the work of Ballou and his colleagues and these have recently been reviewed in detail [6, 7, 38, 39]. In general the polysaccharide is made up of a backbone of mannosyl units in α-1,6-linkage to which are attached shorter side chains containing both α-1,2- and α-1,3-linked units. Linkage to the peptide moiety, which makes up 5-10% of the complex, occurs in one of two ways. Mannan of higher molecular weight is linked *via* di-*N*-acetylchitibiose to an asparagine residue in the peptide whereas shorter oligosaccharides, containing in the case of *S. cerevisiae* up to four mannosyl units, are held in direct *O*-glycosidic linkage to serine and threonine residues (Fig. 14.5). The latter linkages are subject to β-elimination in weak alkali. There is no evidence, however, to determine whether these two types of linkage are located on the same or different peptides.

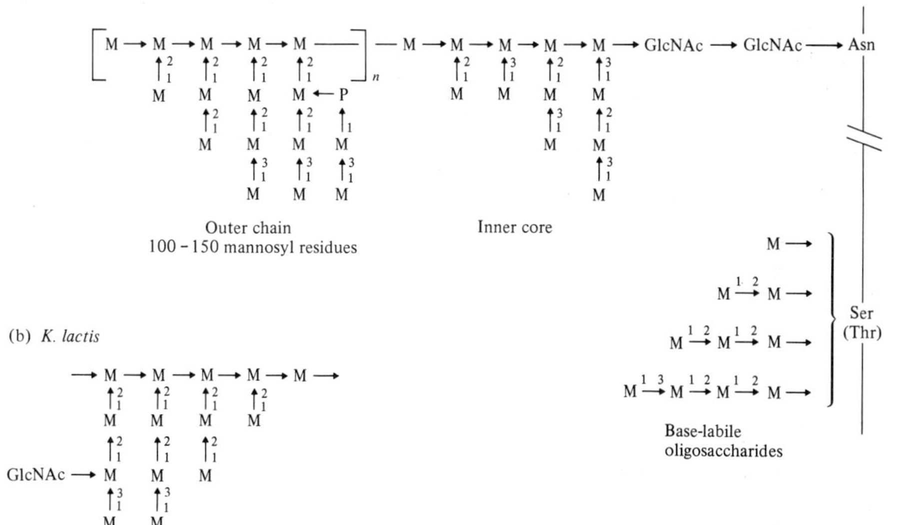

Figure 14.5 The structure of wall-mannan from (a) *Saccharomyces cerevisiae* and (b) *Kluyveromyces lactis.* The structures depicted are those proposed by [87, 88] and [98]. All anomeric linkages have the α-configuration with the exception of those between the reducing terminal mannose and the two *N*-acetylglucosamine resides which are β (1,4) linkages. The configuration of the mannose attached directly to serine or threonine has not been established.

The earlier studies of biosynthesis investigated only the carbohydrate portion of the mannan. Thus, particulate enzyme preparations, obtained from lysed protoplasts of *S. carlsbergensis*, catalysed the synthesis of mannan from GDP-mannose, in the absence of added primer [1, 12]. The synthesis was specific for GDP-mannose and showed an almost absolute requirement for Mn^{2+}. Moreover, degradation of the newly-synthesized material by acetolysis revealed that a branched mannan, similar in structure to that found *in vivo* was being synthesized.

The involvement of lipid-linked intermediates in mannan biosynthesis was first reported by Tanner [117]. A particulate enzyme preparation, obtained from alumina-ground *S. cerevisiae*, transferred mannosyl residues from the nucleotide sugar precursor to a lipophilic intermediate. Evidence for the intermediate role of the mannosyl-lipid came from a study of the effects of cations, particularly Mg^{2+} and Mn^{2+}, on the synthesis of both this lipid and the mannan. Whereas in agreement with the findings of Behrens and Cabib [12], mannan synthesis required the presence of Mn^{2+}, the formation of the mannosyl-lipid occurred also in the presence of Mg^{2+}. Enzyme preparations deficient in Mn^{2+} but containing Mg^{2+} appeared to accumulate radioactivity as the mannosyl-lipid. If, however, Mn^{2+} was subsequently added to this system then the rate of mannan synthesis increased and the amount of radioactivity that could be isolated as the mannosyl-lipid declined. Following these observations evidence accumulated that the lipophilic moiety of the intermediate was a polyprenyl phosphate [21, 109, 118] and the intermediate was finally identified as dolichyl-monophosphate-mannose (DMP-mannose) by Jung and Tanner [54] (Fig. 14.6). *Aspergillus niger* contains a similar polyprenyl phosphate, identified as *exo*-methylene-hexahydroprenyl-phosphate [8]. This was shown to act in mannan biosynthesis when particulate enzyme preparations from *A. niger* catalysed the transfer of mannosyl residues from GDP-mannose to both mannan and a manno-lipid [9]. The latter product was isolated and characterized as a mannosylated derivative of the polyprenyl phosphate and it is therefore analogous to DMP-mannose in *S. cerevisiae.*

More recently the role of DMP-mannose in the mannosylation of glycoproteins of yeast has been further clarified; the biosynthesis of *N*-glycosidically linked polymer is discussed below. In those oligosaccharides linked *O*-glycosidically to

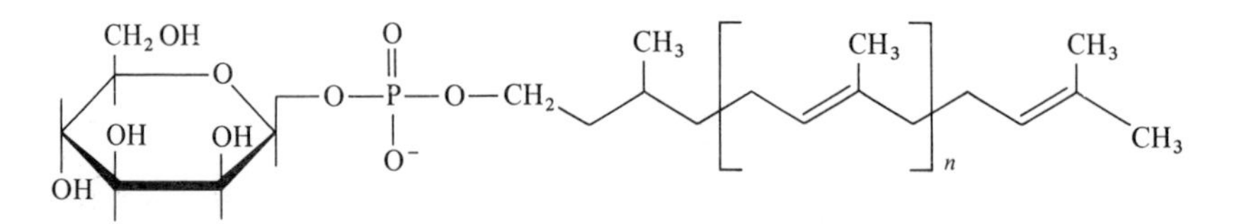

Figure 14.6 The structure of dolichyl monophosphate mannose. In yeast n = 12 to 16. The structure shown is that of dolichyl-β-*D*-mannopyranosyl phosphate although the anomeric configuration of the intermediate involved in yeast glycoprotein biosynthesis is not yet known.

serine or threonine residues of the peptide, only the mannosyl residue involved directly in the linkage was incorporated from the lipid intermediate. Addition of the subsequent mannosyl units, forming oligosaccharide, occurred as a direct glycosylation of the preceding mannose residue [3, 20, 112] (Fig. 14.7). In each of these various investigations the nature of the major product being synthesized was established after release of the radioactive material from the particulate enzyme preparation by β-elimination. However, in neither case did this procedure result in the complete release of the radioactivity [20, 112]. Thus the possibility remained that high molecular weight branched mannan, of the type linked to the peptide through *N*-acetylglucosaminyl–asparagine bonds, was also being synthesized. In *S. cerevisiae* the non-diffusible material remaining after β-elimination of the radioactive product synthesized from DMP–mannose, became diffusible after treatment with pronase [112].

More recently a membrane-bound mannosyl transferase of *S. cerevisiae* has been shown to catalyse the transfer of mannose from the nucleotide sugar to mannose, mannobiose and mannotriose with the formation of the corresponding higher homologues [67]. Addition of the mannosyl residues which was Mn^{2+}-dependent occurred with the release of GDP and the synthesis of α-1,2-linkages. The same enzyme preparation also catalysed the transfer of mannose onto acid-treated mannan-protein to give a product from which almost all the mannose could be released by mild alkali treatment. This observation together with the type of linkage being synthesized suggests strongly that these enzymes are involved only in the synthesis of *O*-glycosidically linked mannan. Earlier a particulate enzyme preparation from *Cryptococcus laurentii*, was shown to contain four distinct mannosyl transferases catalysing the synthesis of α-1,2-, α-1,3- and α-1,6-mannosyl–mannose linkages, and a mannosyl–xylose linkage [108]. Two of these linkages, and possibly all four, are known to occur in wall polysaccharides of this organism. Such evidence strongly suggests the occurrence in other organisms of multiple enzyme activities having the same general function, ie mannosyl transferases in the example described above, but which are involved in the synthesis of linkages having a quite different specificity. As discussed in a later section this has largely been confirmed in *S. cerevisiae.*

14.3.1 *The biosynthesis of N-glycosidically-linked mannan*

The possible participation of a polyisoprenyl-*N*-acetylglucosamine intermediate in the synthesis of the high molecular weight mannan was suggested by the observations of Lampen and his colleagues [63, 64, 120], who found that the antibiotic tunicamycin (Chapter 9.5.2) inhibits the synthesis by yeast protoplasts of mannan-containing glycoproteins such as invertase and acid phosphatase and of mannan itself. On the other hand, tunicamycin is without effect except at very high concentrations (see p. 315), on the synthesis of other wall polysaccharides including chitin and glucan [63, 64], neither does it inhibit the transport by the protoplasts of mannose or glucosamine nor the synthesis of UDP-*N*-acetylglucosamine.

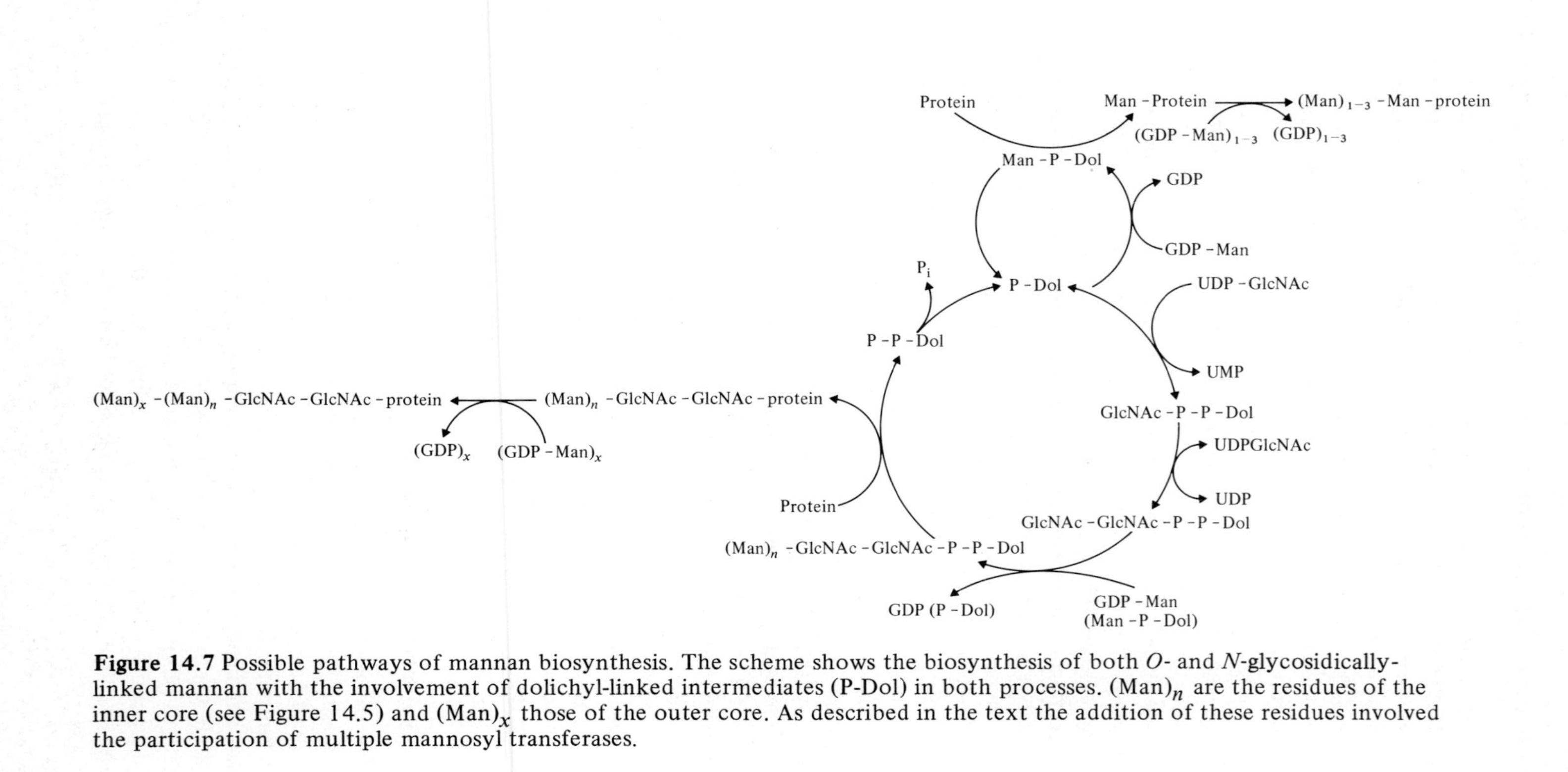

Figure 14.7 Possible pathways of mannan biosynthesis. The scheme shows the biosynthesis of both *O*- and *N*-glycosidically-linked mannan with the involvement of dolichyl-linked intermediates (P-Dol) in both processes. (Man)$_n$ are the residues of the inner core (see Figure 14.5) and (Man)$_x$ those of the outer core. As described in the text the addition of these residues involved the participation of multiple mannosyl transferases.

However, in several bacteria and a microsomal preparation from calf-liver [120] the antibiotic has been shown to inhibit the formation of polyisoprenyl-intermediates containing *N*-acetylglucosamine (see Chapter 9.5.2). The synthesis of DMP–mannose by the microsomes was unaffected. Thus it seemed likely that tunicamycin would act in an analogous manner in yeast glycoprotein synthesis by preventing the formation of some intermediate linked to a polyisoprenoid carrier. Evidence largely confirming this supposition has come with the isolation from membrane preparations of *S. cerevisiae* of dolichyl-pyrophosphoryl-*N*-acetylglucosamine and dolichyl-pyrophosphoryl-di-*N*-acetylchitobiose [68]. Moreover, radioactivity originally present in D-P-P–*N*-acetylglucosamine could be 'chased' into the di-*N*-acetylchitobiose derivative by continued incubation in the presence of unlabelled UDP–*N*-acetylglucosamine (Table 14.3). At this time the nature of the lipid carrier was inferred from the observation that biosynthesis of these two intermediates could be dramatically stimulated by the addition of dolichyl monophosphate to the incubation mixtures. More recently, dolichyl-P-P-*N*-acetylglucosamine from *S. cerevisiae* has been isolated and fully characterized [99]. The same membrane preparation also catalysed the transfer of the mannosyl residues to D-P-P–di-*N*-acetylchitobiose to yield a lipid-linked tetrasaccharide [68]. The isolation of lipid-linked trisaccharides has also been reported [90]. Although these short oligomers and indeed di-*N*-acetylchitobiose can be transferred to protein they do not then undergo any further mannosylation [69, 90]. Yeast membranes also synthesize lipid-linked oligosaccharides containing 10 to 15 mannose units in addition to the di-*N*-acetylchitibiose and it is these that are the precursors of at least the core region of *N*-glycosidically linked mannan [70, 93–5]; whether DMP–mannose is the precursor of only some or all of the mannose residues remains unclear. As described above an involvement of this lipid intermediate had earlier been inferred from the observation that non-diffusible material remaining after β-elimination of the product synthesized from DMP–mannose became diffusible after digestion with pronase [68].

Table 14.3 The formation of lipid-bound oligosaccharide by a membrane preparation from *S. cerevisiae* (from [68]).

Additions	*Radioactive products (cpm) obtained after acid hydrolysis of the glycolipid*		
	N-*acetylglucosamine*	*Di-*N-*acetylchitobiose*	*Oligosaccharide*
^{14}C-labelled glycolipid* (zero time control)	2392	3304	0
^{14}C-labelled glycolipid + GDP-Man	2560	0	3496
^{14}C-labelled glycolipid + UDP-Glc*N*Ac + GDP-Man	1256	0	4864

*The ^{14}C-labelled glycolipid contained a mixture of dolichyl-P-P-GlcNAc and dolichyl-P-P-GlcNAc-GlcNAc.

Once the oligosaccharide, which may represent all or only part of the core, has been *N*-glycosidically linked to the protein acceptor then further mannosylation takes place by the direct transfer of single residues from GDP-mannose. The formation of the various bonds in the outer chain and the addition of side chain residues is undoubtedly controlled by specific mannosyl transferases. Parodi [95] has recently shown that the groups of enzymes involved in the synthesis of the inner core and outer chain can be distinguished by their cation requirements, stability at low temperatures and by ion exchange chromatography. Evidence for the specificity of these enzymes has come from studies of the addition of mannose to manno-oligosaccharides of known structure [39, 89] and from studies of mannan-deficient mutants [88]. These are discussed in more detail below.

14.3.2 *Mannosyl transferases and mutants in mannan biosynthesis*

A series of mutants defective in various aspects of mannan biosynthesis have been obtained from both *S. cerevisiae* [7, 55, 88] and *Kluyveromyces lactis* [116]. These were selected as a result of their being non-agglutinated by specific antisera raised against mannan oligosaccharides of known composition. In *S. cerevisiae* six classes of mutant have now been described and the mannan synthesized by each of these mutants has been characterized chemically. From a study of these mutants and of the addition of mannose to short oligosaccharides of known structure, largely the work of Ballou and his colleagues [6, 7, 55, 88, 89] it has become clear that multiple mannosyl transferases are involved in mannan biosynthesis. The phenotypes of four of the mutants (*mnn* -1, -2, -3 and -4) have been ascribed to defects in specific transferases although an examination of the effects of these mutations has shown the complexity of the biosynthetic process. The synthesis of one of the *O*-glycosidically linked oligomers and the side chains of *N*-glycosidically linked mannan are affected. For example, mutants lacking α-1,3-mannosyl transferase (*mnn* -1) were defective in the addition of terminal α-1,3-linked units to both the outer chains (Fig. 14.8a) and to one of the oligomers attached directly to the hydroxyamino acids. Thus the transferase appears to participate in the synthesis of both fractions. This was not the case with a mutant lacking α-1,2-mannosyl-transferase (*mnn* -2). In this organism the mannan synthesized lacked all the side chains on the outer chain (Fig. 14.8b) whereas synthesis of a branched inner core and of the oligosaccharide fraction remained unaffected implying the existence of one or more additional α-1,2-mannosyltransferases. Presumably the enzyme participating in the synthesis of the oligosaccharide fraction is among those studied by Lehle and Tanner [67] and described in the previous section. A second α-1,2-linkage is also found in the side chains of the outer chain (Fig. 14.5a) and a third class of mutants (*mnn* -3) do not synthesize these linkages (Fig. 14.8c). The enzyme involved has been designated α-1,2-mannosyltransferase II and its relationship if any to the enzymes discussed above remains unclear. Finally the fourth class of mutants (*mnn* -4) synthesize mannan lacking mannosylphosphate substituents of the mannotriose side chains (Fig. 14.8d) and have recently been shown to be

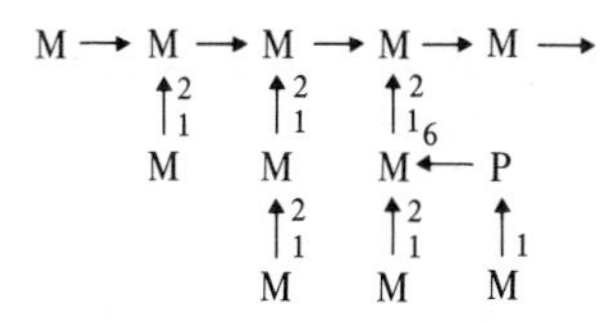

(a) *mnn* - 1 (α - 1, 3 - mannosyltransferase⁻)

M → M → M → M → M →

(b) *mnn* - 2 (α - 1, 2 - mannosyltransferase I⁻)

→ M → M →
↑ 2 1
M

(c) *mnn* - 3 (α - 1, 2 - mannosyltransferase II⁻)

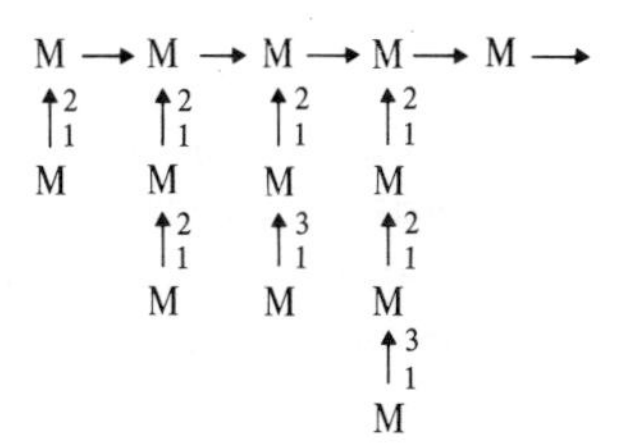

(d) *mnn* - 4 (mannosyl - phosphate transferase⁻)

Figure 14.8 Structures of the outer-chain mannan in the various mutants isolated in *Saccharomyces cerevisiae.* All the linkages have an α- configuration those of the backbone being α-1,6.

defective in mannosylphosphate transferase. These mutants appear to be regulatory in nature and the suggestion has been made that other recently isolated mutants (*mnn* -6) which have the same phenotype are structural mutants in the mannosylphosphate transferase. In fact the structure of wall mannan discussed in detail by Nakajima and Ballou [87, 88] requires the participation of at least three other mannosyltransferases. These would synthesize the α-1,6-linkages of the mannan backbone where separate enzymes may be required for the inner core and the outer chain, the β-1,4-linkage between the first mannosyl residue and the di-*N*-acetylchitobiose linkage unit and the α-1,3-linkages formed directly to the mannosyl residues of the back-bone in the inner core.

The incorporation of phosphate into mannan has been studied using particulate preparations from both *Hansenula holstii* [18, 60] and *H. capsulata* [78]. In both organisms, GDP-mannose was the precursor of both phosphate and mannose residues. Mild acid hydrolysis of the radioactively-labelled product resulted in the release of the newly incorporated mannose and the incorporated phosphate becoming available for release by phosphomonoesterase, whereas normal acid hydrolysis resulted in the release of labelled mannose-6-phosphate. Thus it was concluded that incorporation occurred through the formation of 1,6-phosphodiester linkages between two mannose residues. The enzymes catalysing these reactions would be analogous to the mannosylphosphate transferase in *S. cerevisiae* described above. Membranes of *H. holstii* also catalyse the formation of both *O*- and *N*-glycosidically linked mannan using DMP–mannose as one of the precursors, [19–21].

In *K. lactis* two classes of mutants lacking the *N*-acetylglucosamine immunodeterminant (Fig. 14.5b) have been isolated by use of the techniques outlined above [116]. One class (*mnn* -1) was defective in α-1,3-mannosyltransferase and consequently synthesized only a trisaccharide rather than a tetrasaccharide side chain. The presence of the α-1,3-linked mannosyl residue is a requirement for the subsequent addition of *N*-acetylglucosamine and the mutant has a normal level of this *N*-acetylglucosaminyl transferase. Particulate preparations from protoplasts of *K. lactis* catalysed not only the transfer of *N*-acetylglucosamine, from UDP-*N*-acetylglucosamine, to endogenous mannotetraose side chains but also to a number of exogenous acceptors which terminated in an α-1,3-linked mannosyl residue. Using α-(1,3)-mannosylmannose as the acceptor *N*-acetylglucosamine was transferred to the reducing mannosyl residue in agreement with the known structure.

The second class of mutants (*mnn* -2) lacked only the *N*-acetylglucosamine determinant. Two examples of this type were isolated; one was defective in the specific transferase whereas the second type had a normal amount of this enzyme and the reason for the absence of the *N*-acetylglucosamine remains unknown.

More recently Douglas and Ballou [31a] have reported the purification of the *N*-acetylglucosamine transferase of *K. lactis* and the isolation of a third class of mutants forming incomplete mannan. The enzyme catalysed the transfer of *N*-acetylglucosamine from the nucleotide precursor to the penultimate mannosyl residue of mannotetraose. The reaction is analogous to the one occurring *in vivo* the acceptor having the configuration αM-1,3-αM-1,2-αM-1,2αM. Using fluorescein-labelled wheatgerm agglutinin, a lectin which recognizes *N*-acetylglucosaminyl substituents, Douglas and Ballou found that wild-type *K. lactis* binds approximately three times as much of the lectin as do mannan-defective mutants. In this way they were able to screen for such mutants and among those isolated found a class which synthesized only chains of unsubstituted α-1,6-mannose. This class of mutant designated *K. lactis mnn* -3 is therefore analogous to *S. cerevisiae mnn* -2 described above.

14.3.3 *Effect of cycloheximide and 2-deoxyglucose on mannan biosynthesis*

As described earlier, the initial investigations of mannan biosynthesis were concerned primarily with the carbohydrate portion of the polymer. However, the requirement for the concomitant synthesis of protein and mannan had been established through the use of cycloheximide. This antibiotic inhibits protein synthesis in certain eukaryotes by preventing the transfer of aminoacyl-*t*RNA from the acceptor site to the donor (peptidyl) site of the ribosomes [81]. Clearly, inhibition of mannan synthesis could result either from a lack of the appropriate acceptor molecules, ie the peptide moiety of the completed polymer, or as a result of a decay of the biosynthetic enzymes. In either case inhibition of protein synthesis would cause mannan synthesis to cease. The former mechanism would be expected to act more rapidly since, as the existing acceptor peptides became glycosylated, they would not be replaced, whereas, the decay of the synthetic enzymes would probably result in a much slower decay in the rate of mannan synthesis. The former explanation appears to fit the experimental facts. Measuring mannan biosynthesis as the incorporation of radioactive mannose from GDP-mannose into particulate preparations from *S. cerevisiae* pretreated for varying periods of time with cycloheximide, Elorza and Sentandreu [35] found three separate phases in the decrease in the rate of mannan synthesis: first a brief period without apparent effect, secondly a rapid decrease in activity and finally a further slower decrease which was proportional to the time of pretreatment with the antibiotic. After pretreatment for 30 min the synthetic activity observed was only some 28–33% of that found in untreated cells. Moreover, the bulk of the incorporated radioactivity was extractable as mannolipids. Thus, the inhibition of mannan synthesis is considerably greater than is immediately obvious from the observed decrease. On the basis of these results it was concluded that cycloheximide inhibits the synthesis of the acceptor peptides whereas the glycosylation of pre-existing acceptor molecules remains unaffected, hence the absence of any decrease in synthetic activity during the first 10 min of pretreatment with the antibiotic. The rapid decrease in activity observed in the second phase represents depletion of acceptor molecules from the biosynthetic system followed by the final slow decrease resulting from a permanent loss of activity by the mannosyl transferases themselves.

Formation of both the glucan, discussed in a later section, and mannan is inhibited by 2-deoxyglucose. This non-fermentable analogue of glucose and mannose has been utilized particularly by the groups of Bauer and Lampen to investigate the biosynthesis of both wall mannan and the mannan-containing enzymes such as invertase and acid phosphatase. As a result of these studies two possible mechanisms of action for 2-deoxyglucose have been suggested. In the first the analogue becomes incorporated into polymeric material leading to modification of the carbohydrate moieties. Alternatively, accumulated 2-deoxyglucose-6-phosphate may interfere with the isomerization of glucose and mannose by hexosephosphate isomerase leading to a shortage of the substrates necessary for glycosylation to occur.

Growth of *S. cerevisiae* on medium containing 2-deoxyglucose (glucose to 2-deoxyglucose ratio of 40:1) resulted in the incorporation of the analogue into *N*-glycosidically-linked mannan throughout the growth cycle. However, 2-deoxy-glucose was not incorporated into mannan oligosaccharides released by mild alkali treatment [13, 15]. This finding contrasts with *in vitro* studies where membranes of *S. cerevisiae* catalysed the formation of dolichyl monophosphate 2-deoxyglucose from GDP–deoxyglucose and transferred the analogue into *O*-glycosidically linked oligosaccharides [66]. Once incorporated the analogue blocked further extension of the chain. In the earlier investigation [13, 15] examination of the isolated *N*-glycosidically linked mannan which contained approximately 5% deoxyglucose revealed that 90% of the analogue was incorporated into the side chains of the outer chain. The majority (76%) of these deoxyglucosyl residues were not futher substituted. The absence of significant incorporation into the main α-1,6-linked backbone of the polymer may also reflect the inability of the α-1,6-mannosyltransferase to link mannose to deoxyglucose. At the relatively low levels of total incorporation measured, the lack of extension of a few main chains would probably go unobserved. Whether 2-deoxyglucose can interfere with the synthesis of the inner core remains unknown, although the formation of DMP–deoxyglucose certainly makes this possible.

More recently Krathy *et al.* [61] have shown that the inhibitory effects of deoxyglucose on glycoprotein biosynthesis in *S. cerevisiae* are markedly influenced by the growth medium employed. Thus, as found earlier [13], the greatest degree of inhibition of mannan biosynthesis, measured as mannan present in the isolated wall, was found in organisms grown in glucose-containing medium. In contrast glucan biosynthesis was suppressed in mannose-containing medium. On the basis of these results it was concluded that the proportion of mannan and glucan found in the wall of these organisms was controlled at the level of the phosphohexose isomerases, ie the interconversion of mannose-6-phosphate and glucose-6-phosphate. These observations provide confirmation for the earlier study of Kuo and Lampen [62] who presented evidence to show that the primary site of inhibition by deoxyglucose was at this same point. Again, in the presence of the inhibitor at a hexose to deoxyglucose ratio of 40:1, the organisms were shown to accumulate 2-deoxyglucose-6-phosphate. Surprisingly, in view of the results described above, accumulation of nucleotide sugars and incorporation of deoxyglucose into polysaccharides was not observed. 2-Deoxyglucose-6-phosphate was, however, shown to inhibit the enzymes phosphoglucose isomerase and phosphomannose isomerase and as a consequence the interconversion of the 6-phosphates of glucose, fructose and mannose. Moreover, an inhibitory effect on the transport of glucose and fructose was also observed, whereas under the particular conditions employed protein synthesis was only slightly inhibited. This finding argues against any decrease in energy production being responsible for the observed inhibition of polysaccharide biosynthesis. On the other hand, the incorporation of deoxyglucose into the wall polysaccharides of *S. cerevisiae* found by Biely *et al.* [15] implies the formation of both GDP–2-deoxyglucose and UDP–2-deoxyglucose; nucleotide sugars which were not found to accumulate by Kuo and Lampen [62]. The reason for this difference is not immediately apparent since both investigations were carried out using similar hexose to 2-deoxyglucose ratios. However, on the basis of the experimental evidence described it appears that inhibition of mannan and glucan biosynthesis is a complex process involving both the mechanisms outlined above.

14.3.4 *Localization of mannan biosynthesis*

As described in the preceding section, continued mannan synthesis requires the concomitant synthesis of protein. Ruiz-Herrera and Sentandreu [103] have investigated the initial glycosylation reactions involved in mannan synthesis in *S. cerevisiae.* Radioactivity, incorporated from ^{14}C-mannose given in short pulses, was first isolated in a sub-cellular fraction identified as polysomes; characterization of the incorporated radioactivity revealed it to be predominantly mannose although small amounts of glucose and glucosamine were also found. Additional evidence for the presence of glucosylated peptides was provided by the finding that incubation of the labelled polysomes at pH 10, or with the antibiotic puromycin, treatments previously shown to result in release of the nascent polypeptides, also caused the release of radioactivity [103]. Kosinova *et al.* [59] have also investigated these early stages of mannan biosynthesis. In this case pulse-chase experiments, followed by autoradiography of thin sections revealed that accumulation of labelled material occurred in the cytoplasm before incorporation into the wall. The specificity of labelling with ^{3}H-mannose had already been established [38]. Growth of *S. cerevisiae* on galactose, at a concentration some 600-fold in excess of that of the radioactive mannose, resulted in the mannose being incorporated exclusively into mannan.

In view of the general agreement between these two observations it appeared likely that mannan biosynthesis commences intracellularly rather than at the cytoplasmic membrane as is the case with chitin. However, when compared with peptidoglycan or lipopolysaccharide biosynthesis in bacteria, the established participation of both DMP-mannose and dolichyl-pyrophosphoryl-di-*N*-acetyl-chitobiose strongly suggests that membranes will have a major role in the biosynthetic process. These could be either the endoplasmic reticulum or the main cytoplasmic membrane, the plasmalemma. In this context it is worth noting that density gradient centrifugation of particulate preparations from *S. cerevisiae* revealed 'mannan synthetase' activity to be located predominantly in a light membrane fraction, presumed to contain fragments of the endoplasmic reticulum, and to a lesser extent in the plasmalemma. 'Synthetase' activity was absent from the isolated wall fraction [29, 59].

These earlier studies measured only total mannose incorporation. More recently Lehle *et al.* [65] have investigated the activity of mannosyltransferases and the formation of DMP–mannose in the endoplasmic reticulum, Golgi-like vesicles and plasmalemma of *S. cerevisiae.* The first mannosyl residue of *O*-glycosidically-linked mannan is preferentially incorporated by the endoplasmic reticulum. Subsequent residues are added with greatest efficiency by the Golgi-like vesicles and the plasmalemma. In the synthesis of *N*-glycosidically-linked material mannosyl residues were transferred with equal efficiency by all three membrane preparations and steps involving lipid intermediates could not be distinguished. On the basis of their own and other studies Lehle *et al.* have suggested the scheme shown in Fig. 14.9 for the subcellular sites for the biosynthesis of yeast mannoproteins.

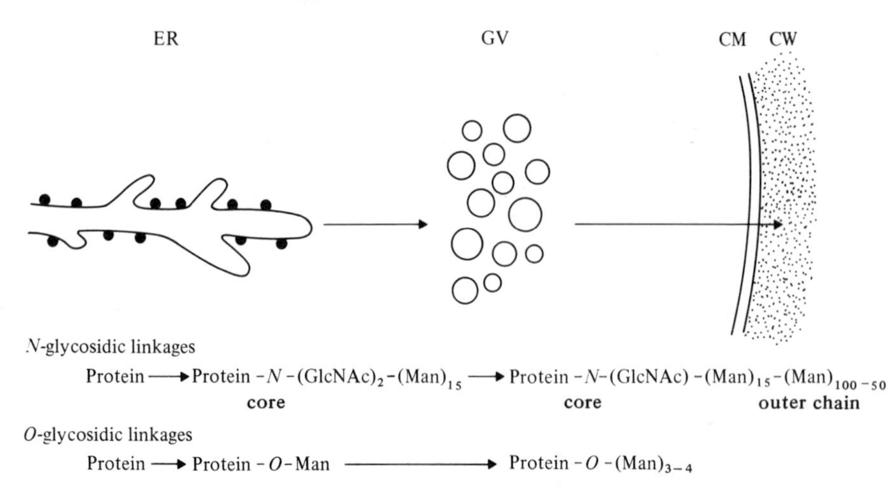

Figure 14.9 A hypothetical scheme for the formation of wall-mannan. (After [65].) ER, endoplasmic reticulum; GV, Golgi-like vesicles; CM, cytoplasmic membrane (plasmalemma); CW, cell wall.

Basically those steps in which lipid intermediates participate, the initial stages of *O*-glycosidically-linked oligosaccharide biosynthesis and the formation of the inner-core of *N*-glycosidically-linked mannan, are catalysed by enzymes of the endoplasmic reticulum. Additional mannosyl residues of both oligosaccharides and the outer chain are subsequently added by enzymes in the Golgi-like vesicles and the plasmalemma before the completed mannan becomes incorporated into the wall or, as in the case of invertase and other enzymes, excreted into the medium.

Insertion of the newly-synthesized mannan into the wall has also been studied in *S. cerevisiae* [38, 119]. Tkacz and Lampen [119] labelled the mannan already present in the wall with fluorescein-conjugated concanavalin A. The organisms were then grown in the absence of the lectin and the sites of growth observed as non-fluorescent areas which were easily distinguishable from the pre-existing wall material. Two main areas of incorporation were observed, the developing bud in which the apex appeared to be the major site of insertion and an area associated with the junction of the mother and older daughter cells presumably resulting from synthesis of the bud scar. Moreover, little if any of the mannan located in the bud wall was derived from the mother cell. Essentially similar results have been obtained by using high-resolution autoradiography of isolated walls previously labelled with ^{3}H-mannose [38]. During the early stages of bud formation, incorporation of newly synthesized material occurred uniformly over the surface, whereas during intermediate stages apical growth predominated. In older buds incorporation again appeared to be occurring over the whole surface. Thus the authors concluded that the ellipsoid shape of *S. cerevisiae* resulted from a combination of both apical growth and spherical extension.

In much the same way that some mechanism must exist to direct the vesicles containing chitin synthase activating factor to the appropriate site so must there be a similar mechanism directing the deposition into the wall of both mannan and glucan. For, as described below, much of the evidence concerning the insertion of newly-synthesized mannan into the yeast bud wall is equally applicable to the insertion of glucan. The participation of vesicles in this process is inferred from numerous ultrastructural observations which show an increase in the number of vesicles during the earlier stages of bud formation [80, 85, 111]. Whether these vesicles contain wall material, perhaps in the case of mannan, pre-synthesized in the endoplasmic reticulum or enzymes, involved either directly in the biosynthesis of the wall polysaccharides or present as activating factors for other biosynthetic enzymes already located in the plasmalemma, is open to speculation. On the other hand, there is some evidence for the presence of a glucanase activity in a vesicle fraction isolated by density gradient centrifugation from *S. cerevisiae* [30]. Perhaps, the action of such lytic enzymes in bringing about a partial dissolution of the wall provides the stimulus necessary to initiate growth at a particular site. Clearly, further information on the composition and enzymic content of these vesicles is required before any definite conclusion can be reached about their possible role in the initiation of new wall synthesis.

14.4 Biosynthesis of glucan

In marked contrast to the detailed information on the biosynthesis of both chitin and mannan, there is little information concerning the glucans. Included under this heading are at least two distinct glucose-containing polysaccharides present in the yeast wall, one or both of which appear to be responsible for the maintenance of shape. Evidence for this latter point comes from the finding that removal of the other wall polymers, ie chitin and mannan together with some soluble glucan, by treatment with alkali and enzymic digestion, did not appear to affect the shape of either whole organisms or isolated walls [4].

The main glucan, isolated from the wall of *S. cerevisiae* by chemical extraction and enzymic digestion (Fig. 14.10) is made up of β-1,3 chains containing some 3% β-1,6-D-glucosidic inter-chain linkages [75] (Fig. 14.11). The minor component of lower molecular weight is a highly branched structure containing mainly β-1,6-linkages but with some 19% β-1,3-linkages [76]. Selective enzymic digestion suggests that the two polysaccharides are organized within the yeast wall such that the major β-1,3-linked polymer provides an inner fibrillar layer on which the β-1,6-linked glucan and other polymers are located [58]. However, it seems likely that the structure should be regarded as an association of the other polymers with a glucan matrix rather than as a series of layers containing discrete components.

Surprisingly perhaps, in view of their structural importance, little is known about the biosynthesis of glucans in either yeast or the filamentous fungi. Until relatively recently, the only study was that of Wang and Bartnicki-Garcia [123]

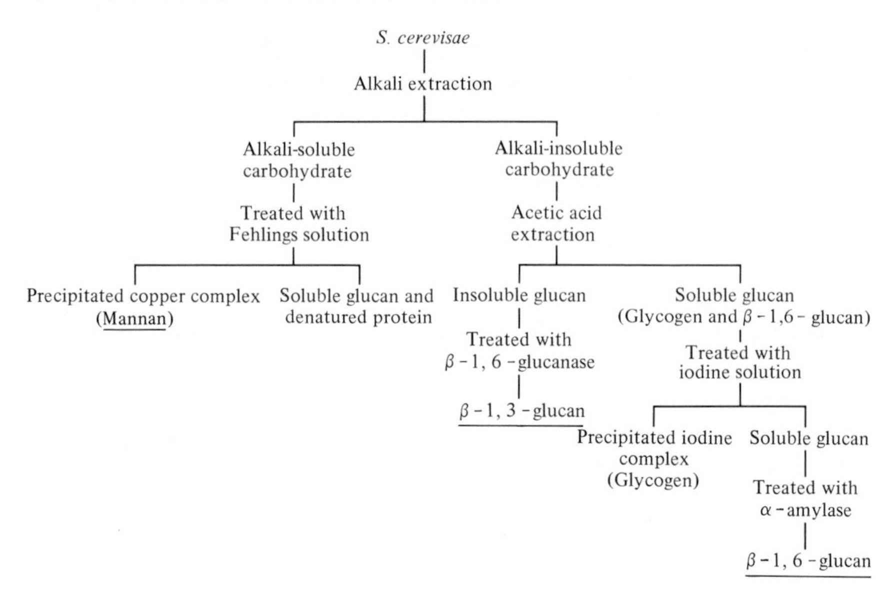

Figure 14.10 Fractionation of *Saccharomyces cervisiae* to yield purified glucans. Data from [75].)

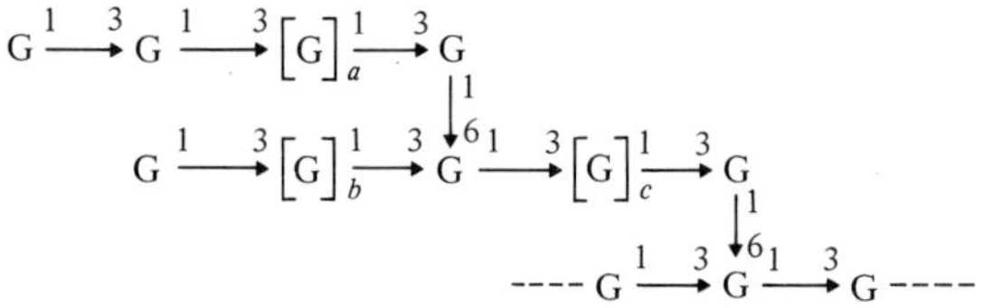

$$G \xrightarrow{1 \quad 3} G \xrightarrow{1 \quad 3} [G]_a \xrightarrow{1 \quad 3} G$$

$$\downarrow {}^{1}_{6}$$

$$G \xrightarrow{1 \quad 3} [G]_b \xrightarrow{1 \quad 3} G \xrightarrow{1 \quad 3} [G]_c \xrightarrow{1 \quad 3} G$$

$$\downarrow {}^{1}_{6}$$

$$---- G \xrightarrow{1 \quad 3} G \xrightarrow{1 \quad 3} G ----$$

Figure 14.11 Partial structure of β-1,3-glucan. All the linkages have the β-configuration. The degree of branching observed was approximately 3%, the interchain linkages being β-1,6. Although the exact lengths are unknown $a + b + c$ is approximately 60 glucose residues [75].

who investigated the biosynthesis of glucan by *Phytophthora cinnamoni*, a plant pathogenic fungus in which glucans compose some 90% of the hyphal wall. Washed hyphal wall fragments incorporated glucose from UDP-glucose, in the presence of Mg^{2+}, into alkali-insoluble material identified by chemical degradation and enzymic digestion as glucan containing both β-1,3- and β-1,6-linkages. Subsequently the same authors [124] used as enzyme a mixed membrane fraction prepared from organisms grown in a glucose-rich medium (in order to minimize the formation of proteases). This preparation catalyses the synthesis of large amounts of β-1,3-glucan microfibrils from UDP–glucose. Surprisingly in view of the earlier studies β-1,6-linkages were not detected. Limited proteolysis with trypsin caused a two-fold stimulation of glucan synthase activity, suggesting that some of the enzyme may be present as a zymogen. Stimulation of activity was also obtained upon the addition of cellobiose which may act as an acceptor in the cell-free system.

Glucan containing both β-1,3- and β-1,6-linkages was synthesized by cell-free preparations of *Cochliobolas miyabeanus* [91, 92].

Cells of *S. cerevisiae* treated with toluene–ethanol have been shown to catalyse the biosynthesis of glucan [110]. Similar results to those described above have been obtained, in that glucose was incorporated from UDP-glucose into material identified as β-1,3-glucan. Again Mg^{2+} was required for maximal activity. In addition the toluene-treated cells also incorporated mannose from GDP–mannose into acid-precipitable material including the lipid intermediate, DMP–mannose. On the other hand, the participation of a lipid intermediate in glucan biosynthesis could not be established although some 60% of the incorporated radioactivity was located in particulate, presumably membrane, fractions. These observations led to the hypothesis that glucan units were synthesized by intracytoplasmic membranes perhaps located in vesicles and subsequently transferred to the wall on fusion of the vesicles with the plasmalemma. As discussed earlier, although vesicles have been described which are thought to contain glucanases [30], it seems unlikely that the organisms would confine both enzyme and substrate in the same vesicle. The participation of other, ie non-glucanase containing vesicles, would clearly obviate this difficulty but, on the other hand, provide an additional complication when considering the biosynthesis of the wall polysaccharides particularly in the ordered direction of vesicles towards the sites of insertion of new material. Membrane preparations of *S. cerevisiae* have also been shown to synthesize glucan from UDP–glucose [5, 73, 113] and GDP–glucose. In the preparation utilizing both nucleotide sugars as substrates, two independent glucosyl-transferases appeared to be present [5]. The glucan synthesized contained both β-1,3- and β-1,6-linkages, the product from UDP–glucose containing the higher proportion of the 1,3 type and that from GDP–glucose more 1,6-linkages. These products may represent the synthesis of both the major and minor components of the yeast wall. More recently Cabib and his colleagues [113, 114] have described the activation of β-1,3-glucan synthase by either ATP or GTP. Activation by ATP can be abolished by treatment of the enzyme with EDTA or pretreatment with alkaline phosphatase. Isolation of glucan synthase in the presence of high concentrations of EDTA results in the enzyme having activity in the absence of either nucleotide triphosphate. Subsequent incubation of the enzyme with Mg^{2+} results in inactivation which can be overcome by treatment with either ATP or GTP. These observations have been interpreted to show a regulatory mechanism for the enzyme which could operate *in vivo*.

Similar results to those described in the preceding section have been obtained for the localized incorporation of glucan into the wall of a number of yeasts including *S. cerevisiae* and *Schizosaccharomyces pombe* [50, 52, 53]. Thus, the incorporation of ^{3}H-glucose into alkali-insoluble polysaccharide was observed by autoradiography. Insertion of newly synthesized material occurred predominantly at the apex of growing buds and at the primary growing end of the fission yeast *S. pombe*. In non-budding cells of *S. cerevisiae* either a more uniform distribution of labelling was observed or, in some cases, a strong localization interpreted as the commencement of bud formation, although the bud itself was not clearly differentiated. In subsequent studies [51] lysis induced by 2-deoxyglucose was also found to coincide with the sites of glucan insertion observed by autoradiography. Some of the information available on the effects of 2-deoxyglucose on glucan synthesis has been discussed earlier in conjunction with the effects of the analogue on mannan biosynthesis. Incorporation of 2-deoxyglucose into the wall glucan has been described although not in as great detail as in the case of mannan [61]. Lysis of *S. cerevisiae* also occurs when 2-fluoro-2-deoxyglucose is present in the growth medium although in this case incorporation of the analogue into wall fractions has not been demonstrated [14]. In each case, disruption of the balance between synthesis of new wall material and the removal of pre-existing material by autolytic enzymes may underlie the response to the sugar analogues. On the other hand, it should be pointed out that Sentandreu and his colleagues [110] were unable to detect appreciable turnover of either β-glucan or wall mannan in *S. cerevisiae.*

References

1. Algranate, I. D., Carminatti, H. and Cabib, E. (1963) *Biochem. biophys. Res. Commun.* **12**, 504–9.
2. Archer, D. B. (1977) *Biochem. J.* **164**, 653–8.
3. Babczinski, P. and Tanner, W. (1973) *Biochem. biophys. Res. Commun.* **54**, 1119–24.
4. Bacon, J. S. D., Farmer, V. C., Jones, D. and Taylor, I. F. (1969) *Biochem. J.*, **114**, 557–67.
5. Balint, S., Farkas, V. and Bauer, S. (1976) *FEBS Letts.* **64**, 44–7.
6. Ballou, C. E. (1974) *Adv. Enzymol.* **40**, 239–70.
7. Ballou, C. E. (1976) *Adv. Microbiol. Physiol.* **14**, 93–158.
8. Barr, R. M. and Hemming, F. W. (1972) *Biochem. J.* **126**, 1193–202.
9. Barr, R. M. and Hemming, F. W. (1972) *Biochem. J.* **126**, 1203–208.
10. Bartnicki-Garcia, S., Bracker, C. E., Reyes, E. and Ruiz-Herrera, J. (1978) *Exp. Mycol.* **2**, 173–92.
11. Bartnicki-Garcia, S., and Lippman, E. (1972) *J. gen. Microbiol.* **71**, 301–9.
12. Behrens, N. H. and Cabib, E. (1968) *J. biol. Chem.* **243**, 502–9.
13. Biely, P., Kratky, Z. and Bauer, S. (1972) *Biochim. biophys. Acta* **255**, 631–9.
14. Biely, P., Kratky, Z. and Bauer, S. (1973) *J. Bact.* **115**, 1108–20.
15. Biely, P., Kratky, Z. and Bauer, S. (1974) *Biochim. biophys. Acta* **352**, 268–74.
16. Bowers, B., Levin, G. and Cabib, E. (1974) *J. Bact.* **119**, 564–75.
17. Braun, P. C. and Calderone, R. A. (1978) *J. Bact.* **133**, 1472–7.
18. Bretthauer, R. K., Kozak, L. P. and Irwin, W. E. (1969) *Biochem. biophys. Res. Commun.* **37**, 820–7.
19. Bretthauer, R. K. and Tsay, G. C. (1974) *Arch. Biochem. Biophys.* **164**, 118–26.
20. Bretthauer, R. K. and Wu, S. (1975) *Arch. Biochem. Biophys.* **167**, 151–60.
21. Bretthauer, R. K., Wu, S. and Irwin, W. E. (1973) *Biochim. biophys. Acta* **304**, 736–47.
22. Cabib, E. (1975) *A. Rev. Microbiol.* **29**, 191–214.
23. Cabib. E. and Bowers, B. (1971) *J. biol. Chem.* **242**, 152–9.

24. Cabib E. and Farkas, V. (1971) *Proc. Natn. Acad. Sci. U.S.A.* **68**, 2052–6.
25. Cabib, E. and Keller, F. A. (1971) *J. biol. Chem.* **242**, 167–73.
26. Cabib, E. and Ulane, R. (1973) *Biochem. biophys. Res. Commun.* **50**, 186–91.
27. Cabib, E., Ulane, R. and Blowers, B. (1973) *J. biol. Chem.* **248**, 1451–8.
28. Camargo, E. P., Dietrich, C. P., Sonneborn, D. and Strominger, J. L. (1967) *J. biol. Chem.* **242**, 3121–8.
29. Cortat, M., Matile, P. and Kopp, F. (1973) *Biochem. biophys Res. Commun.* **53**, 482–9.
30. Cortat, M., Matile, P. and Wienken, A. (1972) *Arch. Mikrobiol.* **82**, 189–205.
31. De Rousset-Hall, A. and Gooday, G. W. (1975) *J. gen. Microbiol.* **89**, 146–54.

31a. Douglas, R. H. and Ballou, C. E. (1979) *Abs. Amer. chem. Soc. biol. Chem. Section* 39.

32. Duran, A., Blowers, B. and Cabib, E. (1975) *Proc. Natn. Acad. Sci. U.S.A.* **72**, 3952–5.
33. Duran, A. and Cabib, E. (1978) *J. biol. Chem.* **253**, 4419–25.
34. Duran, A., Cabib, E. and Blowers, B. (1979) *Science* **203**, 363–5.
35. Elorza, M. V. and Sentandreu, R. (1973) In *Yeast, Mould and Plant Protoplasts* eds. Villanueva, J. R., Garcia-Archa, I., Gascon, S. and Uruburu, F., pp. 205–9, London and New York: Academic Press.
36. Endo, A., Kakaki, K. and Misato, T. (1970) *J. Bact.* **104**, 189–96.
37. Farkaš, V. (1979) *Microbiol. Rev.* **43**, 117–44.
38. Farkas, V., Kovařík, J., Košinova, A. and Bauer, S. (1974) *J. Bact.* **117**, 265–9.
39. Farkas, V., Vagabov, V. M. and Bauer, S. (1976) *Biochim. biophys. Acta* **428**, 573–82.
40. Glaser, L. and Brown, D. H. (1957) *J. biol. Chem.* **228**, 729–42.
41. Gooday, G. W. (1971) *J. gen. Microbiol.* **67**, 125–33.
42. Gooday, G. W. (1977) *J. gen. Microbiol.* **99**, 1–11.
43. Gooday, G. W. and de Rousset-Hall, A. (1975) *J. gen. Microbiol.* **89**, 137–45.
44. Gooday, G. W. and Hardy, J. C. (1978) *Abstracts XII Int. Cong. Microbiol. Munich 1978*, 202.

44a. Gooday, G. W. and Trinci, A. P. J. (1980) In *The Eukaryotic Microbial Cell: 30th Symposium of the Society for General Microbiology*, ed. Gooday, G. W., Lloyd, D. and Trinci, A. J. P., pp. 207–51. Cambridge University Press.

45. Grove, S. N. and Bracker, C. E. (1970) *J. Bact.* **104**, 989–1009.
46. Hasilik, A. (1974) *Arch. Mikrobiol.* **101**, 295–301.
47. Hasilik, A. and Holzer, H. (1973) *Biochem. biophys. Res. Commun.* **53**, 552–9.
48. Isono, K., Asahi, K. and Suzuki, S. (1969) *J. Am. Chem. Soc.* **91**, 7490–505.
49. Jan, Y. N. (1974) *J. biol. Chem.* **249**, 1973–9.
50. Johnson, B. F. (1965) *Exptl. Cell Res.* **39**, 613–24.
51. Johnson, B. F. (1968) *J. Bact.* **95**, 1169–72.
52. Johnson, B. F. and Gibson, E. J. (1966) *Exptl. Cell Res.* **41**, 297–306.
53. Johnson, B. F. and Gibson, E. J. (1966) *Exptl. Cell Res.* **41**, 580–91.
54. Jung, P. and Tanner, W. (1973) *Eur. J. Biochem.* **37**, 1–6.
55. Karson, E. M. and Ballou, C. E. (1978) *J. Biol. Chem.* **253**, 6484–92.
56. Katz, D. and Rosenberger, R. F. (1971) *J. Bact.* **108**, 184–90.
57. Keller, F. A. and Cabib, E. (1971) *J. biol. Chem.* **246**, 160–6.
58. Kopecka, M., Phaff, H. J. and Fleet, G. H. (1974) *J. Cell. Biol.* **62**, 66–76.
59. Kosinova, A., Farkas, V., Machala, S. and Bauer, S. (1974) *Arch. Mikrobiol.* **99**, 255–63.
60. Kozak, L. P. and Bretthauer, R. K. (1970) *Biochemistry* **9** 1115–22.
61. Kratky, Z., Biely, P. and Bauer, S. (1975) *Eur. J. Biochem.* **54**, 459–67.
62. Kuo, S. C. and Lampen, J. O. (1972) *J. Bact.* **111**, 419–29.

63. Kuo, S. C. and Lampen, J. O. (1974) *Biochem. biophys. Res. Commun.* **58**, 287–95.
64. Kuo, S. C. and Lampen, J. O. (1976) *Archs. Biochem. Biophys.* **172**, 574–81.
65. Lehle, L., Bauer, F. and Tanner, W. (1977) *Arch. Microbiol.* **114,** 77–81.
66. Lehle, L., and Schwarz, R. T. (1976) *Eur. J. Biochem.* **67**, 239–45.
67. Lehle, L. and Tanner, W. (1974) *Biochem. biophys. Acta* **350**, 225–35.
68. Lehle, L. and Tanner, W. (1975) *Biochim. biophys. Acta* **399**, 364–74.
69. Lehle, L. and Tanner, W. (1978) *Eur. J. Biochem.* **83**, 563–70.
70. Lehle, L. and Tanner, W. (1978) *Biochim. biophys. Acta* **539**, 218–29.
71. Lenney, J. F. and Dalbec, J. M. (1969) *Archs. Biochem. Biophys.* **129**, 407–9.
72. Lopez-Romero, E. and Ruiz-Herrera, J. (1976) *Antonie van Leeuwenhoek J. Microbiol. Serol.* **42**, 261–76.
73. Lopez-Romero, E. and Ruiz-Herrera, J. (1977) *Biochim. biophys. Acta* **500**, 372–84.
74. Lopez-Romero, E., Ruiz-Herrera, J., and Bartnicki-Garcia, S. (1978) *Biochim. biophys. Acta* **525**, 338–45.
75. Manners, D. J., Masson, A. J. and Patterson, J. C. (1973) *Biochem. J.* **135**, 19–30.
76. Manners, D. J., Masson, A. J., Patterson, J. C., Bjorndas, H. and Lindberg, B. (1973) *Biochem. J.* **135**, 31–6.
77. Marchant, R. and Smith, D. G. (1968) *Arch. Mikrobiol.* **63**, 85–94.
78. Mayer, R. M. (1971) *Biochim. Biophys. Acta* **252**, 39–47.
79. McClure, W. K., Park, D. and Robinson, P. M. (1968) *J. gen. Microbiol.* **50**, 177–82.
80. McCully, E. K. and Bracker, C. E. (1972) *J. Bact.* **109**, 922–6.
81. McKeehan, W. and Hardesty, B. (1969) *Biochem. biophys. Res. Commun.* **36**, 625–30.
82. McMurrough, I. and Bartnicki-Garcia, S. (1971) *J. biol. Chem.* **246**, 4008–16.
83. McMurrough, I. and Bartnicki-Garcia, S. (1973) *Archs. Biochem. Biophys.* **158**, 812–16.
84. McMurrough, I., Flores-Carreon, A. and Bartnicki-Garcia, S. (1971) *J. biol. Chem.* **246**, 3999–4007.
85. Moor, H. (1967) *Arch. Mikrobiol.* **57**, 135–46.
86. Moore, P. M. and Peberdy, J. F. (1978) *Microbios,* **12**, 29–39.
87. Nakajima, T. and Ballou, C. E. (1974) *J. biol. Chem.* **249**, 7679–84.
88. Nakajima, T. and Ballou, C. E. (1974) *J. biol. Chem.* **249**, 7685–94.
89. Nakajima, T. and Ballou, C. E. (1975) *Proc. Natn. Acad. Sci. U.S.A.* **72**, 3912–16.
90. Nakayama, K., Araki, Y. and Ito, E. (1976) *FEBS Letts.* **72**, 287–90.
91. Namba, H. and Kuroda, H. (1974) *Chem. Pharm. Bull., Japan* **22**, 610–16.
92. Namba, H. and Kuroda, H. (1974) *Chem. Pharm. Bull., Japan* **22,** 1895–1901.
93. Parodi, A. J. (1977) *Eur. J. Biochem.* **75**, 171–80.
94. Parodi, A. J. (1978) *Eur. J. Biochem.* **83**, 253–9.
95. Parodi, A. J. (1979) *J. biol. Chem.* **254**, 8343–52.
96. Peberdy, J. F. and Moore, P. M. (1975) *J. gen. Microbiol.* **90**, 228–36.
97. Porter, C. A. and Jaworski, E. G. (1966) *Biochemistry* **5**, 1149–54.
98. Raschke, W. C. and Ballou, C. E. (1972) *Biochemistry* **11**, 3807–16.
99. Reuvers, F., Boer, P. and Hemming, F. W. (1978) *Biochem. J.* **169**, 505–8.
100. Robertson, N. F. (1965) In *The Fungi, Vol. I,* eds. Ainsworth, G. C. and Sussman, A. S., pp. 613–23, London and New York: Academic Press.
101. Ruiz-Herrera, J. and Bartnicki-Garcia, S. (1974) *Science* **186**, 357–9.
102. Ruiz-Herrera, J., Lopez-Romero, E. and Bartnicki-Garcia, S. (1977) *J. biol. Chem.* **252**, 3338–43.

103. Ruiz-Herrera, J. and Sentandreu, R. (1975) *J. Bact.* **124**, 127–33.
104. Ruiz-Herrera, J., Sing, V. O., van der Woude, W. J. and Partnicki-Garcia, S. (1975) *Proc. natn. Acad. Sci. U.S.A.* **72**, 2706–10.
105. Ryder, N. S. and Peberdy, J. F. (1977) *J. gen. Microbiol.* **99**, 69–76.
106. Ryder, N. S. and Peberdy, J. F. (1977) *FEMS Microbiol Letts.* **2**, 199–201.
107. Schölt, E. H. and Holzer, H. (1974) *Eur. J. Biochem.* **42**, 62–6.
108. Schutzbach, J. S. and Aukel, H. (1971) *J. biol. Chem.* **246**, 2187–94.
109. Sentandreu, R. and Elorza, M. V. (1973) In *Yeast, Mould and Plant Protoplasts*, eds. Villanueva, J. R., Garcia-Acha, I., Gascon, S. and Uruburu, F., pp. 187–204. London and New York: Academic Press.
110. Sentandreu, R., Elorza, M. V. and Villanueva, J. R. (1975) *J. gen. Microbiol.* **90**, 13–20.
111. Sentandreu, R. and Northcote, D. H. (1969) *J. gen. Microbiol.* **55**, 393–8.
112. Sharma, C. B., Babczinski, P., Lehle, L. and Tanner, W. (1974) *Eur. J. Biochem.* **46**, 35–41.
113. Shematek, E. M., Braatz, J. and Cabib, E. (1978) *Fedn. Proc. Fedn. Am. Socs. exp. Biol.* **37**, 1394.
114. Shematek, E. M., and Cabib, E. (1979) *Abstracts IX Inst. Cong. Biochem. Toronto, 1979,* 512.
115. Siepen, D., Yu, P. H. and Kula, H. R. (1975) *Eur. J. Biochem.* **56**, 271–81.
116. Smith, W. L., Nakajima, T. and Ballou, C. E. (1975) *J. biol. Chem.* **250**, 3426–35.
117. Tanner, W. (1969) *Biochem. biophys. Res. Commun.* **35**, 144–50.
118. Tanner, W., Jung, P. and Behrens, N. H. (1971) *FEBS Letts.* **14**, 109–113.
119. Tkacz, J. S. and Lampen, J. O. (1972) *J. gen. Microbiol.* **72**, 243–7.
120. Tkacz, J. S. and Lampen, J. O. (1975) *Biochem. biophys. Res. Commun.* **65**, 248–57.
121. Ulane, R. and Cabib, E. (1974) *J. biol. Chem.* **249**, 3418–22.
122. Van Laere, A. J. and Carlier, A. R. (1978) *Arch. Mikrobiol.* **116**, 181–4.
123. Wang, M. C. and Bartnicki-Garcia, S. (1966) *Biochem. biophys. Res. Commun.* **24**, 832–7.
124. Wang, M. C. and Bartnicki-Garcia, S. (1976) *Archs. Biochem. Biophys.* **175**, 351–4.
125. Yu, P. H., Siepen, D., Kula, M. R. and Tsai, H. (1974) *FEBS Letts.* **42**, 227–30.
126. Zubenko, G. S., Mitchell, A. P. and Jones, E. W. (1979) *Proc. Natn. Acad. Sci. U.S.A.* **76**, 2395–9.

15
The cell wall in the growth and cell division of bacteria

15.1 Introduction

An essential feature of the growth of all microbes is that the individual cells must increase in volume to approximately twice their original size and divide to form two closely similar units. Even small continuous departures in one direction or another from this regulated process, would ultimately lead to grossly abnormal cells. Mechanically strong walls are not the only way of achieving precision in shape replication, as can be seen by the distinct forms adopted by different species of mycoplasmas [88] or, for that matter, by the recognizable different cell types in mammalian tissues. In the growth of some other organisms, however, not having mechanically strong cell walls external to their membranes, variation in size and shape at each division process seems to occur but these variations are, presumably, random since statistically the population still maintains its general morphology and number of living units per unit mass of cells. Wall-less L-forms of bacteria are an example of this and illustrate the importance of the proper formation and growth of bacterial walls to maintain regularity in the replication of shape. L-forms differ from the bacteria from which they are derived by their inability to form peptidoglycan, usually either because of the presence of specific wall-inhibitors in the growth medium or because of genetic lesions in the biosynthetic processes for the formation of this polymer [117].

It is, thus, important with bacteria to study the growth of the wall and the nature of the increase in surface area that lead to the formation of two biological units from one. At the same time, it has to be clearly borne in mind that the growth of the walls depends not only, or even principally perhaps, on the biosynthesis of peptidoglycan, but also upon the growth of the membranes which contain the enzymes involved in the biosynthesis of this polymer. This in turn involves all the complex machinery for protein biosynthesis and the necessary codings by the cell genome. The very complexity of the whole process has led to quite separate approaches which are still very far from synthesized. These approaches may be enumerated as follows:

(1) Attempts to visualize the topology of the growth of the wall by the use of immunofluorescent antibodies, phage absorption, or autoradiography.
(2) Studies of wall growth by qualitative and quantitative electron microscopy.
(3) Measurements of cell dimensions and numbers using light microscopy under

conditions varying from growth on solid media, to growth of steady state or synchronous cultures in liquid media.
(4) Enumeration of bacteria in relation to the initiation, replication and termination of DNA synthesis using synchronous and steady state cultures.
(5) The study of mutants disturbed in aspects of division and cell growth.
(6) The creation of mathematical models for cell growth, and the testing of these particularly by methods (3) and (4) above.

Studies have often involved more than one of these approaches. The present book is clearly not the place for a comprehensive treatment of the whole of this complicated and highly sophisticated subject of bacterial cell growth and division. The text will be limited to a few examples in which the behaviour of the wall and membrane during cell growth and division have been studied in some detail. No serious attempt will be made to review literature primarily concerned with the relations between the replication of the genome and the compartmentalization of two new bacteria so that methods 3, 4 and 6 will not be considered. For those wishing to pursue these aspects of the subject, they have been extensively discussed in a recent review [104].

15.2 Growth of streptococcal cell walls

Information about the growth of envelopes of bacteria provides an illustration of the uncertainty of choosing satisfactory models in advance. Knowledge of the chemistry of the walls of many organisms such as staphylococci and bacilli advanced much more rapidly than that for those of streptococci; genetic systems, on the other hand, for *E. coli* and *B. subtilis* are well established, whereas none exists for streptococci. Our knowledge of the interrelations between DNA replication, surface growth and cell division has been obtained by exploiting the former two organisms. Streptococci on the other hand have yielded the clearest most consistent evidence about the growth of the wall. Some possible reasons for this will be discussed later.

In the first recorded experiment to demonstrate wall growth, the insertion of new wall was visualized by the use of labelled antibody produced against whole streptococci and coupled with fluorescein isothiocyanate [27]. The experiment showed that new wall was being inserted into the area at which the organisms were dividing and into the region where the next division would occur, that is at the septal regions of the cell. The antibody was first used by a so-called direct method and results then confirmed by a reverse method. In the direct method, *Streptococcus pyogenes* was allowed to grow in medium containing the fluorescent labelled antibody which was removed, and the organisms washed and then grown in medium containing unlabelled antibody. When examined under ultraviolet light (see Fig. 15.1), the new wall appeared as bands of non-fluorescent material. The reverse method yielded the converse results. Similar results were obtained with *Strep. faecalis* (Cole quoted by [50, 24]) and *Diplococcus pneumoniae* [116].

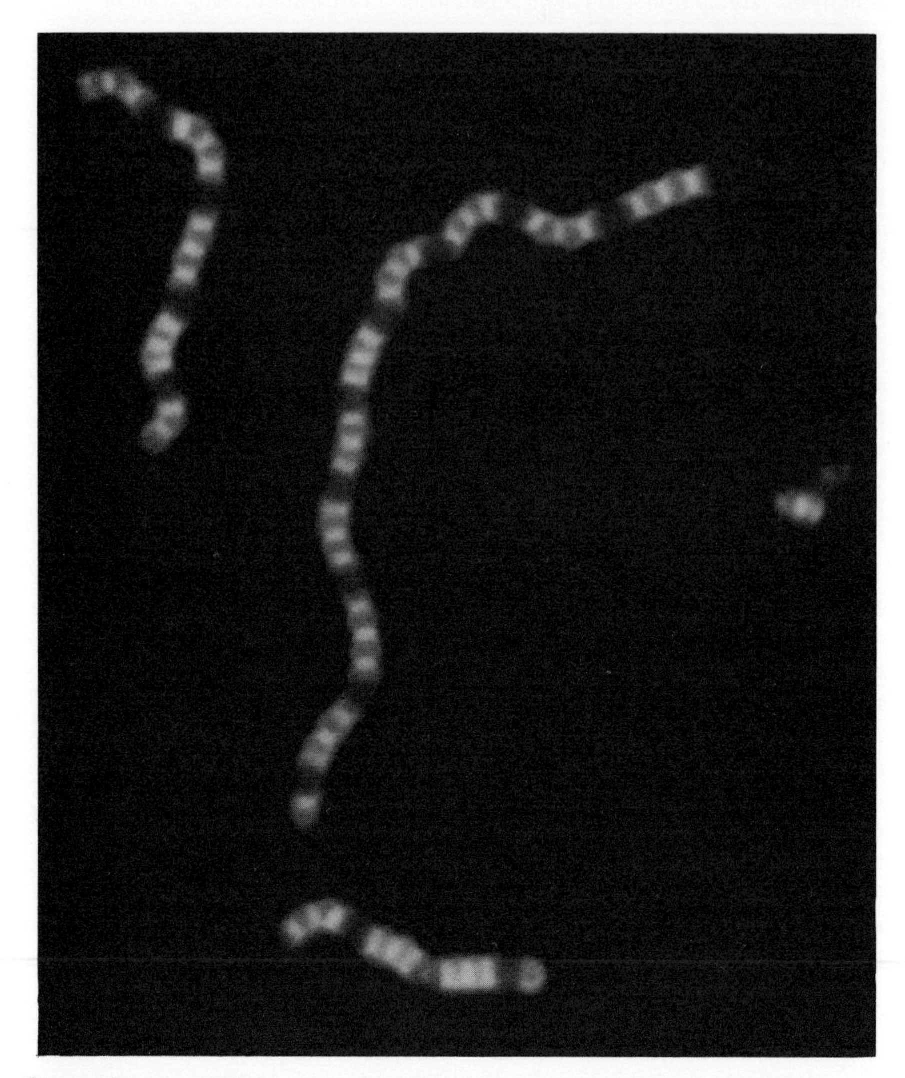

Figure 15.1 Use of fluorescent antibodies to show the growth of the surface of *Streptococcus pyogenes.* The organism shown has been growing for 60 min in unlabelled anitbody using the reverse method of Cole and Hahn [27]. The dark bands show the position of wall that has been newly formed during this 60 min. (The original slide from which this print was made was kindly given to the authors by Dr. Roger Cole of the National Institutes of Health, Bethesda, U.S.A.).

Autoradiography using radioactively labelled wall compounds with the latter organism gave similar results [18]. Attempts to apply the immunofluorescence method to other Gram-positive organisms. *Bacillus cereus, B. megaterium* [25] *B. licheniformis his lyt* [57] or Gram-negative species *Salmonella typhimurium*

[74] *Salmonella typhosa* [26] *E. coli* [8], with one exception led, either to indecisive results, or to evidence for the insertion of new material randomly over the whole cell surface. Of these results the one that was most similar to that for streptococci, was for the autolytic deficient strain of *B. licheniformis* [57] but the fluorescent bands were not as impressively clear. Some possible reasons for the failure of such an elegant method to give clear cut results with most of these other organisms will be discussed later.

More detailed evidence about the growth of the wall of streptococci, in particular of *S. faecalis,* has come from a combination of physiological and ultra-structural studies [50-54, 107]. When bacteria growing exponentially are suspended in 0.01 M phosphate buffer (pH 6.5) at 37° C and samples are removed at intervals between 15 and 60 min, fixed in 4% glutaraldehyde to stop further autolytic action, postfixed with osmium tetroxide and sectioned, autolytic action can be seen as partial wall removal. In early samples this removal is confined to the leading edge of the cross-wall and a central core of this structure. As the time of autolysis increases this action gradually spreads backwards to the peripheral wall. At the time this work was done a hypothesis favouring a necessary role for autolysins in extending the peptidoglycan network was popular (see Chapter 11). The autolysin present in *Strep. faecalis* is an *N*-acetylmuramidase (like lysozyme) and thus, it liberates free-reducing groups of *N*-acetylmuramic acid from the glycan chains. At the level of existing knowledge about peptidoglycan synthesis it was reasonable to suppose that the autolysin nicked the glycan chains, allowing the insertion of new units into the 'bag-shaped molecule' of peptidoglycan but subsequent evidence has cast severe doubts on this hypothesis. In order to test the behaviour of newly added wall compared with that of the old wall, exponentially growing cultures were either pulse-labelled with ^{14}C-lysine, or labelled continuously with this amino acid for at least ten generations followed by a pulse of ^{12}C-lysine for 0.8 or 0.1 of a generation. The walls were isolated from the cells and allowed to autolyse. The conclusion drawn from the rates of liberation of radioactivity relative to the loss in turbidity was that the newly-synthesized wall was always the first to be solubilized by the autolysin. Since the autolysin had been shown to be localized in the region of the cross-wall, it followed that new wall was likely also to be deposited there. This conclusion is discussed elsewhere [97] and in Chapters 11.4 and 11.5.1. The third line of evidence that wall growth takes place in the region of the cross-wall eventually forming two cells out of one, is very direct. It is well known [50, 67, 106] that, when protein synthesis is inhibited, wall synthesis continues and that the walls of a number of organisms become thicker (see Chapter 1.2). Thus, if protein synthesis is first shut off either by the omission of essential amino acids, or by inhibition with antibiotics such as chloramphenicol, and subsequently re-initiated, wall that had already been formed at the time protein synthesis was re-initiated would be expected to be thicker than the new wall formed subsequently. A further marker was required, however, to distinguish between new and old wall on the surface of the cell. A small blip as seen in sections or a ring of raised wall as seen by the scanning electron microscope [2] was found

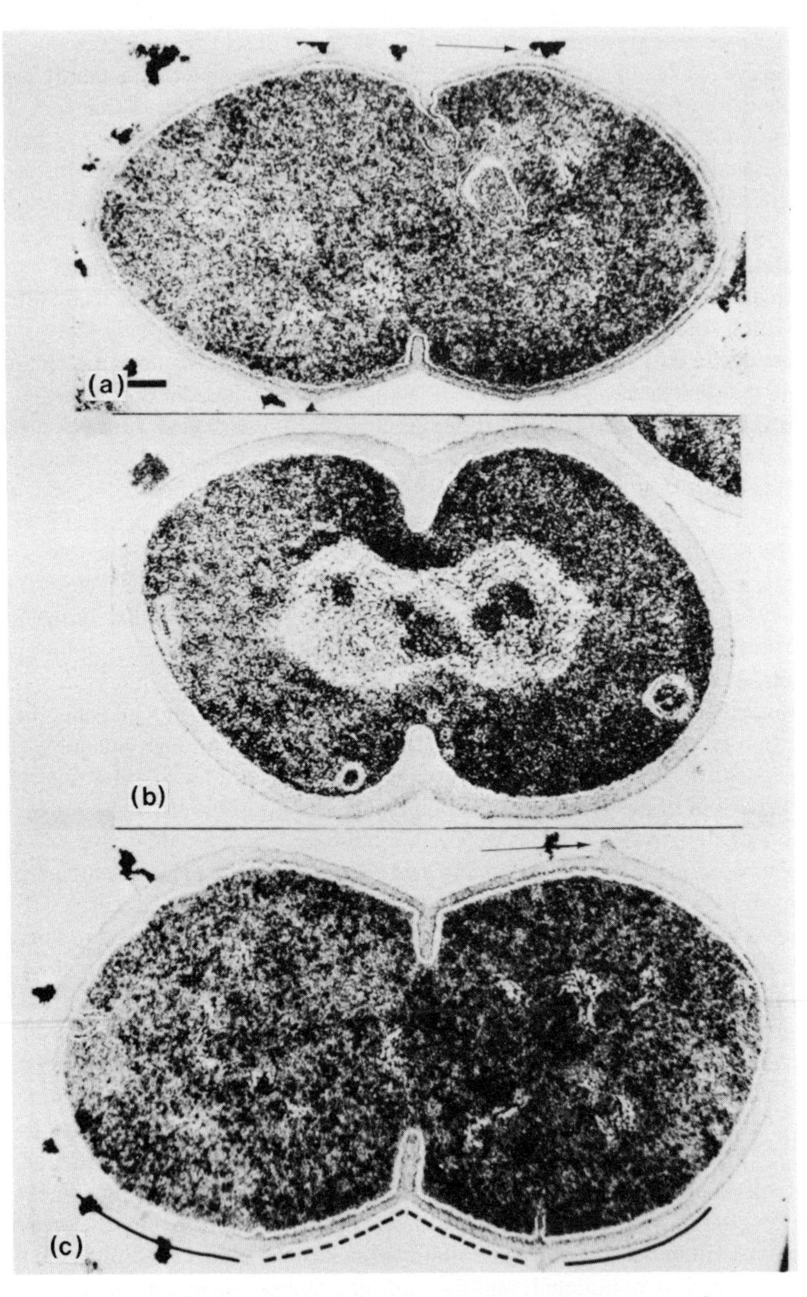

Figure 15.2 Longitudinal sections of *Streptococcus faecalis* recovering from treatment with chloramphenicol. Thus the new wall is thin (see broken line on c) and the old wall thick (solid line). These are separated by the external marker band seen in section here as raised blips. (Reproduction from slide, kindly provided by Professor G. D. Shockman.)

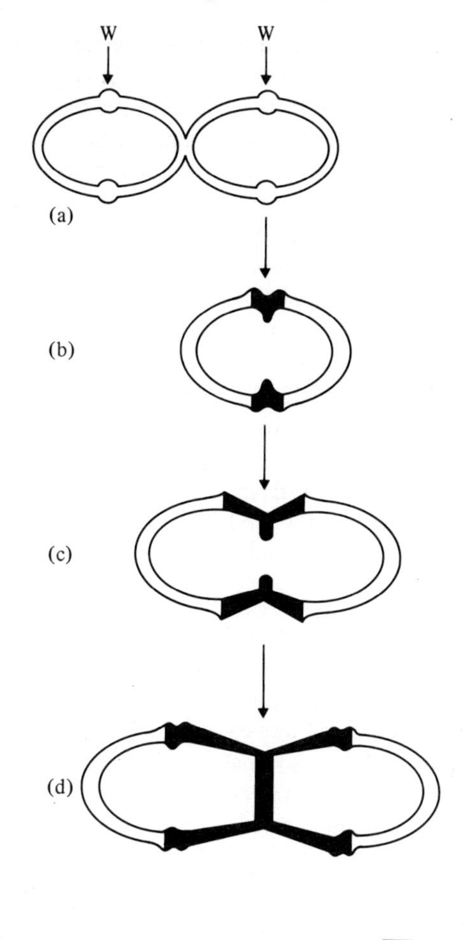

Figure 15.3 A diagram of a cycle of wall growth in *Streptococcus faecalis* based on the work of Professor G. D. Shockman and his colleagues. (a) The potential site for enlargement of the cell surface is located under the raised wall band, (W) at the equator (see also Figure 15.2); the cycle of growth (b–d) is shown in only one of the pair of cocci. The wall band marker splits (b) and new wall is bilaterally fed out from the septal region. When the septum has closed new wall enlargement sites are initiated at the centre of the daughter cocci (from [97]).

to remain either side of the septum. These rings moved apart during the extension of the wall that occurred when protein synthesis was restored. The sort of pictures obtained are shown in Fig. 15.2. Growth of streptococci may then be shown diagrammatically as in Fig. 15.3. Again a picture was obtained of the wall being fed out from the region at which cell division occurs. Although elegant, this wall-thickening technique has disadvantages and attempts [14, 41] to apply it to other organisms such as bacilli might be expected to fail (see Section 11.4). The streptococci had to be deprived of the amino acid necessary for protein synthesis during 10 h. incubation at 37° C in order to obtain sufficient wall thickening. When the amino acid was restored, the exponential growth rate characteristic of the medium was not re-established for some time and a proportion of the cells failed to grow normally, but formed septa at odd angles. The success of the technique is undoubtedly due partly to the special behaviour of the autolytic system of *S. faecalis*, when the streptococci are incubated in the absence of protein synthesis and partly to the absence of wall-turnover in this organism [14]. With bacilli, for example, which are much more lytic and have active wall-turnover, total cell lysis often occurs when protein synthesis is shut off and wild-type strains normally only continue to add wall material for relatively short times [58]. Under these conditions, the net result is that any thickened wall built up is removed by turnover even if frank lysis of the bacteria does not occur.

Returning to wall growth in the cells from exponential cultures of *S. faecalis*, all the above evidence is consistent with the model proposed by Higgins and Shockman [50], the essence of which can be conveyed in a quotation from their own paper:

> both peripheral wall elongation and centripetal cross-wall extension result from biosynthetic activity in the vicinity of the leading edge of the cross-wall. Enzymatic activity in this area of the cell results in (i) initiation of cross-wall penetration, (ii) external wall notching and peeling apart of the new cross-wall into peripheral wall, (iii) peripheral wall extension so that new wall bands are pushed to subequatorial positions and finally, (iv) centripetal closure of the cross-wall. This sequence of interrelated events probably occurs in a more continuous than stepwise fashion.

A more sophisticated approach, however, to the whole problem of surface growth and morphological change during division and replication of cells is now available [49] and has been applied to the growth of the wall of *S. faecalis* [52]. The principles of the technique are fairly simple. Exactly axial and longitudinal sections of suitable fixed cells are photographed and enlarged. These photographs are then marked and the *x* and *y* co-ordinates transferred to tape or traced round by a probe activated to record its positions in terms of them [22]. The co-ordinates are fed to a computer programmed to rotate the shape, around its central axis. A symmetrical three-dimensional form is generated from which volumes and areas of the whole cell wall or chosen parts of it can be extracted. In the very careful paper introducing this method, a number of the more obvious hazards in interpreting the results were examined. Firstly, it was established that unlike *E. coli* no significant shrinkage of the cell occurs during fixing with glutaraldehyde and osmium tetroxide, with the size of the freeze-fractured cell being used as a standard. It was observed, however, that the walls of the organisms are about 33% thinner in the fixed sections than in the freeze-etched preparations. It was argued, despite this, that since the increases of thickness that occur correlate well with increases in chemically determined peptidoglycan, it is reasonable to use, with caution, the calculated value of the volume of wall formed. Emphasis must be placed perhaps on the phrase 'with caution' until adequate density measurements have shown that the shrinkage of the walls is uniform over the whole surface.

Secondly, it was shown that providing the wall appeared as clearly tribanded throughout (see Section 1.2), sections had a very high probability of being radial and longitudinal. Thirdly, it was established that the closing septum or side wall, pictured as analogous to an iris diaphragm, closes in a symmetrical manner. Clearly this is essential if meaningful results are to be extracted for the surface area and

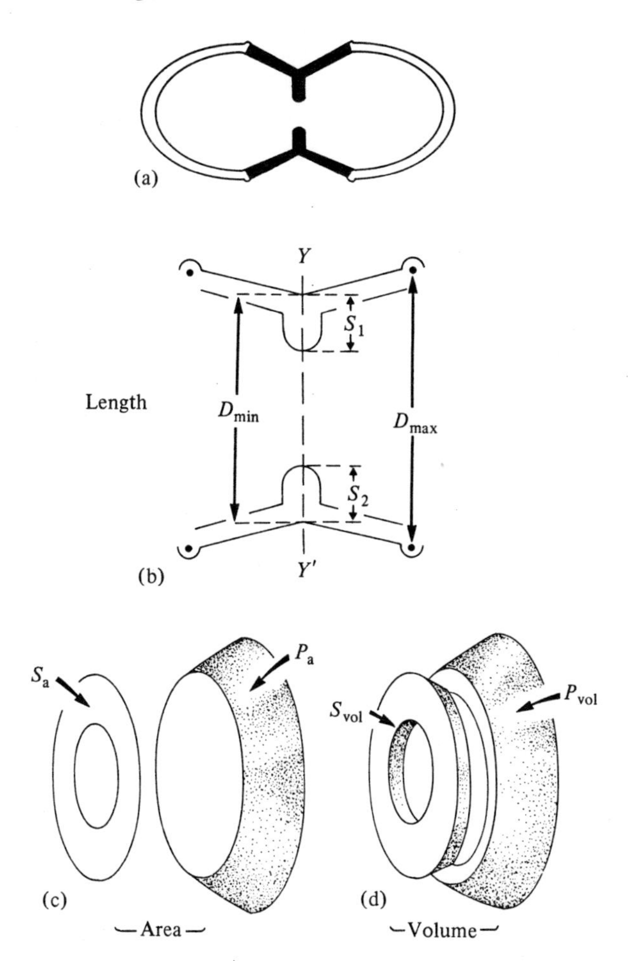

Figure 15.4 Analysis of the growth zone of organisms, in particular *Streptococcus faecalis* by the section rotation technique. In the diagram S_1 and S_2 are the lengths of the cross-walls, D_{min} is the diameter of the cells at the base of the septum, whilst D_{max} is the diameter at the centres of the wall band markers. P_a and P_{vol} are the surface area and volume of the wall in the newly formed polar material, and S_a and S_{vol} are the area and volume of wall to be deduced from the plane passing through the centre of the cross-wall (from [52] and [97]).

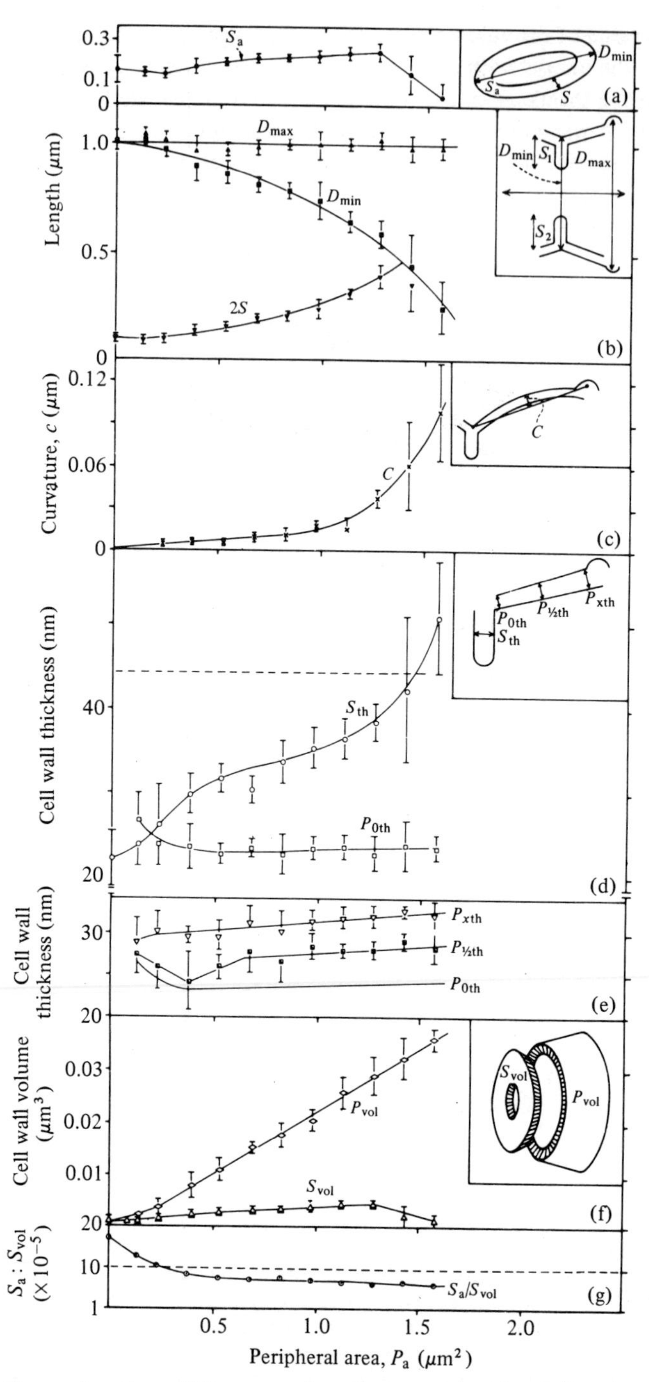

Figure 15.5 Reconstruction of the geometry of a nascent pole of *Streptococcus faecalis*, during a round of envelope synthesis. All parameters are plotted as

volume of the septum from the rotation about the axes of two-dimensional representations of the cells.

As we have seen, previous work had already shown that growth of the peripheral wall was occurring from the septal region. It was therefore reasonable to concentrate attention on this region in analysing the data. The sections were analysed in terms of trapezoids as shown in Fig. 15.4. Inasmuch as the peripheral walls are fed out as frustra of cones, this method of analysis is impeccable and by making the trapezoids progressively thinner, curvature of the walls can be accommodated. The error involved using trapezoids of a thickness that could be conveniently digitized was checked by dividing a circle in a similar way and comparing the volume obtained upon rotation with that for the sphere obtained by rotating the circle itself. The estimated error was less than 4%. A comparison, by this method, of the surface area of the septal wall (Sa) with the peripheral wall during growth and division of the cell is shown in Fig. 15.5. For the majority of the time that the peripheral wall is being formed the septum remains open and the septal area constant. Sa then begins to decrease rather sharply. During this latter phase, previous thinking suggested that the peripheral wall resulted entirely from a double layer of septal wall pealing apart, one layer forming the pole of each of the two new cells. The present more exact analysis, however, shows that more peripheral wall is formed than can be accounted for by such a straightforward process. Rather complex changes were found in the thickness of the wall which are also not consistent with a septum of twice the thickness of the new peripheral walls. The wall grows thicker as it moves away from the growth zone. To account for this, the original model for growth (Fig. 15.3) was modified to allow for wall deposition at points further along the growing wall (Fig. 15.6). A number of other aspects of cell growth were quantified in this first paper using the new method. Clearly it is too early to begin to think of molecular interpretations to account for the results. So far the obvious deficiency of the method is the lack of a time scale of cell division against which to set the morphological changes. Analysis of this sort is undoubtedly a powerful tool that will greatly contribute to our understanding of the processes of growth and division of bacterial and other cells.

---

functions of P_a (the surface area of the nascent pole; the bars represent ± one standard deviation). (a) shows that the area of the septal annulus remains nearly constant until two-thirds of the cycle is completed. Closure occurs at a value of 1.3 μm^2 for P_a. (b) shows how the value of D_{min} is gradually reduced during the cycle and the intersection of this curve with that plotted for the value of $S_1 + S_2$ (ie $2S$) the area of the septal annulus, occurs when the septum closes. (c) shows the curvature of the new peripheral wall, (d and e) show the alterations in thickness of the peripheral wall at the cross-wall (S_{th}), and at three points (P_{0th}, $P_{1/2th}$ and P_{xth}), along the gradient of its thickness. (f and g) show the alterations in volume of the newly formed wall (from [52]).

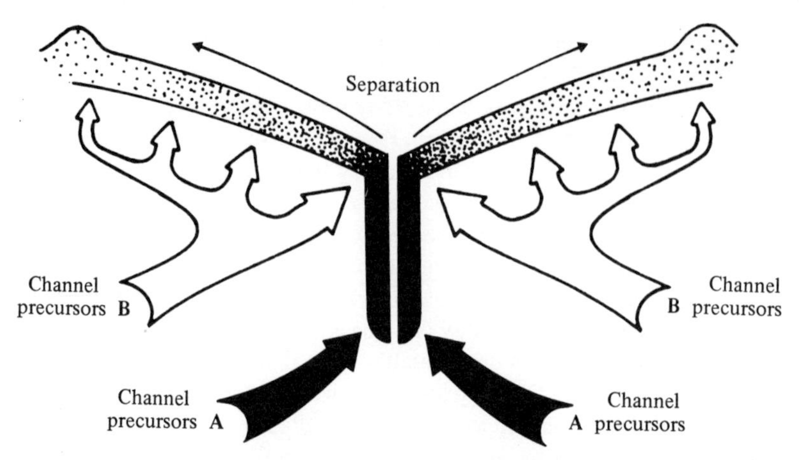

Figure 15.6 Diagram for the suggested additions of newly synthesized wall to allow for the results obtained by Higgins and Shockman [52]. Channel A would be involved in the continued growth of the cross-wall into peripheral wall, whilst Channel B would thicken and expand the wall as it is pealed outwards (from [52]).

15.3 Growth of the walls of Gram-positive rod-shaped bacteria

Knowledge about the growth of the walls of other Gram-positive organisms is in no way so complete, nor is there yet much degree of accord between different workers studying the problem. What follows must, therefore, to some extent, be measured against the likely prejudices of the authors.

Attention has been almost entirely concentrated on *B. megaterium* and *B. subtilis.* Apart from these two species among the rod-shaped Gram-positive bacteria, only *B. cereus* and *B. licheniformis* have received occasional attention. It may be easier for the reader if we start with the more recent work on *B. subtilis* and then look back to earlier observations and try to assess them in its light.

Measurements of the growth of *B. subtilis* cells by the section-rotation method of Higgins [49] became possible [22] partly because blips could be seen in sections of carefully fixed bacteria. The fixation method used [23] was found to inactivate autolysins rapidly and this may be highly relevant to the preservation of the blips. These appeared analogous to those in sections of *S. faecalis* (see p. 512). Moreover, in pictures of the organism taken by the high-resolution scanning electron microscope, raised rings can also be seen [2]. These rings occur near the poles of the cell. During division, they arise in the longitudinal mid-point of the rods and move apart as growth in length takes place. If the increasing region between the septum and the nearest raised ring is analysed by procedures similar to those used for *S. faecalis,* pole formation in the bacillus can be specified. Whereas the peripheral wall of the streptococci is fed out by the layers pealing apart from the open septum, the walls of the poles of *B. subtilis* are formed during and after septal closure. Plots of 2S (the length of the septal wall, see Fig. 15.4) and D_{min} (the

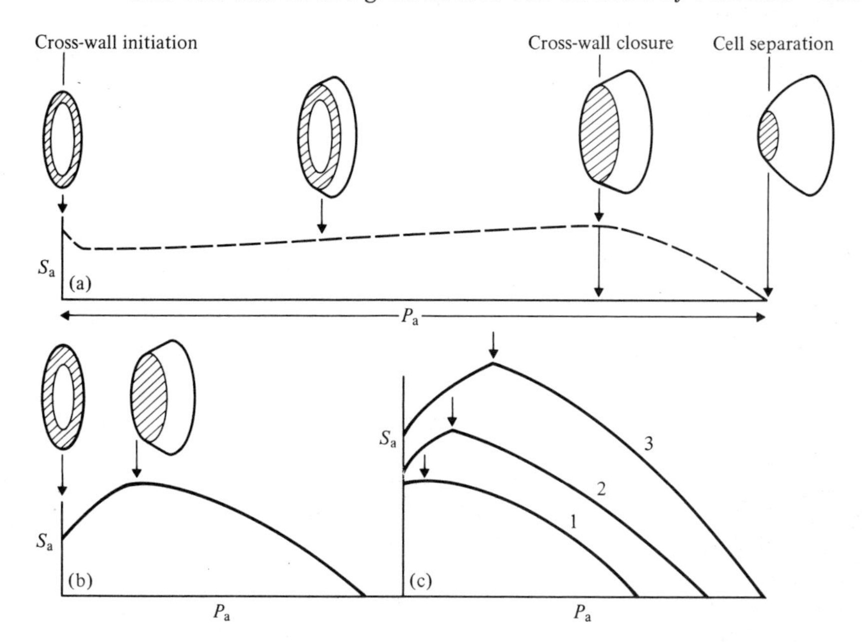

Figure 15.7 A comparison of the growth of the poles of *Streptococcus faecalis* (a) and of *Bacillus subtilis* (b) in terms of the increase in areas of peripheral (P_a) and septal wall (S_a). The septum in (b) closes much earlier in the growth cycle than in (a), and the S_a and P_a curve no longer shows the very flat form of the curve seen in (a). (c) shows curves illustrating the effect that would be caused by closure of the cross-wall at progressively later times in the cycle (from [97]).

minimum diameter of the separating cross-wall) intersect, when only about half the peripheral wall of the pole of a bacillus has been made [22], whereas about 75% of the streptococcal wall has already been formed when this happens (see Fig. 15.7). As in the streptococcus, the peripheral wall of the bacillus grows thicker as it is fed out, and again the total volume of polar wall is much greater than present in a flat disc. Thus, the formation of the poles of *B. subtilis* not only has some characteristics in common with peripheral wall formation by *S. faecalis*, but also demonstrates clear differences. The poles formed from the septal area of the bacillus account for only about 15% of the total wall volume, whereas all the peripheral wall of a streptococcus is thus formed. A plot of the formation of the whole wall of the rod shaped organisms would look as shown in Fig. 15.8 taken from Rogers *et al.* [97]. This shows that the majority of extension in the length of the rod occurs in a region not connected with the forming septum. We know little so far about the wall growth that accounts for the majority of the extension in length of the rod. Until this experiment was done, it was possible to suppose that the whole of wall formation by rod-shaped Gram-positive organisms was by feeding out material from the region of the developing septum, but that the areas of conservation thus implied, were disguised by some secondary rearrangements. Localized formation of

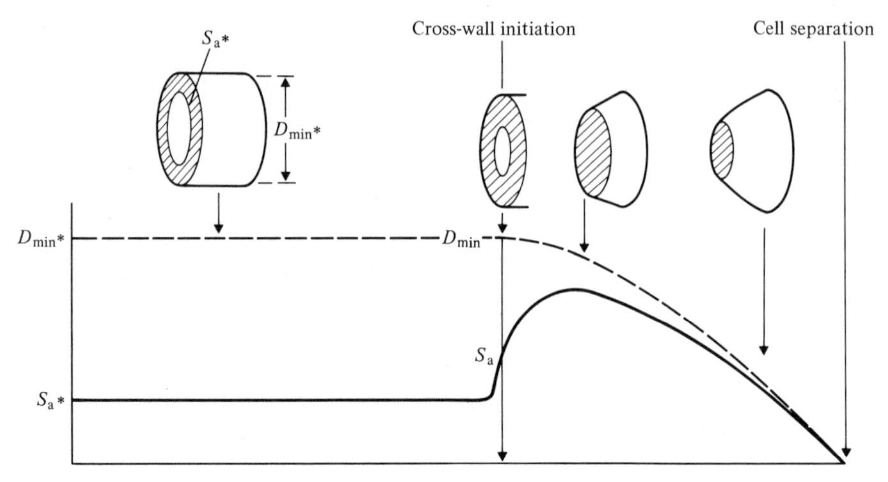

Figure 15.8 Diagramatic representation of the probable growth sequence for the whole surface (including the cylindrical part) of *B. subtilis.* It is assumed that longitudinal growth of the cylindrical part of the cell occurs by a zone or zones within this part of the wall and that while the majority of this peripheral wall is being formed, the septal wall does not appear. The cross-wall would then commence to be formed and the nascent poles arise by separation at its base as described in Fig. 15.7 (b) (from [97]).

wall in the region of the next division of the rod is not ruled out, of course. This would be at a place between the points at which a septum can first be seen to be developing, but if it happens it must occur before any trace of wall irregularity can be seen in median sections of the bacilli.

Returning for a moment to the experiment of Hughes and Stokes [57] (see p. 512) it will be remembered that this was done with a mutant strain of *B. licheniformis* that expressed very poor autolytic activity. Such strains as we now know show little ability to turnover their wall polymers (see Chapter 11.5.3). *S. faecalis* also fails to show wall turnover [14] yet *Lactobacillus acidophilus* which also has a muramidase as its only autolytic enzyme does so. *S. faecalis* and *B. licheniformis* Lyt$^-$ [57] show marked wall segregation when examined by the fluorescent labelled antibody technique but wild-type bacilli and *L. acidophilus* do not. One is, therefore, naturally tempted to wonder whether wall turnover alone may not disguise segregation in wild-type strains of bacilli. Examination of poorly autolytic mutants of *B. subtilis* [86] by autoradiography gives further support to this view. The mutant used, Ni15, had already been shown [84, 85] to be deficient in wall turnover. The walls of *B. subtilis* strain 168 Ni15 were labelled to a constant specific activity by growing the organism for 5–6 generations with ^{3}H-*N*-acetylglucosamine and then filtered off and transferred to unlabelled medium. The distribution of radioactive grains was measured after 0, 1.5, 3.0 and 5.0 generations of subsequent growth. The first and clearest conclusion to be drawn is that the wall material already present in the poles, when the bacteria were transferred to new medium, is conserved and not redistributed along the length of the cells. This result is to be expected from the electron microscopic work already described and is in excellent quantitative agreement with the foregoing since the conserved radioactivity accounted for 13% of the total wall present when the septa were formed; it will be remembered that polar wall accounts for 15% of the total cell wall. The distribution of grains along the longitudinal walls became increasingly non-Poissonian during the 5 generations of growth in unlabelled medium. In other words some degree of conservation of longitudinal wall seems to occur. After 5 generations of growth the probability of the distribution of grains being Poissonian was <0.05. Again, when a pulse of radioactivity was applied and chased through the wall, the probability of random distribution was low after even 1.5 and 3.0 generations of growth, providing only short cells in the population were considered. This was true for the wild-type as well as for the autolytic-deficient mutant. It would thus appear likely that wall deposition on the cylinder of the *B. subtilis* cells does not occur in a dispersed random fashion. That it took a number of generations of growth to demonstrate the non-random insertion of wall may be connected with the spreading of the old wall, as previously suggested to occur during cell growth [85]. The authors ([86] and Schlaepi J-M and Karamata, D unpublished work) conclude that insertion of new wall is localized and occurs at a small number of sites per cell, related to DNA segregation.

De Chástellier *et al.* [29] also studying the distribution of radioactively labelled new wall during the exponential growth of bacillary cultures, came to the reverse conclusion from that of Pooley *et al.* [86]. Before summarizing this work, however, it is important to point out the differences between the approaches in the two studies. Firstly, this work was undertaken with a different bacterial species namely a strain of *B. megaterium,* auxotrophic for lysine and 2,6-diaminopimelic acid. Secondly, very short pulses of radioactivity were used equivalent to 0.026 or 0.05 of a generation time, and no period of chase was applied. Thus, only differences in rates of formation of peptidoglycan will affect the amounts of radioactivity appearing in the wall. Sites where new chains of peptidoglycan are being initiated should be identified by this method and to this extent wall growth zones would be identified providing growth in length depends primarily on the initiation of new peptidoglycan chains. The distribution of radioactivity in the wall, however, is also likely to depend not only on incorporation into initiation sites, but also into partially completed chains. If biosynthesis proceeds by the addition of new units at the reducing ends of chains which are associated with the cytoplasmic membrane (see Section 8.5), then their labelling may spread throughout the apparent growth zone. If peptidoglycan chain extension takes place by addition by some unknown means in the depth of the wall, then more dispersion might be expected but much would depend on the orientation of the chains, the labelling of which might add to a random distribution of radioactivity. The bacterial strain used was not an autolytic deficient mutant and therefore the possible effects of wall turnover also have to be considered.

One of the surprising findings [29] was that 40% of the total number of grains in the autoradiographs was concentrated in narrow zones around the sites of cross-wall formation. The fully formed poles on the other hand were very poorly labelled. This latter result is easy to understand in the light of the result of Pooley *et al.* [86] and other work to be described later. If the mature cell poles are conserved and it is remembered that De Chastellier *et al.* [29] were measuring the *rate* of wall formation, then uptake of radioactivity into them from a pulse would be expected to be poor since they had been formed in the unlabelled medium and so needed no further wall synthesis, except possibly that required to make good turnover losses [36]. The very high incorporation into the region of the forming cross-wall, however, is another matter and difficult to understand in terms of our knowledge in the growth of the wall of *B. subtilis.* If only some 13–15% of the surface of the whole bacterium is formed by growth at the septal region, then to account for the accumulation of 30–40% of the radioactivity into this region, the local rate of peptidoglycan synthesis would have to be very high indeed compared with that into the longitudinal wall of the rod. In *B. subtilis* grown under the conditions used by Burdett and Higgins [22] the septum closed when its area was about 0.09 μm^2. This occurred about half way through formation of the pole which signalled separation of the fully extended bacterium. The ratio of the average rate of formation of the cross-wall, relative to that of the longitudinal wall, in this organism would then be between 7 and 151. Although one would expect more grains from a very short pulse to be deposited in the septum, the very high value found for *B. megaterium* is not easily explained in terms of observations made on *B. subtilis.* Examination of the distribution of radioactivity along the wall of the longitudinal rod of *B. megaterium* led the authors to find that there was no evidence for any localized deposition. Certainly the histograms presented no very obvious clearly demarcated zones of deposition. It is less clear, however, whether the distribution of grains is Poissonian or, as Pooley *et al.* [86] find for *B. subtilis,* non-Poissonian. Moreover it must be remembered that non-Poissonian distribution was found in *B. subtilis* only after chasing the radioactivity through the wall for more than one generation of growth. This was irrespective of whether the label was a general one throughout the wall or had been applied to the bacteria as a pulse. However, no evidence for conservation of labelled wall or non-Poissonian distribution of grains in the walls of *B. megaterium* during growth of cells initially labelled with ^{3}H-2,6-diaminopimelic acid in unlabelled medium, was reported in another experiment [71].

Another quite different approach to the study of the topology of growth of Gram-positive rods has been that involving the use of bacteriophages. Certain of the latter have the glucose residues of wall teichoic acids in *B. subtilis* as important components of their receptor sites [43, 118]. Therefore, the distribution of teichoic acids, presumably at the outer surface of the walls, can be demonstrated by the pattern of bacteriophage adherence as seen under the electron microscope.

When some organisms are grown in continuous culture, the composition of the walls can be manipulated by the nature of the nutrient limitation. Thus of many Gram-positive organisms grown under phosphate limitation, for example, the teichoic acids in walls are replaced by uronic acid containing phosphate-free polymers [32, 33, 39] which do not fix the bacteriophages. Such a change in one group of wall polymers can also be used to indicate behaviour of the peptidoglycan because during such changes, the newly appearing polymer (eg during phosphate limitation, the teichuronic acid) is covalently attached only to new strands of peptidoglycan and not to old existing strands [70]. Using *B. subtilis* strain W23 and bacteriophage SP50, Archibald [3, 4] made use of these observations to study wall growth. In accordance with the ideas of Pooley [84], there was a considerable lag after switching the nutritional limitation before the phage binding changed. For example, a generation of growth had to occur after changing from phosphate to potassium limitation, before maximum phage binding capacity was expressed. There was little or no lag, on the other hand, in the appearance of teichoic acid in bulk preparations of the wall. It is reasonable to suppose that teichoic acid is first deposited on the inner layers of the wall, and that the lag represents the time required for this material to grow outwards through the thickness of the wall. Switching the limitation in the opposite direction gave even more impressive results since about 6 generations had to elapse before the phage binding sites had disappeared, despite the fact that 90% of the teichoic acid had disappeared from the wall in 2–3 generations. The change from 43% of the maximum phage binding capacity by the cells to <5% involved the loss of only a further 4% of the total maximum wall teichoic acid. As would be expected, walls from bacteria examined soon after the change from phosphate to potassium limitation, fixed phage only on their inner edges and vice-versa. The distribution of phage particles along the length of the walls both on their inside and outside layers did not show marked non-random distribution. However, very marked conservation of the material covering the poles of the cell, or sometimes one pole only, was clearly demonstrated, and presumably accounted for the very slow disappearance of phage fixation sites mentioned above. For example, complete coverage of the older poles of pairs of cells was still present about 3–4 generations after applying phosphate limiting conditions. Thus, these experiments confirm the conservation of polar material but do not identify localized zones of growth. Once again wild-type strains were used in which active wall turnover would be expected. The extent of randomness of the particles fixed on the longitudinal walls was not examined statistically. Whether or not cross-wall formation is particularly rapid could not be explored, of course, because this structure is inaccessible to the bacteriophage until it is revealed as a mature cell pole.

We can now afford to return to earlier attempts to examine the growth of the walls of Gram-positive rod-shaped bacteria by the methods so successfully applied to the cocci. The wall thickening method (see p. 512) has been applied to *B. subtilis*, *B. megaterium* [41] and to *Lactobacillus acidophilus* [14]. In all three cases the attempt to look for the growth of unthickened wall was confused by extensive damage along the cylinder of the wall presumably caused by autolytic enzymes. The poles of the cells in all three examples, however, appeared thicker and relatively undamaged. These observations are explained by the relative insusceptibility to the action of autolytic enzymes of the poles of some bacilli compared with the side walls [34, 35]. However, it would appear that the poles of *B. subtilis* cells turnover as well as the side walls [36]. How far the preservation of the poles in wall thickening experiments is, therefore, related to their conservation during labelling is problematical.

Labelling of bacilli with fluorescent antibodies by the method successfully used for streptococci has also been studied by Chung *et al.* [25] as well as by Hughes and Stokes [57]. The results obtained with *B. cereus* appeared to show well demarcated localized zones of growth; even more distinct than those later found in the autolytic deficient strain of *B. licheniformis* [57]. However, in order to obtain good fluorescent labelling the authors had first to grow their strains with trypsin and ribonuclease. It is, therefore, possible that the final results reflected the formation of some capsular or other extracellular material during growth in the absence of these enzymes rather than giving a true picture of wall growth. Further study is necessary. Another attempt to study wall growth by using the wall teichoic acids relied upon identifying their presence on the surface with concanavalin A [31]. This polypeptide combines with glucose when linked by α-glycosidic bonds. If the concanavalin is linked to a fluorescent dye or ferritin, its distribution on the cell surface can be identified. The experiments were done with a phosphoglucomutase negative mutant of *B. subtilis.* Such mutants have two important relevant characteristics. Firstly, when grown on media not containing galactose they cannot glycosylate polymers in the wall, being unable to make the essential intermediate glucose-1-phosphate. If, however, the organisms are grown on an assimilable carbon source such as glycerol, and galactose is also present, the phosphoglucomutase lesion is partially circumvented and some glycosylation of wall polymers occurs [12, 38, 39, 42, 118]. The second relevant characteristic of such mutants is that in the absence of galactose in the medium they are deficient in autolysin [37, 38]. Doyle *et al.* [31] followed the appearance of glycosylated teichoic acid after the addition of galactose to cultures by the application of fluorescein-linked concanavalin A. No evidence of distinct growth zones was found, along the cylinder of the cells. Neither were zones present in the converse experiment when galactose was removed from the cultures. In the latter experiment the autolytic enzyme action and therefore wall turnover would be greatly reduced.

In summary, it would seem that the walls of rod-shaped bacillary cells do not extend from a single growth zone, as does the peripheral wall of streptococci. Nevertheless, the very careful experiment by Pooley *et al.* [86] suggests that extension by diffuse intercalation of new material randomly all over the walls is also unlikely. That positive identification of localized growth zones has been obtained in experiments with autolytic deficient strains with low wall turnover and not in others shows how important turnover may be. Clearly, more work is required to examine just how far wall turnover and the great thickness of the walls of Gram-positive organisms confuse the pictures so far obtained.

15.4 Growth of the Gram-negative cell wall

The problems involved in looking at the topology of the growth of Gram-negative rod-shaped bacteria are necessarily somewhat different from those met in studying the Gram-positive forms. As will be remembered (see Chapter 1.3), in the former the structure of the envelope is more complicated with an outer membrane having the lipopolysaccharides inserted in it. Underlying this is the thin layer of peptidoglycan with its covalently attached lipoprotein which anchors the outer membrane to it. This structure has advantages and disadvantages for the interpretation of the results from the various methods for examining the topology of wall growth. Most of the general problems have already been met in studying the Gram-positive bacilli. Large molecules or entities such as antibodies or bacteriophage can be used to label the lipopolysaccharides and outer membrane and only the problems associated with possible mobility of membrane components might be thought to arise (see Chapter 3.2). On the other hand, the peptidoglycan layer is thin, certainly consisting of no more than 2-3 layers, so that methods such as autoradiography after radioactive labelling are less liable to confusion. Also, the peptidoglycan turns over only very slowly if at all.

15.4.1 *The outer membrane*

Early experiments employing fluorescent antibodies [26, 72] with *E. coli* and *S. typhimurium* showed only a diffuse spread of the label during growth with no evidence for conserved areas. Until recently it was usual to disregard these early experiments because of the possible mobility of the antibody molecules when attached to the surface lipopolysaccharides. However, it is now clear [78] that the mobility of LPS molecules in the outer membrane is only low. The distribution of receptor sites for bacteriophage λ gives a different picture [101, 103]. In this work, use was made of the observation that *E. coli* only fixes significant numbers of phage λ particles when the organisms are grown in the presence of maltose and *c*AMP. This appears to be because the gene *lamB* specifying the λ receptor protein in the outer membrane happens to be located in one of the maltose operons [17, 56] (Chapter 4.10.3), so that when maltose and cAMP are added to glucose grown cultures, λ receptors appear linearly with time after a lag of 4-5 min. Uninduced bacteria fixed only a very small number of phage particles, whereas after full induction the surfaces were almost completely covered. The bacteria were then induced for 9 min which represents 0.07 of a generation, allowing for the lag in receptor synthesis of 5 min and a doubling time of 60 min. Two classes of organisms were found:
(1) Larger cells which had a concentration of particles spreading out from the septal areas. These were recognized by the indentations that appear midway along the cells. About 30% of the surface of these organisms was covered.
(2) A class of small organisms on which the bacteriophages were fixed only to the polar areas. Presumptive evidence was produced to suggest that cells of the second

type arose from the first by the process illustrated in Fig. 15.9. The implication of these results is that the receptor sites in the outer membrane only become operational in a period towards the end of the division cycle. Attempts to show that, once formed, the receptors could migrate by lateral diffusion through the membrane were unsuccessful, as is supported by the work of Mühlradt *et al.* and Nikaido [77-79]. The authors, therefore, suppose that the formation (and/or export through the cytoplasmic membrane) of the receptor protein takes place only during the latter part of the division cycle in the region of the forming septum. The receptors then preferentially fill up the outer membrane in the region, diffusing in the periplasmic space. When longer times of induction are used, they eventually cover the whole surface of the cells. Accepting this picture it has to be suggested that different components of the outer membranes of Gram-negative bacteria may be added in different ways. Lipopolysaccharide, capsular polysaccharides [7], and the porin proteins [108] for example, appear to be formed around the adhesion sites between cytoplasmic membrane, peptidoglycan and outer membrane that were first recognized by Bayer [5, 6]. These are randomly distributed over the surfaces. The lipopolysaccharides move out from such centres [77, 78] to cover the surface. This is probably unlikely to be by lateral diffusion alone because of the relatively slow rate of this process in outer membranes of Gram-negative bacteria [9].

Other experiments to follow outer membrane growth by the use of bacteriophages have been done with T6 [10, 11, 64]. The results obtained by Begg and Donachie [10, 11] need particular comment. These authors used a strain of *E. coli* carrying two lesions one in *tsx* that controls the formation of the receptor for phage T6 (see Chapter 4.10.3) and a temperature-sensitive one that can suppress this gene. Thus, by simply shifting the temperature of growth, they could either stop or start receptor synthesis. A factor which may complicate interpretation of this work is that the authors added penicillin at the time of the temperature shift in order to

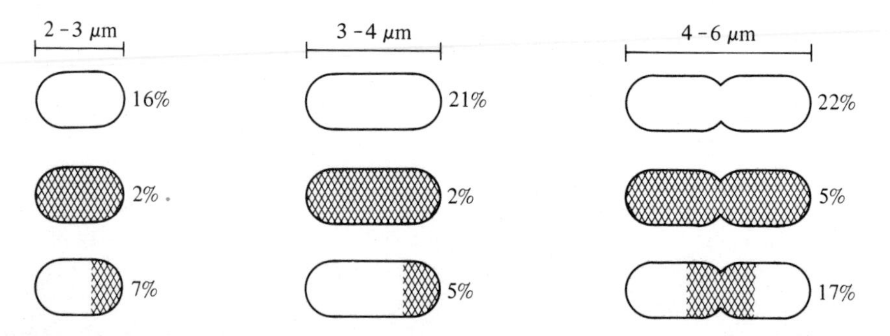

Figure 15.9 The distribution of various classes of bacteria after 9 min of λ receptor induction. The bacteria were classified according to their size and a total of 360 were examined. Unshaded bacteria carry less than 50 phage particles whilst fully shaded ones carry more than 50. The bottom line reading right to left shows the likely origin of the small cells carrying phage particles at one pole (from [103]).

stop cell division. Although the authors say such treatment did not affect 'the growth in mass or length', this may be a rather over-simplified attitude to the effects of penicillin upon Gram-negative cells. A further problem in the work is the apparent inefficiency of the temperature-sensitive suppressor. Cells grown at 42° C fixed a mean of 2 ± 3 phage particles, whilst at 30° C they fixed only 19 ± 13. The authors found that, when the synthesis of receptor sites was switched off, the distribution of T6 particles fixed became non-random with a strong bias against the cell poles. It should be noted, however, that even after growth at 30° C, there was a tendency for there to be phage particles fixed to the poles. The results were taken as evidence for the authors' own hypothesis [30] that cells of *E. coli* grow from their poles. They considered that the insertion of the receptor sites was random over the whole membrane and that sites once inserted did not diffuse. Thus, when synthesis was switched off, the new membrane at the growing poles of the cell did not have receptor sites inserted. This work suggesting polar growth of the membrane has been strongly criticized by Smit and Nikaido [108] who examined the insertion of one of the major outer membrane proteins; a porin (see Chapter 3.4.2) in *Salmonella typhimurium.* They could find no evidence for polar growth of the bacteria. The formation of the outer membrane protein of a molecular weight of 35 000 was very sensitive to the ionic strength of the medium. When a mutant strain unable to form two of the major proteins (those with molecular weights of 34 000 and 36 000) was grown in a yeast peptone medium containing 0.5 M NaCl, no protein of 35 000 mol. wt. could be detected by polyacrylamide gel electrophoresis. By switching the cells to and from media with and without added NaCl, the topology of formation was examined using ferritin-labelled antibody made against the purified 35 000 mol. wt. protein. The mutant strain used for these experiments also had to be defective in lipopolysaccharide formation. The conclusion was that as with lipopolysaccharide and capsular polysaccharides, the 35 000 mol. wt. protein was inserted into the outer membrane via the adhesion sites [5, 6] in the envelope. These are randomly distributed over the whole surface of the cells. No evidence could be found for preferential insertion of proteins into the poles. They suggest that possibly the results of Begg and Donachie [11] might be explained by the normal distribution of T6 receptors themselves. Begg [9], in a later experiment, examined the distribution of radioactivity in 'ghost' preparations containing only matrix (porin) proteins prepared from *E. coli* cultures which had been pulsed with ^{3}H-histidine. The distribution of grains was random whether or not penicillin had been present. No part of the outer envelope had been conserved.

The general impression is that, in fact, the outer membrane of the Gram-negative envelope may well be built up by diffuse intercalation over its surface with material being exported via adhesion sites. This leaves, difficulties in explaining the observed [103] septal distribution of λ receptor sites. One is inclined to agree with Ryter *et al.* [103] that this experiment may reflect either the site or timing of synthesis of the receptors under the conditions of induction, or the availability of space allowing their insertion, rather than sites of synthesis of the outer membrane as a whole. By hindsight, it is salutary to reflect that if the outer membrane is formed by diffuse intercalation, the interpretation of the results obtained by Cole *et al.* [28] with fluorescent antibody, available nearly 20 years ago, was quite straightforward and uncomplicated.

15.4.2 *Growth of the peptidoglycan layer*

Problems concerned with the growth of this part of the envelope structure are clearly likely to be different from those for the outer or cytoplasmic membrane. Unlike the membranes, the peptidoglycan, just like that in the walls of Gram-positive bacteria, is a covalently constructed relatively immobile structure. There would seem to be no necessity for two very different structures (ie the sacculus and the outer membrane) to expand by the same mechanisms. There would be nothing incompatible with the growth of the peptidoglycan from localized areas and that of the outer membrane by diffuse intercalation.

Three recent sets of experiments have been done to examine the growth of the peptidoglycan or sacculus of *E. coli* [62, 102, 105]. All three involved labelling with ^{3}H-2,6-diaminopimelic acid followed by preparation of the sacculi and autoradiography. In the experiments by Ryter *et al.* [102] a narrow band of intense radioactivity was found at the division plane after pulse-labelling either fast or slow growing cultures. A second more diffuse band was present, particularly in fast growing bacteria, approximately mid-way between this and the poles of the two future cells. These results were very well supported by the subsequent study of Koppes *et al.* [62]. In the first experiments [102], a chase was carried out by washing the pulsed-cells and regrowing them in media not containing the radioactive label. A surprisingly complete randomization of the label took place without loss from the sacculi. Considering the nature of peptidoglycan and the fact that the spreading was complete in as little as 0.25 of a generation of time, the only obvious possibility is that a considerable amount of the labelled 2,6-diaminopimelic acid remained in the pool in the cells, despite the washing procedures, and continued to be incorporated during the chase period. Examination of temperature-sensitive filamentous mutants at the restrictive temperature showed, as was to be expected, a reduction in incorporation in the region of septa. The different strains were surprisngly variable however, and the results were not easy to interpret on the basis of a simple inhibition of septum formation.

The relationship of these more recent results, which would seem to support localized growth of the peptidoglycan layer in *E. coli*, to earlier studies [65, 115] is unclear. In the older work, *E. coli* auxotrophic for diaminopimelic acid was heavily labelled by growing it with ^{3}H-diaminopimelic acid and then the organisms were grown for a number of generations in unlabelled medium. There was no loss of label from the cells and it was concluded that the original labelled material was diluted without conservation of wall, as the organisms grew and divided. Such a dilution could only result by insertion of new material at a large number of sites: a conclusion apparently at variance with the more modern work.

Both Ryter *et al.* [102, 105] and Koppes *et al.* [62] concluded that incorporation of radioactivity into peptidoglycan accelerated towards the end of the cell cycle. Koppes also examined DNA replication and suggested that the emergence of the lateral areas (that is, areas of incorporation about mid-way between the developing septum and the poles of the cell) was related in time to DNA segregation. Other workers [46, 55] have found that the incorporation of proteins and lipids into the wall of *E. coli* as well as murein synthesis accelerated 15–20 min before cell division. The conclusion reached by Koppes *et al.* [62] is that 'genome doubling and the new spatial orientation of the genomes with respect to the cell envelope could then be the trigger for new growth areas (beside the central one)'. This conclusion is similar to the general ideas developed by Sargent [104] working with *B. subtilis*, although it has only recently been possible to see localized zones of growth in the bacillary wall (see p. 522). Such ideas have obvious affinities with the replicon hypothesis [59] but they, in effect, stand the original hypothesis on its head. The original hypothesis said that the segregation of the nuclei took place as a result of growth of the envelope. According to present ideas, wall growth zones appear as a result of segregation. The authors are well aware that one category of experiments has been excluded from this consideration of the growth of the surfaces of Gram-negative species of bacteria: namely those that use the location of penicillin-sensitive sites as markers of new wall synthesis. The exclusion is deliberate since it is felt that when penicillin is applied to the organisms the situation is too complicated for ready interpretation and the sensitive site may mark the site of action of autolytic enzymes rather than the site of envelope growth (see Section 11.4).

15.5 Growth of cytoplasmic membranes

An understanding of the topology of the growth of the outermost layers of bacteria is clearly an important way to approach cellular growth itself. Nevertheless, it is important to realize that wall growth could be thought of simply as the accretion of insoluble material on to the surface of the cytoplasmic membrane. Important enzymes concerned with the later stages of peptidoglycan synthesis are firmly fixed in this membrane (see Sections 8.3–8.6). Their surface distribution and localized regulation could well dictate the orientation of the wall and thus decide between length extension or division of rod-shaped organisms. According to such thinking, the prime mover in cell growth and division would then be the behaviour of the cytoplasmic membrane. Unfortunately, if the topologies of the expansion of the walls of some Gram-positive organisms and the outer membrane and sacculus of Gram-negative species have proved difficult to define unambiguously, those of cytoplasmic membranes have so far been impossible. One need not look far for the reason for this difficulty since it is well known that the component molecules are likely to be capable of rather rapid lateral movement (see Chapter 3.2). The lipids move very fast and the proteins at quite appreciable rates. Attempts then to look for conserved entities in the membranes are likely to be vitiated by randomization. In a number of experiments, no clear evidence for the conservation of membranes has been found using labelled lipid components to demark 'new' from 'old' material [40, 45, 65, 76, 114]. It is reasonable to assume that these difficulties are due to lateral movement of the lipids not only after membrane isolation but also *in vivo.* Sargent [104] points out that we know rather little about the dynamics of the membrane lipids in growing cells where the situation may be very different from that in artificial membranes or even in membranes isolated from cells.

The observation [45] of a large number of discrete growth sites in the membranes of *E. coli* B15 remains difficult to understand. When *E. coli* was uniformly labelled with ^{3}H-2-glycerol and grown in unlabelled medium, the distribution of radioactivity in the cell population remained Poissonian for about 6 generations. Departures from random distribution however, then occurred and a class of unlabelled cells appeared. Analysis showed that the rate of appearance of this departure corresponded to about 2^7 to 2^8 sites for membrane formation. The relationship of this to the non-Poissonian distribution of wall label [86] needs consideration. Attempts to show a clearer picture of conservation in membrane synthesis using either mini cells, assuming that these represented poles of cells (or more accurately one old pole and one special septum) and filamentous mutants, failed. The same negative results were obtained when the bacteria were first pulsed with ^{3}H-aminolaevulinic acid to label the haem in the cytochromes.

The only evidence for conservation of membrane comes from an ingenious study of the inheritance of permeases [60]. In these experiments, *E. coli*, was induced for β-galactosidase with thiomethylgalactoside for a number of generations, washed with EDTA to remove any pool of inducer and then grown in lactose medium containing 250 units/ml of benzylpenicillin. Under these circumstances, only those bacteria carrying an induced permease would be able to grow and be killed by the penicillin, which only kills 'growing' bacteria (see Chapter 11.5.6). The authors then measured the rates of lysis of the bacteria. They found that after a time a step occurred in the curves (see Fig. 15.10). They interpreted this as showing that the culture was, by this time, a mixture of induced and uninduced cells: a result explicable if the membrane including the permease was conserved and not shared between daughter cells. They obtained similar results using a number of other inducers although not all. When the bacteria were labelled with ^{3}H-2-glycerol and ^{14}C-stearate the glycerol label was conserved during deinduction whereas the fatty acid was not. Such a result for glycerol incorporation conflicts with much other evidence and, in view of the likely mobility of lipids in the cytoplasmic membrane, is difficult to understand. It must also be noted that other, less often quoted, studies of the cellular inheritance of permeases [61] came to contrary conclusions about conservation of the membrane site. Other early studies of the kinetics of induction of enzymes such as β-galactosidase using sub-maximal concentrations of gratuitous inducers [81] are sometimes used as further evidence against membrane conservation but alternative explanations are probably applicable. Evidence for membrane conservation that is not easy to dismiss, however, is an

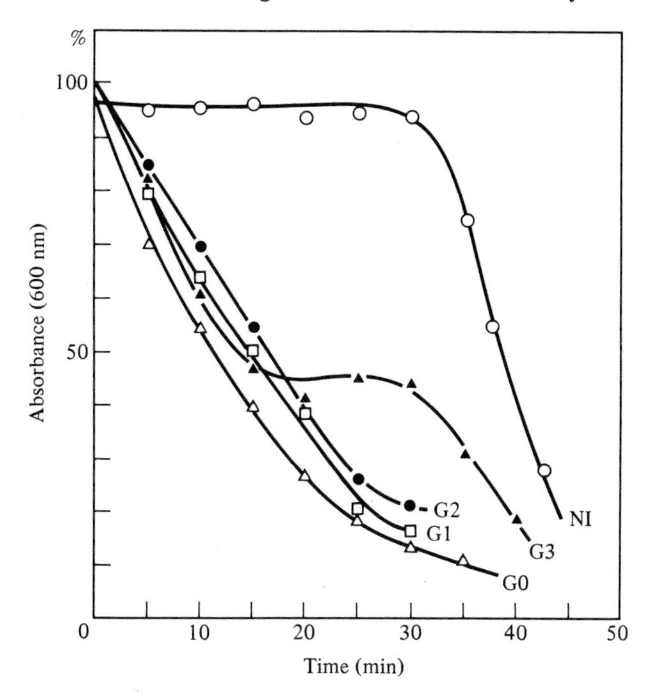

Figure 15.10 Lysis of *E. coli* by 250 units/ml of benzyl penicillin during growth on lactose containing medium after pre-induction of β-galactosidase and subsequent deinduction. The pre-induced bacteria were deinduced by 0, 1, 2 and 3 generations (G0, G1, G2 and G3) of growth in a glucose containing medium not containing inducer. Fully induced populations (G0) lyse immediately whilst non-induced bacteria (N1) do not lyse for 30 min. After 3 generations of growth in the deinducing medium (G3) the population behaves as if it were heterogeneous. About half the bacteria grow and lyse rapidly whereas half lyse later. This was thought to be the explanation of the plateau in the curve G3 and to be explained by a non-random distribution of permease molecules in the bacterial population (from [60]).

early study of the distribution of tellurite crystals in the cytoplasmic membrane of *B. subtilis* [99]. When potassium tellurite is supplied to exponentially growing bacteria, it becomes reduced by membrane enzymes to form tellurium crystals which can be seen under the electron microscope fixed to the membrane. If potassium tellurite is supplied as a 'pulse' to non-aerated cultures, and then aerobic growth allowed to occur in its absence, only the old membrane is labelled with tellurium crystals. The distribution of the crystals was followed for up to 3 generations of growth and suggested a strongly conservative growth with new membrane synthesized in the nuclear region, whilst that at the poles was conserved. One possible vitiating circumstance to this experiment would be toxicity of the tellurium crystals, forcing membrane synthesis into regions of low concentration. However, there is no evidence for or against this hypothesis. The topology of membrane growth supported the replicon hypothesis of Jacob *et al.* [59].

Another very different type of experiment that similarly supports localized growth of either the wall, the cytoplasmic membrane or of both in *B. subtilis* is with a mutant conditionally affected in flagella formation [100]. This strain formed flagellae when grown at 30° C but not at 42° C. When the cells were grown first at 30° C and then at the higher temperature they became bare of flagella in the nuclear region suggesting that this was the situation for the formation of new cell surface. A generalized conclusion that growth of rod-shaped bacteria occurs from near the new pole formed by a previous division was drawn from a study of the Gram-negative *S. typhimurium* and *Proteus vulgaris* temperature-sensitive for flagellum formation [13]. However this study was made by the examination of individuals in micro-colonies growing on collodion. A study [87] of individual cells of *S. typhimurium* isolated by micromanipulation came to the conclusion that each flagellum was segregated into a separate daughter cell, consistent with non-conservative generalized growth of the surface. The exact relationship between the distribution of flagella and the growth of the surfaces of the bacteria is not clear and more evidence is needed about the specification of the location of flagella on the surface.

Thus even in the morphologically simple rod-shaped organisms, evidence about the topology of the growth of the cytoplasmic membranes is so conflicting that no dogmatic conclusions can be drawn.

15.6 Mutants with disturbed surface growth

A variety of mutants of bacteria have been described that grow with shapes markedly different from that of the wild-type and some of these are conditional. Wild-type strains of *Arthrobacter* species likewise change their morphology according to nutritional circumstances. Reference has already been made (p. 529) to the experiments examining the growth of the peptidoglycan sacculus of conditional filamentous mutants of *E. coli*. Little work has been done on the surface growth of filamentous mutants of Gram-positive rods.

Considerable attention has been paid to other types of conditional mutants that have been isolated from both Gram-positive and negative species. These, like Arthrobacter grow as either cocci or rods according to the conditions. Conditional so-called *rod* mutants were first isolated from a strain of *B. subtilis* [94, 95]. A non-conditional spherical mutant of *E. coli* was simultaneously isolated [1]. Subsequent to this a variety of *rod* mutants has been isolated from these and other species (see Table 15.1).

The nature of the growth of the surface of the rod and coccal forms of *B. subtilis rod B* has been examined [20, 21]. These mutants can be changed in morphology by alterations in the Mg^{2+} and anion concentrations in the growth medium [96] or by a shift of temperature [91]. They have the advantage that when cultivated in a suitable minimal-salts-glucose medium [96] they grow as single cells, allowing measurements of cell dimensions. Temperature-conditional *B. subtilis rod A* mutants, on the other hand, grow as large groups of unseparated

Table 15.1 Morphological mutants of various species of rod-shaped bacteria that can grow as cocci (ie *rod* mutants)

Bacterial species	*Genotype or phenotype*	*Growth conditions*		*Reference*
		rods	*cocci*	
Bacillus subtilis	*rod* A	30° C	45° C	90
	rod B	high Mg^{2+}	low Mg^{2+}	
		halide ions	high temperature	
		NO_3^-		
		high glutamate	no glutamate	
		low temperature	low ionic strength	
	rod C	0.8 M NaCl	low ionic strength	
	gta C	normal media	PO_4^{3-} limitation	
	glu 8332	high glutamate	glutamate limitation	
Escherichia coli	*env*B	spherical and not conditional		80
	Rod⁻	minimal media	complex media	82
		complex media + D-alanine		
	rod	30° C	42° C	48
	rod A	spherical and non-conditional		68
	rod⁻	spherical and non-conditional		1
	mecillinam resistant	spheroid and non-conditional		69
	Rod⁻	changed from *rod* to sphere by carbon source		63
Agrobacterium tumefaciens	1) *Rod*⁻	27° C	37° C	42a
	2) *Rod*⁻	spheroid and non-conditional		
Klebsiella pneumoniae	*Rod*⁻	pH 5.5	pH 7.0	73a

cocci at 42° C [15, 28] presumably because autolytic activity is switched off [16, 19] and this is known to lead to difficulties in bacterial cell separation (see Chapter 11.5.2). It will be remembered that in wild-type *B. subtilis* only about 15% of the surface area of the rod-shaped organism is derived from extension from the septal area (see p. 520). As the *rod B* mutant was converted to a coccus increasing proportions of the peripheral wall had this origin. When finally growing as a coccus >90% of the new wall was thus derived. Thus, in this sense, growth of the coccal form of *rod B* mutants closely resembles that of *S. faecalis.* However, whereas the septum of the streptococcus closed only after most of the peripheral wall had been made (see p. 518) in the *rod B* coccal form, the septum closed much earlier. This, of course, gives a quite different form to plots of the area of the septal region against that of the peripheral wall (see Fig. 15.11).

The complex biochemical phenotype of *rod A* mutants includes a derepression of peptidoglycan synthesis [93, 98] and a complete suppression of the activity of the enzyme forming the *N*-acetylgalactosamine-containing wall teichoic acid [47]. Exactly how these changes are related to the morphological phenotype is unclear. Certainly depriving the walls of some *Bacilli* of their negatively charged polymers (teichoic and teichuronic acids) leads to their growth as cocci instead of rods [38, 39].

A number of spherical mutants of *E. coli* have been described [1, 48, 63, 68, 69, 80, 82]. Examination of the temperature-conditional mutant isolated by Henning *et al.* [48] showed that the diameter of the rod roughly doubled with no increase in length and that this change took place in the presence of nalidixic acid and therefore presumably did not require concomitant DNA synthesis [44]. Many of the mutants might be suspected of having abnormal outer membranes from their increased sensitivity, for example, to neutral detergents [79] whilst the structure of the peptidoglycans, where examined, has proved similar to that of the wild-type. The morphology of two of the mutants [63, 82] could be corrected by rather high concentrations of D-alanine, a universal peptidoglycan constituent (see Section 6.1).

An interesting story relates to the penicillin antibiotic mecillinam (FL1060). When acting on *E. coli* it produces ovoid cells [66, 73] but does not inhibit the transpeptidase or carboxypeptidase [83]. Spratt and Pardee [110, 112] examined the kinetics of fixation of a variety of penicillins to five binding proteins in the cytoplasmic membrane of *E. coli* (see Section 9.7.2) and found that mecillinam bound with a rather low affinity constant to protein 2. Spratt [110] therefore considered that in some way this protein was shape-determining for *E. coli.* Spherical mutants resistant to the antibiotic [69] might be expected either to have lost penicillin binding protein 2 or have it modified. Examination of such mutants indeed showed this protein was no longer functional in binding antibiotic. Moreover a temperature-sensitive rod mutant [48] had a modified penicillin-binding protein 2 such that antibiotic was bound at 30° C but not 42° C (111). Thus, it would appear that at least one class of *rod* mutants of *E. coli* have a changed cytoplasmic membrane protein.

Evidence, however, shows that disturbances to the formation of the outer

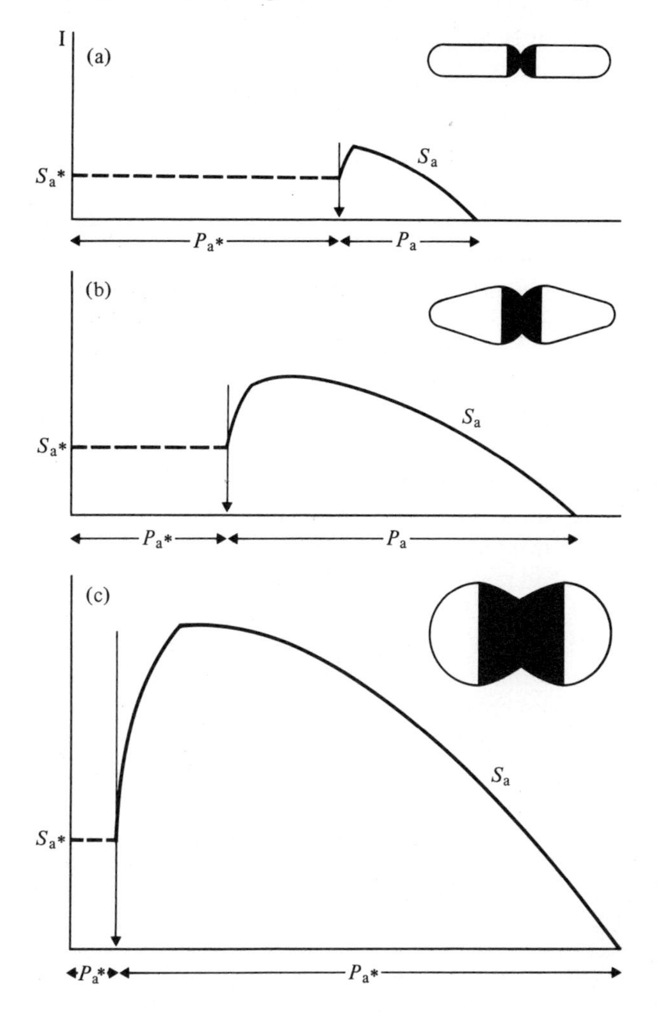

Figure 15.11 The sequence of growth of the wall of *rod* mutants of *B. subtilis* during their conversion from rods to cocci. In (a) growth is as for the normal rod and has been illustrated already in Fig. 15.8. After the shift in growth conditions there is an increase in the width of the cell and an increasing proportion of the peripheral wall (P_a) is fed out from the base of the septum until in (c) when the bacteria are coccal in shape most or all of the new peripheral wall has this origin. In the inset diagrams the new wall is shown in black. The curves show the plots of P_a against S_a as in Figs. 15.5a, 15.7 and 15.8 (from [97]).

membrane proteins and lipoprotein can also lead to spherical shaped cells. Mutants with all the outer membrane proteins missing except the lipoprotein grow with normal morphology, but if the lipoprotein is missing as well as the *tolG* protein or Protein II (see Chapter 3.4.2), then the bacteria have a spherical form [109]. The outer membrane was clearly dissociated from the peptidoglycan layer as would be expected in mutants lacking the lipoprotein (see Chapters 1 and 10) and very interestingly these mutants were hypersensitive to a variety of substances such as deoxycholate, actinomycin and dodecylsulphate. They also required abnormally high concentrations of divalent cations in the medium to allow maximal growth rates and to prevent lysis. In these characteristics they resembled a number of the other rod sphere mutants of *E. coli.* Examination of the penicillin binding proteins in these mutants showed no differences from the parents. Thus, it would seem at present, that coccal morphology in *E. coli* can arise from quite different sorts of disturbances. However, much more work is yet desirable in this fascinating field of endeavour. A more complete review of the subject has recently been published [90].

15.7 Helical growth of bacteria

Helical growth and the arrangement of macromolecules in the form of helices is a common feature of nature, to be found in the growth of a number of trailing plants, in the arrangement of cellulose and of course in DNA. Growth of bacteria as helices, however, had received little attention until Mendelson [74] studied helical growth in *B. subtilis.* His strain carried a *div* IV-B1 mutation [89] disturbing septation and therefore grew as filaments. He assumed that, since this mutation was obtained after treatment with nitrosoguanidine, other mutations leading to helical growth were also present. In order to obtain the helical filaments, growth had to start from germinating spores. The double helices of the filaments obtained were very regular (see Fig. 15.12) and their angle was 70°. Subsequently [75]

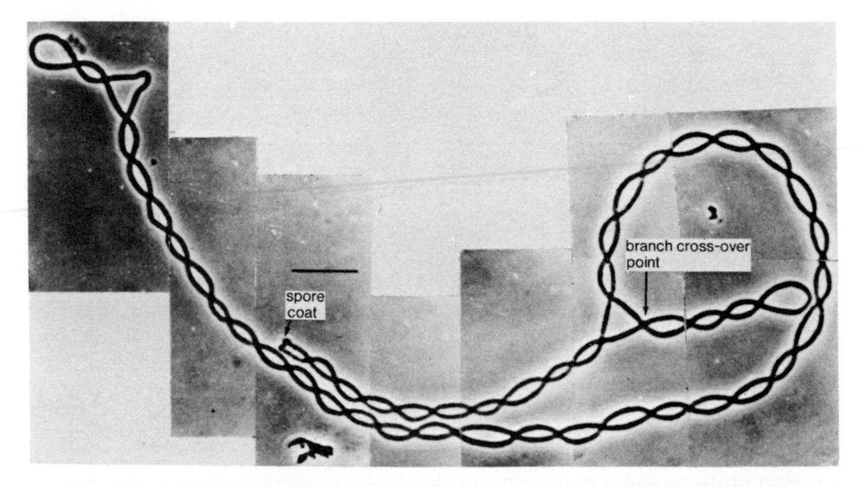

Figure 15.12 The regular double helices formed by strains of *Bacillus subtilis* carrying mutations either interfering with cell division or cell separation and which have the origins of the filament fixed together. In this example the bacilli had germinated from spores and the residual coat can be seen at the site of initiation (from [74]).

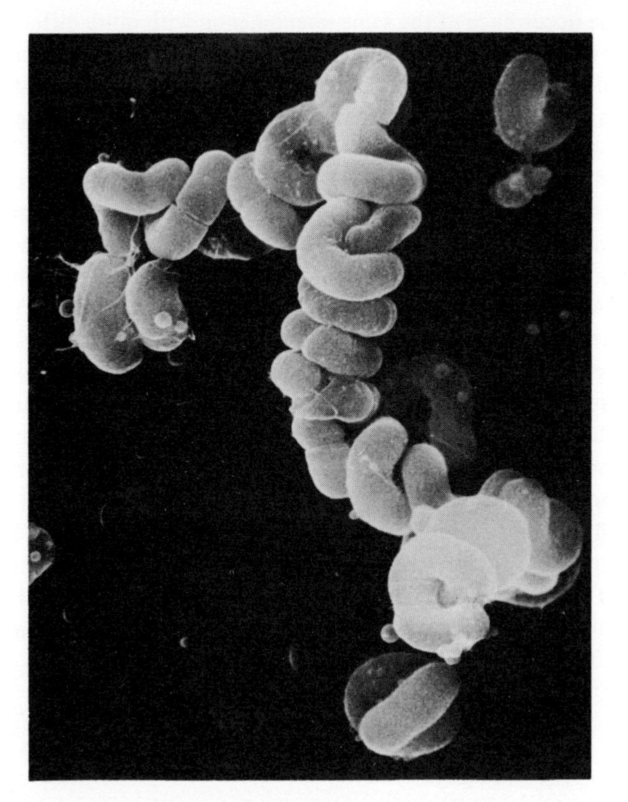

Figure 15.13 Tight corkscrew-like helices also seen in cultures of *B. subtilis* bearing unknown mutations or occasionally in double *rod lyt* mutations (from [113]).

complex multi-helix organizations were formed. Mendelson's hypothesis was that a helical surface organization of the bacteria was present as the phenotype of a mutation leading to rotation of the ends of cells in opposite directions and that when the origins of the growing filament were laterally fixed, a torque developed leading to the production of helices. An 'artificial' system can be set up [92] that supports the major part of this hypothesis except that it makes more likely the arrangement in helical form of the wall polymers in the wild-type. When the cultural conditions of double mutants of *B. subtilis rod A lyt* or *rod B lyt* are appropriately changed so that the bacteria cease to grow as unseparated cocci, they form unseparated rod-shaped cells – the *lyt* gene ensures that the bacteria do not separate (see Chapter 11.5.2). About 60–70% of the filaments from the *rod B lyt* strain formed helices very similar to those of Mendelson [74]. The simplest explanation of this observation is that the ends of the developing filaments were fixed together in the unseparated coccal forms, as were Mendelson's by the germinating spore coats. Inhibition of septation is not necessary providing the individual bacteria cannot separate. This again would be expected from Mendelson's hypothesis. The occurrence of the same phenomenon with quite different strains and mutations suggests that special mutations may not be necessary for helical arrangement of polymers in the wall but may already exist in the wild-type from which mutants are made. A different type of helical growth can also be seen in cultures of these double mutants of a type that had been described by Tilby [113]. These occurred in cultures of his wild-type strain subjected to various treatments such as the presence of oral anaesthetics or low concentrations of benzylpenicillin or even while restarting growth from the resting phase. In cultures of a Triton X-100 resistant strain, however, >95% grew as helices but they were not double helices like Mendelson's, but appeared as tightly wound springs (see Fig. 15.13). Tilby suggested that such growth could result from the presence of left-handed and right-handed helically arranged wall polymers under different degrees of extension or compression forces. The distances separating Tilby and Mendelson's ideas would not seem to be great since if the walls of a rod-shaped bacterium contain helically arranged components in the way suggested by Tilby, internal pressure (turgor pressure) known to be present in bacterial cells would be expected to lead to rotation of the ends of the cell in opposite directions as suggested by Mendelson. The appearance from photographs of Tilby's mutants (see Fig. 15.13) would suggest the presence of a number of unseparated cells in each helix, as in the *lyt* mutants. Measurements of cultures of these mutants show them to be autolysin-defective strains (H. J. Rogers and C. Taylor, unpublished work).

References

1. Adler, H. I., Terry, C. E. and Hardigree, A. A. (1968) *J. Bact.* **95**, 139–42.
2. Amako, K. and Umeda, A. (1977) *J. gen. Microbiol.* **98**, 297–9.
3. Archibald, A. R. (1976) *J. Bact.* **127**, 956–60.
4. Archibald, A. R. and Coapes, H. E. (1976) *J. Bact.* **125**, 1195–206.
5. Bayer, M. E. (1968a) *J. gen. Microbiol.* **53**, 395–404.
6. Bayer, M. (1968b) *J. Virol.* **2**, 346–56.
7. Bayer, M. and Thurow, H. (1977) *J. Bact.* **130**, 911–36.
8. Beachey, E. H. and Cole, R. M. (1966) *J. Bact.* **92**, 1245–51.
9. Begg, K. J. (1978) *J. Bact.* **135**, 307–10.
10. Begg, K. J. and Donachie, W. D. (1973) *Nature, Lond. New Biol.* **245**, 38–9.
11. Begg, K. J. and Donachie, W. D. (1977) *J. Bact.* **129**, 1524–36.
12. Birdsell, D. C. and Young, F. E. (1970) *Bact. Proc.* 39.
13. Bisset, K. A. and Pease, P. (1957) *J. gen. Microbiol.* **16**, 382–4.
14. Boothby, D., Daneo-Moore, L., Higgins, M. L., Coyette, J. and Shockman, G. D. (1973) *J. biol. Chem.* **248**, 2161–9.
15. Boylan, R. J. and Mendelson, N. H. (1969) *J. Bact.* **100**, 1316–21.
16. Boylan, R. J., Mendelson, N. H., Brooks, D. and Young, F. E. (1972) *J. Bact.* **110**, 281–90.
17. Braun, V. (1978) In *Relations Between Structure and Function in the Prokaryotic Cell, 28th Symp. Soc. Gen. Microbiol.*, eds. Stanier, R. Y., Rogers, H. J. and Ward, J. B., pp. 111–38, Cambridge University Press.
18. Briles, E. B. and Tomasz, A. (1970) *J. Cell Biol.* **47**, 786–90.

19. Brown, W. C., Wilson, C. R., Lukeheart, S., Young, F. E. and Shiflet, M. A. (1976) *J. Bact.* **125**, 166–73.
20. Burdett, I. D. J. (1979a) *J. Bact.* **137**, 1395–405.
21. Burdett, I. D. J. (1980) *J. gen. Microbiol.* (in press).
22. Burdett, I. D. J. and Higgins, M. L. (1978) *J. Bact.* **133**, 959–71.
23. Burdett, I. D. J. and Murray, R. G. E. (1976) *J. Bact.* **119**, 303–24.
24. Chung, K. and Hawirko, R. Z. and Isaac, P. K. (1964) *Can. J. Microbiol.* **10**, 473–82.
25. Chung, K., Hawirko, R. Z. and Isaac, P. K. (1964) *Can. J. Microbiol.* **10**, 43–8.
26. Cole, R. M. (1964) *Science* **143**, 820–2.
27. Cole, R. M. and Hahn, J. J. (1962) *Science* **135**, 722–4.
28. Cole, R. M., Popkin, T. J., Boylan, R. J. and Mendelson, N. H. (1970) *J. Bact.* **103**, 793–810.
29. De Chastellier, C., Hellio, R. and Ryter, A. (1975) *J. Bact.* **123**, 1184–96.
30. Donachie, W. D. and Begg, K. J. (1970) *Nature, Lond.* **227**, 1220–4.
31. Doyle, R. J., Streips, U. N. and Helman, J. R. (1977) In *Microbiology 1977*, ed. Schlessinger, D., pp. 44–9, Washington D.C: Am. Soc. Microbiol.
32. Ellwood, D. C. and Tempest, D. W. (1969) *Biochem. J.* **111**, 1–5.
33. Ellwood, D. C. and Tempest, D. W. (1972) *Adv. Microbiol. Physiol.* **7**, 83–117.
34. Fan, D. P. and Beckman, B. E. (1973) *J. Bact.* **114**, 790–97.
35. Fan, D. P., Petit, M. C. and Cunningham, W. P. (1972) *J. Bact.* **109**, 1226–72.
36. Fan, D. P., Beckman, B. E. and Beckman, M. M. (1974) *J. Bact.* **117**, 1330–4.
37. Forsberg, C. W. and Rogers, H. J. (1971) *Nature, Lond.* **229**, 272–3.
38. Forsberg, C. W. and Rogers, H. J. (1974) *J. Bact.* **118**, 358–68.
39. Forsberg, C. W., Wyrick, P. B., Ward, J. B. and Rogers, H. J. (1973) *J. Bact.* **113**, 969–84.
40. Fox, C. F. (1972) In *Membrane Molecular Biology*, eds. Fox, C. F. and Keith, A., p. 345, Stamford: Sinauer Ass. Inc.
41. Frehel, C. A., Beaufils, A. M. and Ryter, A. (1971) *A. Inst. Pasteur* **121**, 139–48.
42. Fukasawa, T. K., Jokura, K. and Murahashi, K. (1962) *Biochem. biophys. Res. Commun.* **7**, 121–5.
42a. Fujiwara, T. and Fukui, S. (1973) *J. Bact.* **110**, 743–6.
43. Glaser, L., Ionesco, H. and Schaeffer, P. (1966) *Biochim. biophys. Acta* **124**, 415–17.
44. Goodell, E. M. and Schwarz, U. (1975) *J. gen. Microbiol.* **86**, 201–9.
45. Green, E. A. and Schaechter, M. (1972) *Proc. Natn. Acad. Sci. U.S.A.* **69**, 2312–16.
46. Hakenbeck, R. and Messer, W. (1977) *J. Bact.* **129**, 1234–8.
47. Hayes, M. V., Ward, J. B. and Rogers, H. J. (1977) *Proc. Soc. Microbiol.* **4**, 85–6.
48. Henning, U., Rehn, K., Braun, V. and Hohn, B. (1972) *Eur. J. Biochem.* **26**, 670–6.
49. Higgins, M. L. (1976) *J. Bact.* **127**, 1337–45.
50. Higgins, M. L. and Shockman, G. D. (1970) *J. Bact.* **101**, 643–8.
51. Higgins, M. L. and Shockman, G. D. (1970) *J. Bact.* **103**, 244–54.
52. Higgins, M. L. and Shockman, G. D. (1976) *J. Bact.* **127**, 1346–58.
53. Higgins, M. L., Pooley, H. M. and Shockman, G. D. (1970) *J. Bact.* **103**, 504–12.
54. Higgins, M. L., Pooley, H. M. and Shockman, G. D. (1971) *J. Bact.* **105**, 1175–83.
55. Hoffman, B., Messer, W. and Schwarz, U. (1972) *J. Supramolec. Struct.* **1**, 29–37.

56. Hofnung, M. (1974) *Genetics,* **76,** 169–84.
57. Hughes, R. C. and Stokes, E. (1971) *J. Bact.* **106**, 694–6.
58. Hughes, R. C., Tanner, P. J. and Stokes, E. (1970) *Biochem. J.* **120**, 159–70.
59. Jacob, F., Brenner, S. and Cuzin, F. (1963) *Cold Spring Harbour Symp. Quant. Biol.* **28**, 329–48.
60. Kepes, A. and Autissier, F. (1972) *Biochim. biophys. Acta* **265**, 443–69.
61. Koch, A. L. and Boniface, J. (1971) *Biochim. biophys. Acta* **225**, 239–47.
62. Koppes, L. J. H., Overbeeke, N. and Nanninga, N. (1978) *J. Bact.* **133**, 1053–61.
63. Lazdunski, C. and Shapiro, B. M. (1972) *J. Bact.* **111**, 498–509.
64. Leal, J. and Marcoutch, H. (1975) *Mol. Gen. Genet.* **139**, 203–12.
65. Lin, E. C., Hirota, Y. and Jacob, F. (1971) *J. Bact.* **108**, 375–85.
66. Lund, F. and Tybring, L. (1972) *Nature, Lond. New Biol.* **41**, 419–29.
67. Mandelstam, J. and Rogers, H. J. (1958) *Nature, Lond.* **181**, 956–7.
68. Matsuzawa, H., Hayakawa, K., Sato, T. and Imahori, K. (1973) *J. Bact.* **125**, 436–42.
69. Matsuhashi, S., Kamiryo, T., Blumberg, P. M., Linnett, P., Willoughby, E. and Strominger, J. L. (1974) *J. Bact.* **117**, 578–87.
70. Mauck, J. and Glaser, L. (1972) *J. biol. Chem.* **247,** 1180–7.
71. Mauck, J., Chan, L., Glaser, L. and Williamson, J. (1972) *J. Bact.* **109,** 373–8.
72. May, J. W., (1963) *Exp. Cell Res.* **31**, 217–20.
73. Melchior, N. H., Blom, J., Tybring, L. and Birch-Anderson, A. (1973) *Acta Microbiol. Scand. Ser. B.* **81,** 393–407.
73a. Meloni, G. A. and Monti-Bragadise, C. (1962) *Ann. Sclavo.* **4**, 143–52.
74. Mendelson, N. H. (1976) *Proc. Natn. Acad. Sci. U.S.A.* **73**, 1740–8.
75. Mendelson, N. H. (1978) *Proc. Natn. Acad. Sci. U.S.A.* **75**, 2478–82.
76. Mindich, L. and Dales, S. (1972) *J. Cell Biol.* **55**, 32–41.
77. Muhlradt, P. F. (1973) *Eur. J. Biochem.* **35**, 471–81.
78. Muhlradt, P. F., Menzel, J. R., Golecki, J. R. and Speth, V. (1974) *Eur. J. Biochem.* **43**, 530–9.
79. Nikaido, H., Takkenschi, Y., Ohnishi, S. and Nakai, T. (1977) *Biochim. biophys. Acta* **465**, 152–64.
80. Normark, S. (1969) *J. Bact.* **98**, 1274–7.
81. Novick, A. and Weiner, M. (1957) *Proc. Natn. Acad. Sci. U.S.A.* **43**, 553–66.
82. Olden, K. and Wilson, T. M. (1973) *Fed. Proc.* **32,** Abst. 1464.
83. Park, J. T. and Burman, L. (1973) *Biochem. biophys. Res. Commun.* **51**, 863–74.
84. Pooley, H. M. (1976a) *J. Bact.* **125**, 1127–38.
85. Pooley, H. M. (1976b) *J. Bact.* **125**, 139–47.
86. Pooley, H. M., Schlaeppi, J–M. and Karamata, D. (1978) *Nature, Lond.* **274**, 264–6.
87. Quadling, C. (1958) *J. gen. Microbiol.* **18**, 227–37.
88. Razin, S. (1978) *Microbiol. Rev.* **42**, 414–70.
89. Reeve, J. N., Mendelson, N. H., Coyne, S. I., Hallock, L. L. and Cole, R. M. (1973) *J. Bact.* **114**, 860–73.
90. Rogers, H. J. (1979) *Adv. Microbiol Physiol.* **19,** 1–62.
91. Rogers, H. J. and Thurman, P. F. (1978a) *J. Bact.* **133**, 298–305.
92. Rogers, H. J. and Thurman, P. F. (1978b) *J. Bact.* **133**, 1508–9.
93. Rogers, H. J. and Taylor, C. (1979) *J. Bact.* **135,** 1032–42.
94. Rogers, H. J., McConnell, M. and Burdett, I. D. J. (1968) *Nature, Lond.* **213**, 31–3.
95. Rogers, H. J., McConnell, M. and Burdett, I. D. J. (1970) *J. gen. Microbiol.* **61**, 155–71.

96. Rogers, H. J., Thurman, P. F. and Buxton, R. S. (1976) *J. Bact.* **125**, 556–64.
97. Rogers, H. J., Ward, J. B. and Burdett, I. D. J. (1978) In *Relations between structure and function in the prokaryotic cell, 28th Symp. Soc. Gen. Microbiol.* eds. Stainer, R. Y., Rogers, H. J. and Ward, J. B., pp. 139–76, Cambridge University Press.
98. Rogers, H. J., Thurman, P. F., Taylor, C. and Reeve, J. N. (1974) *J. gen. Microbiol.* **85**, 335–50.
99. Ryter, A. (1967) *Folia microbiol.* **12**, 283–90.
100. Ryter, A. (1971) *Ann. Inst. Past.* **121**, 271–88.
101. Ryter, A. (1974) *Annls microbiologie* **125B**, 167–80.
102. Ryter, A., Hirota, Y. and Schwarz, U. (1973) *J. molec. Biol.* **78**, 185–95.
103. Ryter, A., Schuman, H. and Schwarz, M. (1975) *J. Bact.* **122**, 295–301.
104. Sargent, M. G. (1979) *Adv. Microbiol. Physiol.* **18**, 139–76.
105. Schwarz, U., Ryter, A., Ramtach, A., Hellio, R. and Hirota, Y. (1975) *J. molec. Biol.* **98**, 749–60.
106. Shockman, G. D. (1965) *Bact. Rev.* **29**, 345–58.
107. Shockman, G. D., Pooley, H. M. and Thompson, J. S. (1967) *J. Bact.* **94**, 1525–30.
108. Smit, M. and Nikaido, H. (1978) *J. Bact.* **135**, 687–702.
109. Sonntag, I., Schwarz, H., Hirota, Y., and Henning, U. (1978) *J. Bact.* **136**, 280–5.
110. Spratt, B. G. (1975) *Proc. Natn. Acad. Sci. U.S.A.* **72**, 2999–3003.
111. Spratt, B. G. (1977) *Microbiology 1977*, ed. Schlessinger, D., p. 182–90. Washington D.C: Am. Soc. Microbiol.
112. Spratt, B. G. and Pardee, A. B. (1975) *Nature, Lond.* **254**, 516–7.
113. Tilby, M. J. (1977) *Nature, Lond.* **266**, 450–52.
114. Tsukoyoshi, N., Fielding, P. and Fox, C. F. (1971) *Biochem. biophys. Res. Commun.* **44**, 497.
115. Van Tubergen, R. P. and Setlow, R. B. (1961) *Biophys. J.* **1**, 589–625.
116. Wagner, M. (1964) *Zentbl. Bakt. Parasitkde.* **195**, 87–93.
117. Ward, J. B. (1975) *J. Bact.* **124**, 668–78.
118. Young, F. E. (1967) *Proc. Natn. Acad. Sci. U.S.A.* **58**, 2377–84.

Index

Main references are in bold type; references to figures and tables are in italic.